AF180988

SQUID AS EXPERIMENTAL ANIMALS

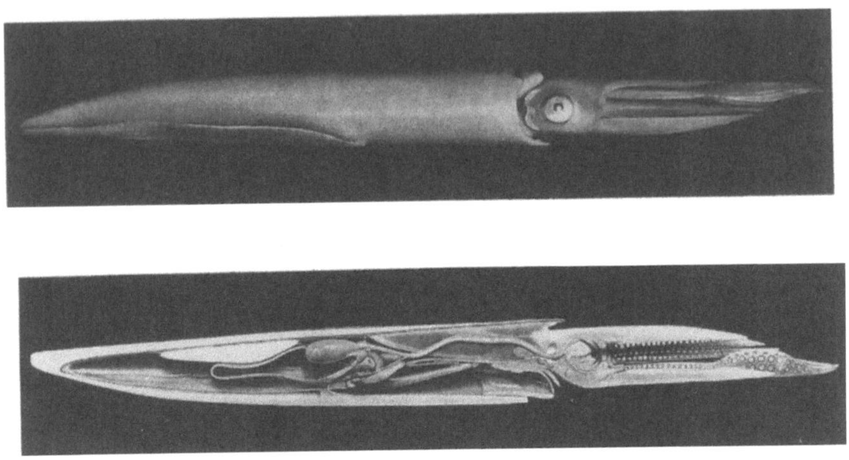

From L. W. Williams, 1909, *The Anatomy of the Common Squid, Loligo pealii, Lesueur,* E. J. Brill, Leiden, Holland. In the illustration by Williams, the top figure of the whole squid is oriented with the physiological dorsal surface facing downwards and the animal appears to be swimming upside-down.

SQUID AS EXPERIMENTAL ANIMALS

Edited by

Daniel L. Gilbert and William J. Adelman, Jr.
National Institutes of Health
Bethesda, Maryland
and
John M. Arnold
University of Hawaii
Honolulu, Hawaii

SPRINGER SCIENCE+BUSINESS MEDIA, LLC

Library of Congress Cataloging-in-Publication Data

Squid as experimental animals / edited by Daniel L. Gilbert, William J. Adelman, Jr., and John M. Arnold.
p. cm.
Includes bibliographical references.
ISBN 978-1-4899-2491-9 ISBN 978-1-4899-2489-6 (eBook)
DOI 10.1007/978-1-4899-2489-6
1. Squids as laboratory animals. 2. Nervous system--Mollusks.
3. Squids--Cytology. I. Gilbert, Daniel L. II. Adelman, William J., 1928- . III. Arnold, John M.
QL430.2.S66 1990
594'.58--dc20 90-6849
CIP

© 1990 Springer Science+Business Media New York
Originally published by Plenum Press, New York in 1990
Softcover reprint of the hardcover 1st edition 1990

All rights reserved

No part of this book may be reproduced, stored in a retrieval system, or transmitted in any form or by any means, electronic, mechanical, photocopying, microfilming, recording, or otherwise, without written permission from the Publisher

Contributors

William J. Adelman, Jr. Laboratory of Biophysics, NINDS, National Institutes of Health, Bethesda, MD 20892

Mario Alberghina Marine Biological Laboratory, Woods Hole, MA 02543 and Institute of Biochemistry, Faculty of Medicine, University of Catania, Viale Andrea Doria 6, 95125 Catania, Italy

Frederick A. Aldrich Ocean Studies Task Force and Department of Biology, Memorial University of Newfoundland, St. John's, Newfoundland A1C 5S7, Canada

John M. Arnold Pacific Biomedical Research Center, Cephalopod Biology Laboratory, 209A Snyder Hall, 2538 The Mall, University of Hawaii at Manoa, Honolulu, Hawaii 96822

Francisco Bezanilla Department of Physiology, Ahmanson Laboratory of Neurobiology and Jerry Lewis Neuromuscular Center, University of California at Los Angeles, Los Angeles, California 90024

F. J. Brinley, Jr. Neurological Disorders Program, National Institute of Neurological Disorders and Stroke, Federal Building, Room 814, National Institutes of Health, Bethesda, MD 20892

Anthony Brown Bio-architectonics Center, School of Medicine, Case Western Reserve University, Cleveland, Ohio 44106

Bernd U. Budelmann Marine Biomedical Institute and Department of Otolaryngology, University of Texas Medical Branch, Galveston, Texas 77550

Lawrence B. Cohen Department of Cellular and Molecular Physiology, Yale University School of Medicine, 333 Cedar Street, New Haven, Connecticut 06510

Rochelle S. Cohen Department of Anatomy and Cell Biology, University of Illinois at Chicago, Chicago, Illinois 60612

Harold Gainer Laboratory of Neurochemistry, NINDS, National Institutes of Health, Bethesda, MD 20892

Daniel L. Gilbert Laboratory of Biophysics, NINDS, National Institutes of Health, Bethesda, MD 20892

Robert M. Gould Marine Biological Laboratory, Woods Hole, MA 02543 and New York State Institute for Basic Research in Developmental Disabilities, 1050 Forest Hill Rd., Staten Island, NY 10314

Roger T. Hanlon Marine Biomedical Institute and Department of Psychiatry and Behavioral Sciences, University of Texas Medical Branch, Galveston, Texas 77550-2772

Francis C. G. Hoskin Biology Department, Illinois Institute of Technology, Chicago, IL 60616

David Landowne Department of Physiology and Biophysics, University of Miami School of Medicine, Miami, Florida 33101

George M. Langford Marine Biological Laboratory, Woods Hole, MA 02543 and Department of Physiology, School of Medicine, University of North Carolina, Chapel Hill, NC 27599

Raymond J. Lasek Bio-architectonics Center, School of Medicine, Case Western Reserve University, Cleveland, Ohio 44106

Charlotte P. Mangum Dept. of Biology, College of William and Mary, Williamsburg, Virginia 23185

I. A. Meinertzhagen Life Sciences Centre, Dalhousie University, Halifax, Nova Scotia, Canada B3H 4J1

Monica A. Meyer Marine Biological Laboratory, Woods Hole, MA 02543

RuthAnne Mueller Laboratory of Biophysics, NINDS, NIH, Bethesda, MD. 20892

Lorin J. Mullins Department of Biophysics, Medical School, University of Maryland, 660 W. Redwood Street, Baltimore, MD 21201

Ron O'Dor Biology Department, Dalhousie University, Halifax, Nova Scotia, Canada B3H 4J1

Harish C. Pant Laboratory of Neurochemistry, NINDS, National Institutes of Health, Bethesda, MD 20892

H. O. Pörtner Institut für Zoologie IV, Universität Düsseldorf, Universitätsstrasse 1 D-4000 Düsseldorf 1, F. R. Germany

Robert V. Rice Department of Biological Sciences, Carnegie-Mellon University, Pittsburgh, PA 15213

Helen R. Saibil Department of Zoology, Oxford University, Oxford OX1 3PS, England. Present address: Department of Crystallography, University of London Birkbeck College, Malet Street, London WC1E 7HX, England

Brian M. Salzberg Department of Physiology, University of Pennsylvania School of Medicine, Philadelphia, Pennsylvania 19104

R. E. Shadwick Biology Department, University of Calgary, Calgary, Alberta, Canada T2N 1N4

Elis F. Stanley Laboratory of Biophysics, NINDS, National Institutes of Health, Bethesda, MD 20892

William C. Summers Huxley College of Environmental Studies and Shannon Point Marine Center, Western Washington University, Bellingham, Washington 98225

Carol Vandenberg Department of Biological Sciences, University of California at Santa Barbara, Santa Barbara, California, 93106

Dieter G. Weiss Marine Biological Laboratory, Woods Hole, MA 02543 and Institut für Zoologie, Technische Universtät München, D-8046 Garching, Fed. Rep. Germany

Preface

The predecessor to this book was *A Guide to the Laboratory Use of the Squid Loligo pealei* published by the Marine Biological Laboratory, Woods Hole, Massachusetts in 1974. The revision of this long out of date guide, with the approval of the Marine Biological Laboratory, is an attempt to introduce students and researchers to the cephalopods and particularly the squid as an object of biological research. Therefore, we have decided to expand on its original theme, which was to present important practical aspects for using the squid as experimental animals. There are twenty two chapters instead of the original eight. The material in the original eight chapters has been completely revised. Since more than one method can be used for accomplishing a given task, some duplication of methods was considered desirable in the various chapters. Thus, the methodology can be chosen which is best suited for each reader's requirements. Each subject also contains a mini-review which can serve as an introduction to the various topics. Thus, the volume is not just a laboratory manual, but can also be used as an introduction to squid biology. The book is intended for laboratory technicians, advanced undergraduate students, graduate students, researchers, and all others who want to learn the purpose, methods, and techniques of using squid as experimental animals. This is the reason why the name has been changed to its present title. Preceding the chapters is a list of many of the abbreviations, prefixes, and suffixes used in this volume.

Cephalopods possess the most advanced nervous system of all invertebrates. For this reason, there are a predominance of chapters on the components of the nervous systems. Part I deals with evolution, history, and maintenance. The first chapter is on evolution of intelligence. Then chapters follow on squid in its natural habitat, the discovery of *Loligo*, squid maintenance and rearing. Part 2 contains two chapters on squid mating and embryology. Part 3 includes the neural membranes. Chapters in this part are concerned with electrophysiology of the squid axon, internal dialysis in the squid axon, the cut-open axon, optical measurements on squid axons, and the squid giant synapse. Part 4 contains chapters on cell biology. These are on tissue culture techniques, squid optic lobe synaptosomes, the cytoskeleton of the giant axon, axoplasmic transport using video microscopy, and lipid metabolism in the nervous system. Part 5 has some chapters on the sensory systems, which are the squid eye, the development of the visual system, and the statocysts of squid. Finally, Part 6 concludes with integrated systems. Chapters in this part deal with the squid as a whole. Blood oxygen and carbon dioxide gas transport to and from the tissues is the topic of the first chapter in this part. This is followed by a chapter on a detoxifying enzyme unique to the squid. The final chapter presents the integration of all the squid systems as a whole for the functioning of the squid in its natural habitat. Due to lack of space, other aspects of squid biology, such as physiology of the central nervous system, digestion, and excretion, are not included.

The frontispiece is taken from the frontispiece in Leonard Worcester Williams classic, *The Anatomy of the Common Squid, Loligo pealii, Lesueur*, published in 1909 by E. J. Brill, Leiden, Holland.

ACKNOWLEDGEMENTS: Thanks are given to Dr. Claire Gilbert for her editorial assistance and to the staff of Plenum Press, especially Mary P. Born, Senior Editor, John Matzka, Managing Editor and his assistant, Gregory Safford. We acknowledge the intramural support of the Basic Neurosciences Program of the National Institute of Neurological Disorders and Stroke, National Institutes of Health, Bethesda, Maryland.

DEDICATION: This book is dedicated to the memory of two distinguished scientists, who recently passed away: Kenneth S. Cole, who did the pioneering studies on the squid giant axon, and Gilbert L. Voss, who was a world renowned authority on squid.

The Editors
Marine Biological Laboratory
Woods Hole, Massachusetts 02543
September, 1989

Contents

Chapter 3
Lol-i-go and Far Away: A Consideration of the Establishment of the Species Designation *Loligo pealei*

FREDERICK A. ALDRICH

Chapter 4
Maintenance, Rearing, and Culture of Teuthoid and Sepioid Squids

ROGER T. HANLON

PART II. MATING BEHAVIOR AND EMBRYOLOGY

Chapter 5
Squid Mating Behavior

JOHN M. ARNOLD

Chapter 6
Embryonic Development of the Squid

JOHN M. ARNOLD

PART III. NEURAL MEMBRANES

Chapter 7
Electrophysiology and Biophysics of the Squid Giant Axon

WILLIAM J. ADELMAN, JR. and DANIEL L. GILBERT

Chapter 9
The Cut-Open Axon Technique

FRANCISCO BEZANILLA and CAROL VANDENBERG

Chapter 10
Optical Measurements on Squid Axons

LAWRENCE B. COHEN, DAVID LANDOWNE, and BRIAN M. SALZBERG

Chapter 11
The Preparation of the Squid Giant Synapse for Electrophysiological Investigation

ELIS F. STANLEY

PART IV. CELL BIOLOGY

Chapter 12
Tissue Culture of Squid Neurons, Glia, and Muscle Cells

ROBERT V. RICE, RUTHANNE MUELLER, and
WILLIAM J. ADELMAN, JR.

Chapter 13
Squid Optic Lobe Synaptosomes: Structure and Function of Isolated Synapses

ROCHELLE S. COHEN, HARISH C. PANT, and HAROLD GAINER

Chapter 14
The Cytoskeleton of the Squid Giant Axon

ANTHONY BROWN and RAYMOND J. LASEK

Chapter 15
Studying Axoplasmic Transport by Video Microscopy and Using the Squid Giant Axon as a Model System

DIETER G. WEISS, MONICA A. MEYER, and GEORGE M. LANGFORD

Chapter 16
Lipid Metabolism In The Squid Nervous System

ROBERT M. GOULD and MARIO ALBERGHINA

PART V. SENSORY SYSTEMS

Chapter 17
Structure and Function of the Squid Eye

HELEN R. SAIBIL

Chapter 18
Development of the Squid's Visual System

I. A. MEINERTZHAGEN

Chapter 19
The Statocysts of Squid

BERND U. BUDELMANN

PART VI. INTEGRATED SYSTEMS

Chapter 20
Gas Transport in the Blood

CHARLOTTE P. MANGUM

Chapter 21
An Organophosphorus Detoxifying Enzyme Unique to Squid

FRANCIS C. G. HOSKIN

Chapter 22
Squid as Elite Athletes: Locomotory, Respiratory, and Circulatory Integration

RON O'DOR, H. O. PÖRTNER, and R. E. SHADWICK

Abbreviations, Prefixes, and Suffixes

α	absorption coefficient or solubility of CO_2 in aqueous media
γ^{32}-P-ATP	ATP labeled with P^{32} in the γ position
ΔH	enthalpy
$\Delta \log P_{50}/\Delta pH$	Bohr coefficient (used as a measure of the Bohr shift)
1,10-ϕ	1,10-phenanthroline
5-HT	5-hydroxytryptamine
8-OHQ-5SA	8-hydroxyquinoline-5-sulfonate
a	refers to postbranchial
A/D	analog to digital
AChE	acetylcholinesterase
ADP	adenosine 5'-diphosphate
AMP	adenosine 5'-monophosphate
AMP-PNP	5'-adenylyl imidodiphosphate
ArgP	arginine phosphate
ASW	artificial sea water
ATCh	acetylthiocholine
ATP	adenosine-5'-triphosphate
ATP-γ-S	ATP with S in the γ position
AVEC-DIC	Allen video-enhanced contrast differential interference contrast microscope
BES	N,N-bis(2-hydroxyethyl)-2-aminoethanesulfonic acid
BIS-TRIS	bis(hydroxethyl)imino-tris(hydroxymethyl)-methane
BIS-TRIS-propane	1,3-bis(tris(hydroxymethyl)methylamino)-propane
C	pressure wave velocity
$CaCO_2$	postbranchial CO_2 concentration
CaMgF	calcium and magnesium free
CaNP	calcium-dependent neutral protease (calpain)
CaO_2	postbranchial O_2 concentration
CD	circular dichroism
CDP	cytidine 5'-diphosphate
CDTA	trans-1,2-diaminocyclohexane-N,N,N',N'-tetraacetic acid
cGMP	cyclic guanosine monophosphate
Chol	cholesterol
Ci	curie
CL	cardiolipin
CL	crista longitudinalis
CNS	central nervous system
CS	common stellate
Cs EGTA	Cs ethylene glycol-bis(2-amino ethyl ether)N,N,N',N'-tetraacetic acid
CTA	crista transversalis anterior

CTP	crista transversalis posterior
CTP	cytidine 5'-triphosphate
CV	crista verticalis
$CvCO_2$	prebranchial CO_2 concentration
CvO_2	prebranchial O_2 concentration
d	blood density
D	Dalton
D/A	digital to analog
DC	direct current
DEAE	diethylamino ethyl
deoxyHc	deoxygenated form of Hc
DFP	diisopropyl phosphorofluoridate
DIC	differential interference contrast
DMSO	dimethylsulfoxide
DNA	desoxynucleic acid
dpm	disintegrations per minute
DTNB	5,5'-dithiobis-(2-nitrobenzoic acid)
E	elastic modulus of the vessel wall
EbO_2	efficiency of O_2 uptake at the gill
EDTA	ethylenediamine tetraacetic acid
EGTA	ethylene glycol-bis(2-amino ethyl ether)N,N,N',N'-tetraacetic acid
EIM	excitability inducing material
EKG	electrocardiogram
EM	electron microscopy
EMG	electromyogram
EPSC	excitatory postsynaptic current
EPSP	excitatory postsynaptic potential
ERG	electroretinogram
ERP	early receptor potential
$ExtrCO_2$	CO_2 extraction from blood
$ExtrO_2$	O_2 extraction from blood
FB	fixation buffer
FBS	fetal bovine serum
FCCP	p-trifluoro methoxy carbonyl cyanide phenyl hydrazine
FET	field effect transistor
FFT	fast Fourier transform
FITC	fluorescein isothiocyanate
g	acceleration due to gravity at sea level or 9.80 m/sec^2
G.A. I	first order giant axon
G.A. II	second order giant axon
G.A. III	third order giant axon
GFAP	glial fibrillary acid protein
GFL	giant fiber lobe
GTP	guanosine 5'-triphosphate
h	vessel wall thickness
H-H equations	Hodgkin-Huxley equations
Hc	hemocyanin
HcO_2	oxygenated Hc
HEPES	N-2-hydroxyethylpiperazine-N'-2-ethanesulfonic acid

HPLC	high pressure liquid chromatography (high performance liquid chromatography)
HRP	horseradish peroxidase
Hz	cycles/sec
i	refers to inhalent
IgG	immunoglobin G
IM	inverted microscope
IR	current times resistance
IR	infrared
IU	international units
KEP	kinetic equilibration paradigm
LPT	linear position transducer
MAO	monoamine oxidase
MAP	microtubule-associated protein
MEM	minimal essential medium
MEPC	miniature excitatory postsynaptic current
mepp	miniature end plate potential
MEPP	miniature excitatory postsynaptic potential
MES	sodium 2-(N-morpholino)ethane sulfonate
ML	mantle length
MMEM	mostly modified Eagle media
MNI	macula neglecta inferior
MNS	macula neglecta superior
MOPS	3-(N-morpholino)propanesulfonic acid
MSP	macula statica princeps
n	Hill coefficient
N	Newtons
N. A.	numerical aperature
NAD	nictinamide adenine dinucleotide
NADH	reduced nicotinamide adenine dinucleotide
NADP	nicotinamide adenine dinucleotide 3'-phosphate
OPA	organophosphorus acid
ORD	optical rotatory dispersion (optical rotation as a function of wavelength)
OT	obligate thermophilic
oxyHc	oxygenated form of Hc
Pa	Pascal
Paraoxon	diethyl p-nitrophenyl phosphate
PC	phosphatidylcholine
PCMB	p-hydroxymercuribenzoate (p-chloromercuribenzoate)
PCO_2	partial pressure of carbon dioxide
PE	phosphatidylethanolamine
PE	polyethylene
PEM	photoelastic modulator
PEP	phospho(enol)pyruvate
PI	phosphatidylinositol
PIP	phosphatidylinositol phosphate
PIP_2	phosphatidylinositol bisphosphate
PIPES	piperazine-N,N'-bis(2-ethanesulfonic acid)

pK_a	constant describing the equilibrium between the dissociated and undissociated forms of an acid
pK'	constant describing the equilibrium between the components of the CO_2 system in aqueous media
PL	phospholipid
PMSF	phenyl-methylsulfonylfluoride
pN	pH at neutrality
PO_2	partial pressure of oxygen
ppt	parts/thousand
PS	phosphatidylserine
PSD	postsynaptic density
psi	pounds/in^2 gauge pressure
P*x*	oxygen pressure in torr when Hc is *x*% oxygenated
Q_{10}	ratio of reaction rates 10 °C apart
R	internal radius
REO	recordable erasable optical
Rf	distance of migration of the solute relative to the solvent front
rpm	revolution/min
S	Siemens
Sarin	isopropyl methylphosphonofluoridate
SDS	sodium dodecyl sulfate
SDS PAGE	SDS polyacrylamide gel electrophoresis
SEM	scanning electron microscopy
SML	sucrose monolaurate
Soman	2,3,3-trimethylpropyl methylphosphonofluoridate
SPM	sphingomyelin
SPM	synaptosomal plasma membrane
SPME	0.6 M sucrose, 0.1 M potassium phosphate, 10 mM magnesium chloride, 1 mM EGTA, pH 7.1
spp.	species
STX	saxitoxin
Tabun	ethyl N,N-dimethylphosphoramidocyanidate
TAME	tosyl L-arginine methyl ester
TC	tissue culture
TCA	trichloroacetic acid
TEA	tetraethyl ammonium chloride
TES	N-tris (hydroxymethyl)-methyl-2-aminoethanesulfonic acid
TLC	thin layer chromatography
TRIS	tris (hydroxymethyl) aminomethane
Tris maleate	mono[tris(hydroxymethyl)-aminomethane] maleate
TRIZMA	tris (hydroxymethyl) aminomethane-HCl (Tris-HCl)
TS	Tris-HCl, 0.25 M sucrose solution
TTX	tetrodotoxin
TV	television
UTP	uridine 5'-triphosphate
UV	ultraviolet
v	refers to prebranchial
VCR	video cassette recorder
VHS	video home system
WORM	write once read many

WW 375	5-[(1-γ-triethylammonium sulfopropyl-4(1H)-quinolylidene)-2-butenylidene]-3-ethylrhodamine
Z	hydraulic impedance to blood flow

Prefixes

Symbol	*Name*	*Factor**	*Examples*
p	pico	10^{-12}	pico Newtons (pN)
n	nano	10^{-9}	nano Siemens (nS)
μ	micro	10^{-6}	micro curie (μCi)
m	milli	10^{-3}	milli curie (mCi)
k	kilo	10^{3}	kiloDalton (kD); kilo Hertz (kHz); kilo Pascal (kPa)
M	Mega	10^{6}	MegaHertz (MHz)
G	giga	10^{9}	gigaohm

* factor by which quantity is multiplied

Suffixes

ase	refers to an enzyme which breaks the substrate designated by the prefix. Examples are ATPase and GTPase.

Part I

EVOLUTION, HISTORY, AND MAINTENANCE

Chapter 1

Evolution and Intelligence of the Cephalopods

JOHN M. ARNOLD

1. Introduction

Aristotle (ca. 330 B.C.) in his observations pointed out that the cephalopods were capable of investigating and perceiving their environment and *learning* from experience. The use of the term *animal intelligence* always raises questions and controversy. The problem seems to separate instinctive behavior (or genetically inherited behavior patterns) from behavior patterns that have been gained with experience by the animal in its environment. The former is a species wide behavior while the latter is exclusive to the individual. This exclusivity implies that sensory mechanisms are utilized to perceive the environment, that data is interpreted in a positive or negative fashion, and that the data is stored for future use when similar situations are presented. This requires certain physical and behavioral characteristics (Wells, 1962, 1978; Young, 1988).

1.1. Intelligence and behavior

First, an animal must have a nervous system capable of perceiving and interpreting its surroundings. Very simple systems will be inadequate. For example, several protozoa avoid bright light while photosynthetic protists are attracted to it. This involves only a light sensitive organ and cilia or flagella control and does not involve information storage. More complicated nervous systems are necessary for true *intelligence and behavior* and the question posed by the scarecrow in the novel *The Wonderful Wizard of Oz* (Baum, 1900) of the results of *only having a brain* are not of trivial importance. Formation of a brain usually requires cephalization of several of the ganglia in an anterior area of the animal. This new compound structure has the combined functions of motor, sensory, interpretation, and information storage. Generally, animal intelligence, both vertebrate and invertebrate, is measured by sensory trial and recall of reward or punishment. This is called *learning* and the speed of learning or number of trials to achieve a defined goal is used as a measure of *intelligence*. This has limited usefulness because humans can only relate to animal intelligence through our own basic interpretation of our own environment. For example,

J. M. ARNOLD • Pacific Biomedical Research Center, Cephalopod Biology Laboratory, 209A Snyder Hall, 2538 The Mall, University of Hawaii at Manoa, Honolulu, Hawaii 96822.

TABLE I. Class Cephalopoda

Infraclass Ectocochia =	Nautiloids and Ammonites
Subclass Ammonites =	All extinct. Distantly related to the Nautiloids
Subclass Nautiloids =	Modern Nautilus and many fossil forms. Some of the earliest cephalopod fossils.
Infraclass Endocochlia =	Coleoidea of coleoids. Shell-less forms or with internal shells only. [cuttlefish, *Spirula* shell, pen or gladius of squids, styletts and skull of octopus].
Order Octopoda:	Eight armed and mainly benthic. Ciratta and Inciratta. (hooks present or lacking).
Order Vampyroteuthis:	Rare octopus-like form with fins. May be between octopods and squids.
Order Decapoda:	Eight arms and two tentacles, an incorrect lumping now generally divided as follows:
Order Sepiodea:	Small squids, Sepioids.
Order Sepiida:	Cuttlefishes.
Order Teuthida:	True squids.
Suborder Oegopsida:	Oceanic squid in general, *Architeuthis, Illex, Dosidicus*.
Suborder Myopside:	Inshore squid including *Loligo*. Many have giant axons.

domestic dogs are easily trained while more intelligent domestic cats are hard to train (Hearne, 1987). It might be that the *trainers* don't know how to appeal to the *trainees*.

The cephalopods definitely qualify as intelligent animals and extensive work has been done on the structure of their nervous system. (For example, Young, 1936, 1963, 1976, 1983; Boycott, 1961; Boycott and Young, 1957; Wells, 1978; Boyle, 1986). In addition, there are several references referring to components of the nervous and sensory systems (Adelman and Gilbert, 1990; Mullins and Brinley, 1990; Bezanilla and Vandenberg, 1990; Cohen, Landowne, and Salzberg, 1990; Rice *et al.*, 1990; Cohen, Pant, and Gainer, 1990; Stanley, 1990; Brown and Lasek, 1990; Weiss, Meyer, and Langford, 1990; Gould and Alberghina, 1990; Hoskin, 1990; Saibil, 1990; Meinertzhagen, 1990; Budelmann, 1990).

There is a vast literature on the behavior and learning ability of the cephalopods and with the possible exception of the primitive and seemingly unintelligent Nautilus, all cephalopods tested have displayed some level of intelligent behavior. Both the cuttlefishes (Boycott, 1961; Messenger, 1973) and octopuses (Wells, 1978) seem to be quite intelligent and capable of learning rapidly. This is not surprising since their brain size in relation to body weight is higher than most fishes and ranks mid-way among the birds and mammals (Packard, 1972).

2. Evolution

The evolution of this intelligence has taken time, and time has been amply available to the cephalopods. The fossil record extends back into the Lower Cambrian era (more or less 570 million years, Teichert, 1988) making the early cephalopods some of the earliest known animals with hard skeletons capable of being well fossilized. There are as many schemes of evolution and classification of the cephalopods as there are workers active in the field and it is hard to choose one that is more right than others. Some of the ideas have been around a long time and are still used although everyone generally acknowledges them to be incorrect. For example, the Decapoda (cuttlefish and squids) and Octopoda (octopods) were recognized since Aristotle (330 B.C.) and Owen (1832) divided the Class Cephalopoda into the Subclass Tetrabranchiata and Dibranchiata with two pairs or one pair of gills, respectively. The ammonites were incorrectly assumed to be tetrabranches because of their resemblance to Nautiloids. This classification is generally considered to be incorrect but the terms Dibranchiata and Teterabranchiata are still used interchangeably instead of the more correct Endococlia (internal or no shell) and Ectocohlia (external shell). Table I is based on several authors (Teichert, 1988; Clarke, 1988 and Lehmann 1981) and my own bias.

2.1. Evolution and competition

The evolution of the cephalopods spans from the Lower and Middle Cambrian eras until the present with periodic explosive expansions of the number of species and sometimes disastrous mass extinctions and near extinctions (Lehmann, 1981). The point has been made that one of the major influences on the evolution of the cephalopods has been the appearance of vertebrate competitors (Packard, 1972). In the Ordovician and Siluarian eras, before vertebrates invaded the marine environment from fresh water, the cephalopods already evolved into sophisticated free swimming animals of considerable size and diversity. The vertebrates therefore not only became competitors but also prey. Probably, until this time cephalopods were mainly predatory upon arthropods, other molluscs and undoubtedly on other cephalopods. The cephalopods were the dominant life form in the oceans and their evolution was greatly modified by the arrival of the vertebrates and in turn, modified the evolution of the fishes.

2.2. Evolution of form

It is generally recognized that the oldest cephalopod fossils are from the Cambrian era where small chambered shells have been found (Fig. 1). There are problematically cephalopods and the single genus *Plectonoceras* is usually given the distinction of being the cephalopod common ancesterial type. The shell is a simple, slightly curved cone which is divided into several chambers which increase in size from the apex to the aperture. These chambers are connected to each other by a tube, the siphuncle, located at the posterior edge of the shell. The chambers were most likely filled with gas which gave the shell upward buoyancy and allowed the animals greater mobility as it crept along the bottom. The foot near the head is drawn as divided into tentacles or arms and the soft parts were attached to the shell by a muscular band. Retractor muscles could pull the animal into its shell. The mantle cavity contained the gills and the intestine probably emptied into this same space.

2.3. Evolution of function

Hypothetically, this simple bottom dweller evolved into a more active bottom feeder similar to *Sciponoceras* in which the shell had become longer with more

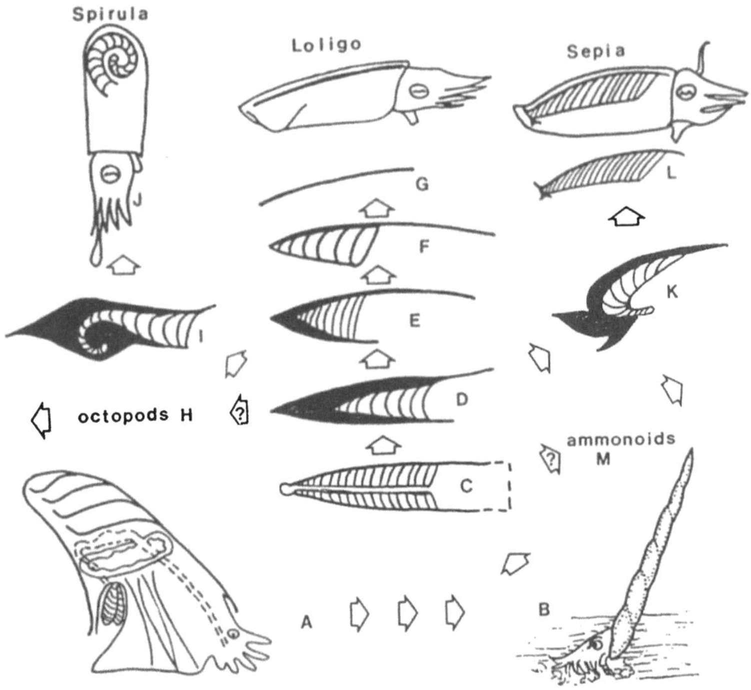

Figure 1. Diagrammatic scheme for the evolution of the Cephalopods. *Plectronoceras*-like animal (*A*) arose from a monoplacophoran and possibly evolved into a bottom feeder like *Sciponoceras*. Chambers in both of these shells were gas filled and connected to the living chamber by a sipuncle. From the *Scopono-ceras*-like creatures evolved the Othroceroids (*C*) which swam Othroceroids (*C*) which swam horizontally as an active predator. The Orthroceroids may have given rise to the nautilid-oncocerid line and the ammonoids (*m*) but did give rise to Belemnites (*d*). The Octopods probably arose from the Belemnites. *Spirula* most likely arose through *Spirulirostra* (*I* and *J*). *Belosepia* (*K*) gave rise to the modern *Sepia*. By reduction of the calcareous shell *Belemnoteuthis* (*E*) gave rise to *Conoteuthis* (*F*) and eventually to the unchambered pen of the oeposids and the myopsids such as Loligo (*G*). There are probably as many evolutionary schemes for the cephalopods as there are people working on cephalopod evolution. This scheme is adapted from Shrock and Twenhofel (1953), Teichert (1967, 1988), and my own bias.

chambers. The arms were better developed for prey capture and the mantle was capable of stirring up the sediments for feeding purposes. From this, it seems simple to leave the constraints of the bottom and assume a horizontal orientation in the ocean rather than have a vertical or oblique orientation. This re-orientation would provide greater motility and increase swimming speed which would have increased the efficiency of predation and escape. The price for this increased mobility involved the development of buoyancy control and maintenance of poise in the water. To offset the tendency to float apex upward, weight had to be added to the apical end of the shell either by additional cameral deposits or by thickening of the shell. These cameral deposits were quite elaborate and involved secondary deposits in the chambers (e.g., in *Orthoceras*), in filling the tip of the siphuncle with dense shell known as encocones (e.g., in *Endoceras*), raylike deposits (e.g., in *Actinoceras*), filling the lower parts of the chambers with dense materials (e.g., in *Protocycloceras*), or by simply breaking off the terminal chamber as did *Ascoceras*.

The evolution of the internal shell (endocochilid) was a major development which led directly to the evolution of the fast swimming squid-like animals. By internalizing the skeleton, additional shell could be added to the outside posterior on the shell, or more importantly, the shell could be reduced to a thinner, lighter structure which eventually was not calcified at all. This was *the* major step in coleioid evolution. The skeleton of the teuthids became reduced to a thin chitinous pen which was even eventually lost or in some octopods is represented only by stylets and skull. The Sepiidia retained the shell as a light buoyancy mechanism (Denton and Gilpeth-Brown, 1961). *Spirula* retained a spiral chambered shell, and internalized it, but it still floats apex upward in the mid-waters. With the lightening and reduction of the shell ever more rapid locomotion was possible and some squids are able to leave the water and *fly* for considerable distances (Cole and Gilbert, 1970).

According to Cannon's *fight or flight* hypothesis (Cannon, 1929), animals can generally be divided into two basic groups: those that fight and those that flee. *Loligo* belong to the group that can flee. The nervous and sensory systems have adapted to this condition. The eyes are very big in relation to the size of the animal. The axons which innervate the mantle make it possible for the squid to swim by jet-action (Adelman and Gilbert, 1990). The larger the axon diameter, the faster the conduction rate of the nerve. In order for jet action to occur, the impulses from the nerves must arrive to the mantle muscles at the same time. The squid accomplishes this function, by having the longest fibers possessing the larger diameters (Purchon, 1977). Of course, increasing the size of a cell increases the problem of getting necessary food and other ingredients to the cell's interior and removing waste and other ingredients from the cell's interior. Interestingly, vertebrates have accomplished this problem of intracellular transport by having a myelin insulation wrapped around the nerve with breaks in the insulation called *nodes of Ranvier.* Nerve conduction can proceed from node to node in myelinated nerve fibers, which is much faster than in unmyelinated fibers. Thus, vertebrates have small fibers.

3. Conclusion

The major trends of cephalopod evolution have been simplified. Still the basic evolutionary pressures have led to the development of rapid locomotion for predation and escape which leads to the need for efficient jet propulsion. The coordination of this synchronous mantle contraction placed demands upon the nervous system which *must have resulted in the evolution of the giant axon* so favored by the neurophysiologist. It is this giant axon that has provided most of our knowledge of nervous transmission and membrane biophysics.

References

Adelman, W. J., Jr. and Gilbert, D. L., 1990, Electrophysiology and biophysics of the squid axon, *this volume.*

Aristotle, ca. 330 B.C., *The Works of Aristotle,* (J. A. Smith and W. D. Ross, eds.), *Volume 4. Historia Animalium,* (D. W. Thompson, translator and editor), 1910), Oxford at the Clarendon Press, Oxford.

Baum, L. F., 1900, *The Wonderful Wizard of Oz*, from: Avon Camelot Books, 1967, *The Wizard of Oz*, Avon Books, New York.
Bezanilla, F. and Vandenberg, C., 1990, The cut-open axon technique, *this volume*.
Boycott, B. B., 1961, The functional organization of the brain of the cuttlefish, *Sepia officinalis*, *Proc. Roy. Soc. B* **153**:503–534.
Boycott, B. B. and Young, J. Z., 1957, Effects of interference with the vertical lobe on visual discriminating in *Octopus vulgaris* Lamarck, *Proc. Roy. Soc. B* **146**:439–459.
Boyle, P. R., 1986, Neural control of cephalopod behavior, in: *The Mollusca. Vol. 9. Neurobiology and Behavior, Part 2*, (A. O. D. Willows, ed.), pp. 1–99, Academic Press, New York.
Brown, A. and Lasek, R. J., 1990, The cytoskeleton of the squid giant axon, *this volume*.
Budelmann, B. U., 1990, The statocysts of squid, *this volume*.
Cannon, W. B., 1929, *Bodily Changes in Pain, Hunger, Fear and Rage*, 2nd Ed., Appleton-Century, New York (reprinted in 1963 by Harper & Row, New York).
Cohen, L. B., Landowne, D., and Salzberg, B. M., 1990, Optical measurements on squid axons, *this volume*.
Cohen, R. S., Pant, H., and Gainer, H., 1990, Squid optic lobe synaptosomes: Structure and function of isolated synapses, *this volume*.
Cole, K. S. and Gilbert, D. L., 1970, Jet propulsion of squid, *Biol. Bull.* **138**:245–246.
Denton, E. J. and Gilpin-Brown, 1961, The buoyancy of the cuttlefish, *J. Marine Biol. Assoc. U. K.* **46**:723–759.
Gould, R. M. and Alberghina, M., 1990, Lipid metabolism in the squid nervous system, *this volume*.
Hearne, V., 1987, *Adam's Task. Calling Animals by Name*, Alfred A. Knopf, New York.
Hoskin, F. C. G., 1990, An organophosphorus detoxifying enzyme unique to squid, *this volume*.
Lehmann, U., 1981, *The Ammonites: Their Life and their World*, Cambridge Univ. Press, Cambridge, U.K.
Meinertzhagen, I. A., 1990, Development of the squid's visual system, *this volume*.
Messenger, J. B., 1973, *Learning performance and brain structure: a study in development*, *Brain Res.* **58**:519–523.
Mullins, L. J. and Brinley, F. J., Jr., 1990, Internal dialysis in the squid axon, *this volume*.
Owen, R., 1932, *Memoir on the pearly Nautilus (Nautilus pompilius)*, Linn. Roy. College of Surgeons, London.
Packard, A., 1972, Cephalopods and fish: the limits of convergence, *Biol. Rev.* **47**:241–307.
Purchon, R. D., 1977, *The Biology of the Mollusca. Second Ed.*, Pergamon Press, New York.
Rice, R. V., Mueller, R. A., and Adelman, W. J., Jr., 1990, Tissue culture of squid neurons, glia, and muscle cells, *this volume*.
Saibil, H., 1990, Structure and function of the squid eye, *this volume*.
Shrock, R. R. and Twenhafel, W. H., 1953, *Principles of Invertebrate Paleontology*, McGraw Hill, New York.
Stanley, E., 1990, The preparation of the squid giant synapse for electrophysiological investigation, *this volume*.
Teichert, C, 1988, Main features of cephalopod evolution, in: *The Mollusca. Volume 12. Paleontology and Neonatology of Cephalopods*, (M. R. Clarke and E. R. Trueman, eds.), pp. 11–79, Academic Press, New York.
Weiss, G. G., Meyer, M. A., and Langford, G. M., 1990, Studying axoplasmic transport by video microscopy and using the squid giant as a model system, *this volume*.
Wells, M. J., 1962, *Brain and Behavior in Cephalopods*, Stanford Univ. Press, Stanford, California.
Wells, M. J., 1978, *Octopus: Physiology and Behavior of an Advanced Invertebrate*, Chapman and Hall, London.
Young, J. Z., 1936, The giant nerve fibres and epistellar body of cephalopods, *Quart. J. Micro. Sci.* **78**:367–386.
Young, J. Z., 1963, Some essentials of neural memory systems. Paired centres that regulate and address the signals of results of actions, *Nature* **198**:626–630.

Young, J. Z., 1976, Central nervous system of Loligo. II. Subesophageal centres, *Phil. Trans. Roy. Soc. London B* **174**:101–167.
Young, J. Z., 1983, The distributed tactile memory system of Octopus, *Proc. Roy. Soc. London B* **218**:135–176.
Young, J. Z., 1988, Evolution of the cephalopod brain, in: *The Mollusca. Vol. 12. Paleontology and Neonatology of Cephalopods*, (M. R. Clarke and E. R. Trueman, eds.), pp. 215–228, Academic Press, New York.

Chapter 2

Natural History and Collection

WILLIAM C. SUMMERS

1. Introduction

This chapter is intended to deal with the lives of squids in their natural setting – the oceans of the world. Many entertaining stories about squid are given by Lane (1960). Squid have been popularized also by Voss and Sisson (1967) and Cousteau and Diolé (1973). There are difficulties in approaching this subject owing to the vast extent of the world's ocean, the numbers of species, their different life stages and the physical limitations on direct observations. Squids are poorly adapted to the laboratory setting because of their size, fragility and general activity (Hanlon, 1987) and there is no simple way to study their organismic biology. At present, their natural history results from an assembly of fragmentary sources including chance observations in the field and laboratory determinations at the physiological or cellular level (Boyle, 1987; Clarke and Trueman, 1988). In a few fortunate cases, long term maintenance and cultivation has allowed laboratory study, especially for cuttlefish and the smaller squids with large eggs. Direct underwater observations have so far been of limited value.

1.1. The ecological dichotomy

The modern predilection to categorize studies of organisms in their natural habitats under the heading of ecology (McIntosh, 1985) is appropriately applied to squids only if it is appreciated that its fit is approximate (Summers, 1983). The traditional study of ecology of terrestrial plants and non-domesticated birds and mammals benefits from a dynamic interplay between direct field access, manipulation and laboratory measurements under highly simplified conditions. These approaches come from the formative roots of ecology, natural history and laboratory physiology as they existed late in the 19th century. Where terrestrial ecologists have spoken convincingly of populations, communities, ecosystems and tropho-dynamics, the study of squid ecology lacks these elements and thus exemplifies a gap in marine ecology.

Quantitative studies of sea life tend to be of two types; (1) community descriptions of sessile organisms as collected by bottom dredging, and (2) single-species yield estimates gathered from fisheries landings (Longhurst, 1981; McIntosh, 1985). The first

W. C. SUMMERS • Huxley College of Environmental Studies and Shannon Point Marine Center, Western Washington University, Bellingham, Washington 98225.

can be traced to the British naturalist, Edward Forbes, who was given credit for the *azoic theory* based on an extrapolation from his brief work, 150 years ago. The latter may be attributed to the German physiologist, Victor Hensen, who undertook careful plankton studies in order to address the problem of variability in fisheries, 100 years ago. It is ironic that Ernst Haeckel who coined the word *ecology* in 1866 (as a part of physiology), strongly criticized his fellow countryman for a quantification of plankton because Hensen was a physiologist and, hence, couldn't appreciate the problems of distribution through statistics. The gap in the ecological approach to the study of squid is partly due to the fact that squids are unlikely to be major parts of the catch taken by either dredges or plankton nets; thus they have been largely omitted from studies directed toward the early life stages and dynamics of valuable fish species.

To complicate matters a bit more, there are at least two further camps from which to be heard. Oceanography, a contemporary of ecology, has belatedly given an emphasis to ocean biology, and then largely at the lowest trophic level and in functional dimensions (Schlee, 1973). Fisheries science, a creature of the 20th century (Cushing, 1975), has been largely driven by commercial and/or political concerns. However, it has discovered that fishes at the market have come through a dynamic life history which often interacts with squid. Fortunately, these have begun to merge in the last two decades so that squid ecologists are welcome to participate in discussions of pelagic trophic communities, adaptations to oceanographic provinces and the exploitation of shallow benthic areas. There are even significant squid fisheries, if that justification is needed. Of course before fish, cephalopods were the top trophic level in the ocean. A cautious heading for the biological overview might be the natural history of squid. The title simply means an inquiry into the subject to create some order and it draws authority from Aristotle (ca. 330 B.C., Book IV) and Buffon (1788–1804).

There is the problem of subject. In many cases, large predatory animals appear to dominate the immediate food resource; they do this by schooling, movement, threat and coordination of activity. This favors a uniformly large size and may hold a cohort together, or favor a continuous reaggregation in a small population. The result is that particular functional roles, especially feeding, are often filled by a single species within a particular area at any one time (Packard, 1972). Reliable sampling might give the impression that only one species exists at that location and time; hence, a single-species, or autecological, view is justified. Changes over time and as a result of dynamic interactions present a more complex view of the community, with the result that the ecological description typically loses more value than it gains. A start on sorting out the added evidence is to view short-term interactions and particular functional levels alone. There is little justification for a theoretical or experimental squid ecology at present.

1.2. Terminology

We need to define our terms. The word squid is a good English term which refers collectively to any naked, swimming cephalopods, particularly if they have eight arms and two tentacles, lateral fins, a funnel, and especially if they are elongated. There are internal features and some additional characteristics best observed in living specimens which further define the sense of the word. Fortunately, it is easy to compare the general appearance with that of a standard or reference squid, say one of the many species of *Loligo*, to define the word. Apparently *squid* refers to the act of squirting and may deliberately include *quid* as a pun or rhyme, to describe the material expelled when squirting. It is debatable whether the squid includes the cuttlefish (which have an internal, calcareous cuttlebone) and/or the generally smaller sepiolids – or other cephalopods. It is likely that the term is fluid and general and should be accepted for nothing more.

It should be noted that squid is both singular and plural. The language allows gatherers (hunters and fishermen) the option to refer to any number of the same sort by the singular term – squid. Where one species is concerned and there is a large-scale operation, we have no trouble with the exception – squid is plural. But the option is based on the perception of the reporter and how narrow a distinction that person wishes to make. *Squids* implies a collection of more than one kind, though it should be considered that the apparent plural may refer to species, numbers of one species, or even sizes (if these have different market value). Scientific literature in English will usually reserve the plural term squids for multiple species; both words are used for several examples of the same species.

It is interesting to compare the Germanic and Celtic root of quid (*that which is chewed*) and another derivative, cud. The similarity with cuttle (=cuttlefish) is unavoidable, but may be trivial if the latter takes its name from cod, the Old English word for bag or husk used here in reference to the body shape or the ink sac.

European languages use many different words for squid, often in a more restricted sense. Even in the same language, local dialects contain words for foods which are often different depending upon the local availability and preparation of these foods. These words are listed in Table I by more formal taxonomic grouping, in no particular order. There are many roots from the Romance languages in this table. Countries bordering the Mediterrean Sea use these marine organisms for food. The word for squid is *calamar* in Spanish and is *calmar* in French, both of which are Romance languages. These words are derived from the Latin *calamus* and the Greek *kalamos* meaning a reed. Reeds were commonly used as writing pens, and this root obviously refers to the pen of the squid. This listing is far from complete and may include errors of transliteration or transcription of which I am unaware. Roper *et al.* (1984) give many of these words. I have not extended this listing beyond the European literature because of my own limitations. It is my understanding that the Japanese refer to all of these as *ika* and separate them with modifying words when being more specific. The Chinese differentiate between the cuttlefish and typical squids, which are called yú yü.

We take the name for the *true* squids, teuthids, and the scientific study of squid (often extended to all cephalopods), teuthology, from a transliteration of the classical Greek word, *teuthides* (singular: *teuthis*). In his categorization of the animal world, Aristotle used this word to describe the common and smaller squid, most likely *Loligo vulgaris*. He referred to a larger squid, *teuthos*, possibly *Todarodes sagittatus*; as well as the octopus, *polypous* and cuttlefish, *sepia* (Aristotle, ca. 330 B.C.). He cites the above as examples of the *softies*, or mollusks, and refers to them collectively as the tribe of cuttlefish.

1.2.1. Terminology of the squid, *Loligo pealei*

Loligo pealei was named by Lesueur in 1821 most likely in honor of Rubens Peale, manager of the Philadelphia Museum in Philadelphia, Pennsylvania. Two of his other brothers were the professional artists, Rembrandt and Raphaelle. His half brother was Titian Ramsay Peale II (Titian Ramsay Peale I was a half brother, who died before

TABLE I. Common Names for Squid and Cuttlefish in European Languages[1]

Squid	*Cuttlefish*	*Sepiolid Squid*
teuthids	sepiids	sepiolids
	seiche[F], sècue	
calamar[S], calmar calamare, calamaro, calamary, kalmar		
	sepia[E], soupia, seppia	sépiole, sepiola, seppiolo, sepieta, sepiolid
cornet[F], encornet		
jibia[S]	jibia	
lula[P], lübje, lignja, lurión, luria, enoploluria, gonalura, loutène		
	choco[P], choquito, chopito	chopito, choco, globito
blaek-, bläck-, blekk-, black-[G]		
inkt-, tint-, timten-[g]		
squid[E] -sprutte, -sprut[D]		
kraken[N] casseron[F] chipirones[S] toutenon[F] pota[S]		

[1] See also Roper, Sweeney and Nauen, 1984.
Examples are:
- [D] These suffixes mean sputter or squirt; in Danish, cuttlefish is Blaeksprutte or ink squirter.
- [E] English
- [F] French
- [G] These prefixes mean ink; in German, cuttlefish is Blackfisch or black fish.
- [g] These prefixes also mean ink; in German, squid is Tintenfisch or ink fish.
- [N] Norwegian
- [P] Portuguese
- [S] Spanish

Titian II was born) and their common father was the famous artist, Charles Willson Peale (Summers, 1974; Aldrich, 1990). Titian II, a friend of Lesueur, was a professional naturalist, who traveled widely, and was on the Wilkes expedition to the South Pacific (Sellers, 1980).

A single *i* is preferred as the suffix in the formation of a scientific name under the *Rules of Zoological Nomenclature*; hence the name *Loligo pealei* for this species (Summers, 1974). Previously, an alternate spelling for this species was *Loligo pealii*, attributed by Verrill (1882) to Blainville in 1823. It is interesting to note that Redfield, one of the early experimenters on squid, used the spelling of *Loligo pealei* (Redfield and Goodkin, 1929).

Loligo pealei is sometimes called the *bone squid* by New England fishermen to distinguish it from *Illex illecebrosus*, the *summer squid.* The remains of small *Loligo pealei* caught in the nets of these fisherman are known as *eyes*.

Loligo pealei is termed *yari-ika* by Japanese fisherman; this term means *spear squid*, possibly due to the spear shaped fins.

1.3. A functional taxonomy

There is little debate over the systematic consistency of the cephalopods, or their close affinities with other mollusks. However, there is a lack of consensus on the significance of the next higher subdivisions and the importance of particular characteristics given a fossil record especially favoring shelled forms. There are, roughly, 500 recognized species of living cephalopods; that number has been estimated to double if current trends in revision and new discovery run their course (Roper, 1985). Borrowing from a standard classification (Voss, 1977), one observes that the squid and cuttlefish occupy two-thirds of the genera, easily twice the numbers of the Octopoda. (This does not suggest that the numbers of individuals follows speciation.) Disagreement centers mainly on the relative commonality between the sepiolid squid, the cuttlefish and the typical squids. It also centers on the extent to which the neritic squids (those living on the continental shelf) should be distanced from the generally pelagic (those living in the open ocean) ones.

Clarke (1988) has reviewed the erratic associations of sepiolid squids, first with the neritic squids (Myopsida) after 1845 and then with cuttlefish in 1921. He proposes three orders: typical squids (Teuthoidea), sepiolids (Sepioloidea), and cuttlefish (Sepioidea), the last including the planktonic *Spirula*. Separation of the sepiolids at the level of order was also proposed by Fioroni (1981) based on a monographic comparison (Sepiolioidea). Those of us who have experience with live sepiolids and Myopsids are struck by functional commonalities and some clear distinctions. The same can be said in pairwise contrasts with live cuttlefish, noting only that the latter possesses a buoyant calcareous shell. Convergence and conservation of features draw the systematics of the squids into a narrow range of options served by diverse solutions. The 25 or more systematic families attest to that.

From a functional point of view, I prefer to view the systematics on the basis of distribution in the sea. The purely pelagic or open deepwater species are a diverse group of Oegopsid squids ranging from strong swimmers to those apparently more sedentary, partially buoyant forms (Arnold, 1979). Moving onto the continental shelf, one observes increasing association with the bottom as depth decreases and a change from Oegopsid to Myopsid squid, which have a closed eye structure, and then to sepiolids and cuttlefish; the last does not occur in the Western Hemisphere. This gradient is accompanied by an increase in egg size and tendency toward direct development, or a tendency to reduce planktonic stages (Mangold, 1987; Sweeney *et al.*, 1990). The higher systematic diversity occurs on the continental shelf; the species diversity occurs largely through the depths and to a lesser extent, expanse of the open ocean. If Clarke's revised classification is used and one arbitrarily separates squid between the open ocean and continental shelf by dividing the Oegopsids from all others, respectively, the former are composed of 26 families and 69 genera and the latter, 3 orders, 7 families and 26 genera. Oceanic habitats are not equitably distributed, of course, and the squids which populate them need only to be adapted well enough to survive. In this light, the taxonomic problem becomes one of identifying unique ecological spaces which balance efficiency or specialization and adaptability; the open ocean appears to favor the former and the continental shelf the latter insofar as squids are concerned. At this point, I fall on the side of the taxonomic lumpers.

1.4. Biological strategies

There are many problems in suggesting a set of biological strategies which typify modern cephalopods or their progression from a simpler, ancestral form. Not least of which is the fact that the living forms have a surprising diversity of patterns when compared with superficially similar morphologies. Adaptability and behavioral characteristics carry cephalopods through changes in their immediate environment, although this may not be apparent from their forms.

The recent literature has compared the role of squids with that of fishes (Packard, 1972; O'Dor and Webber, 1986), noting that the fishes are relative newcomers in marine ecology. Shell-less squids may also be of relatively recent origin; marine ecosystems have changed through time, in part because of the arrival of both groups. A single date for the origin of squids is not adequate in itself to explain the present role of squid. Neither is an attempt to synthesize ancient life strategies based on shell features (see Ward and Bandel, 1987).

Looking more directly at the ecological role of marine predators (all squids are carnivorous, but not all fishes), we can observe constraints which have shaped all of them. First, marine food chains are generally longer and more branched than those described on land. There have to be higher trophic efficiencies at some of the initial levels to support so many levels. Furthermore, the branching food web is the result of adaptability in the form of opportunism and, importantly, there are changes in upper trophic levels with growth. Fish may pass through two or three trophic levels (often beginning with eggs about 1 mm in diameter)(Cushing,1975; Pitcher and Hart, 1982). Squids likely do not exceed two steps, but may hatch more fully developed from eggs as much as 1 cm in diameter (Mangold, 1987). The smallest hatchlings of both groups are planktonic and feed extensively on crustaceans. The largest squids are raptorial and feed on smaller fishes. Large fish feed on all but the largest sizes of squid and on other fishes. Diving birds take both squid and fish at smaller sizes, and marine mammals feed on all sizes of squid and fishes. By dominating the resources, fishes also have become the target for predation.

Squids have certain distinctive advantages (Boyle, 1986). They have excellent and very specialized eyes, an ability to change colors and patterns quickly, and a defensive inking capacity. They swim easily in both directions and change direction or turn rapidly. Notable in comparison with fishes is the fact that squids contact their prey at a distance – usually first by the pair of tentacles, then by the eight arms.

Feeding is extended away from the body and not limited to the gape of the mouth. Here many smooth or serrated suckers, or hooks (for more fleshy prey), and several arms provide a nearly inescapable food gathering mechanism. Several small prey items can be held at the same time and these can be subdued by bites from the beak and salivary poisons. Food processing may also be externalized by sorting and elimination of some portions without ingestion.

Apparently, squids are limited to a protein metabolism (O'Dor and Webber, 1986; O'Dor and Wells, 1987) and they do not store lipids except in the family Gonatidae where small amounts may be used for buoyancy (Clarke, 1988). Their range of metabolic activities may also be narrow, and never as efficient for locomotion as that of equivalent sized fishes.

The higher trophic levels in the ocean are supported by a strategy which favors rapid growth in size, with the commitment to search ever larger areas for proportionately larger prey. It should be noted that swimming speeds as a rule are directly related to size. Sometime during the course of evolution, squids came in second in this competition or they opted to compete in another contest. There are fish predators for every squid, or at least a marine mammal if vertebrates are allowed to substitute for one another. Squids have known physiological inefficiencies relative to fishes (O'Dor and Webber, 1986) and they are short-lived, capping eventual sizes even if they do grow rapidly. Their ecological role appears to be in penultimate productivity which emphasizes reproduction, the short game and adaptability in uneven environments. They do not risk the end game which involves preparation to endure production cycles up to a decade long. Because fish and squid are commonly associated in particular communities, we can assume that the ecological relationship is stable in a dynamic sense: the intervention of human fishing pressure often favors the proportion of squid in a reduced, total system. Given that there are about 100 times as many species of fish as squid, the prize for persistence belongs to the squid by way of reproduction and adaptability. Squids must make good use of the nearby environment, be conservative in long searches for food, and avoid missed opportunities within striking distance. This suggests ambush hunting whenever possible.

The ecological model for squid is a subordinate one; where the dominant group, fish, effectively owns the available resources, we might say that squid either have to deal on a black market for resources or operate as a distinct minority in the open market; they probably do both. They use their special skills, but do not succeed in usurping resource dominance except briefly in the most dynamic ecosystems.

When ancestral squids gave up the chambered shell, they not only lost the passive armor that provided, but also took on the problem of counteracting negative buoyancy. It was a case of sink and swim. None of the modern squids have replaced the armor, but ten oceanic families have neutralized their buoyancy by storing ammonium rich fluids in the coelom or in other, specific locations in the body (Clarke, 1988). Roughly speaking, equal volumes of ammoniacal fluid and dense muscle are required for neutral buoyancy, so many of these squids have reduced musculature as well (see Voss, 1988). Not surprisingly, they live deep in the open ocean, in a dark world where they are less vulnerable to all but whale predation (Clarke, 1987). Many squid either counteract the lack of buoyancy by swimming steadily as do the remaining oceanic squids, or by swimming part of the day and resting on, or gliding close to, the bottom at other times as do the loliginids, or by swimming only occasionally and otherwise resting on or burrowing into the bottom as do the sepiolids. Cuttlefish follow this last pattern even if they are neutrally buoyant. The jet propulsion of squids has been documented rather well (Trueman, 1983); but gets poor ratings on efficiency as size increases (O'Dor and

Wells, 1987). The analyses have largely avoided the contribution of the fins, which are used in the normal swimming of all squids and which take on a variety of sizes and shapes. The energetics of jet swimming suggest that it was, or is, a mistake for squids to grow much beyond the size of macroplankton and to live in temperate or colder oceans. Trophic opportunities favor growth, and the uneven nature of oceanic productivity favors parts of the high latitude oceans (Longhurst, 1981).

Fisheries biologists regularly use a migrational scheme to describe long-lived fish (Harden Jones, quoted by Cushing, 1975). It assumes that there are more or less identifiable areas for the spawning, nursery and feeding of any well adapted and resource dominating stock. Most fish drift planktonically as larvae and/or eggs from the spawning area to a nursery site where initial growth occurs. At some point, small fish *recruit* to the adult feeding population and begin repeated seasonal migrations to the spawning area and back again to the feeding site.

We note some interesting variations when applying this scheme to squids. First, squid would make the circuit only once if they don't have to breed in a particular locality (Mangold, 1987). Next, predation on squid eggs is not known to occur to any significant degree. The oegopsids are the only squids which have planktonic eggs. Loliginids and sepiolids lay in relatively shallow, benthic sites. The oegopsids and, for a limited time loliginids have planktonic young; sepiolids hatch as miniature adults. Finally, recruitment is not a cephalopod issue, because the adults die around the time of spawning.

Thus, the life cycle loop for an oegopsid is planktonic from the spawned egg mass through a small planktonic stage. The nursery function of protection and suitable food resources is provided by access to rich, warm surface waters with ontogenetic descent over time. The adult feeding occurs as an even aged group, perhaps at the edge of a current system. At the appropriate season, they probably migrate upstream to the spawning area and die. There is much risk and a high population variability in this scheme which is offset by the potential for large numbers of small eggs. Loliginids are more linear and confined to coastal currents. Their eggs are moderate in number, and the larger planktonic young drop out after a few weeks or months, returning to the spawning sites a year later. Planktonic stages may be regulated by the same processes that cause variations in sessile coastal populations, including relocations of current systems immediately offshore (Roughgarden, *et al.*, 1988). There may be an offshore and inshore component in the adult movement, especially just before spawning. Sepiolids work around a point, by comparison. The eggs are few and large, the hatchlings not planktonic, all life stages eat the same food (Bergström and Summers, 1983; Summers, 1985) and dispersal may be more important than migratory movements.

The above highly simplified schema, runs from the largest to the smallest adult sizes and the reverse in egg or hatchling sizes. It goes from squids which are planktonic and pelagic all of their lives to those which swim only part of the time and electively. It is not known how they aggregate or what advantages that confers. They occur in all of the productive oceans.

1.5. Distribution and fisheries

Loligo pealei, the so-called *common squid*, is abundant along the eastern coast of North America from New Brunswick, Canada to the Caribbean (Summers, 1974). The Marine Biological Laboratory at Woods Hole, Massachusetts supplies this species in large numbers for biological research. These squid are not generally found in the ocean below 8 °C (Summers, 1969); in fact, their behavior is aberrant below 8 °C (Summers and McMahon, 1974).

Dosidicus gigas, the large squid found along the west coast of the Americas, has also been used for biological research, especially in Chile. However, for the past couple of decades, they have disappeared along the coast of Chile.

Squid fisheries differ in only a few points from modern technological fisheries directed toward various fin fishes (Brandt, 1984; Rathjen, 1984). Industrial requirements are best served by adapting existing gear to many uses, maintaining the ability to quickly relocate and change the target species when market opportunities exist. Squid have become one of many varieties sought by the larger vessels and fishing fleets.

In recent years, the recorded world catch of marine organisms of all sorts has been about 70 million metric tons per year, at least one million of which is cephalopods (including Octopus) and, in recent years, most of that catch is attributed to Japan (Worms, 1983; Rathjen and Voss, 1987). The same sources indicate that cuttlefish and Octopus each comprise about one-sixth of the cephalopod total. The oceanic squids of the family Ommastrephidae make up one-third to a half of the cephalopod total; the neritic squids of the family Loliginidae, though highly valued, often run slightly behind the catches of Octopus and cuttlefish.

It should be cautioned that major new fisheries for squid have recently begun in the Southern Hemisphere (Csirke, 1987) and that the reliability of catch data varies considerably, so that these figures are only indicators of the fishery. In particular, squid generally require more careful handling and prompt preservation when compared with fishes. Therefore, landings in less developed areas which are consumed immediately, and those which are used for bait, may not appear in the statistics.

Operationally, those squid species targeted tend to be found close to shore or near the surface, they are typically ones which aggregate in migration or when spawning. Squids which live in the deep ocean are unexploited and little known. Benthic species distributed on the continental shelf are often avoided, except as a mixed catch or unless they present particular behavioral vulnerabilities. Artisanal fisheries are less restricted, but not necessarily more reliable.

Squid are difficult to locate visually because they generally avoid brightly illuminated areas where they presumably would be vulnerable to predators and also because of their cryptic coloration. They may be identified through the activities of birds and fishes. Squid produce relatively weak sonar signals because they lack sonic contrast. This may also be an adaptive advantage where they are hunted by toothed whales. The most modern fish finding sonar devices detect squid through high output power and multiple (especially higher) frequency transmissions. Fisheries utilizing search equipment will deploy bottom trawls in the daytime (if the depths and bottom surface permit it) or a surround net floated around a school at night - a purse seine or lampara net. The latter requires moderate sea conditions and ample space to avoid entanglement. Midwater trawling is possible, but is usually restricted to sophisticated, larger vessels or experienced paired vessels working closely together.

When a regular migration route is known, set nets may be deployed to intercept them. If they are following coastal contours, this may take the form of a fishtrap set perpendicular to the shore. This was the main source of squid in the Eastern United States previous to banning of traps for management reasons. In this decade, high sea gillnets of immense size have been deployed for squid, particularly in the North Pacific (Fiscus and Mercer, 1982; Sloan, 1984). These are indiscriminant and highly controversial; they are mostly used by Asian nations, partly in response to the 200 mile fisheries conservation zones established by the United Nations.

Fisheries for loliginid squids are normally directed at a few, highly localized spawning sites on known dates. Squid are particularly reckless at these times and often aggregate in dense concentrations – they can literally be scooped up by any net dragged through the water. At other times, loliginids may be trawled as a part of the productive undersea community.

Feeding squids are taken by jigging with a peculiar lure which they mistake for prey. The lure is usually cylindrical or tapered toward both ends and have one or two circles of barbless hooks surrounding the lower end. When used singly, the jigger is weighted and jerked upwards frequently. The jiggers may also be connected at intervals by monofilament line and fished from an eccentric reel cranked by hand or driven mechanically in combination with electronic sensing devices. There are bizarre combinations of colors, lights, textures, articulations and underwater whistles (sic!) built into these lures, and they are typically mixed along an individual string and fished at night with bright lights. Jiggers are most effective with the oegopsid squids and may be highly erratic in their results. They also appear to favor capture of sexually immature animals. It should be noted that the traditional curved fish hook is useless for squid because they do not swallow the bait directly. Likewise, squid are usually too small and/or too soft bodied to warrant attempts to capture them with harpoons or other impaling devices.

Cuttlefish and sepiolid squids are usually taken in bottom trawls along with various fishes and associated bottom fauna. I am not aware of any successful traps for catching them selectively excepting brush materials set out to attract spawning cuttlefish. Cuttlefish also respond to an ploy known since the time of Aristotle – males will seize a female (or good facsimile) weighted and dragged along the bottom.

The use of gillnets for the capture of squid is a puzzling matter. In the first place, squid usually have no hard or rough surfaces and are generally flexible in the dimensions which might fetch up when attempting to pass through net meshes. Secondly, they are adept at reaching most or all of their surfaces with the arms. Finally, being visually acute, it seems unlikely that they would approach a panel of netting without being aware of it unless swimming blindly or in panic. Nevertheless, oceanic squids do become tangled in drifting gillnets and cuttlefish in sinking gillnets.

1.6. Selection and handling

The collection of squid for laboratory or experimental purposes has a very different focus than the commercial fishery, though it probably shares the same technology. Assuming that intact, if not live, specimens are desired, the scale of operations and sequence of events may have to be carefully planned. Unless experiments are to be conducted at the collecting site or aboard the vessel, one is usually constrained to work as close to the laboratory as possible to minimize the time interval between collection and study aquaria. It is not practical to anesthetize squid on capture. Chilling the seawater used to transfer them may be desirable to reduce their activity but temperature shocks also need to be avoided.

Perhaps the least traumatic capture is in fishtraps if any exist or are permitted. Squid therein are often healthy and undamaged if not too crowded or in the presence of predators, and if they are removed promptly. Traps are unpredictable and one never knows when squid will venture into them. Fishtraps are subject to significant permit restrictions and are site specific in design. They do not scale down in size. Thus, arranging to collect from an existing fishtrap, or a restored one, is highly desirable. Lacking a precedent – including catch records of squid – makes them impractical for immediate laboratory purposes.

Perhaps the best way to remove squid from a trap is to raise the bottom of the impoundment slowly in a section where the squid are isolated in order to dipnet them into a transfer tank. The dipnets should be smooth (knotless netting), blunt and not so long that squid easily get caught behind the rim when the net is turned over. The ring diameter should exceed the long dimension of the squid, if possible, and the meshes should be fine enough to prevent entanglement by the arms. One has to work quickly and one at a time in this case.

Squid jigging is an attractive alternative for collection. It is relatively new in the Western Hemisphere (excepting Newfoundland) and has the potential to allow very selective fishing, albeit mainly at night, probably from vessels equipped with sonar, and at the risk of experimenting with lots of gear. The extent of damage caused by the hooks is not documented; in the short-term, it appears to be minimal, but this is a consistent injury and may require sorting or treatment before experimental uses. Then too, sexually mature squid may not be vulnerable to this gear and the different species respond individually. There is good opportunity to try out squid jigging through short-term vessel charters, or simply by jigging by hand or with a mechanical jigger as possible. This should be directed toward oceanic species and those which migrate inshore when they are not yet in spawning condition (i.e., early in the inshore season).

Jigging is closely associated with the use of lights and raises the question of direct light attraction in combination with other gear. In fact, torches and other lights have long been used with the surround net and brailing fisheries in California and the lampara net takes its name from the Italian word for the light which was used. As a practical matter, the fishermen use lights at night to handle their gear and/or see squid to catch. We know that many invertebrates come to lights, but do squid come in the same sense? One view is that squid are attracted to the feed provided by crustaceans, but do not respond mechanically themselves. Another is that squid as highly visual animals are put at a disadvantage in the region of sharply contrasting illumination around a bright light at night. (These two views are not mutually exclusive.) As noted by Munz (1977), the cephalopod eye quickly accommodates to increased illumination by closure of the iris; decreased illumination is only accommodated over a period of minutes and that can be countered by fright. Squid may venture into unnaturally illuminated areas only to find themselves unable to see well enough to leave. They seem to strike at jiggers from adjacent shadowed zones.

A small surround net, such as a lampara net, may be useful for laboratory collection with or without lights. Instruction on applicability should be sought in the regional fishery along with gear and experienced personnel. Because this is a mobile variation of the fishtrap, an arrangement for access to squid at the earliest stage of a commercial haul is highly desirable, though many fishermen will be reluctant to slow operations once underway. This fishery is dependent upon locating the squid and making an accurate set on an aggregation. For these reasons, small nets and random sites will be unproductive. The lampara net fishery in Monterey Bay, California, is an example of the successful use of modest gear and unspecialized small boats in a very restricted area of known spawning aggregations.

The major means for collecting squid for all purposes is through the use of bottom trawls. Where a suitable vessel is available and the bottom allows trawling, this is practical and can be conducted even with a relatively small trawl (so long as it has an adequate vertical opening). It is also very damaging to the squid. Trawls serve laboratory needs only through redundancy and there are limitations to improvements which promote live capture. There is no reason to look for living specimens in commercial operations because the quantities sought kill them all. Trawls are effective in the daytime and cover lots of ground; they appear to catch representative squid of those species which move to the bottom daily. Location of the squid is important and more difficult because they are probably dispersed on the bottom; skillful gear handling is necessary.

To improve live capture in trawls, a number of modifications can be employed. Foremost among these is establishment of an operational discipline to make short tows and even to reduce these if the sonar shows good signals. If only a few squid are caught, chances are good that they will be in a lively condition when they are promptly brought to the surface. The cod end of the net (where the catch collects) can be improved several ways. First, a means to haul it aboard first by a (floating) line attached to one of the trawl doors or wing ends speeds the repeated short haul problem introduced above. This is sometimes called a splitting strap. Next, lining the cod end with knotless mesh netting reduces abrasion. A fine mesh is desirable, but this may be regulated by licensing agencies. A hoop between the outer cod end and liner helps to prevent crushing and a quick release closure hastens handling. The collecting trawl can be made from much lighter twine than a commercial trawl which has several operational advantages, but does make it fragile. In addition, the trawl should be rigged to *tend bottom* (i.e., scrape the sediment surface) only as heavily as needed. For loliginids this can be very light because they readily leave bottom when disturbed. Sepiolids and cuttlefish require a firmer touch because they burrow in the bottom. The net may be semipelagic for some species, that is, not in contact with the bottom beyond the trawl doors. These several suggestions directed toward lightness of gear and delicacy on the bottom have manifold advantages. They allow a more species directed catch and less rough surfaced bycatch, a larger net, and ease of handling. Cuttlefish and sepiolid squid are taken quite adequately in more conventional nets such as shrimp trawls, but also benefit from smaller total catch and the cod end modifications. They may be more available to trawls at night close to shore because of their tendency to burrow in the daylight.

It is useful to empty the cod end directly into a shallow tank on deck frequently filled with fresh seawater. Live squid can be dipnetted from this and placed in small numbers in transfer tanks as desired. These last may be filled with chilled water and/or be equipped with lids or barriers to prevent surging when the vessel rolls. Translucent material which provides visual isolation seems desirable in these tanks as well as smooth surfaces and a lack of sharp corners. See Hanlon (1990) for additional remarks on tankage, transfer and handling.

Open ocean and deepwater squid are not the subjects of intensive fisheries. Sampling done there is not consistent and has failed to give a clear picture of the species present and/or their representative distributions. Comparisons between catches of larger midwater trawl nets (projected mouth areas ranging from 9 m^2 to 700 m^2) and the stomach contents of sperm whales (large, far ranging, specialized squid predators) differ widely in family proportions and sizes collected (Clarke, 1987). What is impressive is the potential numbers and biomasses of squids which can be extrapolated worldwide based on these data – easily more than the present world catch of all fisheries. These whales, of course, are not the only predators of squid, though probably some of the most effective.

1.7. Opportunities

There are several areas in the life history and whole animal biology of squids which need additional work. One hopes the practical questions relating to fisheries management will receive immediate attention (see NAFO, 1985). We need to find better ways to observe squid under normal conditions. Is it practical to put camera equipment in locations where squids live? Ideally, these could be remotely controlled or even part of a submersible program (Vecchione and Gaston, 1985). However, if human participants are involved, studies in large aquaria such as those at Dalhousie University or containment areas may be more practical than underwater observatories. The visual and spatial orientation of squids bears considerable opportunity for work. Is there language associated with the coloration, pattern, texture and posture of squids as proposed by Moynihan (1985)? Or is this cryptic behavior and physiological response? It is particularly appropriate to assess the effectiveness of inking and color changes in avoiding predators.

What are the population dynamics of squids? For that matter, is it possible to identify races or stocks using natural markings, tagging or electrophoresis as it is with certain fishes (Ryman and Utter, 1986; Augustyn and Grant, 1988)? Are squids universally semelparous? It would be desirable to perfect the techniques for aging squid which depend on statoliths (Clarke and Maddock, 1988). There is much need from an ecological viewpoint to determine the predator-prey relationships of squid. As relatively abundant and fast growing animals, they not only have large food requirements themselves, but also provide a resource for larger or more able predators (Summers, 1983). How do they function as links in the marine food chain?

The migratory species must be particularly adapted to the seasons, current systems, and productivity events which occur along their route. These links are poorly described, but very important. What cues promote spawning? Sepiolids (Summers, 1985) may have an elective hatching time which could be of ecological benefit.

References

Aldrich, F. A., 1990, Lol-i-go and far away: A consideration of the establishment of the species designation *Loligo pealei, this volume.*

Aristotle, ca. 330 B.C., *Historia Animalium. Books IV–VI*, (Peck, A. L., ed., 1970), Harvard Univer. Press, Cambridge, MA.

Augustyn, C. J. and Grant, W. S., 1988, Biochemical and morphological systematics of *Loligo vulgaris* and *Loligo vulgaris* reynaudi: D'Orbigny *Nov. Comb.* (Cephalopoda: Myopsida, Malacologia), 29(1):215–233.

Arnold, G. P., 1979, Squid a review of their biology and fisheries, Ministry of Agriculture Fisheries and Food Directorate of Fisheries Research, Lowestoft Laboratory Leaflet No. 48.

Bergström, B. L. and Summers, W. C., 1987, *Sepietta oweniana*, in: *Cephalopod Life Cycles*, vol. I (P. R. Boyle, ed.), pp. 75–91, Academic Press, London.

Boyle, P. R., 1987, *Cephalopod Life Cycles*, vol. II., Academic Press, London.

Boyle, P. R., 1986, Neural control of cephalopod behavior, in: *The Mollusca. Vol. 9. Neurobiology and Behavior, Part 2*, (A. O. D. Willows, ed.), pp. 1–99, Academic Press, New York.

Brandt, A. von, 1984, *Fish catching methods of the world*, 3rd ed., London Fishing News, London.

Buffon, G. L. L., 1788-1804, *Histoire Naturelle des Animaux.*

Clarke, M. R., 1988, Evolution of buoyancy and locomotion in recent cephalopods, in: *The Mollusca. Vol. 12. Paleontology and Neontology of Cephalopods*, (M. R. Clarke and E. R. Trueman, eds.), pp. 203–214, Academic Press, New York.

Clarke, M. R., 1988, Evolution of recent cephalopods – A brief review, in: *The Mollusca. Vol. 12. Paleontology and Neontology of Cephalopods*, (M. R. Clarke and E. R. Trueman, ed.), Academic Press, New York, pp. 331–340.

Clarke, M. R., 1987, Cephalopod biomass-estimation from predation, in: *Cephalopod Life Cycles*, vol. II (P. R. Boyle, ed.), pp. 221–237, Academic Press, London.

Clarke, M. R. and Maddock, L., 1988, Statoliths from living species of cephalopods and evolution, in: *The Mollusca. Vol. 12. Paleontology and Neontology of Cephalopods*, (M. R. Clarke and E. R. Trueman, ed.), pp. 169–184, Academic Press, New York.

Clarke, M. R. and Trueman, E. R., eds., 1988, Paleontology and neontology of cephalopods, in: *The Mollusca. Vol. 12. Paleontology and Neontology of Cephalopods*, (K. M. Wilbur, gen. ed.), Academic Press, New York.

Cousteau, J., and Diolé, P., 1973, *Octopus and Squid. The Soft Intelligence*, (J. F. Bernard, translator from French), Doubleday, Garden City, New York.

Csirke, J., 1987, FAO Fish. Tech. Paper No. 286, pp. 1–75.

Cushing, D. H., 1975, *Marine Ecology and Fisheries*, Cambridge University Press, Cambridge.

Fioroni, P., 1981, The exceptional position of sepiolids, a comparison of the Orders of recent cephalopods (in German), *Zool. Jahrb. Syst.* **108**:178–228.

Fiscus, C. H. and Mercer, R. W., 1982, NOAA Technical Memorandum NMFS F/NWC-28, pp. 1–32.

Hanlon, R. T., 1987, Mariculture, in: *Cephalopod Life Cycles*, vol. II (P. R. Boyle, ed.), pp. 291–305, Academic Press, London.

Hanlon, R. T., 1990, Maintenance, rearing, and culture of teuthoid and sepioid squids, *this volume*.

Lane, F. W., 1960, *Kingdom of the Octopus. The Life History of the Cephalopoda*, Sheridan House, New York.

Longhurst, A. R., 1981, *Analysis of Marine Ecosystems*, Academic Press, New York.

Mangold, K. M., 1987, Reproduction, in: *Cephalopod Life Cycles*, vol. II (P. R. Boyle, ed.), pp. 157–200, Academic Press, London.

McIntosh, R. P., 1985, *The Background of Ecology: Concept and Theory*, Cambridge University Press, Cambridge.

Moynihan, M., 1985, *Communication and Noncommunication by Cephalopods*, Indiana University Press, Bloomington.

Muntz, W. R. A., 1977, Pupillary response of cephalopods, in: *The Biology of Cephalopods*, (M. Nixon and J. B. Messenger, eds.), pp. 277–285, Academic Press, London.

NAFO (Northwest Atlantic Fisheries Organization) Scientific Council Studies No. 9, Sept. 1984, pp. 1–175.

O'Dor, R. K. and D. M. Webber, 1986, The constraints on cephalopods: why squid aren't fish, *Can. J. Zool.* **64**:1591–1605.

O'Dor, R. K. and M. J. Wells, 1987, Energy and nutrient flow, in: *Cephalopod Life Cycles*, vol. II (P. R. Boyle, ed.), pp. 109–133, Academic Press, London.

Packard, A., 1972, Cephalopods and fish: the limits of convergence, *Biol. Rev.* **47**:241–307.

Pitcher, T. J. and Hart, P. J. B., 1982, *Fisheries Ecology*, AVI Publishing, Westport, Connecticutt.

Rathjen, W. F., 1984, *Squid Fishing Techniques*, Gulf and South Atlantic Fisheries Development Foundation, prepared by: Nat. Mar. Fish. Serv., Gloucester, Massachusetts, pp. 1–15.

Rathjen, W. F. and G. L. Voss, 1987, the cephalopod fisheries: a review, in: *Cephalopod Life Cycles*, vol. II (P. R. Boyle, ed.), pp. 253–275, Academic Press, London.

Redfield, A. C. and Goodkind, R., 1929, The significance of the Bohr effect in the respiration and asphyxiation of the squid, *Loligo pealei*, *J. Exp. Biol.* **6**:340–349.

Roper, C. F. E., 1985, Is there a squid in your future: perspectives for new research, *American Malacological Bulletin, special edition* **1**:93–100.

Roper, C. F. E., Sweeney, M. J., and Nauen, C. E., 1984, FAO species catalogue, vol. 3, Cephalopods of the World, FAO Fish. Synop., (125).

Roughgarden, J., Gaines, S., and Possingham, H., 1988, Recruitment dynamics in complex life cycles, *Science* **241**:1460–1466.

Ryman, N. and Utter, F., eds., 1986, *Population Genetics and Fisheries Management*, Univ. of Washington Press, Seattle.

Schlee, S., 1973, *The Edge of an Unfamiliar World*, E. P. Dutton, New York.

Sloan, N. A., 1983, Canadian-Japanese experimental fishery for oceanic squid off British Columbia, Summer 1983, *Canadian Industry Report of Fisheries and Aquatic Sciences* No. **152**.

Sellers, C. C., 1980, *Mr. Peale's Museum. Charles Willson Peale and the First Popular Museum of Natural Science and Art*, W. W. Norton, New York.

Summers, W. C., 1969, Winter population of *Loligo pealei* in the mid-Atlantic bight, *Biol. Bull.* **137**:202–216.

Summers, W. C., 1974, Natural history of the squid, in: *A Guide to Laboratory Use of the Squid Loligo pealei*, (J. M. Arnold, W. C. Summers, D. L. Gilbert, R. S. Manalis, N. W. Daw, and R. J. Lasek, eds.), pp. 5–8, Marine Biological Laboratory, Woods Hole, Massachusetts.

Summers, W. C., 1983, Physiological and trophic ecology of cephalopods, in: *The Mollusca. Vol. 6. Ecology*, (W. D. Russell-Hunter, ed.), pp. 261–279, Academic Press, New York.

Summers, W. C., 1985, Ecological implications of life stage timing determined from the cultivation of *Rossia pacifica* (Mollusca: cephalopoda), *Vie Milieu* **35**:249–254.

Summers, W. C., and McMahon, J. J., 1974, Studies on the maintenance of adult squid (*Loligo pealei*), I. Factorial survey, *Biol. Bull.* **146**:279–290.

Sweeney, M. J., Roper, C. F. E., Mangold, K. M., Clarke, M. R., and Boletzky, S. v., eds., 1990, *"Larval" and juvenile cephalopods; a manual for their identification, Smithsonian Contr. Zool., in press.*

Trueman, E. R., 1983, Locomotion in molluscs, in: *The Mollusca. Vol. 4. Physiology, Part 1*, (A. S. M. Saleuddin and K. M. Wilbur, eds.), pp. 155–198, Academic Press, New York.

Vecchione, M. and Gaston, G. R., 1985, In-situ observations on the small-scale distribution of juvenile squids (Cephalopoda, Loliginidae), *Vie Milieu* **35**:231–235.

Verrill, A. E., 1882, Report on the cephalopods of the northeastern coast of America, *Rep. U. U. S. Comm. Fish.* **1879** (7):211-455.

Voss, G. L., and Sisson, R. F., 1967, Squids. Jet-powered torpedoes of the deep, *Nat. Geograph.* **131**:386–411.

Voss, G. L., 1977, Classification of recent cephalopods, in: *The Biology of Cephalopods*, (M. Nixon and J. B. Messenger, eds.), pp. 575–579, Academic Press, London.

Voss, N. A., 1988, Evolution of the cephalopod family Cranchiidae (Oegopsida), in: *The Mollusca. Vol. 12. Paleontology and Neontology of Cephalopods*, (M. R. Clarke and E. R. Trueman, eds.), pp. 293–314, Academic Press, New York.

Ward, P. and Bandel, K., 1987, Life history strategies in fossil cephalopods, in: *Cephalopod Life Cycles*, vol. II (P. R. Boyle, ed.), pp. 329–350, Academic Press, London.

Worms, J., 1983, World fisheries for cephalopods: a synoptic overview, *FAO Fish. Tech. Paper* **231**:1–19.

Chapter 3

Lol-i-go and Far Away: A Consideration of the Establishment of the Species Designation *Loligo pealei*

FREDERICK A. ALDRICH

The object of the Code is to promote stability and universality in the scientific names of animals, and to ensure that each name is unique and distinct.
-- Preamble, International Code of Zoological Nomenclature (1961)

1. Introduction

Long ago and far away [from Newfoundland], this story begins. The designation *Loligo pealei* first appears in a paper read at a meeting of the Academy of Natural Sciences of Philadelphia on the evening of March 20th, 1821, by the then curator Charles Alexandre Lesueur (1778–1846). Born in LeHavre, France, he early became noted for his talents at sketching, drawing and painting natural history specimens. He had a most successful four years (1800–1804) with François Péron on the **GEOGRAPHIE** Expedition to Australia and Tasmania, which saw the collection of some 100,000 zoological specimens, from which 2,500 new species were described and ultimately published. Only the first volume of results from these collections was published before Péron's death. Péron bequeathed all his unfinished manuscripts "to the most intimate friend, to the faithful companion of (my) work and research in natural history, to the good and faithful M. Lesueur."

As history has recorded, it fell to others to author subsequent publications from the GEOGRAPHIE material studied by Péron. Lesueur was not much of a writer. As I have written elsewhere (Aldrich, 1981), "Throughout Lesueur's career there is a story of incomplete notes, missing information, unfinished papers, but always a wealth of sketches, drawings, water colours – some hurried, some more detailed, others completely finished – but always graphic representations of what he saw."

Largely through his having met William Maclure, an eminent Scottish-American geologist and philanthropist, in Paris in 1815, Lesueur eventually settled at the Academy of Natural Sciences of Philadelphia, an institution of which Maclure was a founding member (1812), a benefactor and subsequently, president.

F. A. ALDRICH • Ocean Studies Task Force and Department of Biology, Memorial University of Newfoundland, St. John's, Newfoundland A1C 5S7, Canada.

And so to the squid. Maclure and Lesueur, upon their arrival in the United States, undertook a lengthy collecting trip in the northeast. On a date of uncertain designation, but in the year 1816, they found themselves in a place Lesueur noted as merely *Sandy Bay*. While there, Lesueur's efforts at securing specimens of squid were well rewarded. As a consequence, he was able to read a paper at that March 20th, 1821 meeting of the Academy, entitled "**Descriptions of several new species of Cuttle-fish**". The paper was subsequently published in the second volume of the Academy's Journal. It reported, in part, on the squid collected at Sandy Bay, as well as other new species he therein described.

To quote from that paper: "When Mr. Maclure and myself were at Sandy Bay in 1816 . . . the beautiful color with which ("Loligos") were ornamented, induced me to take a drawing of one immediately, but not then having leisure to complete it, I took a specimen with me to finish the drawing at my leisure."

One version has it that a sudden rainstorm was the cause of the interruption. But to continue Lesueur's own account: *"But recently upon comparing this specimen with my drawing, I was much surprised to perceive that I had brought with me a very distinct species from that which I had observed."*

The drawing was of a "*loligo*" which Lesueur named *Loligo illecebrosus*, properly assigned later to *Illex illecebrosus* by Steenstrup (1880). Both parts of the name refer to the use of the squid as bait. The drawing ("the cause of the brevity of the following description") proved to be the lectotype of the species, until Lu (1973) in his review of the genus *Illex* established a neotype. It proved to be a mature male specimen from Old Scantum, New Hampshire, some 34 km northwest of Sandy Bay, north of Cape Ann, Massachusetts. Lu could find no justification to place the neotype locality in maritime Canada or Newfoundland, particularly since Hamy (1968) identified the Sandy Bay as that in Massachusetts. However, Lesueur was indeed in Canada in the period under discussion. Perhaps "*Loligo*" was not so far away, after all, with the Marine Biological Laboratory at Woods Hole, Massachusetts being *ground central*. But it is far away from the *squid jigging grounds* of Newfoundland of which Lesueur wrote. But this is not the appropriate place to continue along this line of investigation.

Several other species were first described in Lesueur, 1821. There was *Loligo bartramii*, *Loligo bartlingii*, *Loligo pavo* (= *Taonius pavo*) and *Loligo pealeii*, along with discussions of other *cuttle-fish*. With regard to *Loligo pealeii*, note the spelling of the trivial component of the name, the *eii*, which in more modern usage is reduced to *ei* (see Summers, 1990).

Lu (1973) and others apparently considered that the specimen which Lesueur removed from his Sandy Bay drawing site along with his drawing of *Illex illecebrosus*, was *Taonis pavo*. Lesueur clearly noted that it was taken in 1816 from Sandy Bay, and thus Lu identified Old Scantum as the type locality for *Taonis pavo*, as a result of his resolution of the type locality for *Illex illecebrosus*.

Loligo pealei presents a different set of problems, and to place them in perspective I resort again to Lesueur's own words: "*Loligo pealei: This species, which appertains to the fine collection of the Philadelphia Museum, was largely confided to my care, for examination, by the Manager of that interesting and superb establishment, Mr. R. Peale.*" Hence the species designation, which was a patronymically inspired one, in honor of Mr. Reubens Peale, Manager of the Museum and son of the eminent portrait artist Charles Willson Peale. The elder Peale's artistic skills have passed along to us well-known likenesses of such personages of the period as George Washington, the Marquis de Lafayette and Benjamin Franklin, to name but a few. In 1818 he wrote, "*I have put into the museum a portrait of M. Lesueur who perhaps has the most knowledge of natural history of any man living. He is an excellent draftsman."* The portrait now hangs in the Library of the Academy.

The museum to which Peale referred was the Philadelphia Museum, of which he was the founder, and among its collections was displayed the first American specimen of mastodon, the excavation of which was the subject of one of his larger canvases.

But with respect to the new species, *Loligo pealei*, Lesueur (1821) indicated with considerable doubt that the collection site was the *coast of South Carolina*? Summers (1974) speculated that the description could have been based on "one lot of preserved squid (perhaps one specimen)". It is clear that Lesueur's report indicated that R. Peale had loaned him "this species", rather than a squid. The animal(s) in question had been preserved in ways that would have fallen far short of the requirements as set forth by Roper and Sweeney (1983) and the description of the taxon fails in almost all respect to meet the ideal minimal requirements for taxonomic descriptions as advocated for universal adoption in cephalopod taxonomy by Roper and Voss (1983).

At any rate, Voss (1962) reviewed the reported types and species of cephalopods in the collections of the Academy of Natural Sciences (the oldest such assemblage in the United States) in 1961. He lamented the fact that the collection had fallen into such a poor state, with several of the historically more important type specimens apparently having been lost, and Lu (1973) reported that a similar search failed to uncover Lesueur's original (lectotype) drawing of *Illex illecebrosus*. Among the lost type specimens was Lesueur's *Loligo pealei*, so in effect the author's original drawings as reproduced in JANSP 2(1) could be properly considered a lectotype (Figure 1). However, it is possible that Voss was mistaken in presuming that the *Loligo pealei*, be it a lot or a single specimen, was ever deposited in the cabinets of the Academy of Natural Sciences. If indeed Reubens Peale had only loaned "the species" to Lesueur for examination and subsequent publication through the Academy, it may be assumed that it was returned to the Philadelphia Museum.

But there is one other possible, indeed probable, explanation. It arises from a re-reading of Lesueur's text yet again. "Thus I regarded the species described in this paper, from the collection of the Academy and that from the Philadelphia Museum, as specifically the same, with one of which I made a drawing at Sandy Bay; but upon comparing them with each other, they all proved distinct."

The above excerpt is far from clear or edifying, a truism of much of Lesueur's writings. I suggest that unless Lesueur was referring to other species with which his paper dealt, specimens of *Loligo pealei* may indeed have been in the Academy's collections at the time. He may have chosen to honor Peale and use only the borrowed specimen(s) from the Philadelphia Museum in his description. Or, the nameless specimen which he tells us he brought back from Sandy Bay along with his drawing of *Illex illecebrosus*, considered by Lu (1973) and others to be *Taonius pavo*, could indeed have been *Loligo pealei*. Frankly, the presence of the latter along with *Illex illecebrosus* in a commercial bait catch seems to me to be more likely than *Taonius pavo*. In Figure 2 the two primary bait squids of the Atlantic coast of what was once *British North America* are shown side by side.

Of course, if Sandy Bay should prove – admittedly highly unlikely – to be the type locality of *Loligo pealei*, or if Academy-housed specimens of that taxon were involved, then the loss of *the type* may once again be laid against the practices of Lesueur and his Academy-based successors, such as myself. But more of these matters later.

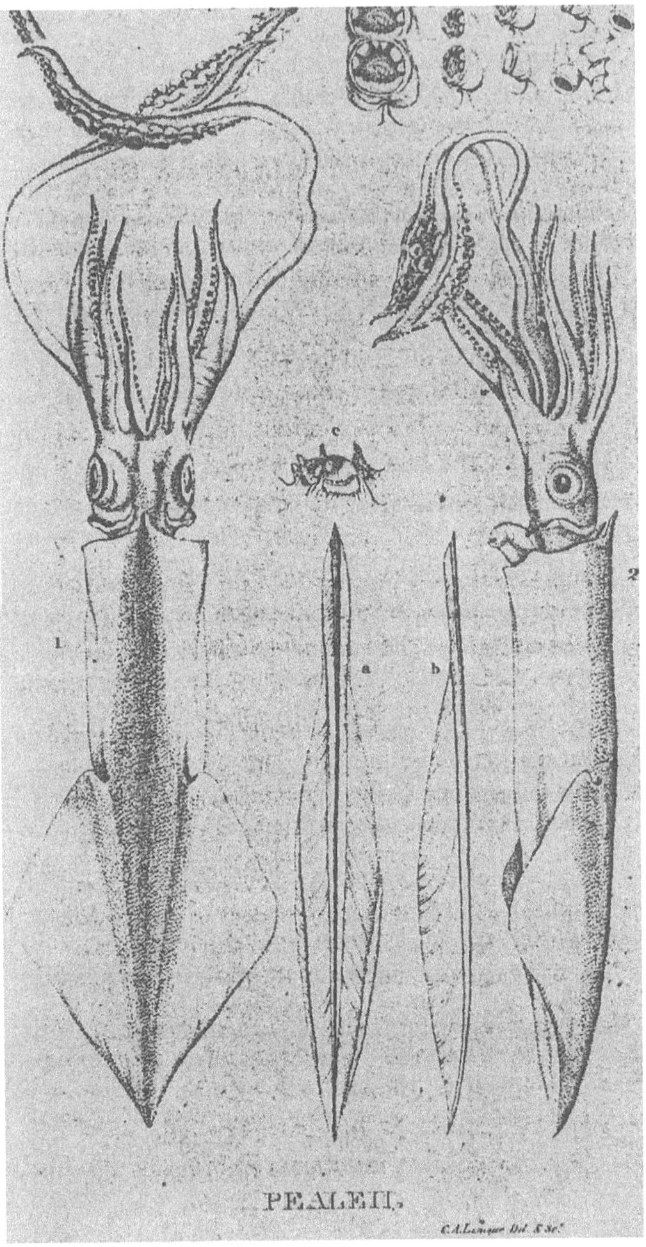

Figure 1. C. A. Lesueur's drawing of the holotype of *Loligo pealei* from *Journal of Academy of Natural Sciences of Philadelphia*. Vol. **2** (1). 1821.

Lesueur's drawings became of considerable zoological importance long before the concept of lectotypes was even contemplated or there was any suggestion of an International Code of Zoological Nomenclature. If there is no holotype, *Article 74* of the Code, *Lectotypes*, calls for "*(b) Designation by means of a figure. Designation of a figure as lectotype is to be treated a designation of the specimen represented by the*

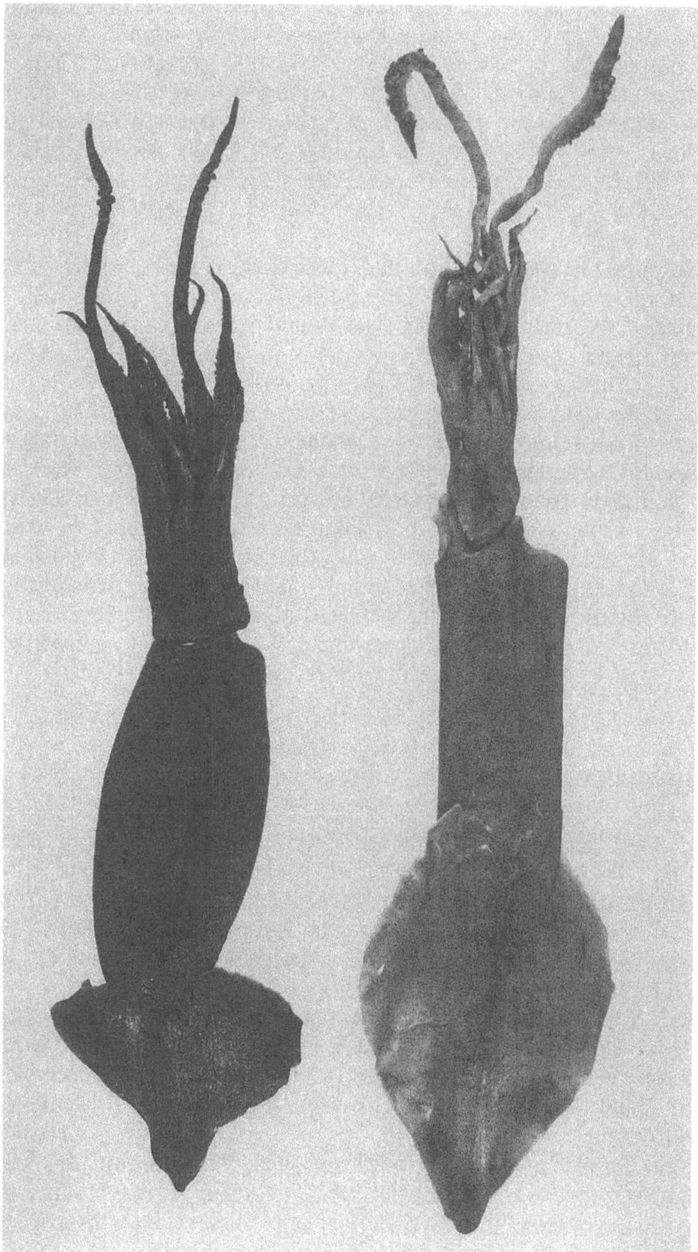

Figure 2. Two species of squid described in 1816 by C. A. Lesueur. Left: *Illex illecebrosus* (Lesueur 1821). Right: *Loligo pealei* Lesueur 1821. (Photo by Roy Fickin, Department of Biology, MUN)

figure ...". Fortunately, Lesueur's artistic skills served our science well, and in most respects far better than did his writings.

Many experimental scientists become exasperated concerning excessive talk of such taxonomic issues as types, the Code, synonymy and priority, etc. Schmidt (1950), a taxonomist himself, put it all in proper perspective when he observed that "elaborate argument over rule and validity of names has been a disgrace to zoology and has contributed more than any other single factor to the low repute of systematics among zoologists as a whole." Many agree with this self indictment, and Schmidt's was not the only sane voice among the systematists. The original International Code of Zoological Nomenclature (1901) and the Rules that enabled its effective use do not include any directives concerning *types*. Provision for generic types was first adopted in 1907, and it was some years before attention was directed to the need to legislate species designations. When all is said and done, we, as zoologists, should take considerable comfort in the fact that, as Mayr (1969) put it, "A type is always a zoological object, never a name".

Therefore, as you pursue experimentation and researches in some aspect of the biology of *Loligo pealei* or whatever squid, whether involving axons, muscles, cells or embryological manipulation, be sure which species of squid it is and from whence it came. Results between species need not be comparable nor compatible, and indeed lead to time-consuming, confusing, and wasteful obfuscation of the facts.

This exhortation is not necessary to many. However, I remember my own experiences. I once attended a seminar given by a biochemist concerning some aspect of echinoderm biology which interested me. The work was doubtlessly of merit and in one of his slides and throughout his presentation the experimental animal was referred to as *the starfish*. In the question period I inquired as to which species of sea star he had worked on. He not only didn't know, but failed to see that it would make any difference. This response to my innocent enough inquiry caused many in his audience to immediately question the relevance of what, beforehand, had appeared to be a significant elucidation of an aspect of asteroidian biology. May cephalopod research fare better.

2. Vernacular nomenclature

The name *squid* itself has an interesting story, so permit me a word about vernacular nomenclature. To Linnaeus, apparently, all squid-like Cephalopods in his grouping which he designated *Vermes mollusca* were loligos, or *cuttlefish* (1758). Sixty-three years later, Lesueur still used the vernacular designation *cuttlefish* in the very work that established several new species, some erroneously, of the genus *Loligo*, as witnessed by both title and text.

Lesueur observed "A great number of Loligos collected by the fishermen, and held in reserve as bait for Cod-fish, which they catch in great numbers on the banks of Newfoundland." Later, in his 1821 presentation, he commented, "This species is known by the name of Squid at Sandy Bay, and is made use of by the fishermen as bait in the Cod-fishery."

The above is from Lesueur's brief description of, and comments on, his new species *Loligo illecebrosus*, properly *Illex illecebrosus*. It is clear, however, that the vernacular designation *squid* entered the English language here, from where I am writing, in Newfoundland, and dealt originally with the ommastrephid *Illex illecebrosus*. Lesueur's comments are reminiscent of the entry in the journal of Captain George Cartwright (1798) of Labrador in which the bait role of *squid, the inkfish* is recorded. For the record, the word squid is reported in common usage in Newfoundland in at least two earlier sources, namely, Richard Hakluyt in 1578 ("...and another..., called a Squid"), and in 1620 John Mason's reference to "Squides, a rare kinde of fish, at his mouth squirting mattere forth like Inke."

Summers (1974) cites *bone squid* as a commonly used designation for *Loligo pealei* by New England fishermen. *Bone* apparently refers to the prominent gladius or internal organic shell characteristic of this species.

This may all seem unimportant, but it is in conformity with our full understanding that squid are not cuttlefish, and with all the physiological attributes which that designation implies in light of latter-gained knowledge. Perhaps this brief investigation of the vernacular form is as good an example of the significance of proper nomenclature as any I could cite.

In conclusion, as has been noted above, Peale (1818) considered Lesueur to be an excellent draftsman. But more to the truth of the situation is what Chinard (1949) wrote, "Its presentation of sea life (was) made by an artist whose eyes seems to have been able to see through marine animals and to distinguish their internal organs as they came from the sea and were still alive." High praise. But I would hasten to point out that it is for the physiologist, of whatever specialty, to complete what Lesueur started with his artistic talents. It is for the zoologist to tell of the workings of internal organs. With respect to *Loligo pealei*, it all began in the sketchpad of an itinerant French artist from LeHavre, sometime between 1816 and 1821. Correctly, Vail (1938), Breton (1978), and others, have referred to Lesueur as "*the Raphael of natural science.*"

3. An afterword

For seven years following graduation with my Ph.D. in those days before post-docs, I served as an Associate Curator in the venerable Academy of Natural Sciences of Philadelphia, on Logan Circle, at 19th and the Benjamin Franklin Parkway. For a brief time, one of my responsibilities was the alcoholic invertebrate collections in the days when Henry Pilsbry, and then R. Tucker Abbott, chaired the Department of Malacology. One day I had a visitor who inquired about the final resting place of the naturalist and one-time recipient of the position I then held, namely, Charles Alexandre Lesueur. I did not know where the mortal remains of C. A. Lesueur reposed, so I contacted the National Historical Burial Commission, or an august body of similar name, and passed along the request. In sepulchral tones I received the reply "*Dr. Lesueur is not with us*". That was my first real introduction to the person whose name appears as the authority on a myriad of taxa with which I was acquainted, not the least of which was *Loligo pealei*. It is strange how "our paths" were to cross again so many years later, here in Newfoundland.

ACKNOWLEDGEMENTS. I wish to thank the following: Mr. F. Burnham Gill for assistance in gathering information on the Lesueur presence in Newfoundland, and Mrs. Virginia Lesueur of Meadville, Pennsylvania, for sharing results of her researches on the history of the family whose name she bears. Special thanks are due and gratefully extended to Miss Sylvia Baker (Librarian) and Mrs. Carol Spawn (Manuscript Librarian), both of the ANSP for granting me access to the Lesueur letters, manuscripts, drawings and paintings in their collections. Mr. Roy Ficken of our Department of Biology prepared the illustrations herein. Thanks, too, to my faculty colleague, Dr. Raymond Clarke, regarding classical roots of scientific taxa. Last but far from least I wish to acknowledge the help and support of my wife Marguerite and, as always, Lillian Sullivan.

References

Aldrich, F. A., 1981, Whither the illecebrosus one?: Charles Alexandre Lesueur and Newfoundland, *Centennial presentation to the Newfoundland Historical Society*. 29 pp.

Breton, G., 1978, A l'occasion du bicentenaire de la naissance de Charles-Alexandre Lesueur, la Ville du Havre, *Bulletin Trimestriel de la Societé Geologique de Normandie et des Amis du Museum du Havre* **55 Suppl(1)**:23 pp.

Cartwright, G., 1798, *A Journal of Transactions and Events During a Residence of Nearly Sixteen Years on the Coast of Labrador. 3 volumes*, Allin and Ridge, Newark, England.

Chinard, G., 1949, The American Sketchbook of Charles Alexandre Lesueur, *Proc. Amer. Philosoph. Soc.* **93(2)**:114–118.

Hakluyt, R., 1578, Letter to Parkhurst in the Original Writings and Correspondence of the Two Richard Hakluyts. Vol. 1, 1935, (E. G. R. Taylor, ed.), The Hakluyt Society, London.

Hamy, E. T., 1968, *The Travels of the Naturalist Charles A. Lesueur in North America, 1815–1837*, (H. F. Raup, ed., M. Haber, translator), The Kent State University Press.

Lesueur, C. A., 1821, Descriptions of several new species of cuttle-fish, *J. Acad. Nat. Sci. of Philad.* **2(1)**:86–101.

Linnaeus, C., 1758, *Systema Naturae per regna tria naturae, secundum classes, ordinis, genera, species, cum characteribus, differentics, synonymis, locis, Facsimilie edition, 1956*, British Museum (Natural History), London.

Lu, C. C., 1973, *Systematics and Zoogeography of the Squid Genus Illex (Oegopsida: Cephalopoda)*, Doctoral thesis, Memorial Univ. Newfoundland, St. John's, Newfoundland.

Mason, J., 1620, *A Briefe Discourse of the New-found-land, with the situation, temperature, and commodities thereof* ..., Andro Hart, Printers, Edinburgh.

Mayr, E., 1969, *Principles of Systematic Zoology*, McGraw-Hill, New York.

Peale, C. W., 1818, Letter to the Rev. Burgess Allison dated February 14th, 1818, in the Peale papers, *Library of the American Philosophical Society*, Philadelphia.

Roper, C. F. E. and Sweeney, M. J., 1983, Techniques for fixation, preservation and curation of cephalopods, *Memoirs of the National Museum of Victoria* **44**:29–48.

Roper, C. F. E. and Voss, G. L., 1983, Guidelines for taxonomic descriptions of cephalopod species, *Memoirs of the National Museum of Victoria* **44**:49–63.

Schmidt, K. P., 1950, More on zoological nomenclature, *Science* **111**:235–236.

Steenstrup, J. S., 1880, Orientering i de Ommatostrephagtige Blaeksprutters indbyrdes Forhold, *Overs. danske Vidensk. Selsk. Forh.* 1880–81:73–110, [The Interrelationships of the Ommatostrephes-like Cephalopods, An Orientation. in: *The Cephalopod Papers of Japetus Steenstrup*, 1962], (A. Volsøe, J. Knudsen and W. Rees, English translators), pp. 52–82.

Summers, W. C., 1974, Natural history of the squid, in: *A Guide to Laboratory Use of the squid Loligo pealei*, (J. M. Arnold, W. C. Summers, D. L. Gilbert, R. S. Manalis, N. W. Daw, and R. J. Lasek, eds.), pp. 5–8, Marine Biological Laboratory, Woods Hole, Massachusetts.

Summers, W. C., 1990, Natural history of the squid, *this volume*.

Vail, R. W. G., 1938, Sketchbooks of a French Naturalist 1816–1837: a description of the Charles Alexandre Lesueur collection with a brief account of the artist, *Proc. Amer. Antiquarian Soc.* April 1938.

Voss, G. L., 1962, List of the types and species of cephalopods in the collections of the Academy of Natural Sciences of Philadelphia, *Notulae Naturae (ANSP)* **No. 356**: 1–7.

Chapter 4

Maintenance, Rearing, and Culture of Teuthoid and Sepioid Squids

ROGER T. HANLON

1. Introduction

The challenges of collecting, transporting, and maintaining *healthy* wild squids have been nearly as formidable as those of determining the structure and function of ion channels in squid giant axons. The differences are a matter of scale, not complexity. Careful perusal of the literature indicates clearly that failures have been the rule, especially with squids of the genus *Loligo*. It is an unfortunate fact that the vast majority of biomedical research on *Loligo* has been done on animals that were moribund when they were prepared for experiments. This was due in most cases to skin damage incurred during capture and transport from the field to the laboratory, and to the secondary infections that invariably followed immediately. Furthermore, the availability of live squids for research is limited seasonally because of the migratory behavior of these fast-swimming invertebrates. Thus any improvements in maintenance, rearing or culture will lead not only to greater availability to meet the demand from researchers, but hopefully to the provision of healthy squids for a wide range of experimentation.

This chapter is not a review, since lengthy reviews have been published as recently as 1987 (Grimpe, 1928; Boletzky, 1974; Shevtsova, 1977; Boletzky and Hanlon, 1983; Hanlon, 1987). Rather, I wish to (1) highlight the salient features of squids that require special attention when these animals are kept in captivity, (2) present some guidelines for choosing seawater type and system design as well as methods of capture and transport of eggs, juveniles or adults and (3) describe recent successes and data in culturing squids and cuttlefishes through the life cycle in captivity. *For the first time, mariculture offers a realistic alternative as a means of providing squids for the research community.* Some scientific and financial obstacles remain before squid mariculture can become convenient and affordable in many locations, but enough information is available to improve existing capture and maintenance facilities. Most importantly, in laboratories such as ours that are committed to culture, the technology has been developed to culture certain species *consistently* throughout the year without major mortality and with the production of first filial generation progeny. This is a major and very recent advance.

R. T. HANLON • Marine Biomedical Institute and Department of Psychiatry and Behavioral Sciences, University of Texas Medical Branch, Galveston, Texas 77550-2772.

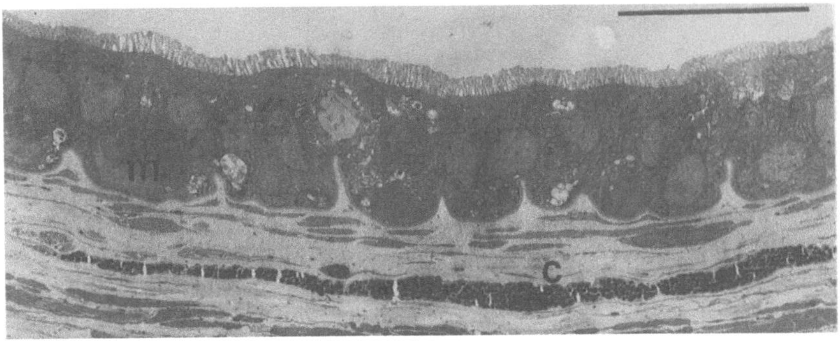

Figure 1. Typical cephalopod skin, illustrating (with *TEM*) the thin epithelium covered with microvilli (top) that are in turn covered with mucus (note mucus-secreting cells, *m*). One of the main features of the dermis is a chromatophore organ (see pigment granules, *c*). Scale equals 20 μm.

To the individual squid *users*, I would hope to redress a prevailing misconception (anchored in outdated literature and heresay) that squids are inordinately difficult animals to care for. It is more important to gain some appreciation for the needs of the squids in captivity and for the unusual organ systems that demand specific consideration. Squids are schooling, social creatures with large brains and complex behavior rivaling that of vertebrates. They are intrinsically fascinating and deserve care and attention when brought to the unnatural confines of the laboratory.

2. Terminology

Confusion persists over terms related to keeping cephalopods in captivity, thus I will reiterate definitions of standardized terms introduced first for plankton culture by Paffenhofer and Harris (1979) and later for cephalopods by Boletzky and Hanlon (1983): *Maintenance*: holding wild-caught late juvenile or adult stages at the same approximate developmental stage for varying periods, with no direct intention of growing them to a more advanced stage. *Rearing*: growing a cephalopod over a certain period of time without achieving a second generation; specifically it refers to any attempt to grow hatchlings or young juveniles to full size and sexual maturity. *Culture (sensu stricto)*: growing a cephalopod at least from hatching, through the complete life cycle (juvenile and adult stages, sexual maturity, mating and egg laying), to hatching of viable young of the first filial (F_1) generation. The term culture also may be used in the sense of collectively referring to maintenance, rearing and culture.

The term squid usually refers to members of the Order Teuthoidea, which includes the nearshore suborder Myopsida (e.g. *Loligo* and *Sepioteuthis*) and the oceanic squids of the suborder Oegopsida (e.g. *Illex, Dosidicus, Todarodes*, etc.). However, the Order Sepioidea contains squid-like animals such as the common cuttlefish of Europe and the Mediterranean, *Sepia officinalis*. There is some disagreement about these systematic arrangements and for the sake of stressing their similarities and usefulness to researchers I will often refer to cuttlefishes as sepioid squids.

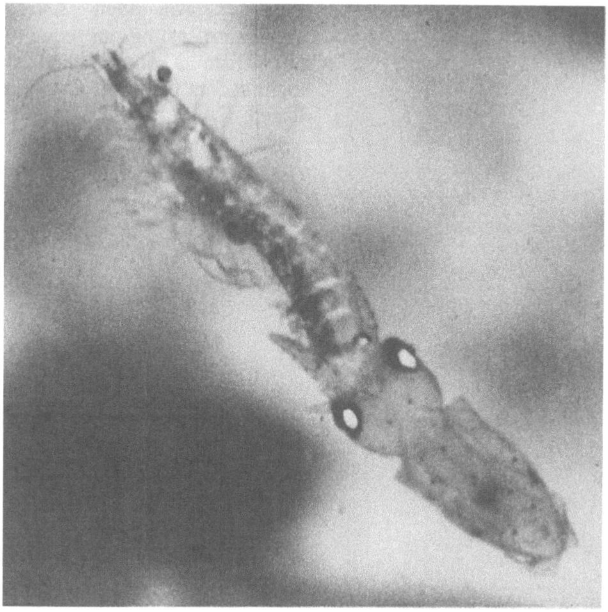

Figure 2. Hatchling *Loligo vulgaris* (3.3 mm mantle length) eating a large mysid shrimp that it has just captured. After Turk *et al.* (1986).

3. Anatomical and behavioral traits relevant to laboratory handling

3.1. Skin

The delicate skin is characterized by a single layer of columnar epithelial cells with a microvillar surface (Fig. 1). Mucus cells are interspersed throughout and provide a mucus covering over the surface of the animal; this is the squid's external barrier to seawater and the many microbes that are commonly present in it. The underlying dermis contains the conspicuous chromatophores and iridescent cells. *Thus it is possible for a squid to have relatively massive damage to its invisible epidermis while the chromatophores are intact and the animal looks normal to a human observer.* Apparently it takes little physical action to alter the mucus and epidermis (net damage, bumping into tank walls, etc.), with the result that the animal begins to bleed and the bacteria in seawater, e.g. *Vibrio* species (spp.), gain access to the wound. Furthermore, there is no vertebrate-like clotting mechanism in cephalopod blood so that bleeding may be persistent in larger wounds; small wounds can be closed by constriction of the vessels or the skin. It is necessary, therefore, that great care always be taken to minimize the amount of physical contact that a squid accrues in every step of handling, including capture, transport, transfer from tank to tank, etc.

Figure 3. An adult *Loligo opalescens* eating an equally large fish. Photographed at night at 5 m depth off Catalina Island, CA.

3.2. Locomotor habits

Squids of the Order Teuthoidea swim constantly except in very rare cases in which they rest in a sitting posture on the substrate. *Loligo pealei* is a species that will sit calmly on the bottom if a suitable substrate is provided. Most squids have terminal fins that constitute only 1/4 to 1/2 the length of the mantle; thus they are not very good at hovering in one spot, but rather tend to sway back and forth. Most squids forage over large distances daily searching for food and for this reason they tend to move about tanks quite a bit. Squids in nature do not have any vertical barriers to deal with. It is useful to paint the walls with a high-contrast pattern and to provide some type of bottom substrate over which they will tend to orient (i.e. a patch of seaweed or even a darkly painted bottom) as some species do in nature. LaRoe (1971), Matsumoto (1976) and Hanlon *et al.* (1983) have all found this helpful. When frightened, all squids use a jet-propulsed escape (either forward or backward) and this results in considerable skin damage when performed in tanks. They can also jet completely out of their tanks. It is useful to provide good tank design (see 4.6.1.) and a quiet atmosphere around the tanks to minimize this sort of problem. Fortunately, some species such as the teuthoid squids *Sepioteuthis sepioidea*, *Sepioteuthis lessoniana* and the sepioid squid (or cuttlefish) *Sepia officinalis* have fins around the entire periphery of the mantle and they can hover very well, thus minimizing wall contact when in captivity. These species are closely associated with the substrates they live near in nature. *Sepia officinalis* normally sits on the substrate or burrows into it and this is conducive to laboratory maintenance. Squids will also ink during or after a jet escape, depending upon the degree of alarm, and this can adversely affect water quality and should be avoided.

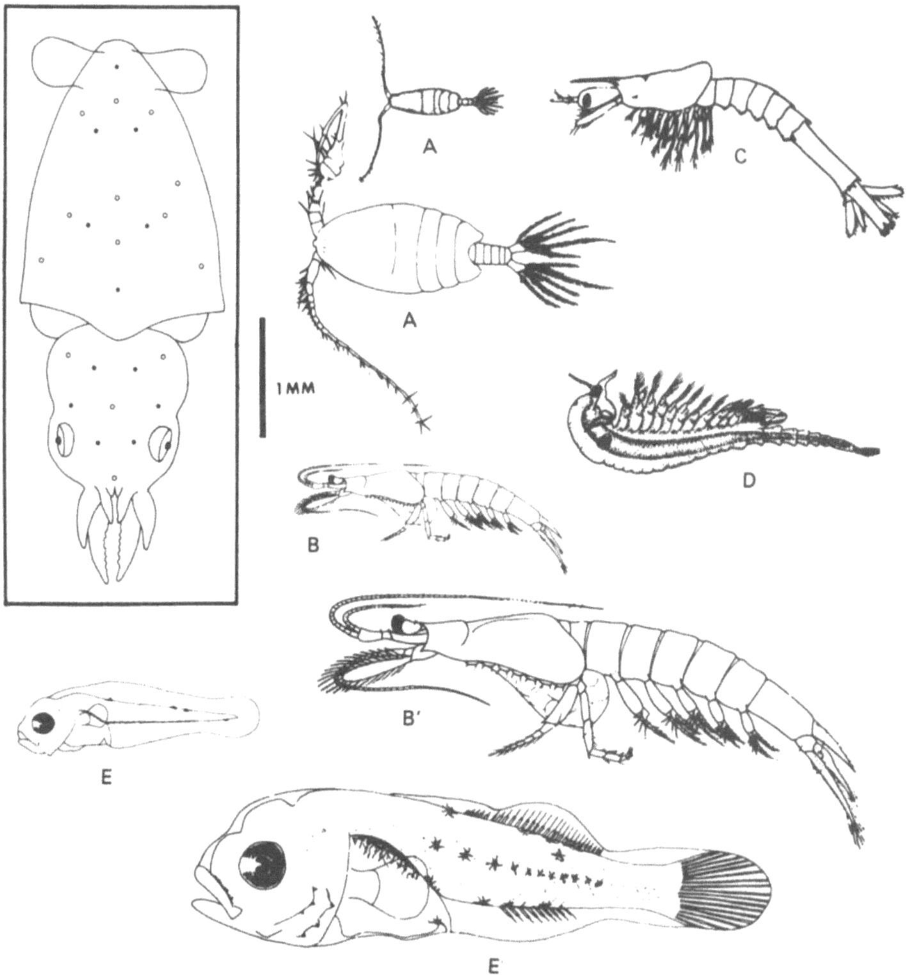

Figure 4. Size relationship of hatchling *Loligo opalescens* and various food organisms fed for the first 30 days posthatching. *A*. copepod *Acartia tonsa*. *A'*. copepod *Labidocera aestiva*. *B*. hatchling mysid shrimp *Mysidopsis almyra*. *B'*. adult *Mysidopsis almyra*. *C*. mysis stage of the shrimp *Penaeus* sp. *D*. adult brine shrimp *Artemia salina*. *E*. 1-day-old larva of the red drum *Sciaenops ocellatus*. *E'*. 13-day-old larva of *Sciaenops ocellatus*. From Yang *et al*. (1986).

3.3. Sensory systems

The squid eye is well developed and squids have excellent vision. The majority of their behavior is mediated by visual input; they have huge optic lobes for processing this information. They are capable of seeing movement outside the confines of their

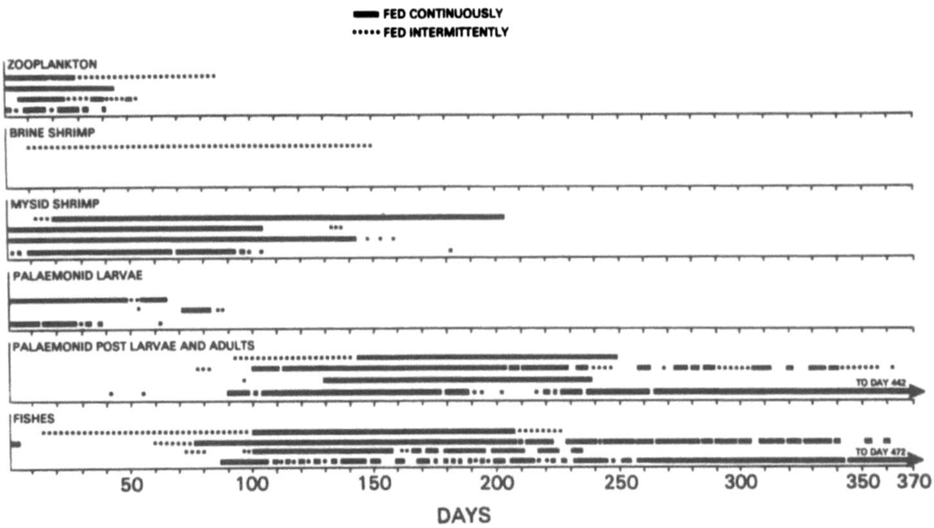

Figure 5. Typical sequence of food types provided to *Loligo* spp. during four major culture experiments. Example types and sizes of food organisms are given in Figure 4. The top line is *Loligo opalescens* (from Yang *et al.*, 1986) and the bottom three lines are *Loligo forbesi* (from Hanlon *et al.*, 1989a).

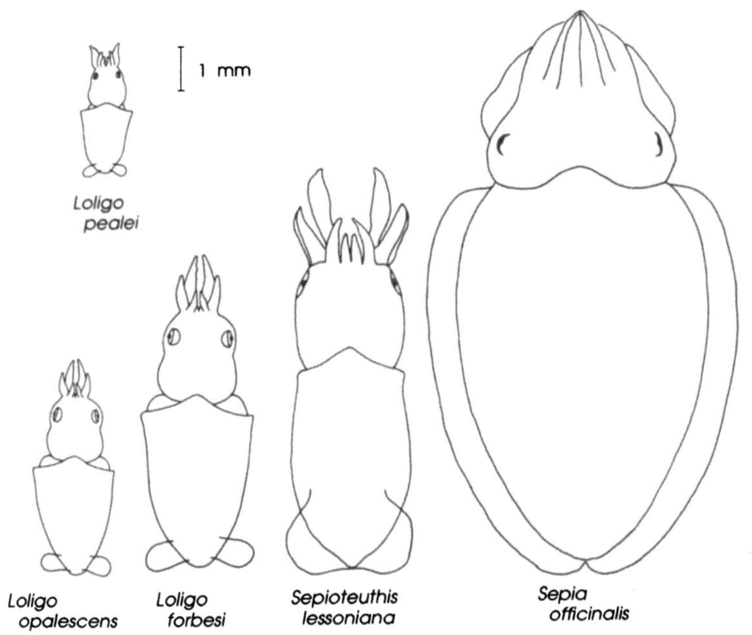

Figure 6. Sizes at hatching of various species, all drawn to the same scale.

tank and light levels can influence their behavior. Intraspecific behavior is expressed by chromatophore body patterns and postures, all of which are visually mediated. Chemoreception is accomplished through the numerous suckers and probably elsewhere, since they can react quickly to noxious chemicals in the water. The normal response is to swim away, but of course this causes problems in the confines of the tank. Squids also detect and respond to vibrations (summary in Hanlon and Budelmann, 1987) and a lateral line analogue has recently been demonstrated (Budelmann and Bleckmann, 1988). With such well developed nervous systems, teuthoid squids are prone to be rather easy to stress in captivity if conditions are not ideal, and deaths have been attributed to *nervous excitability*.

3.4. Mode of feeding

Squids have a muscular buccal mass and a powerful beak that enables them to bite through fish scales and crustacean carapaces. They capture their prey with a tentacular strike and then seize and maneuver the prey with their strong and agile arms. They then kill the prey by biting through the head area or injecting a neurotoxin from the posterior salivary gland. The significant consequence is that, from hatching onwards, they can consume a variety of prey of a very wide size range, often their own size or larger (Figs. 2 – 5).

More importantly, feeding in squids is stimulated largely by the behavior of the prey. Squids are attracted to an erratic swimming motion in prey, such as the quick jerky movements of copepods and shrimp or the wriggling darts of fishes. Due to their high metabolism, squids need food often (at least once daily) during all stages of their life cycle.

3.5. Egg size and hatchling behavior

Cephalopods do not have true larval stages (Young and Harmon, 1988) but some species have larger hatchlings that are more advanced in their capabilities to swim, feed, change body patterns and avoid predators. These species have proved easier to rear or culture (see review in Boletzky and Hanlon, 1983). For example, there is an enormous difference in egg size among *Loligo* spp. (e.g. 1.5 – 3.0 mm egg diameter), *Sepioteuthis* spp. (6 mm) and the sepioid *Sepia* spp. (20 mm) and the hatchlings of each are correspondingly easier to culture. Even among *Loligo* spp. there is a significant size difference in hatchlings (Fig. 6) and the largest ones (*Loligo forbesi* at 4.0 mm mantle length, ML) are far stronger swimmers than the small ones (*Loligo pealei* at 1.8 mm ML) and can feed on larger food organisms immediately after hatching (Fig. 4).

3.6. Social behavior

Squids begin schooling as soon as they can swim strongly enough, at about 5 – 10 mm ML in *Loligo opalescens* (Yang *et al.*, 1983) and *Loligo forbesi* (Hanlon *et al.*, 1989a); this size corresponds to as young as 40 days in *Loligo forbesi* but takes longer in *Loligo opalescens* because they hatch at a much smaller size (Fig. 6). Hatchling

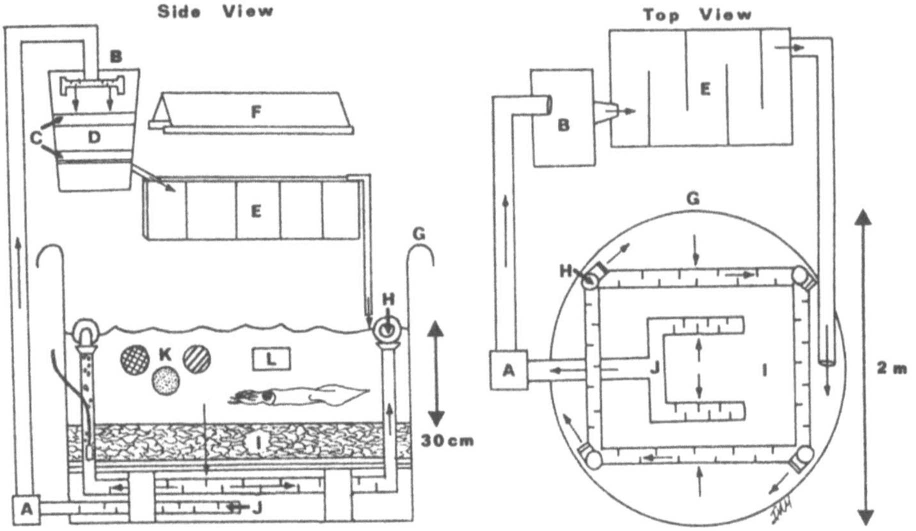

Figure 7. A 2 m circular tank system for squid maintenance. Eleven *Loligo plei* are swimming in this closed system. A pump (*A*) pushes water to an auxiliary filter (*B*), where it then flows by gravity through two layers of polyester fiber (*C*) and granular activated carbon (*D*) into an

algal filter (*E*) that is under continuous illumination (*F*) and back into the squid holding tank (*G*). Water circulation in *G* moves in a clockwise direction that is caused primarily by the flow from air-lift pumps (*H*). Water is drawn through the filter bed (*I*) into the perforated subsurface pipes of the air lifts. Water is also drawn into another set of subsurface pipes (*J*) by the pump (A). Painted walls (*K*) make the sides more visible to the squids. Viewing ports (*L*) are for behavioral observations. From Hanlon *et al.* (1983).

Sepioteuthis lessoniana are larger and begin to swim against a current 5 - 10 days posthatching; schooling has been observed between 11 and 23 days posthatching. Squid behavior becomes erratic when they are separated from the school and thus for laboratory purposes it is wise to keep several animals together for a calming effect. Squid behavior also changes when they are put in small tanks and this should be avoided. Since squids are cannibalistic, it is also advisable to keep animals of similar size together and to keep feeding levels on the order of 10 - 15% of squid body weight, otherwise smaller animals will be cannibalized (cf. Hanlon *et al.*, 1983).

Social behavior related to courtship and mating can also contribute to skin damage and mortality. Male squids engage in mock battles to achieve and maintain dominance (Arnold, 1965; Hurley, 1977; Hanlon, 1982; Moynihan and Rodaniche, 1982) for the presumed purpose of gaining better access to females. During these agonistic bouts, feeding is disrupted significantly and subordinate males in particular often bump into tank walls while attempting to jet away from dominant males; one palliative method is to segregate males from females and to keep squids of the same size together (Hanlon *et al.*, 1983).

4. Water quality and closed vs. open seawater systems

Teuthoid squids require high-quality water. Most importantly, pH must be kept above 7.6 and dissolved oxygen levels must remain high to accommodate the metabolic demands of constant swimming and fast locomotion. *Loligo* spp. are especially sensitive to high levels of ammonia-nitrogen (> 0.17 mg/liter) and nitrite-nitrogen (> 0.20 mg/liter) that accumulate in closed systems; they can withstand relatively high levels of nitrate-nitrogen (50 mg/liter) if other parameters such as pH, temperature and dissolved oxygen are at acceptable levels. Salinity generally has to be above 30 parts/thousand (ppt) except for *Lolliguncula brevis*, an estuarine species that can withstand salinities down to 16 ppt (Hendrix *et al.*, 1981). Ultraviolet sterilization is necessary to keep the bacterial flora minimized in the water column; this is a major problem in open seawater systems with minimal physical filtration of incoming water, which has a fairly rich microflora. Juveniles and adults have good tolerance to temperature differentials, even when subjected to quite severe shocks incurred during transport or transfer (Hanlon *et al.*, 1983). However, as a rule, embryos are intolerant of temperature change and generally require high-quality water throughout development.

Sepioid squids such as *Sepia officinalis* are far more tolerant of nitrogenous waste levels and salinity. In Galveston, these cuttlefishes have been cultured through four generations and levels have reached as high as 1.0 mg/liter ammonia, 1.0 mg/liter nitrite and 125 mg/liter nitrate. However, at about 80 mg/liter nitrate, *Sepia* become agitated and ink often. Boletzky (1983) and Guerra and Castro (1988) reported that this species can tolerate salinities as low as 20 ppt. This is a hardy species that is currently being reared at public aquaria in America and Europe.

Figure 8. The 10,000 liter raceway, a closed system. Seventeen *Loligo plei* are swimming over a white fiberglass sheet (put in for the photograph only). Air lift pumps are turned off for the photograph. Water leaves the raceway (*A*) via a siphon to a pump (*B*) that pushes the water to a rotating biodisc tank (*C*) that is under continuous illumination (*D*). The water then flows by gravity first into the auxiliary filter (*E*) then through the main filter (*F*, a circular tank system) and then back to the raceway. Water within the raceway is circulated in a clockwise direction

by the discharge from the main filter and by the air lifts. Two air blowers (*I*) drive the air lifts. A foam fractionator (*J*) is mounted in the raceway. From Hanlon *et al.* (1983).

Natural and artificial seawater have been used successfully for cephalopod culture. One major problem emerged. Embryos incubated in Instant Ocean® brand seawater developed a severe behavioral deficit in which hatchlings swam in a spinning trajectory and soon died because they could not capture food (Colmers *et al.*, 1984). It was found later that the element strontium was missing or in low quantities and that this prevented the statolith from forming within the statocysts, rendering the hatchlings incapable of detecting gravity; adding strontium chloride to achieve a level of 8 mg/l overcame this problem (Hanlon *et al.*, 1989b). However, this incident should alert future workers of the potential difficulties of using artificial seawater mixes because few, if any, contain all constituents of natural seawater. Nevertheless, artificial seawater is an adequate medium.

Open seawater systems are proven and reliable but require a coastal location with good water quality. There is no problem with buildup of nitrogenous wastes but there is little or no control over temperature, salinity, disease organisms, turbidity and pollutants.

A closed or recirculating system can provide a stable, controllable and reproducible environment once it is established, and with artificial seawater it can be used anywhere. The main problem is the buildup of nitrogenous wastes and thus great attention must be paid to the biological filtration system. Example systems that have proved reliable for squid maintenance, rearing and culture are illustrated in Figures 7 – 10.

5. Capture and transport of eggs, juveniles and adults

5.1. Egg care

Eggs require the utmost care at *all* stages of development and this cannot be overemphasized. Egg capsules may be retrieved from spawning grounds in nature (preferably by SCUBA diving) or from the laboratory since cephalopods commonly spawn in captivity. Egg laying can be stimulated by placing egg-like artifacts in the tank (Arnold, 1962). The best guarantee for normal hatching is to keep all physical parameters constant, to include temperature, pH, salinity, nitrogen wastes, daily light levels and cycles, and no mechanical disturbance of the egg masses. If temperature is to be adjusted (e.g. to meet production schedules) then it should be changed no more than 1 °C per 24 hrs. *Premature hatching must be avoided* or mass mortality will result within a few days posthatching; any mild shift in the parameters mentioned above will cause premature hatching. In late stages, embryos are metabolizing at a high rate and it is important not to place egg capsules in buckets or small volumes of water because of ammonia production and oxygen depletion that can occur within a few hours.

When shipping eggs long distances a difficult decision arises: whether to ship eggs at early stages because their metabolic rate is lower and mechanical disturbance cannot cause premature hatching, or to ship embryos late in development to hopefully ensure normally developed hatchlings. We have had successes and failures with both concepts. Eggs early in development do not need as much water during shipping and are thus less expensive to ship; however, temperature shifts in these early stages can result in major structural abnormalities of the hatchlings. Eggs in late stages (e.g. stages 26 – 30; Arnold, 1965; Segawa *et al.*, 1988) require much more water volume per egg capsule and occasionally hatch prematurely because of deteriorating water quality and/or physical jarring during postal handling.

A generalized packing method for *Loligo* spp. is to place five egg capsules (stage < 15) in 10 *liters* of seawater and an equal volume of oxygen in two heavy-duty plastic bags, each of which is tied securely, and to place this in a heavily insulated shipping box. One must take into account the ambient weather conditions locally and at the destination and consider packing ice accordingly (i.e. in tropics more ice is needed to maintain temperatures in the box, in winter in temperate zones very little or no ice is needed). *Never place the ice in direct contact with the plastic bag with eggs, but pack it within layers of newspaper.* In any case, time in transport should not exceed 24 hrs. Finally, eggs should be acclimated very slowly to the new tank water at destination, especially if there are differences in any aspect of water quality. Egg capsules should receive good water circulation and should not be placed on the bottom of the tank but suspended in the water column to prevent detritus and benthic organisms from collecting on the eggs.

5.2. Juveniles and adults

Wild-caught juvenile or adult squids *must be captured with atraumatic methods.* Bottom trawling is one of the worst methods because it inflicts much skin damage both from the net material and from compression in the cod end during trawling, retrieval and dumping the catch on board ship. In rare cases, trawling can be successful if the water is very shallow (< 5 m), bottom time is short (< 10 min), the catch is lowered carefully into water on deck and the net is modified with knotless mesh to reduce skin abrasion and metal hoops in the cod end to reduce compression. Furthermore, success depends upon the bycatch (e.g. jellyfishes, crustaceans and spiny fishes damage squids) and in any case only a fraction of the squids will make it into onboard live wells with only minor skin damage. In Galveston we have been able to accomplish this with *Lolliguncula brevis* because this species inhabits shallow water nearshore (Hanlon *et al.*, 1983). Lipinski (1985) also achieved some success trawling small *Alloteuthis subulata* in very shallow water (5 m) off Plymouth, England. Cuttlefishes have tougher skin and can also be obtained by trawling. Static pound nets or encirclement nets (e.g. purse seines, lampara nets) are good capture methods because squids are herded slowly into one spot and can be dipped out with soft hand nets. Light attraction and dipnetting are equally satisfactory. Many squids will take squid jigs and this is an excellent method because the small arm wounds heal quickly; unfortunately many loliginid squids are not attracted well to jigs. Other details of capture and transport can be found in Tardent (1962), Summers and McMahon (1970), Flores *et al.* (1976), Matsumoto (1976), O'Dor *et al.* (1977) and Hurley (1978). In Naples, Italy, adult cuttlefish are captured by tethering a female and drawing her slowly to the surface as males follow in attempts to court her.

Shipboard transport and laboratory transfer are crucial steps. Historically, very little attention has been paid to this step of the collection process. Onboard ship, the squids should be kept in large tanks supplied with seawater of good quality (usually through the hull) and care must be taken to shut off this supply when approaching shore or harbors due to different or polluted water. It is useful to insulate the tanks well to avoid temperature changes. Oxygen or air should be bubbled into the tank until docking. Squids should never be crowded and the tanks should be as wide as possible and be designed to retard sloshing of water. Some representative tank designs are illustrated in Hanlon *et al.* (1983), who concluded that transporting 10 adult *Loligo* per 580 liter tank (1 squid/58 liters) was the proper density. Matsumoto (1976) transported 15 adult *Doryteuthis bleekeri* in a 1 × 1 × 1 m tank (1 squid/66 liters) for 3 – 5 hrs with no mortalities. Both studies represent the best survival achieved during long transport and verify that squids cannot be crowded if laboratory survival is expected.

For transfer from shipboard to the laboratory or for shipping squids from one laboratory to another, squids should be caught with soft dipnets and placed gently (with moist hands) into clear plastic bags (approx. 38 × 80 cm for medium-sized squids) filled 2/3 with seawater and 1/3 dissolved oxygen. Care must be taken to avoid inking by the squids or the process must be repeated. The bags should be laid *horizontally* so that the 1 – 3 squids can swim gently and not rub against the bags often. Several bags can be placed horizontally in insulated boxes and transferred gently to the laboratory where the bags are placed gently into the tank and left to float for 5 – 30 min to let temperatures equilibrate and allow the squids to see their new environment. The bags should be opened in the horizontal position, allowing the water types to mix and the squids to swim slowly into the tank. Water-air-water transfers are not recommended, *nor are buckets of any sort.* Flores *et al.* (1976), Hanlon *et al.* (1983) and Lipinski (1985) employed similar methods that proved effective. This process is laborious but essential if maintenance beyond several days is expected.

For long-distance shipping, no more than one large (e.g. 170 mm ML) or two small squids should be packed per bag. Buffering the seawater with approximately 1 g/liter of tris (hydroxymethyl) aminomethane (TRIS) buffer will help maintain pH and extend successful shipping time by nearly a factor of two. It is clear that the squids will succumb first to pH drop because when pH is maintained during shipment they can survive levels of 10 mg/l NH_4-N (ammonia-nitrogen), which is 100 times the recommended levels of Spotte (1979) (Hanlon *et al.*, 1983). As a rule of thumb, about one liter of water per 20 g of squid should be packed per bag, and it is advisable to starve the squids for 12 – 15 hrs prior to shipment to reduce excretion (and thus nitrogen buildup) during shipment. Cooling the animals with ice packed separately (as above) will also slow metabolism and extend shipping time. Using these guidelines, we in Galveston have shipped *Lolliguncula brevis* successfully all over the United States and to southern Canada; the maximum shipping time is about 30 hrs with this technique, provided that the shipping container does not warm up. It is mandatory to use well constructed and insulated shipping boxes. The more tolerant *Sepia officinalis* can be shipped by the following guideline: 20 cuttlefish (each 30 – 40 mm ML) per six liters seawater. The limitations to squid shipment are: (1) the shipping cost of the water that is required to keep the squids alive, (2) the condition of the squids and (3) the gentleness with which the squids are handled during packing and unpacking.

Overall, young squids are more delicate and harder to capture, transport and transfer. All squids should be fed as soon as possible in the new tank systems; those that feed are more likely to survive the trauma of the process. Most mortalities occur within 48 hrs of transfer or shipment and it would be a good practice to use only squids that are surviving well for experimentation.

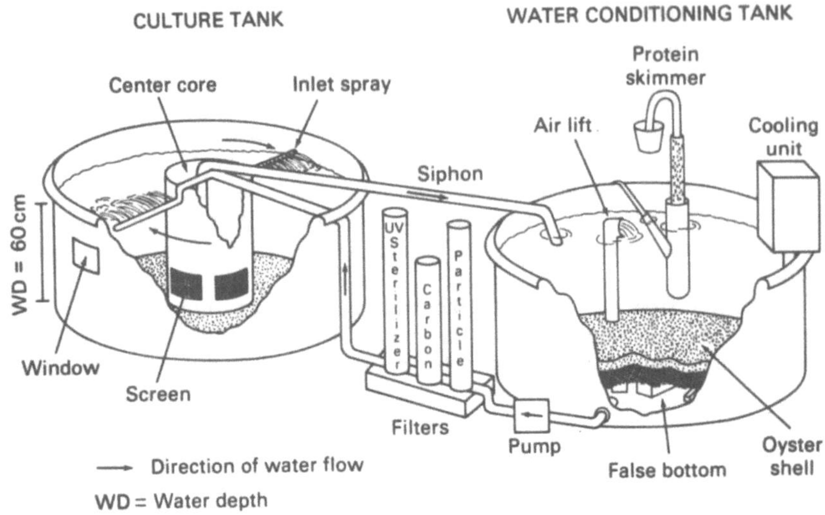

Figure 9. Sectional view of the circular tank system for egg incubation and hatchling rearing. The circulation and flow patterns are indicated by arrows. From Yang *et al.* (1989).

6. Maintenance of wild-caught juveniles and adults

The inherent practical difficulties in maintenance lie in avoiding skin damage (especially during capture, transport and transfer) and providing sufficient horizontal space in the tanks for the squids to swim and behave normally. Other important factors include quality of the seawater, species-specific behavior, stocking densities, segregation of squids, food types, feeding schedules and light levels.

Table I compares laboratory maintenance of wild-caught squids. Note that in all successful attempts the capture method was *not* by trawl, except in the cases of *Lolliguncula brevis* and *Alloteuthis subulata* (explained above).

6.1. Tank configurations

The simplest guidelines are to use tanks with (1) no corners, (2) no filtration or other objects in the same tank with squids and (3) the widest possible horizontal dimensions to accommodate swimming and schooling. Regardless of the delivery system, water quality must remain good and relatively steady. In practice, seawater systems are usually taxed with fluctuating animal biomass. This is not a problem in open systems but it is of great significance in closed systems, which must maintain biological filtration capabilities parallel with total animal biomass.

In Galveston we developed two simple tank systems that have proved reliable for maintaining squids: a 2 m circular tank system (Fig. 7) and a 10,000 liter raceway (Fig. 8). Both provided the essential criteria, i.e. the capability to sustain high-quality

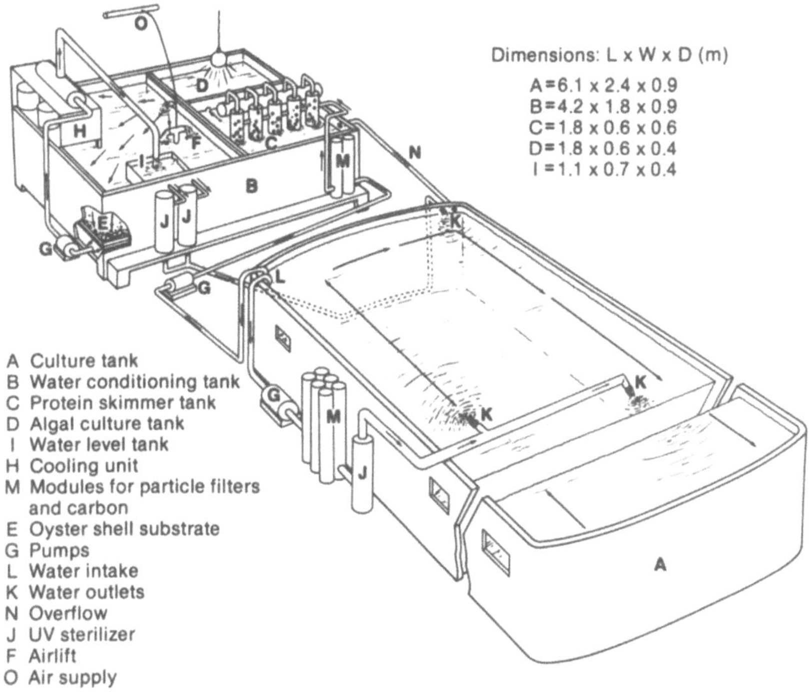

Figure 10. Raceway system for grow-out of juvenile squids. Arrows indicate circulation and flow pattern. From Yang *et al.* (1989).

water and the physical dimensions to accommodate the movements and behavior of the squids (see below). The advantages of these closed systems are (1) independence from natural seawater supply and hence reproducibility at inland locations, (2) rapid and efficient filtration, (3) large volume and wide horizontal dimensions for distribution of squids, (4) accessibility to and observation of live animals, (5) simple construction and (6) low cost.

The 2 m circular system is basic and modifiable, but the essential requirement *in this closed system* was not to exceed the carrying capacity of the filtration system. Hanlon *et al.* (1983) determined that this small system could maintain the following numbers of adult squids in a healthy state for several weeks: 10 – 15 Loligo spp. (150 – 250 mm ML) or 25 *Lolliguncula brevis* (40 – 80 mm ML). This 1500 liter system contained approximately 360 kg crushed oyster shell in the 6 cm-deep substrate; this acted as the primary biological filtrant on which lived bacteria for reducing levels of nitrogenous waste products. *New systems must be biologically* ***conditioned*** for approximately three weeks to allow the bacterial flora to grow and stabilize (Spotte, 1979). The algal tank illustrated in Figure 7 is optional and of questionable value; some recent studies indicate that algae compete directly with bacteria for ammonia, whereas traditionally algae were thought to assimilate only nitrate. In any case, monoculture of algae is nearly impossible and it is important to crop the algae regularly and attempt to minimize the range of organisms that will thrive under the bright light;

furthermore, nitrate tends to accumulate in closed systems and generally has to be removed by water changes (see also Yang *et al.*, 1989). A similar successful tank size and design was published by Matsumoto (1976) and Matsumoto and Shimada (1980). Open seawater systems with similar tank size are described by Tardent (1962), LaRoe (1971), Neill (1971), Soichi (1976), Hurley (1978) and Lipinski (1985).

Painting the tank walls a high-contrast pattern (Matsumoto, 1976; Hanlon *et al.*, 1983) apparently helps the squids see the vertical barriers and stay away from them, but this seems to benefit injured squids more than healthy ones. The same is true of complicated bumper systems on the walls (e.g., Summers *et al.*, 1974; Lipinski, 1985) and these are not recommended unless one deals exclusively with damaged animals.

The 10,000 liter raceway (Fig. 8) is a design borrowed from shrimp mariculturists and used widely for aquaculture (Mock *et al.*, 1977). The long horizontal dimensions are especially conducive to squid behavior; in general, depth is not so critical for loliginid squid schooling behavior. This system was 10 m × 2 m but raceways are made in a variety of sizes by several manufacturers in the U. S. For closed systems, the biological filter can be the size of the 2 m circular tank or upgraded as in similar culture tanks described by Yang *et al.* (1989). In an open system, this tank could hold perhaps 100 – 200 squids, while in the closed system the number would depend upon the filtration capabilities built into the system; Hanlon *et al.* (1983) kept 50 adult *Loligo* for two months. However, in recent experiments with better filtration, a similar raceway held adult *Sepioteuthis lessoniana* for months with a squid biomass eight times that of the 50 *Loligo*.

The ultimate squid maintenance tank is the Aquatron facility at Dalhousie University (O'Dor *et al.*, 1977). This open system is 15 m circular and holds about 580,000 liters. This huge system affords a much more natural environment for schooling squids but its expense is prohibitive in most cases.

6.2. Behavior, feeding and growth

The behavior of squids provides the best evaluation of tank design and filtration. Aspects of behavior are the true limiting factors to survival and growth of wild-caught squids in captivity. Survival has clearly been better in wide round tanks, all other factors being equal. Squids *require* a minimum horizontal space to behave normally; thus it is unreasonable to expect to keep squids in small rectangular tanks or any tank less than about 2 m in diameter; this has been shown clearly in the experiments of Summers and McMahon (1970, 1974) and Summers *et al.* (1974) (see Table I).

Squids school naturally so that it is advantageous to keep several squids together to enhance survival. However, it is better to keep squids of the same size and sex in each tank to reduce competition for food, cannibalism and courtship activity. All teuthoid squids appear to be cannibalistic, especially during food shortage, but smaller or injured squids are cannibalized even during times of food availability. During laboratory maintenance, loliginid squids tend to mature precociously. In mixed schools, males begin to court females with specific behavioral displays, but they also engage in intraspecific aggression directed towards other males competing for females (Hanlon *et al.*, 1983). All of this leads to disruption of feeding and increased fin and skin damage. Careful observations of squids can lead to the correlation of specific body patterns of chromatophores and postures with specific types of behavior (e.g., calmness, stress, aggressiveness, etc.) and these may be used as indicators of how well the animals are doing in captivity (Hanlon *et al.*, 1983). Most of these body patterns are

Table I. Comparisons of Laboratory Maintenance of Wild-caught Squids.

Species	*Mean Survival (days)*	*Max. Survival (days)*	*No. of Squids Evaluated*	*Capture Method*
Suborder Myopsida				
Loligo plei	≈7	38	25	Night Light[A]
"	11	84	453	Night Light[B]
Loligo pealei	2	5	241	Trawl[C]
"	5	17	–	Fish Trap[D]
"	≈4	≈7	≈250	Trawls & Fish Traps[E]
"	4	19	468	Trawls[F]
"	7	39	246	Trawls, Seine[G]
"	7	83	≈500	Trawls, Jigs[H]
"	28	71	37	Night Light, Jigs[B]
"	≈4	≈5	≈46	Fish Trap[I]
Loligo vulgaris	–	60	–	Jigs[J]
"	≈14	30	–	Jigs[K]
Lolliguncula brevis	19	125	313	Trawl[B]
Alloteuthis subulata	16	57	51	Trawl[L]
Doryteuthis bleekeri	≈17	–	12	Jigs[M]
"	≈14	–	53	Jigs[N]
"	5	–	32	Jigs[N]
"	≈43	66	10	Jigs[N]
Sepioteuthis lessoniana	75*	75	150	Jigs[O]
Suborder Oegopsida				
Illex illecebrosus	≈30	41	15	Floating Box Trap[P]
"	≈30	82	35	Floating Box Trap[P]
"	13	24	35	Floating Box Trap[P]
Todarodes pacificus	≈29	35	63	Cast Net[Q]
"	≈7	11	7	Jigs[R]
"	≈10	25	9	Jigs[S]
"	≈30	50	10	Jigs[S]
"	≈20	45	44	Jigs[T]

*held in floating cages

[A] LaRoe, 1971
[B] Hanlon *et al.*, 1983
[C] Brinley and Mullins, 1964
[D] Arnold, 1962
[E] Summers and McMahon, 1970
[F] Summers and McMahon, 1974
[G] Summers *et al.*, 1974
[H] Macy, 1980
[I] Chabala *et al.*, 1986
[J] Tardent, 1962
[K] Neill, 1971
[L] Lipinski, 1985
[M] Matumoto, 1976
[N] Matsumoto and Shimada, 1980
[O] Matsumoto, 1975
[P] O'Dor *et al.*, 1977
[Q] Mikulich and Kozak, 1971
[R] Flores *et al.*, 1976
[S] Flores *et al.*, 1977
[T] Soichi, 1976

species-specific (op.cit.; Hanlon, 1988) and require considerable behavioral observation time, but they *can lead to important findings such as the fact that adult Loligo pealei will sit calmly on the bottom if the substrate is suitable and they are healthy squids.*

Light levels and cycles affect squids but the effects are fairly species-specific and not well documented. Many researchers have kept squids in constant light, a few in a day/night cycle and one (Matsumoto, 1976) in constant low-light levels. Our experience in Galveston was that during a night cycle many squids jetted out of the tanks, indicating increased activity either from their natural hunting behavior at night or increased intraspecific aggression; survival was considerably higher with constant light (Hanlon *et al.*, 1983).

Squids must be fed in captivity; they have a high metabolism, they consume and digest their meals within hours and they have few energy stores in their bodies (Boucher-Rodoni and Mangold, 1987; O'Dor and Wells, 1987). Furthermore, they have usually been injured to some extent during capture and transport and need energy to combat blood loss and infection and to reduce stress and cannibalism. Many past researchers have not fed their animals, mainly because it is an inconvenience. Squids easily consume 15% of their body weight per day in live fishes or shrimps and can grow 20 – 40 mm ML per month on this diet (e.g., Hanlon *et al.*, 1983; Hirtle *et al.*, 1981). Freshwater fish are an acceptable food (Matsumoto, 1976) and some cephalopods will accept frozen or raw fish fillets, although little is known about how well squids survive or grow on these diets (DeRusha *et al.*, 1989).

7. Rearing and culture

Numerous attempts to rear or culture teuthoid squids have been made (Boletzky and Hanlon, 1983) but only recently have significant gains been achieved. Table II summarizes successful culture efforts with teuthoid and sepioid squids. In general, *Loligo* spp. have been very difficult to culture, *Sepioteuthis* spp. have been relatively easier and *Sepia* spp. have proved rather easy to culture. With the smaller hatchlings of *Loligo* and *Sepioteuthis* (Fig. 6), the breakthroughs have come with the realization that the very young squids have specialized requirements for feeding and they require high-quality water.

7.1. *Loligo* spp.

There is a long history of failed attempts to culture *Loligo*. In our laboratory we have concentrated from 1978 – 1985 on culturing *Loligo*, culminating first with success with *Loligo opalescens* (Yang *et al.*, 1983; Yang *et al.*, 1986) and later with the largest species of *Loligo*, *Loligo forbesi* (Hanlon *et al.*, 1989a). However, the results have been disappointing; mortality has been high and growth has been slow because these are cold-water species. Furthermore, attempts to culture *Loligo vulgaris* and *Loligo pealei* through the life cycle have been restricted to short-term rearing experiments (Turk *et al.*, 1986; Hanlon *et al.*, 1987). Vast details are available in these citations, but this section will focus on the most recent and successful culture results with *Loligo forbesi* (Hanlon *et al.*, 1989a).

Loligo forbesi eggs were obtained from England or the Azores Islands and air-shipped to Galveston. Hatching success was 72 – 87% and the best survival during

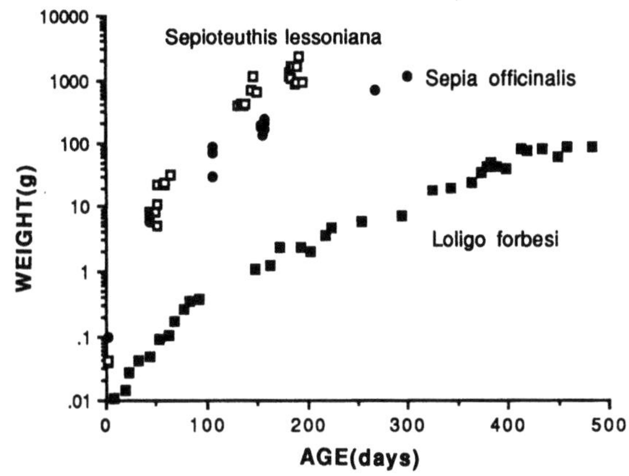

Figure 11. Comparison of growth from hatching to adult size. Note the log scale. Maximum size for *Loligo forbesi* was 124 g, *Sepia officinalis* was 1400 g and *Sepioteuthis lessoniana* was 2200 g.

the first critical 75 days was 15% (more commonly it was 2 – 8%). Figure 5 lists the sequence of food types provided during 3 major experiments and Figure 4 gives an impression of the range of sizes that the hatchlings would feed upon. Considerable effort was directed toward determining the relationship of squid density and food density and how it affected survival during the first 30 days. However, no clear results were obtained. Nevertheless, we noticed that squids fed more vigorously when a variety of small plankton was available: copepods 1 – 3 mm long, palaemonid shrimp larvae 1 – 4 mm long and small mysid shrimp 2 – 4 mm long. Larval fishes 3 – 5 mm long were also actively attacked, and in nature they are probably an important constituent of the diet. It is important to provide large quantities of food nearly round the clock for good survival and growth of the youngest squids. The recommendations of Hanlon *et al.* (1989a) were: (1) to provide 15 – 20 copepods per squid from days 1 – 20, (2) provide small mysid shrimps from day 4 or 5 onwards, increasing mysid size as squids grow, (3) provide a variety of other foods to promote feeding and (4) slowly raise temperatures 2 – 4 °C within 2 weeks post-hatching to enhance feeding rates and growth.

Mean daily feeding rate of juvenile squids from days 100 to 369 was estimated at approximately 14% of squid body weight; this ranged from squids of 10 g feeding at 20% and declining until squids at 60 g were feeding at a rate of 4.6%.

Growth was analyzed carefully by Forsythe and Hanlon (1989). The hatchlings are typically 4.2 mm ML and 5 – 9 mg (wet weight) and the largest squids cultured were a male 155 mm ML and 124 g in 413 days and a female 135 mm ML and 108 g at 434 days. Growth was exponential for the first three months at rates of about 5% body weight gain per day; thereafter growth slowed to a logarithmic form at decreasing rates of 3 – 1% (Fig. 11).

Survival was always poor in the first few weeks (mentioned previously) and thereafter there was a fairly typical slow attrition for months until maturation approached. Even in the best experiment only 40% of the squids transferred to raceways for growout were alive 200 days later. Maturation and spawning were achieved only with *Loligo opalescens* (Yang *et al.*, 1986) but not with *Loligo forbesi* in four major experiments, although the largest squids had mature gonads.

7.2. *Sepioteuthis lessoniana*

Preliminary rearing trials with this species began long ago in Korea and Japan (e.g. Choe and Ohshima, 1961) but focused upon different parts of the life cycle and met with limited success. In 1971, LaRoe reared *Sepioteuthis sepioidea* from the Caribbean Sea to adult size, then Segawa (1987) reared *Sepioteuthis lessoniana* to adult size. Shortly after Dr. Segawa's postdoctoral fellowship in Galveston, we first cultured *Sepioteuthis lessoniana* and have since then achieved highly successful and reproducible culture results. Seven major experiments have been completed and the squids have been cultured three times to the second generation.

In two of the experiments, eggs of late developmental stages were imported from Japan (early stages are preferable) and resulted in high mortality immediately posthatching – 85% and 26% within five days, due to the poor water quality from the long shipment time in a small amount of water. In both cases hatching rate was reasonable – 83% and 86% – but many of these were premature hatches, which contributed greatly to mortality. The hatchlings were huge by comparison with *Loligo* (Fig. 6): about 6 mm ML and 44 mg, or roughly seven times the weight of a *Loligo forbesi* hatchling (it takes *Loligo forbesi* about 45 – 60 days to attain this weight). In all experiments the young squids were fed mysid shrimp (*Mysidopsis* spp.) of 2 – 4 mm length, with the occasional addition of small larval palaemonid shrimp 1 – 3 mm long. These shrimp were added in large quantities (often 80 – 200 per squid) during the first weeks to stimulate feeding. In general, feeding was first observed four days posthatching and then proceeded at a high rate. As the squids grew, they were fed primarily penaeid shrimp and fishes (mainly mullet) and secondarily smaller estuarine fishes such as cyprinidonts and silversides. Cannibalism resulted when food shortages occurred. These squids had prodigious appetites, consuming as high as 35% of their body weight per day.

Growth and survival after the first week were phenomenal. At a mean temperature of 25 °C, the young squids grew at rates of 7 – 12% body weight gain per day; overall growth rates were approximately 7% throughout the life cycle. This is exceptionally fast, even for a cephalopod (Forsythe and Van Heukelem, 1987). Maximum size attained was 360 mm ML and 2200 g in 184 days (Fig. 11). In all experiments to date, squids attained weights of 1 kg in about five or six months. Furthermore, survival has been exceptional after the initial mortality of the first few weeks, which was usually only 5% or so until reproduction, spawning and senescence. The numbers of adult squids cultured to full size have ranged from 30 to 131, respectively.

Reproduction and spawning have occurred in each experiment, with variable success. First matings have been observed as early as 130 – 160 days posthatching and spawning as early as 153 days. Although numerous egg strands (600 – 1000) have been laid in the experiments, viability of the eggs has been low (< 10%) due to poor fertilization and disintegration of the embryos before hatching. These problems are being studied. However, first filial generation hatchlings have just been cultured through the life cycle with good results and indicate clearly the potential for providing semi-continuous laboratory culture of this species.

Table II. Squids That Have Been Cultured at Least to The First Filial (F_1) Generation in the Laboratory.

Species	*Mean ML at Hatching (mm)*	*Approx. Adult Size at Reproduction (mm ML)*	*Maximum Time in Culture (days)*	*Number of Generations*	*Type of Seawater System*
Order Sepioidea*					
Sepia officinalis	8.0	70 – 250	420	2	closed[A]
"	8.0	70 – 250	500	"several"	open[B]
"	8.0	70 – 250	730	1	open[C]
"	8.0	70 – 250	609	3	open[D]
"	8.0	70 – 250	1800	5	open[E]
"	8.0	70 – 250	1161	4	closed
Sepiola affinis	3.2	20	250	1	open[F]
Sepiola robusta	2.2	20	220	1	open[G]
Sepiola rondeleti	3.7	20	150	1	open[G]
Sepietta oweniana	≈5.0	30 – 40	≈300	1	open[H]
Order Teuthoidea†					
Loligo opalescens	2.7	85	233	1	closed[I]
"	3.0	75	234	1	closed[J]
Sepioteuthis lessoniana	5.0	360	575	1	closed

* Mode of life of young animals is benthic.
† Mode of life of young animals is planktonic.

[A] Schröder, 1966
[B] Richard, 1966, 1975
[C] Boletzky, 1979
[D] Pascual, 1978
[E] Zahn, 1979; 1989
[F] Boltezky, 1975
[G] Boletzky *et al.*, 1971
[H] Summers and Bergström, 1981
[I] Yang *et al.*, 1980, 1983
[J] Yang *et al.*, 1983

7.3. *Sepia officinalis*

The common cuttlefish of Europe and the Mediterranean has been a popular and important research model for many decades. This cephalopod has proved to be easy to rear and was first cultured through two generations in closed systems in the Berlin Aquarium by Schröder in 1966, and Zahn (1979) cultured them through five generations in Düsseldorf (Table II). Many aspects of the life cycle of *Sepia officinalis* have been reviewed by Boletzky (1983). Four important advantages of culturing these sepioid squids are: (1) the hatchlings are extremely large (Fig. 6), (2) they sit on or bury into the substrate, (3) they tolerate crowding well and (4) they are relatively catholic in food acceptability. In Galveston, we now have cultured four consecutive generations of this species with little problem.

Eggs are very large (ca. 1.3 cm wide and 2 – 3 cm long) and are laid individually by the female. Our original genetic stock came from England via the Monterey Bay Aquarium in California. Hatching success is normally excellent (> 90%) and the hatchlings are fully formed, sophisticated cuttlefish with nearly all the capabilities of adults (Hanlon and Messenger, 1988). These large hatchlings are usually 6 mm ML and 950 mg (Fig. 6). They prefer to bury in a sand or fine gravel substrate, but they do well with no particulate substrate and will either sit on the bottom or hover in shoals (by regulating buoyancy in the cuttlebone). The young animals feed vigorously, with a preference for mysid shrimp (Wells, 1958); however, various authors have reported that they will eat other small shrimp and amphipods (Boletzky and Hanlon, 1983) and recently they have been reared for the first 6 weeks on high concentrations of adult brine shrimp (*Artemia*) with 95% survival (DeRusha *et al.*, 1989). After only two weeks on mysids, larger shrimp such as *Palaemonetes* spp. or small fishes will be fed upon and eventually only the size of the shrimp or fish need be increased to culture these cuttlefishes.

DeRusha *et al.* (1989) have evaluated some alternative diets that worked fairly well with *Sepia officinalis*. Hatchlings accepted and grew not only on brine shrimp but also fresh-water guppies, although growth rate was only 3.7% per day compared to 7.6% per day on live mysids. Juveniles (4 – 40 g) were reared for two months on frozen shrimp and showed very similar feeding rates (≈8%) and growth rates (≈3.5%) compared to squids on live diets. Larger juveniles (30 – 160 g) grew reasonably well on frozen marine fish (≈2.7% growth rate) with feeding rates on the order of ≈6.5%.

In each of the four generations approximately 50 cuttlefish have been grown to full adult spawning size, although many more animals were culled out at smaller sizes for experiments. At 16 – 18 °C it took about 14 months per generation, while at 21 – 24 °C it took only 10 months; *Sepia officinalis* tolerated temperatures between 13 and 25 °C. In most cases the average adult male was 800 g and the females about 600 g; the largest animal reached 1400 g. These growth rates compare favorably with those reported by others both in the laboratory and in the sea (cf. Pascual, 1978; Boletzky, 1983; Guerra and Castro, 1988). Survival in our tanks has been excellent up to 100 g but variable thereafter because of fluctuations in water quality due to uncommonly good survival. Unexpectedly high biomasses taxed the filtration capacities of available culture systems and demonstrated to us that disease incidence increases when water quality decreases and food is heavily competed for. However, during periods when water quality and food were adequate we found practically no mortality, and this must be considered the normal case.

It is noteworthy that this species tolerates crowding *throughout* its life cycle. As many as 300 cuttlefishes 3 – 5 weeks old have been reared in a water tray 60 cm × 240 cm. In fact, the limiting factors are water quality and food, not the behavior of the cuttlefish. Furthermore, this species tolerates the shipping procedure extremely well, and we have been successful in sending cuttlefish of all sizes around the U. S. and southern Canada.

Reproduction in captivity has been good. Mating and spawning commence regularly near the end of the life cycle and thousands of eggs have been produced. The only problem encountered thus far has been that only half or so of the eggs have been fertilized. The fertile eggs developed and hatched normally, with high viability. For all of these reasons we believe this is an ideal laboratory animal.

8. Mortality and disease

Embryonic mortality in teuthoid squids is often high with imported eggs in closed seawater systems. Egg capsules sometimes become infected with a pink bacterium and it is suspected that other epizootic organisms (benthic copepods, etc.) may destroy the jelly matrix that supports the individual embryos in loliginid squids. Recent experiments in Galveston indicate that dipping the egg strands in solutions of iodine cuts down this problem somewhat, but this deserves future experimentation. Nevertheless, hatching success is usually greater than 75%.

Mortality in all squids is always highest within two weeks posthatching and this is presumably due to starvation when the squids must make the transition from yolk feeding to active predation. Sample mortalities are: *Loligo pealei* 99% in five days; *Loligo vulgaris* 90% in 10 days; *Loligo opalescens* 50% in 15 days; *Loligo forbesi* 50% in seven days; *Sepioteuthis lessoniana* 40 – 70% in seven days; and *Sepia officinalis* 0 – 5% in 14 days (Hanlon *et al.*, 1987; Turk *et al.*, 1986; Yang *et al.*, 1986; Hanlon *et al.*, 1989a; Segawa, 1987 and Hanlon this chapter; Boletzky, 1983, respectively).

Juvenile and adult teuthoid squids reared from eggs most often slowly accrue skin damage that destroys the fins and impairs feeding (Yang *et al.*, 1986; Hanlon *et al.*, 1989a). Occasionally females laden with eggs swim along the substrate and wear a hole in the ventral mantle, which becomes badly infected (op.cit.). Bacterial abscesses sometimes occur in the posterior mantle or the eyes and are eventually lethal. Intraspecific aggression related to reproductive behavior also results in some squids (usually smaller ones) jetting out of the tank.

Wild-caught juvenile and adult teuthoid squids suffer similar maladies (Leibovitz *et al.*, 1977; Hulet *et al.*, 1979; Hanlon *et al.*, 1983). Typical survival of loliginid squids is poor. The high initial mortality is due to injuries incurred during capture, transport and transfer and also to the fact that all squids captured alive (regardless of size or condition) were included in this study (Hanlon *et al.*, 1983); culling out small or injured squids during transport and transfer is advisable and would lead to higher survival. A rare but unusually devastating skin lesion has been observed in *Lolliguncula brevis* in which the dorsal mantle splits open and exposes the gladius (Ford *et al.*, 1986). Cannibalism, exacerbated by crowding or low food levels, also can cause considerable mortality. Jones and O'Dor (1983) reported mortality in the oegopsid squid *Illex illecebrosus* from gill infection by thraustochytrid fungi. Without question, skin damage from hitting walls is the major contributor to mortality. Females tend to mature precociously in captivity and this leads to spawning and subsequent death (Hanlon *et al.*, 1983).

Sepioteuthis lessoniana and the sepioid *Sepia officinalis* are hardier and less susceptible to disease than the other teuthoid squids; the reasons are unknown but one obvious reason is the larger fin around the mantle that enables greater swimming maneuverability. Furthermore, both species live near or on the substrate on coral or rock reefs and are apparently better adapted to swimming amidst vertical structures than open-water squids. The eggs of both are very large compared to other squids and the well-developed hatchlings are better equipped to swim and feed. It is possible that these species have tougher skin, more active mucus-secreting cells in the epidermis or stronger internal defense mechanisms, but this remains speculative.

A review of cephalopod diseases has recently been assembled (Hanlon and Forsythe, 1989).

9. Summary

Squids have several distinct anatomical and behavioral traits that require specific attention when maintenance, rearing or culture are attempted. Furthermore, different species of squids and cuttlefish have significantly different attributes that render them easier or more difficult to handle in the laboratory and these should be considered before a species is chosen for a particular project.

Many reliable and reproducible techniques and tank systems are now available for handling wild-caught eggs or squids.

Sepioteuthis lessoniana should be considered as a viable alternative to *Loligo* spp. in the near future because: (1) it has essentially the same anatomical features that have made *Loligo* spp. useful and (2) unlike *Loligo* spp., it can be cultured throughout the life cycle rapidly and with predictability.

The cuttlefish *Sepia officinalis* is easy to rear and has many anatomical and behavioral attributes that complement the types of research that are being conducted on *Loligo* spp.; furthermore, this species may be shipped nationwide. Thus far in the U. S., this species is beginning to be used for neurobiological studies of the giant axon, stellate ganglia, oculo-motor system, vestibular apparatus, chromatophores and epidermal mechanoreceptors. Other studies include the morphology of the cuttlebone, statoliths, muscle and hair cells as well as the biochemistry of the retina and ink gland.

10. Future considerations

The principal limitations to expanded use of squids in the laboratory are the costs involved with providing live food to animals of all sizes (particularly in a long-term mariculture experiment) and the current difficulty and expense of shipping large squids around the country. In Galveston, we are currently testing alternative live and frozen diets for cephalopods (DeRusha et al, 1989) and we are attempting to develop an artificial diet. Success in this endeavor would not only reduce the costs of mariculture but would enable researchers away from the coast to maintain or rear squids. The shipping issue is mainly one of cost; large animals could be shipped now because we know the water quality and volume requirements, but the associated cost of the water is too great to be considered feasible in most cases. Future work on the use of anaesthetics during shipping to reduce the need for so much water may be worthwhile.

ACKNOWLEDGMENTS. I am most grateful to my colleagues in Galveston who have contributed to our success in culturing cephalopods. In particular I thank Won Tack Yang, Phillip G. Lee, Philip E. Turk and John W. Forsythe for helping bring together some of the data in this chapter. I am especially thankful for support from the Division of Research Resources, National Institutes of Health, on grants DHHS RR01024 and DHHS RR 04226. Laura Koppe was very helpful in typing and assembling the manuscript.

References

Arnold, J. M., 1962, Mating behavior and social structure in *Loligo pealii*, *Biol. Bull.* **123**:53–57.

Arnold, J. M., 1965, Observations on the mating behavior of the squid *Sepioteuthis sepioidea*, *Bull. Mar. Sci.* **15**:216–222.

Boletzky, S. v., 1974, Élevage de céphalopodes en aquarium, *Vie Milieu* **24**:309–340.

Boletzky, S. v., 1975, The reproductive cycle of Sepiolidae (Mollusca, Cephalopoda), *Pubbl. Staz. Zool. Napoli Suppl.* **39**:84–95.

Boletzky, S. v., 1979, Growth and life-span of *Sepia officinalis* under artificial conditions (Mollusca, Cephalopoda), *Rapp. Comm. int. Mer Médit* **25/26**:159–168.

Boletzky, S. v., 1983, *Sepia officinalis*, in: Cephalopod Life Cycles, Volume I: Species Accounts (P. R. Boyle, ed.), pp. 31–52, Academic Press, New York.

Boletzky, S. v. and Hanlon, R. T., 1983, A review of the laboratory maintenance, rearing and culture of cephalopod molluscs, *Mem. Natl. Mus. Victoria* **44**:147–187.

Boletzky, S. v., Boletzky, M. V., Frösch, D., and Gätzi, V., 1971, Laboratory rearing of Sepiolinae (Mollusca: Cephalopoda), *Mar. Biol.* **8**(1):82–87.

Boucher-Rodoni, R. and Mangold, K., 1987, Feeding and digestion, in: *Cephalopod Life Cycles, II. Comparative Reviews* (P. R. Boyle, ed.), pp. 85–108, Academic Press, London.

Brinley, Jr., F. J. and Mullins, L. J., 1964, The collection of squid for use at an inland laboratory, 6 pp., *unpublished manuscript.*

Budelmann, B. U. and Bleckmann, H., 1988, A lateral line analogue in cephalopods: water waves generate microphonic potentials in the epidermal head lines of *Sepia* and *Lolliguncula, J. Comp. Physiol. A* **164**:1–5.

Chabala, L. D., Morello, R. S., Busath, D., Danko, M., Smith-Maxwell, C. J., and Begenisich, T., 1986, Capture, transport, and maintenance of live squid (*Loligo pealei*) for electrophysiological studies, *Pflügers Arch.* **407**:105–108.

Choe, S. and Ohshima, Y., 1961, On the embryonal development and the growth of the squid, *Sepioteuthis lessoniana* Lesson, *Venus* **21**(4):462–477.

Colmers, W. F., Hixon, R. F., Hanlon, R. T., Forsythe, J. W., Ackerson, M. V., Wiederhold, M. L., and Hulet, W. H., 1984, "Spinner" cephalopods: defects of statocyst suprastructures in an invertebrate analogue of the vestibular apparatus, *Cell Tissue Res.* **236**: 505–515.

DeRusha, R. H., Forsythe, J. W., DiMarco, F. P., and Hanlon, R. T., 1989, Alternative diets for maintaining and rearing cephalopods in captivity, *Lab. Anim. Sci.* **39**:306–312.

Flores, E. E. C., Igarashi, S., Mikami, T., and Kobayashi, K., 1976, Studies on squid behavior in relation to fishing. I. On the handling of squid, *Todarodes pacificus* Steenstrup, for behavioral study, *Bull. Fac. Fish. Hokkaido Univ.* **27**:145–151.

Flores, E. E. C., Igarashi, S., and Mikami, T., 1977, Studies on squid behavior in relation to fishing. II. On the survival of squid, *Todarodes pacificus* Steenstrup, in experimental aquarium, *Bull. Fac. Fish. Hokkaido Univ.* **28**:137–142.

Ford, L. A., Alexander, S. K., Cooper, K. M., and Hanlon, R. T., 1986, Bacterial populations of normal and ulcerated mantle tissue of the squid, *Lolliguncula brevis*, *J. Invert. Path.* **48**:13–26.

Forsythe, J. W. and Hanlon, R. T., 1989, Growth of the Eastern Atlantic squid, *Loligo forbesi* Steenstrup (Mollusca: Cephalopoda), *Aquacul. Fish. Manag.* **20**:1–14.

Forsythe, J. W. and Van Heukelem, W. F., 1987, Growth, in: *Cephalopod Life Cycles, II. Comparative Reviews* (P. R. Boyle, ed.), pp. 135–155, Academic Press, London.

Grimpe, G., 1928, Pflege, Behandlung und Zucht der Cephalopoden für zoologische und physiologische Zwecke, *Abderhalden's Handbuch biol. Arbeitsmeth.* **9**:331–412.

Guerra, A. and Castro, B. G., 1988, On the life cycle of *Sepia officinalis* (Cephalopoda, Sepioidea) in the ria de Vigo (NW Spain), *Cah. Biol. Mar.* **29**:395–405.

Hanlon, R. T., 1982, The functional organization of chromatophores and iridescent cells in the body patterning of *Loligo plei* (Cephalopoda: Myopsida), *Malacologia* **23**:89–119.

Hanlon, R. T., 1987, Mariculture, in: *Cephalopod Life Cycles*, Volume II: Comparative Reviews (P. R. Boyle, ed.), pp. 291–305, Academic Press, New York.

Hanlon, R. T., 1988, Behavioral and body patterning characters useful in taxonomy and field identification of cephalopods, *Malacologia* **29**(1):247–264.

Hanlon, R. T. and Budelmann, B. U., 1987, Why cephalopods probably are not "deaf," *Am. Nat.* **129**(2):312–317.

Hanlon, R. T. and Forsythe, J. W., 1989, 1: Diseases of Cephalopods, in: *Diseases of Marine Animals*, Vol. III. Cephalopoda to Urochordata (O. Kinne, ed.), Biologische Anstalt Helgoland, Hamburg, W. Germany, *in press*.

Hanlon, R. T. and Messenger, J. B., 1988, Adaptive coloration in young cuttlefish (*Sepia officinalis* L.): the morphology and development of body patterns and their relation to behaviour, *Phil. Trans. Roy. Soc. Lond.* **320**:437–487.

Hanlon, R. T., Hixon, R. F., Hulet, W. H., 1983, Survival, growth, and behavior of the loliginid squids *Loligo plei*, *Loligo pealei*, and *Lolliguncula brevis* (Mollusca: Cephalopoda) in closed sea water systems, *Biol. Bull.* **165**(3):637–685

Hanlon, R. T., Turk, P. E., Lee, P. G., and Yang, W. T., 1987, Laboratory rearing of the squid *Loligo pealei* to the juvenile stage: growth comparisons with fishery data, *Fish. Bull.* **85**(1):163–167.

Hanlon, R. T., Yang, W. T., Turk, P. E., Lee, P. G., and Hixon, R. F., 1989a, Laboratory culture and estimated life span of the Eastern Atlantic squid *Loligo forbesi* (Mollusca: Cephalopoda), *Aquacul. Fish. Manag.* **20**:15–33.

Hanlon, R. T., Bidwell, J. P., and Tait, R., 1989b, Strontium is required for statolith development and thus normal swimming behaviour of hatchling cephalopods, *J. exp. Biol.* **141**:187–195.

Hendrix, Jr., J. P., Hulet, W. H., and Greenberg, M. J., 1981, Salinity tolerance and the responses to hypoosmotic stress of the bay squid *Lolliguncula brevis*, a euryhaline cephalopod mollusc, *Comp. Biochem. Physiol.* **69A**:641–648.

Hirtle, R. W. M., DeMont, M. E., and O'Dor, R. K., 1981, Feeding growth, and metabolic rates in captive short-finned squid, *Illex illecebrosus*, in relation to the natural population, *J. Shellfish Res.* **1**(2):187–192.

Hulet, W. H., Villoch, M. R., Hixon, R. F., and Hanlon, R. T., 1979, Fin damage in captured and reared squids (Cephalopoda, Myopsida), *Lab. Anim. Sci.* **29**(4):528–533.

Hurley, A. C., 1977, Mating behavior of the squid *Loligo opalescens*, *Mar. Behav. Physiol.* **4**:195–203.

Hurley, A. C., 1978, School structure of the squid *Loligo opalescens*, *Fish. Bull.* **76**(2):433–442.

Jones, G. M. and O'Dor, R. K., 1983, Ultrastructural observations on a thraustochytrid fungus parasitic in the gills of squid (*Illex illecebrosus* Lesueur), *J. Parasitol.* **69**:903–911.

LaRoe, E. T., 1971, The culture and maintenance of the loliginid squids *Sepioteuthis sepioidea* and *Doryteuthis plei*, *Mar. Biol.* **9**(1):9–25.

Leibovitz, L., Meyers, T. R., and Elston, R., 1977, Necrotic exfoliative dermatitis of captive squid (*Loligo pealei*), *J. Invert. Path.* **30**:369–376.

Lipinski, M. R., 1985, Laboratory survival of *Alloteuthis subulata* (Cephalopoda: Loliginidae) from the Plymouth area, *J. Mar. Biol. Ass. U.K.* **65**:845–855.

Macy, III, W. K., 1980, The ecology of the common squid *Loligo pealei* Lesueur, 1821 in Rhode Island waters, Ph.D. dissertation, Univ. Rhode Island, Narragansett.

Matsumoto, F., 1975, Short-term holding culture of Aori-ika (*Sepioteuthis lessoniana*) [in Japanese], *Gyoson* **41**(4):21–22.

Matsumoto, G., 1976, Transportation and maintenance of adult squid (*Doryteuthis bleekeri*) for physiological studies, *Biol. Bull.* **150**:279–285.

Matsumoto, G. and Shimada, J., 1980, Further improvement upon maintenance of adult squid (*Doryteuthis bleekeri*) in a small circular and closed-system aquarium tank, *Biol. Bull.* **159**(2):319–324.

Mikulich, L. V. and Kozak, L. P., 1971, Experimental rearing of Pacific Ocean squid under artificial conditions, *Ékologiya* **3**:94–96.

Mock, C. R., Ross, L. R., and Salser, B. R., 1977, Design and preliminary evaluation of a closed system for shrimp culture, *Proc. 8th Ann. Meet. World Maricul. Soc.*, pp. 335–372.

Moynihan, M. and Rodaniche, A. F., 1982, *The Behavior and Natural History of the Caribbean Reef Squid Sepioteuthis sepioidea With a Consideration of Social, Signal, and Defensive Patterns for Difficult and Dangerous Environments*, Verlag Paul Parey, Berlin and Hamburg.

Neill, S. St. J., 1971, Notes on squid and cuttlefish; keeping, handling and colour-patterns, *Pubbl. Staz. Zool. Napoli* **39**:64–69.

O'Dor, R. K. and Wells, M. J., 1987, Energy and nutrient flow, in: *Cephalopod Life Cycles, II. Comparative Reviews* (P. R. Boyle, ed), pp. 109–133, Academic Press, London.

O'Dor, R. K., Durward, R. D., and Balch, N., 1977, Maintenance and maturation of squid (*Illex illecebrosus*) in a 15 meter circular pool, *Biol. Bull.* **153**:322–335.

Paffenhofer, G. A. and Harris, R. P., 1979, Laboratory culture of marine holozooplankton and its contribution to studies of marine plankton food webs, *Adv. Mar. Biol.* **16**:211–308.

Pascual, E., 1978, Crecimiento y alimentación de tres generaciones de *Sepia officinalis* en cultivo, *Inv. Pesq.* **42**(2):421–442.

Richard, A., 1966, La température, facteur externe essentiel de croissance pour le Céphalopode *Sepia officinalis* L., *C. R. Acad. Sc. Paris* **263**:1138–1141.

Richard, A., 1975, L'élevage de la seiche (*Sepia officinalis* L., Mollusque Céphalopode), *10th Europ. Symp. Mar. Biol., Ostend (Belgium)* **1**:359–380.

Schröder, W., 1966, Beobachtungen bei der Zucht von Tintenfischen (*Sepia officinalis* L.), *Sber. Ges. naturf. Freunde (N.F.)* **6**:101–107.

Segawa, S., 1987, Life history of the oval squid, *Sepioteuthis lessoniana*, in Kominato and adjacent waters of central Honshu, Japan, *J. Tokyo Univ. Fish.* **74**(2):67–105.

Segawa, S., Yang, W. T., Marthy, H.-J., and Hanlon, R. T., 1988, Illustrated embryonic stages of the Eastern Atlantic squid *Loligo forbesi*, *Veliger* **30**(3):230–243.

Shevtsova, V. D., 1977, Cephalopods as a potential object of rearing [in Russian], Centr. Res. Inst. Techn. Econ. Res. & Inform. Fisheries, Moscow. *Review N5 Info. Ser. Fish. Resources World Ocean:* pp. 1–46.

Soichi, M., 1976, On the growth and food quantity of young common squid, *Todarodes pacificus*, in captivity [in Japanese], *Dosuishi* **13**(4):79–82.

Spotte, S., 1979, *Fish and Invertebrate Culture. Water Management in Closed Systems*, 2nd Edition, John Wiley & Sons, New York.

Summers, W. C. and Bergström, B., 1981, Cultivation of the sepiolid squid, *Sepietta oweniana*, and its ecological significance, *Am. Zool.* **20**:927 (abstr).

Summers, W. C. and McMahon, J. J., 1970, Survival of unfed squid, *Loligo pealei*, in an aquarium, *Biol. Bull.* **138**:389–396.

Summers, W. C. and McMahon, J. J., 1974, Studies on the maintenance of adult squid (*Loligo pealei*). I. Factorial survey, *Biol. Bull.* **146**(2):279–290.

Summers, W. C., McMahon, J. J., and Ruppert, G. N. P., 1974, Studies on the maintenance of adult squid (*Loligo pealei*). II. Empirical extensions, *Biol. Bull.* **146**:291–301.

Tardent, P., 1962, Keeping *Loligo vulgaris* L. in the Naples aquarium, *1st Congres International D'Aquariologie A:* pp. 41–45.

Turk, P. E., Hanlon, R. T., Bradford, L. A., and Yang, W. T., 1986, Aspects of feeding, growth and survival of the European squid *Loligo vulgaris* Lamarck, 1799, reared through the early growth stages, *Vie Milieu* **36**(1):9–13.

Wells, M. J., 1958, Factors affecting reactions to *Mysis* by newly hatched *Sepia, Behaviour* **13**:96–111.

Yang, W. T., Hanlon, R. T., Krejci, M. E., Hixon, R. F., and Hulet, W. H., 1980, Culture of California market squid from hatching – first rearing of *Loligo* to subadult stage, *Aquabiology* **2**(6):412–418. [in Japanese]

Yang, W. T., Hanlon, R. T., Krejci, M. E., Hixon, R. F., and Hulet, W. H., 1983, Laboratory rearing of *Loligo opalescens*, the market squid of California, *Aquaculture* **31**:77–88.

Yang, W. T., Hixon, R. F., Turk, P. E., Krejci, M. E., Hulet, W. H., and Hanlon, R. T., 1986, Growth, behavior and sexual maturation of the market squid, *Loligo opalescens*, cultured through the life cycle, *Fish. Bull.* **84**(4):771–798.

Yang, W. T., Hanlon, R. T., Lee, P. G., and Turk, P. E., 1989, Design and function of closed seawater systems for culturing loliginid squids, *Aquacul. Eng.* **8**:47–65.
Young, R. E. and Harmon, R. F., 1988, "Larva","paralarva" and "subadult" in cephalopod terminology, *Malacologia* **29**(1):201–207.
Zahn, M., 1979, *Sepia officinalis* (Sepiidae) – Ruheverhalten, Tarnung und Fortbewegung [in German], Film # 2271 des IWF, Gottingen 1977, *Publ. Wiss. Film, Sekt., Biol.*, Ser. **12**, No. 2/2271, 17 pp.
Zahn, M., 1989, *pers. commun.*

Part II

MATING BEHAVIOR AND EMBRYOLOGY

Chapter 5

Squid Mating Behavior

JOHN M. ARNOLD

1. Introduction

This discussion of the reproductive behavior and anatomy will be limited to a discussion of the teuthoids (Squid, Cuttlefish, Sepiolids, and their kin). For studies on nautiloids and octopods, refer to Arnold (1984), Saunders and Landman (1987), and Boyle (1983).

2. Reproductive anatomy

All of the cephalopods are sexually dimorphic with extreme variation in the size ratios of the sexes in different species. In *Argonauta* the male is about 2% the size of the female and lives independently without the shell-like egg capsule typical of the female. When they mate, the males loses his reproductive arm which becomes wrapped around the opening of the oviduct of the female. This arm, which is approximately the size of the rest of the body of the male is so great, that the zoologist Cuvier mistook it for a parasite and called it the *hectocotylus*, a name that is presently still in use. In a more typical case in the teuthids, the males and females differ slightly in general body proportions; the gonads can be seen through the mantle, so the sexes can easily be distinguished in the living animal. In the loliginids (the group which includes *Loligo pealei*). the male is slightly longer than the female and the females are broader in the posterior region of the mantle. The testis is clearly visible between the fins of the male as an elongate whitish mass; while in the female, the ovary is a larger, diffuse, yellowish area between the fins. When the female is sexually mature, the accessory nidamental gland can be seen through the mantle near the opening of the oviduct (Fig. 1). This gland changes color as the female reaches reproductive maturity, possibly by the development of a symbiotic bacterium within the gland itself (Margolis, 1989).

I will describe the reproductive systems of *Loligo* as typical and mention other species differences only if they represent significant departures.

J. M. ARNOLD • Pacific Biomedical Research Center, Cephalopod Biology Laboratory, 209A Snyder Hall, 2538 The Mall, University of Hawaii at Manoa, Honolulu, Hawaii 96822.

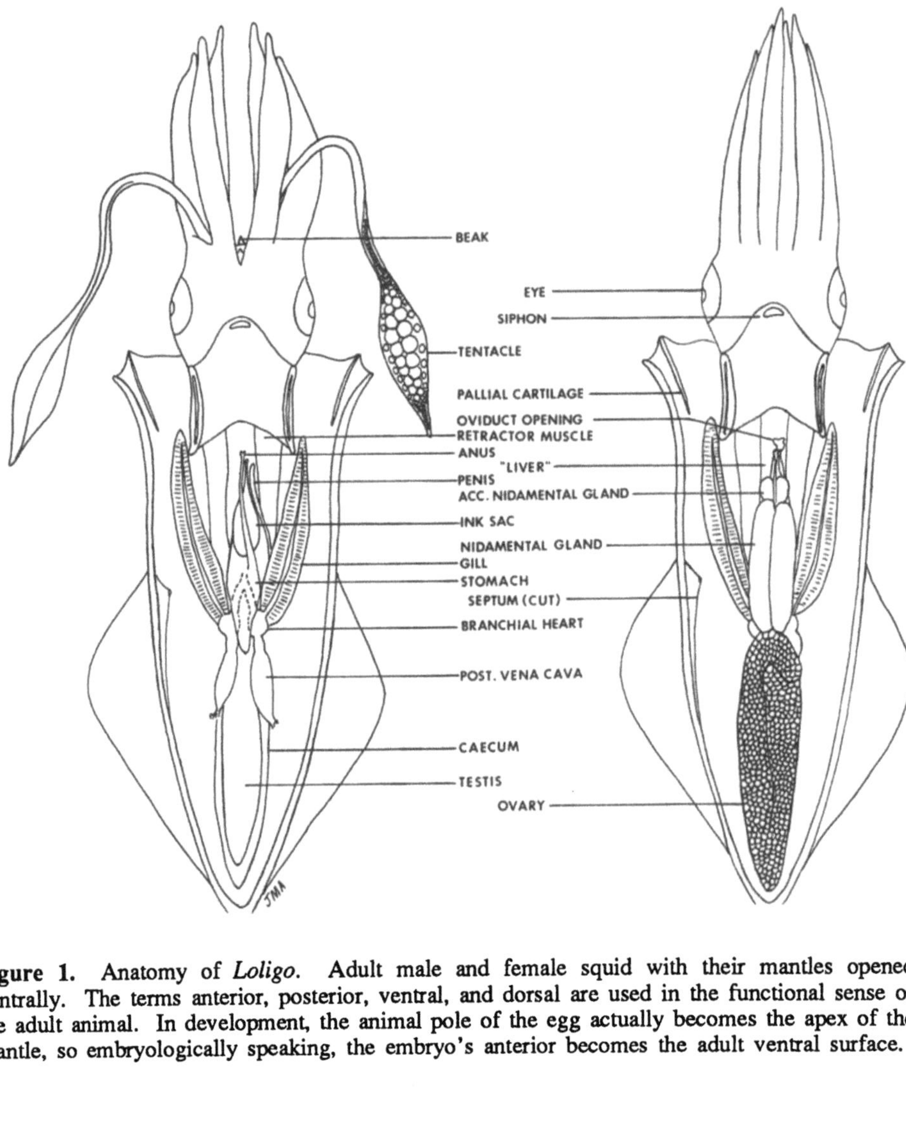

Figure 1. Anatomy of *Loligo*. Adult male and female squid with their mantles opened ventrally. The terms anterior, posterior, ventral, and dorsal are used in the functional sense of the adult animal. In development, the animal pole of the egg actually becomes the apex of the mantle, so embryologically speaking, the embryo's anterior becomes the adult ventral surface.

2.1. Male reproductive system

The male reproductive system (Fig. 1) consists of a large white median testis which lies between the fins and behind the heart. The caecum lies ventral to it and it is surrounded by parts of the renal system. The testis empties into a ciliated funnel at the end of the vas deferens, and the sperm are collected by it after being nominally, but not functionally, released into the coelom. The tubular vas deferens leads into a much specialized, convoluted spermatophoric organ. The sperm are first twisted into a dense spiral and then are covered by various tunics, membranes, and accessory structures to form the spermatophores which are then stored in a seminal vesicle called

Needham's sac. This sac leads directly into the penis from which the spermatophores are gathered. There are essentially three major components: the sperm mass in which the sperm are contained; the tough mucopolysaccharide tunics, membranes, and fluid filled spaces; and the ejaculatory apparatus with its associated cement body. The ejaculatory apparatus has some kind of species characteristic *spacer* to keep open the tubular membranes in order to allow passage of the sperm mass during ejaculation. In *Loligo pealei*, this is a spiral filament which resembles, but does not function as, a coiled spring. The unejaculated spermatozoa are densely packed into a highly oriented array and are highly birefringent. X-ray diffraction has been done on the sperm mass and some of the early information about the structure of desoxyribonucleic acid (DNA) was derived from these studies (Wilkins, 1953). It is estimated in *Loligo pealei* that there are between 7.2×10^6 and 9.6×10^6 spermatozoa per spermatophore (Austin *et al.*, 1964). The spermatophores are 8 to 10 mm long; they are small compared to those measuring 50 mm of *Thysanoteuthis rombus*, 35 cm of *Nautilus belaunsis*, and greater than 1 m of *Octopus doflienai*.

Drew (1909, 1911) and Austin *et al.* (1964) described the structure of the spermatophores in detail and discussed the process of ejaculation. Ejaculation is accomplished by hydrostatic pressure within the tunics of the spermatophore (Mann *et al.*, 1981). Essentially the membranes and sperm mass of the pre-ejaculation spermatophore turn inside out and form the sperm reservoirs which are attached by the ejaculatory apparatus and its associated cement body to the area around the female's oviduct or inside the female's seminal receptacle, if present. The process of ejaculation can be watched by carefully isolating a few spermatophores onto a microscope slide in a little fluid from the mantle cavity and then poking them with a probe or by pulling the cap thread. The sperm are explosively released into the newly formed sperm reservoir from which they are gradually released again and begin to swim actively. The sperm are quite large, highly birefringent, and species characteristic (Arnold and Williams-Arnold, 1978).

2.2. Female reproductive system

The female reproductive tract takes its origin from the large median ovary located in the coelom in the apex of the mantle cavity. The ovary is single and a large oviduct lies directly upon it and at maturity is filled with clear ovate eggs. In several oceanic species, the oviduct is paired and filled with millions of small eggs, 1 mm in diameter or less and often colored pink to red. This color would decrease their visibility in the open ocean since red is preferentially filtered out by water, thus making the eggs appear somewhat darker against the down-welling light. In *Loligo pealei*, the eggs are ovate and very yolky; they measure about 1.6 mm × 1.0 mm in size. They are contained in a clear chorion with a dish-shaped micropyle at the smaller end. At maturity, the eggs are shed from the semi-opaque follicular syncytium (Selman and Arnold, 1977); they are clear, slightly indented on one side and have a small clear blastodisc below the micropyle. This micropyle increases in size as the egg is activated by exposure to sea water. The indentation disappears during activation and the egg appears to decrease slightly in volume. The eggs are then shed from the transparent oviduct into the egg capsule, which is covered with several layers of orangish outer layers presumably arising from the accessory nidamental gland. Each egg capsule (known as *egg fingers* or *egg candles* on the west coast of North America) contains about 180 eggs in *Loligo pealei*. Spent females may lay empty egg capsules. The egg

Figure 2. Reproductive behavior of *Loligo pealei* from observation of captive animals. Part 1. When a male sees an egg mass which has been introduced into a tank of several healthy adults, it responds by swimming toward the egg mass with his arms extended toward it. He then feels the center of the egg mass with his arm tips and gropes the individual egg strings. When he retreats from the egg mass, he may flush it violently with water from his funnel. If the egg mass is *satisfactory*, the male will attempt to establish a pair with a female squid. If the egg mass is old or otherwise *unsatisfactory*, he will withdraw and may or may not seek a mate. Females also show this behavior.

jelly suppresses ciliary action and thereby probably helps to prevent contamination by other invertebrates (Atkinson and Granholm, 1968, Atkinson, 1973). The egg jelly stiffens on contact with sea water and as the embryos develop, their chorions swell so the whole egg capsule increases in size considerably. The egg jelly apparently is a mucopolysaccharide essential to further development of the embryos (Arnold *et al.*, 1974; Arnold and O'Dor, 1989).

The structure of the egg capsules is highly variable between species. In *Loligo*, the egg capsules seem to be most complex with the eggs placed in a spiral pattern around a central common core formed from the jelly folds. In the closely related *Loligunculus*, the eggs are enclosed in a similar capsule but are not in a spiral fold of jelly; instead they accumulate at the distal end of the egg capsule. In *Sepioteuthis*, the eggs are much larger and only a few (2 to 5) eggs are enclosed in the egg capsule in a straight line. In *Illex*, the eggs are deposited in a loose mass of very fluid jelly and development within this free floating egg mass is relatively rapid (O'Dor *et al.*, 1982). *Sepia latimanus* lays its eggs individually among the branches of coral. Each egg is contained within a multilayer capsule which is more or less spherical with a nipple-like point on one end and two extended flaps on the other end which are wrapped around the substrate. Sepiolids appear to lay their eggs singly in multilayered capsules covered by a parchment like layer (Arnold *et al.*, 1974; Singley, 1984; Bergstrom and Summers, 1984). Egg capsules of many cephalopod species have been described and attempts have been made to draw functional correlations with their structure and deposition (Sweeney *et al.*, 1990). Some oceanic squids appear to shed their eggs directly into the ocean with only a thin layer of egg jelly and they float individually in mid water. These squids have small eggs which are laid in vast numbers.

2.3. Mating behavior

The earliest recorded observations on mating behavior in the cephalopods were made by Aristotle (ca. 330 B.C.) on *Sepia*, *Octopus*, and an unidentified squid. His observations suffered from anthropomorphic interpretations; reproductive behavior is often misinterpreted. The following observations apply mainly to *Loligo pealei* which I am most familiar with, but also apply to other teuthids with some modification.

Typically, healthy unmated squid swim about side by side in a small school moving back and forth more or less in synchrony when held in a rectangular tank. In larger circular tanks, the schools tend to swim around and around the tank against the flow of water and are thus less damaged. The males and females move about without any apparent regard to one another. However, if a recently laid mass of egg capsules is introduced into the tank, the squid break formation and orient toward the egg mass, form their arms into a cone and flush the egg mass with spurts of water from their siphons (see Fig. 2). This is followed by individual animals darting up to the egg mass, extending their arms into the egg capsules and groping around among the egg capsules with the tips of their arms. The males display this behavior more readily than the females, but both sexes usually eventually participate. The males become increasingly excited and dart about quickly in the tank challenging other males and attempting to place themselves between particular females and the rest of the school. Apparently, there is some subtle difference in appeal of the females because the males tend to favor particular females. The males challenge one another by special displays, but rarely contact each other; yet they may on occasion grasp each other and bite pieces from the mantle or arms. It appears, therefore, that the stimulus to reproductive behavior is related to the visualization of an existing egg mass or anything that resembles an egg mass. Artificial egg masses made of rubber tubing or water filled rubber gloves will initiate investigation by the males. This generalization seems to apply to many squids. *Sepia bandensis* will investigate egg masses of *Sepioteuthis* and even lay eggs among them (Arnold, 1984). In *Sepioteuthis sepiodea* where the eggs are laid cryptically under conch shells or rocks, neither naturally laid egg strings or artificial models elicited either investigative behavior or mating (Arnold, 1965). Tinbergen (1939) observed similar preferential egg deposition among existing egg clutches by *Sepia officinalis* and concluded that tactile and chemical sensation reinforced the visual stimulus but that the visual stimulus is most important.

When a male has selected a female, he swims beside her raising one of his medial arms upward in a characteristic *S-shaped* curved posture (see Fig. 3). In species other than *Loligo pealei* two median arms may be raised (Arnold, 1965b; Corner and Moore, 1980) and the threat may involve intense color displays. For example, *Sepia bandensis* males turn a grayish purple with light green spots and actively push each other alternatively toward the surface or bottom of the aquarium as shown by Arnold (1984) in a photograph. *Loligo pealei* develops a dark brown patch of varying intensity in front of his eyes by expanding selected chromatophores. If another male approaches, the male attempting mate selection will raise his median arm even higher, intensify his color pattern, and may even dart quickly at the approaching challenger. If the challenger persists, the male will frequently chase him and may, infrequently, seize and bite his competitor. Often they bump their bodies together and several dark spots of expanded chromato-phores appear along the edge of the fin on the side in contact. If the challenger is unsuccessful, he withdraws and attempts to find another female. If another female is found, the challenger may select her and establish a mating pair. If the mated male is displaced by the challenger, he will usually withdraw and lessen his

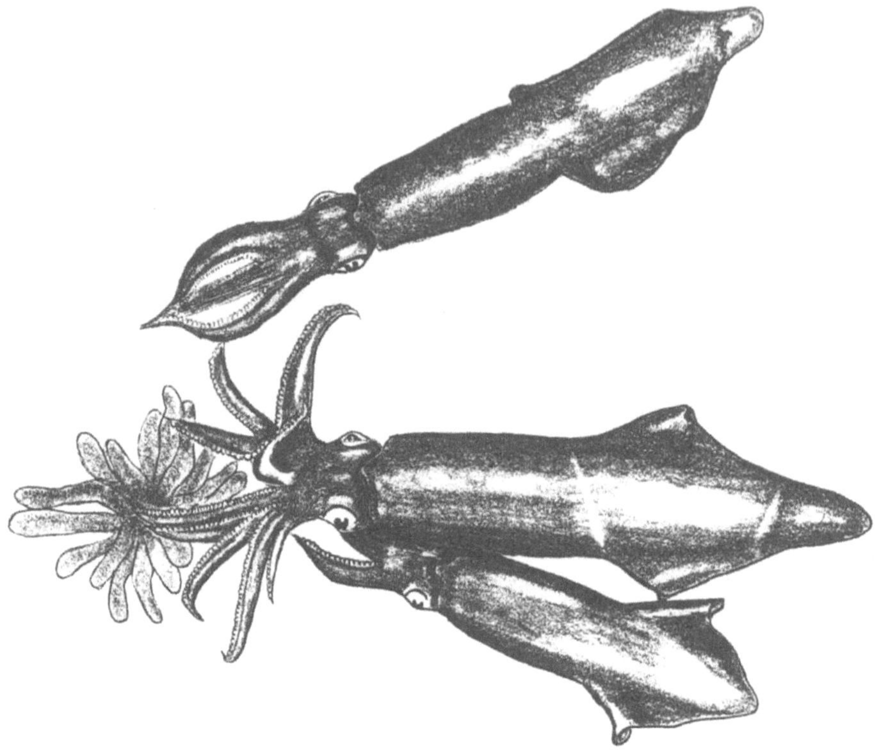

Figure 3. Reproductive behavior of *Loligo pealei* from observation of captive animals. Part 2. After the male has completed investigation, he will attempt to isolate a female from the rest of the school by placing himself between the selected female and other males. He may signal challenging males by raising a median arm and fanning out his other arms. Dark areas of expanded chromatophores appear between his arms and eyes. See text for further details.

display. In this way, a social hierarchy is established with the largest or healthiest males selecting the most desirable females. The features which make one female more desirable than another remain unknown. Once established the hierarchy tends to remain relatively stable but if a mated female high in the *pecking order* is removed, the now unmated male will displace a subordinate male.

During all of the male display and challenging, the females remain passive or attempt to avoid the disturbance by withdrawing from the area. Occasionally, a vigorously pursued female will jump out of the tank in an attempt to avoid an aggressive male. In rectangular tanks, the corners should have a piece of cardboard placed on the top of the tank to avoid the loss of animals and subsequent mess and potential damage to equipment by squirted ink. More typically, however, it appears that the female's acceptance of a male is expressed simply by her not avoiding his advances. This is followed by the female's investigative behavior and subsequent copulation followed by egg laying.

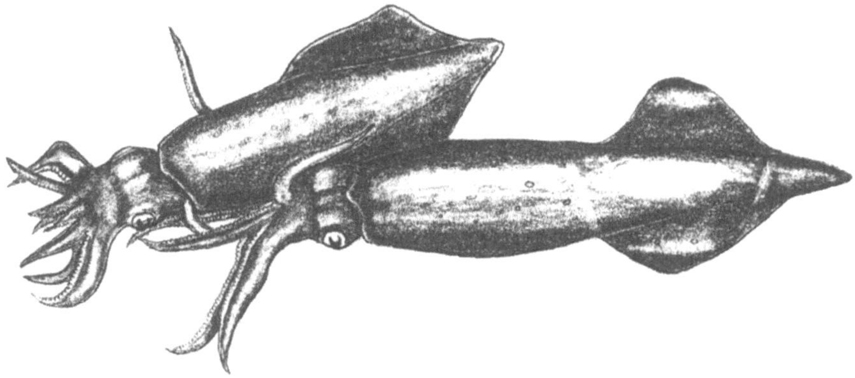

Figure 4. Reproductive behavior of *Loligo pealei* from observation of captive animals. Part 3. After a mated pair is established, the animals copulate. The male grasps the female around the mid-mantle and transfers a mass of spermatophores into her mantle with his modified arm, the hectocotylus. In this illustration, the male is just releasing the female, but the arm tip is still inserted into her mantle cavity.

2.4. Copulation

Copulation and spawning follow the establishment of the pairs and the social hierarchy, but these steps frequently overlap and intermingle so it is hard to distinguish individual steps (see Fig. 4). Before copulation in *Loligo*, the male swims beside and slightly below about his potential mate and flashes his chromatophores. He grasps the female from slightly below about the mid-mantle region and positions himself so his arms are close to the opening of her mantle. He then reaches into his mantle with his hectocotylus and picks up several spermatophores from his penis. In one quick motion, the spermatophores are ejaculated and cemented to the inside of the mantle near the opening of the oviduct or upon it. The male then releases the female and they return to paired swimming. The entire process is consummated in 5 to 20 seconds and usually alternates with egg laying. During copulation, the female remains passive or occasionally will resist the male by attempting to avoid him or pushing his arms away when he grabs the mantle. If the female has an egg capsule in her arms she will frequently raise her median arm and the male will cease copulation attempts. A second position of copulation is less frequently observed in captivity, but must be common in nature since virtually all of the mature females have spermatophores in their sub-buccal seminal receptacles. In this position, the males and females meet face to face and intertwine their arms. The spermatophores are transferred by the hectocotylus and burst in the sac shaped *bursa copulatrix* to form the sperm reservoirs described above. This mode of sperm transfer is referred to as *Aristotle's position* and apparently is the only method used in some species (e.g., *Sepia*). In the inshore ocean, whole schools of squid are stimulated to mass matings and egg layings (McGowan, 1954; Fields, 1965; Arnold, 1962; Arnold and Williams-Arnold, 1977). Massive egg masses are formed which litter the sea bottom and may cover many hectares (McGowan, 1954).

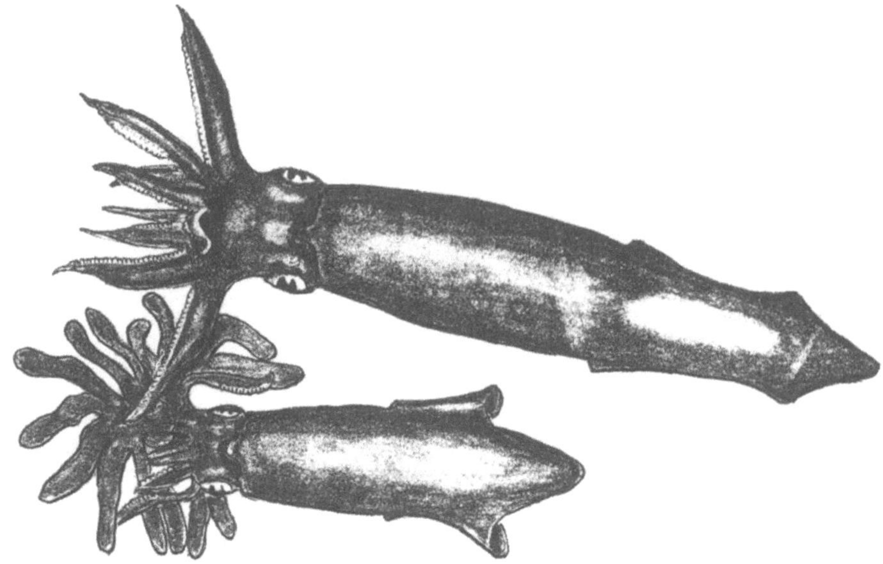

Figure 5. Reproductive behavior of *Loligo pealei* from observation of captive animals. Part 4. After copulation, the female alternates egg laying with copulation. Here she has her arms embedded in the egg mass and is intertwining an egg capsule into the common center of the mass. Copulation and egg laying continue until both of the animals are exhausted and death follows within hours.

These massive matings have been commercially exploited for thousands of years by humans and other marine animals.

In the cuttlefish observed by Aristotle (ca. 330 B.C.) and by Tinbergen (1939), the male approaches the female with his arms together with the tips curled back to form a *love-basket* and the female meets him in a similar fashion and they mate. In *Sepia latimanis*, the male will sometimes stroke the female on the top of her head with his arms and she responds by curling back her arms to form a *basket* (Corner and Moore, 1980; Arnold, 1989) to facilitate copulation.

2.5. Egg deposition

When a female *Loligo* lays eggs, she passes an egg capsule from the oviduct to the funnel and holds it in her arms which are formed into a characteristic cone-shape (see Fig. 5). The egg jelly is penetrated by the actively swimming sperm escaping from the sperm reservoirs near her oviduct and they can be seen swimming in the egg jelly for hours after egg capsule deposition. The sperm congregate at the micropyle of each egg and R. Miller (1989) feels this is chemo-attraction. The actual event of fertilization occurs a few minutes after the egg string has been deposited and can be easily observed with either phase or differential interference contrast (DIC) microscopy. Once the female has the egg capsule in her arms, she attempts to attach it into a pre-existing egg mass or to any object that looks like an egg mass. Frequently, the algae *Fucus* is the center of egg deposition. One end of the egg capsule is devoid of eggs and the jelly there is quite sticky. The female holds the egg capsule, sticky end forward, and attaches it to the egg mass by entwining the end into the center of the pre-existing common egg mass. The female then darts quickly away and flushes the egg mass with sea water from her siphon and rejoins her mate. As the jelly hardens by exposure to sea water, it becomes firmly attached. If the female drops an egg string during her attempt to attach it, she continues as though she still had it in her arms. While she has an egg string in the process of deposition, she will not allow the male to attempt copulation.

In many species, the eggs are laid individually and are contained in many layers of jelly. They may even have parchment like integuments over the softer jellies. These tertiary egg envelopes (i.e., egg jelly and coatings) are laid down by the nidamental and accessory nidamental glands as the eggs are passed down the oviducts. In some of the oegopsid squids, the egg masses are just huge blobs of jelly with the eggs more or less randomly arranged throughout (Hamabe, 1961a, 1961b; O'Dor, 1983). In some species of cuttlefish, such as *Sepia bandensis*, the eggs are squirted with ink as they are laid or covered with ink in passage down the oviduct. These black eggs superficially resemble tunicates and are often overlooked among the coral or other material upon which they are laid.

The copulation and egg laying behavior may be studied as a 5 step process, but the observer must keep in mind that these stages mean nothing to the animals and are completely artificial. They are as follows: 1. Stimulation of mating on an immediate and long term basis. 2. Mate discovery and identification. 3. Copulation and spermatophore transfer. 4. Egg deposition, either individually or in oviduct fabricated egg capsules. 5. Brooding and parental care. In the teuthioids there seems to be little or no parental care but in octopods, it seems to be quite common.

In the laboratory, it is possible to stimulate reproductive behavior and spawning with the appropriate stimulus. With *Loligo pealei*, it is most convenient to tie 10 to 20 egg strings from an egg mass only a few days old to a weight. Large glass stoppers work well as weights and they are non-toxic. The weight can be placed in the center of the tank with mature males and females. The animals will fixate upon the egg mass as soon as they see it and begin to investigate; select partners; copulate; and lay eggs in about an hour. This technique works best in the early evening hours or as the sun comes up in the morning. This stimulus is obviously initially visual, since dialysis tubing, rubber gloves, finger cones, and even human fingers will initiate investigation. An egg mass in a glass jar will also cause the squid to attempt to investigate it. Once copulation is initiated, the females will begin to lay on the egg mass, or lacking a proper substrate, they will begin to pass egg strings through their arms, but then drop them. If enough stray egg strings accumulate together, this may serve as a center of common deposition. In most cases, the egg masses are formed by common efforts of many animals and certainly the very large ones observed in nature are the efforts of hundreds or thousands of individuals.

In many of the cephalopods, the result of mating and egg deposition is death within a short time. The exact hormonal pathways involved have yet to be elucidated and should be studied. Wodinsky (1977) claimed that he could reverse this *aging* process by surgically implanting the optic glands from an unmated *Octopus* into the head of *brooding* females. Attempts to repeat this experiment by others have led to questionable results. It is advisable, therefore, to separate the males and females if long term maintenance is desired. The embryos obtained by the artificial induction of mating and egg laying are best maintained by separating them from the egg mass and hanging them up by fishing line or string in the sea tanks where they will be exposed to a constant flow of running sea water. This way, they will develop uniformly and in synchrony.

References

Aristotle, ca. 330 B.C., *The Works of Aristotle*, (J. A. Smith and W. D. Ross, eds.), *Volume 4. Historia Animalium*, (D. W. Thompson, translator and editor), 1910), Oxford at the Clarendon Press, Oxford.

Arnold, J. M., 1962, Mating behavior and social structure in *Loligo pealii*, *Biol. Bull.* **123**:53–57.

Arnold, J. M., 1965, Observations on the mating behavior of *Sepioteuthis sepioidea*, *Bull. Mar. Sci. of the Gulf and Carib.* **15**:216–222.

Arnold, J. M., 1984, Cephalopods, in: *The Mollusca. Vol. 7. Reproduction*, (A. S. Tompa, N. H. Verdonk, and J. A. M. van den Biggelar, eds.), Academic Press, New York, pp. 419–454.

Arnold, J. M., 1989, *unpublished data*.

Arnold, J. M. and O'Dor, R. K., 1989, *in preparation*.

Arnold, J. M. and Williams-Arnold, L. D., 1977, Cephalopoda: Decapoda, in: *Reproduction of Marine Invertebrates*, Vol. **4**, (A. C. Giese and J. S. Pearse, eds.), Academic Press, New York, pp. 243–290.

Arnold, J. M. and Williams-Arnold, L. D., 1978, Spermiogenesis of *Nautilus pompilus*. I. General Survey, *J. Exp. Zool.* **205**:13–26.

Arnold, J. M., Singley, C. T., and Williams-Arnold, L. D., 1972, Embryonic development and post-hatching survival of the sepiolid squid *Euprymna scolopes* under laboratory conditions, *Veliger* **14**:361–364.

Atkinson, B. G., 1973, Squid nidamental gland extract: isolation of a factor inhibiting ciliary action, *J. Exp. Zool.* **184**:335–340.

Atkinson, B. G. and Ganholm, N. A., 1968, A ciliary activity inhibitor extracted from the nidamental gland of *Loligo pealii*, *Biol. Bull.* **135**:413.

Austin, C. R., Lutwak-Mann, C., and Mann, T., 1964, Spermatophores and spermatozoa of the squid *Loligo pealii*. *Proc. Roy. Soc. Lond. B* **161**:143–152.

Bergstrom, B. and Summers, W. C., 1983, *Sepietta oweniana* in: *Cephalopod Life Cycles*, Vol. **1**, (P. R. Boyle, ed.), Academic Press, London, pp. 69–74.

Boycott, B. B., 1957, An octopus lays eggs at London Zoo, *New Scient.* **2**:12–13.

Boyle, P. R., 1983, *Cephalopod Life Cycles*, Vols. **1**, **2**, Academic Press, London.

Corner, B. D. and Moore, H. T., 1980, Field observations on reproductive behavior of *Sepia latimanus*. *Micronesica* **16**:235–260.

Drew, G. A., 1909, The breeding habits of the squid, *Science N. S.* **29**:436.

Drew, G. A., 1911, Sexual activities of the squid, *Loligo pealii* (Les.). I. Copulation, egg-laying, and fertilization, *J. Morph.* **22**:327–360.

Fields, G., 1965, The structure, development, food relations, reproduction, and life history of the squid *Loligo opalescens*, *Berry State of California, Department of Fish and Game, Fish Bulletin* **131**:1–108.

Fisher, W. K., 1923, Breeding habits of a cephalopod, *Ann. Mus. Nat. Hist.* **9**, **12**:147–149.

Fox, D. L., 1938, An illustrated note on the mating and egg-breeding habits of the two spotted octopus. *Trans. S. Diego Soc. Nat. Hist.* **9**:31–34.

Gravely, F. H., 1908, Notes on the spawning of *Eledone* and on the occurrence of *Eledone* with the suckers in double rows, *Mem. Manchr. Lit. Phil. Soc.* **53**:1–14.

Hamabe, M., 1961a, Experimental studies on breeding habits and development of the squid, *Ommastrephes sloani pacificus* Steenstrup. I. Copulation, *Dobutsugaku Zasshi (Zool. Mag.)* **70**:378–384.

Hamabe, M., 1961b, Experimental studies on breeding habits and development of the squid, *Ommastrephes sloani pacificus* Steenstrup. II. Spawning behavior, *Dobutsugaku Zasshi (Zool. Mag.)* **70**:384–394.

Mann, T., Martin, A. W., and Thiersch, J. B., 1981, Changes in the spermatophore plasma during spermatophore development and during the spermatophore reaction in the giant octopus of the North Pacific *Octopus dofleini martini. Mar. Biol. (Berlin)* **63**:121–128.

Margolis, 1989, *personal commun.*

McGowan, J. A., 1954, Observation on the sexual behavior and spawning of the squid *Loligo opalescens* at La Jolla, California, *Calif. Fish Game* **40**:47–54.

Miller, R., 1989, *personal commun.*

O'Dor, R. K., 1983, *Illex illecebrosus*, in: *Cephalopod Life Cyces*, Vol. **1**, (P. R. Boyle, ed.), Academic Press, London, pp. 69–74.

O'Dor, R. K., Balch, N., and Amaratuna, T., 1982, Laboratory observations on midwater spawning by *Illex illecebrosus, NAFO SCR, Doc 82/VI/5*. pp. 8.

Roper, C. F. E., 1965, A note on egg deposition by *Doryteuthis plei* (Blainville, 1823) and its comparison with other North American Loliginid squids, *Bull. Mar. Sci.* **15**:589–598.

Saunders, W. B. and Landman, N. H. (eds.), 1987, *Nautilus: The Biology and Paleobiology of a Living Fossil*, Plenum Press, New York.

Selman, K. and Arnold, J. M., 1977, An ultrastructural and cytochemical analysis of oogenesis in the squid, *Loligo pealii. J. Morphol.* **152**:381–400.

Singley, C. T., 1984, Chapter 5, *Euprymna scolopes* in: *Cephalopod Life Cycles*, (P. R. Boyle, ed.), Academic Press, London, pp. 69–74.

Sweeney, M. J., Roper, C. F. E., Mangold, K. M., Clarke, M. R., and Boletzky, S. v., eds., 1990, *"Larval" and juvenile cephalopods; a manual for their identification, Smithsonian Contr. Zool., in press.*

Tinbergen, L., 1939, Zur Fortpflanzungscthologie von *Sepial officinalis*. L., *Arch. Neerl. Zool.* **3**:323–364.

Wilkins, M. H. F. and Randall, J. T., 1953, Crytallinity in sperm heads: Molecular structure of nucleoprotein *in vivo, Biophys. Biochim. Acta* **10**:192–193.

Wodinsky, J., 1977, Hormonal inhibition of feeding and death in *Octopus*: Control by optic gland secretion, *Science* **198**:948–951.

Chapter 6

Embryonic Development of the Squid

JOHN M. ARNOLD

1. Introduction

There has been such a considerable expansion of our knowledge of squid embryology since the first edition of this book was written that a complete review of the literature is impossible. The scientific importance and the commercial value of squids have prompted an increased interest in all aspects of squid biology; embryology has followed this trend. The reader is particularly referred to the reviews of Arnold (1974), Arnold and Williams-Arnold (1976), Fioroni and Meister (1974), Boletzky (1987, 1988), Segmüller, M. and Marthy (1989), and the comments of Marthy (1982) for an idea of the status of experimental investigations of the cephalopods and theories on the regulation of ontogeny, as well as some of the outstanding questions concerning development. What follows here is a brief description of the embryonic development of the squid; no survey of the vast literature is attempted. See Arnold (1974) for a fairly extensive list of references for major sources of embryonic information.

2. Handling techniques

Over the years it has become necessary to develop handling techniques unique to cephalopod embryos. Many of the manipulative techniques used for similar sized eggs such as amphibian embryos or fish eggs can be applied to squid embryos in the 0.5 mm to 3.0 mm diameter range; some techniques are, of necessity, exclusive to the squid. Some of the less obvious of these are given here, but the investigator should be prepared to be creative in approaching these embryos. What follows here are some of the *tricks of the trade* used by this author. Other cephalopod embryologists will have their own techniques and skills and none is really superior to the other.

Depending upon the type of egg mass the species under consideration produces the handling of the egg will vary. In all instances it is absolutely necessary for the eggs to be maintained in well aerated sea water, preferably running sea water. If the eggs are individually laid, it may be enough to store the eggs in a basket-like container with a gentle current flowing by them. If the eggs are laid in a large mass like *Illex*, it may be necessary to separate some of the embryos from their very loose egg jelly

J. M. ARNOLD • Pacific Biomedical Research Center, Cephalopod Biology Laboratory, 209A Snyder Hall, 2538 The Mall, University of Hawaii at Manoa, Honolulu, Hawaii 96822.

and keep them in a finger bowl or watch glass; but they will not withstand crowding. It is best to have a monolayer of embryos that covers less that one half of the bottom of the dish. If the eggs are laid in egg strings as are *Loligo* eggs, the egg strings should be separated from the egg mass when the embryos are in early stages, as the center of the egg mass tends to become fairly anaerobic; it will eventually rot and development will become retarded or abnormal. Hanging the egg strings individually by a thread or piece of fishing line in a gentle stream of running water will insure adequate aeration. If well maintained, all the eggs will develop synchronously; since they are all siblings, they represent a homogeneous developing population that can be repeatedly sampled throughout their development.

To make observations on the early stages a freshly laid egg mass is necessary. With *Loligo pealei* the eggs are typically laid at night or at sunrise; early stages can be obtained simply by sorting through a new egg mass. It is much more convenient, however, to artificially stimulate the adults to mate and lay so the eggs can be had on demand. Eggs surgically removed from the oviduct can be fertilized in shallow dishes by breaking spermatophores into a dish of sea water. Only eggs that are fully mature should be used. Mature eggs are translucent and free of the follicular syncytium and have a shallow indentation in their side. Several hundred eggs should be placed in a large finger bowl with enough sea water to cover them by a few mm. Spermatophores removed from Needham's sac of the male should be broken open among the eggs. By swirling the bowl, the eggs and shedding spermatophores will be concentrated in the center of the dish where fertilization will occur. After 15 to 30 minutes of sperm contact, the eggs should be washed with several changes of fresh sea water and checked with the dissecting microscope and debris removed. First cleavage will occur approximately three hours after successful fertilization. This technique usually results in 10 to 20% of the eggs being fertilized. Although the embryos will cease development at about stage 11 or 12 (the egg jelly is necessary for the chorions to lift away from the embryo), early stages can be observed without the frustrating inconvenience of the egg jelly. It is possible to mechanically wrap individual eggs in egg jelly and subsequent development will proceed normally. By using a lyophilized preparation of the nidamental gland, Arnold and O'Dor (1989) have been able to get not only artificial fertilization of oceanic species whose embryos were previously unknown but also to raise some of these embryos to hatching.

To observe the cleavage and gastrulation stages, it is convenient to position the embryo upright in a depression made in a sheet of dental impression wax fixed to the bottom of a suitable dish. Holes the diameter of the egg can be punched in the wax with a pipette drawn to the appropriate diameter. Simple perfusion chambers can also be made from these wax bottom dishes or slides by inserting hypodermic needles into either side of the chamber to function as an inlet and drain. Wax bottom dishes can also be carved into operating dishes in which the embryos can be held in place by pushing wax against the embryo to clamp it into position for regional treatments or surgery.

To observe later stages of development, it is necessary to remove the egg capsule membranes which become covered with dirt and discolored. With *Loligo* and similar egg capsules it is possible to strip off the outer membranes by gently pinching and pulling them free by turning them inside out. By pushing one blade of a scissors into the center of the egg string, it can be opened up and laid out flat in a dish. The egg jelly is the bane of cephalopod embryologists, but it can be either teased off with forceps or partially dissolved by agitation in sea water at pH 5. As much as possible of the jelly should be removed mechanically before the eggs are gently shaken in a glass stoppered graduated cylinder for a few minutes. Several changes of pH 5 sea water may be necessary. As soon as possible the embryos should be rinsed with several changes of sea water because they will die if held at low pH. Mechanical removal is preferable in most instances. With cuttlefish embryos and sepiolids as well as other large singly laid eggs it is possible to *wind up* the jelly with a small dry cotton swab and remove most of it fairly cleanly. Since the jelly is necessary for development it is advisable to leave at least some of it in contact with the chorion if the embryo is to be followed in subsequent development.

The chorions can be removed in several ways. In the cleavage stages it is best to leave the chorion intact since these eggs have little intrinsic strength and will collapse without support. However, holes can be easily cut through the chorion with iridectomy scissors or by tearing out pieces with watchmaker's forceps. After microsurgery these holes can be resealed by pressing the edges of the tear together as the cut surfaces are sticky for a short while after cutting. It should be remembered that there is a fertilization membrane on the egg if surgery is attempted. In the older embryos where the chorion has swollen away from the embryo it can simply be torn open with sharpened forceps. If large numbers of dechorionated older embryos (stage 19 and older) are desired, the chorions can be removed in mass with a sharp scissors. A membrane-free egg capsule is placed in a small finger bowl and about double its volume of sea water is added. The finger bowl is tipped so that the egg string goes to one side and the egg string is *snipped* at with the dissecting scissors. After several dozen *snips* many of the chorions will have been cut and the embryos will fall out into the sea water. Some embryos will be damaged by this procedure but they can be discarded. Hatching at stage 29 and 30 can be stimulated by simply rubbing the egg capsule or removing its outer membrane. This mechanical stimulation causes the embryo to attach their hatching glands (Hoyle organ) to the inside of the chorion and to digest their way out. At first the newly hatched paralarva are attracted to light and can be concentrated in this way.

Later stages are easily observed in their chorion either in a watch glass or on a slide with the coverslip supported by wax or a toothpick. Prolonged temperatures above 22 °C are lethal to *Loligo pealei*. The first indication of this is a discoloration of the yolk from translucent white to opaque yellow to brown in the center of the yolk mass. If caught early enough this can be reversed by returning the embryos to cooler sea water. Rapidly moving embryos can be stilled by chilling them or by using chloretone dissolved in sea water. To observe organ primordia at stages 17 or 18 careful substage illumination is helpful and ring dark field illumination is best.

In my experience dechorionated *Loligo pealei* embryos will not continue to develop in sea water alone. Marthy (1973), in studying *Loligo vulgaris*, used pasteurized sea water and had success. It has been possible to get complete normal development from stage 16 onward in a simple culture medium made of 10% house serum, sterile sea water, and penicillin-streptomycin. Sterile technique is necessary. Frozen and thawed whole squid blood may also be used and it has the advantage of having bacteriostatic properties. The embryos should be removed from their egg capsules with a minimum of jelly and passed through several changes of sterile sea water before they are dechorionated in sterile sea water. They then should be passed through at least three more changes of sterile sea water and placed into the culture medium. Dishes with small embryo-sized depressions in the bottom work best because the early embryos tend to flatten on flat surfaces. The early embryo cultures should be incubated at no more than 22 °C, but cultures derived from older embryonic organs will survive higher temperatures. Rice *et al.* (1990) discuss tissue cultures of squid

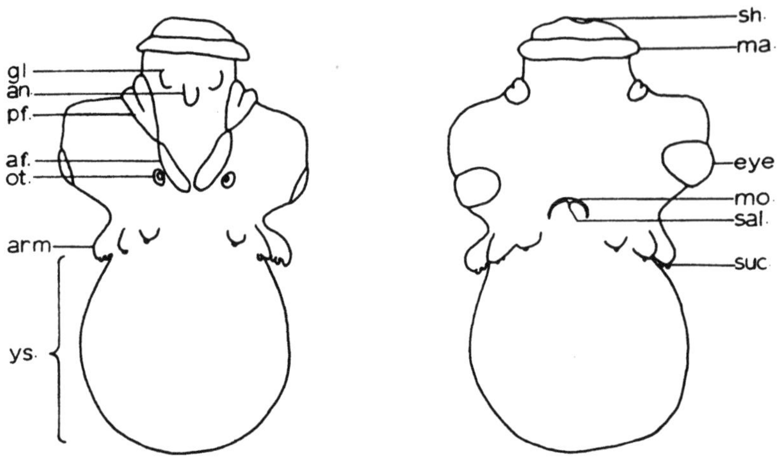

Figure 1. Key to the organ primordia of an embryo at about stage 21: af. = anterior funnel fold; *an.* = anal papilla; *gl.* = gill primordium; *ma.* = mantle; *mo.* = mouth; *ot.* = otocyst; *pf.* = posterior funnel fold; *sal.* = salivary gland invagination; *sh.* = shell gland invagination; *suc.* = sucker primordium.

cells. They have been successful in growing neurons, glial cells and muscle cells. Szabo and Arnold (1963) cultured embryonic tissues and adult ink sac in osmotically modified media 199 or squid serum for ten months and the ink gland cells produced ink until the cultures were terminated. Marthy and Aroles (1987) have had success with a calf serum based medium in growing organ fragments and neural tissues into substantial masses.

Fixation of the embryos presents problems associated with the yolk mass which becomes brittle and shrinks after fixation and subsequent dehydration. The best fixative I have found is 2.6% glutaraldehyde in sea water buffered to pH 7.2 – 7.4 with s-collidin HCl. After fixation of at least one hour at room temperature (longer for larger embryos, and overnight doesn't seem to do any harm) the embryos were rinsed with buffered filtered sea water for one or more hours then post-fixed in 1% OsO_4/sea water for one hour. All material for the fixatives and rinses should be Millipore™ filtered to keep the material clean if scanning microscopy is planned. The OsO_4 will keep best if made up as a 2% solution in distilled water which is added to an equal volume of sea water which has been reduced to 70% of its original volume by boiling. These fixatives have the approximate osmolarity of sea water. I call this fixative *hotel* since it is so simple it can be made up in hotel bathrooms. Some of the most interesting cephalopod embryos occur in areas of the world where laboratories do not occur. The embryos are changed directly from the OsO_4 into three rapid changes of 50% ethanol, soaking for 15 minutes in 50%, then through 70% ethanol and storage in 95% ethanol for subsequent examination. I find that this is adequate for all but the most delicate work at high resolution. For scanning electron microscopy, the embryos can be stored in 95% ethanol for long periods without obvious damage; but for thin sectioning it is best to embed the tissues as soon as an adequate laboratory is reached. For light microscopy a single fixation in the buffered glutaldehyde is adequate and does not make the tissues opaque as does Bouin's fixative. With glutaldehyde followed by

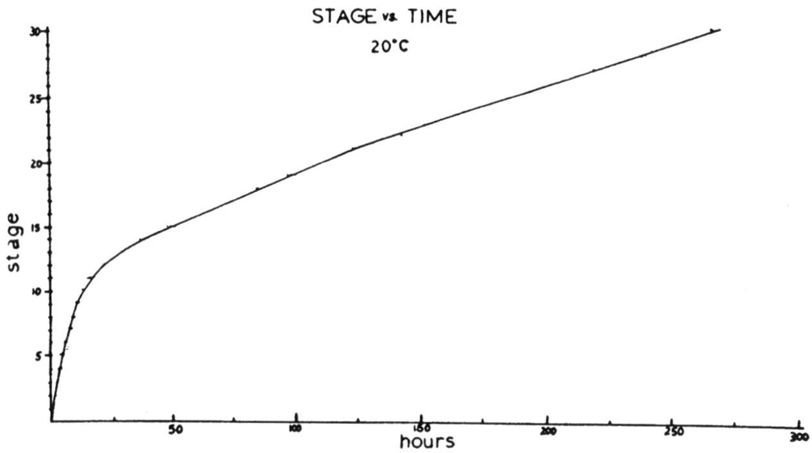

Figure 2. Graph of developmental stage versus time at 20 °C.

ethanol storage the embryos remain semi-translucent so some internal details can be seen which are critical for truly adequate staging. The embryos can be cleared in methyl salicylate or another oil if necessary.

Sectioning squid embryos, particularly the yolky earlier stages, presents special problems. I find avoiding Bouin's fixative in favor of buffered glutaraldehyde or formaldehyde is highly desirable. The glutaraldehyde should be from sealed glass ampules with nitrogen rather than from a bottle where oxidation can take place. If at all possible, the formaldehyde should be made fresh from paraformaldehyde to avoid the methanol in bottled formalin. Embedding for light microscopy should be in water soluble methacrylate or in waxes specially prepared for the histology of hard tissues. For transmission electron microscopy embedding in medium hard epoxy resins has been most the successful method in my laboratory. Although the plastic techniques are tedious it is well worth the extra effort, because the results from standard paraffin techniques are often unsatisfactory and even misleading.

3. Stages of normal cephalopod development

Since the original stages of *Loligo pealei* development were published in 1965 there have been a number of *staging papers* written on the development of various teuthid species (eg. Lemaire (1970), *Sepia officinalis*; Arnold, Singley, and Williams-Arnold (1972), on *Euprymna scolopes*; Yamamoto (1982) on *Sepiella japonica*; Segawa (1987), on *Sepioteuthis lessoniana*; Yamamoto (1988) on *Idiosepius pygmaeus paradoxus*; and Segawa *et al.* (1988) on *Loligo forbesi*. In addition, Natsukari (1970), has provided an excellent series of photographs of the developing embryo of *Idiosepius* but did not assign them stages, as Yamamoto did later. Arnold and O'Dor (1989) have provided photographs of a few selected staged embryos of the oceanic squids *Abralia*, *Abraliopsis*, *Stenoteuthis*, and *Thysanoteuthis*. All of the above have used a 30 stage scheme which begins either with fertilization or first cleavage and goes to hatching and

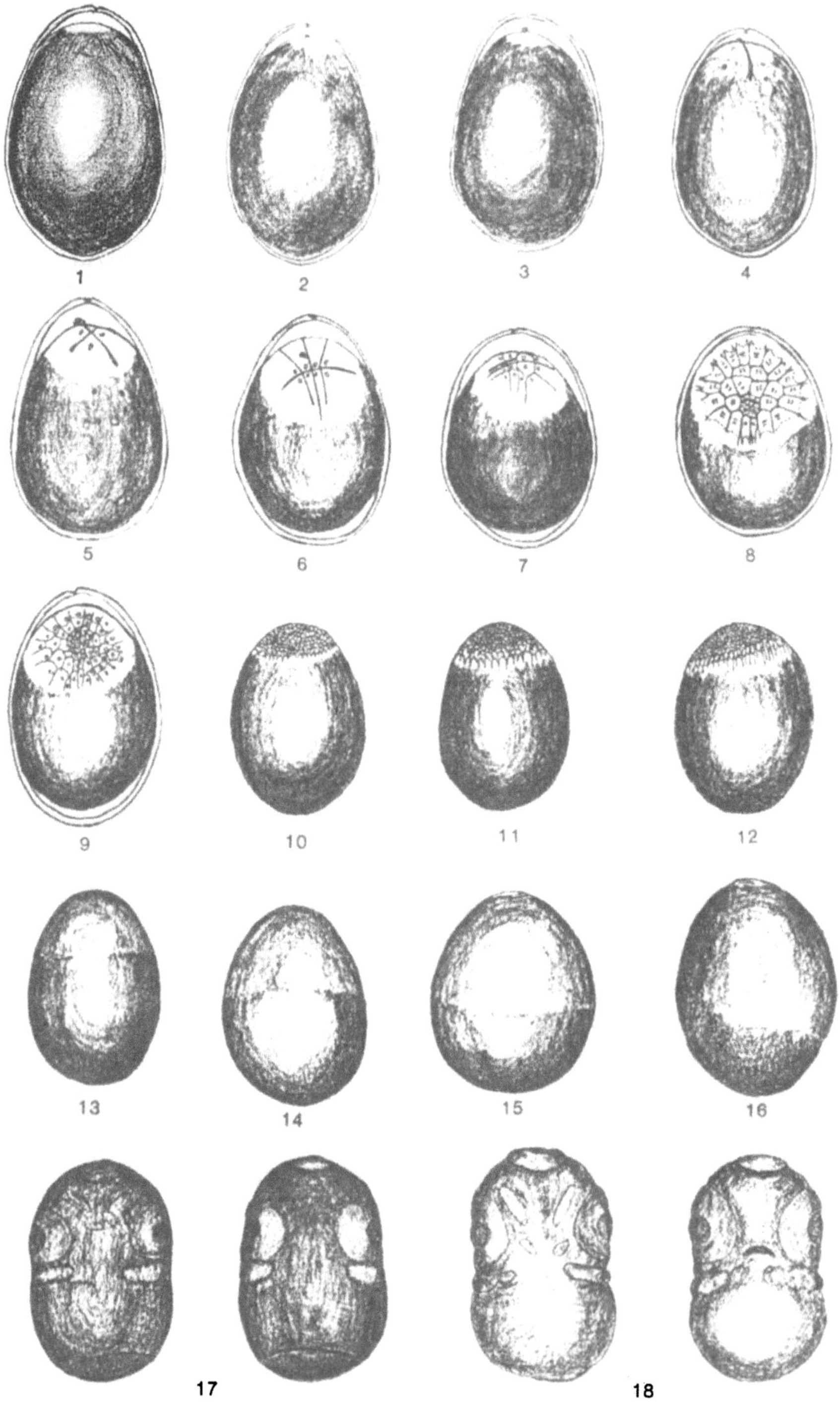
1
2
3
4
5
6
7
8
9
10
11
12
13
14
15
16
17
18

loss of the external yolk sac (if present). Naef (1928a, 1928b) used a scheme of twenty stages which included everything from fertilization through cleavage to early gastrulation in stage 1. Although his drawings are accurate and classically beautiful, and most criteria are well thought out, his staging series are unworkable for most *developmental experimentation.* It has been amply demonstrated that early events in cephalopod embryos are instrumental in subsequent differentiation and morphogenesis (Arnold, 1965b, 1968a; Arnold and Williams-Arnold, 1970, 1976; Arnold, Singley, and Williams-Arnold, 1972; Marthy, 1973, 1982), so it is necessary to have these early stages closely defined to make meaningful developmental analysis possible. Although it has been suggested by Boletzky (1987) that the stages of Arnold be used for early stages and the later stages of Naef be used in combination, such a combined scheme is unfeasible and untenable for cohesive experimentation. Priority and nostalgia notwithstanding, adequacy and efficiency should determine scientific utilization. This is not to degrade the monumental and comprehensive nature of Naef's monograph (1928a) which is an aesthetic joy and covers the teuthids *Loligo vulgaris*, *Sepia officinalis*, *Sepiola rondeleti*, and Oegopsid (either *Calliteuthis* or *Histioteuthis*) and an unknown ommatosrephid as well as several octopods.

What should be devised is an integrative set of developmental stages using morphological criteria applicable to all cephalopod embryos regardless of the amount or placement of the yolk or heterosynchrony of some of the less significant organs. The size of the egg mass may cause major differences in total appearance and shape of the embryos, but it has trivial significance in the relationship of organ placement or sequence of developmental events. The basic cleavage pattern of all cephalopods is uniform and directly comparable despite taxonomic separation. Similarly, gastrulation appears to be uniform in its major features and the amount of yolk only effects the placement of the margin of celluation when it occurs. Some events such as cleavage number and pattern, gastrulation, the ontogeny of the arms, timing of the appearance and invagination of the mouth and anus, and especially the eyes and visual system, can be used as a basis of a combined universal cephalopod embryonic staging system which would transcend the taxonomic differences and developmental heterochrony. Until that ideal is reached the following staging of *Loligo pealei* is offered. Although the drawings are of a single species the criteria used in drawings can be easily generalized (and have been by the above mentioned authors) to fit any species, including Nautilus, thus far observed.

As is the case with all *stages* of any developmental process, the stages given here are arbitrary steps in a continuous process and should be considered only as landmarks in a successive orderly sequence of ongoing events. These stages are characterized by morphological events that are relatively constant and easily recognizable in the living and preserved embryo with a low power dissecting microscope. The organ primordia are labeled in Fig. 1 and an approximate time scale of development is given in Fig. 2. The illustrations of the embryonic stages (Figs. 3, 4, and 5) have been redrawn and slightly revised according to new information gained through additional years of experience with this species and other cephalopods. Stage 10 which includes segregation of the germinal layers has been extensively studied by Singley (1974, 1977) and subdivided into three sub-stages. Stage 8 has been reoriented to better show the arrangement of the cleavage pattern and stage 22 redrawn

Figure 3. Developmental stages 1 to 18.

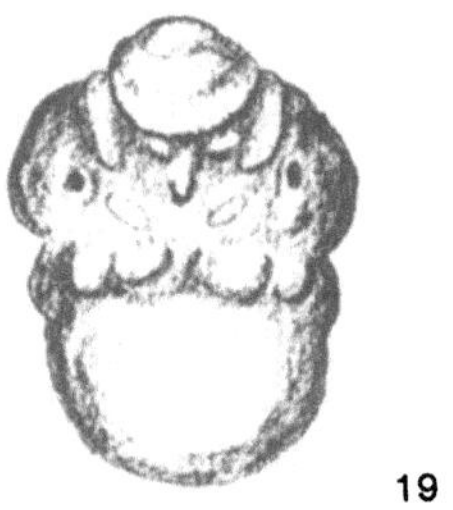

19

20

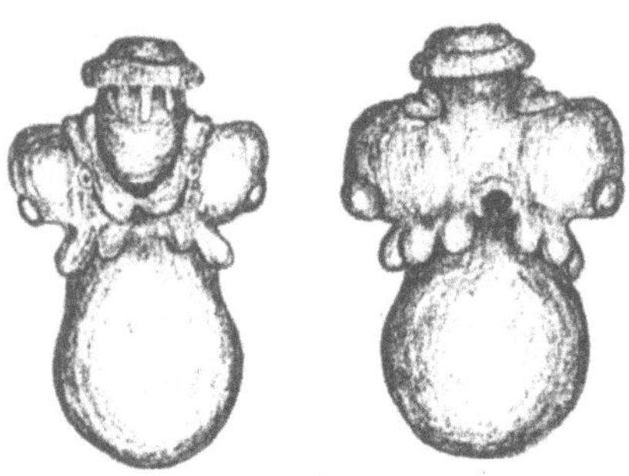

21

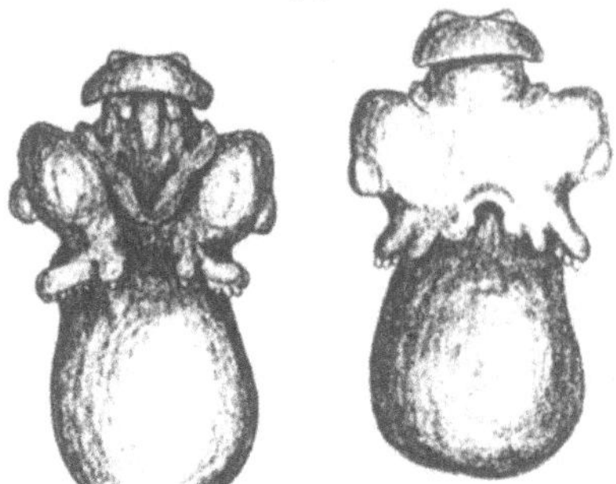

22

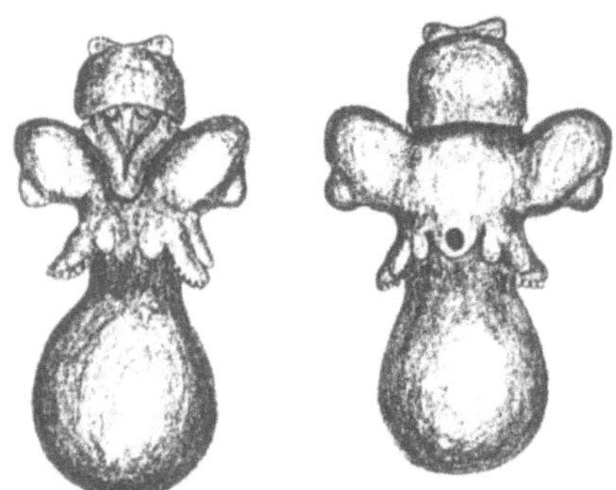

23

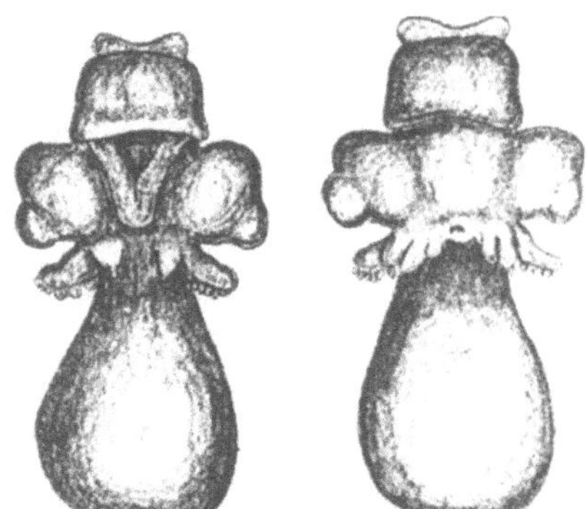

24

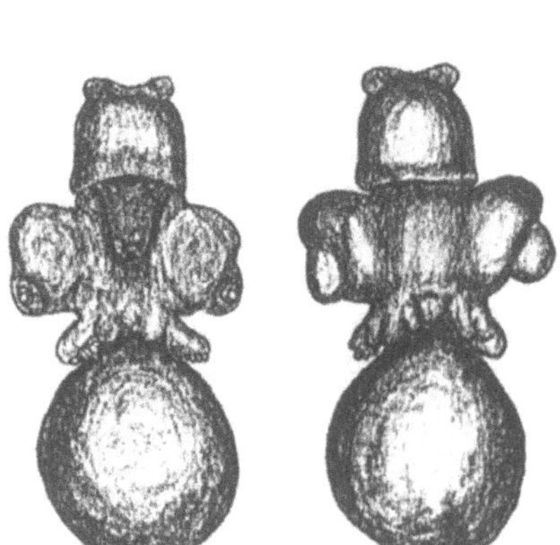

25

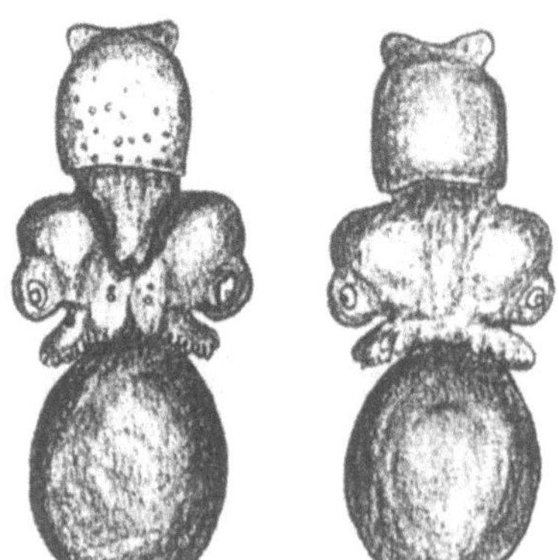

26

to better reflect the average embryo rather than the specific individual from which it was drawn. The original drawings (Arnold, 1965a) were made from randomly chosen individuals that had been carefully chosen from a group of a specific stage. Because the embryos vary in size due to differences in yolk volume (small females have smaller eggs) this led to some variation in the size of the camera lucida drawings. These illustrations are presented with the animal pole upwards as is standard with drawings of all embryos. The tendency of authors of cephalopod illustrations to go against this convention is only a mistake of history which should be abandoned. Scanning electron microscope images which give greater detail of some selected stages have been published in the following references: Arnold, (1984a, 1984b), Arnold and Williams-Arnold (1976, 1980), Arnold, Williams-Arnold, and Peters (1978). Many papers by Arnold (1974, 1984a, 1984b), Boletzky (1988), and Marthy (1973, 1982) have light micrographs of whole embryos which show details of the mantle organs and the development of the eye by transmitted light. Singley (1977) has made excellent light micrographs of cleavage and gastrulation.

3.1. Fertilization and meiosis

Developmental stages 1 to 3 occur during fertilization and meiosis (Fig. 3).

Stage 1: Fertilized but without polar bodies. The blastodisc is forming at the animal pole (small end) and frequently sperm can be seen swimming inside the chorion. There usually is a substantial indentation in the side of the egg which disappears as the egg rounds up slightly following fertilization.

Stage 2: First maturation division. The blastodisc is increasing in size by ooplasmic flow though pathways on the sides of the egg. The meiotic events can be observed in the clear cytoplasm of the blastodisc.

Stage 3: Second maturation division. The blastodisc continues to increase in size. The polar bodies may continue to divide until 6 or 7 are present.

3.2. Cleavage

Developmental stages 4 to 8 occur during cleavage (Fig. 3).

Stage 4. First cleavage begins below the polar bodies and divides the future right and left halves of the embryo. The future embryonic anterior is evident because that side of the egg is sloped at an oblique angle to the animal-vegetal axis of the egg. This sometimes is evident before cleavage.

Stage 5. Second cleavage occurs centrally and slightly subapically near the position of the polar bodies.

Stage 6. Third cleavage. The cells divide asynchronously and are asymmetrical, the smaller ones being in the future posterior and the larger ones being in the future anterior of the embryo. The furrows of the smaller cells are formed first and displace the second furrow toward them temporarily. When the anterior third furrows form, the second furrow is pulled into a jog at the junction of the third and second furrow.

Figure 4. Developmental stages 19 to 26.

Stage 7. Fourth cleavage. A population of inner cells is established while the surrounding cells remain continuous at their outer margins. The formation of the anterior fourth furrow changes the shape of the inner anterior cells from trapezoidal to pentagonal.

Stage 8. Fifth cleavage. The blastoderm is somewhat asymmetrical and division is less synchronous. The smallest cells divide slower than the larger cells.

Stage 9. Sixth through eighth cleavage. Very synchronous division is typical and brief stages are seen with 48, 56, and 64 cells. The central blastoderm is synchronous with the peripheral cells dividing synchronously. A central group of cell segregate which will eventually give rise to the shell gland primordium is shown.

3.3. Establishment of the germinal layers

Developmental stages 10 to 12 occur during the establishment of the germinal layers (Fig. 3).

Stage 10. Ninth cleavage through eleventh nuclear division. After ninth cleavage a yolk syncytium is established which is continuous around the central blastoderm. The germ layers are established in the following substages. This is the modified form of gastrulation without massive tissue migrations or establishment of a true blastopore.

Substage 10 early. Early layer segregation of germ layers. A ring shaped group of cells begin to centrifugally overlap the row cells distal to them. These overlapped cells form the presumptive middle layer of the embryo.

Substage 10 middle. The overlapped cells undergo mitosis to form a ring-shaped middle layer between the now covering cells and the developing yolk syncytium. The yolk syncytium is established by the immigration of the blastocone nuclei into the yolk cytoplasmic layer. The yolk syncytium nuclei divide twice more but the furrows between them are transitory.

Substage 10 late. The now bilayered blastoderm is completely superimposed over the yolk syncytium nuclei. The increase of middle-layer cells causes a depression in the yolk and forms a central yolk papilla above which the blastoderm consists of only the outer layer of cells.

Stage 11. Middle layered cells flatten and spread over the central yolk papillae so that it disappears and the middle layer thins. The edge of the blastoderm completely covers the yolk syncytium nuclei.

Stage 12. The middle layer completes the separation of the outer layer of cells from the yolk syncytium. The peripheral cells are greatly flattened, polygonal in outline, and eventually give rise to the external yolk sac cells.

3.4. Completion of the cellulation of the egg surface

The rate of cellulation is dependent upon the size of the yolk mass. In some other species cellulation is not completed until organogenesis is well under way. Developmental stages 13 to 15 occur during this time (Fig. 3).

Stage 13. About one-third of the egg has been covered by the growth of the blastoderm over the embryo's surface.

Stage 14. About two-fifths of the surface is cellulated.

Stage 15. About three-fifths of the surface is cellulated.

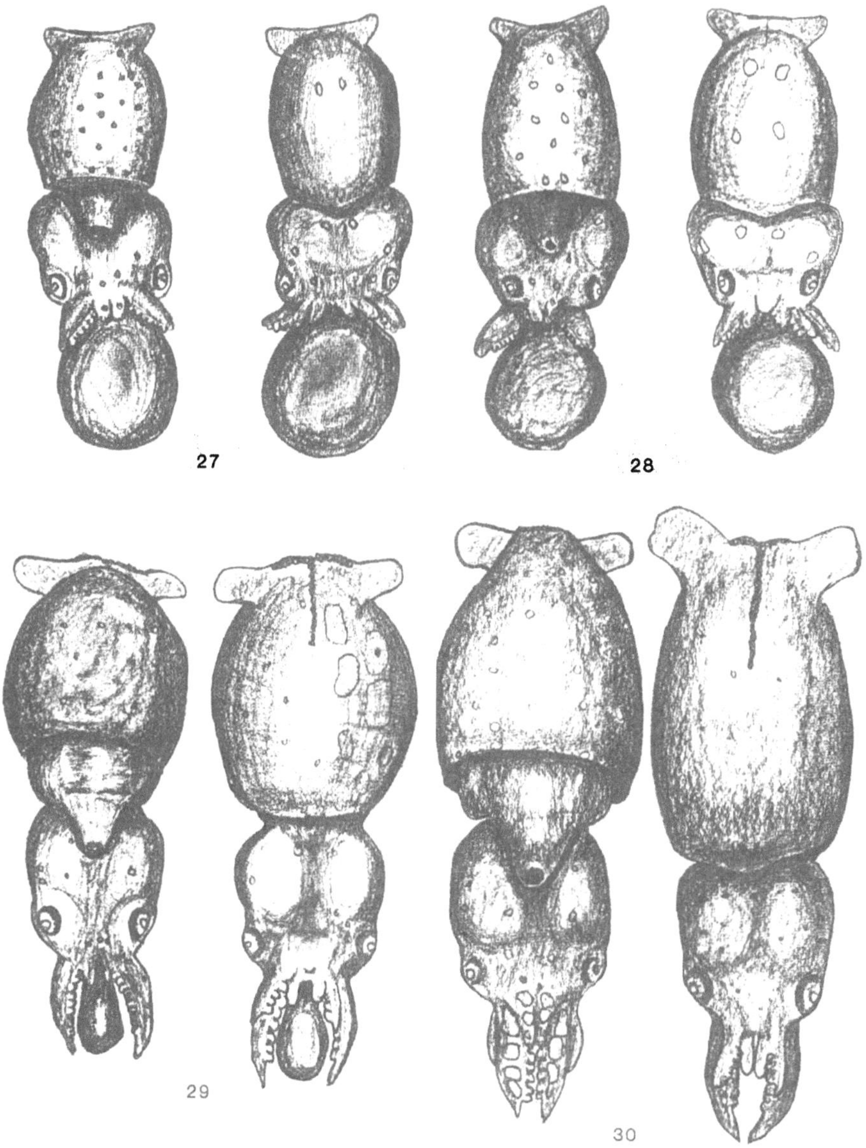

Figure 5. Developmental stages 27 to 30.

3.5. Organogenesis

Developmental stages 16 to 30 occur during organogenesis (Figs. 3, 4, and 5).

Stage 16 (Fig. 3). About two-thirds of the egg is cellulated. The shell gland primordium is evident as a thickening at the animal pole. This is the direct descendent of the central small cells derived in early cleavage.

Stage 17 (Fig. 3). Beginnings of the major organ primordia appear as thickened placodes. About five-sixth of the surface is now cellulated.

Stage 18 (Fig. 3). Completion of cellulation. An equatorial constriction separates the yolk sac from the future embryonic body. All the major organ primordia are evident.

Stage 19 (Fig. 4). Elevation of the organ primordia. The eyes, mantle, and arms are easily seen. The mouth, eye and statocyst invagination begin.

Stage 20 (Fig. 4). Gills, arms, funnel folds and mantle are prominent. Invagination of the salivary pit is evident in the mouth. The shell gland invagination closes.

Stage 21 (Fig. 4). Optic stalks are prominent and the invagination of the optic vesicle is completed. The optic lobes of the brain are evident. The anterior funnel folds fuse at their medial margin. The sucker primordia first appear as small ridges grading into a double line of bulbs along the second arms.

Stage 22 (Fig. 4). The mantle begins downward growth and the fins appear at the animal end of the embryo (future posterior end of the embryo). The lens primordium just becomes evident as a short rod of uniform diameter inside the optic vesicle. The median edges of the funnel folds begin to bend toward the mid-line in the center and the posterior funnel folds fuse to the anterior ones. The salivary pit is prominent inside the mouth.

Stage 23 (Fig. 4). Circulation is established in the external yolk sac and pulsations of this hemal organ are obvious and active. The gills are about one-half covered by the mantle and the iris folds are formed on the surface of the eyeball.

Stage 24 (Fig. 4). The anal papilla and the gills are completely covered by the mantle, but the funnel muscles are visible. Median margins of the funnel fold fuse during this stage. The mouth is completely circular and the branchial hearts are beating at the base of the gills.

Stage 25 (Fig. 4). The posterior portion of the funnel fold is covered by the down growing mantle and only a small triangular opening remains. The lens is quite evident in the eye and is flask-shaped and refractile. Pigmentation is appearing in the iris. The fin primordia are fusing at their median margin and individual gill filaments are barely visible.

Stage 26 (Fig. 4). The mantle completely covers the posterior margin of the funnel. The individual gill filaments are prominent. Pigmentation of the chromatophores begins with a row at the margin of the ventral mantle. The iris is prominent with a ring of pigmentation at its margin.

Stage 27 (Fig. 5). The *secondary* cornea covers the eye and an orbit is formed in which the eye freely moves. The lens is spherical. Dorsal chromatophore are pigmented. The mantle actively contracts.

Stage 28 (Fig. 5). The eye is completely covered by the cornea so the head is trapezoidal. The yolk sac is approximately the same size as the head and the Hoyle organ cells are first evident between the fins. Four dorsal chromatophores are active and prominent.

Stage 29 (Fig. 5). Pigmentation appears in the ink sac and the arms are approximately equal in length to the yolk sac. Hatching may occur at this stage if the embryos are agitated. The retina is deeply pigmented. The arms move actively and the suckers are obviously hollow.

Stage 30 (Fig. 5). The external yolk sac is shed when the embryo accomplishes hatching by attaching the Hoyle organ to the inside of the chorion. Active swimming can occur inside the chorion but the animal is generally quiescent until hatched.

References

Arnold, J. M., 1965a, Normal embryonic stages of the squid *Loligo pealei* (Lesueur), *Biol. Bull.* **128**:24–32.

Arnold, J. M., 1965b, The inductive role of the yolk epithelium in the development of the squid, *Loligo pealei* (Lesueur), *Biol. Bull.* **129**:72–78.

Arnold, J. M., 1968a, The role of the egg cortex in cephalopod development, *Devel. Biol.* **18**:180–197.

Arnold, J. M., 1968b, Formation of the first cleavage furrow in a telolecithal egg (*Loligo pealei*), *Biol. Bull.* **135**:408–409.

Arnold, J. M., 1974, Embryonic development, in: Arnold, J. M., Summers, W. D., Gilbert, D. L., Manalis, R. S., Daw, N. W., and Lasek, R. J. (eds.), *A Laboratory Guide to the Squid Loligo pealei*, Marine Biological Laboratory, Woods Hole, Massachusetts, pp. 24–44.

Arnold, J. M., 1984a, Closure of the squid cornea: A muscular basis for embryonic tissue movement, *J. Exp. Zool.* **232**:187–195.

Arnold, J. M., 1984b, Cephalopods, in: *The Mollusca. Vol. 7. Reproduction*, (A. S. Tompa, N. H. Verdonk, and J. A. M. van den Biggelaar, eds.), pp. 419–454, Academic Press, New York.

Arnold, J. M. and O'Dor, R., 1989, *In vitro* artificial fertilization and embryonic development of oceanic squid for developmental studies. Paralarval identification and mariculture, *in prep.*

Arnold, J. M. and Singley, C. T., 1989, Structure and function of the Hoyle organ (hatching gland) of *Loligo pealei* embryos, *J. Ceph. Biol.* **1**:1–10.

Arnold, J. M., Singley, C. T., and Williams-Arnold, L. D., 1972, Embryonic development and post hatching survival of the Sepiolid squid *Euprymna scolopes* under laboratory conditions, *The Veliger* **14**:361–365.

Arnold, J. M. and Williams-Arnold, L. D., 1970, The effects of Cytochalasin B. on cytoplasmic movement, cleavage and subsequent development of the squid embryo *Loligo pealei, Biol. Bull.* 139:413.

Arnold, J. M. and Williams-Arnold, L. D., 1976, The egg cortex problem as seen through the squid eye, *Am. Zool.* **16**:421–446.

Arnold, J. M. and Williams-Arnold, L. D., 1980, Development of the ciliature pattern on the embryo of the squid *Loligo pealei*: A scanning electron microscope study, *Biol. Bull.* **159**:102–116.

Arnold, J. M., Williams-Arnold, L. D., and Peters, V., 1978, Fusion of tissue masses in embryogenesis. A scanning electron microscope study of funnel development in the squid *Loligo pealei, Develop. Biol.* **65**:155–170.

Boletzky, S. v., 1987, Encapsulation of cephalopod embryos: a review, *Am. Malacol. Bull.* **4**, *in press.*

Boletzky, S. v., 1988, Characteristics of cephalopod embryogenesis, in: *Cephalopods – Present and Past*, (J. Wiedmann and J. Kullmann, eds.), pp. 167–179, Schweizerbatsche verlagsbuchhandlung, Stuttgart.

Fioroni, P. and Meister, G., 1974, *Embryologie von Loligo vulgaris Lam. Gemeiner Kalmar, Grosses Zoologisces Prakikunn, No. 16 c/2*, Gustav Fisher Verlag, Stuttgart.

Lemaire, J., 1970, Table de developpement embryonnaire de *Sepia officinalis* L. (Mollusque: Cephalopode), *Bull. Soc. Zool. Fr.* **95**:773–782.

Marthy, H. J., 1973. An experimental study of eye development in the cephalopod *Loligo vulgaris*: determination and regulation during formation to the primary optic vesicle, *J. Emb. Exp. Morph.* **29**:347–361.

Marthy, H. J., 1982, The cephalopod egg, a suitable material for cell and tissue interaction studies, in: *Embryonic Development. Part B: Cellular Aspects*, pp. 223–233, Alan R. Liss, New York.

Marthy, H. J. and Aroles, L., 1987, *In vitro* culture of embryonic organ and tissue fragments of the squid *Loligo vulgaris* with special reference to the establishment of a long term culture of ganglion-derived nerve cells, *Zool. Jahrb. Physiol.* **91**:189–209.

Naef, A., 1928a, Die Cephalopoden, *Fauna Flora Golf. Neapel.* **35**:1–357.

Naef, A., 1928b, *Die Cephalopoden*, Monographia 35 Fauna e Flora del Golfo di Nàpoli, Teil I, Band 2, Embryologie.

Natsukari, Y., 1970, Egg-laying behavior, embryonic development and hatched larva of the pygmy cuttlefish, *Idiosepius pygmaeus paradoxas, Ortmann. Bull. Faculty of Fisheries* **30**:15–29.

Rice, R. V., Mueller, R. A., and Adelman, W. J., Jr., 1990, Tissue culture of squid neurons, glia, and muscle cells, *this volume*.

Segawa, S., 1987, Life history of the oval squid. *Sepioteuthis lessoniana* in Kominato and adjacent waters Central Honshu, Japan, *J. Tokyo Univ. Fish.* **74**:67–105.

Segawa, S., Yang, W. T., Marthy, H. J., and Hanbu, R. T., 1988, Illustrated embryonic stages of the Eastern Atlantic Squid, *Loligo forbesi*, *The Veliger* **30**:230–243.

Segmüller, M. and Marthy, H.-J., 1989, Individual migration of mesentodermal cells in the early embryo of the squid *Loligo vulgaris*: *in vivo* recording combined with observations with TEM and SEM, *Internat. J. Develop. Biol.* **33**:287–296.

Singley, C. T., 1974, Analysis of gastrulation in *Loligo pealei. Symp. Biology of the Cephalopoda. 55th Ann. Western Society of Naturalists. Abstracts* pp. 22–23.

Singley, C. T., 1977, *An Analysis of Gastrulation in Loligo pealei*, Ph.D. Thesis, 163 pp., University of Hawaii, Honolulu.

Szabo, G. and Arnold, J. M., 1963, Studies of melanin biosynthesis in the ink sac of the squid (*Loligo pealei*). III. *In vitro* culture of tissues and isolated cells of the adult ink sac, *Biol. Bull.* **125**:393–394.

Yamamoto, M., 1982, Normal stages in the development of the cuttlefish, *Sepiella japonica Sasaki*, *Zool. Mag.* **91**:146–157.

Yamamoto, M., 1988, Normal embryonic stages of the pygmy cuttlefish *Idiosepius pygmaeus paradoxas* Ortmann, *Zool. Sci.* **5**:989–998.

Part III

NEURAL MEMBRANES

Chapter 7

Electrophysiology and Biophysics of the Squid Giant Axon

WILLIAM J. ADELMAN, JR. and DANIEL L. GILBERT

1. Introduction

The giant axon of the squid, *Loligo pealei*, is a valuable research preparation currently in use to solve several physiological problems. Many investigators first became acquainted with squid either from the giant axon literature, or from direct involvement in squid axon electrophysiological experiments. Several books have featured work on the electrophysiology of the squid giant axon (Hodgkin, 1964; Katz, 1966; Cole, 1968; Tasaki, 1968; Adelman, 1971a; Baker, 1984b). In addition, there have been several reviews on this subject (Hodgkin, 1951; Baker *el al.*, 1976; Rosenberg, 1973, 1981; Fishman, 1982; Bezanilla *et al.*, 1982; Baker, 1984a; Meech, 1986). Because of their large diameters, giant axons have been excellent preparations for the measurement of whole membrane bioelectrical phenomena. In fact, much of the basic knowledge of the mechanism of nerve fiber or axon impulse conduction has been obtained from the squid giant axon (Adelman and French, 1976).

Originally, Williams (1909) noticed and described giant axons in the squid, but the significance of these as experimental preparations was not recognized until J. Z. Young announced his rediscovery in 1936 at a Cold Spring Harbor Symposium (Young, 1936, 1939). Young urged Kenneth S. Cole to study these giant axons. Cole then remembered that when he was on a fellowship at Harvard in 1928, he rented a room from the widow of Williams (Cole, 1979). From the mid 1930s until the involvement of the United States in World War II, Kenneth S. Cole and his collaborators, working at the Marine Biological Laboratory in Woods Hole, Massachusetts, opened a new era in neurophysiology by subjecting the squid giant axon to a wide variety of physical experiments. An extensive history of these developments and an intimate memoir of the details of these experiments is given in Cole (1968). From its first use until the present, the giant axon preparation has been used in many novel ways. Electrophysiological experiments performed on the axon include intracellular recording of resting and action potentials, membrane impedance measurements, membrane current and voltage clamping, membrane and channel noise determinations, and ion channel gating current measurements. Single channel recording using patch clamp techniques

W. J. ADELMAN, JR. and D. L. GILBERT • Laboratory of Biophysics, NINDS, National Institutes of Health, Bethesda, MD 20892.

(Bezanilla and Vandenberg, 1990) as well as optical measurements on the axon have also been performed (Cohen *et al.*, 1990).

1.1. Overall view of nerve function

The function of the nerve is to conduct information by producing action potentials. There are three processes which under normal conditions give rise to the action potential (Hodgkin and Huxley, 1952a, 1952b, 1952c, 1952d). The *sodium on* (sodium activation) process occurs first in time, and permits the sodium ions to move from the external medium, where the sodium ion concentration is high, across the cell membrane to the internal medium where the sodium ion concentration is low. The entry of these positive ions causes the membrane potential to change its value from about –65 millivolts to about +55 millivolts. The second process, called the *sodium off* (sodium inactivation) process, then stops this ion movement. A third slower process permits the potassium ions from the internal cellular environment (where the potassium ion concentration is high) to move across the cell membrane to the external medium where the potassium ion concentration is low. This permits the membrane to return electrically to its resting state.

After these electrical events have occurred, there is a slow return of the original ionic environment within the cell by active metabolic energy consuming processes, so that the cell is ready chemically and electrically for another action potential (Adelman, 1971a; Hodgkin, 1964). Dean (1941; 1987) was the first to recognize that there had to be an energy consuming process in the cell membrane, the so-called *sodium pump*, for removing sodium ions from the cell. Independently, Krogh (1946) also arrived at this same conclusion. Thus, the energy for the action potential is derived eventually from the active transport processes. Several investigators have measured fluxes using the dialysis technique (Mullins and Brinley, 1990) for studying active transport mechanisms in the squid axon.

Gating or non-ionic currents precede the sodium and potassium current changes in response to a change in membrane potential. Membrane channel conformational changes give rise to these gating currents. Gating currents are studied by inhibiting the ionic currents using chemicals.

2. The giant axon

Squid are free swimming mollusks. *Loligo pealei* live in the western regions of the North Atlantic. The life history and habits of this species has been described (Arnold, 1990a, 1990b, 1990c; Summers, 1990; Aldrich, 1990; Hanlon, 1990). Needless to say, swimming and jet propulsion are important modes of movement for the squid through its environment (see Fig. 1). Swimming is brought about by a rippling undulating motion of the tail fins. Jet propulsion is accomplished by the squirting of a volume of sea water previously taken into the mantle cavity through the molluskan siphon apparatus. This squirting depends upon the rapid contraction of the muscles of the mantle so that a rapidly flowing jet stream may be directed outwardly through the siphon. Activation of mantle muscle contraction is initiated through giant nerve fiber or axon innervation of these muscles. The nerve cells of these giant axons receive

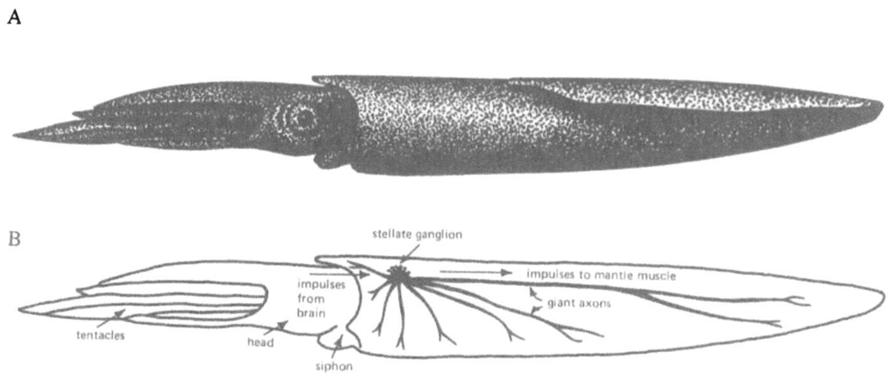

Figure 1. *Loligo pealei,* the common squid along the coast of North America. A. Side view of the left side of the squid. B. Representation of the squid giant axon system in the left side of the mantle.

neural messages from higher nerve centers (ganglia or *brain*) of the squid nervous system (see Stanley, 1990 for details of the giant synapse).

The muscle fibers throughout the mantle must receive nerve impulses almost simultaneously so that by mantle cavity pressurization sufficient momentum can be imparted to the water flow through the siphon in order that the animal be propelled through the sea with a velocity rapid enough that escape from predators is brought about. Through evolution, a modification of the sizes of the axons innervating these mantle muscles came about so that this end could be achieved. In unmyelinated nerve fibers (such as in the squid), conduction velocity is directly proportional to axon diameter. Axons that innervate remote muscle fibers in the mantle are larger in diameter than those that innervate muscle fibers near to the two stellate ganglia. The stellate ganglia are where these innervating nerve cells have their origin and where they are activated through synaptic connections by presynaptic neurons. Thus, all muscle fibers can be signaled to contract simultaneously regardless of distance from the stellate ganglia, because the relation between fiber diameter and distance has been genetically made proportional to the distance the impulses must travel. The largest diameter axon innervating the mantle is found in the medial posterior mantle nerve, and it is this axon that has been the subject of over five decades of intensive research.

3. Electrophysiology of the giant axon

Once techniques had been developed for isolating the giant axon from the squid, the axon preparation was used to measure the transverse impedance of the nerve fiber (Curtis and Cole, 1938). These measurements confirmed and extended earlier observations done on red cells (Fricke, 1923; 1933), marine eggs (Cole, 1928; 1937), and giant plant cells (Curtis and Cole, 1937). These studies taken together established a universal biological constant for cell membranes, namely that the membrane capacity is about 1.0 μfarads per square centimeter of membrane surface. Impedance measurements provide useful information as to the structure and function of cells (DeFelice *et al.*, 1981).

In the late 1930s, Cole and Curtis (1939) were able to combine action potential recording with transverse impedance measurements and show that coincident with the action potential there was a decrease in the transverse resistance of the axon with little or no change in the capacitance. This finding put emphasis on the search for a resistive or conductive element as being responsible for the origin of the nerve impulse. To this day, many membrane studies measure conductance as a function of membrane potential. This chapter will describe in detail systems for making such measurements on the squid giant axon.

3.1. Intracellular recording

In 1939, Howard J. Curtis took the first step into the modern era of electrophysiology by inserting an electrode into the interior of the isolated squid giant axon. Shortly after, he and Kenneth S. Cole in Woods Hole (Curtis and Cole, 1940; 1942), and A. L. Hodgkin and A. F. Huxley in Plymouth, England (Hodgkin and Huxley, 1939; 1945) were able to measure the potential difference across the giant axon membrane between this internal electrode and an external electrode. Composed of a long glass capillary, about 0.1 mm in diameter, and filled with a conducting salt solution, this electrode, once inserted into the axon interior, was able to sense the intracellular potential at its open end and convey this voltage to a reversible metal electrode contacting the electrode solution at the other end. This electrode was connected to the input of a suitable recording device to register the internal potential of the axon. Curtis's electrode became the model for five decades of intracellular electrodes despite repeated difficulties at the onset with its salt makeup, its longitudinal resistance, and its glass wall capacity (Cole, 1968, pp. 145–146). Voltage sensing electrodes still present similar difficulties in design, construction and use, and this chapter will be concerned in part with these problems.

3.2. From current to voltage clamp

At the end of World War II, Cole found himself at the University of Chicago where with the cessation of his role as Director of Atomic Biology for the Manhattan Project, he became Professor of Biophysics and founded the first Department of Biophysics at the University. He and George Marmont began an attack on the membrane conductance problem that led in succession to the invention of the squid axon current clamp (Marmont, 1949) and voltage clamp (Cole, 1949) methods. The current clamp experiments showed that one could record a zero net current action potential implying that the membrane conductance was voltage dependent and not current dependent. The current clamp approach was quickly abandoned, and the squid giant axon experiment was reconfigured as a voltage clamp experiment. The voltage clamp technique permits the experimenter to measure current densities of the cell membrane at pre-set potentials. While the designs and fabrication of the equipment were done in Chicago, the experimental voltage clamp work was performed in Woods Hole, particularly during the summer of 1947. After visiting Cole in Chicago early in 1948, A. L. Hodgkin returned to England, constructed an approximation to Cole's squid axon voltage clamp system and proceeded to voltage clamp *Loligo forbesi* giant axons in Plymouth (Hodgkin, Huxley and Katz, 1949; 1952). This chapter will describe both current and voltage clamp methods as applied to the squid giant axon.

3.3. Microinjection, internal dialysis, and internal perfusion

While the voltage clamp work (Hodgkin and Huxley, 1952a, 1952b, 1952c, and 1952d) had provided the basis for an understanding of nerve excitation and provided the motivation for awarding a Nobel Prize to Hodgkin and Huxley and a National Medal of Science to Kenneth Cole, the squid giant axon subsequently lent itself to several new types of electrophysiological experiments. The first of these involved combining electrical measurements with the microinjection into the axoplasm of a variety of substances. The object of this approach was to control both the external and internal chemical environments so that membrane potentials and currents could be directly related to known chemical gradients across the membrane. Hodgkin and Keynes (1956) developed a microinjector for the squid axon without causing a damaging pressure buildup during the injection. Peter C. Caldwell and collaborators continued these microinjection experiments (Caldwell *et al.*, 1960). The microinjector allowed a volume of solution to be deposited into the interior of the axon and represents an early effort to control the interior environment of the nerve membrane.

However, in 1960 Ichiji Tasaki and coworkers in Woods Hole (Oikawa *et al.*, 1961) and Trevor Shaw and Peter Baker (Baker *et al.*, 1961), working in Plymouth, England, were able to develop methods for replacing the axoplasm of the squid giant axon with flowing artificial media. These methods were so useful that they became standards for most giant axon work for almost three decades. In 1962, Adelman and Gilbert developed a method for voltage clamping the internally perfused squid giant axon (Adelman and Gilbert, 1964; Adelman and Fok, 1964). This chapter will discuss the several methods developed for internal perfusing and voltage clamping axons.

Shortly after the development of internal perfusion methods for the giant axon, Brinley and Mullins (1967) developed a variant using sintered glass capillaries so that they could measure the effects of metabolic substrates on the role of sodium extrusion in membrane processes in the squid giant axon. This method is referred to as internal dialysis (Brinley and Mullins, 1967; Mullins and Brinley, 1969, 1990), and is still used for many types of radioisotope and other experiments where trace or small amounts of substances are to be equilibrated with the cytoplasm in the axon interior, or where internal perfusion would be detrimental to the functions being studied.

3.4. Chemical blockers of specific ionic conductances

In their classical work on the squid giant axon, Hodgkin and Huxley (1952a, 1952b, and 1952c) showed that the currents flowing through the axon membrane in response to a step depolarization in membrane potential could be separated into two major components. The first of these components was an inward current over the voltage range from the resting potential up to the equilibrium potential for the sodium ion gradient across the membrane. The second was predominately an outward current over the same voltage range. The former component rose with time constants in the tenths of milliseconds range and declined with time constants in the millisecond range, being bi-phasic with a definite peak value. The later component was monotonic and achieved steady state values with time constants in the millisecond range. The mathematical description which Hodgkin and Huxley (1952d) used is commonly referred to as the *Hodgkin-Huxley equations* or simply as the *H-H equations*. Hodgkin and Huxley (1952a) called these current components the sodium and potassium currents respectively, since substitution of choline chloride (assumed impermeant) for external sodium reduced the early component in a manner predictable from the theory that the charge carriers for this current were sodium ions (Hodgkin and Huxley, 1952a). In a similar manner but not quite as straightforwardly, they claimed that the late outward current was carried by potassium ions.

For many years it had been the hope of most workers in this field that an independent test of these ideas could be carried out. One approach taken was to search for specific chemical agents which would block either one of these current components without affecting the other. Pharmacological agents were eventually found that could block the sodium current without having an influence on the potassium current. Likewise, the potassium current could be blocked without affecting the sodium current by other agents.

While for hundreds of years there were world wide reports of human poisonings following the eating of certain fish, particularly of the *Tetrodon genus* (puffer fish), it was not until the work of Yoshizumi Tahara (1910) that chemical isolation of the active principle (tetrodotoxin) was started. With isolation methods being perfected and the structure reported by American and Japanese scientists in 1964 at the International Conference on Chemistry of Natural Products held in Kyoto, Japan, it was Toshio Narahashi (Narahashi *et al.*, 1960; Narahashi *et al.*, 1964) who showed that tetrodotoxin (TTX) was a specific blocker of the sodium channel in nerve membranes. TTX has many uses in the study of the excitation properties of ionic channels in the squid membrane, and many electrophysiological techniques make use of TTX. Narahashi (1984) has reviewed the literature and described the techniques and methods of study of sodium channels by means of the blocking action and pharmacological effects of a wide variety of chemical agents and toxins (Narahashi, 1984). Saxitoxin (STX) possesses similar pharmacological effects as TTX (Kao, 1981).

In 1957, Tasaki and Hagiwara reported that injection of tetraethyl ammonium chloride (TEA) ions inside the squid giant axon resulted in prolongation of the action potential, and that such responses to stimulation had plateaus on the falling phases. Upon voltage clamping these axons, Tasaki and Hagiwara found that the potassium currents were greatly reduced below normal. These findings were confirmed by Armstrong and Binstock (1965), and later Armstrong (1969) presented a model of the effects of TEA on the potassium channel.

3.5. Membrane noise

Ion channels are proteins whose configurations enable them to span the lipid bilayer of the cell membrane and form ion conducting pores across the cell membrane. These channels have a water filled central pore connecting the intracellular fluid with the extracellular. These pores fluctuate between closed and open states. The electrical events occurring in the squid axon membrane studied by Cole and Curtis and by Hodgkin and Huxley actually involved thousands of these channels. Such channels at rest exhibit some fluctuations between closed and open states; not all channels open at once. Macroscopic membrane electrical events represent a marshalling in time and space of a population of channels such that a significant number of open states exist simultaneously even though the individual channel openings and closings are probabilistic.

Prior to the mid 1970s estimates of single channel currents were obtained from channel noise measurements. As DeFelice (1981) has pointed out, the first measurements of biological noise were made by Paul Fatt and Bernard Katz (Fatt and Katz, 1950; 1952) at the vertebrate neuromuscular junction, and these events took the form of miniature end plate potentials (mepps). While these were spontaneous and each was probably the result of several thousand acetylcholine molecules acting on the post-synaptic membrane, Fatt and Katz clearly stated that there should be even smaller sub-units or elements contributing to the mepps. When Katz and Miledi (1970, 1971, 1972) found that neuromuscular junctions had a continuous very low amplitude of apparently random electrical noise that increased in frequency when acetylcholine was applied to the junction, they concluded that the activity of individual ionic channels in the post-synaptic membrane was responsible for this new noise. After careful analysis of their records Katz and Miledi (1972) concluded each of these channels was open for about 1 msec, and that during that interval 4.7×10^4 ions moved through the channel. The conductance of an open channel was estimated to be about 0.1 nS (nano Siemens). These numbers were so large that most models involving some form of carriers were abandoned.

Noise measurements were begun on the squid axon membrane in Woods Hole in 1973 by Harvey Fishman and his colleagues (Fishman, 1973a, 1973b) using a patch electrode placed against the inside of the axon membrane and isolated from its surround by a flowing sucrose guard. That same year Emelio Wanke, Franco Conti and Louis DeFelice began noise measurements in Camogli, Italy on the giant axon of the Mediterranean squid, *Loligo vulgaris* (Wanke, DeFelice and Conti, 1974, Conti, DeFelice and Wanke, 1975; DeFelice, Wanke and Conti, 1975).

They were able to obtain noise spectra for sodium and potassium channels and give figures for the conductances of individual channels much as Katz and Miledi had done for the acetylcholine activated post-synaptic channel. They also were able to give estimates of the density of channels per unit membrane area that confirmed previous estimates obtained from tetrodotoxin binding to sodium channels and tetraethylammonium ion binding to potassium channels. For a complete description of methods for membrane and channel noise measurements, the reader is referred to the extensive treatment given in DeFelice's book on membrane noise (DeFelice, 1981).

3.6. Single channel currents

Each of these channels is capable of responding to appropriate stimuli by changing from a closed non-conduction pore state to one in which the pore is selectively open to the conduction of specific ions across the membrane. While it had long been considered that electrical events in excitable membranes were due to an ensemble of single channels, it was not until Ross Bean and coworkers (Bean *et al.*, 1969) were able to record discrete miniature current deflections in artificial membranes doped with Mueller and Rudin's EIM (*excitability inducing material*) (Mueller *et al.*, 1962; Mueller and Rudin, 1967) that the behavior of individual channels could be measured directly. Ehrenstein and Lecar and their colleagues (Ehrenstein *et al.*, 1970; Ehrenstein *et al.*, 1974) were able to describe the kinetics of the closed to open transitions exhibited by these single channel conductances. With the advent of single channel recording using the patch clamp technique (Neher and Sakmann, 1976), it soon became possible to record currents through single channels in the squid (*Loligo vulgaris*) giant axon membrane (Conti and Neher, 1980). Sigworth and Neher (1980) demonstrated that the ensemble of single channels does correspond to the macroscopic current recordings. Bezanilla and colleagues have been able to record from single sodium channels in the cut open squid (*Loligo pealei*) axon preparation by means of the patch clamp technique (Bezanilla, 1987; Bezanilla and Vandenberg, 1990). The single channel patch clamp technique has been extended to potassium channels in the squid giant axon (Llano *et al.*, 1988).

3.7. Ion channel gating currents

As the sodium channel is the primary machine for producing the nerve impulse, its structure underlies the excitable characteristics of nerve and muscle membranes. Because the squid giant axon can be readily perfused with artificial media both internally and externally, thus allowing open channel blockade by specific toxins, agents, and procedures, it has become a standard preparation for the measurement of gating currents.

Hodgkin and Huxley predicted that sodium channels would possess electrically charged structures associated with the flexible parts of the channels that move in accord with changes in the membrane electric field. Such *carrier* movements were envisioned as those changes in channel structure that would eventually lead to ion conduction. These charged structures (or in modern terms, protein sub-groups) would be the channels' gating entities, and their gating movements (or conformational changes) should give rise to a small measurable current which would be non-ionic. For many years, investigators looked unsuccessfully for such *carrier* (or gating) currents (eg. Chandler and Meves, 1965). Gating currents were first measured by Armstrong and Bezanilla (1973, 1974) in squid giant axons and by Schneider and Chandler (1973) in skeletal muscle (Chandler *et al.*, 1976), and they have been shown to exist in many preparations (see review by Almers, 1978). Most importantly, these currents were shown to represent a conservative dielectric charge movement and not to be due to any ionic charge transfer.

Channel gating leads to ionic flow across the membrane. In the open state, channels admit ion flow, and in the closed state ion flow is prohibited. Transitions between resting (closed), activated (open), and inactivated states occur both randomly and in a marshaled or recruited manner as a function of the trans-membrane electric field or the membrane potential. Many investigators believe there are several closed or *hidden* states existing between the resting and the open states (DeFelice, 1981; DeFelice and Kell, 1983). Single channel currents measured by such techniques as patch clamping are open channel currents, and do not measure the kinetics of the closed state transitions, as such.

The gating current is a capacity current having a unique characteristic over a given voltage range. Because this current is larger for a depolarization from rest than for an equivalent hyperpolarization, the gating current is unlike a charging current for a pure capacity, and is often referred to as an asymmetry current.

3.8. Gating currents indicate molecular conformational changes

It soon became apparent to early workers on the asymmetry current that there was no way to predict, from the kinetics of the sodium ionic current, the shape or time course of an expected gating current to be obtained during a pulse clamp experiment. Almers (1978) discusses this problem in terms of a macromolecule undergoing a number of transitions, each causing a change in channel dipole movement with an associated current. The charge transfer needed to reach the open state of the channel molecule would be the vector sum of these charge movements. In the most general sense, the gating current appears to precede the ionic current which is suggestive of the possibility that the gating current represents dipolar movements in the channel molecule voltage sensor making several closed state transitions prior to the closed to open channel transition. This has been the view of many investigators (Armstrong and Bezanilla, 1977; Armstrong and Gilly, 1979; Fohlmeister and Adelman, 1985b; Fohlmeister and Adelman, 1986; Rayner and Starkus, 1989). Gating currents provide a direct insight into the molecular transitions involved in sodium channel activation. Many gating current measurements have been made on the squid giant axon (Keynes and Rojas, 1974, 1976; Meves, 1974; Fohlmeister and Adelman, 1987).

4. The giant axon preparation

For most electrophysiological work involving a large area of membrane to be studied, the use of a large diameter (ca. 400 to 1000 μm) axon having a length of over one centimeter has become a standard. Several species of squid have giant axons that meet these requirements (Table I). Among these are the two most frequently used species, *Loligo pealei* and *Loligo forbesi*. *Loligo pealei* is found from the New England coast southward to the Gulf of Mexico (Summers, 1990). Because in New England waters these squid spawn inshore in estuaries and along the coast in the spring and summer and spend the winters offshore in deeper warmer waters (Gulf stream), it is generally believed that work on the axons of this species is only possible during the late spring and summer months. It is true that the squid catches are larger during this period; the axons of squid caught in May and in the first two weeks in June have very large diameters. However, squid can be obtained in other months as well, as indicated in surveys made by Summers (1967, 1968, and 1969). Adelman collected *Loligo pealei* during the fall, winter, and spring of 1963 to 1971 off the Maryland and Virginia coasts.

Loligo forbesi is found along the Atlantic coasts of the United Kingdom, Ireland, France, Spain and Portugal. Depending on local conditions, these squid are available year round at several marine laboratories.

The giant axons from *Dosidicus gigas* have very large diameters (greater than 1.0 mm). This species has been found in the Pacific Ocean from Chile north to California, and was available for axon research in marine laboratories in Chile and Peru. Because of several disastrous shifts in the Humboldt current due to a phenomenon called El Niño, obtaining these animals for research purposes has become almost impossible for the past couple of decades.

Many researchers in countries on the east coast of South America or in countries bordering the Caribbean Sea have been able to obtain giant axons from tropical squid

TABLE I. Squid Species Having Useful Giant Axons

Squid Species	*Diameter (μm)*	*Reference*
*Architeuthis dux**	137–210	Steele and Aldrich, 1990
Doryteuthis bleekeri	700–900	Tasaki, Watanabe, and Takenaka, 1962
Dosidicus gigas	600–1500	Tasaki and Luxoro, 1964
Loligo forbesi	700–900	Baker, Hodgkin and Shaw, 1962
"	500–1040	Cohen, Hille, and Keynes, 1970
"	985†	O'Dor, 1989
Loligo opalescens	ca. 250	Llinas, 1974
Loligo pealei	300–800‡	Adelman and Gilbert, 1989
"	350–450**	Gilbert, 1971
Loligo plei	290–394	Villegas and Villegas, 1960††
"	400–500	DiPolo *et al.*, 1985, 1989
Loligo vulgaris	530–560	Spyropoulis, 1965
Lolliguncula brevis	75–150	Joyner and DeGroof, 1974
Sepioteuthis sepioidea	250–450	Villegas *et al.*, 1963
Todarodes pacificus‡‡	140–260	Chailaklyan, 1961
Todarodes sagittatus	ca. 400	Mauro, Conti, Dodge, and Schor, 1970

* Hardly *useful* as this species is very difficult to obtain.
† Largest size seen in the Azores, autumn 1989.
‡ Yearly range of giant axon diameters obtainable in Woods Hole, Massachusetts.
** From squid caught by Marine Biological Laboratory vessels during July and August near Woods Hole, Massachusetts.
†† Cited as *Doryteuthis plei.* Animal now known as *Loligo plei.*
‡‡ Cited as *Ommastrephes pacificus sloanei* which is probably *Ommastrephes sloanei pacificus.* According to Voss (1963), *Todarodes pacificus* should be used.

such as *Loligo plei* and *Sepioteuthis sepioidea.* Giant axons obtained from these species have been found to have diameters ranging from 200 to 500 μm.

Much work has been done on giant axons from the Mediterranean squid, *Loligo vulgaris.* For many years, researchers at marine laboratories in Camogli and Naples in Italy were able to prepare giant axons having diameters of about 500 μm. Mauro and coworkers (Mauro *et al.*, 1970) have introduced giant axons obtained from another Mediterranean squid, *Todarodes sagittatus.* Axons of about 400 μm in diameter have been obtained from this species.

In cold waters of the western Pacific, the species *Doryteuthis bleekeri* provides Japanese neurophysiologists with large diameter giant axons. Many physiologists in inland research laboratories in Okazaki and Tsukuba in Japan have been able to keep these animals in laboratory aquaria for many months, thus assuring a convenient and ready supply of giant axons for their research.

4.1. Location of the giant axon in the squid

For the general anatomy of the squid, consult one of the following references: Williams (1909), Pierce (1950), Berman (1961), Bullock and Horridge (1965), Barnes (1968), or Sherman and Sherman (1970). The nervous systems of squid have three levels of organization containing giant nerve fibers (Young, 1939). The anatomical organization of these levels and the origin at the tertiary level of the largest giant axons arising from the stellate ganglia is described by Brown and Lasek (1990). The reader is also referred to articles by Martin (1969) and Villegas and Villegas (1984). The giant axons that have been the objects of study by generations of electrophysiologists are the singularly large diameter fibers in the two hindmost stellar nerves. These nerves also are the most medial of the mantle nerves, and in squid such as *Loligo pealei* (Fig. 1A) are located close to the lateral margins of the cartilaginous pen. Figure 1B shows diagrammatically the general location of the left stellate ganglion and the mantle nerves when viewing a squid from the side. Figure 2 shows their location in a rough diagram of a supine split open eviscerated squid mantle.

4.2. Preparations for dissecting the mantle nerves

The techniques which are described are not the only methods for obtaining viable preparations. The techniques used require concentration and attention to detail as errors rarely can be corrected and usually lead to starting afresh with a new nerve or a new animal.

A small table is used containing a well where the squid mantle can be manipulated, illuminated, bathed in an appropriate physiological medium (usually sea water), and easily viewed with a magnifier or low power dissecting microscope. In some cases this table is mounted on the side of a tank or aquarium containing running sea water. The preparation can be bathed directly with running sea water. Alternatively, the table is perfused with sea water from a small reservoir and the waste perfusate from the well is drained away. If space is at a premium, the dissection can take place in a portable well with a static bath of sea water which can be freshened by removal and replacement of the fluid periodically. Irrespective of the choice, the well has to be deep enough so that a squid mantle once slit open and pinned flat can be covered with physiological solution, but shallow enough that visualization of the preparation is not made difficult. Visualization of the mantle nerves is best achieved by trans-illumination of the mantle from below through a glass or plastic window set in the well. It is preferable to use some form of dark field illumination for doing this dissection. Lamps should be battery powered or run from house current though ground fault interrupters in order to reduce the possibility of electrocution of the dissector.

A table is constructed which has a central well 25 cm wide, 30 cm long, and 3 cm deep. A small window of clear glass measuring 7 cm wide and 11 cm long is set flush in the bottom of the well, and an illuminator is positioned below the window with a movable black panel placed between the lamp and the window so that the light from the illuminator can produce various degrees of dark field illumination depending on the position of the panel. A circle fluorescent lamp of 10 inch or 12 inch diameter or a battery powered industrial flashlight are ideal illuminators. Any form of reflected light from the surface of the preparation occurring as a result of direct overhead illumination or scattered lighting resulting from using a translucent window makes the dissection close to impossible. Investigators in Plymouth, England have used an automobile headlamp powered by a storage battery.

The bottom of the well is coated with a material so that pins can be used to secure the mantle to the floor of the well. It is necessary to leave the area of the window clear so as to transmit light through the squid mantle. One of the most satisfactory coatings is silicon rubber (General Electric RTV-41). Paraffin wax can be used as a substitute for silicon rubber; the paraffin wax is applied to the well bottom

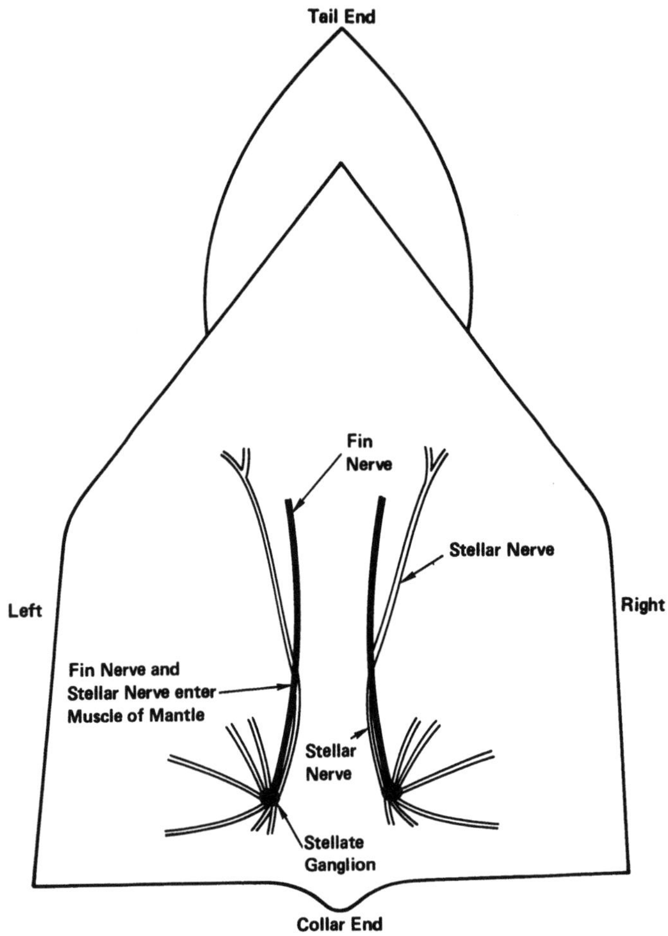

Figure 2. The exposed mantle of the squid.

in a molten state, trimmed, and shaped when solid. Another satisfactory well bottom consists of a sheet of paraffin wax impregnated cork cut to fit the well bottom with a hole over the illumination window. A centimeter scale affixed to the well surface is useful for measuring mantle dimensions.

Dissecting instruments are: a pair of large bandage scissors for beheading the squid and cutting open the mantle; a pair of angular dissecting scissors with 2 cm blades to remove internal organs, muscle and connective tissue overlaying the mantle nerves; two large toothed forceps for handling the mantle and internal organs; two small toothed forceps for handling the muscle and connective tissue near the mantle nerves; two no. 1 Dumont forceps for preparing and removing for storage the medial mantle nerves; silk thread for ligating the cut ends of the mantle nerves and for handling for storage and subsequent dissection (cleaning); a pair of angular Castroviejo eye scissors for removing the mantle nerves from the mantle. The latter item is quite expensive and while desirable is not an absolute necessity. All of these instruments are constructed of stainless steel. If these instruments are sharpened regularly, they can last several years.

4.3. Dissecting the mantle nerves

In netting the squid to be dissected, care should be taken not to agitate the remaining squid in the holding tank or sea water aquarium as this might shorten their lives. For this purpose, a fine mesh nylon bait net (25 cm diameter, 28 cm deep, with a 50 cm long handle) can be used. The squid should be decapitated as quickly as possible, and the time of decapitation should be recorded. Some cautions are deemed necessary at this stage. Bending and squeezing the mantle are best avoided as both of these actions tend to damage the mantle nerves. In addition, the squid is likely to ink if frightened or harmed before decapitation, and inking often makes the dissection more difficult than necessary. The body of the squid (mantle) is then placed on the dissecting table over the trans-illuminated window so that the bottom or ventral (functional use of the term) side is up. The siphon should be facing you and the tail pointing away. The animal is opened by cutting through the mantle along a central longitudinal line on the ventral surface. The mantle is then flattened out (the internal organs should be uppermost and visible), and a record of the mantle length and the circumference of the collar should be obtained, i.e., the height and base of the roughly triangularly shaped mantle. Usually, specimens with a long mantle length and a large collar circumference possess large axon diameters. Female squid have larger diameter axons for a given mantle length than males, but female squid also have a larger collar dimension for a given length of mantle. When squid are in short supply, particularly late in the summer in Woods Hole, one is likely to find that an order for large squid results in a delivery of exclusively male squid.

The mantle is then pinned to the dissecting table so that both stellate ganglia are over the near end of the trans-illuminated window, and several cm of mantle caudal to the two ganglia are centered over the window. There should be sufficient mantle overlap in the non-window area of the dissecting trough for the margins of the mantle to be securely fastened to the table by pinning to the silicon rubber or to the paraffin waxed base. Before removing the viscera, the ink sac is first located. The ink sac is silvery colored and has a tear drop shape, one to two cm long. It can be seen caudal to the siphon and in the center of the mantle. Its ejecting orifice can be seen at the anterior end, the caudal end being the more rounded end of the sac. The narrow end of the sac facing the dissector is then gripped with the toothed forceps pinching off ink ejection and the sac and the digestive tract, heart, and digestive and excretory organs are dissected free from the mantle using the dissecting scissors and a lifting pull on the ink sac. The siphon is then picked up with the large forceps and the muscle and visceral attachments to the mantle are cut with the dissecting scissors. At this point it is possible to see the mantle nerves radiating out from the two stellate ganglia. It is also possible at this stage of the dissection to see the pen, which is a thin structure of *cartilage* running the entire mantle length. While the stellate ganglia appear to be located just lateral to the pen and above the mantle muscle, the hindmost medial mantle nerves can be seen to course along the edge of the pen, then retreat below the pen and eventually enter the musculature of the mantle itself lateral to the edge of the pen. Removal of the pen from the mantle appears to be quite easy, but the dissector is warned that the largest axons are found in the medial mantle nerve which runs in close proximity to the pen. The gills are then removed, and the animal is turned over.

For purposes of visibility, the skin is removed from the outer surface of the mantle. The skin next to the collar is cut from edge to edge, and stripped away. This very thin tissue contains colored chromatophores which can impair the visibility for dissection of the axon both directly by light absorption and indirectly by light scattering. The mantle is then turned back to the position where the interior is again facing upwards, and the mantle is repinned to the dissecting table with the tail facing away from the dissector. In all these procedures, it is important to avoid damaging the mantle muscle as damaged muscle cells lose their impermeability to external calcium ions. When calcium enters these damaged cells the muscle proteins react with the calcium in such a way that the muscles become opaque and scatter light.

So as to provide better access to the medial mantle nerve, the pen is removed by first making a straight line cut through the tissue overlaying the pen. With a blunt tipped forceps, the tissue above the pen, is scraped back from the cut to the margin of the pen. The edge of the pen is then freed from the mantle. At this stage care must be taken to avoid damaging the medial mantle nerve which lies very close to the edge of the pen. The rostral end of the pen is then cut free of the mantle and upon inserting a finger between the pen and the mantle the pen is removed from the mantle by sliding the finger down under the pen for the length of the mantle, while lifting the pen from the mantle.

At this stage of the dissection the right and left stellate ganglia can be easily seen and recognized by the ten or so stellar nerves radiating from each ganglion (see Fig. 2). Each stellar mantle nerve contains a giant post-ganglionic axon (tertiary neuron). These axons are large enough to be identified without magnification. The largest and longest of these axons are located in the two hindmost stellar nerves. Each of these medial mantle nerves is overlaid and joined by another nerve composed exclusively of small diameter fibers. These nerves innervate the tail fins, and hence are called the fin nerves. These nerves have been used from time to time for anatomical and biochemical studies. The retractor capitis muscles were previously cut in removing the pen and these lay above the medial mantle and fin nerves. These muscles have to be removed carefully. The use of fine scissors is best for this purpose as in some places the muscle insertion is only a millimeter away from the giant axon.

At this stage there is a choice open to the dissector. The fin nerve can be stripped away from the underlying medial mantle nerve, and those who have access to a dissecting microscope or use powerful magnifiers often elect to do so at this point in the dissection, as it makes subsequent removal of the mantle nerve somewhat less difficult. However, if adequate visibility is not the case because of lack of magnification or because of excessive light scattering, then the giant axon is at risk of being damaged by the procedure. If the medial mantle nerve is followed from the stellate ganglion caudally, the point at which the mantle and fin nerves dip into a tunnel in the mantle muscle may be readily observed. Further observation will reveal that the tunnel bifurcates into two, the fin nerve tunnel being more medial than the mantle nerve tunnel.

At this point in the dissection, the tunnels may be opened by inserting the tip of the lower blade of a pair of fine scissors into the common tunnel entrance and cutting in a caudal direction opening up the tunnels so that they appear as troughs with the fin and mantle nerves resting in their bottoms. This cutting should be done with an upward lifting motion so that the nerve directly below the scissors blade is not damaged. Once the fin and mantle nerves are properly exposed, the fin nerve can be cut in two at a point where the separation between the fin and mantle nerves is defined clearly. Once the fin nerve is transected, the distal end of the proximal portion can be grasped with the no. 1 forceps. By lifting slightly and cutting under the fin nerve from its cut end back toward the ganglion until a point is reached where the two nerves appear as one, the fin nerve may again be transected and the loose segment of fin nerve removed and discarded. Often in very large squid, these procedures are unnecessary as a sizeable length of mantle nerve lies entirely in the mantle cavity. A length of nerve of 5 cm provides more than adequate lengths of useful giant axon once fine cleaning is accomplished. For much of the summer season at the Marine Biological Laboratory in Woods Hole, it will be necessary to dissect well into the mantle nerve tunnel to obtain a useful length of axon.

Once an adequate length of medial mantle nerve is exposed, the nerve (usually about 5 cm from the stellate ganglion) is ligated with silk thread. The best procedure for performing this task consists of passing the thread under the nerve using fine forceps with curved tips. Once knotted around the nerve, one of the two falls can be cut away leaving a thread end of several cm for handling and mounting. The proximal end of the mantle nerve can then be ligated. Another tie is made by passing the silk thread under the stellate ganglion and tying a firm knot around the ganglion and the overlaying fin nerve. Alternatively, a tie is made just around the nerve close to the ganglion. In either case one of the two thread falls is cut away, the remaining one to be used for handling and mounting. For work involving the giant synapse, the procedure is quite different (Stanley, 1990). These ties are often color coded, so as to distinguish the larger head end from the smaller tail end; i.e., red for head, blue for tail, or vice versa. By picking up the thread end at the ganglion end of the mantle nerve and inserting the fine scissors blades under the nerve it is possible to cut away the nerve from the mantle. The procedure which seems to work best in our hands involves a continuous but slight pull upward on the thread while cutting the two insertions of connective tissue that hold the nerve attached to the interior mantle wall. Looking through a low power dissecting microscope (ca. 5 to 10 ×), or using magnifying loupes can be of help in obtaining routinely viable preparations. Others have found that even this stage of the preparation could be performed well by using their unamplified native vision.

4.4. Storing the mantle nerves

Once the nerve is freed from the squid it should be placed immediately in a storage medium, such as filtered sea water or artificial sea water (see Section 4.6). Although unfiltered natural sea water can be used, the presence of particulate matter, as well as microflora and fauna make storage and subsequent dissection more difficult. The media can be used for both storage and for dissection of the giant axon. A good storage container is a 9 cm Petri dish with a cover. A double width microscope slide can be just fitted inside the bottom of one of these dishes, particularly if one trims off the corners of the slide slightly by pinching with a pair of small needle nosed pliers. Polyethylene tubing can be fitted in the bottom of the Petri dish to rest the slide on and keep the slide off the bottom of the dish. The diameter of the tubing should be just enough to allow the preparation to be adequately covered with medium without the nerve breaking the surface.

Stainless steel clips fasted much like paper clips to the slide can be used for holding the nerve to the slide. The threads at the ends of the nerves can be slid under these clips and be positioned and clamped in place by the spring-like action of the clips. The steel for making these clips must be tempered and should be between 0.5 to 1.0 mm in diameter. Stainless steel fishing leader is a readily available inexpensive material for this purpose. Another form of hold-down makes use of small pieces of rubber stoppers glued to the ends of the slides with Elmer's® contact cement. By cutting V slits into these rubber stopper pieces, the threads tied to the ends of the dissected nerves can be slipped down into these grooves at opposite ends of the slide. It is convenient to record the length of the nerve between ties before removal from the squid so that when the dissected nerve is mounted on the glass slide the same length can be kept. A 4 cm length of nerve fits nicely on a slide without slack or stretching. At this stage, the nerve can be refrigerated in the covered Petri dish while the experimental setup is readied. It is best to mark the date, and time of the dissection on the cover of the Petri dish. Other details such as the squid's sex and whether the nerve is from the right or left side can be added. Keeping a dissection log is helpful and such details can be incorporated in the log. The nerve can be identified by a code referring to the logged information. A log is useful for predicting the best time of year for particular experiments.

4.5. Isolation of the giant axon

In the argot of squid axon physiology, this aspect of the dissection has been called *cleaning*. In essence, a giant axon is stripped or cleaned of its surrounding fibers and connective tissue so that it can be used for certain experiments. In critical experiments where radioisotopes are equilibrated with the axoplasm or where membrane fluxes are to be traced solely to the giant axon membrane, this process may require that all fibers other than the giant be removed. In other cases, removal of fibers and connective tissue might only be limited to reducing the size of the preparation, or making visibility of the inside of the axon possible. For macroscopic whole axon voltage clamping, reducing the material external to the giant axon membrane reduces the external series resistance, and the use of relatively clean axons is recommended. Whenever the choice is between a *clean* weak axon and a *dirty* robust axon the *dirty* one should be chosen.

A dissecting table with a width of 70 cm and length of 90 cm is used for cleaning the giant axon. A cutout of dimensions 12 cm wide and 40 cm long is made in the center of the front, so that the table top has arm supporting wings. Inset into the center of the top of the table is a well with inside dimensions 15 cm square by 3 cm deep. This well has a flange so that it can be set into a hole in the table top which has a supporting 1/8 in deep recess around the margin of the hole such that the well is flush with the table top, but can be easily removed for cleaning. This well is constructed of 1/8 inch thick clear plastic (Lucite or Plexiglass) and is put together to be water tight so that it can be used to contain an ice and water mixture for cooling the preparation during the dissection. A dissecting microscope, with zoom type objective lenses, is mounted on the table so that the center of the well is in the center of the visual field. A good magnification can be obtained by using a 10 × wide field ocular and a variable objective magnification of from 0.7 to 4 ×, giving an overall magnification range of from 7 × to 40 ×. The criteria for choosing a stereo microscope should be used based on the flatness of field, the convenience of use, as well as the sharpness of image. The power of the right and left oculars should be adjustable independently to accommodate differences in the eyes of individual users.

The lighting of the preparation has almost universally been of the dark-field variety. A 9 in diameter circle fluorescent lamp, located below the lucite well of the table, is used for lighting. Hung below this lamp is a metal shelf with an inexpensive mirror attached such that the illumination from the circle lamp is reflected upwards and through the well. A two bladed propeller-like black vane is fastened to a shaft rising from the center of the metal shelf. The overall length of the vane is about 20 cm and

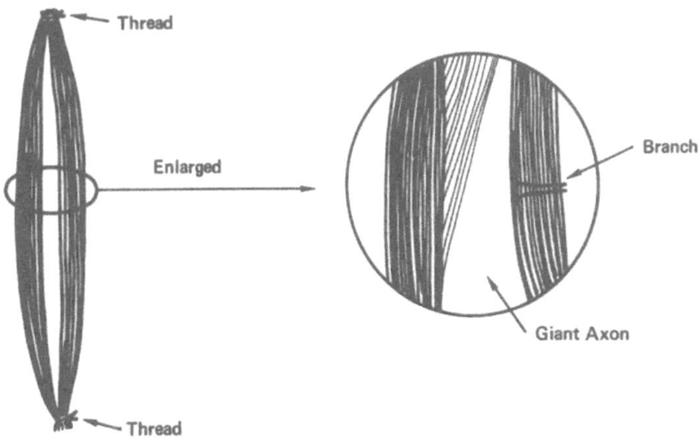

Figure 3. The stellar nerve.

its width is about 10 cm at the tips and about 5 cm at the hub. The vane is so fastened to the shaft that it can be turned. Various dark-field arrangements can be achieved with this vane. Covering the well is a disk of opaque plastic with a hole bored in its center. The disks are 25 cm in diameter, and are composed of 2 to 3 mm thick gray PVC plastic. The holes bored in these disks are 93.4 mm in diameter. This hole diameter was chosen because most 90 mm Petri dishes will fit into this hole and still have a sufficiently larger rim bead to prevent them from falling through the hole. Petri dishes are rejected if they either fall through the hole or do not fit the hole at all. When a Petri dish containing a squid nerve is placed in the PVC disk the dish extends below the surface of the disk and into the coolant in the well below. With the disk centered over the ice-water filled well, the nerve is centered in the microscope visual field and may be moved around in the field by gentle pressure applied to the disk by the heels of the dissector's hands. Such an arrangement leaves the fingers of both hands free to wield dissecting instruments. If the focusing knob of the dissecting microscope is attached to a servo-motor driven by an actuating pedal under the table, the dissector can proceed with the dissection without having to stop for positioning of the preparation or focusing of the microscope.

If the dark-field illumination is arranged so that the light is transverse to the long axis of the mantle nerve (and the giant axon), the outline of the giant axon inside the mantle nerve is brilliantly lit with the axoplasm inside the axon remaining dark (see Fig. 3 in which the dark areas are represented in reverse). When the vane under the well is rotated so that the light is longitudinal with the axis of the mantle nerve the margins of the giant axon fade from view, and other structures either transverse or oblique to the axon are revealed. As the giant axon has branches which innervate the medial musculature of the mantle, some of these branches exist quite close to the stellate ganglion. Cutting these branches off close to the giant axon proper or damaging them by pulling causes leakage of calcium ions into the axoplasm as well as gross loss of cytoplasm to the external medium. Neither of these is good for the viability of the axons. In transverse illumination, damaged regions (cut or pulled branches) appear white, because the usually impermeant calcium ion has entered the axon and precipitated the axoplasm (Hodgkin and Katz, 1949) causing it to go from a transparent ordered state to a light scattering disorganized state, viewed as *white spots*. Using the variable dark-field during dissection is a great aid to seeing branches, and thus is an aid to avoiding pulling or cutting these branches too close to the giant axon

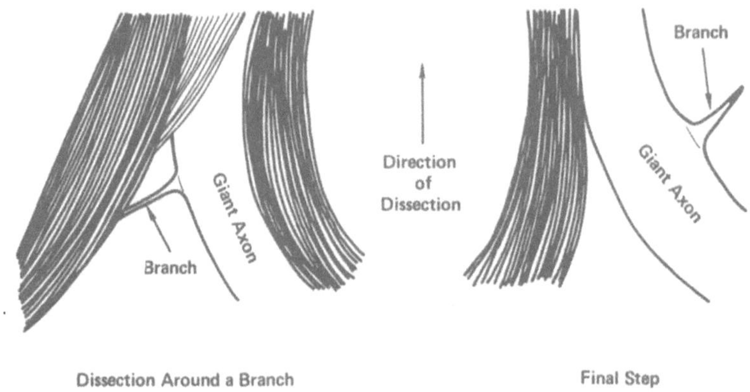

Figure 4. Isolating the giant axon.

proper. Examining the nerve before beginning cleaning is helpful in determining the best approach to the dissection.

The dissection is usually begun by cutting away any muscle tissue adhering to the mantle nerve using angular Castroviejo eye scissors. If the fin nerve is still present, it is cut out. In removing the fin nerve, care is taken to avoid damage to the giant axon in the region where fibers between the fin and mantle nerves are exchanged. This decussation is usually about 2 cm caudal from the ganglion. The remaining preparation consists of the giant axon surrounded by small fibers and connective tissue. For removal of these fibers and connective tissue, a sharp number 4 or 5 Dumont forceps and/or needles made from sharpened insect pins are used. These pins are held in either wooden handles, glass handles made from fine bore thick walled tubing, or pin vises with plastic handles. When necessary, the forceps is sharpened using a very fine Arkansas stone or very fine Emery paper (or Emery cloth). Holding the small fibers around the giant axon with the forceps, enables one to gently pull these fibers back from the giant axon. If a sufficient opening occurs then the eye scissors or a needle can be inserted in the gap. The scissors or the needle can then be used to separate the small fibers from the giant axon by teasing or cutting. This procedure usually requires several passes (see Fig. 4). The extent that these are removed depends on the type of experimental purpose that is planned for the giant axon. *It is necessary that extreme caution be taken so that small branches off the giant axon are not pulled or cut short.*

Generally, it is better to cut a branch by pinching-off than by direct severance with a knife of with scissors. It is sometimes advisable to leave a large amount of tissue around a branch so that it can be easily identified. Nerve fibers are usually found in groups or bundles surrounded by a connective tissue layers or perineuria. Natural cleavages exist between these nerve bundles, and in dissecting the small fibers from the giant axon, separation by bundles using the natural cleavages between these groups is a most successful approach. It is better to proceed with working an entire bundle free before proceeding with another. A few carefully planned passes up and down the full length of the mantle nerve are usually more successful than many small passes. *Branches are visible generally only on the sides of the giant and usually cannot be seen on the upper or lower axon surfaces.*

Once a giant axon has been cleaned to a satisfactory degree, the Petri dish containing the axon should be placed in a refrigerator for at least 15 minutes. At the end of this period it should be possible to see any visible signs of damage (white spots) to the axon caused by pulled or cut branches. If there is a sufficient length of apparently undamaged axon available for experiment, the axon can then be tied off with fine silk thread very much like the nerve was before removing from the squid. The silk thread that anglers use to tie flies is very appropriate for this purpose. At this point the axon has been made ready for an experiment or is available for further preparation.

4.6. Physiological saline for the giant axon

The ionic composition of sea water (Robertson and Webb, 1939) is almost the same as squid hemolymph (Shoukimas, Adelman, and Sege, 1977). These later authors gave the following cation values for *Loligo pealei* hemolymph: 393 mM sodium; 11.7 mM potassium, 45.9 mM magnesium, and 10.4 mM calcium. These concentrations were determined by atomic absorption spectroscopy, and were obtained from both whole hemolymph and protein-free hemolymph. The presence of protein in the hemolymph did not alter significantly the mean concentration values. Robertson (1965) obtained similar values for *Loligo forbesi* with the exception of potassium which was given as 20.6 mM. These values are concentrations and not activities. Given that the total osmolarity of either sea water or squid hemolymph is about 1.0 osmol, then one should expect the activity coefficients to be significantly less than 1.0, and hence the activities to be also lower. Steele and Gilbert (1980) measured the activities of calcium and potassium in squid hemolymph and in an artificial sea water using ion-sensitive electrodes. The composition of the artificial sea water used by Steele and Gilbert was 9.27 mM $CaCl_2$, 9.0 mM KCl, 423 mM NaCl, 22.9 mM $MgCl_2$, and 25.5 mM $MgSO_4$. They found that the activity ratio of the hemolymph Ca^{2+} compared to their sea water Ca^{2+} was 1.05 ± 0.02; and for K^+, this ratio was 1.21 ± 0.04. When natural sea water was used instead of the artificial sea water, the ratios were 0.99 ± 0.02 for Ca^{2+} and 1.26 ± 0.08 for K^+. The hemolymph was taken from fresh squid. When the hemolymph was taken 2 minutes from the squid, the K^+ was elevated by 20 to 50%. A potassium leakage from dying cells could account for this elevation and suggest that high concentration values of K^+ found in the hemolymph could be due to this time factor artifact. Furthermore, the potassium concentration in the hemolymph *Octopus dofleine* is only 10.3 mM (Potts and Todd, 1965). Shoukimas *et al.* (1977) give several reasons for believing that 10 mM is closer to the squid hemolymph potassium concentration value than 20 mM. Most solutions used for electrophysiological experiments are devoid of sulfate since sulfate has little significance for the electrical behavior of the axon. For the practical purpose of preparing solutions, concentrations are used instead of activities. When calculating ionic equilibrium potentials, it is proper to use activities.

Schmitt (1955) showed that squid axons survive longer in an external solution depleted in potassium, and it is often recommended that potassium-free sea water be used for storage of squid nerves or axons.

Howell and Gilbert (1976) have shown that in *Loligo pealei* the pH of hemolymph is 0.4 to 0.8 units above the neutral pH at any temperature. Table II gives the pH of water at neutrality as a function of temperature and shows that the pH at neutrality increases as the temperature decreases. Since solutions are usually made up at room

TABLE II. Temperature Effect on the pH of Water

*pH at Neutrality**	*Temperature in °C*
7.472	0
7.367	5
7.267	10
7.173	15
7.083	20
7.000	24
6.998	25
6.916	30
6.840	35
6.767	40

* Calculated from Weast *et al.* (1988). The pH = −0.5 log K_w.

temperature, the pH of the solution at neutrality increases at the colder temperatures used in most voltage clamp experiments. A good buffer should have the same temperature dependence of its ionization constant, i.e., pK, as that of water, so that the pH above the neutral pH remains constant as the temperature is changed. In addition, since the buffers have a maximum buffer value at its pK, the pK of the buffers should be close to the desired pH. Good *et al.* (1966) synthesized buffers which fit these two criteria. Three buffers that these investigators include are N,N-bis (2-hydroxyethyl)-2-aminoethanesulfonic acid or BES buffer, N-tris (hydroxymethyl)-methyl-2-aminoethanesulfonic acid or TES buffer, and N-2-hydroxyethylpiperazine-N'-2-ethanesulfonic acid or HEPES buffer. Table III gives the pK values as well as their temperature dependence. Included in table III is also another buffer, tris (hydroxymethyl) aminomethane or TRIS buffer. Note that the Good buffers fit the criteria much better than does the TRIS buffer.

The chemical composition of a simple physiological saline is: 430 mM NaCl, 10 mM $CaCl_2$, 50 mM $MgCl_2$, 10 mM KCl, and 10 mM HEPES buffered to pH 7.4 at 20 to 25 °C. To increase the buffer capacity, the HEPES concentration can be increased. The KCl concentration can be less than 10 mM, even zero if desired.

5. Electrophysiological methods

5.1. Whole axon voltage clamp

The technique presented here is applicable to giant axons having diameters of about 450 µm and lengths of about 2 cm. The size of the internal cannulas and electrodes are matched to the axon diameter.

For measuring voltage clamp currents, it is important that the following rules be adhered to (Cole, 1971; Cole and Moore, 1960): 1.) The potential difference across the membrane should be known and uniform over the area of measurement to within a few mv. 2.), The delay in the measurement and control of the potential should be as small

TABLE III. Temperature dependence of buffers

BUFFER	*pK*	*ΔpK/°C*
HOH	7.08*	−0.018†
BES	7.15	−0.016
TES	7.50	−0.020
HEPES	7.55	−0.014
TRIS	8.3	−0.031

pK values are for a temperature of 20 °C.

* Value obtained from Table II.
† Value obtained from performing a least squares fit to data in Table II.

as possible (for the classical squid axon clamp, not more than 10 μsecs). 3.), the current density through the membrane over the area of measurement should be uniform within a few per cent at any instant in time.

Conforming to these simple rules has always been difficult for investigators voltage clamping the membranes of whole axons. The voltage clamp method (Moore and Cole, 1963) requires that the potential across a given area of membrane be forced to conform to a given set of voltage values commanded by the investigator. Using a feedback circuit, the membrane potential is sensed, fed-back to a comparator, and controlled by current flow from the control amplifier through the membrane to set the potential across the membrane equal and opposite to the command. This means that if the clamp works properly, the command forcing function can be of any waveform. For a square wave pulse as the command function, Cole's second rule is applied by comparing the settling time of the membrane potential with the command. For a sinusoidal forcing function, this rule is applied in the frequency domain to the phase shift between command and membrane potential. In essence, all waveforms can be expressed and analyzed in the frequency domain.

5.1.1. Axon preparation for voltage clamping

Giant axons are prepared by the techniques described earlier in this chapter. For beginning most voltage clamp experiments where ionic currents are to be measured, the axons are internally perfused with artificial media with some internal solution. An internal solution which has been used successfully is: 190 mM K glutamate, 550 mM glycine, 40 mM NaF, 10 mM KF, and 40 mM NaHEPES buffered to pH 7.6 at room temperature with KOH or glutamic acid. The high concentration of HEPES is added to ensure that the pH is kept constant. Other internal solutions use isethionate and aspartate as anions (see Mullins and Brinley, 1990). This is done in a chamber in which the temperature is maintained by a Pelletier device driven by current from proportional controller under feed-back from a temperature sensor in the axon chamber. In the present version, the temperature can be dial set to a given value and maintained constant to within 0.1 °C. There are two very different methods for internally perfusing the giant axon. The first method is to remove the axoplasm by rolling out the axoplasm (Baker *et al.* 1961; Narahashi, 1963). The other method is to insert a cannula longitudinally into the axon and by means of capillary action, remove the axoplasm (Oikawa *et al.*, 1961; Tasaki *et al.*, 1962, Adelman and Gilbert, 1963). The removal of axoplasm technique will not be discussed. However, Gould and Alberghini (1990), Brown and Lasek (1990), Weiss *et al.* (1990) have used this technique for obtaining axoplasm.

5.1.2. Voltage clamp chamber and electrodes

The design of the chamber (Clay and Shlesinger, 1983; Fohlmeister and Adelman, 1984) is based upon considerations discussed in detail by Cole (1968). The external guard and current electrodes are fabricated from silver, plated alternately with platinum and silver chloride in accord with the method of Cole and Kishimoto (1962). The central external current measuring electrode is 1.0 mm in length, and each of the guard electrodes is 5 mm in length. The electrodes are separated from each other in the chamber by mylar strips. As the fluid-filled trough is lined on each side of the chamber by this external electrode array and is 2 mm deep and 2 mm wide, current flowing outwardly to the external wall electrodes from an axial wire electrode located in the cross-sectional center of the trough approximates radial flow to a remote concentric cylindrical electrode (Binstock *et al.*, 1975). Therefore, for a giant axon impaled with a 100 μm axial wire, the current from this wire would traverse the axon membrane transversely (normal to the surface), and the density of current would be equally distributed over the membrane area encompassed by the electrode array for a uniform membrane.

Voltage sensing electrodes consist of glass-walled salt bridges filled with half molar KCl solution into which silver-silver chloride metal electrodes are inserted. At the sensing ends (tips) they are about 50 μm in diameter, and at the metal electrode ends they are about 0.7 mm in diameter. The glass is pulled so that a very long gradual taper is produced. Consequently, they are interchangeable and when used for internal voltage sensing take up little volume inside the axon. The electrodes contain a *floating* platinum black shunt wire inserted through the glass lumen to reduce the high frequency impedance of the long salt bridge (Fishman, 1973a). The KCl solution contains sufficient agar (ca. 3%) to just gel the solution. A variation of this electrode, which was suggested by Gen Matsumoto (1989), is to plug the tip of the capillaries with asbestos fibers before filling with KCl solution. The tips are fire polished to close the tips around the fibers and then back filled with KCl solution. These electrodes can be stored in air, do not need agar in the KCl solution, and can be made ready by simply rinsing the tips briefly in distilled water. Either of these type of electrodes are connected electrically to appropriate amplifiers with field effect transistor (FET) input stages.

The voltage clamp electronics were designed for relatively wide-band operation, direct current (DC) to about 1 megacycle/sec (MHz), consistent with low noise considerations. The measurement of membrane current is accomplished through a current to voltage transducer connected to the central electrode pair. As this electrode pair is only 1 mm in length the maximal sodium ion current (about 5 mamp/cm^2) flow to this electrode from the axon is observed to be stable with no thresholds or notches (Taylor *et al.*, 1960; Cole, 1968) as long as the axial wire is carefully plated with platinum black and the external electrodes are plated as prescribed by Cole and Kishimoto (1962). The clamp circuitry contains a series resistance (R_s) compensation that compensates for series resistance (Adelman *et al.*, 1977) without drawing any current from the current to voltage transducer. This compensatory device adjusts the command voltage as a function of the I_mR_s drop. As I_m varies with time and R_s is a constant, it is a simple matter to build a device that samples I_m and feeds back a voltage equivalent to the error in the true value of the membrane potential induced by the presence of the series resistance. Tests of the clamp on an axon dummy circuit with a variety of suitable R_s values indicated that proper series resistance compensation is accomplished over a R_s range of several ohm cm^2. Measurement of series resistance in the giant axon is described in Binstock *et al.* (1975).

5.1.3. Internal perfusion and electrode placement

For doing a perfusion-voltage clamp experiment, one has to insert a perfusion cannula, a current injecting electrode (axial wire), and a voltage sensing electrode inside the axon. There must also be on the outside of the axon, a current collecting electrode and a voltage sensing electrode. The old methods required a microtip electrode for measuring the internal potential.

5.1.3.1. *Piggy-back* technique

This method combines both internal electrodes into one assembly, commonly called the *piggy-back* assembly. Two cuts are made, one of either side of the axon. A perfusion cannula is placed into one of the cuts. The axoplasm is sucked into the perfusion cannula by applying a slight negative pressure to the cannula as it is passed through the cut on the other side of the axon. Then the axoplasm is removed from the cannula. With the cannula outside of the axon, the *piggy-back* assembly is inserted into the cannula. As the cannula is being slowly withdrawn, the *piggy-back* assembly is slowly inserted at the same rate, so that the assembly remains inside the cannula until the assembly is in position as in figure 5. See Section 5.1.3.2 for details about the insertion, withdrawal, and tying of the cannula to the axon. Care must be taken in making the *piggy-back* assembly so that electrical cross-talk is inhibited between the two electrodes. This is done by making certain that the floating platinum wire does not touch the axial wire. In order to reduce the possibility of this cross-talk, the internal electrode can be made so that the tip is bent at right angles. The two electrodes are glued together with 5 minute epoxy glue. This cross-talk problem is avoided in the two internal electrode technique described in Section 5.1.3.2.

5.1.3.2. Two internal electrode technique

The method which is described here is derived from the original method first developed by Adelman, Kishimoto and Dalton (1960–1961) and improved in 1962 by Adelman and Gilbert (1963; 1964). This technique has been under continuous development since its introduction and incorporates within the perfusion cannula a platinized platinum axial wire for delivering current through the axon membrane for voltage clamping purposes (Gilbert, 1971; Adelman, 1971b; Adelman and French, 1978). The axon voltage chamber is designed to accommodate an axon that is 2.6 cm

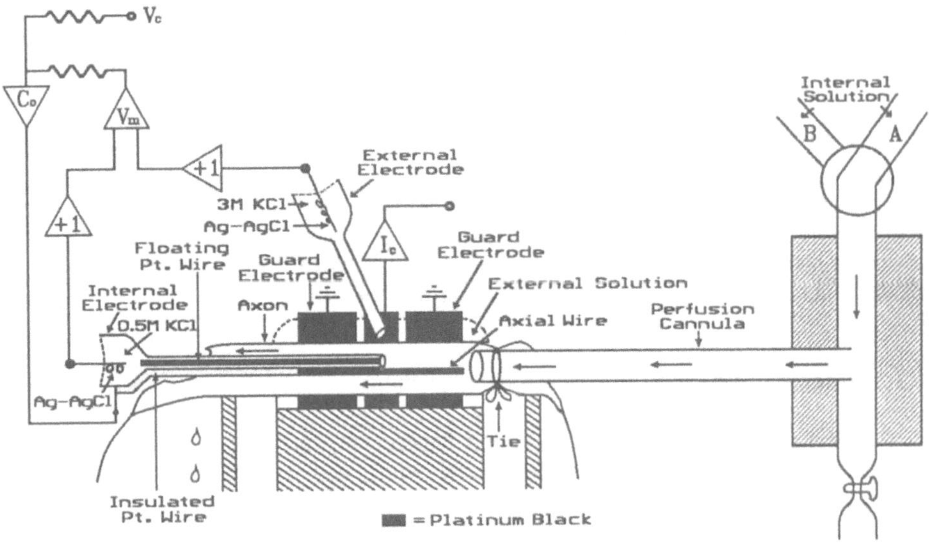

Figure 5. A simplified diagrammatical sketch (not to scale) of a voltage-clamp perfusion technique using the *piggy-back* assembly. Note the guard regions are on either side of the central measuring area. Internal perfusion is accomplished using a valve arrangement for perfusing with either solution *A* or *B*. A stop-cock arrangement permits a quick change of the internal solution. At other times, it is closed so that the solution can pass through the axon. The perfusion cannula is inserted into a plastic mound and is sealed into place using dental wax. For the two internal electrode method, the axial wire is inserted into the perfusion cannula.

The simplified voltage clamp circuitry is also illustrated. The two potential sensing electrodes have a high impedance. For this purpose, they are connected to impedance converters with a gain of 1 and are designated *+1* in the diagram, which converts the high impedance input to a low impedance output. *Vm* is the measured potential across the axon membrane, *Vc* is the control potential to which the membrane is clamped, *Ic* is the measured current, and *Co* refers to the control amplifier. Current is passed through the axial wire into the axon. A computer program can program the control potential through a digital to analog (D/A) converter connected to a computer and record the measured current through an analog to digital (A/D) converter also connected to a computer. Not shown are the grounds attached to the operational amplifiers. The diagram does not give the sign of the potentials and currents. This depends upon the circuit actually used. An outward current is designated as positive. The membrane potential is about −60 mv.

long between ties and that ideally is 460 μm in diameter. This length provides for ample overhangs on the ends for bending the axon so holes may be put in the axon for cannula and electrode insertion into the axon interior with horizontal movements. Once the axon is tied in place the axon should lie flat with no additional bends or kinks that might prevent threading the cannula and electrodes inside the axon between end bends. Once the axon is mounted in the chamber and the external perfusion is deemed to be flowing at an appropriate rate with a temperature of about 7 °C, holes are poked into the ends of axon at the bends using a fine sharp needle. The end into which the internal perfusion cannula will be threaded has a hole made which is just larger than the cannula diameter; the other end into which the internal voltage sensing electrode will be threaded is made quite large (approaching several hundred μs). With practice it soon becomes easy for the investigator to judge the size that this hole should be so that the axon will not collapse and yet present no restriction to internal perfusion fluid flow out of the end of the axon once the internal voltage sensing electrode is threaded inside the outlet hole.

Once the axon is deemed properly mounted in the chamber, the internal perfusion cannula and axial wire assembly is positioned so that the axial wire tip is centered at the inlet hole. The axial wire-cannula assembly is then threaded down inside the axon until all of the exposed portion of the axial wire is outside the axon at the outlet end, and the tip of the cannula is just outside the outlet hole. Perfusion flow is begun and the cannula end is examined to see if a vigorous flow has been achieved and the cannula is cleared of any axoplasm accumulated during penetration. The flow may then be slowed but still maintained significantly, and the cannula tip is then withdrawn back inside the axon for a distance of about 3 mm. Perfusion flow is continued with the cannula in this position until the investigator is certain that the outlet hole and the immediate end of the axon is cleared of all visible axoplasm. The outlet hole can be cleared by picking pieces of axoplasm with a fine forceps. As both axoplasm and the internal perfusion medium are distinctly different in their refractive indices from the external medium, one can clearly see the boundary between static axoplasm and sea water, and between the moving perfusion medium and the external sea water. The boundary between sea water and moving perfusion medium forms a clearly visible edge or *schlieren*. This will be used throughout the procedure as an index of perfusion. The chamber has several outlet holes in the external fluid trough. Strong suction assures that flow of perfusion fluid is immediately taken to waste without flowing back into the main portion of the chamber.

At this stage the cannula is withdrawn slowly back into the axon maintaining internal medium flow. If this is done slowly enough the axoplasm adhering to the inner surface of the axon is eroded away. Depending on the speed of this process, the type of perfusion medium, and the presence or absence of proteolytic enzymes, the axon eventually is internally perfused and the end of the cannula reaches a point when the exposed portion of the axial wire is directly opposite the array of external metal electrodes. If proteolytic enzymes, such as pronase, are used, care must be taken to avoid an excess; these enzymes can block sodium inactivation (Rojas and Armstrong, 1971; Rojas and Rudy, 1976). At this point there should be about 0.5 to 1 cm of cannula still inside the inlet end of the axon. This cannula is tied in place by cinching a fine silk thread down around the axon so that the cannula is patent and the perfusion fluid does not flow backwards out the inlet hole. At this stage passing a small but adequate pulse of current through the membrane from the axial wire to the external electrodes can excite the axon to produce a membrane action potential. This action potential can be recorded from the axial wire. An axon with an 100 mv or greater action potential is considered acceptable for experiment.

If the axon passes muster, the internal and external voltage sensing electrodes are both placed in the external solution (usually artificial sea water). Junction potentials are nulled for each electrode with respect to ground and then with respect to each other. The designated internal voltage sensing electrode is then moved so it can be threaded inside the axon through the perfusion outlet hole in the end of the axon. The designated external voltage sensing electrode is then moved so that its tip as close to the axon's surface as possible in the external solution opposite the mid-point of the 1 mm long chamber current measuring electrode. The internal voltage sensing electrode is threaded into the axon through the outlet hole in one end of the axon. This

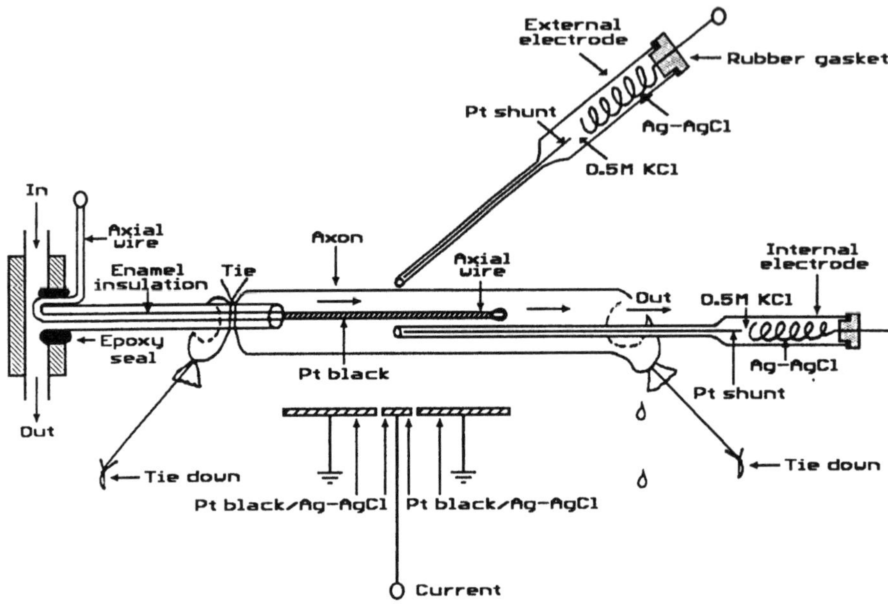

Figure 6. A simplified diagrammatical sketch (not to scale) of a voltage-clamp perfusion technique using the *two-electrode* system. The simplified voltage clamp circuit is not shown, but is given in Figure 5. The perfusion cannula is concentric around the axial wire. The perfusion fluid flows around the axial wire between the cannula wall and the enamel insulation of the wire. The wire is held in place by a single drop of epoxy glue applied to the outside of the cannula where the wire makes a hairpin turn. The internal voltage sensing electrode is inserted into the open cut end. The external electrode contains a floating platinum wire which serves as a shunt, and permits the use of 0.5 M KCl in the electrode. The current collecting electrodes are made of platinum black and silver-silver chloride using the method of Cole and Kishimoto (1962).

electrode is positioned so that it is as physically separated from the axial wire electrode as is possible (Adelman, Palti, and Senft, 1973) consistent with its tip being as close to the interior surface of the axon as is possible. The tip of this electrode is positioned opposite the external voltage sensing electrode's tip (Fig. 6).

The positioning of the two voltage sensing electrodes is important so as to eliminate electrical cross-talk between the axial wire and internal voltage sensing electrode, and to reduce the series resistance to as low a value as possible before electronic compensation. Methods for voltage clamping have been described by Adelman and French (1978), French and Shoukimas (1981), and Fohlmeister *et al.* (1984).

5.2. Forcing functions

The forcing functions that are used to command the feedback circuit controlling the membrane potential can be of almost any form. The forcing functions sequences

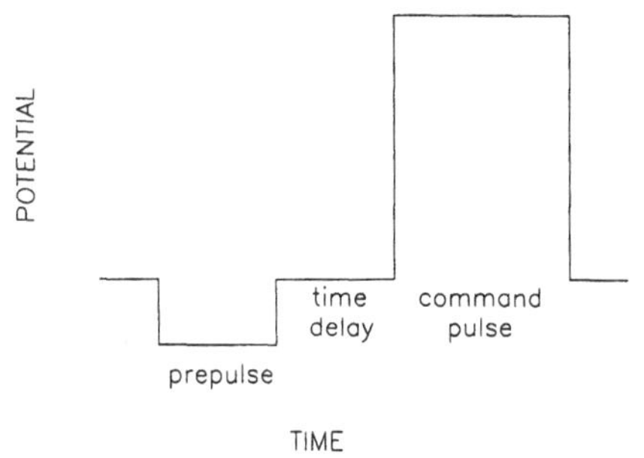

Figure 7. Voltage square pulse protocol. The conditioning first pulse is called the pre-pulse and the second pulse is called the test or command pulse. A time delay separates the two pulses. The first pulse can be modified to be a series of pulses or eliminated. The time delay can also be eliminated.

can be under computer control through a digital to analog (D/A) converter. The main consideration in selecting a particular forcing function for an experiment is determined by the type of information sought and whether the membrane or channel current elicited by the command has some known means for its analysis. Cole (Adelman, 1983) recognized that the machinery for axon membrane conductance gating was voltage dependent and that the way of studying this mechanism was by means of using membrane potential forcing functions.

5.2.1. Square waves

The most commonly used forcing function is the rectangular pulse or square wave (Fig. 7). Rectangular pulses can be applied either singly or in a given sequence arranged to elicit a particular set of data so as to describe a specific channel function. The advantage of using the rectangular pulse depends on the rapidity of its rise to peak and the constancy of its value after the rise or settling time. As the current flowing through the membrane capacity (I_c) will depend on C_m times (dV/dt), whenever dV/dt is zero (V_m is a constant) I_c equals zero. Therefore, once settling occurs, the total membrane current flowing in response to a square pulse in membrane potential will be due solely to resistive membrane elements rather than to capacitative elements.

Pulses are taken from a set value of the steady membrane potential, commonly referred to as the *holding potential*. If the holding potential is set equal to the resting potential of the axon as measured just before clamping, then the current required to hold the membrane at that potential will be zero. This is a convenient index for determining the resting potential of axons that are continuously held in clamp. When the holding current is zero, the holding potential is equal to the resting potential.

Figure 7 shows the general form of the use of square waves as a forcing function. The first pulse is a conditioning pre-pulse which conditions the axon membrane. Following the pre-pulse is a time delay before the command test pulse. For certain protocols, the time delay is zero, constant, or variable. The pre-pulse amplitude can also be zero, constant, or variable. In addition, the duration of the pre-pulse can be constant or variable. The command pulse amplitude can be constant or variable and should be of sufficient duration to obtain the desired currents. Various pulsing protocols are used.

Protocol 1. A current voltage relationship is obtained by maintaining the membrane potential held at a hyperpolarized steady potential, the *holding potential*, of about −120 mv. Then, a train of single command pulses is presented at a pulse rate of less than one per second, with the voltage value of each pulse incrementing by a fixed value (eg. +10 mv) so as to encompass the physiological range. The last pulse is about +220 mv, so that the absolute membrane potential during the command pulse is +100 mv. Many investigators have used a command pulse duration of 50 msec so as to be able to see both the early transient or sodium current and the delayed steady state or potassium current. If only the sodium current is to be studied (by blocking potassium currents with agents such as internal TEA^+ or Cs^+ ions), then a shorter duration pulse may be used. The interval between pulses is several seconds in duration so as to prevent stochastic errors arising from the excitable effects induced by one pulse carrying over to the next pulse. This rest interval is arrived at by experimental trial using some index of response which will reveal the carry over if it exists.

Protocol 2. A variation of protocol 1 is to add a pre-pulse without any time delay. The pre-pulse is given an absolute voltage value of −120 mv with a duration of 50 msec. If the *holding potential* is –60 mv or at a value close to the previously measured resting potential, the holding current is kept low or zero for the majority of the experiment's duration, and resting fast sodium inactivation is overcome by the prepulse hyperpolarization.

Protocol 3. The pre-pulse is of fixed amplitude and of varying duration. Again there is no time delay. The amplitude and duration of the command pulse are constant. Development or recovery of the sodium current from inactivation is measured using this protocol. If positive depolarizing pulses are used, inactivation develops as the pre-pulse duration increases. Similarly, for pre-pulses that are hyperpolarizing, resting sodium channel inactivation is overcome as the duration of the pre-pulse increases.

Protocol 4. The pre-pulse is of varying amplitude and of a long constant duration. There is no time delay. The amplitude of the command pulse is constant and the duration is of short constant duration. If the test pulse by itself is sufficient to elicit a sodium current, then by varying the amplitude of the pre-pulse the influence of membrane potential on inactivation of the sodium channel is measured in the steady state from the amplitude of the sodium current as a function of the conditioning pulse voltage. Steady state in this case depends on the selection of the duration of the pre-pulse. Adelman and Palti (1969a, 1969b) have shown that there are both fast and slow inactivation processes with plateau values in the tens of milliseconds range and in the seconds range, respectively. In some cases, Adelman and Palti (1969b) used pre-pulses that were several minutes in duration in order to overcome all resting sodium channel inactivation.

Protocol 5. The prepulse and command pulse are of constant duration amplitude. The time delay is varied. Frequently, the amplitude of the pulses in the pair is chosen to produce a significant sodium current, and the duration of the pulses in the pair is selected so that it is longer than the time to peak of the sodium current elicited. As the first pulse is used to produce some constant level of conditioning, the second pulse measures the relaxation from that level as a function of time after the conditioning voltage is removed (Hodgkin and Huxley, 1952c).

Protocol 6. Trains of sets of pulses are configured so as to subtract capacity and linear leakage current components. Several different pulse protocols fall in this category. The simplest is a pulse pair of the same duration but of equal but opposite amplitude. If one of these elicits an appropriate response and the other does not, then adding together the currents corresponding to each of these two voltage pulses subtracts from the summed current all linear well-behaved components such as the capacity current and leaves as a remainder the asymmetry current. As one usually holds the membrane at some value that does not inactivate the conductance mechanism, the pulses in the hyperpolarization direction achieve very large values of absolute membrane potential with electric field strengths approaching dielectric breakdown. Bezanilla and Armstrong (1977) and Adrian and Almers (1976) introduced pulse protocols in which for every one depolarizing pulse of amplitude, *P*, there are four identical pulses of amplitude *P/−4*. The *−4* indicates that the four pulses are hyperpolarizing. The *P/4* procedure indicates that the four pulses are depolarizing. In the *P/−4* procedure, the five currents are summed to obtain the asymmetry current. In the *P/4* procedure, the four currents corresponding to the quarter pulses are summed and then subtracted from the current corresponding to the single larger depolarizing pulse. This later protocol only works as a measure of unidirectional asymmetry currents if the four small depolarizing pulses do not induce any significant non-linear currents on their own, i.e., they are out of the voltage range of the asymmetry current of interest. To achieve this one might have to use some larger number of smaller fractional pulses for the summation (eg. *P/−8* or *P/8* procedures).

5.2.2. Sinusoidal forcing functions

While sine waves, and other waveforms having a periodic character are not as popular as square waves or rectangular pulse commands in voltage or patch clamp experimental designs, they can serve very useful roles in obtaining information about certain membrane (Fishman and Cole, 1969; Palti and Adelman, 1969; Poussart *et al.*, 1977; DeFelice *et al.*, 1981; Fishman *et al.*, 1981) or channel (Fohlmeister and Adelman, 1985a) functions. As the membrane or channel response to a sine wave voltage clamp is inherently periodic, the analytical approach most frequently taken has involved frequency domain techniques (Fohlmeister and Adelman, 1985b), such as the fast Fourier transform (FFT). When the forcing functions are non-sinusoidal but none the less are highly periodic (Palti and Adelman, 1969), combinations of analytical methods are used which lean on techniques derived from both rectangular and sinusoidal wave analysis. Many investigators have used mathematical models of membrane channels both to predict the form of current that is elicited by a given voltage waveform (Palti and Adelman, 1969; FitzHugh, 1981; Fohlmeister and Adelman, 1985b; 1986), and to select the best method for analyzing the records (Fohlmeister and Adelman, 1985b).

The membrane potential during a sine wave voltage clamp is described by the following relation:

$$E_M(t) = E_H + E_0[\sin(2\pi ft)] \tag{1}$$

where f is the frequency of the wave
t is the time
$E_M(t)$ is the membrane potential as a function of time
E_H is the holding potential
E_0 is the potential measured from E_H to the waveform peak

The membrane potential during a triangular wave voltage clamp is described by the following relation:

$$E_M(t) = E_H + At \tag{2}$$

where A is a slope coefficient (sign with each half cycle)

Systems such as the squid giant axon membrane can be considered as being composed of linear and non-linear elements. For the most part, the membrane capacity as derived from the dielectric properties of the membrane can be thought of as linear. A linear system responds to a sine wave with a sinusoidal response of the same frequency as the forcing waveform function. However, in most instances, this response is shifted in phase with respect to the forcing waveform, and has an amplitude that changes proportionally in a linear fashion with changes in the amplitude of the forcing function (DeFelice *et al.*, 1981). If the nerve membrane were purely linear, then the application of a sine wave voltage clamp would result in membrane currents that are pure sine waves of the same frequency as the voltage waveform. The amplitude ratio of the current to voltage waves would define the membrane admittance at each forcing frequency. However, most membrane excitable properties are considered to be non-linear.

While impedance measurements are limited by the requirement of linearity (Chandler *et al.*, 1962; FitzHugh, 1981), most voltage clamped sine wave forcing functions usually give rise to distorted membrane currents which contain several higher order harmonics of the forcing function's fundamental frequency. This has meant that the membrane impedance can not be measured directly with any degree of accuracy. In order to examine the non-linear characteristics of the membrane, an approach was taken which examined the effect of sinusoidal forcing functions on membrane or channel model systems (eg. Hodgkin and Huxley, 1952d; Adelman and FitzHugh, 1975) to simulate the non-linear current response (FitzHugh, 1981; Palti and Adelman, 1969; DeFelice *et al.*, 1981; Fohlmeister and Adelman, 1985a, 1985b). By making comparisons between theoretical predictions and experimental results for the same set of forcing functions, qualitative agreement could be obtained between theory and experiment (DeFelice *et al.*, 1981).

In order to determine the contributions of higher order harmonics to the recorded membrane currents in response to a sine wave voltage, a fast Fourier transform has been employed to extract the frequency components from these distorted membrane current waves. For membrane currents in squid axon membranes with conducting ionic channels, small signal amplitudes (ca. 1 to 2 mv peak to peak) were used (DeFelice *et al.*, 1981; Taylor *et al.*, 1981). For asymmetry currents (channel gating currents) in membranes with channel ionic conduction blocked, moderately large sine wave amplitudes (ca. 50 to 70 mv peak to peak) were used (Fohlmeister and Adelman, 1985a). In the earlier work of Palti and Adelman (1969) on membrane capacity and ionic conductances, large amplitude sine waves were employed, but the recorded current waves were analyzed simply by direct comparison with computed current waves obtained upon voltage clamping the models without recourse to the Fourier transform.

5.2.3. Other forms of forcing functions

Other types include the so-called *ric-rac* (named by Richard FitzHugh) pulse sequence (Adelman, 1979). Fohlmeister and Adelman (1982) used this sequence to measure the slope conductance of potassium channels. The fast ramp or saw-tooth clamp has been used for obtaining a rapid description of the current-voltage relationship for the sodium current (Fishman, 1970). Gilbert (1974) analyzed this type of voltage clamp in terms of the *H-H equations.*

Harvey Fishman and his colleagues have pioneered in the use of *noise* and pseudorandom signals as forcing functions in voltage clamp experiments (Fishman,

1975; Fishman, Poussart *et al.*, 1977; Poussart *et al.*, 1977; Fishman, Moore *et al.*, 1977). In order to determine impedance of the membrane it has been a standard practice to limit the amplitude of the random and pseudorandom forcing functions to very low values so as to be in the range of linearity (Taylor *et al.*, 1981).

5.3. Data acquisition

The voltage clamp system in use at present is a low impedance system capable of producing a step change in membrane potential with a rise time (10 – 90 per cent) of about 1 μsec and a settling time of 5 – 10 μsec. The device has means for setting the steady membrane (or holding) potential, setting the value in ohm cm^2 of the compensation for the membrane series resistance, and normalizing the measurement of membrane current to a milliamp/cm^2 scale for a range of axon diameters from 300 to 800 μm. The membrane current output is filtered with an 8 pole Bessel low-pass filter in order to avoid *aliasing*, a phenomenon of a high frequency input masquerading as a low frequency phenomenon when the current is digitized. *Aliasing* (Robinson and Silvia, 1978; Foster, 1981) occurs when the current waveform is undersampled. The digitizing rate must be at least twice the highest frequency (*Nyquist frequency* or *folding frequency*) present in the current sample. However, it is recommended that the digitizing rate be at least ten times the *Nyquist frequency*. Note that the *Nyquist frequency* refers to the highest frequency *present* in the input signal and not necessarily to the highest frequency of interest in the input signal. Incorporated in the device is a means for delivering controlled currents of varying amplitudes and durations to the axon in open loop mode or in current clamp mode. These stimulus currents are used primarily for testing the viability of the axon, but they can also be used as commands for classical current clamp experiments (Marmont, 1949).

The input of the current to voltage transducer is maintained at virtual ground. Usually the low pass anti-aliasing filter is set for a cut-off at 100 kHz when digitizing at 200 kHz, which also reduces high frequency noise. Before digitizing, the filtered output is amplified so as to present to the analog to digital (A/D) converter a signal of sufficient amplitude that conversion is optimized without saturating the converter. A fixed gain amplifier coupled with an variable attenuator is used to select the correct overall gain.

Numbers in digital instruments, such as A/D converters, are stored as digits of 0 or 1, known as binary numbers. Each binary digit is known as a *bit*. If a binary number is composed of *n bits*, then the binary number can represent 2 to the power of *n* different values. Thus, a 2 *bit* binary number can have 4 different values, i.e., 0, 1, 2, or 3. Similarly, a 10 *bit* binary number can have 1024 different values and a 12 *bit* binary number can have 4096 different values. The resolution of a binary number is represented in *bits*. Thus, the digitizing rate times the binary resolution equals the rate of receiving information. Generally, the fastest converters have the lowest resolution. However, it is possible to achieve digitization at 1 MHz, i.e., 1 μsec intervals, at 10 *bit* resolution. Digital oscilloscopes commonly have a resolution of 12 *bits* and a digitizing rate of 1 MHz. Depending upon the type of experiment, a digitizing rate of 200 kHz, i.e., 5 μsec intervals, at a resolution of 10 bits can be used.

All of our voltage clamp systems have been under computer control, both as to the generation of sequences of programmed command pulses and the synchronized collection of digitized data. Thus, the generation of voltage clamp forcing functions, synchronization of all operations, the setting of digitizing rates, and the archiving and analysis of data are accomplished through on-line laboratory computers. It is crucial to trigger all operations from the same clock as any disparity between input and output is reflected in the digitized records as a phase shift relative to the input.

Mostly the processors making up the core of these devices have been PDP-11 type mini-computers. Some investigators have used micro-computers, mostly of the IBM® AT® type using the Intel® 80286 chip with 80287 math co-processor. At present there is an effort to make use of Motorola 68020 chip based computers such as the Macintosh II and Intel® 80386 based computers such as the Compaq® and IBM® PS/2 for data acquisition and to perform on-line analysis. A complete system for data acquisition has been described (Waltz, Tyndale, and Adelman, 1983). The described system utilized an earlier type processor whose functions were similar to those of more modern processors; therefore the description of this design can be used to understand the general logic that was considered.

5.3.1. Storage of voltage clamp data

The data acquired in a conventional voltage clamp experiment is dependent on the form and timing of a given pulse sequence. This sequence results in an electrophysiological event which consists of a voltage-dependent time-variant current. If this current is precisely monitored through a current to voltage transducer and converted to digital form by an analog to digital converter, the results can be transmitted to a host computer for processing, archiving and analysis. For one experimental run, as many as 60 pulse sequences, consisting of from 2048 to 4096 10 *bit* microprocessor numbers of input per sequence, can be recorded. Between any two pulse sequences there is a rest period of several seconds (typically 5 seconds, but often longer) to allow the axon to recover. An experiment usually contains 15 runs. Between runs there are periods of time to allow for changes in solutions, alterations in protocol, latency of treatment effect, and recovery from test effects. These times and the actual times required for the individual runs mean that a typical experiment takes several hours to perform. It became apparent that such a mass of data would require special computer techniques for analysis.

Most of the analytical work is performed on data that exists on some archival medium such as magnetic tape, floppy disk, hard disk or optical disk such as a WORM (write once read many) or the recently introduced erasable laser disk or REO (recordable erasable optical).

6. Conclusion

As has been described earlier in this chapter, sodium channels in the membrane of the squid giant axon have been studied extensively for over half a century, and a larger body of knowledge exists as to the macroscopic behavior of these channels, in ensemble, than exists for any other preparation. The squid giant axon preparation is particularly useful in measuring gating currents, and seems to continue to have utility in the area of channel ensemble kinetics. A recent development has also made it possible to patch clamp this preparation using the cut-open axon technique. Thus, it is possible to obtain macroscopic as well as microscopic currents on the same preparation.

References

Adelman, W. J., Jr., ed., 1971a, *Biophysics and Physiology of Excitable Membranes*, Van Nostrand Reinhold Co., New York, 527 pp.

Adelman, W. J., Jr., 1971b, Electrical studies of internally perfused axons, in: *Biophysics and Physiology of Excitable Membranes*, (W. J. Adelman, Jr., ed.), pp. 274–319, Van Nostrand and Reinhold, New York.

Adelman, W. J. Jr., 1979, Tests of a simple method for continuously monitoring membrane slope conductance during a voltage clamp, *Biophys. J.* **25**:14a.

Adelman, W. J., Jr., 1983, An introduction to membrane conductances, in: *Structure and Function in Excitable Cells* (D. C. Chang, I. Tasaki, W. J. Adelman, Jr., and H. R. Leuchtag, eds.), pp. 145–158, Plenum Press, New York.

Adelman, W. J., Jr. and FitzHugh, R., 1975, Solutions of the Hodgkin-Huxley equations modified for potassium accumulation in a periaxonal space, *Fed. Proc.* **34**:1322–1329.

Adelman, W. J., Jr. and Fok, Y. B., 1964, Internally perfused squid axons studied under voltage clamp conditions, II. Results. The effects of internal potassium and sodium on membrane electrical characteristics, *J. Cell. Comp. Physiol.* **64**:429–444.

Adelman, W. J., Jr. and French, R. J., 1976, The squid giant axon, *Oceanus* **19, No.2**:6–16.

Adelman, W. J., Jr. and French, R. J., 1978, Blocking of the squid axon potassium channel by external caesium ions, *J. Physiol. (Lond.)* **276**:13–25.

Adelman, W. J., Jr. and Gilbert, D. L., 1963, Voltage clamp studies on internally perfused squid axons. *Biophys. Soc. Abst.*, Seventh annual meeting, WC 4.

Adelman, W. J., Jr. and Gilbert, D. L., 1964, Internally perfused squid axons studied under voltage clamp conditions. I. Method, *J. Cell. Comp. Physiol.* **64**:423–428.

Adelman, W. J., Jr. and Gilbert, D. L., 1989, *unpublished data.*

Adelman, W. J., Jr., Kishimoto, U., and Dalton, J., 1960–1961, *unpublished data.*

Adelman, W. J., Moses, J., and Rice, R. V., 1977, An anatomical basis for the resistance and capacitance in series with the excitable membrane of the squid giant axon, *J. Neurocytol.* **6**:621–646.

Adelman, W. J., Jr. and Palti, Y., 1969a, The effects of external potassium on the inactivation of sodium currents in the giant axon of the squid, *Loligo pealei*, *J. Gen. Physiol.* **53**:685–703.

Adelman, W. J., Jr. and Palti, Y., 1969b, The effects of external potassium and long duration voltage conditioning on the amplitude of sodium currents in the giant axon of the squid, *Loligo pealei*, *J. Gen. Physiol.* **54**:589–606.

Adelman, W. J., Jr., Palti, Y., and Senft, J. P., 1973, Potassium ion accumulation in a periaxonal space and its effect on the measurement of a membrane potassium ion conductance, *J. Memb. Biol.* **13**:387–410.

Adrian, R. H. and Almers, W., 1976, Charge movement in the membrane of striated muscle, *J. Physiol. (Lond.)* **254**:339–360.

Aldrich, F. A., 1990, Lol-i-go and far away: A consideration of the species designation *Loligo pealei*, *this volume.*

Almers, W., 1978, Gating currents and charge movements in excitable membranes, *Rev. Physiol. Biochem. Pharmacol.* **82**:96–190.

Armstrong, C. M., 1969, Inactivation of the potassium conductance and related phenomena caused by quaternary ammonium ion injection in squid axons, *J. Gen. Physiol.* **54**:553–575.

Armstrong, C. M. and Bezanilla, F., 1973, Currents related to movement of the gating particles of the sodium channels, *Nature (Lond.)* **242**:459–461.

Armstrong, C. M. and Bezanilla, F., 1974, Charge movement associated with the opening and closing of the activation gates of the Na channels, *J. Gen. Physiol.* **63**:533–522.

Armstrong, C. M. and Bezanilla, F., 1977, Inactivation of the sodium channel. II. Gating current experiments, *J. Gen. Physiol.* **70**:567–590.
Armstrong, C. M. and Binstock, L., 1965, Anomalous rectification in the squid axon injected with tetraethylammonium chloride, *J. Gen. Physiol.* **48**:859–872.
Armstrong, C. M. and Gilly, W. F., 1979, Fast and slow steps in the activation of sodium channels, *J. Gen. Physiol.* **74**:691–711.
Arnold, J. M., 1990a, Evolution and intelligence of the cephalopods: The alternative form of intelligent life on the planet earth, *this volume.*
Arnold, J. M., 1990b, Squid mating behavior, *this volume.*
Arnold, J. M., 1990c, Embryonic development of the squid, *this volume.*
Baker, P. F., 1984a, Internal perfusion and dialysis: Application to large nerve and muscle cells, in: *Internal perfusion of excitable cells,* (P. G. Kostyuk and O. A. Krishtal, eds.), pp. 1–17, John Wiley, New York.
Baker, P. F., ed., 1984b, *The Squid Axon. Current Topics in Membranes and Transport,* **Volume 22**, Academic Press, New York.
Baker, P. F., Gilpin-Brown, J. B., and Meves, H., 1976, The squid and its giant nerve fibre, *J. Physiol.* **263**:99P.
Baker, P. F., Hodgkin, A. L., and Shaw, T. I., 1961, Replacement of the protoplasm of a giant nerve fibre with artificial solutions, *Nature* **190**:885–887.
Baker, P. F., Hodgkin, A. L., and Shaw, T. I., 1962, Replacement of the axoplasm of giant nerve fibres with artificial solutions, *J. Physiol. (Lond.)* **164**:330–354.
Barnes, R. D., 1968, *Invertebrate Zoology.* Second Edition, W. B. Saunders Co., Philadelphia, 743 pp.
Bean, R. C., Shepherd, W. C., Chan, H., and Eichner, J. T., 1969, Discrete conductance fluctuations in lipid bilayer protein membranes, *J. Gen. Physiol.* **53**:741–757.
Berman, W., 1961, *How to Dissect. Exploring With Probe and Scalpel,* Sentinerl Book Pub., Inc., New York, 127 pp.
Bezanilla, F., 1987, Single sodium channels from the squid giant axon, *Biophys. J.* **52**:1087–1090.
Bezanilla, F. and Armstrong, C. M., 1977, Inactivation of the sodium channel. I. Sodium current experiments, *J. Gen. Physiol.* **70**:549–566.
Bezanilla, F. and Vandenberg, C., 1990, The cut-open axon technique, *this volume.*
Bezanilla, F., Vergara, J., and Taylor, R. E, 1982, Voltage clamping of excitable membranes, in: *Methods of Experimental Physics:.* **Volume 20. Biophysics**, (G. Ehrenstein and H. Lecar, eds.), pp. 445–511, Academic Press, New York.
Binstock, L., Adelman, W. J., Jr., Senft, J., and Lecar, H., 1975, Determination of the resistance in series with the membranes of giant axons, *J. Memb. Biol.* **21**:25–47.
Brinley, F. J. and Mullins, L. J., 1967, Sodium extrusion by internally dialyzed squid axons, *J. Gen. Physiol.* **50**:2303–2331.
Brown, A. and Lasek, R. J., 1990, The cytoskeleton of the squid giant axon, *this volume.*
Bullock, T. H. and Horridge, G. A., 1965, *Structure and Function in the Nervous System of Invertebrates.* Vols. I and II. W. H. Freeman & Co., San Francisco, 1719 pp.
Caldwell, P. C., Hodgkin, A. L., Keynes, R. D., and Shaw, T. I., 1960, The effects of injecting "energy-rich" phosphate compounds on the active transport of ions in the giant axons of *Loligo, J. Physiol. (Lond.)* **152**:561–590.
Chailaklyan, L. M., 1961, Measurement of the resting and action potentials of the giant fibre of the squid in various conditions of recording, *Biophysics* **6**:344–353.
Chandler, W. K., FitzHugh, R., and Cole, K. S., 1962, Theoretical stability properties of a space clamped axon, *Biophys. J.* **2**:105–127.
Chandler, W. K. and Meves, H., 1965, Voltage clamp experiments on internally perfused giant axons, *J. Physiol. (Lond.)* **180**:788–820.
Chandler, W. K., Rakowski, R. F., and Schneider, M. F., 1976, A nonlinear voltage dependent charge movement in frog skeletal muscle, *J. Physiol. (Lond.)* **254**:245–283.
Clay, J. R. and Shlesinger, M. F., 1983, Effects of external cesium and rubidium on outward potassium currents in squid axons, *Biophys. J.* **42**:43–53.

Cohen, L. B., Hille, B., and Keynes, R. D., 1970, Changes in axon birefringence during the action potential, *J. Physiol. (Lond.)* **211**:495–515.
Cohen, L. B., Landowne, D., and Salzberg, B. M., 1990, Optical measurements on squid axons, *this volume.*
Cole, K. S., 1928, Electrical impedance of suspensions of Arbacia eggs, *J. Gen. Physiol.* **12**:37–54.
Cole, K. S., 1937, Electrical impedance of marine egg membranes, *Trans. Faraday Soc.* **33**:966–972.
Cole, K. S., 1949, Dynamic electrical characteristics of the squid axon membrane, *Arch. Sci. Physiol.* **3**:253–258.
Cole, K. S., 1968, *Membranes, Ions and Impulses*, 569 pp., U. Calif. Press, Berkeley.
Cole, K. S., 1971, Some aspects of electrical studies of the squid giant axon membrane, in: *Biophysics and Physiology of Excitable Membranes* (W. J. Adelman, Jr., ed.), Van Nostrand Reinhold Co., New York, pp. 125–142.
Cole, K. S., 1979, Mostly membranes, *Ann. Rev. Physiol.* **41**:1-24.
Cole, K. S. and Curtis, H. J., 1939, Electric impedance of the squid giant axon during activity, *J. Gen. Physiol.* **22**:649–670.
Cole, K. S. and Kishimoto, U., 1962, Platinized silver chloride electrode, *Science* **136**:381–382.
Cole, K. S. and Moore, J. W., 1960, Ionic current measurements in the squid giant axon membrane, *J. Gen. Physiol.* **44**:123–167.
Cole, K. S., 1979, Mostly membranes, *Ann. Rev. Physiol.* 41:1–24.
Conti, F., DeFelice, L. J., and Wanke, E., 1975, Potassium and sodium ion current noise in the membrane of the squid giant axon, *J. Physiol. (Lond.)* **248**:45–82.
Conti, F. and Neher, E., 1980, Single channel recordings of K^+ currents in squid axons, *Nature* **285**:140–143.
Curtis, H. J. and Cole, K. S., 1937, Transverse electric impedance of Nitella, *J. Gen. Physiol.* **21**:189–201.
Curtis, H. J. and Cole, K. S., 1938, Transverse electric impedance of the squid giant axon, *J. Gen. Physiol.* **21**:757–765.
Curtis, H. J. and Cole, K. S., 1940, Membrane action potentials from the squid giant axon, *J. Cell. Comp. Physiol.* **15**:147–157.
Curtis, H. J. and Cole, K. S., 1942, Membrane resting and action potentials from the squid giant axon, *J. Cell. Comp. Physiol.* **19**:135–144.
Dean, R. B., 1941, Theories of electrolyte equilibrium in muscle, *Biol. Symp.* **3**:331–348.
Dean, R. B., 1987, Reminiscences on the sodium pump, *TINS* **10**:451–454.
DeFelice, L. J., 1981, *Introduction to Membrane Noise*, Plenum Press, New York, 500 pp.
DeFelice, L. J., Adelman, W. J. Jr., Clapham, D. E., and Mauro, A., 1981, Second-order admittance in squid axon, in: *The Biophysical Approach to Excitable Systems* (W. J. Adelman, Jr. and D. E. Goldman, eds.), Plenum Press, New York, 258 pp.
DeFelice, L. J. and Kell, M. J., 1983, Noise, impedance and single-channel currents, in: *Structure and Function in Excitable Cells* (D. Chang, I. Tasaki, W. J. Adelman, Jr., and H. R. Leuchtag, eds.), Plenum Press, New York, pp. 197–209.
DeFelice, L. J., Wanke, E., and Conti, F., 1975, Potassium and sodium current noise from squid axon membranes, *Fed. Proc.* **34**:1338–1342.
DiPolo, R., 1989, *personal commun.*
DiPolo, R., Bezanilla, F., Caputo, C., and Rojas, H., 1985, Voltage dependence of the Na/Ca exchange in voltage-clamped, dialyzed squid axons, *J. Gen. Physiol.* **86**:457–478.
Ehrenstein, G., Blumenthal, R., Latorre, R., and Lecar, H., 1974, Kinetics of the opening and closing of individual EIM channels in a lipid bilayer, *J. Gen. Physiol.* **63**:707–721.
Ehrenstein, G., Lecar, H., and Nossal, R., 1970, The nature of the negative resistance in bimolecular lipid membranes containing excitability inducing material, *J. Gen. Physiol.* **55**:119–133.
Fatt, P. and Katz, B., 1950, Some observations on biological noise, *Nature* **166**:597–598.
Fatt, P. and Katz, B., 1952, Spontaneous subthreshold activity at motor nerve endings, *J. Physiol.* **117**:109–128.

Fishman, H. M., 1970, Direct and rapid description of the individual ionic currents of squid axon membrane by ramp potential control, *Biophys. J.* **10**:799–817.
Fishman, H. M., 1973a, Relaxation spectra of potassium channel noise from squid axon membranes, *Proc. Nat. Acad. Sci. USA* **70**:876–879.
Fishman, H. M., 1973b, Low impedance capillary electrode for wide-band recording of membrane potential in large axons, *IEEE Trans. Bio-Med. Electron.* **20**:380–382.
Fishman, H. M., 1975, Axon membrane noise measurements, *Fed. Proc.* **34**:1330–1337.
Fishman, H. M., 1982, Current and voltage clamp techniques. P119, *Techniques in Cellular Physiology. Part 2*, pp. 1–42.
Fishman, H. M. and Cole, K. S., 1969, On line measurement of squid axon current potential characteristics, *Fed. Proc.* **28**:333.
Fishman, H. M., Moore, L. E., and Poussart, D. J. M., 1977, Asymmetry currents and admittance in squid axons, *Biophys. J.* **19**:177–183.
Fishman, H. M., Moore, L. E., and Poussart, D., 1981, Squid axon K conduction: Admittance and noise during short- versus long-duration step changes, in: *The Biophysical Approach to Excitable Systems* (W. J. Adelman, Jr. and D. E. Goldman, eds.), Plenum Press, New York, pp. 65–95.
Fishman, H. M., Poussart, D. J. M., Moore, L. E., and Siebenga, E., 1977, K^+ conduction description from the low frequency impedance and admittance of squid axon, *J. Memb. Biol.* **32**:255–290.
FitzHugh, R., 1981, Nonlinear sinusoidal currents in the Hodgkin-Huxley model, in: *The Biophysical Approach to Excitable Systems* (W. J. Adelman, Jr. and D. E. Goldman, eds.), Plenum Press, New York, pp. 25–36.
Fohlmeister, J. F. and Adelman, W. J., Jr., 1982, Anomalous potassium channel-gating rates as functions of calcium and potassium ion concentrations, *Biophys. J.* **37**:427–431.
Fohlmeister, J. F. and Adelman, W. J., Jr., 1984, Rapid sodium channel conductance changes during voltage steps in squid giant axons, *Biophys. J.* **45**:513–521.
Fohlmeister, J. F. and Adelman, W. J., Jr., 1985a, Gating current harmonics I: sodium channel activation gating in dynamic steady states, *Biophys. J.* **48**:375–390.
Fohlmeister, J. F. and Adelman, W. J., Jr., 1985b, Gating current harmonics II: model simulations of axonal gating currents, *Biophys. J.* **48**:391–400.
Fohlmeister, J. F. and Adelman, W. J., Jr., 1986, Gating current harmonics III: dynamic transients and steady states with intact sodium inactivation gating, *Biophys. J.* **50**:489–502.
Fohlmeister, J. F. and Adelman, W. J., Jr., 1987, Gating current harmonics IV: dynamic properties of secondary activation kinetics in sodium channel gating, *Biophys. J.* **51**:335–338.
Fohlmeister, J. F., Adelman, W. J., Jr., and Brennan, J. J., 1984, Excitable channel currents and gating times in the presence of anticonvulsants ethosuximide and valproate, *J. Pharm. Exp. Therap.* **230**:75–81.
Foster, C. C., 1981, *Real Time Programming. Neglected Topics*, Addison-Wesley Pub., Reading, Massachusetts.
French, R. J. and Shoukimas, J. J., 1981, Blockage of squid axon potassium conductance by internal tetra-n-alkylammonium ions of various sizes, *Biophys. J.* **34**:271–291.
Fricke, H., 1923, The electrical capacity of cell suspensions, *Physic. Rev.* **21**:708–709.
Fricke, H., 1933, The electrical impedance of suspensions of biological cells, *Cold Spring Harbor Symp. Quant. Biol.* **1**:117–124.
Gilbert, D. L., 1971, Internal perfusion of squid axons: technical considerations, in: *Biophysics and Physiology of Excitable Membranes* (W. J. Adelman, Jr., ed.), Van Nostrand Reinhold Co., New York, pp. 264–273.
Gilbert, D. L., 1974, Potential and time constant effects in the fast ramp voltage clamp, *Math. Biosciences* **20**:67–74.
Good, N. E., Winget, G. D., Winter, W., Connolly, T. N., Izawa, S., and Singh, R. M. M., 1966, Hydrogen ion buffers for biological research, *Biochemistry* **5**:467–477.
Gould, R. M. and Alberghina, M., 1990, Lipid metabolism in the squid nervous system, *this volume.*

Hanlon, R. T., 1990, Maintenance, rearing, and culture of teuthoid and sepioid squids, *this volume.*
Hodgkin, A. L., 1951, The ionic basis of electrical activity in nerve and muscle, *Biol. Rev.* **26**:339–409.
Hodgkin, A. L., 1964, *The Conduction of the Nervous Impulse*, Charles C Thomas, Springfield, Ill., 108 pp.
Hodgkin, A. L. and Huxley, A. F., 1939, Action potentials recorded from inside a nerve fibre, *Nature (Lond.)* **144**:710–711.
Hodgkin, A. L. and Huxley, A. F., 1945, Resting and action potentials in single nerve fibres, *J. Physiol. (Lond.)* **104**:176–195.
Hodgkin, A. L. and Huxley, A. F., 1952a, Currents carried by sodium and potassium ions through the membrane of the giant axon of *Loligo, J. Physiol. (Lond.)* **116**:449–472.
Hodgkin, A. L. and Huxley, A. F., 1952b, The components of membrane conductance in the giant axon of Loligo, *J. Physiol. (Lond.)* **116**:473–496.
Hodgkin, A. L. and Huxley, A. F., 1952c, The dual effect of membrane potential on sodium conductance in the giant axon of Loligo, *J. Physiol. (Lond.)* **116**:497–506.
Hodgkin, A. L. and Huxley, A. F., 1952d, A quantitative description of membrane current and its application to conduction and excitation in nerve, *J. Physiol. (Lond.)* **117**:500–544.
Hodgkin, A. L., Huxley A. F., and Katz, B., 1949, Ionic currents underlying activity in the giant axon of the squid, *Arch. Sci. Physiol.* **3**:129–150.
Hodgkin, A. L., Huxley, A. F., and Katz, B., 1952, Measurement of current-voltage relations in the membrane of the giant axon of *Loligo, J. Physiol. (Lond.)* **116**:424–448.
Hodgkin, A. L. and Katz, B., 1949, The effect of calcium on the axoplasm of giant nerve fibres, *J. Exp. Biol.* **26**:292–294, Plate 8.
Hodgkin, A. L. and Keynes, R. D., 1956, Experiments on the injection of substances into squid giant axons by means of a microsyringe, *J. Physiol. (Lond.)* **131**:592–616.
Howell, B. J. and Gilbert, D. L., 1976, pH-Temperature dependence of the hemolymph of the squid, *Loligo pealei, Comp. Biochem. Physiol.* **55A**:287–289.
Joyner, R. W. and DeGroof, R. G., 1974, *personal commun.*
Kao, C. Y., 1981, Tetrodotoxin, saxitoxin, chiriquitoxin: new perspectives on ionic channels, *Fed. Proc.* **40**:30–35.
Katz, B., 1966, *Nerve, Muscle and Synapse*, McGraw Hill Book Co., New York, 193 pp.
Katz, B. and Miledi, R., 1970, Membrane noise produced by acetylcholine, *Nature* **226**:962–963.
Katz, B. and Miledi, R., 1971, Further observations on acetylcholine noise, *Nature New Biol.* **232**:124–126.
Katz, B. and Miledi, R., 1972, The statistical nature of the acetylcholine potential and its molecular components, *J. Physiol. (Lond.)* **224**:665–699.
Keynes, R. D. and Rojas, E., 1974, Kinetics and steady state properties of the charged system controlling sodium conductance in the squid giant axon, *J. Physiol. (Lond.)* **239**:393–434.
Keynes, R. D. and Rojas, E., 1976, The temporal and steady state relationships between activation of the sodium conductance and movement of the gating particles in the squid giant axon, *J. Physiol. (Lond.)* **255**:157–189.
Krogh, A., 1946, The active and passive exchanges of inorganic ions through the surfaces of living cells and through living membranes generally. Croonian Lecture, *Proc. Roy. Soc. B* **133**:140–200.
Llano, I., Webb, C. K., and Bezanilla, F., 1988, Potassium conductance of the squid giant axon. Single channel studies, *J. Gen. Physiol.* **92**:179–196.
Llinas, R., 1974, *personal commun.*
Marmont, G., 1949, Studies on the axon membrane, *J. Cell. Comp. Physiol.* **34**:351–382.
Martin, R., 1969, The structural organization of the intracerebral giant fiber system of cephalopods. The chiasma of the first order giant axons, *Z. Zellforsch.* **97**:50–68.
Matsumoto, G., 1989, *personal commun.*
Mauro, A. F., Conti, F., Dodge, F., and Schor, R., 1970, Subthreshold behavior and phenomenological impedance of the squid giant axon, *J. Gen. Physiol.* **55**:497–523.

Meech, R. W., 1986, Membranes, gates, and channels, in: *The Mollusca. Vol. 9. Neurobiology and Behavior, Part 2*, (A. O. D. Willows, ed.), pp. 189–277, Academic Press, New York.
Meves, H., 1974, The effect of holding potential on the asymmetry currents in squid giant axons, *J. Physiol. (Lond.)* **243**:847–867.
Moore, J. W. and Cole, K. S., 1963, Voltage clamp techniques, in: *Physical Techniques in Biological Techniques* (W. L. Nastuk, ed.), Vol. 6, Academic Press, New York, pp. 263–321.
Mueller, P. and Rudin, D. O., 1967, Action potential phenomena in experimental bimolecular lipid membranes, *Nature* **213**:603–604.
Mueller, P., Rudin, D. O., Tien, H. T., and Westcott, 1962, Reconstitution of all membrane structure *in vitro* and its transformation into an excitable system, *Nature* **194**:979–981.
Mullins, L. J. and Brinley, F. J., 1969, Potassium fluxes in dialyzed squid axons, *J. Gen. Physiol.* **53**:704–740.
Mullins, L. J. and Brinley, F. J., Jr., 1990, Internal dialysis in the squid giant axon, *this volume*.
Narahashi, T., 1963, Dependence of resting and action potentials on internal potassium in perfused squid giant axons, *J. Physiol. (Lond.)* **169**:91–115.
Narahashi, T., 1984, Pharmacology of nerve membrane sodium channels, *Curr. Topics Membs. Trans.* **22**:483–516.
Narahashi, T., Deguchi, T., Urakawa, N., and Ohkubo, Y., 1960, Stabilization and rectification of muscle fiber membrane by tetrodotoxin, *Am. J. Physiol.* **198**:934–938.
Narahashi, T., Moore, J. W., and Scott, W.R., 1964, Tetrodotoxin blockage of sodium conductance increase in lobster giant axons, *J. Gen. Physiol.* **47**:965–974.
Neher, E. and Sakmann, B., 1976, Single-channel currents recorded from membrane of denervated frog muscle fibers, *Nature* **260**:799–802.
O'Dor, R., 1989, *personal commun.*
Oikawa, T., Spyropoulis, C. S., Tasaki, I., and Teorell, T., 1961, Methods for perfusing the giant axon of *Loligo pealei*, *Acta Physiol. Scand.* **52**:195–196.
Palti, Y. and Adelman, W. J., Jr., 1969, Measurement of axonal membrane conductances and capacity by means of a varying potential control voltage clamp, *J. Memb. Biol.* **1**:431–458.
Pierce, M. E., 1950, *Loligo pealeii*, in: *Selected Invertebrate Types*, (F. A. Brown, Jr., ed.), pp. 347–357, John Wiley and Sons, Inc., New York.
Potts, W. T. W. and Todd, M., 1965, Kidney function in the octopus, *Comp. Biochem. Physiol.* **16**:479–489.
Poussart, D., Moore, L.E., and Fishman, H. M., 1977, Ion movements and kinetics in squid axon. I. Complex admittance, *Ann. N. Y. Acad. Sci.* **303**:355–379.
Rayner, M. D. and Starkus, J. G., 1989, Steady-state distribution of gating charge in crayfish giant axons, *Biophys. J.* **55**:1–19.
Robertson, J. D., 1965, Studies on the chemical composition of muscle tissue. III. The mantle muscle of cephalopod molluscs, *J. Exp. Biol.* **42**:153–175.
Robertson, J. D. and Webb, D. A., 1939, The micro estimate of sodium, potassium, calcium, magnesium, chloride and sulphate in seawater and body fluids of marine animals, *J. Exp. Biol.* **16**:155–177.
Robinson, E. A. and Silvia, M. T., 1978, *Digital Signal Processing and Time Series Analysis, Pilot Edition*, Holden-Day, Inc., San Francisco.
Rojas, E. and Armstrong, C. M., 1971, Sodium conductance activation without inactivation in pronase-treated axons, *Nature (New Biology)* **229**:177–178.
Rojas, E. and Rudy, B., 1976, Destruction of the sodium conductance inactivation by a specific protease in perfused nerve fibres from *Loligo*, *J. Physiol. (Lond.)* **262**:501–531.
Rosenberg, P., 1973, The giant axon of the squid: A useful preparation for neurochemical and pharmacological studies, in: *Methods of Neurochemistry*, **Volume 4**, (R. Fried, ed.), Marcel Dekker, Inc., New York, pp. 97–160.
Rosenberg, P., 1981, The squid giant axon: Methods and applications, in: *Methods in Neurobiology. Volume 1*, (R. Lahue, ed.), pp. 1–133, Plenum Press, New York.

Schmitt, O. H., 1955, Dynamic negative admittance components in statically stable membranes, in: *Electrochemistry in Biology and Medicine* (T. Shedlovsky, ed.), John Wiley and Sons, New York, pp. 91–120.

Schneider, M. F. and Chandler, W. K., 1973, Voltage-dependent charge movement in skeletal muscle: a possible step in excitation-contraction coupling, *Nature (Lond.)* **242**:244–246.

Sherman, I. W. and Sherman, V. G., 1970, *The Invertebrates: Function and Form. A Laboratory Guide*, Macmillan, Toronto, 304 pp.

Shoukimas, J., Adelman, W. J., Jr., and Sege, V., 1977, Cation concentrations in the hemolymph of *Loligo pealei*, *Biophys. J.* **18**:231–234.

Sigworth, F. J. and Neher, I., 1980, Single Na^+ channel currents observed in cultured rat muscle cells, *Nature* **287**:447–449.

Spyropoulis, C. S., 1965, The role of temperature, potassium and divalent ions in the current-voltage characteristics of nerve membranes, *J. Gen. Physiol.* **48**, No. 5, Part 2:49–53.

Stanley, E., 1990, The preparation of the squid giant synapse for electrophysiological investigation, *this volume.*

Steele, R. E. and Gilbert, D. L., 1980, Calcium and potassium activities in the hemolymph of the squid, *Loligo pealei*, *Biol. Bull.* **159**:492.

Steele, V. J. and Aldrich, F. A., 1990, The giant axon of the squid *Architeuthis dux*, in: *Monograph of the Genus Architeuthidae, in press.*

Summers, W. C., 1967, Winter distribution of *Loligo pealei* determined by exploratory trawling, *Biol. Bull.* **133**:489.

Summers, W. C., 1968, The growth and size distribution of current year class *Loligo pealei*, *Biol. Bull.* **135**:356–377.

Summers, W. C., 1969, Winter population of *Loligo pealei* in the mid-Atlantic bight, *Biol. Bull.* **137**:202–216.

Summers, W. C., 1990, Natural history of the squid, *this volume.*

Tahara, Y., 1910, Ueber das tetrodongift, *Biochem. Z.* **30**:255–275.

Tasaki, I., 1968, *Nerve Excitation. A Macromolecular Approach*, Charles C. Thomas, Springfield, Ill., 201 pp.

Tasaki, I. and Hagiwara, S., 1957, Demonstration of two stable potential states in the squid giant axon under tetraethylammonium chloride, *J. Gen. Physiol.* **40**:859–885.

Tasaki, I. and Luxoro, M., 1964, Intracellular perfusion of Chilean giant squid axons, *Science* **345**:1313–1315.

Tasaki, I., Watanabe, A., and Takenaka, T., 1962, Resting and action potential of intracellularly perfused squid giant axon, *Proc. Nat. Acad. Sci. (U. S. A.)* **48**:1177–1184.

Taylor, R. E, Fernandez, J. M., and Bezanilla, F., 1981, Squid axon membrane low-frequency dielectric properties, in: *The Biophysical Approach to Excitable Systems* (W. J. Adelman, Jr. and D. E. Goldman, eds.), Plenum Press, New York, pp. 97–106.

Taylor, R. E, Moore, J. W., and Cole, K. S., 1960, Analysis of certain errors in squid axon voltage clamp measurements, *Biophys. J.* **1**:161–182.

Villegas, G. M. and Villegas, R., 1960, The ultrastructure of the giant nerve fiber of the squid: Axon-Schwann cell relationship, *J. Ultrastruct. Res.* **3**:362–373.

Villegas, G. M. and Villegas, R., 1984, Squid axon ultrastructure, in: *Current Topics in Membranes and Transport* **22**:3–37.

Villegas, R., Villegas, L., Gimenez, M., and Villegas, G. M., 1963, Schwann cell and axon electrical potential differences. Squid nerve structure and excitable membrane location, *J. Gen. Physiol.* **46**:1047–1067.

Voss, G. L., 1963, Cephalopods of the Philippine Islands, *U. S. Nat. Mus. Bull.* **234**:1–180.

Waltz, R., Tyndale, C., and Adelman, W. J., Jr., 1983, A solution to data acquisition in a multi-user environment, *Proc. DECUS*, **May**:129–140

Wanke, E., DeFelice, L. J., and Conti, F., 1974, Voltage noise, current noise and impedance in space clamped squid giant axon, *Pflügers Arch.* **347**:63–74.

Weast, R. C., Astle, M. J., and Beyer, W. H., eds., 1988, *CRC Handbook of Chemistry and Physics*, **69**th edition 1988–1989, CRC Press, Boca Raton, Florida.

Weiss, D. G., Meyer, M. A., and Langford, G. M., 1990, Studying axoplasmic transport by video microscopy and using the squid giant axon as a model system, *this volume.*

Williams, L. W., 1909, *The Anatomy of the Common Squid, Loligo pealii, Lesueur*, E. J. Brill, Leiden.

Young, J. Z., 1936, Structure of nerve fibres and synapses in some invertebrates, *Cold Spring Harbor Symp. Quant. Biol.* **4**:1–6.

Young, J. Z., 1939, Fused neurons and synaptic contacts in the giant nerve fibers of cephalopods, *Phil. Trans. Roy. Soc. Ser. B* **229**:465–503.

Chapter 8

Internal Dialysis in the Squid Giant Axon

LORIN J. MULLINS and F. J. BRINLEY, JR.

1. Historical

An axon used for *in vitro* transport studies necessarily has cut ends so that methods used for the measurement of flux must be designed to cope with the inevitable leakage in such regions of the substance under measurement. Nerves consisting of a large number of small fibers have been used successfully in transport studies but they offer certain complications. Their extracellular space may change with time, and if the fibers are quite small the resulting high surface/volume ratio may lead to such rapid equilibration that these kinetics are difficult to separate from the equilibration of the extracellular space. Hence, both for electrical studies and for transport measurement, emphasis has been on the use of single-fiber preparations of a size sufficient for the introduction of capillaries inside the fiber and for adequate volumes of axoplasm to carry out the various analytical measurements.

Progress in understanding mechanisms of transport has relied almost entirely on the ability of the experimenter to measure unidirectional fluxes, a measurement that is only possible with isotopes. Early isotope work with squid and *Sepia* axons relied on the quite simple technique of immersing an isolated axon in a radioactive solution of a particular isotope and a short time later removing the axon, rinsing it briefly in non-radioactive solution, and counting the fiber. An important improvement to this technique was the extrusion of axoplasm from such a fiber in order to avoid radioactive contamination from the inclusion of connective tissue and Schwann cells in the sample to be counted. Such a method for influx measurement avoids external contamination and is in use at the present time. It does have the disadvantage that only one influx measurement can be made on each axon and hence it is difficult to apply to influx measurements where a substantial difference is to be expected from axon to axon as a result of dissection methods.

The immersion of an axon in a radioactive solution for several hours will allow the accumulation of sufficient radioactivity in the axoplasm for efflux measurements to be made. The fractional equilibration of the isotope with intracellular K^+, for example, is small (of the order of 10 – 20%), but when the axon is placed in seawater, sufficient counts emerge to make efflux measurements on samples of seawater collected

L. J. MULLINS • Department of Biophysics, Medical School, University of Maryland, 660 W. Redwood Street, Baltimore, MD 21201. F. J. BRINLEY, JR. • Neurological Disorders Program, National Institute of Neurological Disorders and Stroke, Federal Building, Room 814, National Institutes of Health, Bethesda, MD 20892.

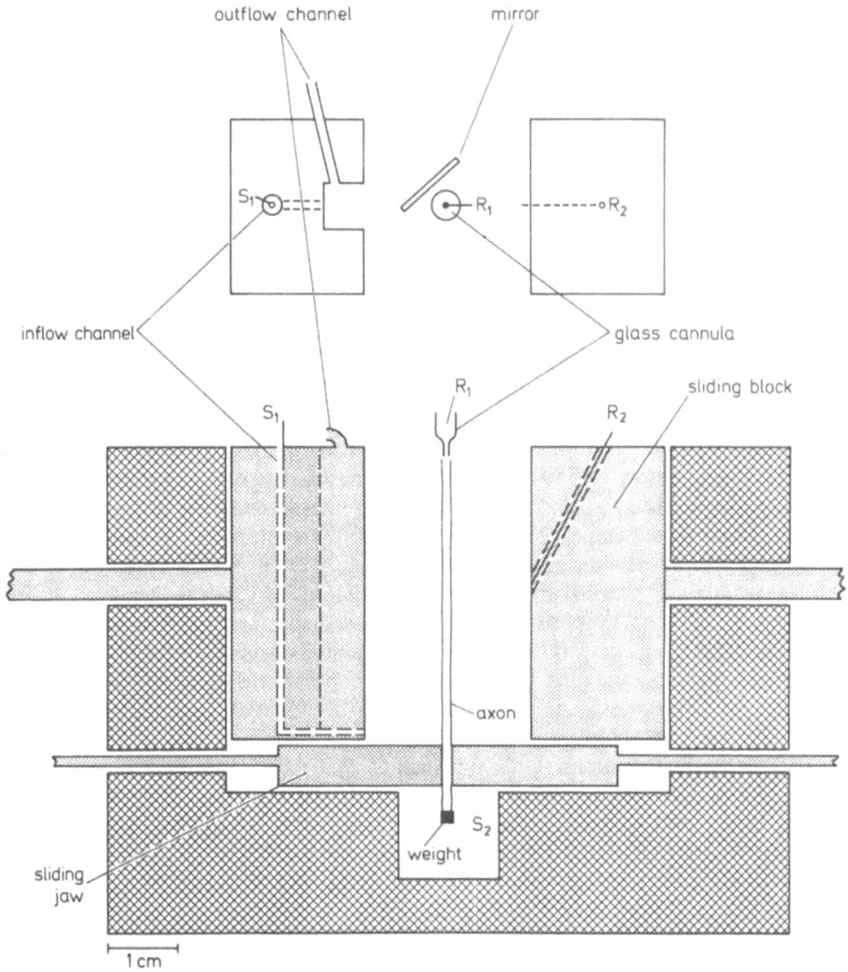

Figure 1. Injection apparatus modified from design of Hodgkin and Keynes. Features include: axon tied to glass cannula at one end and gold weight at the other. Sliding jaws provide electrical insulation between well at bottom of apparatus and top chamber, making stimulating currents (from S_1, S_2) flow through membrane. Action potentials recorded from R_1, R_2. Sliding blocks necessary to steer glass cannula through axon to site of injection; both direct observation and mirror used to keep capillary centered. Manipulators move chamber to keep injection capillary centered. Sliding blocks then are brought together, forming small efflux collection chamber. Mechanical pumps move seawater into chamber and collect solution emerging from top into fraction collector. Hamilton syringes of 1 μl capacity are connected to injection capillary (usually 100 μm O.D.). (Reproduced from Brinley and Mullins, 1965).

at 10-minute intervals. Efflux measurements for Na^+ are probably reasonably accurate with this method, because the exchange time for Na^+ in extracellular compartments is relatively brief; for K^+ efflux it has been shown (Caldwell and Keynes, 1960; Sjodin and Mullins, 1967) that the specific activity of the counts emerging from an axon falls over a period of two hours, so that flux measurements are inaccurate. The mechanism of this effect is reasonably attributed to a fairly rapid equilibration of ^{42}K with the Schwann cells and a much slower rate of equilibration with the axoplasm.

The difficulties indicated with respect to efflux measurements of potassium can be obviated by a method developed by Hodgkin and Keynes (1956), whereby it is possible to inject substances into squid axons rather than having to equilibrate them from the outside. The principle of the method is illustrated in Figure 1 and essentially involves the insertion of a glass capillary into the axoplasm and the coupling of this capillary to a micro-syringe. The internal diameter of the capillary and the volumetric displacement of the micro-syringe are coupled in such a way that equivalent volumes of material are injected as the capillary is removed from the axoplasm. In effect, this produces a line of injected solute which diffuses rather uniformly throughout the axoplasm within a period of 15 seconds or so unless the material injected is a substance that is bound. The technique is useful not only for the introduction of isotopes for flux measurements, but also for the injection of substrates in connection with studies of active transport, and can also be used to inject current or to measure membrane potential.

It is important to recognize that injection can only add substrates or isotopes to the axoplasm; it is not a technique that allows a reduction of ion concentrations, nor does it permit the removal of products of metabolism if these are relatively nondiffusible. It does permit the introduction of enzymes (DeWeer, 1968) that may hydrolyze existing substrates or catalyze transphosphorylations that do not normally take place.

In 1962, Baker *et al.* showed that it was possible to remove a substantial fraction of the axoplasm from a squid axon, replace it with a flowing solution of isosmotic potassium sulfate and still find the axon electrically excitable. The arrangement used is shown in Figure 2. The technique of internal perfusion is a very useful one in examining the dependence of the electrical properties of the membrane on internal solutions. Experimentation with a variety of anions to accompany potassium as an internal perfusion medium showed that potassium fluoride gave the best results, although potassium glutamate has also been used. Measurements of transport fluxes associated with bioelectric activity using the internal perfusion technique have only been really successful when potassium fluoride was the internal solute, and under these conditions a sodium influx of the order of 20 – 30 pmol/cm^{2}s has been observed (Atwater *et al.*, 1969). Since F^- irreversibly inactivates membrane adenosine-5'-triphosphatase (ATPase) and presumably other proteins of interest in connection with active transport studies, attempts were made (Shaw, 1966; Baker *et al.*, 1971) to use glutamate and other nontoxic anions in an internal perfusion medium that would allow active transport to proceed. The results of such studies have not been very successful, either because there is a large Na^+:Na^+ exchange component to the flux and the Na^+ efflux itself is extremely high, or because it was not possible to observe any ouabain-sensitive Na^+ efflux or any real dependence on a substrate such as adenosine-5'-triphosphate (ATP). Some success with short-term internal perfusion (ca. 60 min) has been reported (Rojas and Canessa-Fischer, 1968; Canessa-Fischer *et al.*, 1968) but it is not judged feasible for experiments of some hours duration.

Causes for the anomalous Na^+ and K^+ fluxes are not easily discerned, but may be related to the fact that when axoplasm is washed away by an internal perfusion system, proteolytic and other kinds of enzymes are released from a bound form and can act on the membrane transport mechanisms to bring about Na^+:Na^+ exchanges that are not energized by ATP.

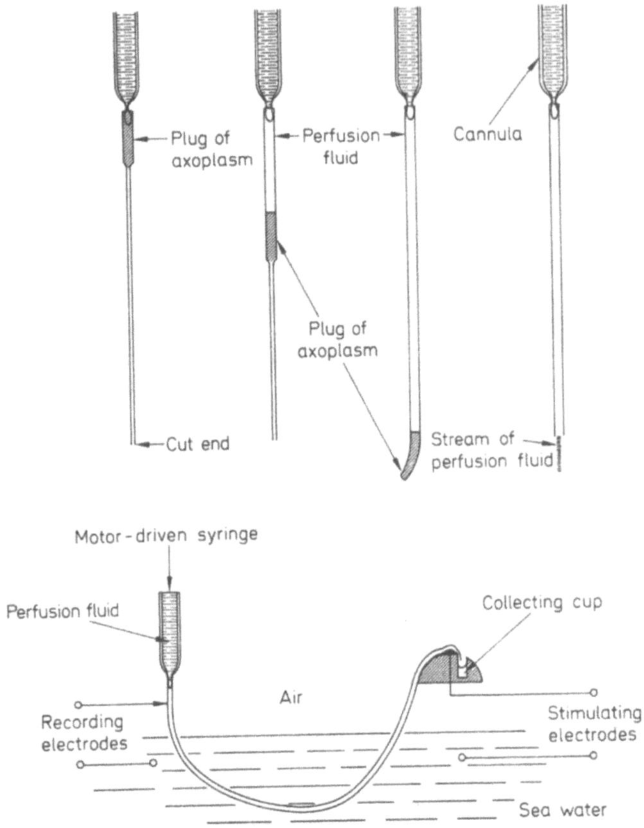

Figure 2. Top section illustrates re-inflation of extruded axon with perfusion fluid. Fluid was driven into the cannula by motor-driven *Agla* syringe at about 6 μl min^{-1}. Bottom section shows standard arrangement for recording external action potentials from perfused axons. (Reproduced from Baker *et al.*, 1962).

A somewhat different approach to internal solute control is the technique of internal dialysis (Brinley and Mullins, 1967) in which a capillary with a porous length of about 15 mm and an overall length of 12 cm is steered through the axon and fluid passed through the capillary at a rate of about 1 μl min^{-1}. The porous region of the capillary (originally porous glass, later porous plastic) allows an interchange of solutes with molecular weights of 1000 or less but is impermeable to protein. The time constant for the exchange of diffusible solutes such as Na^+ is of the order of 5 minutes in a 600 μm axon and Na^+ and K^+ fluxes as measured with this technique are in good agreement with values obtained from microinjected axons. The general arrangement for the internal dialysis technique is shown in Figure 3.

With suitable modifications, the technique can measure the influx of solutes as well, and can do so continuously rather than relying on the single measurement of influx that can be obtained from the extrusion of axoplasm after a predetermined time of treatment with radioisotope.

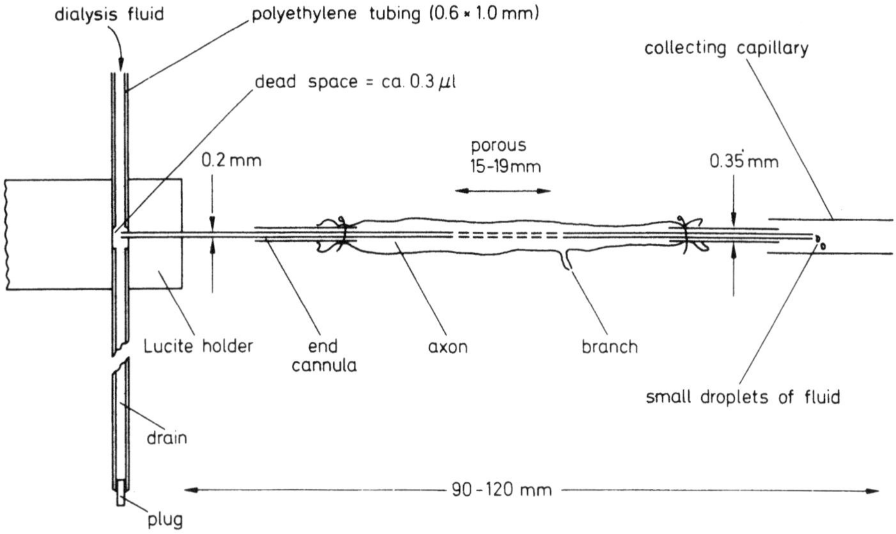

Figure 3. Diagram (not to scale) of lucite holder for porous capillary is shown on left. Lucite holder mounted on micromanipulator. Center: porous capillary inserted in a squid axon. Right-hand side: capillary that collects the effluent fluid being pumped through porous capillary by motor-driven syringe. (Reproduced from Brinley and Mullins, 1967)

Internal dialysis is capable of removing all substrates from an axon, although for substances such as ATP, that may be internally compartmentalized, this may take an hour or more to reach μM levels of concentration. It can readily control levels of Na^+, K^+ and other nonmetabolizable substances. With ATP, however, such control is less effective, since instead of local generation of this substrate by mitochondria, one substitutes a system that delivers ATP on the axis of the fiber. Diffusion from its site of delivery to the membrane can result in hydrolysis.

2. Materials for internal dialysis

The first attempt to develop an internal dialysis capillary involved a stainless steel tube approximately 150 μm in diameter in which a number of 10 μm holes had been drilled over a 1 cm length of tubing. The non-perforated regions of the stainless steel capillary were coated with a thin layer of varnish as insulation. A platinum-iridium wire was then passed through the lumen of the tube, and the entire assembly dipped in a hot agar solution to coat the center region and plug the 10 μm holes with agar. The wire was then withdrawn, hopefully leaving the agar plugs filling the holes intact. It was intended that the plugged agar holes would allow a diffusion pathway for ions and other small molecules to diffuse outward from an internal dialysis solution passed through the interior of the capillary. Unfortunately, this technique was never successful because it proved impossible to position the tube inside an axon without dislodging some of the agar plugs, which allowed bulk flow of solution through the 10 μm holes to distend the axon and cause irreversible damage. Furthermore, it was soon realized that the mechanical strength of the capillary would not permit a sufficient number of holes to be drilled in the central region to achieve a significant porous region even if the agar plugs remained intact.

After some weeks of fruitless experimentation, the problem was discussed with Dr. P. W. Davies of the Physiology Department at Johns Hopkins who recalled that Dr. M. E. Nordberg had published a description of a type of porous glass manufactured by Corning Glass Works. Dr. Nordberg and his associate, Mr. Thomas Elmer, were approached for more information about porous glass, and the possibility of fabricating some small porous glass capillaries for testing.

The porous glass actually was one stage in the manufacturing of Vycor. The silica used for the manufacture of Vycor ware was initially doped with a proprietary mixture of heavy metals and other materials to lower the melting point. This allowed silica ware of various shapes to be formed at lower temperature than would be possible with quartz. The doping materials were subsequently removed by prolonged leaching in strong acid at elevated temperatures. This resulted in the creation of a nearly pure silica matrix with a number of submicrometer channels in the glass membranes resulting in a porous wall. Controlled heating of the porous silica caused fusion of the holes resulting in loss of porosity and the formation of commercial Vycor. From our standpoint, this doping of the silica meant that it was relatively easy to fabricate thin walled glass capillaries which were subsequently acid leached to make them porous. The ends of the capillary could then be easily rendered non-porous simply by controlled heating of both ends, leaving an unheated, porous region of the desired length in the center of the capillary.

Our initial *in vitro* tests showed that these capillaries indeed were porous to small ions such as sodium, potassium, chloride, and to a lesser extent to ATP. However, diffusion of this latter material, crucial for fueling the sodium and potassium pumps, was marginal. Fortunately, we were able to increase the porosity of the central region several fold by prolonged treatment of the capillaries with one normal nitric acid at 90 °C. This procedure caused considerable loss of the mechanical strength of the capillaries and many capillaries were lost by breakage. Furthermore, many of our initial experiments with squid axons failed when the axons rapidly lost excitability. Inspection of the dialysis capillary in these cases showed coagulation of the axoplasm on the surface of the porous region. When this problem was discussed with Dr. Nordberg and Mr. Elmer, they pointed out that the standard acid leaching procedure did not remove all of the heavy metals used in the initial doping of the silica melt. They concluded that our additional extensive acid treatment probably exposed additional toxic heavy metals. We then modified our capillary preparation protocol to follow the acid treatment with a prolonged soak in 50 mM ethylenediamine tetraacetic acid (EDTA) for about 48 hours to chelate the heavy metals. The parameters used in our leaching and chelating procedures were critical because deviation from a pH of 6.5 – 7 markedly accelerated the dissolution of the porous region as did higher concentrations of EDTA.

Due to the very small size and thin walls of these porous capillaries, we found that they had to be stored in water because when allowed to dry, the surface tension generated by drying frequently caused the capillaries to break. However, storing the capillaries in distilled water apparently allowed additional materials in the porous wall to dissolve, because such capillaries gradually became toxic. Therefore, capillaries were routinely stored in weak solutions of EDTA in ethanol at neutral pH.

2.1. Evaluation of porosity to low molecular weight substances

Our estimate of the amount of porous area in the central regions indicated that about 10 – 20% of the geometrical area was actually water filled channels. This result agreed with the estimate of porous area obtained by the manufacturer on the basis of electron micrographs of porous glass.

By measuring the rate at which isotope passed across the wall of the capillary, we estimated that for small molecules the diffusion coefficient for sodium and potassium was about 50% of that in bulk solution. We also checked the porosity of the capillaries to proteins such as hemoglobin and aldolase and concluded that the diffusion coefficient was very much less than 1/10% of that of free solution.

The results of this type of measurement indicated that small molecules passed across the porous wall primarily by diffusion, although at high flow rates a certain amount of bulk water flow occurred. After considerable experimentation we found that a flow rate of 0.5 μl/min through the lumen of the capillary which corresponded to hydraulic pressure of 40 cm of water at the porous region for porous capillaries of the order of 125 μm outside diameter, and a wall thickness of about 25 μm did not produce any measurable change in the diameter of the axon.

2.1.1. Porous acetate capillaries

Although the porous glass capillaries allowed effective control of the concentration of small molecules of squid axoplasm, they were mechanically fragile, and very difficult to fabricate. Many capillaries had to be rejected because of small bends or kinks. Furthermore, as the experiments progressed, it became clear that electrical measurements needed to be made, and it proved impossible to place wires inside the glass capillaries. Therefore, a search for other porous materials ensued.

Fabric Research Laboratory in Dedham, Massachusetts proved to be a reliable source of cellulose acetate capillary tubing of various dimensions. Appropriate regions of this capillary could be made porous simply by exposure to 50 mM sodium hydroxide at room temperature to hydrolyze the acetate residues. Sodium hydroxide concentrations in excess of 50 mM destroyed the cellulose matrix, whereas lower concentrations seemed to act too slowly.

The absolute diffusion coefficient for sodium across the wall of these capillaries was about 10% of that of free solution. However, the porosity to calcium and ATP appeared to be rather variable. DiPolo (1977) made a more complete evaluation of the porosity of these capillaries to calcium and concluded that the porosity was markedly reduced if the cellulose was allowed to dry thoroughly, although porosity could be maintained simply by keeping the capillary damp when it was exposed to air. Furthermore, there was some evidence that exposure of the capillaries to distilled water irreversibly reduced porosity. Therefore, capillaries, once permeabilized, were kept in

0.1 M KCl. Despite all of these precautions, however, the porosity of hydrolyzed section gradually decreased with time. Therefore, capillaries needed to be freshly prepared.

2.1.2. Hollow cellulose fibers

Recently regenerated cellulose hollow fibers with molecular weight cut off between 6000 and 9000 have become available from Spectrum Medical Industries, Los Angeles, CA. These fibers are porous over their entire length; however, the walls can be easily coated with varnish or nail polish with minimal overall increase in diameter. Fibers currently available have wall thicknesses of 9 to 18 μm and an overall diameter of 170 to 250 μm. As with the cellulose acetate capillaries available from Fabric Research, the porosity is reduced if the fibers are allowed to dry, and, in fact, the manufacturer recommends that the relative humidity around the fiber not be less than 50%.

The availability of these relatively inexpensive hollow cellulose fibers has greatly increased the usefulness of the dialysis capillary technique and, with the use of voltage clamp techniques, has allowed the membrane potential as well as the internal concentration of diffusible materials to be controlled.

3. Solutions for internal dialysis

Initial trials of the method of internal dialysis led to sufficient problems with the toxicity of glass porous capillaries and with the survival of axons with such large internal structures as a dialysis capillary inserted in them so that our approach to internal solutions for dialysis was a conservative one. It seemed best to have a solution that was a close approximation to the composition of the squid giant axon diffusible solutes as measured by biochemical analysis. These analyses showed that a) there was a substantial non-electrolyte component to the internal osmotic pressure (mostly taurine) and b) that the internal anions were mostly isethionate, a rare sort of internal anion with minor contributions from aspartate and to a lesser extent glutamate (see Table I).

After achieving some degree of confidence in the ability of this sort of solution to support internal solute control and a normal cation flux across the membrane, it seemed possible to explore changes in internal dialysis solution with changes of a variety of sorts. These changes can be grouped into a) changes in internal substrate for transport, b) in the internal cation of axoplasm, i.e., the exchange of internal K for Na or other cations, and c) a change in the loss of electrical polarization caused not as in b) but by increasing external K to levels approaching those of internal K.

Previous studies of Na exchange in a wide variety of cells had established that ATP or some substrate readily transformed into ATP was the source of energy for Na extrusion. Internal dialysis allowed us to test singly ATP, and arginine phosphate (ArgP) along with a large number of high energy phosphate compounds such as guanosine 5'-triphosphate (GTP), uridine 5'-triphosphate (UTP), cytidine 5'-triphosphate (CTP) as well as intermediates such as phospho(enol)pyruvate (PEP) and similar substances. One might anticipate that depriving the membrane of a metabolic substrate might cause leaks in the ionic pathways and might have other irreversible effects. No such findings were ever noted; indeed, the membrane could be kept with internal levels of ATP of 1 μM for several hours without any suggestion of irreversible effects since

Table I. Constituents of *Loligo* axoplasm

Constituent	*Axoplasm** *mM*	*Corrected to* $Na^+ + K^+ = 373$ *mM*
Potassium	344	314
Sodium	65	59
Calcium	3.5	3.2
Magnesium	10.0	9.1
Arginine, Lysine, Ornithine	8.8	8.0
Sulfate	7.5	6.8
Phosphate	17.8	16.2
Chlorine	151	137
Glutamic acid	21.2	19.2
Aspartic acid	79.0	71.9
Isethionic acid	165	150
Hoemarine	20.4	18.6
Betaine	73.7	67.1
Taurine	107	97
Cysteic acid amide	4.9	4.5
Alanine	8.6	7.8
Glycine	11.6	10.5
Other amino acids and sugars	27.8	25.3
Totals	1127	1025
Σ Cations (meq)	444.8	
Σ Anions (meq)	455.2	

* Taken from Deffner (1961)

the restoration of ATP always produced a Na efflux appropriate to the experimental conditions employed.

Actual substrates tested included PEP, ArgP, XTP where X is G, U, or C, hexose phosphates, and acetyl phosphate. None of these were capable of energizing more than a trivial amount of Na extrusion and even this small amount was likely due to ATP contaminants in the preparations as analyzed for by the firefly flash reaction. Similarly, adenosine 5'-diphosphate (ADP) and adenosine 5'-monophosphate (AMP) were without effect when allowance was made for the fact that they can be converted to ATP via kinase reactions. The only substrate that showed genuine ability to induce Na efflux was deoxyATP.

The composition of a number of dialysis solutions used for various purposes is given in Table II where in addition to the constituents of axoplasm indicated above, N-tris (hydroxymethyl)-methyl-2-aminoethanesulfonic acid (TES) was used as a pH buffer, and ethylene glycol-bis(2-amino ethyl ether)N,N,N',N'-tetraacetic acid (EGTA) or EDTA as Ca and heavy metal complexing agents were also employed. Some reagents are heavily contaminated with Ca (above the 100 μM level) and hence require purification before use. In addition, glass capillaries sometimes had heavy metal

Table II. Composition and identification of solutions used*

Ion	*ASW‡*	*Li-ASW*	*A'*	*E'*	*F*	*G*	*H*	*M*	*N*	*O*	*O'*
	mM	*mM*	*mM*	*mM*	*mM*	*mM*	*mM*	*mM*	*mM*	*mM*	*mM*
Na^+	429	–	80	80	–	80	–	231	–	80	80
Li^+	–	429	–	–	–	–	–	–	80	–	–
K^+	9	9	304	305	385	303	383	153	304	304	304
Mg^{++}	48	48	4	4	4	4	4	4	4	4	4
Ca^{++}	9	9	–	–	–	–	–	–	–	–	–
Cl^-	496	496	88	88	88	88	88	88	88	88	88
$Isethionate^-$	–	–	151	151	151	151	151	151	151	151	151
$L\text{-}Aspartate^-$	–	–	151	151	151	151	151	151	151	–	–
$D\text{-}Aspartate^-$	–	–	–	–	–	–	–	–	–	151	151
CN^-	–	–	2	2	2	–	–	–	–	–	–
$SO_4^=$	25	25	–	–	–	–	–	–	–	–	–
$EGTA^{\blacksquare}$	–	–	–	–	–	–	–	–	–	–	0.1
$EDTA^{\blacksquare}$	0.1	0.1	0.1	0.1	0.1	0.1	0.1	0.1	0.1	0.1	–
TES^-	4	4	–	1	1	1	1	2	2	2	2
Taurine	–	–	275	275	275	275	275	275	275	275	275

* pH of external solutions 7.5, of internal solutions 6.9 – 7.1.
‡ CN-ASW formulated by addition of 1 – 4 mM NaCN, buffered to pH 7.5.

contamination that chelators readily removed. Modification of the dialysis solution such that a single anion (either aspartate or glutamate) was present or a chelating anion such as trans-1,2-diaminocyclohexane-N,N,N',N'-tetraacetic acid (CDTA) or EGTA led to axons with entirely stable fluxes over many hours of flux measurement. Hence, it can be concluded that internal dialysis solutions do not have a critical requirement for a specific anion. This is in marked contrast with perfused squid axons where in the absence of fluoride as a substantial constituent of the internal perfusion fluid, a rapid run down of the axon is to be expected. Aspartate and glutamate help but still are inferior to fluoride; still, Na fluxes appear unstable and presumably are heavily contaminated with ion exchange fluxes.

Since the method of internal dialysis necessarily involves a filtration of dialysis fluid into the axon and hence to a volume increase in the axon, dialysis solutions were sometimes made hypotonic so that water would flow outward to compensate for the filtration that tended to increase the volume of the axon. With a seawater adjusted to 1025 mosmole/liter, dialysis solutions were made 925 mosmole/liter. Because the solutions contained many uncommon reagents, a dew point osmometer was indispensable to the preparation of internal dialysis fluids.

4. Theoretical analysis of diffusion within the porous capillary and axoplasm

4.1. Efflux experiments

Although preliminary experiments with radioactive sodium clearly indicated that porous dialysis capillaries delivered radioactive isotope to the axoplasm, and that a steady state of isotope efflux was obtained within five to ten minutes, there was no direct way to evaluate the adequacy of diffusion control of multivalent high energy substrates such as phosphoarginine, or ATP. Another matter of concern in evaluating the usefulness of the dialysis technique for measuring efflux, was the effect of longitudinal diffusion of isotope out of the central region of the porous capillary into the unperfused ends of the axon. Although it seemed intuitively reasonable that using a long central porous region would reduce the contribution to tracer efflux crossing the membrane radially in the unperfused regions following axial diffusion, the magnitude of the end effect for any given experimental condition was not obvious.

For this reason a numerical integration of the diffusion equation (Brinley and Mullins, 1967), for one particular set of experimental parameters was obtained, for the composite system: dialysis capillary, axoplasm, axolemma. For simplicity, the only initial condition considered was an instantaneous change of concentration of isotope at the inner face of the dialysis capillary at time zero, i.e., the effect of a finite axial laminar flow in the interior of the dialysis capillary was ignored.

Subsequently, Horn (1983) considered the more general case of finite laminar flow in the lumen of the dialysis capillary, as well as the effect of a wide range of capillary and axon geometries, diffusion coefficient membrane permeabilities, etc. This analysis permitted a more accurate estimation of the effects of diffusion limitation within the wall of the porous capillary and of edge effects.

The theoretical calculations performed by Brinley and Mullins (1967) showed that monovalent ions with diffusion coefficient comparable to those for sodium and potassium (10^{-6} cm^2/sec) reached diffusional equilibrium with a half time of two-three minutes. This computed half-time was significantly smaller than the experimentally observed half-times for equilibration which were of the order of five to ten minutes. At the time these calculations were presented, it was suspected that the simple assumption of instantaneous change of concentration at the interior surface of dialysis capillary was responsible for the discrepancy. The analysis of Horn confirmed this suspicion, because theoretical half-times estimated with the more sophisticated boundary conditions were indeed of the order of five to ten minutes.

The theoretical calculation also showed that diffusion equilibrium for a molecule such as ATP, being consumed uniformly throughout the axoplasm at the experimentally observed rate (of 0.6 μmole per millimole ATP concentration per gram of axoplasm per minute at 15 °C), was reached with only a very slightly increased half-time. Although ATP was hydrolyzed during radial diffusion, the steady state concentration at the surface of the axolemma was calculated to be almost 85% of that delivered at the inner surface of the dialysis capillary. Horn (1983) confirmed this conclusion for high concentrations of ATP, i.e., 1 – 4 mM. However, his analysis indicated that at low concentrations of ATP, i.e., 50 μmol, most would be broken down by the time it reached the membrane. Fortunately, and fortuitously, when Brinley and Mullins (1968) measured the sodium efflux under low ATP conditions, they had added phosphoarginine as a codiffusing high energy substrate which presumably would allow for rephosphorylation of any ATP broken down. Although they did this principally with the thought of maintaining a relatively constant and high ATP/ADP ratio, their protocol also ensured that the ATP concentration in the dialysis fluid was maintained during radial diffusion to the axolemma.

4.1.1. End effects

In an attempt to estimate the magnitude of *end effects* at the interface between porous and non-porous regions, Brinley and Mullins (1967) considered the somewhat artificial case of axial diffusion into a semi-infinite cylinder with one face maintained at a constant concentration C_o. Their qualitative conclusion was that the outward radial diffusion of isotope across the axolemma in the unperfused regions of the axon could become significant compared to the steady state radial diffusion in the porous region. However, this effect developed only slowly over a period of many hours. A more comprehensive analysis by Horn (1983) confirmed this initial conclusion, but showed that this circumstance would occur only if the ratio of porous length to axon radius was less than 10.

Since, in most dialysis experiments performed in the early experiments, the length to radius ratio was greater than 20, probably the end effects were much less important than we originally thought. For that reason the rather elaborate guard region which we used in the initial experiments was probably unnecessary. However, the development of this guard system proved to be very useful for other purposes, i.e., it did protect the central region from contamination by efflux from damaged end regions, and it also provided experience with fractionation with flows in the central slot which turned out to be critical in calcium influx experiments performed during the seventies.

4.2. Influx experiments

Although the internal dialysis technique was originally developed to measure the effect of internal concentrations of various substances on ion efflux, it was soon realized that it was possible to use the technique to measure influx by placing radioactive isotope outside the axon and dialyzing the interior of the axon with non-radioactive medium. In the original dialysis chamber it was not possible to confine extracellular radioactive solution only to that region of the axon containing the porous capillary. It was realized that any isotope entering the axon from the unperfused end regions would be able to diffuse axially towards the porous region where it would be picked up by the capillary and counted as influx.

A theoretical analysis of the problem (Brinley and Mullins, 1967) using unrealistic boundary conditions (assuming uniform influx into a very long end region all of which diffused longitudinally towards the porous region), indicated that although a steady state diffusion into the ends of the porous region could contribute as much as 50% of the total influx, the steady state was approached with a half-life of many hours. We anticipated that this edge effect would not be serious for relatively short-term influx experiments for sodium and potassium. This expectation was borne out in pilot experiments where the half-time for steady state influx of sodium and potassium isotope was about 10 minutes.

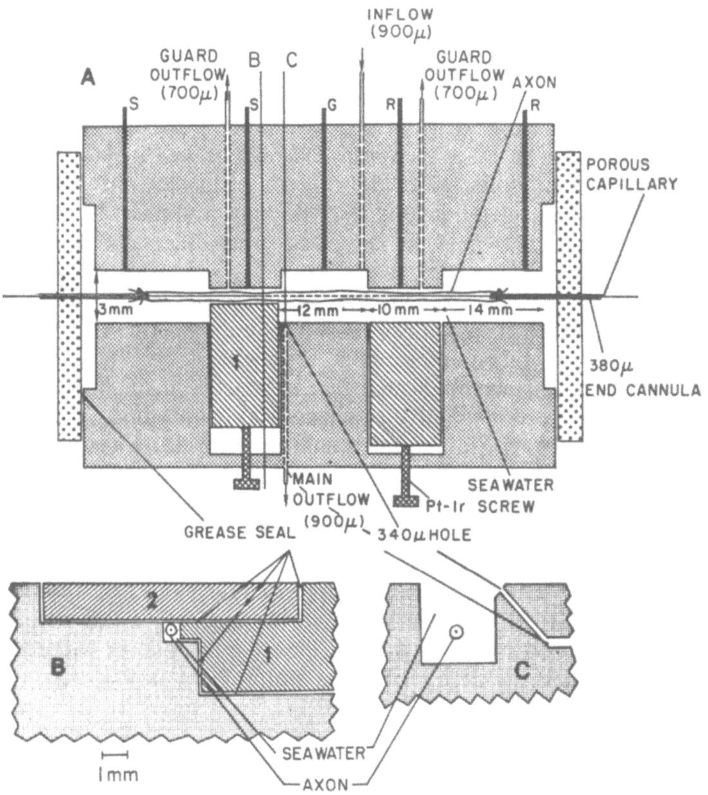

Figure 4. Schematic drawing of apparatus used to hold the cannulated axon during dialysis. A is a top view (not to scale); B and C are cross-sections (to scale) at the indicated levels. Labels *S*, *R*, and *G* refer to stimulating, recording, and ground leads, respectively, which are recessed slightly to avoid contact with the axon. The cannulae, to which the ends of the axon are tied, are seated in the shaded blocks which are attached to micromanipulators to allow positioning of the axon in the chamber or in the slot. After the axon has been lowered into the slot, the Lucite pieces labelled *1* are advanced with the thumbscrew to form part of the guard chamber. The piece labelled *2* is then placed on top to complete the guard compartment. The clearance between the axon and any side of the guard compartments is 50 – 100 μm. (Figure and Legend from Brinley and Mullins, 1967).

Horn (1983) examined the effect of a finite porous length upon influx in a much more comprehensive manner and showed that the influx error, defined as the percent of total influx contributed by diffusion from the unperfused end regions, could be virtually eliminated if the radioactive isotope in the external medium did not extend beyond the porous non-porous junction. Furthermore, the error was not serious even if the area of exposed axon extended as much as 0.5 to 1 millimeter beyond the non-perfused region.

Horn (1983) also considered a situation of experimental interest, in which the influx is suddenly reduced from its initial value. Such condition would occur experimentally if one is determining the effect of ATP removal, for example, upon potassium influx. In this case the influx in the dialyzed region would be reduced, whereas influx in the unperfused end regions would continue at a high level (because ATP has not been reduced in the end regions), and therefore axial diffusion into the porous region would be unaffected. Therefore, the radioactive isotope collected by the porous capillary will reflect not only the reduced radial diffusion in the central porous region, but a relatively higher contribution from the end regions. The initial analysis of Brinley and Mullins (1967) took no account of this situation. However, the analysis of Horn (1983) showed that the effect could be substantial if the length of axon exposed to isotope extended more than 0.5 to 1 millimeter beyond the porous non-porous junction. Furthermore, if isotope is removed from the fluid bathing the central part of the axon during the performance of a wash out experiment, the axial diffusion from isotope already in the axoplasm in the unperfused end regions will make a very substantial contribution to the isotope collected in the dialysis effluent. The general conclusions from these theoretical considerations are that the time resolution of influx experiments is substantially less than would be expected from the same system operating in the efflux mode.

Unfortunately, these theoretical calculations are not readily applied to the situation of calcium influx because a great deal of isotope can be stored in the internal buffering systems such as mitochondria and other intracellular organelles. In fact, early experiments with the conventional chamber indicated that radioactive calcium influx, after a period of dialysis, was relatively insensitive to changes in extracellular isotope.

In order to manage this problem, two different techniques were developed. One used by Brinley *et al.* (1975) involved two coiled porous capillaries. One capillary had a central porous region, while the other had two regions of porosity located in such a manner that when the two porous capillaries were lined up these porous regions collected isotope inwardly diffusing from the end regions and therefore prevented longitudinal diffusion into the central region. A second approach developed by DiPolo (1979) for use in the smaller axons of the tropical squid, *Loligo plei*, involved an arrangement of guard regions such that the end regions of the axon were never exposed to radioactive isotope. Further details on these techniques are provided in the sections describing dialysis chambers.

5. Design of dialysis chambers

5.1. Efflux chamber

A schematic diagram of the apparatus first used for efflux measurements is shown in Figure 4 (Brinley and Mullins, 1967). The squid fiber is cannulated at both ends and mounted horizontally as shown. The end cannulae are mounted on lucite blocks which in turn are attached to micromanipulators. The porous capillary, mounted on a third micromanipulator is advanced into the interior of the fiber. Positioning of the capillary in the axon during insertion is facilitated by adjusting the position of the fiber by moving the two lucite porous blocks. Since the original experimenters were both left-handed, the porous capillary was mounted to the left of the dialysis chamber so that the capillary could be passed from left to right through the fiber. The porous region is located in the center block of the dialysis chamber by reference to vernier marks on the micromanipulator scale and by noting the obvious change in the physical properties of the glass at the junction between porous and non-porous regions. It is also quite acceptable to stain the porous regions with phenol red before inserting the capillary in the axon. Sufficient dye remains in the wall of the glass to mark the porous region.

After the capillary has been positioned inside the squid axon, the fiber and capillary assembly is positioned inside the dialysis slot. This was accomplished in the earlier models by simultaneously lowering the two manipulators holding the axon and

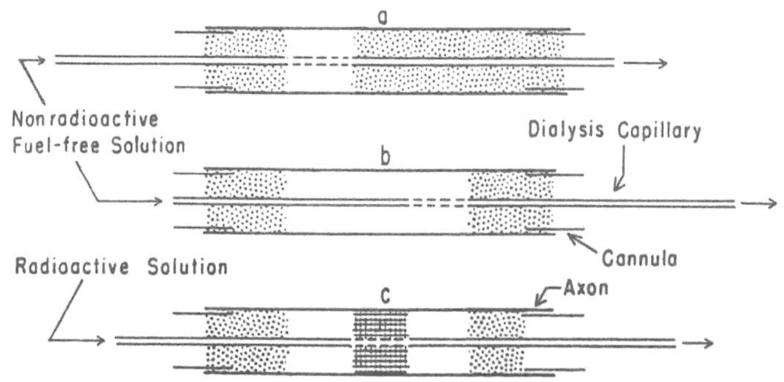

Figure 5. The method for removing substrates from the end regions of the axon is shown in this figure. The axon is cannulated and the porous capillary is inserted and brought to the position shown in a. The loss of stippling by the axoplasm indicates the removal of substances not contained in the nonradioactive dialysis fluid. The dialysis capillary is advanced toward the right at a rate of 1 mm/min until the porous region reaches the position shown in b. The result of this manipulation is to remove substrates from the central part of the axon. The direction of movement of the capillary is reversed and brought to the location shown in c; the fluid flowing through the dialysis capillary is changed to a medium containing ^{24}Na and this is shown as cross-hatching in the axoplasm. The region of flux measurement is thus bordered on each side by axoplasm that has been washed free of substrates such as ADP. (Figure and Legend from Mullins and Brinley, 1967).

the one holding the porous capillary. In later experiments it proved more convenient to attach a manipulator to the main dialysis chamber block and simply raise it. The dimensions of the center and guard regions shown in the diagram, i.e., 12 mm and 10 mm, respectively, proved convenient for axons isolated from *Loligo pealei*. The volume of the guard slot is minimized by moving sliding blocks into close apposition with the squid axon as shown in Panel B, Figure 5. The dimensions shown in the diagram are approximate. The critical point is that the length of the porous capillary should extend one to two millimeters beyond the end of the central collecting pool (see section 4 for a theoretical discussion).

The principle of the guard chamber is that rated fluid entering the central collecting region is slightly in excess of the amount withdrawn from the central region. The excess is withdrawn through the guard outflow pipes creating a net flow of fluid from the central region outward into each guard region. During dialysis any isotope diffusing across the membrane near the junction between porous and non-porous will be swept out by the current of flow in the guard chamber and will not pass into the central region. In this way efflux is collected only from that portion of the axon exhibiting radial diffusion and not from those end regions of the axon where dialysis control of the internal solutes was not adequate. With the dimension given it was found that a flow rate into the center compartment of about 1.2 ml per minute and withdrawal though each guard region of approximately 0.07 ml per minute was adequate to provide a rapid wash out at the central chamber (about 1 - 2 minutes) while simultaneously preventing more than 95% of the isotope diffusing across the axolemma in the guard region from reaching the central slot.

The arrangement of inflow, outflow and guard flow ports proved not to be critical as long as the outflow was near the extreme lateral end of the guard region. Later models of the dialysis chamber used by the authors and others, generally shortened the length of the guard region to five or six millimeters and added an extra inflow port into the center region to facilitate mixing of solutions in the center compartment. DiPolo (1972) introduced a useful modification of the procedure by sealing the outside of the

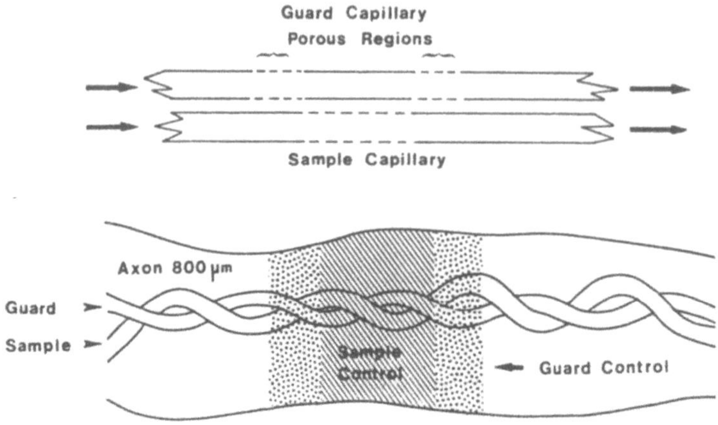

Figure 6. Schematic diagrams illustrating the principle of influx measurement are shown below. The porous region of the sample capillary is 10 – 12 mm long and the guard region 5 – 7 mm. The separation between guard and sample region was 0 – 1 mm. The lower diagram illustrates the actual assembly with the two capillaries wound around one another with a pitch of about one turn per 5 – 10 mm. Inside an axon, this assembly controls the central region (hatched area) by introducing regions of solute control (dotted areas) on each side of the sample region. (Figure and Legend from Brinley and Mullins, 1975).

guard region with a bit of vaseline, and then withdrawing solution from the lateral regions of the chamber. This produced drying of the fiber in the end regions and tended to reduce the depolarization of axons. Furthermore, when this arrangement was used for barnacle muscle fibers, the drying of the terminal portions of the barnacle muscle fiber tended to reduce the development of contractures during dialysis.

5.2. Special precautions to remove ATP from axoplasm

Although preliminary experiments indicated that dialysis adequately controlled the internal concentrations of sodium and potassium, in initial experiments the sodium outflux appeared insensitive to the presence or absence of ATP in the dialysis fluid. Analysis of the ATP concentration in the dialysis fluid after it had passed through the porous region of the axoplasm indicated that it was not greatly reduced when the axon is dialyzed with ATP free solution. The obvious explanation was that ATP was being generated sufficiently rapidly by the mitochondria to provide a source of ATP in the axoplasm. Longitudinal diffusion of ATP from the end regions which were not dialyzed apparently made a significant contribution to ATP in the central region of the axon. This problem was overcome by treating the sea water bathing the axon with cyanide to block phosphorylation and also by using the porous capillary to *end wash the axoplasm.*

The principle of end washing is indicated by Figure 6 (Mullins and Brinley, 1967). The capillary was stopped in its forward advance when the porous region was still to the left of the selected central region. The axon was then dialyzed with ATP-free non-radioactive media for 10 – 15 minutes to remove ATP from the left-hand end region. The capillary was then advanced beyond the central region and positioned with the porous region towards the right end of the axon and the end washing procedure repeated. The porous capillary was then finally withdrawn and repositioned in the central region. Although this end washing added between 20 – 30 minutes to the preparation, in our hands it proved essential to reduce ATP to levels a few micromolar or less. There may, however, be some species variation in the ability of axoplasm to supply ATP in the presence of ATP free dialysis, because DiPolo and Beauge (1989) were able to achieve very low ATP concentrations without the necessity for such extensive end washing in experiments performed with *Loligo plei.*

5.3. Influx chamber

The approximate theoretical calculations described in Section 4, suggested that influx measurements would not be seriously affected by edge effects due to poor dialysis control at the junction between porous and non-porous capillary. The predictions were generally borne out when sodium and potassium influx were measured (Brinley and Mullins, 1968). Relatively stable baselines of influx were obtained, which showed the expected variations when internal ion or ATP concentrations were changed.

However, the simple technique used for sodium and potassium influx was completely inadequate for calcium influx measurement. Early measurements of calcium influx using the original glass capillaries or a single porous plastic capillary showed that radioactive calcium influx did not reach a steady state level but increased steadily during the course of the long influx measurement. Furthermore, at the end of the experiment, removal of extracellular radioactive calcium had little effect on the amount of tracer collected by the dialysis capillary.

There appeared to be two reasons why calcium influx cannot be measured accurately by simple dialysis, 1) longitudinal calcium diffusion from the unperfused end regions, and 2) calcium sequestration in mitochondrial or other intracellular buffers with slow release into the axoplasm. The problem of longitudinal diffusion from unperfused regions was addressed in two ways. Brinley *et al.* (1975) devised a double capillary technique illustrated in Figure 6; this has already been described in the section on capillaries. DiPolo (1979) improved the technique by providing a more elaborate guard system with continual flushing of the end regions so that there was always a continual circulation of fluid from guard region to the end region. The flow rates and geometry were such that radioactive isotope was strictly confined to the length of axon slightly less than the length of the porous region of the internal capillary. The apparatus is diagrammed in Figure 7.

Although the end effects proved troublesome, probably the most serious experimental obstacle to an accurate measurement of calcium influx was the tremendous buffering power of mitochondria and other intracellular calcium buffers. When the double guard system was used, isotope washout from the interior of the axon was much slower than would have been expected from simple diffusion. Even when EGTA was added in relatively high concentrations in the dialysis solution (1 millimole) (Brinley *et al.*, 1975), this buffering capacity was still not sufficient to remove calcium from the mitochondrial buffer systems, unless the mitochondria were inactivated by combinations of p-trifluoro methoxy carbonyl cyanide phenyl hydrazine (FCCP), cyanide and oligomycin (DiPolo, 1979).

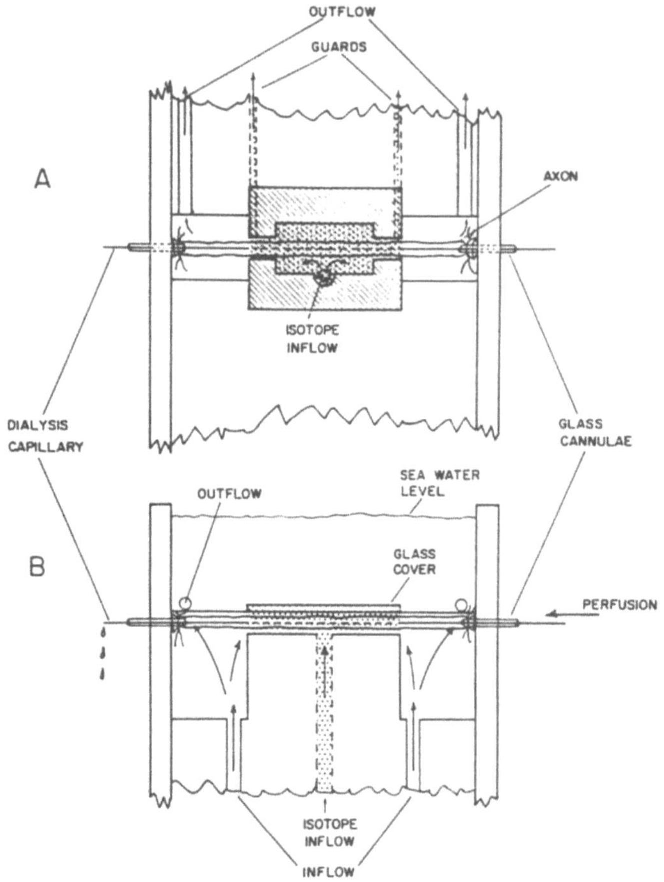

Figure 7. Diagrammatic view of the experimental apparatus. (A) Top view; (B) cross section through the middle of the central slot. The chamber is not drawn to scale and is meant primarily to illustrate the positions of the axon, porous capillary, and chamber compartments. The narrow regions of the chamber in the top view are the *guard* regions. The isotope entering the central compartment (black dots) exited through the guard outflows. The lateral compartments were kept at infinite dilution by the inflow-outflow system. (Figure and Legend from DiPolo, 1979).

6. Hardware for control of internal dialysis

Micromanipulators of the Prior design have been generally used for the movement of end cannulae, and the adjustment of chamber height. Some special requirements were noted in connection with the unconventional nature of internal dialysis itself. These involved the rather long horizontal distances that were necessary to insert a glass capillary into an end cannula and have it penetrate perhaps 1 cm of cannula, the axoplasm which was usually at least 5 cm, followed by another 1 cm of end cannula before the capillary would emerge into air. This requirement for long horizontal distances made it necessary to design a manipulator with at least 10 cm of horizontal travel and it proved convenient to have electric drive of the manipulator since advancing was a slow process. The requirement was met by mounting Prior manipulators on a base with a sliding drive of at least 10 – 15 cm, and with an electric drive.

Another problem involved the need to do *end washing*, that is, the positioning of the porous region somewhat further along the axon than its ultimate position in order to wash out substrates in this region followed by a retreat to a position 3 – 4 mm less than that ultimately desired where again washing was carried out. This procedure was straightforward, but it was noticed early that glass capillaries tended to bond to axoplasm so that subsequent efforts to move the capillary led to a destruction of the axon. This led to the development of a manipulator with an electric drive and one that included a *dither* where the capillary upon reaching its position for end washing was moved backward and forward over a 2 – 3 mm range by continuous movement provided by the electric drive. Such an arrangement led to no binding of the axoplasm to the glass and it was possible to control end washing in a quite satisfactory way.

In addition to manipulators, it was also necessary to provide for a variety of flows by syringe-driven pumps. In the case of internal dialysis fluid flow, this rather critical operation required that very close control over internal fluid flow be held to values of the order of 1 μl/min, a value that would be difficult to achieve with conventional syringe-driven pumps because of the tendency of syringes to leak. Fortunately, Hamilton had come out with teflon-tipped gas tight syringes in the range of 10 – 100 μl and these proved to be ideally suited to delivery of the requisite volume of internal dialysis fluid given Harvard Apparatus Co. pumps operating at a level of about 1 revolution/min (rpm). It was also necessary to provide some mechanism for the rapid control of changes in internal solute so that time was not wasted in moving solutions to the dialysis capillary. This was originally made possible by having a 4-port lucite fitting into which the internal dialysis solution was flowing and another port in which a new solution was allowed to vent to waste. A change could then be made by removing a pin from one solute line and replacing it in another. An improvement was to use teflon valves developed for chromatography where the dead space was of the order of 3 μl and simply make the change in internal solute by moving the valve. A final improvement was to use electrically operated Hamilton valves with similar characteristics which allowed such changes from a central control panel.

External solution flows were initially effected by large (100 ml) syringes driven by a Harvard Apparatus Co. pump. These gave steady flows but were inconvenient since the syringes delivering the flow needed refilling during the experiment (at 1 ml/min the maximum duration of flow was 100 min.). Attempts to use peristaltic pumps were quite unsuccessful owing to the large pressure transients in such pumps until LKB pumps were properly developed. These became a standard for dialysis in a very short time, both for the delivery of external solutions to the slot where fluxes were being measured, and also for the operation of guard systems.

References

Atwater, I., Bezanilla, F., and Rojas, E., 1969, Sodium influxes in internally perfused squid giant axon during voltage clamp, *J. Physiol. (Lond.)* **201**:657–664.

Baker, P. F., Hodgkin, A. L., and Shaw, T. I., 1962, Replacement of the axoplasm of giant nerve fibers with artificial solutions, *J. Physiol. (Lond.)* **164**:330–354.

Baker, P. F., Foster, R. F., Gilbert, D. S., and Shaw, T. I., 1971, Sodium transport by perfused giant axons of *Loligo*, *J. Physiol. (Lond.)* **219**:487–506.

Brinley, F. J., Jr. and L. J. Mullins., 1965, Ion fluxes and transference numbers in squid axons, *J. Neurophysiol.* **28**:526–544.

Brinley, F. J., Jr. and Mullins, L. J., 1967, Sodium extrusion by internally dialyzed squid axons, *J. Gen. Physiol.* **50**:2303–2331.
Brinley, F. J., Jr. and Mullins, L. J., 1968, Sodium fluxes in internally dialyzed squid axons, *J. Gen. Physiol.* **52**(2):181–211.
Brinley, F. J., Jr. and Mullins, L. J., 1974, Effects of membrane potential on sodium and potassium fluxes in squid axons, *Annals New York Academy of Science* **242**:406–433.
Brinley, F. J., Jr., Spangler, S. G., and Mullins, L. J., 1975, Calcium and EDTA fluxes in dialyzed squid axons, *J. Gen. Physiol.* **66**:223–250.
Brinley, F. J., Jr., Scarpa, A., and Tiffert, T., 1977, The concentration of ionized magnesium in barnacle muscle fibres, *J. Physiol.* **266**:545–565.
Caldwell, P. C. and Keynes, R. D., 1960, The permeability of the squid giant axon to radioactive potassium and chloride ions, *J. Physiol. (Lond.)* **154**:177–189.
Canessa-Fischer, M., Zambrano, F., and Rojas, E., 1968, The loss and recovery of the sodium pump in perfused giant axons, *J. Gen. Physiol.* **51**(5, Pt 2), 162s–171s.
Deffner, G. G. J., 1961, The dialyzable free organic constituents of squid blood; a comparison with nerve axoplasm. *Biochim. Biophys. Acta* **47**:378–388.
DeWeer, P., 1968, Restoration of a potassium-requiring sodium pump in squid giant axons poisoned with CN and depleted of arginine, *Nature* **219**:730–731.
DiPolo, R., 1972, Chloride fluxes in isolated dialyzed barnacle muscle fibers, *J. Gen. Physiol.* **60**:471–497.
DiPolo, R., 1977, Characterization of the ATP-dependent calcium efflux in dialyzed squid giant axons, *J. Gen. Physiol.* **69**:795–813.
DiPolo, R., 1979, Calcium influx in internally dialyzed squid giant axons, *J. Gen. Physiol.* **73**:91–113.
DiPolo, R. and Beauge, L. B., 1989, *unpublished.*
DiPolo, R., Bezanilla, F., Caputo, C., and Rojas, H., 1985, Voltage dependence of the Na/Ca exchange in voltage-clamped, dialyzed squid axons. Na-dependent Ca efflux, *J. Gen. Physiol.* **86**:457–478.
Hodgkin, A. L. and Keynes, R. D., 1956, Experiments on the injection of substances into squid giant axons by means of a micro-syringe, *J. Physiol. (Lond.)* **131**:592–616.
Horn, L., 1983, Numerical analysis of the method of internal dialysis of giant axons, *Biophys. J.* **42**:243–254.
Mullins, L. J. and Brinley, F. J., Jr., 1967, Some factors influencing sodium extrusion by internally dialyzed squid axons, *J. Gen. Physiol.* **50**(10):2333–2355.
Mullins, L. J. and Brinley, F. J., Jr., 1969, Potassium fluxes in dialyzed squid axons, *J. Gen. Physiol.* **53**(6):704–740.
Rojas, E. and Canessa-Fischer, M., 1968, Sodium movements in perfused squid giant axons, *J. Gen. Physiol.* **52**:240–257.
Shaw, T. I., 1966, Cation movements in perfused giant axons, *J. Physiol. (Lond.)* **182**:209–216.
Sjodin, R. A. and Mullins, L. J., 1967, Tracer and nontracer potassium fluxes in squid giant axons and the effects of changes in external potassium concentration and membrane potential, *J. Gen. Physiol.* **50**:533–549.

Chapter 9

The Cut-Open Axon Technique

FRANCISCO BEZANILLA and CAROL VANDENBERG

1. Introduction

The giant axon of the squid has been the preparation of choice for the recording of electrical events associated with the opening and closing of the ionic conductances. Its large size allows the introduction of electrodes and the exchange of solutions giving almost complete control of the chemical environment and voltage across the axolemma (Adelman and Gilbert, 1990; Mullins and Brinley, 1990).

The giant axon has provided a wealth of information on the macroscopic ionic currents and gating currents, two of the three electrical expressions of voltage dependent ionic conductances. The third expression, single channel fluctuations, is normally done with the patch clamp technique that requires a clean glass pipette pressed against a clean membrane. This technique has been difficult to apply to the giant axon due to the presence of external connective tissue and the Schwann cell layer. Conti and Neher (1980) introduced a bent pipette inside a perfused axon to record the fluctuations of single potassium currents and this work was the first direct demonstration that the squid potassium current is produced by a large number of discrete channels. However, the technique had a limited frequency response and it was not possible to achieve gigaseals due to the presence of remaining axoplasm. A similar technique was used by Lopez-Barneo *et al.* (1981) to record from a small population of channels in the giant axon.

2. The cut-open axon technique for small population of channels

A different approach was to cut the axon open to have free access to the internal surface of the membrane. The original technique (Llano and Bezanilla, 1980) consisted of isolating a small segment of axon and cutting it open with microscissors over a hole of about 200 µm on a plastic partition separating two compartments. This operation exposed the internal surface which became oriented upwards. The solutions in both compartments could be exchanged and by introducing electrodes it was possible to

F. BEZANILLA • Department of Physiology, Ahmanson Laboratory of Neurobiology and Jerry Lewis Neuromuscular Center, University of California at Los Angeles, Los Angeles, California 90024. C. VANDENBERG • Department of Biological Sciences, University of California at Santa Barbara, Santa Barbara, California, 93106.

voltage clamp the axon segment using a conventional negative feedback voltage clamp circuitry. The seal between the axon and the plastic partition did not exceed 1 Mohm which constituted a short circuit for the normal conductances, eliminating the resting potential and making the distribution of the potential non-uniform in the segment. Even though macroscopic and gating current could be recorded with this technique, the main objective was to isolate small patches of membrane using patch clamp pipettes. The frequency of gigaseals was very low and only a few successful recordings of potassium channels were obtained (Llano and Bezanilla, 1983). Most frequently, seals of tens of megohms were obtained and fluctuation studies of small population of channels were undertaken (Llano and Bezanilla, 1984) using the technique introduced by Sigworth (1980).

Another cut-open axon technique was introduced by Bekkers *et al.* (1986) and described in detail by Forster and Greeff (1988). In their technique the axonal membrane was sandwiched between the two apices of cones machined in plexiglass. This method allowed recording of fluctuations of a large number of channels (ca. 10^6), but was not intended to be used with patch pipettes.

3. The cut-open axon technique for single channel recording

The original cut-open axon technique demonstrated that it was possible to slit the axon open and still have functional channels. During the procedure of cutting the axon open, care was taken not to expose the internal side of the membrane to external solutions such as sea water. It was found later by Levis *et al.* (1984) that actually, the exposure of the internal face of the axonal membrane to sea water did not eliminate all the functional channels and made possible the recording of several thousand to millions of channels with large glass pipettes pressed against an open piece of axon.

This technique gave origin to a method to record single channels from the cut-open axon (Llano and Bezanilla, 1985). The exposure to sea water was beneficial in obtaining gigaohm seals with excellent reproducibility by activating proteases present in the axoplasm which helped in cleaning the internal surface of the axon membrane. This modified cut-open axon technique was first used to record potassium channels (Llano and Bezanilla, 1985; Llano *et al.*, 1988) and then single sodium channels (Bezanilla, 1987), macroscopic currents, and even gating currents with patch pipettes (Vandenberg and Bezanilla, 1988). A similar method has also been applied for the recording of single sodium channels from the cut-open *Myxicola* giant axon (Schauf, 1987).

4. Method and results

4.1. Experimental set-up

The set-up is built around a modified inverted microscope. Some of the modifications are standard for patch clamp recording, such as a movable stage independent from the microscope body and a strong base for the motor driven micromanipulator used to approach the patch pipette. The stage is modified to set an

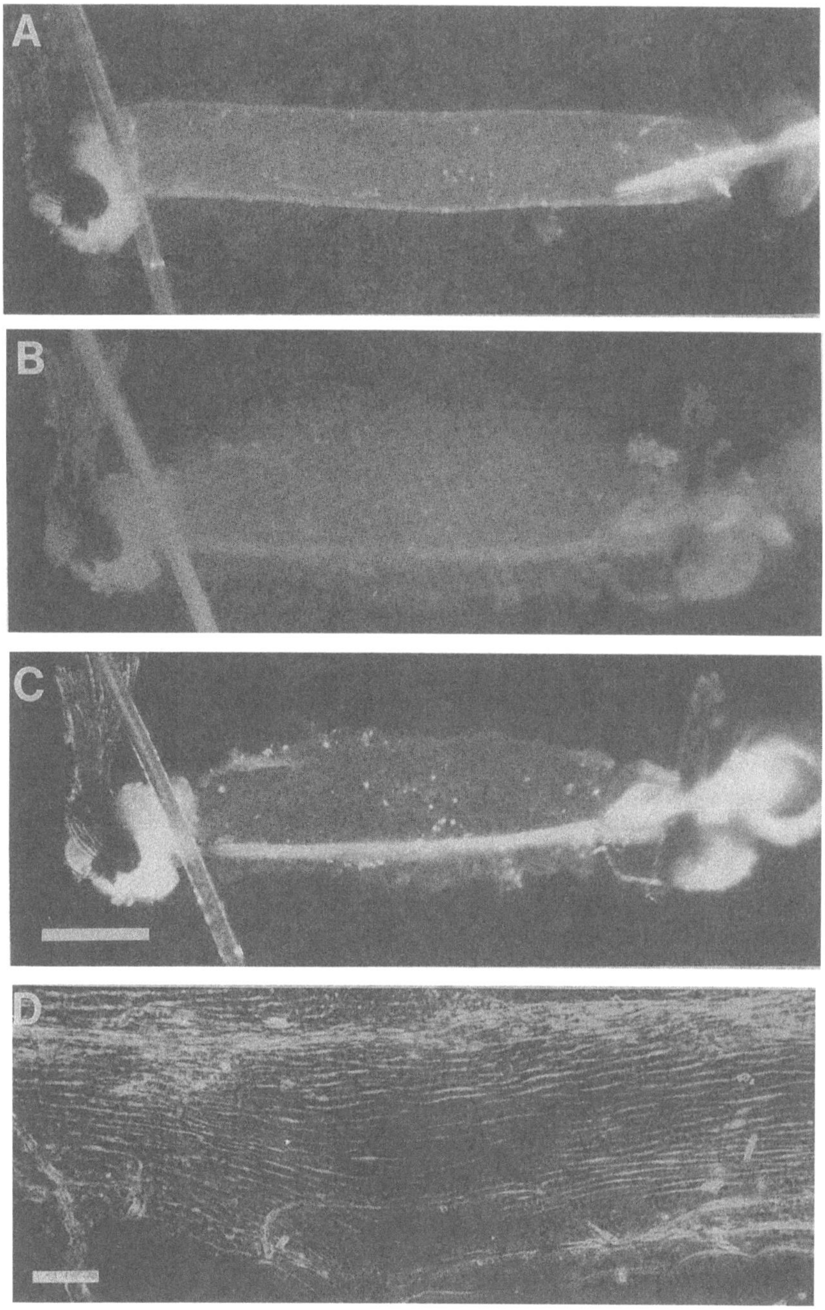

Figure 1. Squid giant axon segment in preparation for patch recording. A) Segment of axon pinned to dish, prior to cutting; B) Axon has been cut open in artificial sea water, and axoplasm is dissolving; C) Axoplasm has been removed by suction, Bar: 500 μm; D) Phase contrast image

of cut-open axon showing regular stripes corresponding to membrane ridges and grooves. Bar: 100 μm.

aluminum block that is in contact with peltier elements connected to a feedback circuit to control the temperature. The temperature is measured in the block for fast and stable control and the peltier elements dissipate heat to the massive stage while cooling the aluminum block. The block has a central cutout where the chamber is press fitted using heat conductive grease.

The chamber is made of very thin plexiglass with a glass coverslip as its bottom (see Fig. 1 of Llano *et al.*, 1988). Two pieces of glass are placed on the inside of the chamber to improve heat conduction and reduce the actual volume of solution to a narrow trough for easy solution exchange which is critical in axoplasm removal (see below). The bottom of the trough is covered with a thin layer of Sylgard™ to allow the positioning of *minutien* pins to hold the cut-open axon.

The condenser of the inverted microscope can be swung out of the way to allow the positioning of a normal dissecting microscope required to perform the cutting operation.

4.2. Experimental procedure

A cleaned segment of giant axon (3 – 5 mm length, 300 – 600 μm diameter) from the squid *Loligo pealei* or *Loligo opalescens* was tied at both ends, and the piece of axon pinned to the bottom of the Sylgard™-coated chamber (Fig. 1A). The axon was bathed in an artificial sea water consisting of 440 mM NaCl, 10mM KCl, 50 mM $MgCl_2$, 10 mM $CaCl_2$, 10 mM N-2-hydroxyethylpiperazine-N'-2-ethanesulfonic acid (HEPES), pH 7.6 at room temperature. Under the dissection microscope and with microdissecting scissors the axon was cut open by a partial cross-sectional cut near each end of the segment, and a longitudinal cut along the upper surface of the axon (Fig. 1B). Incubation in the bathing solution for about 3 minutes allowed the axoplasm to soften and begin to dissolve, forming a cloud over the internal surface of the cut-open axon (Fig. 1B). The axoplasm was removed by flowing solution into the bathing chamber while withdrawing the axoplasm by suction through a fine needle. Care was taken to avoid touching the axonal membrane while vacuuming the axoplasm. The bathing solution was then exchanged to a low divalent ion solution (540 mM NaCl, 10 mM HEPES, pH 7.6, 1000 mOsm) for the recording of sodium channels and cooled to 5 °C. This procedure results in a clean internal surface of the membrane suitable for patch clamping (Fig. 1C). Using an inverted microscope with Hoffman modulation optics, a healthy preparation showed regularly spaced (20 μm) longitudinal grooves and ridges throughout the membrane segment (Fig. 1D).

In a variation of the cut-open axon technique, advantage can be taken of the native enzymes in the axoplasm to alter the resting state of the channels, for example by phosphorylation or dephosphorylation (Perozo *et al.*, 1989). This technique was used to examine single potassium channels under different phosphorylating conditions (Vandenberg *et al.*, 1989). Axons can be predialyzed in various solutions prior to cutting them open, or they can be cut open and preincubated while maintaining the axoplasm intact by using solutions lacking calcium and chloride ions. The axoplasm is then removed following incubation in artificial seawater as above.

Recordings were made using standard patch clamp equipment, and patch pipettes that were pulled from Corning 7052 glass (Garner Glass Co., Claremont, CA) to have

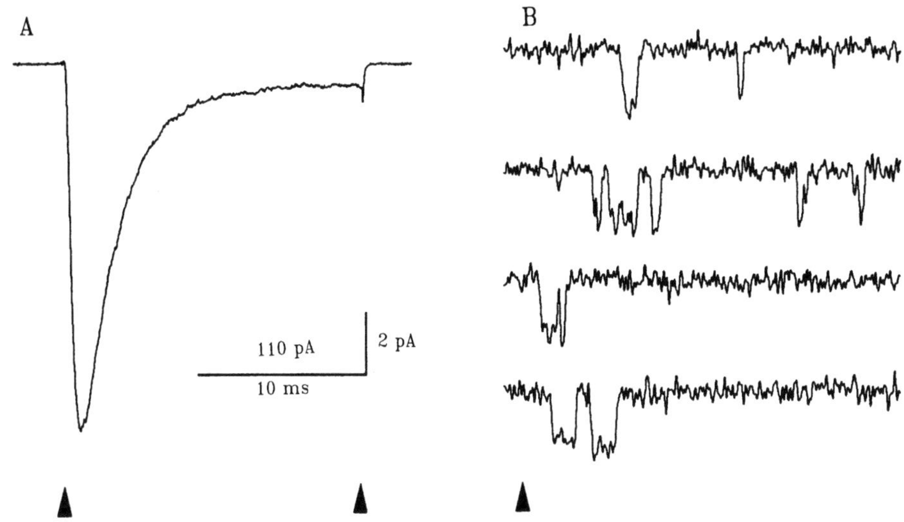

Figure 2. A) Macroscopic sodium current recorded in response to an 18 ms voltage pulse to −8 mV, preceded by a prepulse of 50 ms to −108 mV, from a holding potential of −98 mV. Linear leakage and capacitative currents have been subtracted using the P/−4 procedure (Bezanilla and Armstrong, 1977; Adelman and Gilbert, 1990). Filter was 10 kilocycles/second (kHz), temperature 5 °C, average of 4 records. B) Single-channel sodium current elicited by a 22 ms pulse to −48 mV from a holding potential of −108 mV. Filter was 3 kHz, temperature 5 °C.

apertures of about 1 μm diameter, and filled with 125 mM Cs glutamate, 20 mM CsF, 40 mM Na glutamate, 10 mM NaCl, 5 mM Cs ethylene glycol-bis(2-amino ethyl ether) N,N,N',N'-tetraacetic acid (Cs EGTA), 470 mM sucrose, and 10 mM HEPES, pH 7.3 (950 mOsm), with resistances of 5 – 10 Mohm. Gigaseals were made to the internal surface of the cut-open axon membrane using an inverted microscope with 400 × magnification. The frequency of gigaseal formation was >80%, and stable patches could be maintained for about 2 hours. Patches were allowed to remain attached to the axonal membrane to avoid vesicle formation on the patch pipet, or were excised from the membrane after a delay of about 15 minutes.

4.3. Experimental results

The number of sodium channels in a patch varied substantially, from one channel to several thousand. This variation was present within a cut-open axon segment, with localized regions of very high or low densities.

The channel distribution did not appear to correlate with anatomical features such as the proximity to axonal branches or the location of the patch relative to the longitudinal grooves and ridges in the membrane. The reasons for this variability are not understood, and could represent natural patchiness in channel distribution or may be the result of localized loss of channel activity during axoplasm removal. High densities of channels for gating current or macroscopic ionic current measurement were obtained by forming patches in high current density areas, and by using large diameter patch pipettes. Although the patch lifetime was decreased, 5 μm diameter pipettes with resistances of 1 – 3 Mohm were routinely used for macroscopic measurements. It was thus possible to record single-channel, macroscopic ionic, and gating current measurements from a single preparation.

Figure 2 shows typical sodium channel currents obtained from the cut-open squid giant axon. In the absence of divalent cations which block the sodium channel (Bezanilla, 1987; Vandenberg and Bezanilla, 1988), single-channel currents can be well resolved (Fig. 2B). Additional resolution is possible by increasing the sodium concentration to several molar (Correa *et al.*, 1989). Perhaps because the axon is cut open, the membrane is unusually stable to changes in osmotic strength, and will tolerate solutions of 50 mM to 4 M NaCl as well as osmotic gradients of about 40 fold across the patch membrane (Correa *et al.*, 1989).

Macroscopic sodium currents are shown in Fig. 2A. Peak sodium currents of up to 2.5 nA have been recorded from a membrane patch. This current has been correlated with the current obtained using the axial wire voltage clamp (Bezanilla, 1987; Vandenberg and Bezanilla, 1989). The small membrane area, compared to a typical mammalian cell with similar current amplitude, gives excellent cancellation of the transient capacitative currents and good voltage control without series resistance problems. The use of patch recording also allows macroscopic currents to be recorded in solutions, such as those that contain no divalent ions, that are not possible with axial wire voltage clamp of the axon.

AKNOWLEDGEMENTS. The cut-open axon technique for single channel recording was developed thanks to the enthusiasm and perseverance of Dr. Isabel Llano. She also contributed the picture in Fig 1D. Mr. E. Perozo contributed with Figs. 1A, 1B, and 1C. This work was supported by USPHS grant GM30376.

References

Adelman, W. J., Jr. and Gilbert, D. L., 1990, Electrophysiology and biophysics of the squid axon, *this volume.*

Bekkers, J. M., Greeff, N. G., and Keynes. R. D., 1986, The conductance and density of sodium channels in the cut-open squid giant axon, *J. Physiol.* **377**:463–486.

Bezanilla, F., 1987, Single sodium channels from the squid giant axon, *Biophys. J.* **52**:1087–1090.

Bezanilla, F. and Armstrong, C. M., 1977, Inactivation of the sodium channel. I. Sodium current experiments, *J. Gen. Physiol.* **70**:549–566.

Conti, F. and Neher, E., 1980, Single channel recording of K currents in squid axons, *Nature* **285**:140–143.

Correa, A. M., Latorre, R., and Bezanilla. F., 1989, Na-dependence and temperature effects on BTX-treated sodium channels in the squid giant axon, *Biophys. J.* **55**:403a (Abstr.).

Forster, I. C. and Greeff, N. G., 1988, Technical aspects of voltage-clamping the cut-open squid giant axon, *J. Neurosci. Methods* **26**:151–168.

Levis, R. A., Bezanilla, F., and Torres, R. M., 1984, Estimate of the squid axon sodium channel conductance with improved frequency response, *Biophys. J.* **45**:11a (Abstr.).

Llano, I. and Bezanilla F., 1980, Current recorded from a cut-open giant axon under voltage clamp, *Proc. Nat. Acad. Sci.* **77**:7484–7486.

Llano, I. and Bezanilla F., 1983, Bursting activity of potassium channels in the cut-open axon, *Biophys. J.* **41**:38a. (Abstr.)

Llano, I. and Bezanilla F., 1984, Analysis of sodium current fluctuations in the cut-open squid giant axon, *J. Gen. Physiol.* **83**:133–142.
Llano, I. and Bezanilla F., 1985, Two types of potassium channels in the cut-open squid giant axon, *Biophys. J.* **47**:221a. (Abstr.).
Llano, I., Webb, C. K., and Bezanilla, F., 1988, Potassium conductance of the squid giant axon. Single-channel studies, *J. Gen. Physiol.* **92**:179–196.
Lopez-Barneo, J., Matteson, D. R., and Armstrong C. M., 1981, Currents recorded through small areas of squid axon membrane with an internal virtual ground voltage clamp, *Biophys. J.* **36**:811–816.
Mullins, L. J. and Brinley, F. J., Jr., 1990, Internal dialysis in the squid axon, *this volume.*
Perozo, E., Bezanilla, F., and Dipolo R., 1989, Modulation of K channels in dialyzed squid axons: ATP mediated phosphorylation, *J. Gen. Physiol.* **93**:1195–1218.
Schauf, C. L., 1987, Properties of single Na channels in cut-open Myxicola giant axons, *Can. J. Physiol. Pharm.* **65**:568–573.
Sigworth, F. J., 1980, The variance of sodium current fluctuations at the node of Ranvier, *J. Physiol.* **307**:97–129.
Vandenberg, C. A. and Bezanilla, F., 1988, Single-channel, macroscopic and gating currents from Na channels in squid giant axon, *Biophys. J.* **53**:226a (Abstr.).
Vandenberg, C. A. and Bezanilla, F., 1989, *in prep.*
Vandenberg, C., Perozo, E., and Bezanilla, F., 1989, ATP modulation of the K current in squid giant axon: a single channel study, *Biophys. J.* **55**:586a (Abstr.).

Chapter 10

Optical Measurements on Squid Axons

LAWRENCE B. COHEN, DAVID LANDOWNE, and BRIAN M. SALZBERG

1. Introduction

A combination of two factors makes the squid giant axon a favorable preparation for optical measurements: there is a relatively large membrane area and one can obtain good control of membrane potential. The large membrane area makes for relatively large signal-to-noise ratios in the optical measurements and the ability to control the membrane potential allows one to investigate the physiological origins of the signals in some detail. Using giant axons it is relatively straight-forward to determine if a given optical signal depends on changes in membrane potential, or the membrane currents, or the increases in membrane permeability (Cohen *et al.*, 1968, 1972b).

Optical measurements on squid axons have been made with two different scientific aims. When optical signals were first discovered the main interest in them was the possibility that they would provide information about the structural events that occur in the axon during the action potential. Somewhat later it was recognized that the potential-dependent optical signals might be useful for monitoring changes in membrane potential in a variety of preparations and that the squid giant axon was a useful preparation for screening this kind of optical signal. In these experiments, the aim was to find larger signals with reduced pharmacologic effects and photodynamic damage.

The fractional changes in light intensity, $\Delta I/I$, available from giant axon optical signals are in the range of 10^{-6} to 10^{-2} per trial. Care must be taken to minimize both extraneous and intrinsic sources of noise and, for the smaller signals, averaging must be used. A description of methods to optimize the signal-to-noise ratio in optical measurements has appeared (Cohen and Lesher, 1986).

The use of internal silver-silver chloride current electrodes in voltage clamp experiments where extensive averaging is necessary can lead to silver poisoning of the axons. One way around this difficulty is to use platinized platinum electrodes. These electrodes have the disadvantage that they cannot be used to measure direct current (DC) potentials and thus the resting potential is unknown. Better control of the experiment, including a DC measurement of membrane potential, is provided by the use

L. B. COHEN • Department of Cellular and Molecular Physiology, Yale University School of Medicine, 333 Cedar Street, New Haven, Connecticut 06510. D. LANDOWNE • Department of Physiology and Biophysics, University of Miami School of Medicine, Miami, Florida 33101. B. M. SALZBERG • Department of Physiology, University of Pennsylvania School of Medicine, Philadelphia, Pennsylvania 19104.

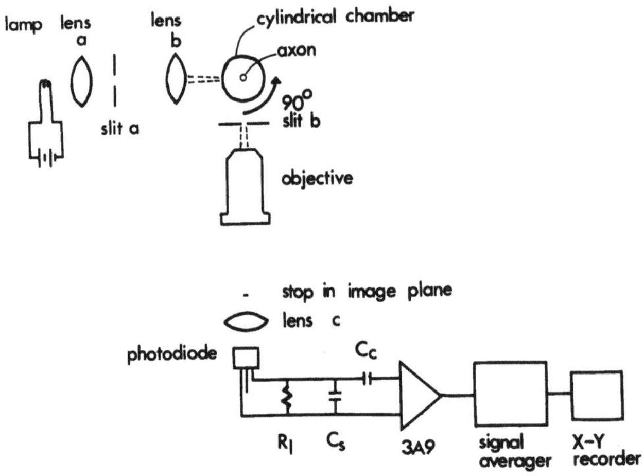

Figure 1. Schematic diagram of the apparatus for measuring light scattering, shown in the position for measuring 90° scattering. Heat filters were placed between lens *a* and slit *a*. Modified from Cohen *et al.* (1972a).

of silver-silver chloride inside a KCl microelectrode for measuring potential and platinized platinum or carbon fiber electrodes for passing current.

Optical measurements on giant axons begin either with cleaned intact axons or with perfused axons. Internal perfusion allows control of the internal ionic milieu. See Adelman and Gilbert (1990) for methods of dissection, cleaning, and perfusion.

2. Optical studies of structural changes in axons

2.1. Light scattering

Because the light scattering of an axon (or any material) is dependent on the angle at which the scattering is measured, the apparatus for scattering measurements from axons needs to have the flexibility to allow measurements at several angles. Figure 1 is a schematic diagram of the apparatus used for measurements of light scattering from giant axons during voltage clamp steps. The light from a tungsten filament bulb was collected by the lens *a*, passed through the slit *a* and focused by the lens *b* onto a 4 mm spot which illuminated a 4 mm length of axon. The axon was perpendicular to the plane of observation (the plane of the paper in the figure) in the center of a cylindrical chamber. A 10 × microscope objective formed an enlarged image of the axon with the scattered light. This light passed through slit *b*. Lens *c* collected the light from the axon onto a silicon photodiode. The photocurrent was measured as the voltage drop across the resistance R_l. The resistor and the two capacitors formed a bandpass filter to reduce the system response to the frequencies of interest.

Slit *a* was used to make the illuminating light more parallel and slit *b* was used to limit the scattering angles over which the measurements were made. The dashed lines in the figure indicate the actual convergence of the incident light and the actual divergence of the scattered light admitted to the detection system in experiments carried out on *Loligo pealei* (Cohen *et al.*, 1972a, 1972b). When the angle between the most

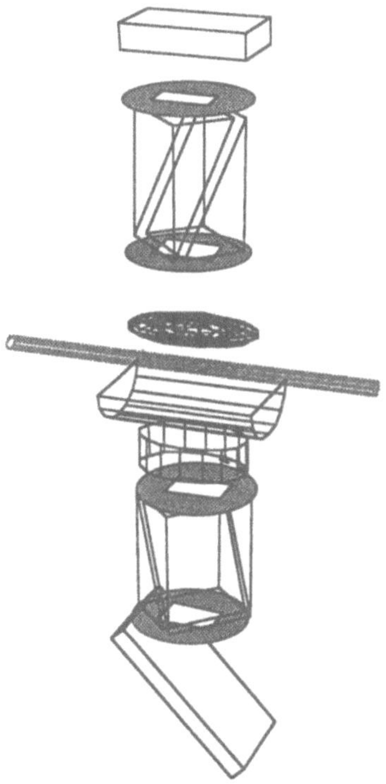

Figure 2. Schematic drawing of the apparatus for measuring axon birefringence.

divergent rays of the incident light and the angle between the most divergent rays of scattered light were added, their sum was between 9° and 14°. No steps were taken to limit the divergence perpendicular to the plane of observation. This divergence was approximately 60°. With this level of angular resolution, it was possible to resolve three kinds of current-dependent light scattering signals as a function of scattering angle (Cohen *et al.*, 1972b). Better angular resolution might allow the detection of additional light scattering signals.

2.2 Birefringence

The basic birefringence experiment is to place an axon between crossed polarizers at 45° to their axes, send a light beam through the system and measure the amount of light coming through the second polarizer (Figure 2).

The major design consideration in these studies is to get the largest number of useful photons detected at the photodetector. In a well designed system, quantal fluctuation of the light is the major noise source so more photons means a greater signal-to-noise ratio. For birefringence experiments this means using calcite crystals for the polarizer and analyzer. (Polaroid sheet works by absorbing the undesired light; at high intensity the plastic base will melt. It also absorbs some of the light polarized in the desired direction. Polaroid sheet does not polarize blue light, requiring additional filters.) The barriers or stops on each end of the calcite crystal are necessary to prevent scattered unpolarized light from the edges of the crystal from reaching the photodetector.

The condenser, the cylindrical lens just below the axon, is a trade-off between more light striking the axon and depolarization of the beam by the curved surface which raises the background. For visualization, Köhler illumination is the preferred arrangement. However, critical illumination puts more light onto the axon and this improves the signal-to-noise ratio. Since the analyzer removes most of the light which has not interacted with the nerve, the situation is not as bad as using critical illumination for ordinary light microscopy. A cylindrical lens will focus the beam onto a line source. It should be focused above the nerve so the axon is in the converging portion of the beam. The convergence keeps the beam off the walls of the chamber which would otherwise scatter depolarized light and raise the background.

The compensator allows measurement of the optical retardation of the axon which permits converting the changes in light intensity into changes in retardation. Quasi-monochromatic light is preferable for measuring the resting retardation. For measuring the changes in retardation it is better to use no filters to have as much light as possible. Because the birefringence change is the same in the visible and the near infra-red (Landowne, 1985), the signals are improved by not using a heat filter.

Biophysical studies are easier when the axon is chilled because the channels react more slowly, allowing more time to collect the light. This brings the additional problem of fogging the windows in the chamber. Using oil between the lower window and the condenser and surrounding the condenser and a few mm column of air with cold metal will prevent condensation at these surfaces. For the upper surface a coverslip (not shown) is used. A slow stream of chilled nitrogen gas flowing across the upper coverslip prevents fogging.

2.3 Optical activity

A change in optical activity (also called optical rotation) associated with nerve impulses has recently been described in crustacean nerve bundles and reported (in an abstract) for squid axons (Watanabe, 1985, 1987). Optical activity can be thought of as circular birefringence; that is, the preparation has two different indices of refraction for right- and left-circularly polarized light. Optical rotatory dispersion (ORD or optical rotation as a function of wavelength) is related to circular dichroism (CD) by Kramers-Kronig transforms. There is a similar relationship between linear birefringence and linear dichroism. The advantage of measuring optical rotation is that it can be done at visible wavelengths where the nerve is not strongly absorbing and therefore not severely affected by the light beam.

Optical rotation is measured with an apparatus similar to that shown in Figure 2 except a photoelastic modulator (PEM) replaces the compensator. The PEM is a crystal compensator whose retardation is varied sinusoidally by piezo-electric compression. In Watanabe's experiment, the output of the PEM alternated between plane polarized light at +45° and at –45° at 50 cycles/sec (Hz). A source of birefringence will rotate both of these beams towards its slow axis, and a source of optical rotation will rotate both beams in the same angular direction. The analyzer passes only the component of these beams which is parallel to the nerve. In the birefrigence case, these will be equal; in the rotation case they will be unequal. The output from the photodetector was high-pass filtered to remove any variation due to changes in absorption and scattering, and then led to a lock-in amplifiier synchronized with the PEM driver. Changes in output of the lock-in amplifier are proportional to changes in optical rotation. Careful alignment of the nerve was necessary to remove the larger birefringence signal. Quasi-

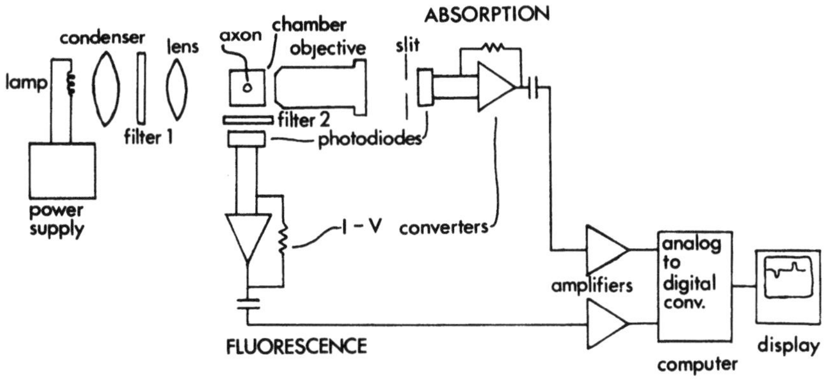

Figure 3. Schematic diagram of the apparatus used for simultaneous measurements of absorption and fluorescence of a squid axon. Filter 1 was an interference filter with a width at half-height of 30 nm. Heat filters were placed between the condenser and filter 1. The barrier filters, filter 2, were obtained from Schott Optical Glass. In absorption, an image of the axon is formed by a 10 × 0.22 numerical aperature (N. A.) objective and a slit in the image plane was used to block light that did not pass through the axon. The field of view of the objective was only 1.5 mm in diameter so that light from only this length reaches the photodetector. The coupling capacitors, C, blocked the DC and low frequency components of the optical signal; to measure the resting light intensity the capacitors were removed from the circuit. Modified from Ross *et al.* (1977).

monochromatic light must be used because the PEP will only produce the two plane polarized beams at a particular selectable wavelength.

3. Screening for larger optical signals for monitoring activity

Most of the effort devoted to finding better signals for use in optical monitoring has involved testing new dyes that transduce changes in membrane potential into changes in light intensity. At present the best dye signals have signal-to-noise ratios that are about 100 times larger than the largest intrinsic optical signals. Many hundreds of dyes have been tested (Cohen *et al.*, 1974; Ross *et al.*, 1977; Gupta *et al.*, 1981).

There are both advantages and disadvantages to using squid giant axons as a screening preparation for larger signals. The advantages include ease of dissection, reproducibility of the preparation, and the fact that membrane potential can be easily measured and controlled. In addition, it is simple to make simultaneous absorption and fluorescence measurements. Finally, the relatively large membrane area means that the signal-to-noise ratio is relatively large. Thus it is not difficult to measure signals that are about 1000 times smaller than the largest available signals. This large signal-to-noise ratio has meant that little effort has gone into improving the apparatus. For example, the use of cylindrical lenses would increase the illumination and collection efficiencies and thus the signal-to-noise ratio but this modification to the apparatus has not been made.

The major disadvantage of the giant axon as a screening preparation is that it is known not to be completely representative of other preparations. There is now considerable experience showing that it will be easier to obtain relatively large signals on giant axons than on other preparations of interest. For example, on squid axons, about 20 different merocyanine-rhodanine analogues gave signal-to-noise ratios within a factor of two of the maximum but, on abdominal ganglia from adult *Aplysia* (> 100g), only a few of those 20 dyes would penetrate through the connective tissue to reach the neuron cell bodies (Zecevic *et al.*, 1989). Similarly, in mammalian cortex, the dyes with the largest signals are considerably less hydrophobic than the dyes which had the largest giant axon signals (Orbach *et al.*, 1985).

Figure 3 illustrates the apparatus for simultaneous measurement of absorption and fluorescence of a voltage clamped axon. If crossed polarizers are placed in the absorption path before and after the axon, then dye related birefringence signals (Gupta *et al.*, 1981) can be measured. Several components of this apparatus are strongly wavelength dependent. The output of the light source and the quantum efficiency of the photodetector both increase monotonically between 350 and 900 nm. Both effects are most steeply wavelength dependent between 350 and 550 nm. These wavelength effects mean that the signal-to-noise ratio for a constant fractional change also increases as a function of wavelength and thus the squid apparatus, and any other apparatus using similar components, is biased toward dyes which absorb and emit at longer wavelengths.

4. Optical determination of the series resistance in *Loligo*

Measurements of extrinsic optical changes (absorption, fluorescence, or birefringence) established (Cohen *et al.*, 1974; Ross *et al.*, 1977; Gupta *et al.*, 1981) that an optical measurement of membrane potential is, to a limited extent, equivalent to an electrical measurement. Under certain conditions, an optical measurement is actually superior to an electrical measurement, because the voltage probe, a dye molecule, has a dimension on the order of 10 Å, and actually resides in the membrane.

Because there is a small resistance (R_{ser}), electrically in series with the membrane capacitance in *Loligo*, arising from the narrow Schwann cell cleft, the true transmembrane potential, V_m, will differ from the value, V_e, recorded between the voltage measuring electrodes, whatever their construction, by an amount $I_m R_{ser}$. Hodgkin *et al.* (1952) introduced the procedure known as compensated feedback, in which a voltage proportional to the membrane current is added at the summing junction of the control amplifier. The proportionality constant depends upon the value of the series resistance, which must be determined accurately, since incorrect series resistance compensation will generate clamp currents that do not reflect the response of the membrane to step changes in voltage, and the inferred kinetics for the membrane conductance changes will be in error. Bezanilla and Salzberg (1983) exploited the fact that potentiometric probes change their optical properties in microseconds (see below) in response to changes in membrane potential, while they are insensitive to changes in membrane conductance or current (Cohen and Salzberg, 1978; Salzberg, 1983). This method has the advantage that it is not essentially an electrical measurement at all, and provides a completely independent determination of the series resistance based solely on Ohm's law and the linearity of the optical response. This is important because, as J. C. Maxwell (Maxwell, 1865; pp. 493–497) clearly implied, it is impossible, in principle, to determine the value of a resistance in series with a lossy dielectric by electrical measurements made on the system as a whole. This difficulty arises from

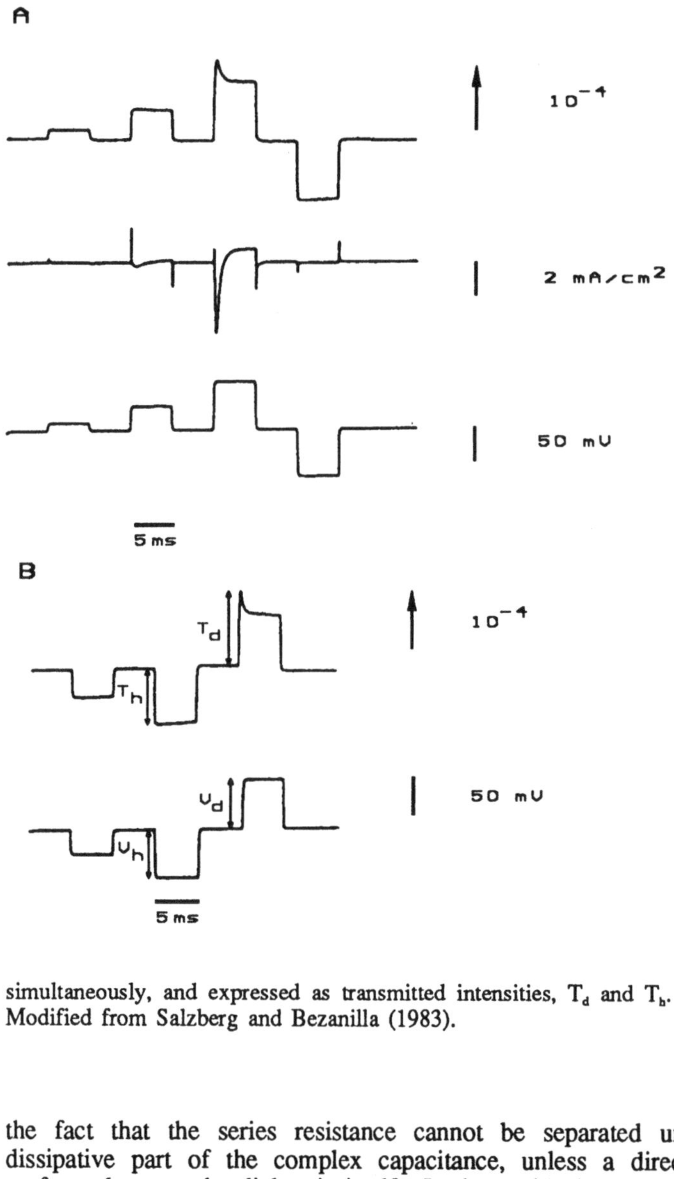

Figure 4. Optical determination of the series resistance in *Loligo*. A. An experiment carried in artificial sea water (ASW) on an axon stained with 10 μg/ml WW 375 for 10 minutes. The upper trace shows the transmitted intensity changes resulting from the imposition of the voltage clamp command steps shown in the bottom trace, which were accompanied by the membrane currents recorded simultaneously in the middle trace. Note that the second depolarizing pulse, in which a non-linear current was barely activated, resulted in a true transmembrane potential change (top trace), which was already contaminated by the uncompensated series resistance. B. Illustration of the method used to determine the value of the series resistance optically. The lower trace shows the depolarizing and hyperpolarizing steps of magnitude V_d and V_h, respectively. The upper trace shows the resulting changes in extrinsic absorption, recorded simultaneously, and expressed as transmitted intensities, T_d and T_h. See equation (3) in text. Modified from Salzberg and Bezanilla (1983).

the fact that the series resistance cannot be separated unambiguously from the dissipative part of the complex capacitance, unless a direct measurement can be performed across the dielectric itself. In the squid giant axon, this would require that one voltage measuring electrode be located in the axoplasm, immediately against the axolemma, with the other electrode in the 150 Å wide periaxonal space, a highly unlikely prospect.

The optical determination of the series resistance in *Loligo* is illustrated in Figure 4. In Panel A, the imposed voltage (lower trace), the membrane current (middle trace), and the transmitted intensity were recorded from a stained axon with 10 μg/ml of a mercyanine-rhodamine dye for 10 min in which the sodium and potassium currents were activated, and result in significant IR (current times resistance) drops across the series resistance. This particular dye is 5-[(1-γ-triethylammonium sulfopropyl-4(1H)-

quinolylidene)-2-butenylidene]-3-ethylrhodamine (WW 375) which was termed dye XVII by Ross *et al.* (1977). During the two smaller depolarizations, the optical record closely resembled the imposed voltage. However, during the third clamp pulse, the optical signal exhibited an extra component having the shape of the inverted membrane current. This is the expected result if the control voltage is imposed across the membrane plus the series resistance and the dye is monitoring the *true* membrane potential, since a part of the clamp voltage proportional to the membrane current drops across the series resistance, as shown in equation (1).

$$V_m = V_c - I_m R_{ser} \tag{1}$$

An approximate value of the series resistance could be determined by varying the feedback compensation until the optical signals during symmetric hyperpolarizing and depolarizing clamp steps had the same square shape.

A more accurate method for determining the value of the series resistance, and the one relied upon by Salzberg and Bezanilla (1983) is illustrated in Fig. 4B. The lower trace shows the depolarizing and hyperpolarizing potential steps of magnitude V_d and V_h, respectively. The upper trace shows the resulting changes in extrinsic absorption, expressed as changes in transmitted intensities, T_d and T_h. Then

$$\frac{T_d}{T_h} = \frac{V_d + I_m R_{ser}}{V_h} \tag{2}$$

Since I_m is measured independently, together with the voltages and transmitted intensities,

$$R_{ser} = \frac{\frac{V_h T_d}{T_h} - V_d}{I_m} \tag{3}$$

yields the value of the total series resistance directly, in ohms per square centimeter. Salzberg and Bezanilla, after making corrections for chamber geometry (Salzberg and Bezanilla, 1983) measured a value of the series resistance in artificial seawater of 3.78 ± 0.95 ohm-cm^2, with the contribution of the Schwann layer estimated to be 2.57 ± 0.89 ohm-cm^2.

5. Fast measurements of potentiometric probe response

Elucidation of the mechanism by which potentiometric probes transduce voltage changes into optical signals may be aided by measured limits on the kinetics of the dye's responses to step changes in membrane potential. For example, electrochromic effects, involving the electronic structure of the dye molecules are expected to be faster than signals that depend upon changes in probe orientation followed by changes in the aggregation state of the dye.

Optical measurements have been made (Salzberg, Bezanilla, and Obaid, 1989) with an apparatus that included very fast voltage clamp circuitry (voltage amplifier settled in less than 1 μsec; full series resistance compensation) modified from that described by Bezanilla *et al.* (1982), and a photodetector circuit with a response time constant of approximately 200 ns. This experimental arrangement permitted a time resolution of 1.2 μsec per point, and, within these limits imposed by the electronics, the response kinetics of the three dyes studied followed the membrane potential without measurable lag (Salzberg, Bezanilla, and Obaid, 1989; Loew *et al.*, 1985).

References

Adelman, W. J., Jr. and Gilbert, D. L., 1990, Electrophysiology and biophysics of the squid axon, *this volume.*

Bezanilla, F., Taylor, R. E., and Fernandez, J. M., 1982, Distribution and kinetics of membrane dielectric depolarization. I. Long-term inactivation of gating currents, *J. Gen. Physiol.* **79**:21–40.

Cohen, L. B. and Lesher, S., 1986, Optical monitoring of membrane potential: methods of multisite optical measurement, in: *Optical Methods in Cell Physiology*, (P. DeWeer and B. M. Salzberg, eds.), *Soc. Gen. Physiol. Ser.* **40**:71–99.

Cohen, L. B. and Salzberg, B. M., 1978, Optical measurement of membrane potential, *Rev. Physiol. Biochem. Pharmacol.* **83**:35–88.

Cohen, L. B., Keynes, R. D., and Hille, B. 1968, Light scattering and birefringence changes during nerve activity, *Nature* **218**:438–441.

Cohen, L. B., Keynes, R. D., and Landowne, D., 1972a, Changes in light scattering that accompany the action potential in squid giant axons: potential-dependent components, *J. Physiol. (London)* **224**:701–725.

Cohen, L. B., Keynes, R. D., and Landowne, D., 1972b, Changes in axon light scattering that accompany the action potential: Current-dependent components, *J. Physiol. (London)* **224**:727–752.

Cohen, L. B., Salzberg, B. M., Davila, H. V., Ross, W. N., Landowne, D., Waggoner, A. S., and Wang, C. H., 1974, Changes in axon fluorescence during activity: molecular probes of membrane potential, *J. Membr. Biol.* **19**:1–36.

Gupta, R. K., Salzberg, B. M., Grinvald, A., Cohen, L. B., Kamino, K., Lesher, S., Boyle, M. B., Waggoner, A. S., and Wang, C. H., 1981, Improvements in optical methods for measuring rapid changes in membrane potential, *J. Membr. Biol.* **58**:123–137.

Hodgkin, A. L., Huxley, A. F., and Katz, B., 1952, Measurement of current-voltage relations in the membrane of the giant axon of *Loligo*, *J. Physiol.* **116**:424–448.

Landowne, D., 1985, Molecular motion underlying activation and inactivation of sodium channels in squid giant axon, *J. Membr. Biol.* **88**:173–185.

Loew, L. M., Cohen, L. B., Salzberg, B. M., Obaid, A. L., and Bezanilla, F., 1985, Charge-shift probes of membrane potential: Characterization of aminostyrylpyridinium dyes on the squid giant axon, *Biophys. J.* **47**:71–77.

Maxwell, J. C., 1865, A dynamical theory of the electromagnetic field, *Phil. Tran. Roy. Soc. London* **155**:459–512.

Orbach, H. S., Cohen, L. B., and Grinvald, A., 1985, Optical mapping of electrical activity in rat somatosensory and visual cortex, *J. Neuroscience*, **5**:1886–1895.

Ross, W. N., Salzberg, B. M., Cohen, L. B., Grinvald, A., Davila, H. V., Waggoner, A. S., and Wang, C. H., 1977, Changes in absorption, fluorescence, dichroism, and birefringence in stained giant axons: optical measurement of membrane potential, *J. Membr. Biol.* **33**:141–183.

Salzberg, B. M., 1983, Optical recording of electrical activity in neurons using molecular probes. in: *Current Methods in Cellular Neurobiology*, (J. L. Barker and J. F. McKelvy, eds.), pp. 139–187, Wiley, New York.

Salzberg, B. M. and Bezanilla, F., 1983, An optical determination of the series resistance in *Loligo*, *J. Gen. Physiol.* **82**:807–817.
Salzberg, B. M., Bezanilla, F., and Obaid, A. L., 1989, *personal commun.*
Watanabe, A., 1985, The optical activity recorded from a squid giant fiber, *J. Physiol. Soc. (Japan)* **47**:402.
Watanabe, A., 1987, Change in optical activity of a lobster nerve associated with excitation, *J. Physiol.* **389**:223–253.
Zecevic, D., Wu, J.-Y., Cohen, L. B., London, J. A., Hopp, H.-P., and Falk, X. C., 1989, Hundreds of neurons in the *Aplysia* abdominal ganglion are active during the gill-withdrawal reflex, *J. Neuroscience* 9, *in press.*

Chapter 11

The Preparation of the Squid Giant Synapse for Electrophysiological Investigation

ELIS F. STANLEY

1. Introduction

The giant axo-axonic synapse in the squid stellate ganglion has advantages as an experimental model of synaptic transmission that can not be equalled by other preparations. These advantages (listed below) all center on the extraordinarily large presynaptic nerve terminal. The object of this chapter is to introduce the experimental techniques that have been used to study the squid stellate ganglion, with particular emphasis on the giant synapse. In addition, the advantages and disadvantages of the giant synapse as an experimental preparation are discussed. This report is not intended as a comprehensive review of scientific findings, but an attempt has been made to make reference to the original technical innovators.

1.1. The stellate ganglion

A nerve trunk, the pallial nerve, exits each of the paired optic ganglia in the head of the squid, traverses the narrow neck of the animal on either side and then descends to the stellate ganglion on the inner surface of the mantle; see the classical work by Williams (1909) for the best general anatomical description and illustration of the dissected squid. The stellate ganglion is located at the point of contact of the pallial nerve with the mantle (Fig. 1), where this nerve divides into two branches. One branch enters the ganglion while the other, the fin nerve, skirts it.

The stellate ganglion is enveloped in a transparent connective tissue capsule and can be seen to be a complex structure with hundreds of neuronal cell bodies and seemingly tangled nerve fibers. The two ganglia are, at least grossly, mirror images of each other and have 10 to 12 nerve trunks radiating away from their centers giving them a *star* like appearance. In addition there is a small inter-stellar connective nerve that joins the ganglia close to the point of entry of the pallial nerve. The ganglion itself can be seen to be composed primarily of nerve fibers on its face away from the mantle and a prominent lobe, composed of neuron cell bodies, on its opposite aspect.

E. F. STANLEY • Laboratory of Biophysics, NINDS, National Institutes of Health, Bethesda, MD 20892.

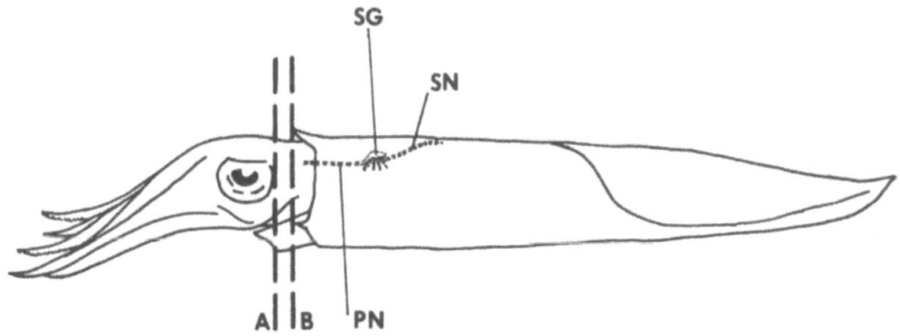

Figure 1. Location of stellate ganglion (*SG*), pallial (*pre*) nerve (*PN*), and last stellar nerve (*SN*) in the intact squid. *A* and *B* indicate levels of decapitation prior to dissection. Section at *A* leaves the presynaptic giant axons intact, while section at *B* cuts the giant axons in the pallial nerve.

1.2. The giant synapse

The giant fiber system in the squid is involved in escape responses and represents a fast conducting pathway between the brain (cerebral ganglia) and the mantle muscles. Excitation of this system culminates in the contraction of the mantle, a forceful expulsion of water from the mantle chamber, and rapid movement of the squid (Young, 1938). The giant fiber system is composed of three segments: the first order giant axon (G.A. I) in the cerebral ganglia, the second order giant axons (G.A. II) that leave the brain and terminate in the stellate ganglion, and the third order giant axons (G.A. III) that innervate the mantle musculature (Young, 1939).

There are in fact a series of third order giant axons, one in each of the radiating stellar nerves. The second order giant in the prenerve sends finger-like branches that synapse near the origin of each of the IIIrd order giant axons in the radiating stellar nerves (Fig. 2). These IIIrd order giant axons increase in diameter, and in the size of their respective synapses, roughly in proportion to their length. Thus, the first stellar nerve, that supplies the nearby mantle, caudal to the ganglion, contains a *giant* axon of relatively modest diameter which is contacted by a small finger of the IInd order giant axon. By contrast, the last stellar nerve, which follows a course that is parallel to the midline of the squid and supplies the most distal mantle musculature, contains the largest giant axon and is contacted by the last, and largest, branch of the IInd order presynaptic giant axon (see Martin and Miledi, 1986).

The synapse between the terminal finger of the IInd order giant axon in the prenerve and the giant axon in the last stellar nerve, i.e., the giant axon routinely used for voltage clamping by Cole, Hodgkin and Huxley, etc. (see Adelman and Gilbert, 1990); this synapse is generally referred to as the giant synapse. One of the best published photographs of the stellate ganglion with dye-labeled pre- and postsynaptic giant axons is presented in an excellent review by Llinas (1984).

Studies on the fine structure of the giant synapse (see Robertson, 1963; Young, 1973; Pumplin and Reese, 1978) show that the presynaptic nerve terminal is a smooth, rather blunt ended structure. By contrast, the postsynaptic axon sends out a dense mat

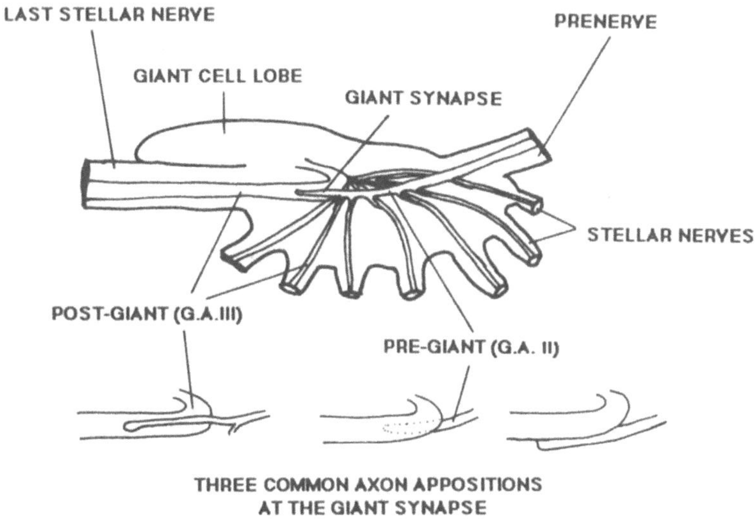

Figure 2. Stellate ganglion showing the giant axon system. The giant axons are not as obvious as indicated in the illustration and appear as pale lines within the neuropil. Hundreds of smaller axons course above and below the giants. The giant cell lobe contains the neuronal cell bodies whose axons fuse to form the IIIrd order giant axons. Other neuronal cell bodies are located across the back of the ganglion and in the fringes between the radiating stellar nerves.

of bulbous projections to contact the presynaptic axon (Young, 1973). The transmitter release sites in the presynaptic axon, identified by accumulations of secretory vesicles, occur opposite the tops of these projections.

2. The giant synapse as an experimental preparation

The advantages of the giant synapse as an experimental preparation have been repeatedly emphasized. This synapse remains unique among chemically conducting synapses as the preparation where intracellular voltage recording, current passing and voltage clamping can be reliably carried out in an identified presynaptic nerve terminal. Unfortunately, this property has often obscured several methodological disadvantages and gaps in knowledge that may make this an inappropriate preparation for certain experimental questions (but also serve as challenges for future research). These disadvantages preclude many of the experimental approaches that may be routine in other synaptic preparations.

2.1. Experimental advantages of the giant synapse preparation

2.1.1. Large size of the pre- and postsynaptic giant axons (Young, 1939).

In ideal squid the post synaptic giant axon is hundreds of microns in diameter and penetration with a micropipette is relatively trivial. Thus, the recording of postsynaptic potentials and ion currents is comparatively easy. The presynaptic giant axon can be of 50 to 100 μm in diameter in the synaptic region and up to a millimeter in length. The penetration of the presynaptic nerve terminal with one or more micropipettes, while not trivial, is possible, allowing the recording of presynaptic action potentials and ion currents. Future use of this preparation will no doubt take advantage of the large size of the presynaptic nerve terminal to directly visualize the points of access of ion currents with the powerful ion sensitive dyes (see for example Smith *et al.*, 1987), and the movements of the secretory and endocytotic vesicles themselves.

2.1.2. *Fast* transmitting synapse

The synapse transmits single action potentials with high speed (Bullock 1948), perhaps as fast as is seen in any other preparation. In this, as in many other properties, the giant synapse is very similar to another important experimental model of synaptic transmission: the neuromuscular junction. The findings from these two preparations are, on the whole, complementary, greatly simplifying the interpretation of some results at the giant synapse.

2.1.3. High release capacity

Under suitable conditions the synapse can transmit many hundreds of impulses without significant deterioration. The release capacity is so large that it is possible that some experimental strategies can be carried out solely in this preparation.

2.1.4. Rapid exchange of external solutions

The advent of the aortic perfusion technique (Stanley and Adelman, 1982, 1984) allows the exchange, within one to two minutes of equilibration, of synapse bathing solutions.

2.1.5. Facilitation and depression

The transmitter release phenomena of facilitation and depression can both be studied at the giant synapse (Takeuchi and Takeuchi, 1962; Miledi and Slater, 1966; Charlton and Bittner, 1978; Charlton *et al.*, 1982; Stanley, 1986).

2.1.6. Analysis of ionic currents in a nerve terminal

At present one of the most attractive physiological features of this preparation is the fact that many aspects of synaptic transmission can ultimately be tracked down to the level of identified ion currents, whether in the pre- (Llinas *et al.*, 1976, 1981) or postsynaptic (Hagiwara and Tasaki, 1958) element.

2.1.7. Single presynaptic input-output relations

Single presynaptic input-output relations have been analyzed (Llinas *et al.*, 1981b; Augustine *et al.*, 1985, 1986). As a result of (5) and (6) above, transmitter release can be analyzed to an impressive level of detail, including the quantity of transmitter released per Ca current influx (although not without some conflicting results).

2.1.8. Nerve terminal capacitance

This subject has been investigated by Gillespie (1979). The release of transmitter has been correlated with an increase in the capacitance of the presynaptic nerve terminal. This can most probably be attributed to the increase in membrane area that is due to the fusion of secretory vesicles at the release sites.

2.2. Experimental disadvantages of the giant synapse

2.2.1. Unidentified transmitter substance

It is not without some irony that at this synapse, which has been analyzed to such detail in terms of transmitter release, the transmitter substance has not as yet been identified. Despite some evidence in its favor, acetylcholine has been essentially ruled out (Stanley, 1984), and glutamate remains the best candidate (Miledi, 1969; Kawai, 1983). Furthermore, it should be noted that the lack of identification of the transmitter, coupled with the generally poorer understanding of drug action in invertebrates, precludes the pharmacological dissection of synaptic function at the giant synapse that is routine in many vertebrate synaptic models.

2.2.2. Miniature excitatory postsynaptic potentials (MEPPs)

Although the release of transmitter in quanta has been confirmed at the giant synapse (Miledi, 1967; Mann and Joyner, 1978), their observation is not trivial. The difficulty in detecting MEPPs is due to the low input resistance of the postsynaptic giant axon. A quantum of transmitter that might be expected to result in one mV or more depolarization of the post synaptic cell of other synapses results in less than 20 uV in the giant synapse. Even in the studies where MEPPs or miniature excitatory postsynaptic currents (MEPCs) have been observed, their amplitude distributions are skewed by a loss of the small events in the baseline noise. Usually the best that can be achieved is to measure postsynaptic membrane noise. It is not generally feasible, therefore, to carry out experiments requiring estimates of quantal size or quantal content. In fact, by analogy with findings obtained at the neuromuscular junction, it is generally assumed that a change in the excitatory postsynaptic potential (EPSP) or current (EPSC) reflects a change in the number of quanta released, rather than the unit size of each quantum. It remains to be determined if this assumption always holds.

2.2.3. More than one presynaptic axon

Young (1939) was the first one to note that the postsynaptic giant axon receives more than one monosynaptic input from the prenerve. An action potential evoked in the postsynaptic giant cannot, therefore, be assumed to originate as a result of excitation of the IInd order giant axon. Other axons in the prenerve form more proximal synapse onto the postsynaptic giant axon (Young, 1939). It is possible to crush these auxiliary axons with fine forceps (Bryant, 1948), but not without the danger of damaging the presynaptic giant – a procedure that is not lightly undertaken after a prolonged dissection. Only when the pre-synaptic giant is stimulated by an intracellular electrode can one be reasonably certain that the transmitter release observed originates from this, and only this, nerve terminal. Even this is not absolutely definitive since there remains the remote possibility of electrical conduction between presynaptic giant axons.

2.2.4. The presynaptic giant projects multiple nerve terminals

Young (1939) noted that the presynaptic giant projects multiple nerve terminals. While it is generally accepted that the IInd order giant axon projects only one presynaptic digit at the giant synapse, it must not be forgotten that this axon also sends similar branches to synapse on each of the IIIrd order giant axons that supply the more proximal stellar nerves. This can be a problem with voltage clamping of the nerve terminal since it cannot be assumed that all the membrane current observed in the nerve terminal originates from the terminal digit of the presynaptic giant axon. Ion currents may also be detected from the neighboring terminal digit.

2.2.5. Diffusion barrier

The diffusion barrier was emphasized by Bryant (1958). The giant synapse can be visualized with incredible clarity and yet repeated studies have shown that the access of bath applied salts or drugs is extremely poor (see table I). This finding should not be surprising since, save for the thin connective tissue sheath, the stellate ganglion is directly exposed to the external sea water in the mantle cavity. Thus, an impenetrable barrier between the blood space in the ganglion and the mantle cavity is essential to conserve blood solutes. This barrier is a major problem for any experiment that requires the change of solutions outside the giant synapse. In many early studies the diffusion barrier was largely ignored, despite the very long times to equilibration (up to and over 1 hr) for small ions such as Ca^{2+} (e.g., Katz and Miledi, 1970). The diffusion barrier is evidently not uniform, however, since some access from the bathing medium is possible: Miledi (1969) noted that iontophoresed glutamate would depolarize the postsynaptic giant axon, but only at very restricted locations. The only method of ensuring the rapid exchange of extracellular medium for the whole synapse is by aortic perfusion of the ganglion (Stanley and Adelman, 1984). Note, therefore, that attempts to superfuse the giant synapse independently of the rest of the ganglion by a closely opposed micropipette (Miledi and Slater, 1966; Augustine *et al.*, 1985) would not be expected to evenly bathe the presynaptic nerve terminal. Instead the superfused medium probably results in a *checkerboard* pattern of high access and low access areas.

2.2.6. Large size of the pre- and postsynaptic giant axons

The large size of the pre- and postsynaptic giant axons presents some electrophysiological problems not encountered in most other preparations. In particular, their low input impedance demands high stimulus currents. In common with other excitable cells with markedly non-spherical structure, voltage clamping of these axons by microelectrodes encounters the acute problem of space clamping (see Section 3.4.5.3).

2.2.7. Variability

There are probably as many configurations for the apposition of the pre- and postsynaptic axons at the giant synapse as there are synapses (see Martin and MIledi, 1986). Virtually every preparation is suitable for postsynaptic recording but the frequency of usable preparations is greatly reduced if electrodes are to be inserted into the presynaptic giant axon (10% on a good day). In analogy with the old real estate adage, there are three factors of primary importance in experimentation on the giant synapse preparation: selection, selection and selection (see Section 3.4.3.1).

2.2.8. Oxygen sensitivity

Early studies reported that the giant synapse has a requirement for high oxygen tensions (Bryant, 1958). Recent tests indicate that the synapse does require oxygen for long term maintenance, but that this requirement is not nearly as acute as first supposed (Colton *et al.*; see Section 3.4.2).

2.2.9. Depletion of transmitter release

Depletion of transmitter release has been demonstrated (Bullock and Young, 1938; Kusano and Landau, 1975). Blood levels of Ca^{2+} in the squid have been determined to be about 10 mM. At this level of external Ca^{2+}, repetitive stimulation of the presynaptic giant axon leads to rapid depression of transmitter release. This difficulty can be a confusing complication in many experiments but is easily avoided by lowering the external Ca^{2+} to 2 to 3 mM, and hence, lowering the quantity of transmitter released per impulse.

3. Dissection, mounting, and experimentation

There are several approaches to the preparation of the functioning stellate ganglion for experimental study. These techniques range in difficulty from morphological fixation to, perhaps the most ambitious, voltage clamping of both the pre- and postsynaptic giant axons with concurrent aortic perfusion. Here, I will describe techniques for: coarse dissection of the ganglion with pre and post nerves, basic electrophysiology, basic aortic perfusion, and aortic perfusion combined with multiple electrode penetration. It is inevitable that in such a description some of the described techniques will include my own technical preferences, but an attempt will be made to point these out so that the reader can refer to alternative approaches. The following discussion will be limited, for simplicity, to the stellate ganglion on the left side of the squid (as viewed in the opened and pinned out mantle), but can be equally well applied to the opposite ganglion with appropriate adjustments. In this description I will make use of the terms: caudal, towards the tip of the tail; rostral, towards the tip of the tentacles; dorsal, the upper surface of the squid when swimming, and ventral.

3.1. Squid

The selection of the ideal squid for giant synapse study depends on the experimental protocol. Generally, it is a balance of obtaining as large a presynaptic terminal as possible, while avoiding the opacity and dense connective tissue that occurs in the older squid. In fact, individual variation is a far larger variable than squid size. The size range within which usable squid occur for electrophysiology-type experiments ranges from a mantle length of 5 cm to 15 cm. Other squid can be used (and often are at times of poor availability) but show a lower success rate. Some experimenters prefer females but good synapses are to be found in both sexes.

Two main dissection techniques are described below. The first is essentially as reported by Bullock (1948) and is adequate for experiments where minimal changes in the external medium of the synapse are planned, i.e., where long waits for equilibration are not a problem. Many highly sophisticated electrophysiological experiments have been successfully carried out with variants of this technique. The second preparative technique, utilizing aortic perfusion, is recommended, however, particularly if the experiment requires any of the following: optimum maintenance of the stellate ganglion, rapid changes in the external medium at the giant synapse, testing the effects of ionic, pharmacologic or toxic agents on the synapse, and any attempts at fixation for morphological studies.

The main problems associated with experiments utilizing aortic perfusion are: increased dissection time (15 to 30 extra minutes), movement of the synapse during perfusion due to *inflation* of the ganglion, and increased opacity. These problems have been largely solved (see below) but may indicate the use of the preparation without perfusion in some circumstances.

3.2. The basic stellate ganglion preparation

3.2.1. Coarse dissection

The squid is held with a firm grip around the mantle, with its ventral surface up, and is decapitated. It is usually sufficient to cut the whole head off (Fig. 1) but this can leave a rather short length of the prenerves, especially in the smaller squid. I prefer to decapitate immediately behind the eyes, cutting through the most caudal cerebral ganglia (Fig. 1), as originally proposed by Bullock (1948). This leaves a maximum length of the prenerves and, avoids damaging these nerves until they can be tied off. A cut at the right level causes the squid to go into a rigid, dark (i.e., melanocyte activated) state, but still able to maintain swimming in a straight line (and to squirt ink!). The remaining ganglia are crushed and the squid loses color and goes flaccid.

The mantle is cut up the ventral midline, opened up to reveal the internal organs, and is rinsed to remove any trapped ink. The skin is then peeled off to improve visualization of the ganglia, and the mantle is pinned out, organelle side up, on the dissecting table (Fig. 2), Fresh running seawater, bubbled with O_2, is continuously passed over the preparation. The oxygenation may be omitted if the ganglion is to be removed very quickly.

The next step of the dissection is to reveal the stellate ganglion and its associated nerves. First, the gills are freed from the mantle. Next the proximal pointed end of the ink sac is grasped with toothed forceps and is stripped away together with the attached gut, reproductive organs, gills etc. The syphon and its attached muscles are next stripped off the mantle, leaving only the yellow digestive gland in its muscular tube. At this point the stellate ganglia can be seen by pushing the digestive gland to either side. They are located close to the midline and are identified by the stellar nerves that radiate out across the mantle (Figs. 2, 3A). The digestive gland muscular tube is cut up either side, taking care not to damage the underlying prenerves, and the muscle above the gland is stripped off to expose the gland. The digestive gland is lifted in its mid portion and gently teased off the underlying muscular tissue and blood vessel, first caudally and then rostrally. Removal of the digestive gland must be carried out with particular care at its most rostral end, where it usually encircles the prenerves, ideally under a dissecting microscope.

At this point the pallial nerves that innervate the stellate ganglia can be seen. The pallial nerve originates from the palliovisceral ganglion in the head, follows the inner surface of the digestive gland tube for about 2 cm, and then descends through a hole in the tube, the *eye*, to enter the ganglion. The prenerve is tied off with a long length of silk thread as close as possible to the remains of the cerebral ganglia. This silk thread will be used to move the prenerve out of the way of the subsequent dissection steps. The nerve is lifted and cut free from the underlying muscle up to the *eye*. The neck muscle is cut parallel with the mantle mid-line, in a straight line from its caudal end to the *eye*, around the prenerve, past the upper surface of the ganglion and along the last stellar nerve. At this point the entire stellate ganglion and attached nerves should be visible (Fig. 2b, c). If the ganglion is not to be perfused through its arterial system the following dissection should be followed, otherwise skip to Section 3.3.

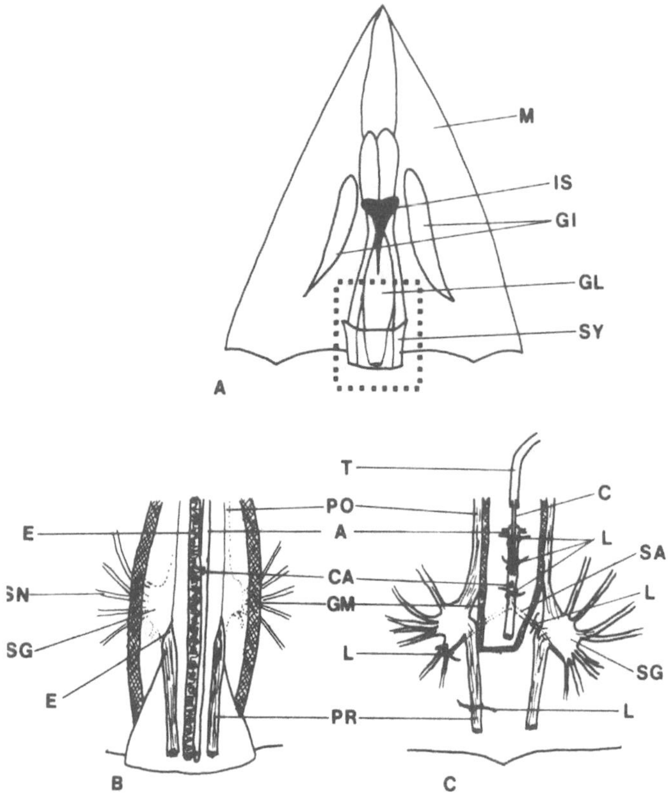

Figure 3. Dissection of the stellate ganglion. A. Diagram of the main organs after cutting the mantle along the ventral midline. B. Exposure of the pallial nerves and aorta after removal of the digestive gland. C. Exposure of the ganglia and location of sutures and cannulation of the aorta for perfusion. *M*, mantle; *IS*, ink sac; *GI*, gills; *GL*, digestive gland; *SY*, syphon; *ES*, esophagus; *SN*, last stellar nerve; *SG* stellate ganglion; *E*, "aye"; *T*, polyethylene tube; *PO* last stellate nerve; *A*, aorta; *CA*, exit of common stellate artery; *GM*, cut digestive gland muscle; *L*, ligature; *PR*, pallial (or pre) nerve; *C*, cannula; *SA*, stellate artery.

3.2.2. Removal of the stellate ganglion

The remaining gland musculature is lifted by its proximal end and, while lifting, is cut free from the stellate ganglion. Once the muscle has been cut beyond the ganglion it can be safely stripped off. The prenerve is lifted by its attached suture and the ganglion is carefully cut free from the underlying mantle. The stellar nerves are cut in turn, leaving a few mm of each to facilitate fixation of the ganglion to the experimental chamber. Approximately 15 mm of the last stellar nerve is ligated and removed with the ganglion. It may be helpful to use different color thread for the two nerves to easily remember which is which. In fact, the stellar nerve can be readily distinguished by the giant axon, which is visible to the naked eye, and by the lobe of ganglion neuron cell bodies that is attached to its stellar end. The ganglion is transferred to an experimental chamber. With practice the entire coarse dissection can be accomplished in less than a minute. For alternative descriptions of the coarse dissection of the stellate ganglion refer to the excellent original report by Bullock (1948), and to that by Miledi (1967).

3.2.3. Fine dissection

The ganglion is placed in a Sylgard™ coated petri dish containing sea water that is continually bubbled with O_2 (I prefer a small, 25 mm dish as it maintains oxygenation better). The ganglion is observed under a 20–40× dissecting microscope with transillumination from a point source of light and pinned out with the prenerve to the right, and the last stellar nerve directly opposite to the left. If the ganglion originated from the left side of the mantle (as viewed from above), the giant fiber lobe is at the top, towards the left and the cut stellate nerves radiate downwards.

The first step of the fine dissection is to remove the *capsule* of connective tissue that encircles the ganglion. The capsule, which is almost transparent, may not be noticed at first; but if the ganglion is observed for a while, it can be seen to be slowly and regularly contracting. With a pair of fine forceps grasp the capsule at a point that is not directly over the giant synapse or the terminations of the prenerve (to avoid inadvertent damage). Lift the connective tissue sheath and, with fine iridectomy scissors, cut it from a mid point on the lower edge of the ganglion to the upper edge. The two sides of the capsule are then grasped with fine forceps and stripped apart and removed.

At this point a nerve branch that approaches the ganglion in the pallial nerve, but then bypasses the ganglion itself along its upper border, will become evident. This nerve, the fin nerve, is best removed by cutting close to where it joins the last stellar nerve and dissecting back to the pallial nerve. Further dissection into the ganglion is difficult, since it usually leads to deterioration of the preparation, and/or to increased opacity of the overlying tissues. A more careful selection of preparations minimizes this danger of tissue damage during dissection in order to gain more direct access to the synapse (see Section 3.4.3.1).

3.3. The stellate ganglion with aortic perfusion

The preparation of the stellate ganglion with aortic perfusion is similar to the above procedure, except in that the aorta must be cannulated, arteries other than that leading to the ganglion must be ligated, and care must be taken not to disturb the blood supply of the ganglion during the fine dissection.

3.3.1. The blood supply of the stellate ganglion

The hearts of the squid are located caudal to the ink gland and send a main vessel, the aorta, towards the head. This artery can be identified easily and it is instructive to inject a dye solution to reveal the structures described below.

The aorta enters the digestive gland muscular tube from its ventral surface at its caudal end, passes through the gland, and then follows the inner surface of the tube dorsal wall towards the head. The artery is distinguished from the esophagus by its smooth, thick walled appearance and by side branches. The aorta can be seen to send pairs of arteries into the mantle along its course. The two stellate ganglia are supplied by a single branch from the aorta, which I term the common stellate (CS) artery (Fig. 3B). This artery penetrates the dorsal wall of the digestive gland and bifurcates to supply each ganglion. The stellate arteries pass the rostral edge of the ganglia and then continue to supply the adjoining mantle. The actual arterial input to the ganglion can vary, but usually includes two branches, one directly behind the ganglion as viewed in the dissection, and the second just as the stellate artery passes the ganglion.

3.3.2. Coarse dissection with cannulation of the artery

Follow the coarse dissection up to the end of Section 3.2.1. Only one ganglion is used in each squid since the artery that is cannulated is the main aorta and is common for both ganglia. I routinely only use the left ganglion, as described here, to avoid the adjustments in the experimental chamber that would have to be possible to accommodate either ganglion.

The dorsal wall of the R cartilaginous/muscular wall of the neck is cut parallel to the midline from the rostral end towards the *eye* of the right pallial nerve past the R ganglion, as was done on the L side. The neck musculature is cut in a straight line between the two *eyes*, just rostral to the descent of the two pallial nerves, cutting the aorta at the same point. This piece of tissue is removed and the aorta is ready for cannulation.

A 30 g syringe needed pushed into polyethylene (PE) tubing. PE 10 tubing is ideal for the cannulation of a wide range of aorta sizes. The cannula should be filled with sea water before insertion. The aorta is cannulated from the caudal side of the exit of the common stellate artery (Fig. 3C). This artery branch leaves the aorta from its dorsal surface, about 15 mm caudal to a line joining the mid points of the two ganglia. It can be seen as a *dimple* in the artery itself, or by pushing the aorta to one side gently. Before cannulation the aorta is ligated rostral to the exit of the CS artery and two loose ligatures are tied caudal to the branch in preparation for the cannula. One ligature is positioned close to the CS ganglion artery (1 mm) and the other about 5 mm caudally. It is useful to include some of the underlying muscle in the ligatures to support the cannula.

The cannula is now inserted into the aorta and positioned to terminate just caudal to the CS artery branch. A useful trick to ease insertion of the cannula is to cut the artery only half way through, so that a little tension on the caudal end of the aorta opens a hole into which the cannula can be inserted and slid into place. The remaining ligatures are now tightened to hold the cannula firmly. The insertion of the cannula with this technique is sturdy enough that the preparation can be moved from one place to another quite safely by the cannula itself.

Two main arteries must be ligated before removal of the ganglion from the mantle. First, expose the stellate artery serving the opposite (R) ganglion. This artery follows the under (dorsal) side of the digestive gland muscle and can be exposed by cutting this muscle close to the R ganglion and parallel to the midline. When the muscle is pulled medially this artery can be seen and ligated (Fig. 3C). Next tie off the L stellate artery just after it has passed the rostral edge of the stellate ganglion.

The preparation is now ready for removal from the mantle. Cut the digestive tube muscle caudal to the canula across the midline to the pen border. Lift the cannula and its attached piece of muscle and cut through the underlying fascia at the midline. With the muscle and attached ganglion reflected to the side, cut through the stellar nerves in order (excepting the last stellar nerve), cutting each as close to the mantle as possible. It is useful at this point to cut along the mantle muscle/last stellar nerve border back towards the ganglion nearly up to the stellate artery. Clean the top surface of the last stellar nerve, cut the nerve free from the underlying mantle and tie a ligature around this nerve about 15 mm from the ganglion. When the last stellar nerve is cut, the cannula, the piece of attached digestive gland muscle, the stellate ganglion, the prenerve, and the last stellar nerve, should be completely free from the mantle.

A variant of the perfusion technique has been described by Augustine *et al.*, (1986), which involves cannulation of the fine stellate artery as it passes the ganglion. At present it is not possible to compare the ease or success rate of this approach with the original method described above.

3.3.3. Fine dissection of the perfused ganglion

The preparation is pinned out in a Sylgard™ coated dish. The cannula is connected to artificial seawater under oxygen at a very gentle pressure, e.g., 1 pound/in^2 gauge pressure (psi), just sufficient to inflate the blood vessels in the ganglion, and is observed through a dissecting microscope. The capsule of connective tissue overlying the ganglion is grasped with a pair of fine forceps and is cut from the lower to the upper midline, as in the description above. In this preparation, however, the capsule is not completely removed since this would strip off the blood vessels on the far side. Instead, the capsule is removed only from the top surface of the ganglion. The perfused ganglion is now ready for electrophysiological study.

3.4. Experimental considerations for studies on the stellate ganglion

The following is a discussion of some of the experimental issues that are, in many respects, unique to this preparation.

3.4.1. Artificial sea water (ASW)

The composition of squid blood has been found to be similar in ionic composition to sea water. In fact, if control of ions is not essential, filtered sea water is an excellent Ringer solution. In most experimental situations, however, it is necessary to have some control over the ionic composition, in particular Ca^{2+}. Some ASW compositions are given in Table I. With regard to the composition of these solutions, note: First, the content of the *major* ions Na, K, Mg, and Ca is very similar. Second, the choice of buffer must be considered. HCO_3^-/CO_2 has for long been used as a buffer in physiological experiments with good tolerance. However, as an H^+ buffer it is very poor in the physiological pH range and should not be relied on alone. Thus, more recent studies have used HEPES, or an alternative, as the main pH buffer (see Adelman and Gilbert, 1990). Charlton *et al.*, (1982) recommend the inclusion of some HCO_3^-/CO_2, which is believed to be important in the buffering of the neuronal axoplasm. Finally, the importance of osmolarity has not been discussed in previous reports. Changes as small as 50 mOs can have a substantial effect on the release of transmitter (Stanley and Ehrenstein, 1989). An increase in external osmolarity results in a delayed enhancement of the EPSP amplitude.

Table I. Artificial Sea Water (ASW) Compositions

Ions					*Buffers*				*Gases*	
Na^+	K^+	Ca^{+2}	Mg^{+2}	SO_4^{-2}	HCO_3^-	*TRIS*	*HEPES*	*pH*	O_2	CO_2
440[A]	9.0	9.0	53.4	–	–	–	–	–	+	–
440[B]	9.0	0.9	53.4	–	–	–	–	–	+	–
425[C]	9.0	9.3	48.0	25.0	2.15	–	–	7.9	+	–
423[D]	9.0	9.3	49.4	25.5	2.15	–	–	7.5*	+	–
466[E]	10.0	11.0	54.0	–	3.00	–	–	–	+	–
450[F]	10.0	40.0	20.0	–	–	–	–	7.2	+	–
424[G]	9.0	9.0	48.0	25.0	2.0	–	–	–	+	–
466[H]	10.0	11.0	54.0	–	3.0	–	10.0	7.2	+	+†
423[I]	8.3	10.0	50.0	–	–	+	–	7.2	+	+†
423[J]	9.0	9.3	48.4	25.5	–	–	5.0	7.4	+	–
470[K]	10.0	11.0	55.0	–	–	10.0	–	7.8	–	–
446[L]	10.0	11.0	54.0	–	–	–	5.0	7.6	+	–

* pH range from 7.5 to 7.8
† CO_2 in trace quantities to aid in cytoplasmic buffering
‡ 0.5% glucose
Cl^- is the missing anion. SO_4^{-2} is added as $MgSO_4$.
Mg^{+2} is added as a mixture of $MgCl_2$ and $MgSO_4$.
References to the solutions are given below.

[A] Bryant, 1958
[B] Miledi & Slater, 1966
[C] Kelly & Gage, 1969
[D] Kusano, 1970
[E] Katz & Miledi, 1967
[F] Llinas, et al., 1976
[G] Erulkar & Weight, 1977
[H] Charlton, Smith, & Zucker, 1982
[I] Llinas, Steinberg, & Walton, 1981
[J] Stanley, 1984
[K] Adams, Takeda, & Umbach, 1985
[L] Westerfield & Joyner, 1982

3.4.2. Oxygen

Since the early study of Bryant (1958) it has remained a truism that the giant synapse is highly dependent on oxygen. We have recently found that the synapse is far less dependent on oxygen than originally supposed. Perfusing with ASW flushed with nitrogen results in a delayed but consistent decline in synapse function that can be reversed by the reintroduction of oxygen (Colton *et al.*). It is clear, however, that if the preparation is perfused through the artery using compressed air to drive the perfusate, high oxygen tensions are no longer necessary. Thus, the high oxygen tensions used in early studies that did not utilize perfusion were probably necessary because of the limited access of the bath oxygen through a diffusion barrier.

3.4.3. Recording from the presynaptic nerve terminal

The giant synapse is of interest as an experimental preparation principally because of the feasibility of recording, current passing, drug injection and visualization directly in a presynaptic nerve terminal. In many respects, however, these techniques remain an art in themselves and prior discussion with someone that has attempted them is invaluable. The following points should prove useful:

3.4.3.1 Selection of synapses

This point cannot be overemphasized. Many hours can be wasted in trying to penetrate non-ideal presynaptic nerve terminals. It is essential that the squid dissecting table is designed to allow the selection of suitable preparations prior to involved dissections. The mantle should be pinned out over a clear glass plate below which a moveable point source of light (a florescent tube is useless) is positioned. Immediately after stripping off the skin and pinning the mantle out, the digestive gland muscle is reflected to one side and the ganglion is observed through a dissecting microscope. The postsynaptic giant axon (post-giant) should be easily observed; if not discard the ganglion. Often the presynaptic nerve terminal that runs along the first mm of the post-giant is also easy to see. However, easy visualization does not necessarily indicate a good presynaptic nerve terminal. Very often the nerve terminal is on the far side of the post-giant, and the latter acts as a magnifying lens, resulting in a pristine nerve terminal that is a terrible temptation! The nerve terminal must be either directly above, or directly on the side (e.g., Fig 2), of the post-giant to be worth any further effort. Furthermore, if the preparation is to be perfused it is necessary that there is are no overlying nerve fibers. Good nerve terminals are often identified by their obvious *three-dimensionality*: they look like tubes rather than clear flat areas.

3.4.3.2. Movement

One of the most troubling aspects of this preparation is the elimination of movement. The ganglion comes complete with convenient points at which it can be fixed to the experimental chamber, the giant fiber lobe and the radiating stellar nerves. However, it is difficult to secure over a flat surface due to its shape with the bulk underneath and nerves on top. Furthermore, it appears to contain muscle fibers that act rather like some vertebrate smooth muscle with continual slow contractions and relaxations. The usual approach to avoiding this problem is to pin the ganglion out with considerable tension between the pre- and post nerves. It is quite remarkable how much tension the preparation will withstand without the synapse deteriorating. Movement problems are further compounded when the preparation is perfused through the arterial blood supply. With pressures greater than about 2 psi, the ganglion tends to blow up like a balloon. Thus, recording from the presynaptic nerve terminal is easiest if the pressure is kept very low. A pressure on the perfusate of between 1–2 psi is sufficient for reasonably rapid changes in the external medium and in addition avoids the increase in opacity that occurs at higher pressures. If, however, the acute effects of a pharmacological agent are to be studied (such as a neurotransmitter candidate), it is necessary to use much higher pressures of up to 15 psi (Stanley, 1984).

3.4.3.3. Micropipette impalement

Whether the object of the micropipette impalement is to carry out an electrophysiological procedure, or to inject a compound, it is necessary to be able to record from the micropipette. Certainty of impalement of the presynaptic digit requires the recording of intracellular resting and action potentials. Thus, it may be necessary to add an electrolyte (e.g., KCl) to the micropipette solution. To impale, first record intracellularly from the post-giant and stimulate the presynaptic axon with external wire electrodes at 3–5 cycles/sec (Hz). Position the tip of the microelectrode over the presynaptic nerve terminal and gradually lower the electrode while monitoring extracellular potentials recorded through the microelectrode on an oscilloscope. As the electrode nears the nerve terminal the extracellular potential gradually grows in amplitude, finally reversing and suddenly increasing in size as the axon is penetrated. Presynaptic recordings are easily distinguished from postsynaptic ones by their short latency and the absence of a step on the rising phase of the action potential. In high Ca (> 4 mM) this step may not always be obvious, but can be exposed with high stimulus frequencies.

3.4.4. Correlation of pre- and postsynaptic events

Electrophysiological experiments at the giant synapse focus in large part on the correlation of effects of treatments on the presynaptic nerve terminal with the release of transmitter. It is useful, therefore, to use the EPSP as a monitor of transmitter release. It is first necessary, however, to reduce the EPSP amplitude below the threshold for an action potential.

An early solution to this problem was to treat the preparation with a marginal dose of tetrodotoxin, since the post-giant was found to be more sensitive to this sodium channel blocker than the pre- (Bloedel *et al.*, 1966). This method is no longer recommended. An alternative, and rather simple, method is to reduce the EPSP amplitude by lowering the external Ca^{2+} ion concentration to 2 or 3 mM. Aortic perfusion is usually required, since the preparation would otherwise take considerable time to equilibrate. The final, and surprisingly easy, method, is to voltage clamp the postsynaptic axon (see Llinas *et al.*, 1974) and determine EPSCs rather than EPSPs. Due to its large size, the insertion of two electrodes into the post-giant is not difficult, but it may be necessary to combine voltage clamping with a lowered external Ca^{2+} ion concentration to effectively control the large currents.

3.4.5. Voltage clamp techniques

A full discussion of these techniques at the giant synapse is beyond the scope of this chapter but the following points may be of use.

3.4.5.1. Equipment

Most commercial voltage clamp units will not deliver the large currents required to effectively control postsynaptic responses through a micropipette. The voltage compliance of the unit should be 100 V or higher.

3.4.5.2. Current passing electrodes

These must have a tip resistance of less than 5 Megohms. While the postsynaptic currents associated with transmitter release can be observed with simple micropipette electrodes, fast and small currents, such as the important presynaptic Ca^{2+} current, require electrically screened current passing electrodes, with driven shields extending close to the pipette tip (see Llinas *et al.*, 1981a).

3.4.5.3. Voltage clamp limitations

It is not as yet possible to totally control ion currents in the presynaptic nerve terminal. The rather complex techniques that have evolved in two main laboratories (Llinas, Steinberg and Walton, 1976, 1981a, 1981b; Charlton and Augustine, 1988) reflect the difficulties of voltage clamp using microelectrodes. The main problem is achieving an adequate space clamping of a tubular structure with a technique that is limited to point recording and current passing. Much of the ion current is inevitably beyond the clamped region. One technique that probably does reflect the accurate (but not total) ion currents at one point of the axon, is three electrode clamping (see Augustine and Eckert, 1984, for an excellent discussion of two and three electrode clamping in this preparation). Unfortunately, voltage clamping of more than one point is required if presynaptic events are to be accurately correlated with postsynaptic ones, since the area of transmitter release in the presynaptic nerve terminal is large. It is for this reason that attempts were made to limit current flow by only providing the permeant ion to one small area of the terminal (Smith *et al.*, 1985). Recently it has been reported that space clamp problems can be greatly reduced by mechanically isolating the nerve terminal by compressing the axon close to the presynaptic digit (Charlton and Augustine, 1988). While it may remain the case that an 'ideal' voltage clamp of the presynaptic nerve terminal will have to await the insertion of an axial wire electrode (very difficult!), it is likely that the results to date are not far off such an hypothetical experiment (see Augustine and Charlton 1986, for perhaps the most ambitious study in terms of technique, combining presynaptic and postsynaptic voltage clamp with arterial perfusion).

3.4.6. Morphology

Arterial perfusion is essential in any attempt to chemically fix the ganglion for morphological studies, as was first carried out by Young (1939). This need not be complicated. Open up the mantle and expose the artery as described above, then with the ganglia remaining in situ clamp the artery rostral to the exit of the common stellar artery. Insert and clamp a syringe needle (30 g for the small squid) in the artery and perfuse the arterial system (Figure 3C). A spasm in the adjoining musculature and a color change in the ganglia and surrounding mantle (at least with a formalin-based fixative) indicates successful perfusion.

3.4.7. Pharmacology

The giant synapse remains one of the most studied synaptic preparations in terms of physiology, but one of the least studied in terms of pharmacology. Part of this

Table II. Bath Applied Pharmacological Agents

Drug	*Concentration (M)*	*Synaptic Transmis. (min)*	*Effect on Pre-Action Potential*	Effect on the *EPSP*
DFP[A]		B(7),R	–	–
d-tubocurarine[B]	14×10^{-3}	N(55)	–	–
Atropine[B]	4×10^{-3}	N(70)	–	–
Tetraethylamm.[B]	143×10^{-3}	B(17)	–	–
3/4 aminopyr.[C]	$1\text{–}30 \times 10^{-3}$	E	PB *IK*	E
Hemicholinium[B]	17×10^{-3}	B(60)	–	–
Acetylcholine[B]	6×10^{-3}	–	–	–
Acetylcholine[D]	50×10^{-3}	B	–	B
Nicotine[B]	15×10^{-3}	PB(41)	–	–
Neostigmine[B]	9×10^{-4}	–	–	–
Physostigmine[B]	1×10^{-3}	–	–	–
Epinephrin[B]	6×10^{-3}	B(35)	–	–
Tyramine[B]	22×10^{-3}	B(40)	–	–
Procaine[B]	25×10^{-5}	B(60)	–	–
Serotonin[B]	57×10^{-4}	PB(60)	–	–
Histamine[B]	27×10^{-3}	B(50)	–	–
Caffeine[B]	31×10^{-3}	B(29)	–	–
Strychnine[B]	9×10^{-3}	B(15)	–	–
Tetracaine[D]	5×10^{-4}	B(15–82)	–	B R
Dibucaine[D]	5×10^{-4}	B(16–65)	–	B R
Pentobarbital[E]	$0.1\text{–}5 \times 10^{-3}$	B	PB I_C^{a}	B
Eserine[D]	75×10^{-3}	B(16–50)	–	B R
Hexamethonium[D]	$40\text{–}60 \times 10^{-3}$	B(60–65)	–	B PR
L-glutamate[F]	$5\text{–}100 \times 10^{-3}$	B	–	B PR
Kainic acid[G]	10×10^{-3}	B	PB of afterhyperpol	B
Tetrodotoxin[H]	1×10^{-6}	B	Block I_{Na}	B
Guanidine[I]	$4\text{–}400 \times 10^{-3}$	E	PB of I_K	E
Joro Toxin[J]	–	B	N	B
Na cyanide[K]	2×10^{-3}	B(60)	N	B
FCCP[K]	$1\text{–}2 \times 10^{-5}$	B(35–40)	N	B
Ruthenium red[K]	1×10^{-5}	B	N	B

B Block
PB Partial block
R Recovery
PR Partial Recovery
N No effect
E Enhanced
– Not tested

References to the drugs are given below.
[A] Bullock, 1948
[B] Bryant, 1958
[C] Llinas, Steinberg, & Walton, 1981b
[D] Webb, 1966
[E] Morgan & Bryant, 1977
[F] Kelly & Gage 1969
[G] de Santis, Eusebi, & Miledi, 1978
[H] Llinas, Steinberg, & Walton, 1976
[I] Kusano 1970
[J] Kawai et al. 1983
[K] Adams, Takeda, & Umbach, 1985

Table III. Aortic Perfusion Pharmacology

Agent	*Concentration (M)*	*Pre*	*EPSP Effect*	*Post*
Acetylcholine[A]	1×10^{-3}	N	N	Depol/anion channel
Carbamylcholine[A]	$1\text{–}10 \times 10^{-4}$	N	N	Depol/anion channel
Glutamate[A]	$1\text{–}10 \times 10^{-4}$	–	B R	Depol/cation channel
d-tubocurarine[A]	1×10^{-3}	–	–	Block cholinergics
Cd^{++}[B]	1×10^{-3}	–	B R	Small depolarization
La^{+++}[B]	1×10^{-3}	–	B	Small depolarization

B Block without recovery
B R Block with recovery
References to the agents are given below.
[A] Stanley, 1986

N No effect
– Not tested

[B] Stanley & Adelman, 1984

Table IV. Pharmacology by Local, External Application.

Agent	*Concentration (M)*	*Pre*	*EPSP Effect*	*Post*
Acetylcholine[A]	0.5 *	–	N	N
L-glutamate[B,C]	2.0 †	–	B	Depolarization

* Applied by diffusion from a micropipette
† Applied by iontophoresis
B Block
N No Effect
References to the agents are given below.
[A] Bryant, 1958
[B] Miledi, 1969

– Not tested

[C] Adams & Gillespie, 1988

deficit is due to the generally poorer understanding of invertebrate than vertebrate pharmacology but, in addition, it is due to the poor access of drugs to the synapse due to the diffusion barrier. The actions of a number of pharmacological substances that have been tested by bath application on this synapse are summarized in Table II. Note the high concentrations and long incubation times required to obtain effects. With a few exceptions drug actions were rather nonspecific, resulting in the decay of the EPSP and block of synaptic transmission. With aortic perfusion pharmacology is improved dramatically but is still not ideal (Table III). More specific actions can be noted (membrane depolarizations, actions within second and minute time scales), but drug concentrations required to see significant effects, while at least ten fold lower, are still on the high side. One disadvantage of aortic perfusion for examining the pharmacology of the giant synapse, is that the whole ganglion is simultaneously exposed to the agent, leading to possible indirect effects. Nonetheless, aortic perfusion remains the most reliable alternative for changing the external medium and is particularly effective for the control of the synapse ionic milieu. Attempts to gain access to the synapse by enzymatic digestion have not as yet succeeded, probably because of a poor understanding of invertebrate catabolic enzymes. The usually preferable techniques of iontophoresis or pressure ejection from a micropipette are possible (Table IV), but are generally unreliable in this preparation.

3.4.8. Temperature

Some control over the temperature of the preparation is necessary in quantitative studies. Transmitter release is quite steeply related to temperature (Charlton and Atwood, 1979) with a Q_{10} (ratio of reaction rates 10 °C apart) of approximately 4.

4. Conclusion

It should be clear from the above discussion that some aspects of the function of the giant synapse have been resolved to a greater degree than in any other synaptic preparation. The detailed understanding of the relationship of ion movement into the nerve terminal and the release of transmitter has led to this preparation becoming a key model in the study of neurosecretion in fast-transmitting synapses. The experimental advantages that make this preparation so attractive to electrophysiologists could equally well be utilized in the study of the biochemical and morphological aspects of neurosecretion. It is perhaps along these lines that the most fruitful future research on the giant synapse will lie. It would seem, however, that before such studies can be undertaken in any detail, the identity of the neurotransmitter substance must be definitively resolved.

References

Adams, D. J., Takeda, K., and Umbach, J. A., 1985, Inhibitors of calcium buffering depress evoked transmitter release at the squid giant synapse, *J. Physiol. (Lond.)* **369**:145–159.

Adelman, W. J., Jr. and Gilbert, D. L., 1990, Electrophysiology and biophysics of the squid axon, *this volume*.

Augustine, G. J. and Charlton, M. P., 1986, Calcium dependence of presynaptic calcium current and post-synaptic response at the squid giant synapse, *J. Physiol. (Lond.)* **381**: 619–640.

Augustine, G. J., Charlton, M. P., and Smith, S. J., 1985, Calcium entry and transmitter release at voltage-clamped nerve terminals of squid, *J. Physiol. (Lond.)* **367**:163–181.

Augustine, G. J. and Eckert, R., 1984, Divalent cations differentially support transmitter release at the squid giant synapse, *J. Physiol. (Lond.)* **346**:257–271.

Bryant, S. H., 1958, Transmission in squid giant synapses. The importance of oxygen and the effects of drugs, *J. Gen. Physiol.* **41**:473–484.

Bryant, S. H., 1959, The function of the proximal synapses of the squid stellate ganglion, *J. Gen. Physiol.* **42**:609–616.

Bullock, T. H., 1948, Properties of a single synapse in the stellate ganglion of squid, *J. Neurophysiol.* **11**:343–364.

Charlton, M. P., 1978, Facilitation of transmitter release at squid synapses, *J. Gen. Physiol.* **72**:471–486.

Charlton, M. P. and Atwood, H. L., 1979, Synaptic transmission: temperature-sensitivity of calcium entry in presynaptic terminals, *Brain Res.* **170**:543–546.
Charlton, M. P. and Augustine, G. J. 1988, Spatial control of membrane potential, a method for the improved voltage clamping of the squid giant synapse, *J. Neurosci. Meth.* **2**:195–202.
Charlton, M. P., Smith, S. J., and Zucker, R. S., 1982, Role of presynaptic calcium ions and channels in synaptic facilitation and depression at the squid giant synapse, *J. Physiol. (Lond.)* **323**:173–193.
Colton, C., Gilbert, D. L., Stanley, E. F., and Colton, J., 1986, Effect of changes in oxygen tensions on the squid giant axon, *Fed. Proc.* **45**:884.
de Santis, A., Eusebi, F., and Miledi, R., 1978, Kainic acid and synaptic transmission in the stellate ganglion of the squid, *Proc. Roy. Soc. Lond. B* **202**:527–532.
Erulkar, S. D. and Weight, F. F., 1977, Extracellular potassium and transmitter release at the giant synapse of squid, *J. Physiol. (Lond.)* **266**:209–218.
Gillespie, J. I., 1979, The effect of repetitive stimulation on the passive electrical properties of the presynaptic terminal of the squid giant synapse, *Proc. Roy. Soc. Lond. B* **206**: 293–306.
Hagiwara, S. and Tasaki, I., 1958, A study on the mechanism of impulse transmission across the giant synapse of the squid, *J. Physiol. (Lond.)* **143**:114–137.
Kelly, J. S. and Gage, P. W., 1969, L-glutamate blockade of transmission at the giant synapse of the squid stellate ganglion, *J. Neurobiol.* **2**:209–219.
Katz, B. and Miledi, R., 1967, A study of synaptic transmission in the absence of nerve impulses, *J. Physiol. (Lond.)* **192**:407–436.
Katz, B. and Miledi, R., 1970, Further study of the role of calcium in synaptic transmission, *J. Physiol. (Lond.)* **207**:789–801.
Kawai, N., Yamagishi, S., Saito, M., and Furuya, K., 1983, Blockade of synaptic transmission in the squid giant synapse by a spider toxin (JSTX), *Brain Res.* **278**:346–349.
Kusano, K., 1970, Effect of guanidine on the squid giant synapse, *J. Neurobiol.* **1**:459–469.
Kusano, K. and Landau, E. M., 1975, Depression and recovery of transmission at the squid giant synapse, *J. Physiol. (Lond.)* **245**:13–22.
Llinas, R. R., 1984, The squid giant synapse, *Curr. Top. in Memb. and Trans. (The Squid Axon)* **22**:519–546.
Llinas, R., Joyner, R. W., and Nicholson, C., 1974, Equilibrium potential for the postsynaptic response in the squid giant synapse, *J. Gen. Physiol.* **64**:519–535.
Llinas, R., Steinberg, I. Z., and Walton, K., 1976, Presynaptic calcium currents and their relation to synaptic transmission: voltage clamp study in squid giant synapse and theoretical model for the calcium gate, *Proc. Natl. Acad. Sci. U.S.A.* **73**:2913–2922.
Llinas, R., Steinberg, I. Z., and Walton, K., 1981a, Presynaptic calcium currents in squid giant synapse, *Biophys. J.* **33**:289–321.
Llinas, R., Steinberg, I. Z., and Walton, K., 1981b, Relationship between presynaptic calcium current and postsynaptic potential in squid giant synapse, *Biophys. J.* **33**:323–351.
Llinas, R., Walton, K., and Bohr, V., 1976, Synaptic transmission in squid giant synapse after potassium conductance blockage with external 3- and 4-aminopyridine, *Biophys. J.* **16**: 83–86.
Mann, D. W. and Joyner, R. W., 1978, Miniature synaptic potentials at the squid giant synapse, *J. Neurobiol.* **9**:329–335.
Martin, R. and Miledi, R., 1986, The form and dimensions of the giant synapse of squids, *Phil. Trans. Roy. Soc. Lond. B* **312**:355–377.
Miledi, R., 1967, Spontaneous synaptic potentials and quantal release of transmitter in the stellate ganglion of the squid, *J. Physiol. (Lond.)* **192**:379–406.
Miledi, R., 1969, Transmitter action in the giant synapse of the squid, *Nature* **223**:1284–1286.
Miledi, R. and Slater, C. R., 1966, The action of calcium on neuronal synapses in the squid, *J. Physiol. (Lond.)* **184**:473–498.
Morgan, K. G. and Bryant, S. H., 1977, Pentobarbital: presynaptic effect in the squid giant synapse, *Experientia* **33**:487–488.

Pumplin, D. W. and Reese, T. S., 1978, Membrane ultrastructure of the giant synapse of the squid *Loligo pealei*, *Neurosci.* **3**:685–696.

Robertson, J. D., 1963, The occurrence of a subunit pattern in the unit membranes of club endings in Mauthner cell synapses in goldfish brains, *J. Cell Biol.* **19**:201–221.

Smith, S. J., Augustine, G. J., and Charlton, M. P., 1985, Transmission at voltage-clamped giant synapse of the squid: evidence for cooperativity of presynaptic calcium action, *Proc. Natl. Acad. Sci. U.S.A.* **82**:622–625.

Smith, S. J., Osses, L. R., and Augustine, G. J., 1987, Fura-2 imaging of localized calcium accumulation within squid 'giant' presynaptic terminals (M-Pos 54), *Biophys. J.* **51**:66a.

Stanley E. F., 1984, The action of cholinergic agonists on the squid stellate ganglion giant synapse, *J. Neurosci.* **4**:1904–1911.

Stanley E. F., 1986, Decline in calcium cooperativity as the basis of facilitation at the squid giant synapse, *J. Neurosci.* **6**:782–789.

Stanley E. F. and Adelman, W. J., Jr., 1984, Direct access of ions to the squid stellate ganglion giant synapse by aortic perfusion: Effects of calcium-free medium, lanthanum, and cadmium, *Biol. Bull.* **167**:467–476.

Stanley, E. F. and Ehrenstein, G., 1989 *personal observ.*

Takeuchi, A. and Takeuchi, N., 1962, Electrical changes in pre- and postsynaptic axons of the giant synapse of *Loligo*, *J. Gen. Physiol.* **45**:1181–1193.

Webb, G. D., Dettbarn, W. D., and Brisin, M., 1966, Biochemical and pharmacological aspects of the synapses of the squid stellate ganglia, *Biochem. Pharmacol.* **15**:1813–1814.

Westerfield, M. and Joyner, R. W., 1982, Postsynaptic factors controlling the shape of potentials at the squid giant synapse, *Neurosci.* **6**:1367–1375.

Williams, L. W., 1909, *The Anatomy of the Common Squid, Loligo pealii,* Brill, Leiden.

Young, J. Z., 1938, The functioning of the giant nerve fibres of the squid, *J. Exp. Biol.* **15**:170–185.

Young, J. Z., 1939, Fused neurons and synaptic contacts in the giant nerve fibres of cephalopods, *Phil. Trans. Roy. Soc. Lond. B* **229**:465–503 (48 plates).

Young, J. Z., 1973, The giant fibre synapse of Loligo, *Brain Res.* **57**:457–460.

Part IV

CELL BIOLOGY

Chapter 12

Tissue Culture of Squid Neurons, Glia, and Muscle Cells

ROBERT V. RICE, RUTHANNE MUELLER, and
WILLIAM J. ADELMAN, JR.

1. Introduction

We first became interested in the culture of squid neurons as an adjunct to our studies of membrane channels in excitable membranes and our investigation of axoplasmic transport in squid giant axons. For several decades the squid giant axon was the quintessential preparation for the study of the properties of sodium and potassium channels in excitable membranes (Hodgkin, 1951; Cole, 1968; Adelman, 1971; Armstrong, 1975; Adelman and French, 1976). With the advent of single channel recording (Bean *et al.*, 1969; Ehrenstein, Lecar and Nossal, 1970) in lipid bilayers and with the patch clamp technique being applied to living membranes (Neher and Sakmann, 1976), it soon became possible to relate the ensemble properties of excitable membranes to the summation of unitary events occurring in single channels in bilayers (Ehrenstein, Lecar, and Nossal, 1970; Ehrenstein *et al.*, 1974) and in living cell membranes (Sigworth and Neher, 1980). Our desire was to obtain clean cultured squid neurons free of glial cells so that we could easily patch clamp single sodium or potassium channels in the membranes of these cultured cells and then predict and verify the ensemble properties of the excitation process from the single channel patch clamp data. The vast archive of information already in existence as to the electrical behavior of the squid giant axon made this cell culture effort seem particularly tantalizing.

As we also were interested in the role of the neuroplasmic lattice (Hodge and Adelman, 1983) in axoplasmic transport in the squid giant axon (Allen *et al.*, 1982), we were interested as well in culturing squid neurons so we could alter the structural chemistry of axoplasmic proteins and observe with various microscopical techniques the effects of such alterations on living vesicle and particle movement (Allen, Metuzals, and Tasaki, 1981; Brady, Lasek, and Allen, 1981; 1982).

R. V. RICE • Department of Biological Sciences, Carnegie-Mellon University, Pittsburgh, PA 15213. R. MUELLER, and W. J. ADELMAN, JR. • Laboratory of Biophysics, NINDS, NIH, Bethesda, MD. 20892.

1.1. Invertebrate tissue culture

Invertebrate tissue culture has been confined largely to insects although cells from a number of mollusks (Bayne, 1976), particularly the snail (Heyneman *et al.*, 1976; Hansen, 1976) and *Aplysia* (Kaczmarek *et al.*, 1979; Dagan and Levitan, 1981; Wong *et al.*, 1981; Schacker and Proshansky, 1983) have been grown *in vitro*. There is a paucity of published results on cephalopods, although Szabo and Arnold (1963) briefly described the culture of adult squid ink glands.

Hansen (1976) reported the successful establishment of a snail cell line by using embryonic tissues, whereas the reports cited above utilized adult cells and organs. Although useful electrophysiological studies can be done on regenerated axons from isolated ganglia and neurons, our stated interest in patch clamping and single channel work necessitated that we adopt more stringent cell culture methods. Single isolated neurons growing on tissue culture surfaces with an easily modified environment should be amenable to electrode attachment to cell membranes for patch clamping and should respond quickly to biochemical alterations of the tissue culture media. As much work on axoplasmic transport involves using high contrast differential interference microcopy, thin cells arrayed on optically advantageous surfaces should improve visualization. For these reasons, we sought to develop cell culture methods for squid neurons. In the process we were also able to develop methods for the culture of squid glial and muscle cells.

1.2 Neurons, glia, and muscle cells

This chapter describes methods for the primary culture of various cells from squid embryos including neurons identified by tetanus toxin binding, glial cells by glial fibrillary acid protein (GFAP) immunofluorescence, and muscle cells by their ability to twitch or contract.

2. Culture strategies

The general strategy adopted assumed that embryonic cells from squid would grow reasonably well when tissue culture procedures commonly used for vertebrates were appropriately modified to conform to the marine environment. The main modifications employed were (1) drastically reducing media carbon sources, (2) adjusting the osmolarity of the media, (3) using an ionic environment similar to that of sea water or squid hemolymph, (4) lowering the incubation temperature, (5) using selected Eagle media constituents, and (6) adding heat inactivated fetal bovine serum to the media.

2.1 Reduced carbon sources

It had been observed in our laboratory that contamination of cultured invertebrate cells was greatly reduced by lowering the carbohydrate concentration in the growth medium. A further study showed that the carbohydrates contained in heat treated embryonic calf serum were sufficient for cell growth. Therefore, we adopted a strategy of utilizing glucose-free media in culturing squid cells so as to both reduce contamination of the cultures and to allow for the addition of other carbohydrate sources to the media in a manner similar to that employed by Hansen (1976) for snail culture. In general, this strategy resulted in culture media which were quite low in carbohydrate concentrations.

2.2. Osmolarity of the media

As squid and other marine invertebrates live in a sea water environment, and as most squid tissue fluids have osmolarities similar to sea water (Shoukimas, Adelman, and Sege, 1977), we assumed that culture would be best served by using a total medium osmolarity close to that of ambient sea water (ca. 1.0 M).

2.3. Ionic composition of the media

By the same token, we assumed that the culture medium should contain the major ions found in squid hemolymph (Shoukimas *et al.*, 1977) in approximately the same concentrations.

2.4. Incubation temperatures

In general, *Loligo pealei* lives in sea water temperatures that range from 8 °C (Summers, 1990) to just below 24 °C. In addition, squid embryos are killed by temperatures above 22 °C (Arnold, 1990), and squid become lethargic and moribund at temperatures below 10 °C. We assumed that culture temperatures just below 22 °C would assure maximum growth without being lethal. In fact, this assumption required modification when we found that squid cell cultures would proceed at temperatures above 22 °C even though the embryos from which the cells were taken would not have survived at these higher temperatures.

2.5. Modified Eagle medium

For thirty years, Eagle (1959) medium has been a standard for tissue and cell culture. We assumed that squid cell culture would be best served by using the Eagle formula as a basic starting point, and that by experiment and observation we could derive a variant of the basic Eagle formulation which would best support growth and maintenance of squid neurons, glia and muscle cells.

2.6. Fetal bovine serum

Our decision to include fetal calf serum in our culture medium in order to stimulate growth was based on the success of Hansen (1976) in employing this material in her culture of embryonic snail cells. We are indebted to her, as her study provided a valuable guide for our efforts.

3. Procedure

The following procedure for the establishment of primary cultures of useful cells for experimental purposes was arrived at by numerous trials and by a series of experimental tests of adopted techniques and preliminary methods. We are sure that improvements in the procedure can be achieved. By the same token, we recommend that those who adopt these methods consider that the procedure is experimental and that variation in the viability and usefulness of the cultures is to be expected.

3.1. Manipulation of eggs

Egg cases of *Loligo pealei* were collected from running sea water tanks at the Marine Biological Laboratory's Marine Resources Department the morning after deposition by female squid. Selected individual egg cases, each containing about 180 eggs (Arnold, 1974), were tied with monofilament nylon line and suspended into a small aquarium supplied with running sea water. The egg cases were monitored daily and their stages of development (Arnold, 1974) were noted. Once it had been determined that an egg case had embryos at an appropriate stage of development for use in tissue culture, the case was removed from the aquarium and readied for embryo manipulation.

The selected egg case was decontaminated by several quick (5 to 10 sec) rinses of 0.5 percent Chlorox in artificial sea water (ASW) followed immediately by flushes of ASW while the egg case rested on the 20 mesh sieve of a Cellector (Carolina Biological Supply, Burlington, NC). The decontaminated egg casings then were stripped of the outer envelope by hand, leaving the embryos in egg jelly.

3.2. Preparation of embryos

Using fine needles, forceps, and scissors under a dissecting microscope (30 ×), the viable embryos were released from their chorions into sterile ASW containing antibiotics. The antibiotic mixture was composed of the following: penicillin, 1000 units/ml; fungizone, 2.5 μg/ml; neomycin, 500 μg/ml; polymyxin B, 5000 μg/ml; linocin, 5000 μg/ml; kanamycin, 1000 μg/ml (Gibco, Grand Island, NY); gentamycin, 1250 μg/ml; chlortetracycline, 50 μg/ml (Schering, Kenilworth, NJ); and streptomycin, 1000 μg/ml (Sigma, St. Louis, MO). A simplified antibiotic solution (Pen Strep Fungizone, Gibco) was also successfully used in some cultures.

The decontamination procedure was arrived at as a compromise adopted in order to eliminate contamination by protozoa yet give high yields of living embryos when monitored for embryonic organ motility under a phase microscope (100 ×). Higher concentrations or prolonged contact with Chlorox killed most embryos and, therefore, these choices were avoided. It was essential to kill the large numbers of native protozoa while the embryos were protected by their chorions and casing membranes.

Freshly released embryos were washed by decantation with 2 to 3 changes of ASW containing antibiotics. The working solution of ASW contained 12.556 g NaCl, 0.3728 g KCl, 0.735 g $CaCl_2$, 5.091 g $MgCl_2$, and 2.383 g N-2-hydroxyethyl-piperazine-N'-2-ethanesulfonic acid (HEPES) buffer dissolved in 500 ml of filtered (0.45 μm pore size) H_2O. All salts were tissue culture grade (Sigma). The pH was adjusted to 7.6; osmolarity was 1024 mOs/Kg.

In addition to embryos collected in the above manner, we also utilized embryos from *in vitro* fertilized *Loligo pealei* eggs following the method of Klein and Jaffe (1984), except that the Agarose cushion and the sea water medium were prepared and used under aseptic conditions.

3.3. Embryo dissection

Mostly, we utilized only dissected portions of embryos for obtaining starting material for the cell cultures. However, embryos survived without appreciable development for more than a month in the tissue culture medium at 22 °C. Dissection was done under aseptic conditions with a 30 × dissecting microscope and fine forceps and needles at room temperature. During dissection the embryos were held in sterile plastic (35 mm or 60 mm) tissue culture dishes bathed with a few cc of sterile ASW containing the antibiotic mixture. Separated pieces of the dissected embryos were transferred to 0.3 or 1 ml micro *V* shaped vials (Wheaton) on ice. Separated tissue from thirty to fifty head regions took about one hour to prepare. Tissues were then washed 2 to 3 times by gravity sedimentation with ASW plus antibiotics. Usually this material was used immediately for further culture preparation, but such material could be stored on ice at this stage and remain viable for as long as five days.

3.4. Cell dispersion and plating out

A number of procedures were used to disperse tissue. The presence of relatively large amounts of yolk seemed to interfere with subsequent cell growth. Avoiding yolk by using later stage embryos (hatchlings at Stage 29 or 30, Arnold, 1974) was fraught with difficulties because of embryo motility even if they could be concentrated in conical ice cold tubes. The desirability of starting with the earliest neurons appearing in the embryo (Stage 17, Martin, 1965) led us to use mostly Stage 18 to 28 embryos. Dispersion of the tissue into individual cells or clumps of a few cells was accomplished both by mechanical and enzymatic means or by a combination of each.

Usually, controlled tryptic digestion was carried out for varying lengths of time depending on the embryo stage. If, after trituration, the suspension was judged too coarse, then sieving was done. Cellector sieves were used with 100 mesh screens. Trypsin (0.25 percent) was dissolved in calcium and magnesium free (CaMgF) ASW and stored frozen in small vials until just before use. The most successful procedure utilized cold infiltration of trypsin.

Following dissociation, storage and cell suspension, cells were plated out on a choice of surfaces. Early in our work we used standard plastic tissue culture dishes. Previous experience in growing chick embryonic neural tissue led us to use the new *Primeria* plastic which appears to inhibit fibroblast growth and thus favors the growth of neurons and muscle cells. Often, cells were first plated out in regular plastic dishes and then, after a period of standing, were transferred to the *Primeria* dishes. The rationale for this procedure assumed that neuronal cells would be enriched in the second permanent dish. However, the simpler approach of using only *Primeria* plastic seemed adequate since a majority of the cells cultured in this manner appeared to be neurons.

3.5. A typical dispersion and plating out protocol:

a. About 50 head pieces in a 1 ml screw cap *V* vial are washed 3 to 5 times with ASW plus antibiotics on ice.

b. The pellet is washed 2 times with freshly thawed 0.25 percent solution of trypsin in CaMgF ASW and then 1 ml of the 0.25 percent trypsin is added to the ice cold pellet.

c. After from 6 to 18 hours infiltration on ice, the excess trypsin solution is removed, and the vial is then incubated at 37 °C for from 5 to 15 minutes. The times are adjusted downward for earlier embryonic stages and upward for the longer infiltration periods used with material obtained from later embryonic stages.

d. Medium (0.3 ml) is added and the pellet is broken up by trituration with a sterile plastic pipette 152 mm long (Fisher #13-711-9B). The number of passes through the pipette is adjusted to give a reasonably homogeneous suspension and usually requires 20 to 60 passes. The medium contains 10 to 40 percent serum which serves to stop tryptic activity.

e. Cells are plated out by adding 0.2 ml of the suspension to 1 to 2 ml of the medium contained in a 60 mm plastic tissue culture dish at room temperature (22 °C).

f. The dish is examined immediately using an inverted phase contrast microscope, and the concentration of cells is adjusted, if necessary. From time to time, hemocytometer counts are performed. We have found that about 10^5 cells/cc are obtained by the procedure. The presence of cell aggregates of about 5 to 10 cells in addition to individual cells has been observed.

g. The dishes then are placed in a water jacketed incubator under an atmosphere of humid air. The temperature is kept at 21 to 23 °C.

h. Usually medium is replaced during the first 24 hours of incubation or not at all.

3.6. Cryopreservation

Dissociated cells were prepared by cryopreservation for use in later months when embryos were unavailable. For cryopreservation the pellet was triturated in medium composed of 20 percent fetal bovine serum (FBS) in mostly modified Eagle (1959) media (MMEM) containing 10 percent glycerol. The cell suspension was put in screw cap cryo-vials (Gibco), and these were left at room temperature for 30 minutes to allow infiltration prior to freezing. The freezing procedure began at a cooling rate of −1 °C/min (22 °C to −80 °C) for 2 hrs; the rate then was increased to −3 °C/min (−80 °C to −176 °C) for 4 hours. The sample then was placed into liquid nitrogen (−196 °C) (Freshney, 1983). Frozen cells were thawed in a 37 °C water bath for 90 sec. Survival was approximately 50 percent.

3.7. Media

Tissue culture (TC) media were MMEM. These contained 430 mM NaCl, 10 mM KCl, 10 mM $CaCl_2$, 50 mM $MgCl_2$, and 20 mM HEPES buffer as the ASW component to which were added the amino acids and vitamins (Microbiological Associates) required for Eagle's composition, giving a measured osmolarity of 985 mOs/kg. Glucose was not included. Heat inactivated fetal bovine serum (Gibco) usually was added to give final concentrations of FBS from 5 to 40 percent. Other components usually added included the antibiotic mixture, phenol red indicator (5 mg/ml), and at first chick embryo extract. The pH was adjusted to 7.6. Typically to prepare 50 ml of 20 percent FBS medium, 10 ml serum and 0.5 ml of 100 × antibiotic mixture were added to 39.5 ml of Eagle's MMEM. This volume was usually sufficient for starting a normal primary culture. The serum and the antibiotic solution were stored frozen, and these were thawed just before use.

4. Cell culture and growth

Cultures were grown in 35 mm and 60 mm diameter plastic tissue culture dishes (Falcon). Both regular plastic dishes and surface treated *Primeria* plastic dishes (Falcon) were used. The latter had their surfaces altered during manufacturing to approximate polylysine coated surfaces. When detailed microscopic examination was planned, polylysine (Sigma) coated glass coverslips were used.

4.1. Effects of media on growth

Several variations of media were tried. All of these included fetal bovine serum with the exception of a medium composed of squid hemolymph in sea water. The hemolymph medium did not support growth of cells. In our early attempts to define an appropriate medium, we included chick embryo extracts along with FBS. Since good growth of cells was obtained with only FBS added to Eagle's MMEM, the inclusion of chick embryo extracts in our media was discontinued. Concentration of FBS was varied from 5 percent to 40 percent with little qualitative difference in the amount of growth. Therefore, 20 percent was chosen as the standard for later cultures.

4.2. Effects of aggregates, explants, and conditioned media on growth

Preliminary experiments indicated that cells grew more quickly than normal if small aggregates of embryonic cells were included in the inoculum. This observation was also made by Hansen (1976). In some of our cultures, we included such aggregates along with the dissociated cells to be cultured. In other trials, we included larger clumps, approaching small explants. In the small explant trials, good growth of individual cells was observed along with prolific outgrowth of cells from the explant margins.

In a few experiments, the medium from a dish containing growing explants was collected and added to a poorly growing culture. To date we have no conclusive evidence that such *conditioned* media contain surviving growth factors. Because we observed that clumps persistently promoted good growth, we avoided changing media after the first change (which was made within 2 to 24 hours of plating).

4.3. Effects of carbohydrates on growth

Our original decision to eliminate glucose from the usual Eagle medium was made so that we could add other carbohydrate sources to the media in the manner utilized by Hansen (1976) for snail cell culture. Inasmuch as good growth was obtained without these additions, we continued to omit carbohydrates from our media. From time to time, cultures failed to grow or were slower to grow than usual. Various media changes were tried when this happened, including adding glucose (500 mg/l), chick embryo extract, or fresh FBS. To date, failure of a culture to grow has only been traced to a serum batch, and not to any specific methodology.

4.4. Effects of temperature on growth

Even though Arnold has reported that squid embryos do not survive above 22 °C (Arnold, 1974), we have found that embryonic cells in culture continue to grow at temperatures over 22 °C up to 28 °C. These cultures will survive for a few hours at temperatures as high as 30 °C. We operated our incubator, for convenience, as close to 22 °C as possible. As expected, at this temperature embryonic squid cells grew more slowly than vertebrate cells at 37 °C. Hansen (1976) made a similar observation for snail cells grown at lower temperatures than that used for vertebrate cell culture.

5. Cell identification

The types of cells that we have been able to culture from squid embryos are similar to those that other workers have been able to culture from vertebrate brain tissue (Antanitus *et al.*, 1975; Booher and Sensenbrenner, 1972; Bottenstein, 1984; Currie, 1980; Gremo *et al.*, 1984; Keller *et al.*, 1985; Raff *et al.*, 1979; Thampy *et al.*, 1983). All of these workers utilizing vertebrate brain tissue were successful in culturing a heterogeneous population of cell morphologies. Many of these studies reported the culture of neurons identified by tetanus toxin, and the culture of glial cells identified by GFAP immunofluorescence.

We were able to identify several squid cell types by direct microscopic observation. Squid cells retained good morphology for up to one month in culture. Cells with the appearance of neurites began to elongate during the first week of incubation, and in a few favorable cases, this elongation began within the first 24 hours after plating. While microscopic observation can give useful suggestions as to the identity of cell types, we chose more definitive techniques to identify the three classes of cells we were most interested in culturing for our biophysical studies. For neurons and glial cells, we used indirect immunofluorescence staining methods for cell identification. For muscle cells, we used muscular contraction (twitching) for cell identification.

5.1. Neuron identification by tetanus toxin binding

The tetanus toxin binding methods of Mirsky *et al.* (1978) were used to mark neurons. The tetanus toxin and the goat anti-toxin that we employed were gifts from Dr. William H. Habig, Office of Biologics Research, FDA, Dept. Health and Human Services. The fluorescein isothiocyanate (FITC) conjugated anti-goat immunoglobin G (IgG) was purchased (Boehringer Mannheim). All personnel were inoculated with tetanus anti-toxin.

The procedure for cell identification began with the rinsing, with fresh medium containing 20 percent heat-inactivated FBS, of cells attached to either plastic or polylysine coated coverslips. These were incubated with 50 μl of tetanus toxin (5 – 10 μg/ml) for 30 minutes at room temperature. Following incubation the cover slips were rinsed ten times in a series of separate beakers containing 20 percent FBS in MMEM. Once rinsed, the preparations were incubated with 50 μl of goat anti-toxin (5 units/ml) FITC-anti-goat IgG (diluted 1:25) for 30 minutes at room temperature. This was followed by a wash with fixative (5 percent acetic acid-95 percent methanol). Finally the coverslips were mounted in Gelvatol (DuPont), and the slides were viewed through a Zeiss inverted microscope (IM) equipped with phase and epifluorescent optics.

5.2. Glial cell identification

Glial cells were identified by glial fibrillary acid protein (GFAP) immunofluorescence (Raff *et al.*, 1979). Human anti-GFAP (Accurate Chemicals) was added after fixation of cells, either separately or subsequent to tetanus toxin tagging. Rhodamine conjugated rabbit IgG (Boehringer Mannheim) was added to the cell media following the above procedures.

Control experiments using conjugated fluorochromes alone were negative for both types of labeling.

5.3. Cell imaging, photography, and video processing

Cells were photographed on 35 mm Plus-X or Tri-X film (Kodak) using an Olympus microscope camera attached to the IM microscope camera ocular tube. Films were developed in Acufine. Many cultures were viewed on a either a Barco color monitor or a Panasonic black and white monitor displaying the output of a Dage Newvicon 65 camera employed for phase images, or a Dage SIT 65 camera for fluorescent images. These video images were recorded directly by a Panasonic VHS NV-8950 video cassette recorder (VCR) using 1/2 inch video tape, or through an analog video processor (Colorado Video) by a Sony U-matic VCR (VO-5800) using 3/4 inch tape.

Photographs of video tape images were made on Plus-X 35 mm film or Polaroid Type 55 film with a Rembrandt Model 3500C Computer Graphics Film Recorder (NISE, Inc., Cerritos, CA) or from single frame images obtained using a frame-grabbing digital video processing system (Imaging Technology, Inc., Woburn, MA), and a PDP-11/73 on-line digital computer. Some video images were converted to digital frames with the help of a time base corrector kindly lent to us by Dr. Shinya Inoué.

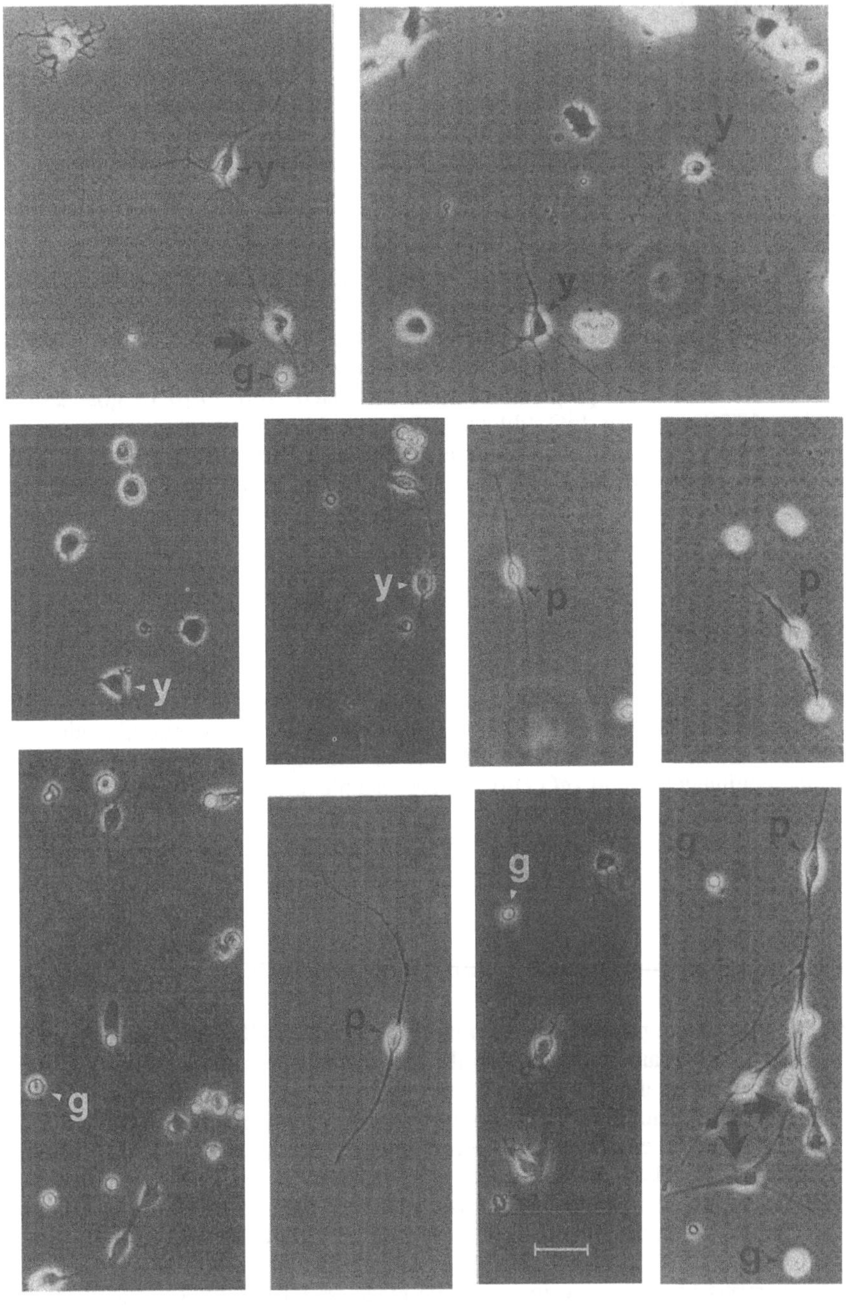

Figure 1. Montage of phase micrographs of cultured embryonic squid cells *in vitro* 3 days. From Stages 22 – 26 embryos. A young culture showing individual cells and short networks. Note bipolar (*p*), granule (*g*), and pyramidal (*y*) cells. This selection underestimates population of granule cells. A few thin cells (arrows) appear to lie underneath presumptive neurons. Not completely representative for other somal morphologies were seen (scale bar = 25 μm).

6. Characteristics of cultured cells

In Primeria plastic culture dishes, the majority of growing cells were positive to tetanus toxin binding. The somal diameters ranged from 5 µm to 40 µm with most soma having a diameter close to 15 µm after 7 to 14 days growth. Long thin processes less than 1 µm wide and reaching several mm in length (some with branches) coursed out from the cell bodies. Some of these processes were tapered; others were not. In most cultures more than a few days old, neurites contacted other processes and cell bodies. Large networks of intermingled processes dominated the morphology of older cell cultures. Some thin processes had many varicosities.

6.1. Morphology of soma

Somal morphologies varied, but most were spheroid, ovoid, fusiform or polygonal in shape when attached to the surface. The ovoid somas usually had long processes arising from each end and, thus, these resembled bipolar cells. Polygonal cells had as many as seven processes arising from different points on the soma surface. The predominant cell morphology consisted of small (7 µm) round phase-bright cells (spheroids) which in early stages of culture were difficult to distinguish from unattached rounded cells.

Figure 1 shows some of the morphologies found in an early culture when shapes of individual cells could be more clearly distinguished. In this early culture, only a few thin (presumptive glial) cells can be seen lying underneath cells with neuronal shapes. Although this montage is representative, it does show all soma morphologies.

6.2. Are fibroblasts present in the cultures?

From the beginning of our work with squid cell cultures, the absence of large numbers of cells with fibroblast-like morphologies was apparent. Although we selected the *Primeria* plastic ware for the purported property of inhibiting fibroblast growth or attachment, we assumed that all cultured cells were not neurons. Indirect immunofluorescence staining with tetanus toxin tagging did show, however, that a large proportion of cells were fluorescing, thus giving definitive evidence that most cells were neurons (Mirsky *et al.*, 1978; Raff *et al.*, 1979).

6.3. Muscle cells

We observed that the cultures were devoid of myotubes, but from time to time spindle-like cells were seen to twitch. Figure 2 is a phase contrast micrograph of cultured cells from the head region of squid embryos in which arrows designate cells that were seen to twitch. This micrograph also shows relatively large numbers of

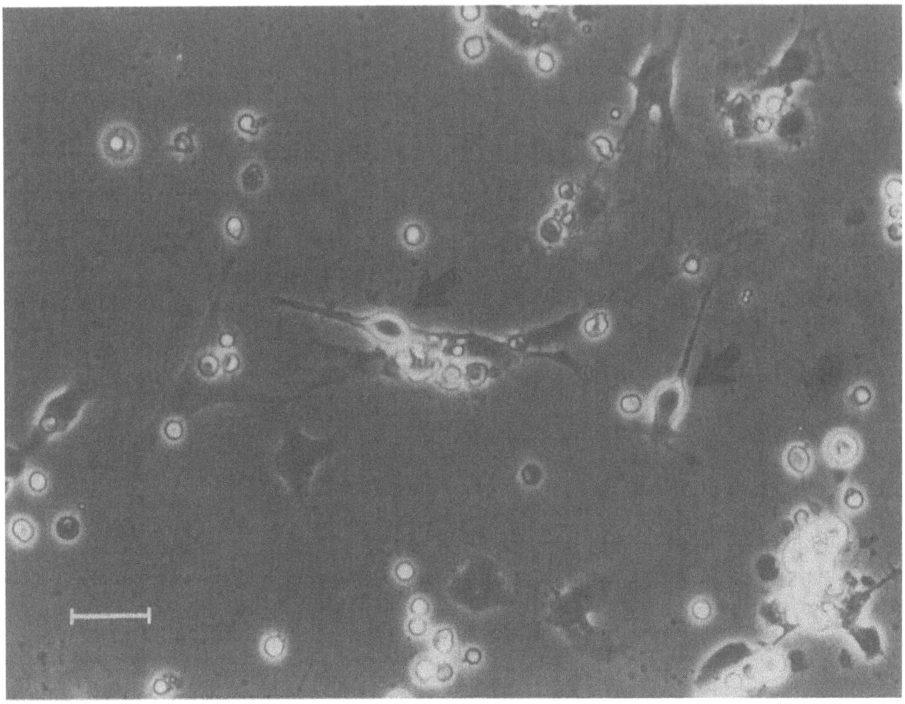

Figure 2. Phase micrograph of cultured embryonic squid cells *in vitro* 7 days. From embyronic Stages 25 – 27. Presumptive neurons are phase bright and presumptive glia are thin and dark. Arrows point to cells within clumps which were observed to twitch (scale bar = 25 μm).

small, spheroidal phase bright cells, and flat dark cells. Large numbers of twitching cells were seen in cultures prepared from separately dissected mantle regions of embryonic squid.

6.4. Distinguishing neurons from glia

Figure 3 is another montage of light micrographs showing the cells viewed by phase contrast and fluorescence microscopy. A large proportion of these cells can be identified as neurons on the basis of tetanus toxin binding fluorescence. In cultures grown on regular Falcon plastic dishes, a much larger proportion of cells did not fluoresce. Some very thin cells could usually be seen in the background, and these became apparent under video contrast differential interference optics. Most of these had neurite processes impinging upon them. Such presumptive glial cells were more apparent in older cultures (> 7 days) or in cultures growing on regular plastic than in freshly plated cultures growing on *Primeria* plastic or polylysine coated coverslips. Double labeling with both tetanus toxin and glial fibrillary acid protein (GFAP) antisera confirmed the purely morphological observations.

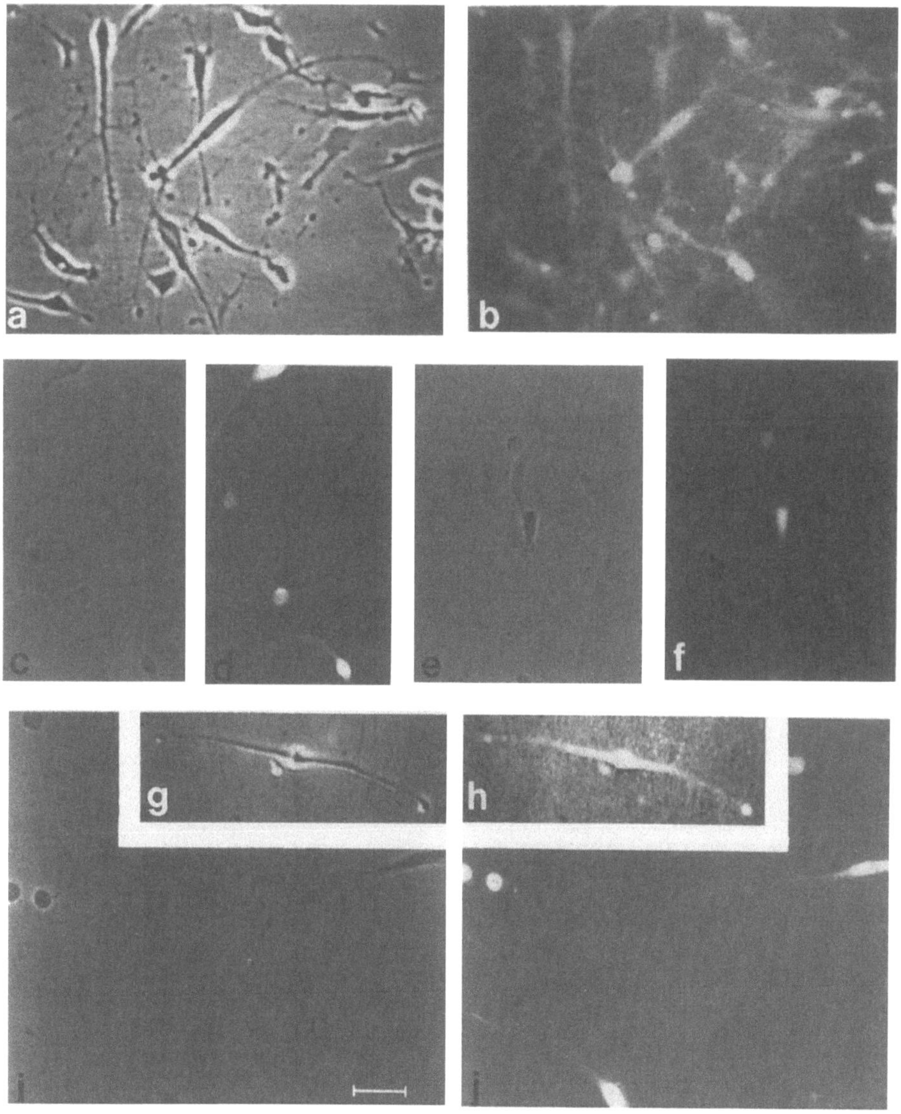

Figure 3. Tetanus toxin labelling of neuronal cell bodies and processes. Light micrographs of cultured embryonic (Stage 27) squid cells comparing phase images (*a*, *c*, *e*, *g*, and *i*) and FITC fluorescence images (*b*, *d*, *f*, *h*, and *j*) of the same cells, *in vitro*, *a* and *b*, 28 days; *c–j*, 5 days. Tetanus toxin, then goat anti-toxin, followed by FITC conjugated anti-goat IgG was added to the cultures with rinsing between additions. Coverslips were fixed in 2% glutaraldehyde and mounted in Gelvatol. *a*, *b*, *g*, and *h* recorded on video tape with SIT camera. Photographed from video screen. *c*, *d*, *e*, *f*, *i*, and *j* photographed on 35 mm Tri-X film. A Zeiss inverted phase microscope with epi-fluorescence optics and a 16 × objective was used. Scale bar = 25 µm. Note in *a* and *b* the extensive neurite processes of this much older culture. Both soma and processes of bipolar and pyramidal cells are labelled, but round, presumptive granule cells lack processes in this figure.

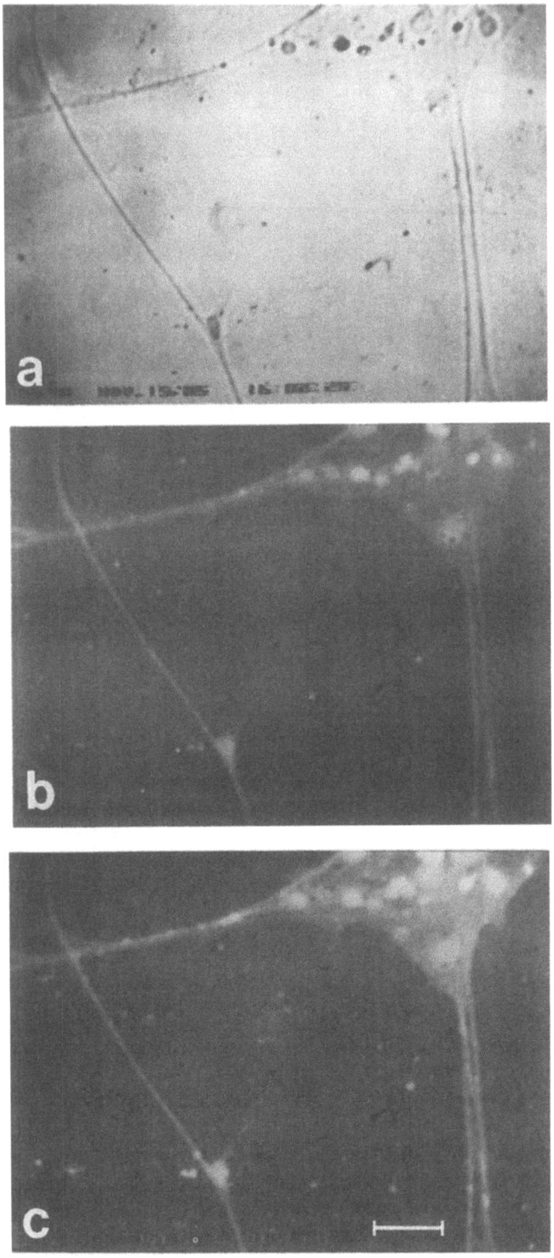

Figure 4. Double labellings of cultured embryonic squid cells (Stage 28). *a*. Bright field. *b*. FITC fluorescence. *c*. Rhodamine fluorescence, *in vitro* 8 days. This preparation was exposed first to tetanus toxin, then in turn to anti-toxin and FITC conjugated antibody in TC medium as in Figure 3. After fixation in anti-glial fibrillary acid protein (GFAP) was added, and then the second rhodamine conjugated antibody was added. Although the flat cell in the upper right can be clearly seen in both the bright field and FITC fluorescence images (*a* and *b*), it is enhanced by the GFAP label suggesting that it is a glial cell.

Figure 4 is a sequence of photographs of the same area of a culture showing several neurites and a portion of a flattened cell in bright field (A), FITC (B), and rhodamine (C) fluorescent optics. The FITC tetanus toxin is seen to enhance the neuronal process and to a limited extent the area of impingement of neurites in the flat cell. The double label clearly enhanced the flat cell viewed with rhodamine filters.

Also, we have noted large numbers of small round phase-bright cells which Currie (1980) identified as granule neurons. Presumably, many of the GFAP positive cells are astrocytes in our squid cell cultures in analogy with the results of Raff *et al.* (1979), and Currie (1980). These workers distinguished two major cell morphologies just as we have: (1) relatively large, flat cells (fibroblastic or glial), and (2) smaller, darker, process bearing cells (neurons). In older cultures, many of the neurites were found on top of flat cells. Both Currie (1980) and Raff *et al.* (1979) used additional markers to closely identify the various types of cells present in embryonic brain cultures. In many cases, their morphological identifications compare favorably to those that we have seen in embryonic squid brain cell cultures.

More recent studies of cultured chick brain embryo cells have revealed a similar heterogeneous populations of neuronal morphologies (Thampy *et al.*, 1983; Gremo *et al.*, 1984). Thampy *et al.* (1983) reported biochemical and electron microscopical results in addition to results on staining for optical microscopy of early (8 day embryos) chick brain cultures. Their work showed that glial cells are essentially absent in cultures grown from that embryonic stage. Their work utilized the results of a previous study (Pettmann *et al.*, 1979). A comparable stage for squid development is not known, but presumably it would occur before stage 18 which is the earliest stage of our embryo cultures. Nevertheless, we have consistently observed a paucity of glial cells from any stage used during the first few days *in vitro.*

6.5. Bipolar and pyramidal neurons

Our observations show that early cultures have large numbers of putative granule neurons along with both bipolar and pyramidal neurons. As the cultures mature, networks are formed primarily among these bipolar and pyramidally shaped cells. Flat glial cells are not easily observed in young cultures solely by phase microscopy, but older (7 days after plating) cultures show many neuronal processes growing on top of them. Video enhanced microscopy clearly showed this association, and the association was even more apparent after GFAP staining. Fetal rat brain cultures have been reported to have a similar sequence of maturation (Keller *et al.*, 1985). These authors reported that *mainly bipolar and pyramidal neurons* were found on glial cells. We have not noticed many, if any, squid granule cells growing on glia or as part of the extensive neurite network. Squid round granule cells do have many active processes. The growth of these processes can be studied best by video enhanced contrast microscopy.

6.6. Characteristics of cultured bipolar cells

In addition to the ubiquitous granule cells, many of the isolated cells in young cultures are bipolar. In well dispersed early cultures, they are the first cells to send out long processes. Thus, these can be quickly identified as putative neurons. The embryonic squid bipolar cells have essentially the same morphology as that reported for chick embryonic retinal cells (Gremo *et al.*, 1984). The eye is one of the most developed neural systems of the embryonic squid head (stages 18 – 28, see Arnold, 1974). We suspect the presence of large numbers of bipolar neurons in the cultures reflects the early maturation of neurons in squid eye tissue.

6.7. Muscle cells

Only a few cultures of cells from the head region of squid embryos exhibit contraction (twitching), but many cultured cells from the separately dissected mantle region contract. Usually these cells are found in relatively large clumps (10 or more cells), but some have been observed moving in clumps comprised of only a few cells. None have been found twitching when completely isolated from other cells. In all cases, neuronal processes are associated with the clumps. Active movement seems to start after heat from the microscope lamp has warmed the culture. In some cases, movement continued for several minutes, but usually the contractions were sporadic. In several instances, the movement was sufficiently violent to detach the cell from the substrate. No cells with striations similar to embryonic striated muscle were observed. It seems probable that these few cells are early stages of smooth muscle, since their morphology is similar to that of smooth muscle cells (Harvey, 1984). Unsuccessful attempts have been made to tag such cells with FITC conjugated tetanus toxin and still observe their twitching. This toxin is reported not to affect the viability of neurons (Mirsky *et al.*, 1978).

7. Conclusions

A method has been described for tissue culture of dissociated cells from molluscan embryos. Dissected head regions of *Loligo pealei* yield high proportions of cultured neurons and glia cells when grown on polylysine coated cover slips, standard plastic or *Primeria* tissue culture dishes. Mantle regions of squid embryos have been shown to be capable of culture to produce twitching muscle cells. All cultured cells grow relatively slowly over several weeks at 22 °C in modified minimal Eagle medium without added carbohydrate sources in the presence of heat inactivated fetal bovine serum (FBS). Neurons were identified by indirect tetanus toxin immunofluorescence, and glia by fibrillary acid protein immunofluorescence. This culture method appears to prevent sepsis from occurring during the culture of cells derived from marine organisms.

ACKNOWLEDGMENTS. We thank Dr. Alan Hodge for much helpful advice and for kindly processing video photographs; Dr. Carl Berg for advice and equipment for cryopreservation; and Dr. John Arnold for advice and assistance in handling embryos. Ms. Lisa Warren and Ms. Melissa Hendricks ably dissected many embryos, and Mrs. Dorothy Martin patiently typed the manuscript.

References

Adelman, W. J., Jr., 1971, Electrical studies of internally perfused axons, in: *Biophysics and Physiology of Excitable Membranes*, (W. J. Adelman, Jr., ed.), pp. 274–319, Van Nostrand and Reinhold, New York.

Adelman, W. J., Jr. and French, R. J., 1978, Blocking of the squid axon potassium channel by external caesium ions, *J. Physiol. (Lond.)* **276**:13–25.

Allen, R. D., Metuzals, J., and Tasaki, I., 1981, Fast axonal transport in the squid giant axon, *Biol. Bull.* **161**:303.
Allen, R. D., Metuzals, J., Tasaki, I., Brady, S. T., and Gilbert, S. P., 1982, Fast axonal transport in squid giant axon, *Science* **218**:1127–1129.
Antanitus, D. S., Choi, B. H., and LaPham, L. W., 1975, Immunofluorescence staining of astrocytes *in vitro* using antiserum to glial fibrillary acidic protein, *Brain Research* **89**:363–367.
Armstrong, C. M., 1975, K pores of nerve and muscle membranes, in: *Membranes: A Series of Advances*, Vol. 3 (G. Eisenman, ed.), pp. 325–358, Marcel Dekker, New York.
Arnold, J. M., 1974, IV Embryonic Development, in: *A Guide to Laboratory Use of the Squid Loligo pealei*, (J. M. Arnold, W. C. Summers, D. L. Gilbert, R. S. Manalis, N. W. Daw, and R. J. Lasek, eds.), pp. 24–44, Marine Biological Laboratory, Woods Hole, MA.
Arnold, J. M., 1990, Embryonic development of the squid, *this volume*.
Bayne, C. J., 1976, Culture of molluscan organs: a review, in: *Invertebrate Tissue Culture*, (K. Maramorosch, ed.), pp. 61–74, Academic Press, New York.
Bean, R. C., Shepherd, W. C., Chan, H., and Eichner, J. T., 1969, Discrete conductance fluctuations in lipid bilayer protein membranes, *J. Gen. Physiol.* **53**:741–747.
Booher, J. and Sensenbrenner, M., 1972, Growth and cultivation of dissociated neurons and glial cells from embryonic chick, rat, and human brain in flask cultures, *Neurobiology* **2**:97–105.
Bottenstein, J. E., 1984, Culture methods for growth of neuronal cell lines in defined media, in: *Methods for serum-free culture of neuronal and lymphoid cells*, (D. W. Barnes, D. A. Sirbasku. and G. H. Sato. eds.), pp. 3–13, A. R. Liss. Inc,. New York.
Brady, S. T., Lasek, R. J., and Allen, R. D., 1981, Fast axonal transport in extruded axoplasm from squid giant axon, *Biol. Bull.* **161**:304.
Brady, S. T., Lasek, R. J., and Allen, R. D., 1982, Fast axonal transport in extruded axoplasm from squid giant axon, *Science* **218**:1129–1131.
Cole, K. S., 1968, *Membranes, Ions and Impulses*, 569 pp., U. Calif. Press, Berkeley.
Currie. D. N., 1980, Identification of cell types by immunofluorescence in defined cell cultures of cerebellum, in: Tissue Culture in Neurobiology, (E. Giacobini *et al.*, eds.), pp. 75–87, Raven Press, New York.
Dagan, D. and Levitan, I. B., 1981, Isolated identified *Aplysia* neurons in cell culture, *J. Neuroscience* **1**:736–740.
Eagle, H., 1959, Amino acid metabolism in mammalian cell cultures. *Science* **130**:432–437.
Ehrenstein, G., Lecar, H., and Nossal, R., 1970, The nature of the negative resistance in bimolecular lipid membranes containing excitability inducing material, *J. Gen. Physiol.* **55**:119–133.
Ehrenstein, G., Blumenthal, R., Latorre, R., and Lecar, H., 1974, Kinetics of the opening and closing of individual EIM channels in a lipid bilayer, *J. Gen. Physiol.* **63**:707–721.
Freshney, R. K., 1983, *Culture of Animal Cells. A Manual of Basic Technique*. P. 191, A. R. Liss, Inc., New York.
Gremo, F., Porru, S., and Vernadakis, A., 1984, Effects of corticosterone on chick embryonic retinal cells in culture, *Dev. Brain Res.* **15**:45–52.

Hansen, E., 1976, A cell line from embryos of *Biomphalaria glabrato (Pulmonata)*, in: *Invertebrate Tissue Culture*, (K. Maramorosch, ed.), pp. 75–99, Academic Press, New York.

Harvey, A. L., 1984, *The pharmacology of nerve and muscle in tissue culture*, P. 161, A. R. Liss. Inc., New York.

Heyneman, D., 1976, Snail tissue culture, in: *Invertebrate Tissue Culture*, (K. Maramorosch, ed.), pp. 57–60, Academic Press, New York.

Hodge, A. J. and Adelman, W. J., Jr., 1983, The neuroplasmic lattice. Structural characteristics in vertebrate and invertebrate axons, in: *Structure and Function in Excitable Cells*, (D. C. Chang, I. Tasaki, W. J. Adelman, Jr., and H. R. Leuchtag, eds.), pp. 75–111, Plenum Press, New York.

Hodgkin, A. L., 1951, The ionic basis of electrical activity in nerve and muscle, *Biol. Rev.* **26**:339–409.

Kaczmarek, L. K., Finbow, M., Revel, J. P., and Strumwasser, P., 1979, The morphology and coupling of *Aplysia* bag cells within the abdominal ganglion and in cell culture, *J. Neurobiology* **10**:535–550.

Keller, M., Jackish, R., Seregi, A., and Hertting, G., 1985, Comparison of prostanoid forming capacity of neuronal and astroglial cells in primary cultures, *Neurochem. Int.* **7**:655–665.

Klein, K. C. and Jaffe, J. A., 1984, Development of *in vitro* fertilized eggs of the squid *Loligo pealei* and techniques for dechorionation and artificial activation, *Biol. Bull.* **167**:518.

Martin, R., 1965, On the structure and embryonic development of the giant fiber system of the squid Loligo vulgaris, *Z. Zellforsch* **67**:77–65.

Mirsky, R., Wendon, L. M. B., Black, P., Stolkin, C., and Bray, D., 1978, Tetanus toxin: a cell surface marker for neurones in culture, *Brain Res.* **148**:251–259.

Neher, E. and Sakmann, B., 1976, Single-channel currents recorded from membrane of denervated frog muscle fibers, *Nature* **260**:799–802.

Pettmann, B., Louis, J. C., and Sensenbrenner, M., 1979, Morphological and biochemical maturation of neurones cultured in the absence of glial cells, *Nature* **281**:378–380.

Raff, M. C., Fields, K. L., Hakomori, S., Mirsky, R., Pruss, R. M., and Winter, J., 1979, Cell-type-specific markers for distinguishing and studying neurons and the major classes of glial cells in culture, *Brain Res.* **174**:283–308.

Schacher, S. and Proshansky, E., 1983, Neurite regeneration by *Aplysia* neurons in dissociated cell culture: modulation by *Aplysia* hemolymph and the presence of the initial axonal segment, *J. Neuroscience* **3**:2403–2413.

Sigworth, F. J. and Neher, I., 1980, Single Na^+ channel currents observed in cultured rat muscle cells, *Nature* **287**:447–449.

Summers, W. C., 1990, Natural history of the squid, *this volume*.

Szabo, G. and Arnold, J. M., 1963, Studies of melanin biosynthesis in the ink sac of the squid (*Loligo pealii*). III. *In vitro* culture of tissues and isolated cells from the adult ink gland, *Biol. Bull.* **125A**:393–394.

Thampy, K. G., Sauls, C. D., Brinkley, B. R., and Barnes, E. M., Jr., 1983, Neurons from chick embryo cerebrum: ultrastructural and biochemical development *in vitro*, *Dev. Brain Res.* **8**:101–110.

Wong, R. G., Hadley, R. D., Kater, S. B., and Hauser, G. C., 1981, Neurite outgrowth in molluscan organ and cell cultures: the role of conditioning factor(s), *J. Neuroscience* **9**:1008–1021.

Chapter 13

Squid Optic Lobe Synaptosomes: Structure and Function of Isolated Synapses

ROCHELLE S. COHEN, HARISH C. PANT, and HAROLD GAINER

1. Introduction

Over the past two decades, the progress made in elucidating some of the cell biological mechanisms involved in synaptic transmission has been due, to a great extent, to the pioneering studies on the isolation of nerve endings, or synaptosomes, by the laboratories of Whittaker and De Robertis (Whittaker, 1959; De Robertis *et al.*, 1961). These isolated nerve endings, formed during tissue homogenization, consist of a presynaptic element which is *pinched off* from the distal end of the axon and a postsynaptic membrane (with its adherent density) which is derived from adjacent membranes surrounding a particular postsynaptic element, such as a dendritic spine, dendrite, soma, or axon (Jones, 1975). The membranes of the presynaptic element seal up, while those of the postsynaptic element may or may not. The two elements appear to be held together by material within the cleft between the pre- and postsynaptic membranes. A multitude of experiments have demonstrated the synaptosomes' ability to retain metabolic properties (e.g., oxygen and glucose utilization) and exhibit membrane phenomena (e.g., active transport of ions and molecules, release of putative neurotransmitters and neuropeptides, and response to depolarizing influences) characteristic of the intact nerve ending (Whittaker and Barker, 1972; De Belleroche and Bradford, 1973). In addition, synaptosomes serve as the starting point for further subfractionation into its component organelles, including synaptosomal plasma membranes, synaptic vesicles, synaptic junctional complexes, and postsynaptic densities (Whittaker and Barker, 1972).

Initial studies on synaptosomes derived from cephalopod head ganglia, in particular the optic lobes, were based on the early finding by Bacq (1935) and Bacq and Mazza (1935) which showed that the cephalopod central nervous system contains much higher levels of acetylcholine than that contained in the mammalian central nervous system. The presence of acetylcholine in squid (*Loligo pealei*) optic lobe synaptosomes at a level about 40 times that of guinea pig cortex (cf. Dowdall *et al.*, 1976, for review), together with a high affinity choline uptake system (Dowdall and Simon, 1973), makes this a central nervous system (CNS) preparation comparable to that derived from

R. S. COHEN • Department of Anatomy and Cell Biology, University of Illinois at Chicago, Chicago, Illinois 60612. H. C. PANT and H. GAINER • Laboratory of Neurochemistry, NINDS, National Institutes of Health, Bethesda, MD 20892

Torpedo electric organ for investigating cholinergic mechanisms of transmission (cf. Whittaker *et al.*, 1972; Dowdall *et al.*, 1976, for reviews).

Several laboratories have now focused their attention on other aspects of the squid optic lobe synaptosome fraction. Morphological investigations range from a correlation between isolated synaptosomes and intact nerve endings in the optic lobe using conventional transmission electron microscopy (Haghighat *et al.*, 1984) to more detailed analyses of the correlations between distributions of diffusible ions and non-diffusible elements and subcellular structure using X-ray microanalysis (Fiori *et al.*, 1988), to the determination of size and dispersity of synaptosomes using laser light scattering (Sattelle *et al.*, 1987). Since the original findings on the high concentration of acetylcholine (Dowdall and Whittaker, 1973) and the presence of a high-affinity choline uptake system in squid optic lobe synaptosomes (Dowdall and Simon, 1973), other putative transmitters have been studied in this fraction, including epinephrine (Matsumura, 1987), L-noradrenaline (Dowdall, 1974; Pollard *et al.*, 1975), 5-hydroxytryptamine (5-HT) (Dowdall, 1974; Pollard *et al.*, 1975), glutamate, aspartate and glycine (Pollard *et al.*, 1975). The neuropeptide somatostatin has been localized in the intact optic lobe (Feldman, 1986), although measurements of this peptide in the synaptosome fraction have not yet been done. The identification of several phosphoproteins (Pant *et al.*, 1979, 1983; Bass *et al.*, 1987), endogenous protein kinases (Unver *et al.*, 1986; Bass *et al.*, 1987), and regulators of phosphorylation, such as calcium (Pant *et al.*, 1983; Unver *et al.*, 1986; Bass *et al.*, 1987), make this fraction a fruitful preparation for studies on the regulation of neurotransmitter release, as well as other events mediated by protein phosphorylation at the nerve terminal.

The special appeal of the squid optic lobe synaptosome fraction is that it represents a subcellular compartment of the neuron that can be studied in relation to other well-characterized neuronal domains of the squid nervous system, including the intact giant axon and extruded axoplasm, the giant synapse, and cell bodies of the stellate ganglion. It is the intent of this chapter to review the current status of knowledge of squid optic lobe synaptosome structure, biochemical composition, and function, and relate this information to comparable data on other neuronal components of the squid nervous system, as well as synaptosome preparations derived from other sources.

2. Subcellular fractionation of squid optic lobe synaptosomes

2.1. Preparation of synaptosomes

The method for the preparation of squid optic lobe synaptosomes depends upon the high osmolarity (i.e., 1200 mOsm/liter) of body fluids of this marine organism (cf. Dowdall *et al.*, 1976, for discussion). Dowdall and Whittaker (1973) developed the original isolation procedure using 0.7 – 0.8 M sucrose for the homogenization medium, since sucrose solutions having the equivalent osmolarity of body fluids are too viscous at 0 °C (Dowdall and Whittaker, 1973). While this solution is hypoosmotic, a large proportion of tissue acetylcholine is bound to the synaptosomes (Dowdall *et al.*, 1976).

The optic lobes are located immediately behind the eye (Fig. 1). After decapitation, the optic lobes are quickly removed, placed in a cold receptacle and weighed. The entire procedure is carried out between 0 °C – 4 °C. According to the

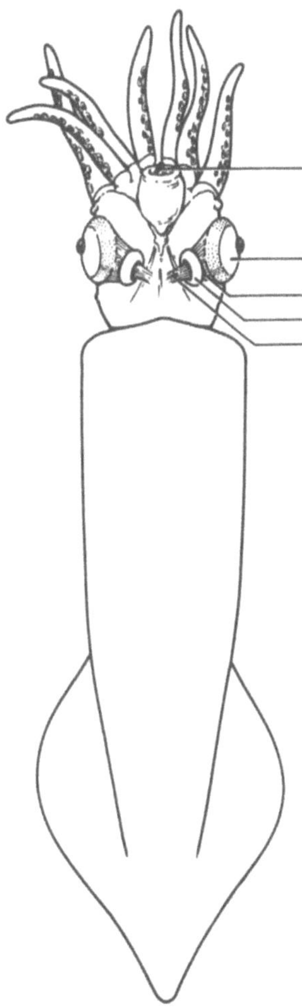

Figure 1. Diagram of dissected squid to show the position of the optic lobes. The optic lobes have an irregular dumb-bell shape and lie with their elongated axis oblique to the eye. Optic nerves from the eye penetrate the lateral surface of the optic lobe. The optic tract emerges from the optic lobe on the opposite side.

original procedure (Dowdall and Whittaker, 1973), homogenates (10% wt/vol) of tissue are prepared in the above sucrose solution using 12 complete strokes of an Aldridge homogenizer (Aldridge *et al.*, 1960) (1% clearance, 840 rev/min.). The resultant homogenate is then centrifuged at 1000 × g (g is the acceleration due to gravity at sea level or 9.80 m/sec^2) for 10 min in a Sorvall preparative centrifuge with an SS 34 rotor, resulting in a pellet and supernatant. The former, consisting of nuclei and cellular debris, can be washed by resuspension in the homogenizing medium and centrifuging again as described above. The supernatants are then pooled and centrifuged at 17,300 × g for 1 hr. This results in essentially three fractions, a small mitochondrial pellet, a supernatant, and a firm, white pellicle, which floats on the surface of the supernatant and which is designated as the synaptosome fraction. The time for this centrifugation can be reduced to 30 min, although a higher yield is obtained at 60 min (Dowdall *et al.*, 1976). Successful removal of the pellicle can be achieved by gently teasing it with the sealed tip of a Pasteur pipette and then carefully decanting it from the centrifuge tube. It is not clear why the synaptosomes float in 0.7 – 0.8 M sucrose since their cytoplasm is relatively hyperosmotic to the homogenizing solution (Dowdall *et al.*,

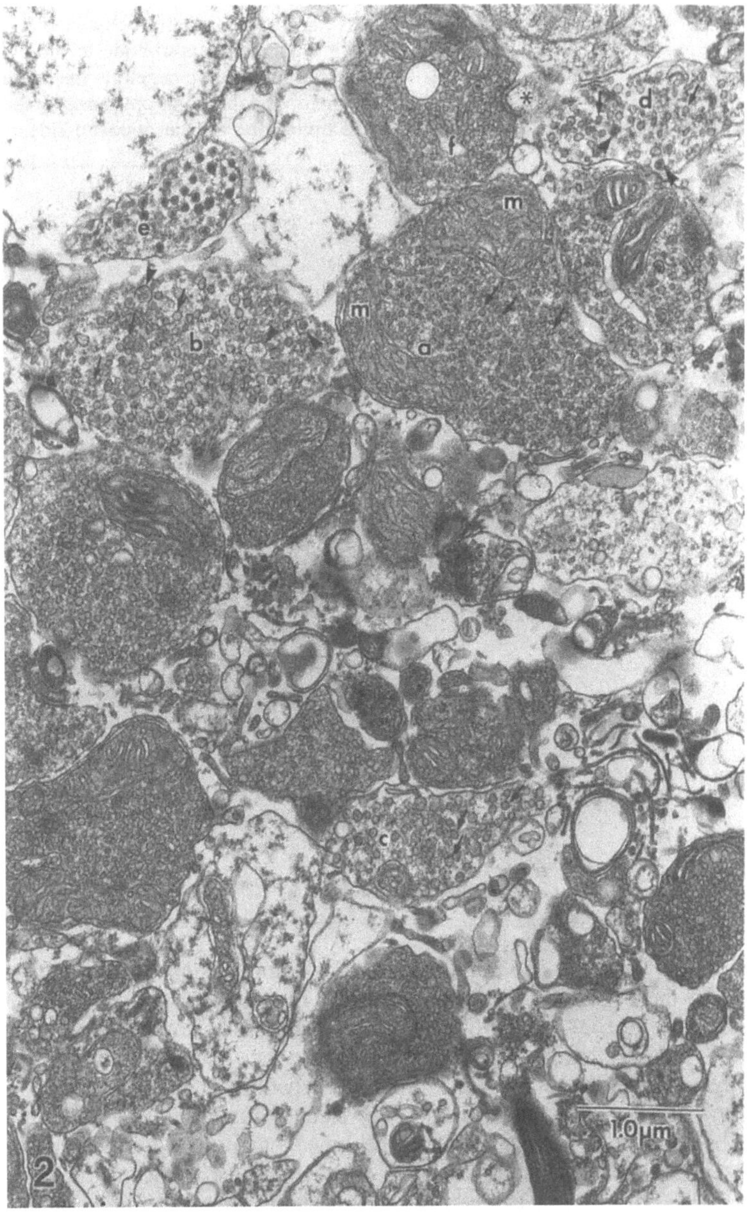

Figure 2. Synaptosome fraction prepared from whole squid optic lobe. The various profiles correspond to endings in the intact optic lobe. Several synaptosomal types can be distinguished. One type (*a*), characterized by many, densely packed clear vesicles (arrows), corresponds to photoreceptor endings. Another type (*b*) contains pleomorphic vesicles (arrowheads) in addition to clear vesicles (arrows) and is also probably derived from a photoreceptor ending. The synaptosome designated (*c*) contains fewer clear vesicles (arrows) than those described above and is likely to have been derived from a chemical synapse. The synaptosome marked (*d*) contains

1976). Dowdall *et al.* (1976) propose that this may be due to the solutes within the synaptosomes which are lower in density than an isosmotic or hypoosmotic concentration of sucrose. Since the squid optic lobe does not contain myelin, the resultant pellicle does not need to be subjected to further subcellular fractionation on sucrose density gradients, as is the case for the crude mitochondrial pellet (which is enriched in synaptosomes) obtained after differential centrifugation of mammalian brain (Dowdall *et al.*, 1976).

A comparable synaptosomal fraction was obtained by a modification of this method by Pollard and Pappas (1979), in which the homogenizing medium is isotonic 1.0 M sucrose. Sixteen squid optic lobes are homogenized as a 20% (wt/vol) solution in 1.0 M sucrose by six complete strokes of a glass-on-glass Ten-broek homogenizer (B pestle). The homogenate is centrifuged at 10,000 × g for 1 hr at 4 °C in a Sorvall preparative centrifuge using an SM-24 rotor. The resultant pellicle is decanted as described above. Figure 2 is an electron micrograph of a representative area of a synaptosome fraction derived in this manner, which is described in detail below.

Dowdall (1974) has shown that freshly prepared squid synaptosomes in artificial sea water (pH 7.4) can take up O_2 in a linear fashion for 30 min. Cyanide inhibited O_2 utilization, which also decreased when synaptosomes were placed in a hypoosmotic solution prior to incubation in sea water. The respiratory rate of synaptosomes can be maintained at 0 °C in 0.7 M sucrose for up to 8 hr following isolation, although a decreased respiratory rate resulted with longer storage in 0.7 M sucrose or shorter storage in sea water.

Synaptosomes derived from squid optic lobes are also capable of a high rate of incorporation of tritiated leucine into synaptosomal proteins (Hernández *et al.*, 1976). Two protein synthetic systems appear to be implicated, i.e., one located in the cytoplasm which is cycloheximide-sensitive and another, associated with mitochondria, which is chloramphenicol-sensitive.

2.2. Preparation of synaptosomal plasma membranes

Synaptosomal plasma membranes (SPMs) can be prepared from squid optic lobe synaptosomes in a manner similar to that described for the isolation of mammalian SPMs (Cohen *et al.*, 1977). The following procedure is based on that described for mammalian SPMs (Cohen *et al.*, 1977) and squid synaptosomal ghosts (Dowdall and Whittaker, 1973). The synaptosomal pellicle, obtained as described above, is resuspended and lysed in distilled water (Dowdall and Whittaker, 1973) or 6 mM tris (hydroxymethyl) aminomethane-HCl (Tris-HCl), pH 8.1 (Cohen *et al.*, 1977) with a glass-on-glass Ten-broek homogenizer. Lysis is carried out for 45 min at 4 °C. The SPMs are spun down at 48,200 × g for 20 min, resuspended and subsequently purified on a sucrose-density gradient consisting of 10 ml each of 0.85 M, 1.0 M, and 1.2 M sucrose solutions, all containing 1.0 mM $NaHCO_3$; 6 ml of the resuspended crude SPM pellet in 0.32 M sucrose – 1 mM $NaHCO_3$ is layered on top. The sucrose density gradients are spun for 2 hrs at 100,000 × g in an SW 27 rotor in a Beckman L565 ultracentrifuge. The SPMs, banding between 1.0 M and 1.2 M sucrose, are recovered

both clear (arrows) and dense-cored (arrowheads) vesicles, while that labeled (*e*) contains predominantly dense-cored vesicles. A portion of a small spine (asterisk) adheres to the synaptosome marked (*f*).

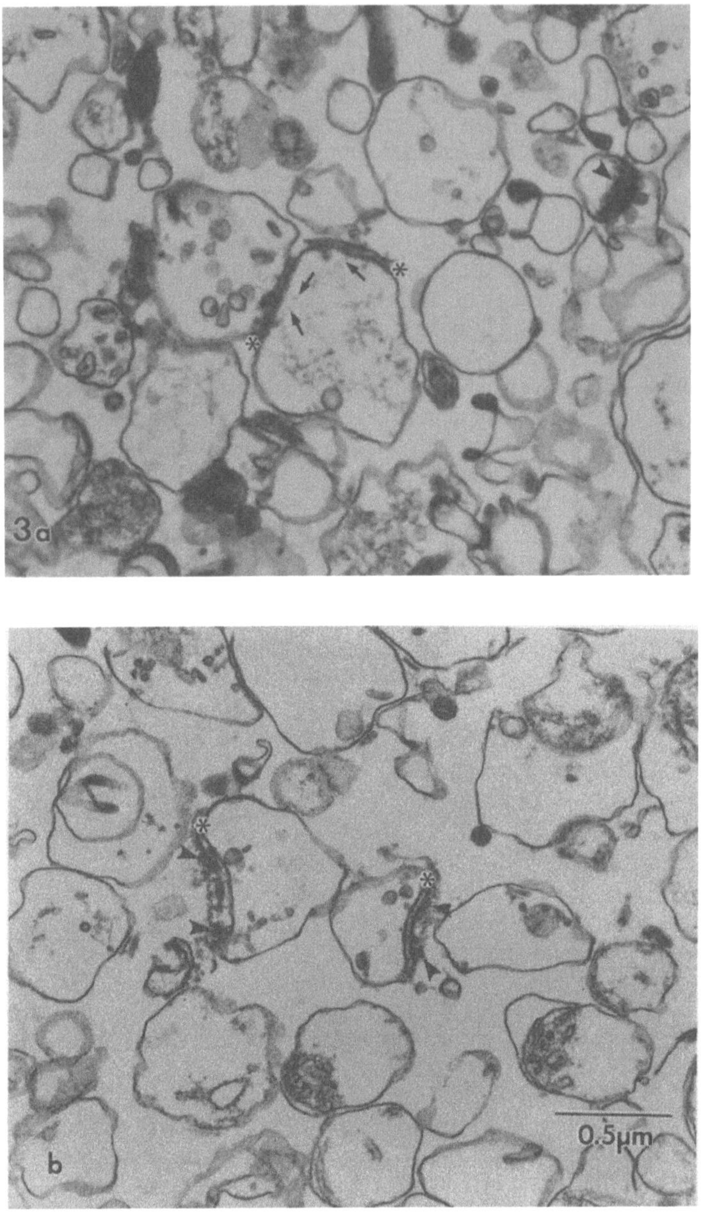

Figure 3. High magnification electron micrographs of the synaptosomal plasma membrane (SPM) fractions derived from squid optic lobe (*a*) and canine cerebral cortex (*b*). Although a postsynaptic density (PSD) (arrowhead) is occasionally associated with the postsynaptic membrane in squid SPMs, usually only a few fine filaments (4 – 6 nm in diameter) (small arrows) are seen to extend from the postsynaptic membrane. In contrast, SPMs derived from canine cerebral cortex usually have a thick PSD (arrowheads) associated with the postsynaptic membrane. Asterisks, synapses.

by aspiration, diluted, and centrifuged at 48,200 × g in a Sorvall RC2B centrifuge using an SS-34 rotor (Cohen *et al.*, 1982). The protease inhibitor phenyl-methylsulfonyl-fluoride (PMSF) can be added to all solutions at a concentration of 1.0 mM to prevent proteolysis of proteins during the isolation procedure.

The ultrastructure of the SPM fractions isolated from canine cerebral cortex and squid optic lobe has been compared (Cohen *et al.*, 1982). Prominent postsynaptic densities (PSDs) are attached to the postsynaptic membranes in canine SPMs (Figure 3b), whereas those derived from squid optic lobe usually lack well-developed PSDs (Figure 3a). The squid SPMs derived by the methods described here are morphologically similar to the synaptosomal ghosts seen in the osmotically ruptured squid optic lobe synaptosome fraction of Dowdall and Whittaker (1973). An enrichment of the enzyme marker, acetylcholinesterase, was seen in the fraction between 1.0 and 1.2 M sucrose when lysed synaptosomes were subjected to sucrose density gradient centrifugation (Dowdall and Whittaker, 1973).

While the squid lobe synaptosome and SPM fraction are highly enriched in their respective subcellular components, it should be noted that they are not completely purified. The isolation of purified fractions from brain is complicated by the heterogeneity of cell types, as well as the problem of isolating pure membrane fragments of unknown cellular origin (cf. review by Jones, 1975). Therefore, brain fractions may show an enrichment of a particular subcellular elements, as measured by enzymatic marker studies, for example (cf. Ueda *et al.*, 1979, for discussion), but should not be regarded as *pure* subcellular components.

2.3. Preparation of synaptic vesicles

The preparation of a synaptic vesicle fraction generally involves lysis of synaptosomes in water or dilute buffer and subsequent density gradient centrifugation (Whittaker *et al.*, 1964). Dowdall and Whittaker (1973) have prepared a synaptic vesicle fraction from squid optic lobe synaptosomes as follows. Synaptosomes are resuspended and subjected to osmotic shock in distilled water (5 ml/g optic lobe). The lysate is then layered onto sucrose density gradients consisting of 5 layers of sucrose (5 ml each of 0.4 M, 0.6 M, 0.8 M, 1.0 M, and 1.2 M). After centrifugation at 53,500 × g for 120 min, synaptic vesicles are recovered at the interface between 0.4 M and 0.6 M sucrose. This fraction contains a large proportion of vesicles similar in diameter to that seen within synaptosomes (approximately 50 nm in diameter). However, membrane vesicles of other sizes, as well as other membrane fragments, were also seen. Acetylcholine was enriched in the following fractions: within the 0.4 M sucrose, between 0.4 M and 0.6 M sucrose, between 0.6 M and 0.8 M sucrose and between 0.8 M and 1.0 M sucrose. With the exception of the fraction located between 0.4 M and 0.6 M sucrose discussed above, electron microscopic examination of the other fractions was not possible due to low yields. Huttner *et al.* (1983) have presented a modification of the procedure of Whittaker *et al.* (1964) and Nagy *et al.* (1976) for the isolation and purification of synaptic vesicles from mammalian brain. This procedure may be superior for the preparation of squid synaptic vesicles. However, it has not yet been tried with squid synaptosomes.

3. Structure of squid optic lobe synaptosomes

3.1. Morphological characterization

The squid optic lobe consists of two distinct regions, an outer cortex, or deep retina, and an inner medulla. The medulla is approximately five to six times thicker than the cortex and separated from that region by a single layer of cells, the mononuclear layer. The cortex is comprised of four main layers, i.e., the outer nuclear layer, the outer plexiform layer, the inner nuclear layer, and the inner plexiform layer. Using silver techniques, the outer plexiform layer can be further subdivided into four radial layers separated by tangential ones (Young, 1974). The cortex receives incoming nerve fibers which expand into conical or carrot-shaped presynaptic endings and terminate on tangential dendrites of second order visual neurons. The inner medulla receives axons of second order visual neurons and sends efferents to the motor centers. Cajal (1917) and Young (1974) distinguished five zones within the medulla, i.e., the palisade layer, the frontier zone, the zone of radial columns, the zone of tangential bundles, and the zone of optic tract bundles. Islands of several neuronal types are also present within this region. Whereas the layers of the cortex are well-delineated, those of the medulla are not clearly defined. The receptive fields of the second-order visual cells located in the plexiform zone of the cortex may function in providing a particular dendritic orientation for the detection of selected characteristics of the visual input, which is then transmitted to the medulla (Young, 1974). The latter then chooses the appropriate response and, in turn, transmits this information to motor centers of the central nervous system (Young, 1974).

The ultrastructure of synaptosomes derived from the optic lobes of the squid (Dowdall and Whittaker, 1973; Haghighat *et al.*, 1984) and the supraesophageal ganglia of octopus (Jones 1967, 1970a, 1970b) have been well-characterized. Five morphological types of synaptosomes are distinguishable (Figure 2) which correspond to synaptic terminals present in the intact squid optic lobe (Cohen, 1973; Haghighat *et al.*, 1984). One type of synaptosome (*a*, Fig. 2), which corresponds to the terminals in the first radial layer of the outer plexiform layer (except the fourth radial and fourth tangential layers), represents the dense photoreceptor terminals present in the plexiform layer, as well as terminals present in the medulla. This type is relatively large; it contains densely staining cytoplasm and is packed with clear vesicles. A second large type (*b*, Fig. 2) also corresponds to photoreceptor terminals but has less dense cytoplasm and fewer clear vesicles but more pleomorphic vesicles than the synaptosomes described above. A third type (*c*, Fig. 2), which is probably derived from a chemical synapse, contains many fewer clear vesicles than the above types. Of the remaining types, one contains both clear and dense-cored vesicles (*d*, Fig. 2) and the other, predominantly dense-cored vesicles (*e*, Fig. 2). The latter three profiles described appear smaller than the photoreceptor-derived terminals and correspond to synaptic terminals found in all layers of the cortex and medulla, except the first radial layer of the outer plexiform layer.

The heterogeneity in synaptosomal size has been investigated by Sattelle *et al.* (1987) using laser light scattering. The mean diameter of the synaptosomes ranged from 0.5 – 2.0 μm. In agreement with the results described above, the fraction essentially consists of two major size classes, i.e., smaller particles, approximately 300 – 700 nm and larger particles, approximately 1500 – 5000 nm.

Portions of the postsynaptic processes adhering to the two photoreceptor-derived types of synaptosomes (for example, see *f*, Figure 2) are derived from postsynaptic processes in the form of small spines (Haghighat *et al.*, 1984). In intact tissue, they are seen in cross-section as small, circular profiles within the photoreceptor terminals (Haghighat *et al.*, 1984). Neither the pre- or postsynaptic densities associated with intact mammalian synapses are well-defined in intact or isolated optic lobe endings, although fine filamentous material appears to extend from the postsynaptic membrane (Haghighat *et al.*, 1984). In fact, paramembranous densities are not obvious in any of the intact or isolated squid optic lobe endings, even when stained with ethanolic phosphotungstic acid which preferentially stains these densities (Haghighat *et al.*, 1984). Postsynaptic densities appear to be greatly reduced in squid optic lobe synaptosomes. In this regard, *Torpedo* synaptosomes (and synaptosomal plasma membranes) completely lack adhering postsynaptic membranes (Morel *et al.*, 1982).

Recently, X-ray maps from freeze-dried cryosections of squid optic lobe synaptosomes have been made (Fiori *et al.*, 1988). This data can provide relevant quantitative analyses about diffusible ions and non-diffusible elements relative to the functional state of various subcellular structures. X-ray maps of the cryosections of a synaptosome suspension indicated the presence of high concentrations of potassium (as well as sulfur), but low internal levels of calcium.

3.2. Preparation of tissue for electron microscopy

The osmolarity of marine organisms' body fluids must be considered when preparing intact tissue and subcellular fractions for electron microscopy. Good preservation of intact optic lobe tissue and synaptosomes has been achieved with the following fixation protocol. Optic lobes are immersed in a cold solution containing the following: 84.0 ml artificial sea water, 16.7 ml 1.0 M sodium cacodylate-HCl buffer (pH 7.2), 20.0 ml 25% glutaraldehyde, 1.7 ml acrolein, 4.1 ml dimethylsulfoxide, 42.0 ml 8% paraformaldehyde, and 0.02 g $CaCl_2$. Because acrolein is flammable and highly reactive, special precautions must be used when handling this reagent. Preparation of the fixative should be done wearing gloves under the hood. Waste acrolein can be neutralized by pouring it into 10% sodium bisulfite (Hayat, 1986). The hazardous effects of acrolein are reviewed elsewhere (Hayat, 1986).

The optic lobes or synaptosomal pellicle can then be divided into small pieces, or portions, respectively, while immersed in the fixative and fixed for approximately 3 1/2 hrs at 4 °C. The piece of tissue or pellicle is then rinsed in 0.1 M cacodylate buffer, pH 7.4, postfixed with 1% OsO_4 in 0.1 M cacodylate buffer, then dehydrated and embedded in an appropriate embedding medium, such as Epon, according to standard electron microscopy procedures (Haghighat *et al.*, 1984). The addition of acrolein is advantageous as it is highly reactive (Hayat, 1986) and has been shown to quickly stabilize proteins *in situ* (Flitney, 1966). King *et al.* (1983) have also demonstrated that a brief perfusion (5 min) of an acrolein solution (5%) preserves antigenicity for immunocytochemical experiments.

Alternatively, subcellular fractions can be fixed at 4 °C while suspended in sucrose solutions. Dowdall and Whittaker (1973) recommend adding 4.5 volumes of ice-cold 2.5% glutaraldehyde in 0.15 M phosphate buffer or 0.1 M phosphate buffer + 0.105 M NaCl. A pellet can then be obtained by centrifugation.

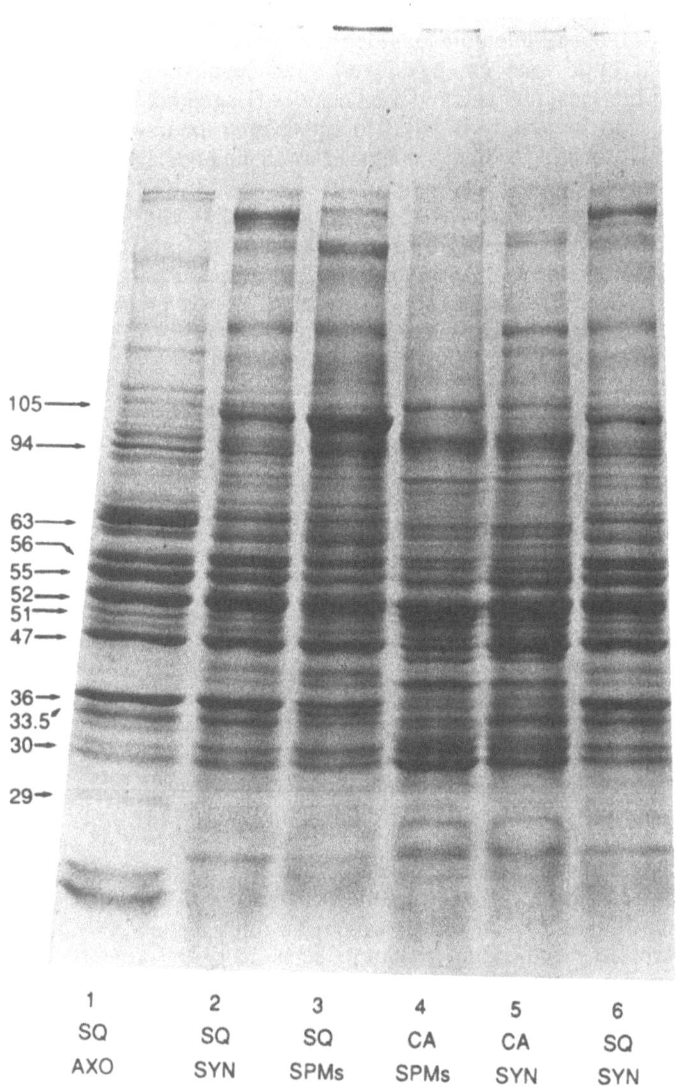

Figure 4. SDS-polyacrylamide gel electrophoretic patterns of fractions derived from squid and canine neural tissues. Major bands are given in descending order of staining density. Slot 1: squid axoplasm (*SQ AXO*). The major bands are those at.52, 36, 47, 63, 56, and about 94 kD; a high molecular weight doublet is also present. Slot 2: squid optic lobe synaptosomes (*SQ SYN*). The most prominent bands are those at 52, 36, 47, 55, 56, 105, and 32 kD and a large polypeptide which could not be extrapolated from the standard curve. Slot 3: squid optic lobe synaptosomal plasma membranes (*SPMs*) (*SQ SPMs*). The major polypeptides include those at 105, 47, 29, 30, 51.5, 55, 56, and 52 kD. Slot 4: canine cerebral cortex *SPMs* (*CA SPMs*). The most predominant bands are those at 51, 29, 47, 84, 36, and 40 kD. Slot 5: canine cerebral cortex synaptosomes (*CA SYN*). Major bands are those at 52, 47, 55, 29, 30, 34, and 97 kD. Slot 6: squid optic lobe synaptosomes (*SQ SYN*). Molecular weights are expressed in kiloDaltons (kD).

4. Proteins in the squid optic lobe synaptosome and synaptosomal plasma membrane fractions

4.1. Polypeptide composition

The protein composition of various fractions derived from squid and mammalian, i.e., canine, central nervous systems were compared by one- and two- dimensional gel electrophoresis (Cohen *et al.*, 1982). Figure 4 shows sodium dodecyl sulfate (SDS)-polyacrylamide gel electrophoretic patterns of synaptosome and SPM fractions derived from squid and canine neural tissues. The major bands in squid axoplasm are also shown for comparison (Fig. 4, slot 1). The most predominant bands in squid synaptosomes (Fig. 4, slots 2 and 6 and in descending order of staining density) include those at: 52, 36, 47, 55, 56, 105, and 32 kiloDalton (kD), and a very large polypeptide which could not be extrapolated from the standard curve, but which differs from the high molecular weight proteins seen in squid axoplasm. Canine synaptosomes (Fig. 4, slot 5) contain only a negligible amount of the 105 kD band, suggesting that this polypeptide may be specific to the squid system. Canine synaptosomes also show a thin band at 51 kD (characteristic of cerebral cortex and midbrain PSDs) which also appears to be present in small amounts in squid synaptosomes.

The polypeptide patterns of squid optic lobe-derived SPMs (Fig. 4, slot 3) were compared with those derived from canine cerebral cortex (Fig. 4, slot 4), as well as with squid optic lobe synaptosomes (Fig. 4, slots 2 and 6). Differences were evident in the protein patterns between the squid and canine SPM fractions. While squid SPMs have only a small amount of the 52 kD band, it is the most salient band in the canine SPM fraction. On the other hand, squid SPMs have a very prominent band at 105 kD, which may be lacking or present in small amounts in canine SPMs. Squid SPMs also lack larger amounts of the 51 kD band seen in canine SPMs and which is the major protein in PSDs derived from canine cerebral cortex and midbrain (Kelly and Montgomery, 1982).

Squid synaptosomes have several bands in common with SPMs derived from them, including similar amounts of the 52, 26, 55 and 56 kD bands. While the 105 kD band is present in the squid synaptosome fraction, it is highly enriched in the SPM fraction. Figure 5 shows the two-dimensional electrophoretic patterns of squid optic lobe synaptosomes and SPMs. Several polypeptides in the synaptosome fraction overlap with those of the SPM fraction and, therefore, may represent the synaptosomal plasma membrane component of synaptosomes. These include polypeptides at about 92.5 kD (see 1, Fig. 5), between 45 and 66 kD (2 – 6), at 45 kD (7), between 25.7 and 45 kD (8 – 14), at slightly below 25.7 kD (15), at slightly above 18.4 kD (16), at around 14.3 kD (17 – 19), and at 12.2 kD (20).

The two-dimensional electrophoretic patterns of squid and canine SPMs are compared in Figure 6. These data reveal some prominent lower molecular weight species in the two fractions which are indistinct in the one dimensional gels and also show the complexity of the individual bands from the one dimensional gels in the region between 45 kD and 66 kD. Most of the polypeptides in the two fractions do not overlap, with the exception of those species at about 45 kD and between 50 and 66 kD.

Several major differences in the polypeptide patterns of SPMs derived from canine cerebral cortex and squid optic lobe were seen. While canine SPMs show a pattern

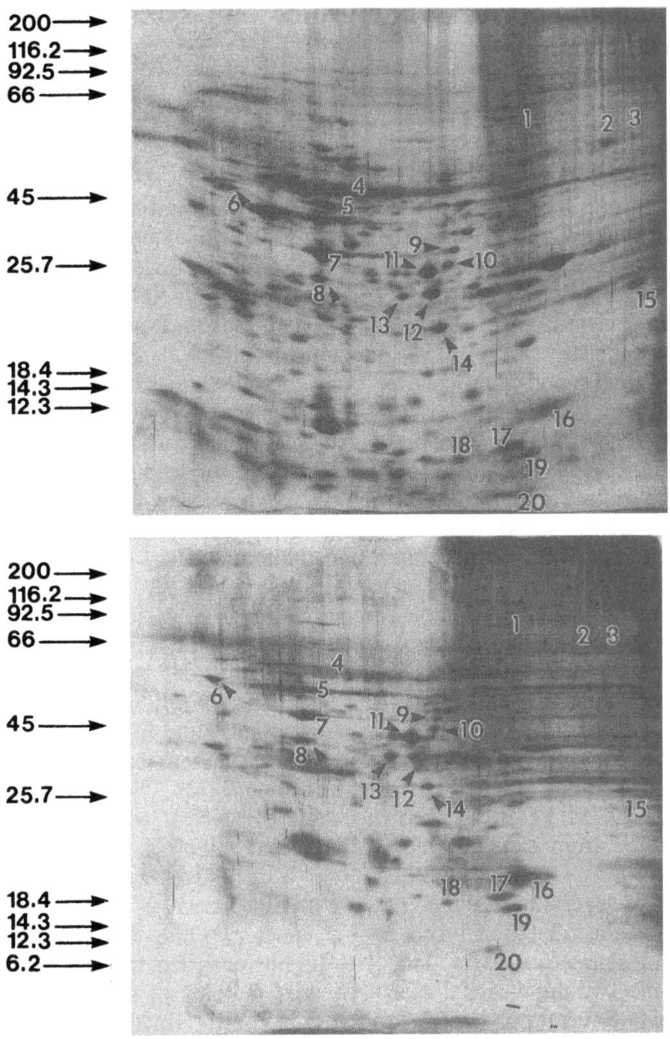

Figure 5. Two-dimensional gels of synaptosomes (a) and synaptosomal plasma membranes (b) derived from squid optic lobes. Several polypeptides (arrowheads) of the synaptosome fractions and SPM fraction overlap. These include a polypeptide around 92.5 kD (1), polypeptides between 45 and 66 kD (2 - 6), one at 45 kD (7), several proteins between 25.7 and 45 kD (8 - 14), one slightly below 25.7 kD (15), one slightly above 18.4 kD (16), three at around 14.3 kD (17 - 19) and one polypeptide at 12.2 kD (20). Molecular weights are expressed in kiloDaltons (kD).

typical of those derived from mammalian brain (Cohen *et al.*, 1977), squid SPMs show a pattern that is comparable to pre- and postsynaptic membranes derived from purely cholinergic endings of *Torpedo* (Morel *et al.*, 1982). The presence of the 105 kD band in squid SPMs and its lack in canine SPMs, and conversely, the smaller amount of the 52 kD band characteristic of mammalian synapses (cf. review by Mahler, 1977) in squid SPMs, denotes the main differences in polypeptide composition between the squid and canine fractions. The larger amounts of the 51 kD polypeptide, which represents a component of a Ca^{2+}/calmodulin-dependent protein kinase (Kelly *et al.*, 1984) and which is designated as a marker for cerebral cortex-derived PSDs (Kelly and Montgomery, 1982), seen in canine SPMs compared to that seen in squid SPMs may underscore the concomitant lack of a prominent PSD in squid optic lobe synapses.

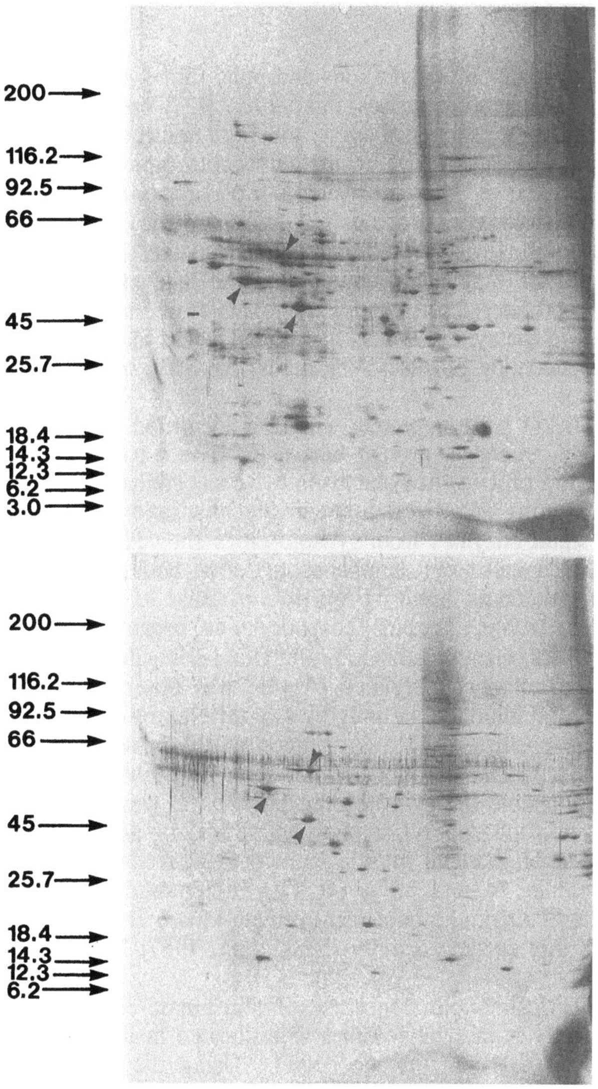

Figure 6. Two-dimensional gels of synaptosomal plasma membranes derived from squid optic lobe (a) and canine cerebral cortex (b). The gel in (a) shows polypeptides in the 18 kD region not detected in the one-dimensional gels and also show the complexity of the bands in the region between 45 and 66 kD. Many differences are seen in the two gels except for a spot at about 45 kD and two spots between 50 and 66 kD (arrowheads).

It should be noted here that comparisons of electrophoretic profiles of proteins from tissues of such diverse species as squid and mammals (as in Figs. 4 and 6) are limited as identifying factors. Evolutionary change can dramatically alter these migration parameters by changes in only a few amino acids. Hence, the absence of co-migration (even on 2D-gels) cannot be taken as evidence of a fundamental (functional) difference in the proteins being examined. This is apparent for the Ca^{2+}/calmodulin-dependent protein kinase II which in mammalian brain migrates as a 51 kD subunit, but migrates 54 kD subunit in the squid (Bass *et al.*, 1987).

4.2. Protein phosphorylation

In a comparison of protein substrates in squid axons and optic lobe synaptosomes which could be phosphorylated by endogenous kinases, Pant *et al.*, (1979) reported that while the major phosphate acceptors in axoplasm were the high molecular weight neurofilament protein subunits (ca., 200 and 400 kD), in the synaptosomes the major substrate was 47-50 kD. They also showed that adenosine-5'-triphosphate (ATP) with P^{32} in the γ position (γ^{32}-P-ATP) was efficacious as a phosphate donor for phosphorylation in *intact* synaptosomes as was labeled inorganic phosphate. In addition, by use of inhibitors of ATP synthesis they demonstrated that the labeled ATP was used directly as a phosphate donor under these conditions. This suggested that the endogenous substrates and kinases in the synaptosomes had direct access to the labeled ATP in the medium, either by translocation of intact ATP across the SPM or by action of ecto-kinases.

The character of this 47 – 50 kD phosphoprotein was later identified by Bass *et al.* (1987) as squid Ca^{2+}/calmodulin-dependent protein kinase II. The activity of this enzyme was 15 times greater (per mg protein) in synaptosomes versus stellate ganglion cell bodies; this enzyme activity in turn was 3 – 5 times greater than axoplasm from the giant axon. Isolation of the squid enzymes from synaptosomes showed it had Mr values of 54, and 58 – 60 kD, that it underwent autophosphorylation both *in situ* and *in vitro*, and that the 54 kD subunit could bind ^{125}I-calmodulin (Bass *et al.*, 1987). These workers also showed that the enzyme, after autophosphorylation, became calcium-independent as has been reported for mammalian Ca^{2+}/calmodulin-dependent protein kinases. Dephosphorylation of the autophosphorylated enzyme was observed in the synaptosome preparation, and could be inhibited partially by 10 mM NaF or completely by the use of ATP with S in the γ position (ATP-γ-S), suggesting the presence of an endogenous phosphatase. This phosphatase activity has not been fully characterized, but using ^{32}P-phosphorylase as a substrate the synaptosome preparation was shown to contain about 2.5 mμ per mg protein of phosphorylase phosphatase activity as compared to a value of 1.1 mμ per mg in rabbit skeletal muscle extracts. Interestingly, squid axoplasm was also shown to contain 54 and 58 – 60 kD ^{125}I-calmodulin binding proteins corresponding to squid Ca^{2+}/calmodulin-dependent protein kinase II (Krinks *et al.*, 1988), and a small amount of this enzyme activity (Bass *et al.*, 1987; Pant *et al.*, 1986).

Approximately 90% of the Ca^{2+}/calmodulin-dependent protein kinase activity and 65% of the protein kinase C-like activity in synaptosomes were located in the 100,000 g particulate fraction, and presumed to be membrane bound. Using the particulate fraction Bass *et al.* (1987) found that the addition of γ^{32}-P-ATP caused three major proteins to be phosphorylated, at Mr values of 54, 58, and 100 – 105 kD. The 54 kD protein was most rapidly phosphorylated to saturation, suggesting that this was the autophosphorylated enzyme, whereas the 100 kD protein was more slowly phosphorylated, presumably as the major substrate. Several other phosphoproteins between 130 – 200 kD are also phosphorylated in synaptosomes and SPMs in a calcium-dependent manner (Pant *et al.*, 1983; Bass *et al.*, 1987) and some of these proteins (at 90, 100, 130 and 180 kD) appear to dramatically increase their incorporation of labeled phosphates in the presence of 4-aminopyridine (Pant *et al.*, 1983).

The efficacy of γ-^{32}P-ATP as a phosphate donor for the Ca^{2+}/calmodulin-dependent protein kinase in intact squid synaptosomes remains curious, since this does not appear to occur in mammalian synaptosomes (H. Pant, 1989). One possible interpretation is that despite the absence of defined PSDs (Figs. 2 and 3), the squid synaptosome preparation could have extensive postsynaptic membranes which are leaky to intact ATP making it accessible to endogenous kinases and substrates. Alternatively, the presynaptic membrane could be leaky to intact ATP. Which of these alternatives, or both, is correct remains to be studied.

4.3. Proteases

Using a ^{14}C-casein degradation assay, Pant *et al.* (1982) found that squid optic lobe synaptosomes were devoid of calcium-dependent neutral protease (CaNP, calpain) activity. The significance of this observation is that, in contrast, giant axons and axoplasm contain abundant CaNP activity which appears to selectively degrade neurofilament proteins in axoplasm. The absence of both neurofilament proteins and CaNP activity in squid synaptosomes, suggests that the neurofilament proteins are degraded by CaNPs in the nerve terminals followed by an autolysis of the CaNPs (Pant *et al.*, 1982).

In a general analysis of protease activity in squid synaptosomes, Gallant *et al.* (1984) reported, using a ^{14}C-casein assay, that there were abundant acid protease activities in synaptosomes versus axoplasm where these activities were very low. The activity at pH 3 was 95% inhibited by leupeptin and p-hydroxymercuribenzoate (p-chloromercuribenzoate) (PCMB), but only 67% inhibited by pepstatin. In contrast, at pH 6 the protease activities were not inhibited by leupeptin and PCMB, but were 92% inhibited by pepstatin. These data suggest that squid synaptosomes contain acid hydrolyases (e.g., cathepsin B and D), but firmer identification of the specific proteases is needed.

5. Squid optic lobe synaptosomes as model cholinergic endings

The original studies on squid optic lobe synaptosomes have identified this fraction as a rich source of cholinergic nerve terminals (Dowdall and Whittaker, 1973; Dowdall and Simon, 1973). Salient features of cholinergic terminals, in general, include the presence of large amounts of acetylcholine, choline acetyltransferase, the enzyme responsible for its synthesis, and acetylcholinesterase, the enzyme which hydrolyzes acetylcholine (Whittaker and Barker, 1972). All of these components of cholinergic terminals are found in synaptosomes derived from squid optic lobes and are described below.

The presence of acetylcholine in squid optic lobe synaptosomes has been demonstrated (Welsch and Dettbarn, 1972; Dowdall and Whittaker, 1973) and has, in fact, found to be present there in concentrations 40 times greater than that found in synaptosomes derived from guinea pig cerebral cortex (Dowdall and Whittaker, 1973). While homogenization of squid optic lobes indicated the presence of acetylcholine in a *free* form, following boiling at acid pH, the *total* levels of acetylcholine increased two-fold. This may reflect the *bound* acetylcholine probably present in synaptosomes (Welsch and Dettbarn, 1972). In general, the acetylcholine content of invertebrate synaptosomes is greater than that of vertebrates. With the exception of synaptosomes derived from octopus head ganglia, those derived from squid optic lobe contain acetylcholine in amounts several orders of magnitude greater than that found in other sources examined, i.e., the lobster nerve cord and dogfish olfactory bulb (cf. Dowdall *et al.*, 1976, for review). Synaptosomes derived from the *Torpedo* electric organ are purely cholinergic (cf. Dowdall *et al.*, 1976, for review) which makes this system an ideal model for studying cholinergic mechanisms of nerve transmission. In terms of comparative studies with mammalian synapses, however, the *Torpedo* electroplaque functions more like a motor discharge and takes at least three days to completely recover after electrical stimulation, whereas squid optic lobe synapses resemble more rapidly adapting mammalian synapses (Zimmerman and Whittaker, 1974).

Choline acetyltransferase was also found in high concentration in homogenates of squid optic lobe tissue (Feldberg *et al.*, 1951; Prince, 1967; Welsch and Dettbarn, 1972). Centrifugation of a homogenate in 1.5 M sucrose resulted in significant amounts of this enzyme in the floating pellicle containing synaptosomes and in the high speed supernatant (Welsch and Dettbarn, 1972).

Welsch and Dettbarn (1972) have demonstrated cholinesterase activity (including that of acetylcholinesterase and butyrylcholinesterase) associated with synaptosomes and subcellular fractions derived from them. In regard to the subfractions the highest activity was located on the fraction floating on 1.0 M sucrose, although following osmotic lysis of synaptosomes and sucrose density centrifugation, relatively high levels of acetylcholinesterase were found in the fraction between 1.0 and 1.2 M sucrose (Dowdall and Whittaker, 1973) where synaptosomal plasma membranes usually band after fractionation of lysed synaptosomes.

High cholinesterase activity has also been demonstrated in squid (*Ommatostrephes sloanei-pacificus*) (Turpaev *et al.*, 1968) using a histochemical technique. When either acetylthiocholine or butyrylthiocholine was used as a substrate, the activity was localized in the areas containing synaptic terminals. Other studies by this group indicate that the squid cholinesterase is different from acetyl- and serum cholinesterases (Turpaev *et al.*, 1968). In regard to substrate specificity, the squid enzyme is similar to acetylcholinesterase. However, in contrast to the latter, the squid enzyme can split butyrylcholine. The activity of the squid optic lobe cholinesterase is higher than that found in the squid cerebral ganglion or stellate ganglion, or in the bovine caudate nucleus, but comparable to that seen in the electric organ of *E. electricus* (cf. Turpaev *et al.*, 1968, for discussion) and *Torpedo* (cf. Dowdall and Whittaker, 1973).

Squid optic lobe synaptosomes also possess a high-affinity choline uptake system, which appears to be specific for cholinergic neurons and which may regulate acetylcholine synthesis (Dowdall and Simon, 1973; Barker *et al.*, 1975). While mammalian synaptosomes can also take up radioactive choline and incorporate it into acetylcholine, the rate of uptake is significantly greater in squid optic lobe synaptosomes (Whittaker *et al.*, 1972; Dowdall *et al.*, 1976). The influx of choline in this system could be resolved into both saturable and non-saturable components. The kinetics of choline uptake in squid synaptosomes indicate the presence of two saturable components which function in choline uptake, i.e., a high-affinity and low-affinity system (Dowdall and Simon, 1973).

Characteristics of both carrier-mediated systems have been studied in detail in squid synaptosomes (Dowdall and Simon, 1973). The uptake is temperature dependent and proceeds at 26 °C; lowering the temperature to 0 °C had an inhibitory effect on the uptake process. These systems are specific since structural analogues of choline, i.e., trimethylamine, trimethylamine oxide, and betaine, do not significantly affect choline uptake. On the other hand, metabolic inhibitors, such as 2,4-dinitrophenol and cyanide, significantly depressed choline uptake (particularly when the synaptosomes were preincubated with these agents) in both the high- and low- affinity systems, demonstrating the dependence of these systems on metabolic energy. Hemicholinium 3, which blocks choline uptake in several tissues and ouabain, an inhibitor of active transport, depresses choline uptake in these synaptosomes. Also, the systems appear to be dependent upon Na^+, as substitution of K^+ or Li^+ for Na^+ depressed choline uptake. Dowdall and Simon (1973) have suggested the high-affinity system is localized in the cholinergic synaptosomes and the low-affinity system is localized in all, or in non-cholinergic, synaptosomes. It has also been suggested that mammalian synaptosomes have a high-affinity uptake system but are undetected because of the great majority of non-cholinergic synaptosomes having a low-affinity system required to take up choline for the synthesis of phospholipids (Whittaker *et al.*, 1973).

Incubation of [^{3}H]choline with squid synaptosomes resulted in a part of the [^{3}H]choline being incorporated into [^{3}H]acetylcholine and [^{3}H]phosphorylcholine (Dowdall and Simon, 1973). In these experiments there was a reciprocal relationship between the length of incubation and the proportion of radioactivity recovered as choline. In squid synaptosomes, the activity of the high-affinity uptake carrier decreased with time concomitant with an increase in acetylcholine. It has been proposed that acetylcholine synthesis is regulated by the high-affinity choline uptake system with the increase in intrasynaptosomal acetylcholine inhibiting the high-affinity choline uptake system (Whittaker *et al.*, 1972; Dowdall and Simon, 1973).

It is curious that despite all the above studies on choline uptake and acetylcholine metabolism that have been done using this preparation, there are no reports in the literature demonstrating stimulus-induced secretion of newly synthesized acetylcholine from squid synaptosomes. Attempts to do this over several summer seasons at the Marine Biological Laboratory in Woods Hole, Massachusetts have failed (Pollard, 1989). In contrast, the secretion of endogenous (non-labeled) acetylcholine from squid and cuttlefish synaptosomes by calcium-dependent depolarization has been achieved (Pollard, 1989; Dowdall, 1989), but these data have not been published. The only published data showing calcium-dependent secretion from squid synaptosomes is for the release of ATP (presumably in cholinergic vesicles) by veratridine stimulation (Pollard and Pappas, 1979). This veratridine-activated release could be blocked by tetrodotoxin.

6. Localization and uptake of other putative neurotransmitters and neuropeptides in squid optic lobe nerve terminals

In addition to acetylcholine, several studies indicate the presence of other neurotransmitters and peptides in the squid optic lobe. As discussed above, dense-cored vesicles are located within the optic lobe and these usually contain monoamines or peptides. Neurotransmitters that have been demonstrated in the cephalopod nervous system include (5-HT) (Juorio, 1971; Tansey, 1979, 1980; Welsh and Moorhead, 1959, 1960; Youdim *et al.*, 1986), dopamine, and norepinephrine (Juorio, 1970, 1971). Catecholamines and 5-HT have also been demonstrated in cephalopod head ganglia using fluorescent techniques (Barlow, 1971; Matus, 1973). The neuropeptide somatostatin has been localized in squid optic lobes, specifically, in the internal granule layer and the medulla as well as in fibers throughout the lobe and in the optic tract (Feldman, 1986).

Squid optic lobe synaptosomes have been shown to possess specific, high-affinity uptake systems for L-noradrenaline and 5-HT (Dowdall, 1974; Pollard *et al.*, 1975). The uptake of these transmitters was inhibited by chlorpromazine, a drug shown to inhibit monoamine uptake in peripheral adrenergic systems (Thoenen *et al.*, 1965). It was found that chlorpromazine selectively and competitively inhibited the uptake of L-noradrenaline and 5-HT. Although aspartic acid, glutamic acid, and glycine were taken up by synaptosomes, the uptake systems appeared to be unaffected by chlorpromazine. The high-affinity uptake system for choline was also resistant to the effects of this drug. All the uptake systems were sensitive to 2,4-dinitrophenol, potassium cyanide, and ouabain at higher concentrations and in a non-competitive fashion (Pollard *et al.*, 1975). It was suggested that chlorpromazine blocks the permeation mechanisms for L-noradrenaline and 5-HT by binding to component common to each of these uptake systems (Pollard *et al.*, 1975).

In regard to 5-HT, recent studies have demonstrated an monoamine oxidase (MAO) activity capable of oxidatively deaminating this transmitter in a synaptosomal fraction (Youdim *et al.*, 1986). It was shown that this activity may be different from the form, or has not yet differentiated into the two forms present in the mammalian CNS. That this enzyme is functional is demonstrated by the increase in 5-HT levels in intact squid optic lobe, as seen with pargyline, an inhibitor of monoamine oxidase.

ACKNOWLEDGEMENT. Supported in part by NIH grant NS 15889 to R.S.C.

References

Aldridge, W. N., Emery, R. C, and Street, B. W. 1960, A tissue homogenizer, *Biochem. J.* 77:326–327.

Bacq, Z. M., 1935, Recherches sur la physiologie et la pharmacologie du système nerveux autonome. *XVII.* Les esters de la choline dans les extraits de tissus des invertébrés, *Arch. int. Physiol.* **42**:24–42.

Bacq, Z. M. and Mazza, F. P., 1935, Recherches sur la physiologie et la pharmacologie du système nerveux autonome. *XVIII.* Isolement de cholaurate d'acétylcholine à partir d'un extrait de cellules nerveuses d'*Octopus vulgaris. Arch. int. Physiol.* **42**:43–46.

Barker, L. A., Dowdall, M. J. and Mittag, T. W., 1975, Comparative studies on synaptosomes: High-affinity uptake and acetylation of N-[Me-^{3}H]choline and N-[Me-^{3}H]N-hydroxyethylpyrrolidinium, *Brain Res.* **86**:343–348.

Barlow, J. J., 1971, The distribution of acetylcholinesterase and catecholamines in the vertical lobe of *Octopus vulgaris, Brain Res.* **35**:304–307.

Bass, M., Pant, H. C., Gainer, H., and Soderling, T. R., 1987, Calcium/calmodulin-dependent protein kinase II in squid synaptosomes, *J. Neurochem.* **49**(4):1116–1123.

Cajal, S. R., 1917, Contribución al conocimiento de la retina y centros ópticos de los cefalópodos, *Trab. Lab. Invest. biol. Univ. Mad.* **15**:1–82.

Cohen, A. I., 1973, An ultrastuctural analysis of the photoreceptors of the squid and their synaptic connections. III. Photoreceptor terminations in the optic lobes, *J. Comp. Neurol.* **147**:399–426.

Cohen, R. S., Blomberg, F., Berzins, K., and Siekevitz, P., 1977, The structure of postsynaptic densities isolated from dog cerebral cortex. I. Overall morphology and protein composition, *J. Cell Biol.* **74**:181–203.

Cohen, R. S., Pollard, H. B., Gainer, H., and Pappas, G. D., 1982, Synaptosomal plasma membranes from cholinergic terminals of brain, *J. Cell Biol.* **95**:138a.

De Belleroche, J. S. and Bradford, H. F., 1973, The synaptosome: An isolated working neuronal compartment, in: *Progress in Neurobiology*, Vol. **1** (G. A. Kerkut, and J. W. Phillis, eds.), Pergamon, Oxford, pp 275–298.

De Robertis, E., Pellegrino De Iraldi, A., Rodriguez, G., and Gomez, C. J., 1961, On the isolation of nerve endings and synaptic vesicles, *J. Biophys. Biochem. Cytol.* **9**:229–235.

Dowdall, M. J., 1974, Biochemical properties of synaptosomes from the optic lobe of squid (*Loligo pealii*), *Biochem. Soc. Trans.* **2**(2):270–272.

Dowdall, M., 1989, *personal commun.*
Dowdall, M. J., Fox, G., Wächtler, K., Whittaker, V. P., and Zimmermann, H., 1976, Recent studies on the comparative biochemistry of the cholinergic neuron, *Cold Spring Harbor Symp. Quant. Biol.* **40**:65–81.
Dowdall, M. J. and Simon, E. J, 1973, Comparative studies on synaptosomes: Uptake of [N-Me-^{3}H]choline by synaptosomes from squid optic lobes, *J. Neurosci.* **21**:969–982.
Dowdall, M. J. and Whittaker, V. P., 1973, Comparative studies in synaptosome formation: The preparation of synaptosomes from the head ganglion of the squid, *Loligo pealii, J. Neurochem.* **20**:921–935.
Feldberg, W., Harris, G. W., and Lin, R. C. Y., 1951, Observation on the presence of cholinergic and non-cholinergic neurones in the central nervous system, *J. Physiol.* **112**:400–404.
Feldman, S. C., 1986, Distribution of immunoreactive somatostatin (ISRIF) in the nervous system of the squid, *Loligo pealei, J. Comp. Neurol.* **245**:238–257.
Fiori, C. E., Leapman, R. D., and Swyt, C. R., 1988, Quantitative X-ray mapping of biological cryosections, *Ultramicroscopy* **24**:237–250.
Flitney, F. W., 1966, The time course of the fixation of albumin by formaldehyde, glutaraldehyde, acrolein and other aldehydes, *J. R. Micros. Soc.* **85**:353–364.
Gallant, P. E., Pant, H. C., and Gainer, H., 1984, Distribution of acid protease activity in the squid nervous system, *J. Neurochem.* **42**(2):590–593.
Haghighat, N., Cohen, R. S., and Pappas, G. D., 1984, Fine structure of squid (*Loligo pealii*) optic lobe synapses, *Neuroscience* **13**(2):527–546.
Hayat, M. A., 1986, *Basic Techniques for Transmission Electron Microscopy*, Academic Press, New York.
Hernández, A. G., Langford, G. M., Martízez, Jr., J. L., and Dowdall, M. J., 1976, Protein synthesis by synaptosomes from the head ganglion of the squid, *Loligo pealei, Acta. Cient. Venezolana* **27**:120–123.
Huttner, W. B., Schiebler, W., Greengard, P., and De Camilli, P., 1983, Synapsin I (Protein I), a nerve terminal-specific phosphoprotein. III. Its association with synaptic vesicles studied in a highly purified synaptic vesicle preparation, *J. Cell Biol.* **96**:1374–1388.
Jones, D. G., 1967, An electron-microscope study of subcellular fractions of *Octopus* brain, *J. Cell Sci.* **2**:573–586.
Jones, D. G., 1970a, A further contribution to the study of the contact region of *Octopus* synaptosomes, *Z. Zellforsch.* **103**:48–60.
Jones, D. G., 1970b, A study of the presynaptic network of *Octopus* synaptosomes, *Brain Res.* **20**:145–158.
Jones, D. G., 1975, *Synapses and Synaptosomes. Morphological Aspects*, John Wiley and Sons, New York.
Juorio, A. V., 1970, The distribution of catecholamines in the nervous tissue of an octopod mollusc, *Brit. J. Pharmac.* **39**:240–242P.
Juorio, A. V., 1971, Catecholamines and 5-hydroxytryptamine in nervous tissue of cephalopods, *J. Physiol.* **216**:213–226.
Kelly, P. T., McGuiness, T. L., and Greengard, P., 1984, Evidence that the major postsynaptic density protein is a component of a Ca^{2+}/calmodulin-dependent protein kinase, *Proc. Natl. Acad. Sci. U.S.A.* **81**:945–949.
Kelly, P. T. and Montgomery, P. R., 1982, Subcellular localization of the 52,000 molecular weight major postsynaptic density protein, *Brain Res.* **233**:265–286.
King, J. C., Lechan, R. M., Kugel, G., and Anthony, E. L. P., 1983, A fixative for immunocytochemical localization of peptides in the central nervous system, *J. Histochem. Cytochem.* **31**(1):62–68.
Krinks, M. H., Klee, C. B., Pant, H. C., and Gainer, H., 1988, Identification and quantification of calcium-binding proteins in squid axoplasm, *J. Neurosci.* **8**:2172–2182.
Mahler, H. R., 1977, Proteins of the synaptic membrane, *Neurochem. Res.* **2**:119–147.
Matsumura, F., 1987, Deltamethrin-induced changes in synaptosomal transport of ^{3}H-epinephrine in squid optic lobes, *Comp. Biochem. Physiol.* **87C**(1):31–35.

Matus, A. I., 1973, Histochemical localization of biogenic monoamines in the cephalic ganglia of *Octopus vulgaris*, *Tissue and Cell* **5**(4):591–601.

Morel, N., Manaranche, R., Israel, M., and Gulid-Krzywicki, T., 1982, Isolation of a presynaptic plasma membrane fraction from *Torpedo* cholinergic synaptosomes: Evidence for a specific protein, *J. Cell Biol.* **93**:349–356.

Nagy, A., Baker, R. R., Morris, S. J., and Whittaker, V. P., 1976, The preparation and characterization of synaptic vesicles of high purity, *Brain Res.* **109**:285–309.

Pant, H. C., 1989, *unpublished.*

Pant, H. C., Pollard, H. B., Pappas, G. D., and Gainer, H., 1979, Phosphorylation of specific distinct proteins in synaptosomes and axons from squid nervous system, *Proc. Natl. Acad. Sci.* **76**(12):6071–6075.

Pant, H. C., Gallant, P. E., Gould, R., and Gainer, H., 1982, Distribution of calcium-activated protease activity and endogenous substrates in the squid nervous system, *J. Neuroscience* **2**(11):1578–1587.

Pant, H. C., Gallant, P. E., Cohen, R., Neary, J. T., and Gainer, H., 1983, Calcium-dependent 4-aminopyridine stimulation of protein phosphorylation in squid optic lobe synaptosomes, *Cellul. Mol. Biol.* **3**(3):223–238.

Pant, H. C., Gallant, P. E., and Gainer, H., 1986, Characterization of a cyclic nucleotide- and calcium-independent neurofilament protein kinase activity in axoplasm from the squid giant axon, *J. Biol. Chem.* **261**:2698–2977.

Pollard, H. B., 1989, *personal commun.*

Pollard, H. B., Barker, J. L. Bohr, W. A., and Dowdall, M. J., 1975, Chlorpromazine: Specific inhibition of L-noradrenaline and 5-hydroxytryptamine uptake in synaptosomes from squid brain, *Brain Res.* **85**:23–31.

Pollard, H. B. and Pappas, G. D., 1979, Veratridine-activated release of adenosine-5'-triphosphate from synaptosomes: Evidence for calcium dependence and blockade by tetrodotoxin, *Biochem. Biophys. Res. Comm.* **88**(4):1315–1321.

Prince, A. K., 1967, Properties of choline acetyltransferase isolated from squid ganglia, *Proc. Natl. Acad. Sci.* **57**:1117–1122.

Sattelle, D. B., Langley, K. H., Obaid, A. L., and Salzberg, B. M., 1987, Laser light scattering determination of size and dispersity of synaptosomes and synaptic vesicles isolated from squid (*Loligo pealii*) optic lobes, *Eur. Biophys. J.* **15**(2):71–76.

Tansey, E. M., 1979, Neurotransmitters in the cephalopod brain: A review, *Comp. Biochem. Physiol.* **C64**:173–183.

Tansey, E. M., 1980, Aminergic fluorescence in the cephalopod brain, *Philos. Trans. R. Soc. Lond. (Biol.)* **291**:127–145.

Thoenen, J. R., Hurlimann, A., and Haefely, W., 1965, On the mode of action of chlorpromazine on peripheral adrenergic mechanisms, *Int. J. Neuropharmacol.* **4**:79–89.

Turpaev, T. M., Abashkina, L. I., Brestkin, A. P., Beick, L. L., Grigorjeve, G. M., Perzner, D. L., Rozengart, V. J., Rozengart, E. V., and Sakharov, D. A., 1968, Cholinesterase of squid ganglia, *Europ. J. Biochem.* **6**:55–59.

Ueda, T., Greengard, P., Berzins, K, Cohen, R. S., Blomberg, F., Grab, D. J., and Siekevitz, P., 1979, Subcellular distribution in cerebral cortex of two proteins phosphorylated by a cAMP-dependent protein kinase, *J. Cell Biol,* **83**:308–319.

Unver, E., Augustine, G. J., Barry, S., and Miljanich, G. P., 1986, Kinase C calcium-phospholipid-dependent protein kinase in squid optic lobes synaptosomes, *Soc. Neurosci. Abs.* **12**(2):821.

Welsch, F. and Dettbarn, W. D., 1972, The subcellular distribution of acetylcholine, cholinesterases and choline acetyltransferase in optic lobes of the squid *Loligo pealii*, *Brain Res.* **39**:467–482.

Welsh, J. H. and Moorhead, M., 1959, Identification and assay of 5-hydroxy-tryptamine in molluscan tissue by fluorescence method, *Science* **129**:1491–1492.

Welsh, J. H. and Moorhead, M., 1960, The quantitative distribution of 5-hydroytryptamine in the invertebrates, especially in their nervous systems, *J. Neurochem.* **6**:146–169.

Whittaker, V. P., 1959, The isolation and characterization of acetylcholine-containing particles from brain, *Biochem. J.* **72**:694–706.

Whittaker, V. P. and Barker, L. A., 1972, The subcellular fractionation of brain tissue with special reference to the preparation of synaptosomes and their component organelles, in: *Methods in Neurochemistry* Vol. **2**, (R. Fried ed.) Marcel Dekker, New York, pp 1–52.

Whittaker, V. P., Dowdall, M. J. and Boyne, A. F., 1972, The storage and release of acetylcholine by cholinergic nerve terminals: Recent results with non-mammalian preparations, *Biochem. Soc. Symp.* **36**:49–68.

Whittaker, V. P., Michaelson, I. A., and Kirkland, R. J. A., 1964, The separation of synaptic vesicles from nerve-ending particles ('synaptosomes'), *Biochem J.* **90**:293–303.

Youdim, M. B. H., Feldman, S. C., Pappas, G. D.,and Pollard, H. B., 1986, Serotonin metabolism and the nature of monoamine oxidase in squid central nervous system, *Brain Res.* **381**:300–304.

Young, J. Z., 1974, The central nervous system of *Loligo*. I. The optic lobe, *Phil. Trans. R. Soc. Ser. B* **267**:263–302.

Zimmerman, H. and Whittaker, V. P., 1974, Different recovery rates of the electrophysiological, biochemical and morphological parameters in the cholinergic synapses of the *Torpedo* electric organ after stimulation, *J. Neurochem.* **22**:1109–1114.

Chapter 14

The Cytoskeleton of the Squid Giant Axon

ANTHONY BROWN and RAYMOND J. LASEK

1. Introduction

The squid giant axon is useful to biologists because of its very large size, but it is the continued applicability of studies on this cell to much smaller diameter mammalian axons that have sustained our interest in this mollusc. For example, in electrophysiology the large size of the squid giant axon has enabled manipulations that would have been difficult or impossible with smaller axons. More recently, new methods, such as patch clamping, have made the detailed study of mammalian cell membranes possible, and these have shown that the basic mechanisms established for the squid giant axon also occur in the excitable cells of other metazoa, including mammals and presumably man.

In addition to its usefulness for studies on electrically excitable membranes, the squid giant axon has also proven useful for many other kinds of cell biological studies. In particular, its large diameter allows axoplasm to be separated from the axonal plasma membrane and its surrounding glial sheath without chemical treatment or mechanical disruption. This accessibility offers cell biologists an unusual opportunity to investigate the properties of living cytoplasm. Many of these opportunities have yet to be exploited and we hope to point the way to some of them in the course of this article.

Since the publication of *A guide to the laboratory use of the squid Loligo pealei* (Arnold *et al.*, 1974), there has been much progress in research on squid axoplasm, and some of this has already been summarized in two reviews by Rosenberg (1981) and Gainer *et al.* (1984). Particular progress has been made during this period in studies on the cytoskeleton of the squid giant axon. In this article we present a practical guide to the special possibilities that the giant axon offers for the study of the cytoskeleton, and we summarize much of our current knowledge about the structure and organization of the proteins in squid axoplasm.

We hope that this article will encourage people to make experimental use of squid axoplasm in their research. However, most cell biologists who might be interested in the information obtained from the giant axon have never had the opportunity to see or use this remarkable preparation. For this large group of scientists, we offer a detailed practical guide to the experimental manipulation of the axon and axoplasm. This partial recreation of the experience of working with the giant axon can not substitute for the actual experience of investigation. Nonetheless, by recreating

A. BROWN and R. J. LASEK • Bio-architectonics Center, School of Medicine, Case Western Reserve University, Cleveland, Ohio 44106.

some aspects of the actual physical experience of experimentation with the giant axon, we hope that investigators who have never had this experience may be better able to understand experimental results obtained with this unique preparation.

This chapter is divided into 22 Sections. Sections 1 and 2 give general aspects of the squid. Sections 3 to 8 deal with the anatomy of the giant nerve fiber and with methods for obtaining axoplasm. In these Sections we present a detailed practical guide to the methods and considerations involved in isolating axoplasm from the giant axon. Next, Sections 9 to 11 cover the composition and physical properties of axoplasm, and Sections 12 and 13 deal with methods for the fixation of axoplasm for electron microscopy of the cytoskeleton, and with methods for analysis of the proteins in axoplasm by gel electrophoresis. Then, in Sections 14 to 19, we describe the structure and organization of the cytoskeleton and its elements. In these Sections we summarize much of the current knowledge on the cytoskeleton of the giant axon. Finally, in Sections 20 and 21, we illustrate the potential of the giant axon for studies on the cytoskeleton by describing two paradigms that we have developed for studying the stability and mechanical properties of the axonal cytoskeleton. These latter studies illustrate new ways in which the squid giant axon can aid our understanding of the cell biology of axons. Section 22 is the summary.

2. The squid

Squid do not survive for very long in captivity and it is generally necessary to use them within a day of capture. A healthy squid has a transparent mantle (body wall) and a lively escape response. Unless otherwise mentioned the contents of this chapter apply to the squid, *Loligo pealei*, which are obtainable at the Marine Biological Laboratory, Woods Hole, Massachusetts. For research on axoplasm, no major differences with other members of the squid family are anticipated.

We can encourage the uninitiated investigator who would like to conduct investigations on the squid giant axon to consider visiting the Marine Biological Laboratory at Woods Hole. The Marine Biological Laboratory has many mechanisms for fostering research on squid. The squid are collected regularly from May through September, and they continue to be reasonably plentiful. Many investigators who work on the giant axon are willing to assist in carrying out a pilot experiment that can be useful in testing the feasibility of a full scale study.

3. The anatomy and development of the giant nerve fiber

To understand how much the squid giant axon can tell us about mammalian axons it is important to understand its anatomy and development. The squid giant axons represent one of the few examples in nature of neurons that are true syncytia. However, unlike other syncytial neurons, such as those found in annelids, all the cell bodies are located at one end of the axon. Thus the squid giant axons share the polarized morphology of other non-syncytial neurons.

There are three orders of giant nerve fibers in squid (Young, 1939). Together, these form part of a motor system that triggers the animal's escape response by causing rapid and synchronous contraction of the body wall. Two first order giant axons

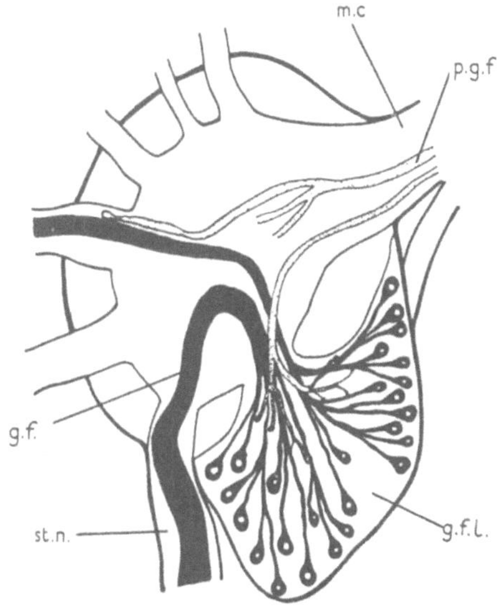

Figure 1. Diagram of the stellate ganglion in the squid *Loligo*. Each stellate nerve contains a single (third order) giant axon that arises from many small cell bodies located in the giant fiber lobe of the stellate ganglion. The second order pre-ganglionic giant axons form giant synapses directly with the giant stellate axons. Abbreviations: mantle connective (*m.c.*); second order or pre-ganglionic fiber (*p.g.f.*); giant fiber lobe (*g.f.l.*); stellate nerve (*st.n.*); giant axon of the giant nerve fiber (*g.f.*). Drawing reproduced with permission from Young (1936).

emerge from the magnocellular lobes of the brain and synapse with the second order giant axons in the palliovisceral lobes. The second order giant axons course through the pallial nerves of the mantle connectives and synapse with the third order giant axons in the two bilateral stellate ganglia, from which the stellate nerves emanate.

In *Loligo* each stellate nerve contains a single third order giant axon that is surrounded by many smaller axons. The giant axons arise from clusters of cell bodies in the giant fiber lobe of the stellate ganglion (see Fig. 1). Apparently each giant axon is formed from a fascicle of ordinary-sized axons that course together in each of the stellate nerves of the developing squid. These axons fuse with each other laterally to form one common axoplasmic mass that is surrounded by a single plasma membrane. The process of fusion is documented in the case of the first order giant fibers (Martin, 1965, 1969). Only the axons fuse; the cell bodies in the stellate ganglia retain their separate identities and each one contributes to the maintenance of the giant axon. Young (1936) estimated that between 300 and 1500 cells fuse to form each giant axon and it is likely that these cells remain transcriptionally active. Thus each giant axon is the product of hundreds of ordinary-sized molluscan neurons that each contribute axoplasmic elements to the single giant axon.

Each giant axon is surrounded by a single layer of adaxonal glial cells (Schwann cells) that are closely apposed to the outer surface of the axolemma (see Fig. 2); a narrow 10 nm periaxonal space separates the axon from the glial cells. The glial cells are extensively folded and interdigitate laterally to form a continuous mosaic sheet that envelops the axon, but a gap of about 10 nm separates adjacent cells forming long tortuous intercellular clefts that provide continuity between the surface of the axon and the extracellular space (Villegas and Villegas, 1984). Diffusion between the periaxonal space and extracellular space is apparently facilitated by a system of tubular trans-glial channels that penetrate the glial cells in a sponge-like fashion (Zwahlen *et al.*, 1988). These channels are about 40 nm in diameter and are continuous with the periaxonal and extracellular spaces. Thus the glial cells do not provide a barrier to the diffusion

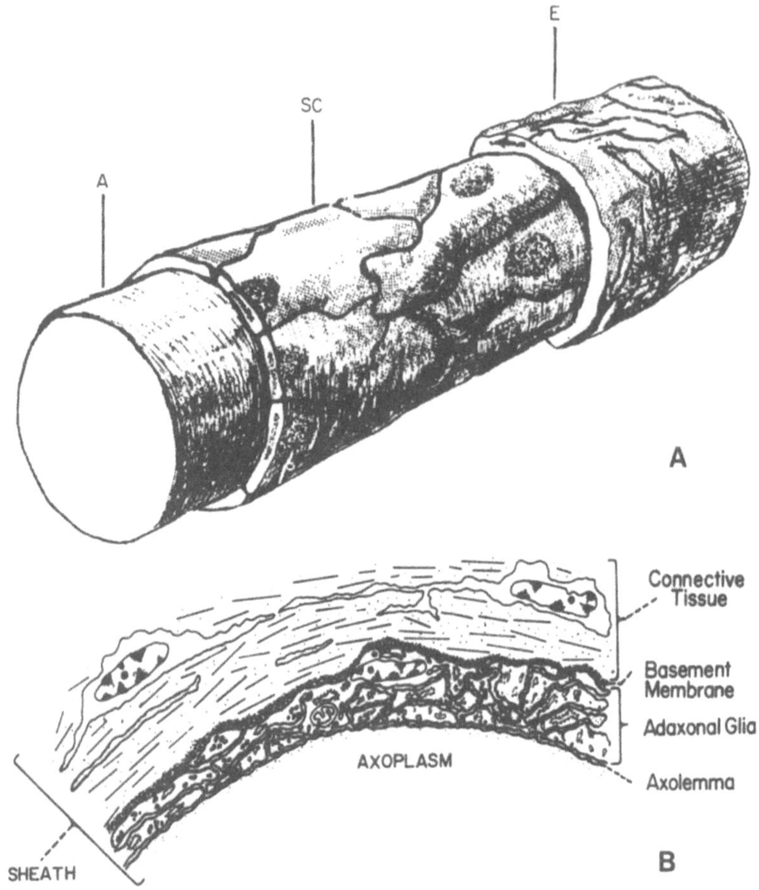

Figure 2. The anatomy of the squid giant nerve fiber. A. Schematic drawing of the giant nerve fiber. Like most other axons, the giant axon (*A*) is surrounded by a cellular sheath, and for the giant axon this sheath is composed of a single layer of Schwann cells (*SC*), and an outer layer of connective tissue (*E*) that corresponds to the endoneurium. The giant axon in *Loligo pealei* can be up to 0.8 mm in diameter. Reproduced from Villegas and Villegas (1984) with permission. B. Schematic drawing of a sector of the giant nerve fiber sheath in cross-section. The outer connective tissue layer contains isolated fibrocytes and is separated from the inner layer of adaxonal glia by a thick basement membrane (the basal lamina). The glial cytoplasm is densely packed with rough endoplasmic reticulum and membranous organelles.

of small molecules from the surroundings. In *Loligo pealei* the Schwann cell layer is about 1 μm thick (Adelman *et al.*, 1977).

Surrounding the layer of glial cells is a basal lamina around which is a layer of loose connective tissue containing isolated fibrocytes. Our observations of axons using video-enhanced differential interference contrast microscopy indicate that the collagen fibers crisscross to produce a lattice-like arrangement that confers mechanical strength and elasticity to the nerve fiber. Many hundreds of smaller nerve fibers (less than 50 μm in diameter) surround the giant nerve fiber, and together these constitute a stellate nerve.

4. Dissection of the giant axons

In general usage the hindmost and medialmost giant axon is referred to as *the squid giant axon*. In each squid there are two of these axons that emerge from the two bilateral stellate ganglia. These giant axons are used because they are the largest and most accessible of the stellate axons. In *Loligo pealei*, they range from 300 to 800 μm (typically 350 to 500 μm) in diameter (Rosenberg, 1981). In larger squid, such as *Loligo forbesi*, which can have mantle lengths of greater than 50 cm, it may also be worthwhile to dissect the next largest of the stellate axons. At the Marine Biological Laboratories in Woods Hole, large specimens of *Loligo pealei* are around 20 to 25 cm in mantle length.

The axons can be dissected from the mantle in natural or artificial seawater using a dissecting tray with illumination from beneath. This is a standard procedure which has been refined over the years by numerous investigators. At the Marine Biological Laboratory in Woods Hole, we use a flowing seawater table. The hindmost giant axons run for a few centimeters along the inner surface of the mantle before they course into the mantle musculature. Generally we dissect the axons up to this point or to the first major branch point within the mantle wall. There are smaller branches more proximal to this point, particularly near the point that the fin nerve and giant axon diverge, and care should be taken in the dissection to cut these, preferably a millimeter or so from the axon, rather than tear them when the axon is excised. The dissection of the giant axon is described in detail by Adelman and Gilbert (1990).

It is preferable to dissect the axons from the mantle immediately after the squid has been killed. Once excised from the mantle the axons can be kept in cold seawater (in the refrigerator at 4 °C, for example) for many hours. At room temperature the survival time is shorter. After excision of axons from the mantle, the subsequent dissections are generally performed in Millipore™-filtered natural seawater or in artificial seawater with or without calcium.

4.1. Fine-cleaning the giant axons

Before excision from the mantle, the ends of the axon are tied with cotton thread to prevent seawater leaking into the axoplasm. If the axon is excised with the stellate ganglion attached, then it is not necessary to ligate the proximal end. The axons are then transferred to a dish containing fresh seawater. For the subsequent dissection we use a large Petri dish containing a layer (about 5 mm thick) of Sylgard™, which is a transparent rubbery resin into which dissecting pins can be secured. For this procedure fine dissecting scissors are required. For most investigations it is necessary to dissect away the surrounding smaller nerve fibers and any small pieces of attached muscle, blood vessels etc. before the axon can be used. For extrusion of axoplasm in the conventional manner (see Section 6.1) this need only be done at the proximal end of the fiber, but for some studies it may be necessary to dissect the whole fiber in this way. In both cases, the dissection is done under a low power binocular microscope with illumination from beneath. We typically use a magnification of 25 ×. Dark-field illumination improves the visibility of the smaller nerve fibers but bright-field illumination provides better visibility of the fine branches of the giant axon. Either of these sources of illumination can be used with success, but we prefer bright-field illumination.

A careful dissection generates a *fine-cleaned* fiber, which is a fiber that has been freed of surrounding small nerves. The fine-cleaned fiber consists of the axon enveloped by a sheath of Schwann cells and an outer layer of loose connective tissue (see Fig. 2). Usually some tissue is left around the branches of the giant axon to minimize the risk of damaging the axon. It is important to realize that the *fine-cleaned* giant fiber includes the Schwann cell sheath, which adheres tightly to the axolemma of the giant axon and cannot be removed by conventional dissection. Removal of the sheath requires a special procedure, which was developed by Metuzals *et al.* (1981b) and is described in detail in Section 13.4. Briefly, they used mild chemical fixation to stabilize the axon and sheath, and osmotic shrinkage to separate the axolemma from the Schwann cell layer. This procedure enables the surface topography of the axon to be viewed by scanning electron microscopy (see Section 14.2).

5. Handling the giant axons

The visualization of the isolated squid giant nerve fiber is a remarkable experience for those used to dealing with smaller axons. The giant axon itself is about the size of a medium size mouse nerve or a small rat nerve, yet it is a single axon that can be seen with the naked eye and can be handled between two lengths of ordinary cotton thread. The connective tissue sheath supports the giant axon sufficiently that it can bear a small amount of stress. Because of this, the giant fiber can be transferred between dissecting dishes or solutions by grasping the attached cotton thread at one end and simply lifting the axon from one solution to the other, but care should be taken not to stretch it too much. When removed from a solution, the fiber conveys a thin covering layer of liquid with it, but it should not be left in air longer than absolutely necessary, in order to avoid drying. When a nerve fiber is transferred into a new solution it often floats at the surface and care should be taken to ensure that it is thoroughly wetted and submerged. If an axon appears more turgid or flaccid after it has been transferred from natural seawater into artificial seawater, this is probably caused by differences in the osmolarity of the two solutions. Axoplasm is in osmotic equilibrium with the external solution (see Section 9) and the axon volume will rapidly shrink, or to some extent swell, in response to changes in the osmolarity of the medium.

6. Methods for obtaining axoplasm from the giant axons

The large diameter of the squid giant axons allows axoplasm to be separated from the axolemma and axonal sheath in milligram quantities. An axon with a diameter of 500 μm contains about 1 μl (1 mg wet weight) of axoplasm for a 5 mm length. Three major factors should be considered in choosing the method for isolation of axoplasm: purity, preservation of structural integrity, and time. In this Section

Table I. Comparison of three methods for isolating axoplasm.

Method	*Advantages*	*Disadvantages*
Extrusion	Quick, essentially free of contamination	Some distortion of cytoskeletal organization
Slitting	Preservation of shape and cytoskeletal organization	Some contamination from sheath tissue
Cannulation	Preservation of plasma membrane, access to cortical cytoskeleton, no contamination from sheath	Time consuming, special equipment and expertise required, axoplasm disrupted

we describe three methods for isolating axoplasm from the giant axon; extrusion, slitting and cannulation. Some advantages and disadvantages of these methods are summarized in Table I.

In all cases the axon should be undamaged along its length or seawater will leak in and produce white spots in the axoplasm. These are most apparent when axons are viewed under dark-field illumination. The white spots may be due to a precipitation of axoplasmic protein in the presence of seawater that results in an increase in light scattering. The precipitation is apparently caused by the high concentration of calcium in seawater, since calcium forms insoluble sticky precipitates with sulfates, which are present in seawater, and with phosphates, which are present in the axoplasm. Damaged regions are most often caused during the dissection by tearing branches at or near their junction with the axon.

6.1. Extrusion

The quickest way to obtain axoplasm is by extrusion. This method is simple and is widely used by many investigators. For this method the giant axon does not need to be fine-cleaned of associated nerve fibers along its entire length, but it is necessary to fine-clean a region of about 1 cm at the proximal end so as to avoid contamination with extracellular fluid or tissue from the sheath. At the proximal end there are fewer branches than more distally and it is possible, with practice, to fine-clean this region in about five minutes without damaging the axon.

It is also possible to extrude axoplasm from the axon without removing the axon from the mantle. This can be done simply by cutting the axon at its proximal end and then applying pressure to the mantle in a distal-to-proximal direction (the axon is widest at its proximal end). Unfortunately this procedure is very messy and it is difficult to avoid contamination of the axoplasm with extracellular fluids during the extrusion process. Thus, we do not recommend this procedure, and in practice it is necessary to dissect out the axon prior to extrusion and to fine-clean its proximal end as described above.

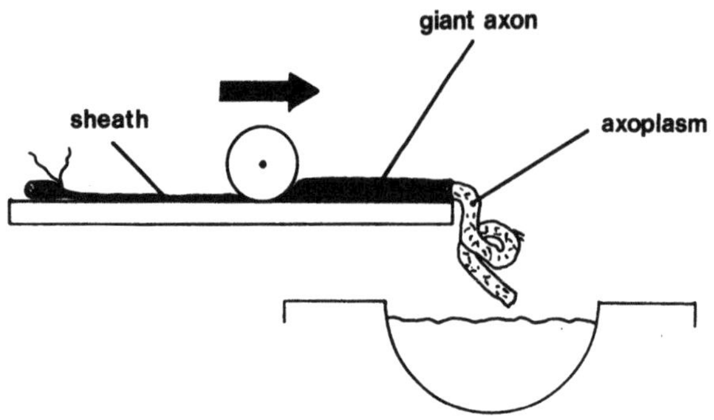

Figure 3. Schematic illustration of axoplasm being extruded from a giant axon. In the method shown here, the giant axon is placed on filter paper on a solid support, such as a glass slide, with the cut end protruding over the edge. For diagrammatic purposes, the axon is shown here as though it has been fine-cleaned along its entire length, though in practice it is only necessary to fine-clean the proximal end. The axoplasm is extruded by gentle pressure in a distal-to-proximal direction (indicated by arrow) towards the cut end. As the axoplasm is extruded it falls away from the cut end. The axoplasm can be collected on a piece of Parafilm™ or, as shown here, in a watch glass containing a small volume of artificial axoplasm solution. In an alternative extrusion method, the cut end of the axon can be placed directly onto a piece of Parafilm™, as described in the text. After extrusion, the flattened sheath is generally discarded. From Morris and Lasek (1982).

For extrusion, the axon is removed from the seawater and excess fluid removed by gently and briefly blotting on a glass coverslip or filter paper. If desired, the axon can be rinsed in artificial axoplasm solution or calcium-free seawater prior to blotting. The axon is then placed on a fresh region of filter paper and cut at its proximal end, leaving at least 5 mm of fine-cleaned axon distal to the cut. For minimal resistance to extrusion and optimal preservation of the structure of the axoplasm, the cut should be made about 5 mm distal to the stellate ganglion, where the axon is at its widest. Nearer the ganglion the axon diameter is smaller and extrusion from this point distorts the axoplasm as it is forced through the narrower region of axon.

After cutting the axon, it can then be lifted by the cotton thread at its distal end, and the fine-cleaned cut end placed onto a small piece of Parafilm™ with the rest of the axon on filter paper. Extrusion is achieved by applying gentle even pressure in a distal-to-proximal direction along the axon. As pressure is applied along the fiber the axoplasm is extruded from the cut end of the axon onto the Parafilm™, like toothpaste from a tube. The filter paper under the axon serves to absorb excess fluid in the sheath as it is compressed. The axon can also be placed on a solid support with the cut end protruding over the edge so that the axoplasm falls away from the cut end as it is extruded (see Fig. 3). Fig. 4 shows a piece of axoplasm that was extruded in this manner and collected in a small volume of artificial axoplasm solution.

In the original method of Bear *et al.* (1937b), the axoplasm was extruded with a pair of bent forceps. Others, such as Baker *et al.* (1962) and Rosenberg (1981), have used a rubber roller. In our hands, a small piece of polyethylene tubing (such as Intramedic PE250 tubing) is preferable, and it is likely to cause less damage to the axon. However, the instrument used is not as important as the protocol employed. For studies where purity is essential, we advise using short lengths of axon and extruding in a single pressure stroke (see Section 7.1). The stroke should not be continued all the way to the end of the axon so that the axoplasm nearest to the site of compression is not extruded. In addition, axoplasm should not be extruded from regions of axon

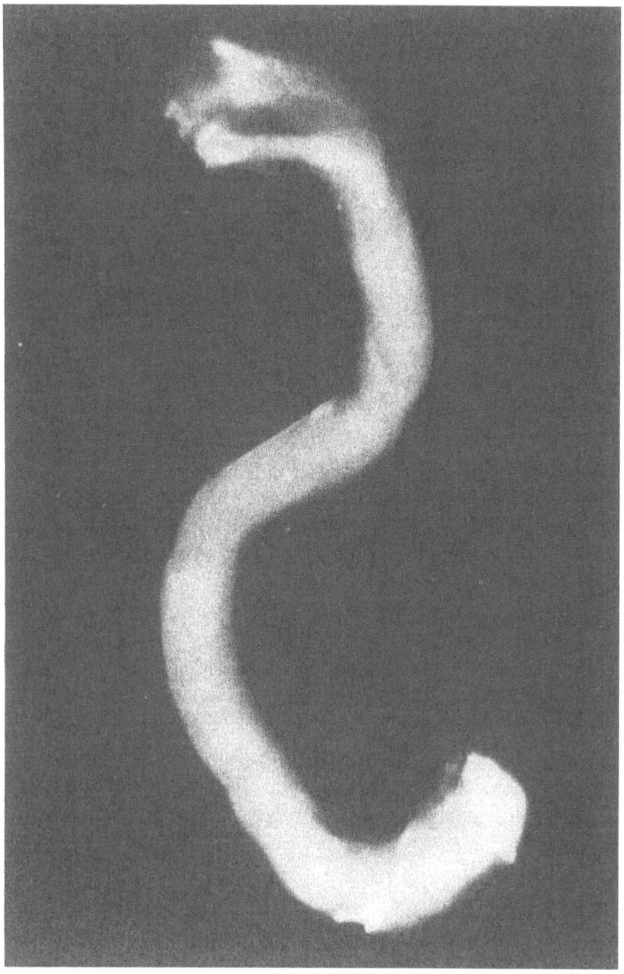

Figure 4. Photograph of a piece of extruded axoplasm. This piece of axoplasm was extruded and immersed in artificial axoplasm solution as shown in Fig. 3, and photographed with dark-field illumination. Axoplasm is almost transparent and therefore it is almost invisible under bright-field illumination. The axoplasm is about 10 mm long and about 0.5 mm in diameter. The *S*-shape of the axoplasm is due to distortion of the longitudinal alignment of the cytoskeletal polymers caused by the compressive forces during extrusion (see text).

within a few millimeters of the knot of cotton thread since these regions may be damaged, thereby producing contamination.

We have observed that when extrusion pressure is applied to the axon, the axoplasm initially resists, and the axon becomes turgid before the axoplasm begins to move. This back pressure on the axoplasm apparently compresses the cytoskeletal elements longitudinally and can distort the cytoskeletal organization of the axoplasm during extrusion. For example, measurements of the length and diameter of the axon and the axoplasmic cylinder indicate that axoplasm extruded onto Parafilm™ is shorter and thicker than the axon from which it came. Maximum change in dimensions is seen if the axoplasm is extruded directly into a liquid rather than onto a surface to which the axoplasm can adhere. In measurements of 36 extrusions directly into artificial axoplasm solution (Brown, 1989), the mean length of the axon segments was 10.2 mm and the mean length of the extruded axoplasms was 5.9 mm. Thus the axoplasms were on average 58% shorter than the original axon (minimum = 26%, maximum = 88%, standard deviation = 14%). Since the volume remains constant, the diameters of the extruded axoplasms were correspondingly larger. Electron microscopy of axoplasm extruded in this way revealed that the cytoskeletal polymers in the center of the extruded axoplasm were orientated obliquely to the long axis of the cylinder of axoplasm. Apparently, the cytoskeletal elements tend to buckle in response to the compressive forces of the extrusion. An indication of the distortion of the extruded cytoskeleton is that the extruded axoplasm often assumes a macroscopic curved or twisted shape (see Fig. 4), which is not seen when axoplasm is isolated without compression (see Section 6.2 and Fig. 5).

For studies that require preservation of the structural integrity of the axoplasm, the distortion of the longitudinal organization of the axoplasmic cytoskeleton can be partially offset by extruding the axoplasm under gentle tension in order to straighten the cytoskeletal polymers as they emerge from the axon. One way to achieve this is by using the natural adhesiveness of the axoplasm to help pull the axoplasm out of the axon. To do this, the roller or plastic tubing is held still and the axon is drawn backwards away from the cut end. As the axoplasm is extruded, it adheres very slightly to the Parafilm™ and with a little practice, this method can yield long unbroken lengths of axoplasm (centimeters in length). Continuous unbroken lengths of axoplasm are necessary for investigations in which the organization and mechanical integrity of the axoplasm is important, such as for our stretching studies on axoplasm (see Section 21) and for the preparation of fibers for X-ray diffraction studies on the cytoskeletal polymers (Wais-Steider *et al.*, 1988).

Extrusion of axoplasm using tension is best achieved on a more adhesive surface such as glass (see Section 8), but in this case the adhesion is so good that the axoplasm tends to maximize its contact with the glass and flatten out. If the axoplasm then needs to be lifted off the glass, further mechanical disruption is inevitable, so extrusion onto glass is not suitable for any studies that require further handling or preservation of the cylindrical shape of the axoplasm. One application in which we have used extrusion onto glass is for obtaining improved longitudinal organization of the cytoskeletal polymers for electron microscopy of extruded axoplasm (Miller and Lasek, 1985).

6.2. Slitting

One of us (A. B.) has experimented with an alternative method for isolating axoplasm that avoids extrusion and the complication of compression. This is the *slitting method*, in which the axon is slit along its length to gain access to the axoplasm. In this procedure, the axon is first fine-cleaned of associated small nerve fibers along its entire length (except around the branches) and rinsed in several changes of 1 – 2 ml artificial axoplasm solution. The axon is then placed in a small Sylgard™ dish containing artificial axoplasm solution and pinned at either end under gentle tension (if required, the dissection can also be done in oil to prevent extraction of diffusible components). Using fine dissecting scissors and a dissecting microscope, it is then possible to carefully slit the sheath open along the length of the axon. Scissors with fine blades and pointed tips have proven to be most effective.

The scissors must be held at a low angle to minimize penetration of the axoplasm. As the sheath is cut it tends to retract so as to expose the axoplasm. The axoplasm can then be gently peeled away from the sheath, while still submerged in solution, and thereby lifted out of the axon (see Fig. 5). The solution can then be removed and replaced with fixative for electron microscopy, or the axoplasm can be used directly for

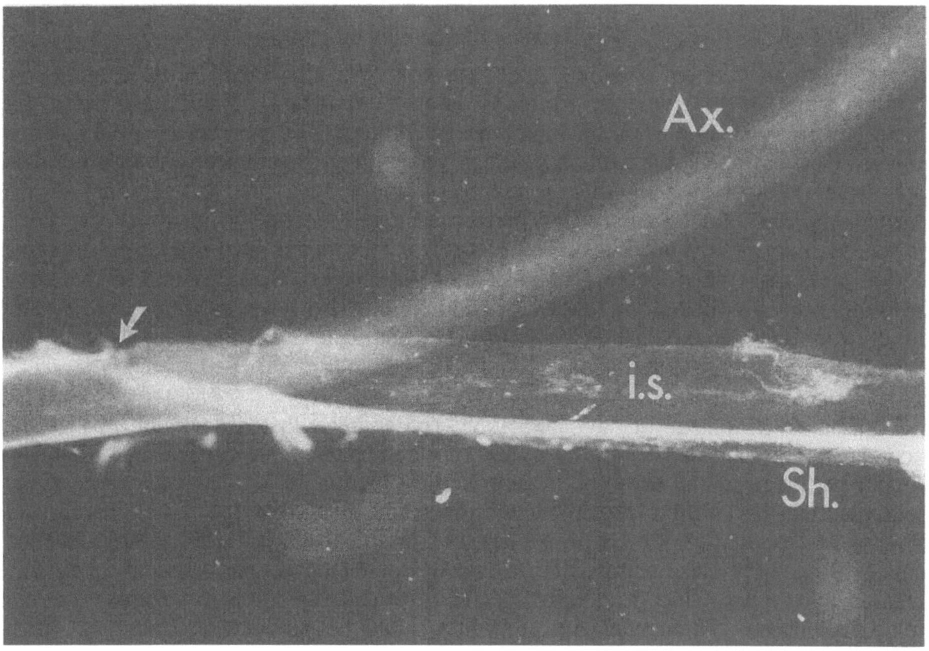

Figure 5. A cylinder of axoplasm being removed from a slit axon. This axon was slit from right to left and the axoplasm was then grasped at one end with fine forceps and gently peeled away from the axon. Note that the axoplasm is removed as a smooth and undistorted cylinder. The white arrow shows the limit of the slit. The axoplasm (*Ax.*), sheath (*Sh.*) and inner surface of the axon (*i.s.*) are labelled. Photograph taken under dark-field illumination. The axoplasm is about 0.5 mm in diameter.

other studies. Some contamination with sheath material is likely in this procedure but contamination with extracellular fluid, which can be destructive to the axoplasm (see Section 11.2), can be avoided by rinsing the fine-cleaned fiber well with artificial axoplasm solution before slitting. In many studies the possibility of slight contamination with extracellular fluid or sheath material is not important.

Axoplasm is almost transparent and, thus, it is almost invisible under bright-field illumination. However, the edges of axoplasm obtained both by extrusion and by slitting appear sharp under the light microscope in dark-field illumination, and this suggests that the axoplasm is removed along a natural annular cleavage plane in the cortex of the axon (see Section 15). Ultrastructural studies have shown that extrusion removes greater than 95% of the axoplasm and leaves behind a 10 μm thick layer of cortical axoplasm. It is likely that the axoplasm peeled from a slit axon also cleaves at this plane, but we have not examined this.

6.3. Cannulation

A third method for obtaining axoplasm from the giant axon is by cannulation. This procedure was developed to allow internal perfusion of the giant axon for electrophysiological studies on the membrane and is described in detail by Adelman and Gilbert (1990). Briefly, the axon is fine-cleaned and then cannulated with tapered glass cannulae at one or both ends. In one method, a long and narrow glass capillary is then inserted into one of the cannulae and advanced axially along the inside of the axon. As the capillary is advanced, a central core of axoplasm is drawn into it by applying suction pressure. When the capillary has been advanced to the other end of the axon, the axoplasm is expelled by applying positive pressure and the capillary is then withdrawn. Because axoplasm adheres to glass, complete expulsion from the capillary requires flow of internal perfusion solution. The axoplasm expelled in this way tends to wash out in fragments.

The main advantage of the cannulation procedure is that it does not damage the axon membrane and thus it affords access to the subaxolemmal cortical cytoplasm by internal perfusion of the axon, as is described in Section 15.2. The intactness of the membrane can be verified throughout the procedure by monitoring the excitability of the axon and by the use of a cell-impermeant extracellular marker such as radio-labelled sucrose or mannose (see Section 7.1). However, cannulation is not practical for routine isolation of axoplasm since it requires special effort and equipment and furnishes a low yield of axoplasm. The cannulation procedure is also not suitable for studies in which preservation of the organization of the cytoskeleton is important since the axoplasm obtained is disrupted and is also diluted with the perfusion solution used to expel it from the core capillary.

7. How pure is axoplasm isolated from the giant axon?

The two possible sources of contamination of axoplasm are the extracellular fluid and the tissue of the sheath.

7.1. Extrusion

Compression of the axon during extrusion may result in damage to the axolemma and sheath cells and this could give rise to contamination of the axoplasm with extracellular fluids or extracellular material such as Schwann cell cytoplasm. One measure of the extent of damage to the axon during extrusion is whether the remaining sheath will propagate a biphasic action potential, though this is not an absolute test since it is known that action potentials can propagate through damaged regions. Baker *et al.* (1962) reported that extrusion of axoplasm using a rubber roller resulted in electrically inexcitable sheaths in 60 out of 207 attempts. Thus, in their hands about one third of the axons were so badly damaged that they became inexcitable. However, in practice the extrusion procedure is satisfactory if these precautions are taken:

1. To avoid contamination with extracellular fluids: fine-clean proximal end of axon, rinse axon in artificial axoplasm solution or calcium-free seawater, and blot well to remove excess fluid before extruding axoplasm.

2. To avoid contamination from sheath: extrude with gentle pressure in a single stroke using shorter lengths where possible, and always end the extrusion stroke before reaching the end of the axon so that the axoplasm in the region of compression is not extruded.

One way to test the amount of contamination from extracellular material is to incubate the giant nerve fiber in an extracellular marker prior to extrusion, and then to measure how much of the marker is present in the extruded axoplasm. For example, Larrabee and Brinley (1968) incubated axons in radioactive inulin and then extruded the axoplasm, taking precautions similar to those listed above. They found that there was a slight contamination of the axoplasm that could be accounted for by the short portion of the axon (near the cut end) that was placed on the plastic film onto which the axoplasm was extruded. Rosenberg and Hoskin (1963) and Hoskin and Rosenberg (1965) tested a variety of other radiolabelled molecules that are also considered to be impermeable to cells, such as sucrose and acetylcholine. Using partially fine-cleaned axons and a slightly different extrusion protocol (see Rosenberg, 1981), they found less than 1% penetration compared to that expected if the plasma membrane of the axon presented no permeability barrier. It is possible that the axon is indeed slightly permeable to these compounds, but nevertheless these determinations place an upper limit on the amount contamination of extruded axoplasm from extracellular fluids.

7.2. Slitting

Contamination with extracellular fluids should not be a problem in the slitting procedure because the fine-cleaned axon can be rinsed in artificial axoplasm solution prior to cutting the axon open. However, slight contamination of the axoplasm with tissue from the sheath tissue is likely.

7.3. Cannulation

In the cannulation procedure, contamination by extracellular fluids and by tissue from the sheath can be avoided, but this procedure is not practical for many studies. One example of its use is in an investigation of the glia-neuron transfer hypothesis by Gainer *et al.* (1977), in which cannulation and internal perfusion of the giant axon was used to provide evidence that glial cells are the sole site of protein synthesis in the giant nerve fiber. The perfused axon preparation has also been used to investigate the composition and properties of the cortical membrane-associated cytoskeleton of the giant axon (see Section 15.2).

8. Handling axoplasm

Axoplasm is gelatinous in appearance and has sufficient structural integrity to be handled gently with flattened or rounded forceps. Nevertheless, axoplasm is, after all, naked cytoplasm, and any handling will locally disrupt its microscopic structure. Thus, it is preferable to reduce the number of times it is handled to the minimum possible. The precautions taken during handling depend on the nature of the study. For biochemical experiments, full recovery of the axoplasm at each stage of handling is the main requirement, while for structural or mechanical investigations, preservation of the organization and shape of the axoplasm is more important.

One consideration in handling axoplasm is its adhesiveness. Axoplasm adheres strongly to glass and tends to spread out and flatten so as to maximize its area of contact when extruded onto a glass surface. We find that better preservation of the cytoskeletal organization, and easier handling of the axoplasm, is afforded by extrusion onto a hydrophobic plastic surface such as a piece of Parafilm™ or Nescofilm™. If necessary, the plastic film can be stretched over a glass slide to provide rigidity and a solid base for extrusion. Handling axoplasm on plastic film also maximizes recovery of the axoplasm when it is subsequently removed, so we also recommend plastic for general handling of axoplasm, such as for weighing or for axoplasm that is being pooled for subsequent treatment.

Axoplasm rapidly dries if left in air. There are a number of ways to avoid or minimize this. The choice of method again depends on what the axoplasm is to be used for. One way is to immediately place the freshly isolated axoplasm in a moist atmosphere on ice. For example, axoplasm can be placed on Parafilm™ over moistened filter paper in a petri dish on ice. We often place axoplasm in a small envelope made by folding a piece of Parafilm™ and pinching together the open sides.

The shape and organization of the axoplasm can also be disrupted by surface tension forces when axoplasm is transferred through an air-water interface. Typically these forces cause the axoplasm to flatten to some extent. In investigations where the structural integrity of the axoplasm is important, we avoid this by extruding the axoplasm directly into a drop of solution placed over the cut end of the axon, or by submerging a fine-cleaned axon in the solution during extrusion. Two examples where we have extruded axoplasm in this manner are in our stretch experiments on axoplasm (see Section 21), in which mechanical integrity is important, and also in our experiments on the assembly dynamics of the cytoskeletal polymers, in which the cylindrical shape of the axon is important for the kinetic analyses (see Section 20).

9. The composition of axoplasm

The accessibility of axoplasm from the giant axon has enabled a detailed analysis of the chemical composition of the axon. Axoplasm contains about 2 – 2.5% (w/w) protein (Huneeus and Davison, 1970; Morris and Lasek, 1984) and 86.5 ± 5% (w/w) water (Koechlin, 1955). The axoplasm is iso-osmotic with seawater (about 980 mOs/kg) and has a pH of about 7.2, with published estimates ranging from 7.0 to 7.35 (Caldwell, 1958; Spyropoulos, 1960; Bicher and Ohki, 1972; Boron and DeWeer, 1976).

9.1. Ions and small molecules

Table II is a comparison of the composition of axoplasm and blood from *Loligo pealei*, and of seawater. The values are given as millimoles/kg axoplasm (i.e. millimolal). Note that since the water content of axoplasm is about 865 g/kg, the actual molar (moles/liter) concentrations of the elements are somewhat greater than their molal (moles/kilogram) concentration. The values for the amino acids, sugars and

Table II. The composition of axoplasm and blood in *Loligo pealei*, and of seawater.

Substance	*Axoplasm*	*Blood*	*Seawater*
Water[5] (g/kg)	865	870	966
Protein[4,6] (g/kg)	25	0.15	–
pH	7.2	7.5	7.9
Inorganic ions[2,3,5,6]			
K	340	22	10
Na	65	460	460
Cl	150	580	540
Ca	3.5	11	10
Mg	10	55	53
Sulfate	–	8.1	28
Phosphate	18	–	–
Amino acids[2]			
Gly	12	2.5	–
Ala	8.6	2.9	–
Val	2.4	2.6	–
Leu/Ile	2.9	4.4	–
Pro	1.1	2.1	–
Ser	4.0	1.3	–
Thr	2.1	1.2	–
Asp	79	1.8	–
Glu	21	2.2	–
Arg	3.5	0.6	–
Lys	2.6	1.0	–
Ornithine	2.0	0.40	–
Others	2.7	7.2	–
Other organic metabolites[2]			
Isethionate	170	1.6	–
Taurine	110	4.6	–
Homarine	20	3.3	–
Betaine	74	4.3	–
Cysteic acid amide	4.9	0.4	–
Carbohydrates[2]			
Glucose	0.24	0.3	–
Sucrose	0.24	–	–
Fructose	0.24	0.5	–
Myo-inositol	7.6	–	–
Glycerol	4.4	–	–
Others[1]			
ATP	0.7–1.7	–	–
Arginine phosphate	1.8–5.7	–	–

Values are given as millimoles/kg unless otherwise noted.
References: [1] Caldwell (1960), [2] Deffner (1961), [3] Hodgkin (1958), [4] Morris and Lasek (1984), [5] Koechlin (1955), and [6] Robertson (1949).

other organic metabolites are taken from the analyses of Deffner (1961), since this is the most comprehensive analysis to date, and accounts for about 95% of the dialyzable (small molecule) constituents of axoplasm. Deffner (1961) also analyzed axoplasm of the giant Chilean squid *Dosidicus gigas* and found a very similar composition, so it is likely that there is little variation between different members of the squid family. The composition of axoplasm has also been analyzed for the giant axon of the marine fanworm *Myxicola infundibulum* (Gilbert, 1975b), and it is similar to that of squid except in the nature of the major organic ions.

If variations between different squid do occur, they are also most likely to be in the relative concentrations of the major organic anions. Because squid axoplasm is iso-osmotic with seawater, it contains much higher concentrations of organic and inorganic ions than the cytoplasm of vertebrates and freshwater invertebrates. The major cations are potassium and sodium and the major anions are normally-occurring organic metabolic intermediates that are present at elevated levels. These include aspartic acid, isethionic acid (2-hydroxyethanesulfonic acid), taurine (2-aminoethanesulfonic acid), betaine (carboxymethyltrimethylammonium) and homarine (1-methyl-2-pyridinium carboxylic acid).

The ionic composition of squid blood and of seawater are very similar. The main difference is the presence of organic metabolites and a small amount of protein in squid blood. This similarity in composition has allowed seawater to be used universally as a medium for dissection, incubation and experimentation on axons. However, for investigations in which the metabolic activity of the axon or Schwann cells is important it would be advisable to include some amino acids and sugars in the seawater and to ensure adequate oxygenation. The ionic composition of squid axoplasm and blood differ markedly. In axoplasm potassium is the major cation, while in blood it is sodium, and in blood the major anion is chloride, but in axoplasm the chloride concentration is low and the balance is provided by organic anions.

9.2. Macromolecules

Axoplasm also contains the macromolecules that make up cytoplasm. The major macromolecule in axoplasm is protein, and the major axoplasmic proteins are those that form the cytoskeleton. The structure and composition of the cytoskeleton is the subject of Sections 14 to 19 of this chapter. In addition, axoplasm contains membranous organelles, including mitochondria, and smooth endoplasmic reticulum, which are composed of proteins and lipids. The lipid composition of axoplasm has been analyzed by McColl and Rossiter (1951). The principal components are phospholipids (0.18 mg/100 mg axoplasm) and cholesterol (0.03 mg/100 mg axoplasm). The study of lipids in squid axoplasm is discussed by Gould and Alberghina (1990). Axoplasm also contains small amounts of RNA, and the interested reader is referred to the review by Gainer *et al.* (1984) for more information.

9.3. Calcium and magnesium in axoplasm

The data in Table II indicate that the total concentrations of calcium and magnesium in axoplasm are 3.5 and 10 mM respectively, but measurements indicate that not all of this is ionized. This is especially so for calcium, and less than 0.1%

of the total calcium in axoplasm is ionized. Baker and Schlaepfer (1978) identified two distinct components of calcium binding in squid axoplasm. One is an energy-dependent process with a high capacity but low affinity for calcium. This component probably represents the mitochondria and endoplasmic reticulum in axoplasm (Baker and Schlaepfer, 1978, Henkart *et al.*, 1978), which can sequester calcium in large amounts (probably as a calcium phosphate or calcium oxalate precipitate). The other component is independent of metabolic energy and has a higher affinity for calcium, but is readily saturated. This binding may be accounted for by proteins in the cytoplasm. For example, squid axoplasm contains calmodulin (Alema *et al.*, 1973; Head and Kaminer, 1980), and it may constitute as much as 1% of the total protein in the axon (Alema *et al.*, 1973), which is about 0.25 mg protein/g axoplasm wet weight. Another likely source of calcium-binding are the peripheral domains of the neurofilament proteins because these domains are known to be highly phosphorylated (see Section 17.3). In support of this, Abercrombie *et al.* (1986) have reported that neurofilament proteins from the giant axon of the marine fanworm *Myxicola infundibulum* bind calcium, and that this can account for most of the energy-independent calcium binding in these axons.

As an aside in this regard, a little-known study by Krishnan and Singer (1974) is also of interest. They used the reagent potassium pyroantimonate, which forms an electron-dense precipitate with sodium, calcium and magnesium, to localize the sites of cation binding in amphibian myelinated nerve. In addition to precipitation of cations in the Schwann cell cytoplasm, precipitate was also observed in the axoplasm. This precipitate was located predominantly on neurofilaments and in a periodic fashion along their length. The periodicity was often regular and measured 45 – 75 nm, which is similar to the axial spacing of the neurofilament side-arms that contain the highly phosphorylated peripheral domains of the neurofilament proteins.

The regulation of the ionized calcium concentration in cells is important because of the large number of cellular processes that are regulated by this ion. One such process in axons is calcium-dependent proteolysis (see Section 17.4). The calcium-dependent protease in squid axoplasm cleaves the cytoskeletal proteins in the presence of millimolar concentrations of calcium causing liquification of the axoplasm, but the low levels of ionized calcium in the axon ensure that the enzyme is not normally active.

Measurements in *Loligo pealei* using the calcium indicators aequorin and arsenazo III (DiPolo *et al.*, 1976), and using calcium-selective electrodes (DiPolo *et al.*, 1983) indicate a resting ionized calcium concentration in freshly dissected axons of about 100 nM (a pCa of 7). In contrast, the calcium concentration in squid blood and seawater is about 10 mM, though the presence of calcium-binding compounds such as sulfate ions result in a free ionized calcium concentration of nearer 4 mM (Blaustein *et al.*, 1974). Thus the ratio of ionized calcium in the extracellular space to that in the axoplasm is about 40,000:1. A similar level of ionized calcium is found in other cells, such as resting muscle, and it seems that under normal conditions the bulk of the calcium inside cells is sequestered. The high calcium concentration in seawater and the biological activity of calcium in axoplasm underscore the importance of avoiding contamination of axoplasm with extracellular fluids during isolation.

For magnesium in axoplasm, a much larger fraction is ionized than for calcium. Measurements indicate that between one third and one half of the total axoplasmic magnesium is free, with estimates varying from 2 – 4 mmoles/kg axoplasm (Baker and Crawford, 1972; Brinley and Scarpa, 1975; DeWeer, 1976). The remainder is probably bound to charged molecules in the axoplasm such as proteins and nucleotides. The large proportion of magnesium that is ionized means that the ratio of free magnesium to calcium in axoplasm is about 30,000:1.

10. Artificial axoplasm solutions

Because the composition of squid axoplasm is known it is possible to design solutions that closely approximate many of the important physical and chemical parameters of this cytoplasm. We consider inorganic molecules, small organic molecules, macromolecules, concentration of protons (pH), ions such as calcium (pCa) and magnesium, ionic strength, and colligative properties such as osmotic strength. In many studies it is not practical nor necessary to approximate all of these parameters. The choice of an appropriate medium therefore depends on defining those factors that are important for the phenomenon being studied. Since we do not know how critically the chemical environment affects many of the processes in cells, we have attempted to simulate the known chemical composition of squid axoplasm in many of our studies, as far as is practical.

10.1. Buffer P

Originally we developed Buffer P, which adheres closely to the known concentrations of small molecules and ions in axoplasm (Morris and Lasek, 1982). Initially Buffer P contained 15 amino acids (ala, ile, leu, met, phe, pro, val, gly, ser, thr, tyr, asp, glu, arg and lys), four carbohydrates (glucose, fructose, myo-inositol and glycerol), five other organic metabolites (isethionate, taurine, citrulline, hypoxanthine and ornithine), and five inorganic ions (K^+, Na^+, Mg^{++}, Cl^- and phosphate). Excess taurine was added to substitute for homarine and cysteic acid amide. Subsequently, in a simplification of this solution, alanine, glycine, aspartic acid and arginine were used in appropriate amounts to substitute for their non-polar, polar, acidic or basic counterparts. Also, citrulline, hypoxanthine, myo-inositol and ornithine were eliminated from the solution. Adenosine-5'-triphosphate (ATP) and guanosine 5'-triphosphate (GTP) were included at their approximate physiologic levels to fulfill the energy requirements of the axoplasm. Ethylene glycol-bis(2-amino ethyl ether)N,N,N',N'-tetraacetic acid (EGTA) and phenyl-methylsulfonyl fluoride (PMSF) were added to chelate calcium and inhibit proteolysis, respectively, but the inclusion of protease inhibitor is not necessary, providing the free calcium concentration is low. The composition of the simplified Buffer P is shown in Table III. Note that in calculating the composition of this artificial axoplasm solution, millimoles/kg axoplasm was taken to be equivalent to millimoles/liter of buffer, though axoplasm is in fact only 86.5% water (w/w) (see Section 9.1).

10.2. Buffer X

A simpler alternative to Buffer P is Buffer X which was developed for our studies on organelle translocation in axoplasm (Brady *et al.*, 1984). In this solution, isethionate is omitted and replaced with aspartate, and alanine and arginine are omitted and

Table III. The composition of two artificial axoplasm solutions.

Substance	*Buffer P*	*Buffer X*
Inorganic ions		
K	344	300–400*
Na	65	–
Cl	147	32
Ca	–	3
Mg	10	13
Phosphate	18	–
Amino acids		
Ala	16	–
Gly	18	50
Asp	100	350
Arg	6	–
Other organic metabolites		
Isethionate	165	–
Taurine	132	130
Betaine	74	70
Carbohydrates		
Glycerol	4	–
Glucose	1	1
Fructose	0.5	–
Others		
ATP	1	1
GTP	0.5	–
EGTA	1	10
PMSF	0.1	–
HEPES	–	20

Values given as millimoles/liter. Composition of Buffer P from Morris and Lasek (1982). Composition of Buffer X from Brady *et al.* (1984). A recipe for Buffer X is given in Table IV.

* The potassium in Buffer X is added as potassium salts of HEPES, EGTA and aspartate, and as potassium hydroxide which is used to adjust the pH (see Table IV). We have not determined the exact final concentration of potassium in Buffer X, but an estimate is given.

replaced with glycine. The other major organic ions are taurine and betaine. Glucose is included as the sugar. An improvement over Buffer P is the use of a mixture of calcium and EGTA (3:10 mole ratio) to buffer the concentration of ionized calcium at

Table IV. Recipe for Buffer X (Developed by Scott T. Brady)

Substance	*Buffer X mM*	M_r	*stock M*	*stock g/100 ml*	*ml of stock/100 ml*
Aspartate	350	171.2	1	17.12	35
Taurine	130	125.1	–	–	1.626 g*
Betaine·H_2O	70	135.2	1	13.52	7
Glycine	50	75.1	1	7.51	5
HEPES	20	238.3	1	23.83	2
$MgCl_2 \cdot 6H_2O$	13	203.3	1	20.33	1.3
EGTA	10	380.4	0.1	3.804	10
$CaCl_2 \cdot 2H_2O$	3	147.03	1	14.703	0.3
Glucose	1	180.16	0.1	1.802	1
ATP	1	553.18	–	–	55 mg*

The table has six columns. These represent (from left to right): the substance, its molarity in Buffer X (mM), its relative molecular mass (M_r), the molarity of the stock solution (M), the amount of substance required to make 100 ml of stock solution (g/100 ml), and the volume of that stock solution required to make 100 ml of Buffer X (ml of stock/100 ml). Stock solutions are prepared for all components, except for taurine and ATP which are added as the solid. The stock solutions are stored frozen and thawed immediately before use. When all the components have been added, the pH of the mixture is adjusted to pH 7.2 with a 20% solution of potassium hydroxide, and then made up to final volume with distilled/deionized water. The Buffer is stored frozen in aliquots at −20 °C. Aspartate is the mono-potassium salt of the L-isomer. The HEPES and EGTA stocks are prepared with the free acids and adjusted to pH 7.2 with potassium hydroxide. ATP is used as the disodium salt. Abbreviations:

HEPES (N-2-hydroxyethlypiperazine-N'-2-ethanesulfonic acid)
EGTA (ethylene glycol-bis(2-amino ethyl ether)N,N,N',N'-tetraacetic acid)
ATP (adenosine-5'-triphosphate)

* Add by weight.

physiological levels. In an artificial axoplasm solution containing 3 mM calcium, 10 mM EGTA, 12.9 mM magnesium, and methane-sulfonate, taurine and glutamate as the major anions, Rubinson and Baker (1979) determined the ionized calcium concentration to be 65 nM, which is within the physiological range (see Section 9.3). Of the magnesium, 2.9 mM was bound to the EGTA, leaving 10 mM ionized magnesium. The composition of Buffer X is shown in Table III, and a recipe for Buffer X that we use in our laboratory is given in Table IV.

Both Buffer P and Buffer X should be iso-osmotic with axoplasm in order to preserve the membranous vesicles, mitochondria and smooth endoplasmic reticulum. These solutions are approximately iso-osmotic with axoplasm but the osmolarity should be checked and, if necessary, the difference should be made up with sucrose rather than salt in order to keep the ionic strength the same. It should be noted, however, that the composition of these artificial axoplasm solutions still differs from axoplasm in at least one very important way in that they contain no macromolecules, and in particular no protein, which is the major macromolecule in axoplasm.

11. The solubility and stability of axoplasm

Axoplasm is normally of gel-like consistency and is strong enough to be handled with forceps. These mechanical properties are due largely to the great length of the neurofilaments and microtubules, which makes them particularly important elements in the axoplasmic architecture. If axoplasm is immersed in physiological solutions that approximate the solute composition of axoplasm, the neurofilaments and many of the microtubules remain stably polymerized, and the axoplasm retains its mechanical integrity (see Section 10 for a description of these solution conditions). However, ionic conditions that differ from the physiological state, such as those produced by increasing the ionic strength or pH, can alter the stability of the cytoskeletal polymers and in this way affect the mechanical properties of the axoplasm.

11.1. Salt and pH

High salt concentrations tend to depolymerize axonal cytoskeletal polymers and thereby destabilize axoplasmic architecture, and comparisons of the effects of different salts on both the mechanical properties of axoplasm and the stability of the cytoskeletal polymers indicate that the chemical nature of the salt is more important than the ionic strength. For example, Baumgold *et al.* (1981) showed, using centrifugation, that the cytoskeletal polymers in squid axoplasm are stable in 360 mM potassium fluoride at pH 7.3 but are completely soluble in 360 mM potassium iodide at the same pH and ionic strength. In potassium fluoride, centrifugation of homogenized axoplasm produced a pellet that was highly enriched in neurofilament proteins, tubulin and actin; these proteins comprised 81% of the sedimentable protein. In potassium iodide, the axoplasm dispersed and none of these proteins sedimented.

In fact, the relative efficacy of different ions for preserving the structure of axoplasm appears to increase in the order of the classical Lyotropic Series of colloid chemistry. In this series, ions are ranked according to their chaotropic effect, which is their ability to destabilize interactions between hydrophilic colloids such as proteins. The chaotropic effect is a consequence of effects on the structure of the water which in turn modify the solvent-macromolecule interactions that are involved in the stabilization of macromolecular assemblies and associations (von Hippel and Wong, 1964; Friefelder, 1985). Physical studies on the interactions between macromolecules have indicated the following anion series:

fluoride > phosphate > aspartate/glutamate > sulfate > citrate > tartrate > acetate > chloride > nitrate > bromide > iodide > thiocyanate

and the following cation series:

rubidium > potassium > ammonium > sodium > lithium

Those ions that are most chaotropic are most effective at solubilizing proteins. Thus solutions that contain fluoride and phosphate salts of potassium are much more favorable for the preservation of the cytoskeleton than those containing iodide or thiocyanate salts of sodium or lithium.

Studies on neurofilaments and microtubules *in vitro* confirm that the stability of these cytoskeletal polymers is dependent on the chaotropic nature of the ionic environment. Apparently, solvent interactions are critical for the stability of these cytoskeletal polymers. For example, in studies on the assembly/disassembly properties of neurofilaments purified from axoplasm and squid brain, we have found that potassium thiocyanate is more effective than potassium chloride at solubilizing filaments, and a more detailed comparison of different potassium halide salts revealed a direct correspondence to the Lyotropic Series (Brown, 1989). In addition, Sakai and Matsumoto (1978) observed a similar relationship in their studies on the assembly of microtubules from squid axoplasm in the presence of various potassium halide salts.

Through their effect on the stability of the protein filaments, chaotropic salts can affect the mechanical properties of the axoplasm. For example, Rubinson and Baker (1979) investigated the effect of various salts on the mechanical properties of axoplasm by measuring the viscous resistance of extruded axoplasm to flow under pressure through a porous cellulose acetate tube. They used an artificial axoplasm solution containing methanesulfonate (0.3 M), taurine (0.15 M) and glutamate (0.1 M) as the major anions. Methanesulfonate was used as a substitute for isethionate (2-hydroxyethanesulfonate), which is the major anion in squid axoplasm (see Section 9.1). Substitution of methanesulfonate in the artificial axoplasm solution with more chaotropic anions had a marked effect on the mechanical properties of the axoplasm; chloride, bromide, iodide and thiocyanate (in order of increasing efficacy) all caused a reversible decrease in the viscosity.

The protein interactions that stabilize the cytoskeletal polymers in axoplasm are also sensitive to pH. The microtubules and neurofilaments in squid axoplasm are optimally stable between pH 6 and 7. These polymers tend to precipitate or aggregate at around pH 5, which is near their isoelectric point, but are readily solubilized at alkaline pH (Brown, 1989).

11.2. Calcium-dependent proteolysis

In addition to their physical effects, ions may also have effects due to their specific chemical or biological activity. This is so for calcium, which has a specific biological action on axoplasm. In early studies on extruded axoplasm it was noted that axoplasm rapidly liquified in seawater. Hodgkin and Katz (1949) showed that this disintegration was caused specifically by calcium at concentrations of 1 mM or more; axoplasm does not disintegrate in calcium-free seawater. These observations were confirmed by the observations of Chambers and Kao (1952) and by the rheological measurements of Rubinson and Baker (1979). It is now known that this effect is due to the action of an endogenous calcium-dependent protease that preferentially cleaves the neurofilament proteins in axoplasm (see Section 17.4). The protease is not active in the axon *in vivo* since the ionized calcium concentrations in axoplasm are very low, and do not exceed micromolar levels due to the strong buffering capacity of the intracellular calcium stores (see Section 9.3).

12. Preparation of axoplasm and sheath for SDS PAGE technique

The protein composition of the axoplasm and sheath of the giant axon can be analyzed by sodium dodecyl sulfate polyacrylamide gel electrophoresis (SDS PAGE). In this technique proteins are dissolved in the ionic detergent, sodium dodecyl sulfate (SDS), in the presence of a sulfhydryl reducing agent, β-mercaptoethanol, which reduces any disulfide bonds. The denatured SDS-polypeptide complexes are then applied to a porous polyacrylamide gel and subjected to an electric field. The proteins move through the gel toward the anode because of the negative charge of the bound SDS. For most proteins, the rate of movement has been found empirically to be related to the size of the denatured polypeptide chain, which in turn is related to its mass. Therefore, this technique has proven useful for the separation of polypeptides of different molecular masses. In addition, SDS PAGE allows an empirical estimation of the molecular mass of a protein by comparison of its mobility with the mobility of proteins of known mass. A comprehensive treatment of the principles and practice of electrophoresis can be found in Andrews (1981).

Axoplasm readily dissolves in a standard sample buffer containing SDS and β-mercaptoethanol, with brief heating in a hot water bath, as originally described by Laemmli (1970). No homogenization is necessary. For standard SDS PAGE with Coomassie Blue (CBBR-250) staining, 1.2 mg axoplasm wet weight (equivalent to about 30 μg protein) gives a good loading on a typical 1 mm thick Studier slab gel (Studier, 1973), and 0.3 mg axoplasm wet weight (equivalent to about 7.5 μg protein) gives a suitable loading for a typical 0.9 mm thick minigel. These loadings will, of course, depend also on the width of the wells on the gel.

In contrast to the axoplasm, the sheath is difficult to solubilize since it contains connective tissues. We recommend homogenization with a small volume glass-glass homogenizer with ground glass surfaces, followed by heating with sample buffer. High speed centrifugation may then be necessary to clarify the sample of insoluble material such as collagen, which can disturb the protein separation in SDS PAGE and cause smearing of the bands. For analysis of the residual cortical axoplasm in the sheath (see Section 15), without extracting glial and extracellular proteins, the sheath can be slit open after axoplasm extrusion and extracted with salt solutions as described by Sakai and Matsumoto (1978). Alternatively, the proteins of the cortical axoplasm can be extracted by the intracellular perfusion technique (see Section 15.2).

13. Preparation of axoplasm for electron microscopy

13.1. A general fixation protocol

For routine preparation of axoplasm for thin sectioning, we use the following protocol (Miller and Lasek, 1985). Axoplasm is fixed for 1 – 2 hours at room temperature in SPME (0.6 M sucrose, 0.1 M potassium phosphate, 10 mM magnesium chloride, 1 mM EGTA, pH 7.1) solution containing 2.5% glutaraldehyde. The SPME solution has an osmolarity of 980 mOs/kg, which is approximately iso-osmotic with axoplasm. The osmolarity of the glutaraldehyde fixative is ignored since it readily crosses membranes during fixation and therefore is not osmotically active (Glauert, 1975). For preservation of membranous organelles, it is better that the fixative solution be slightly hyper-osmotic than hypo-osmotic. The osmolarity should be adjusted with sucrose rather than salt since it is desirable to keep the ionic strength low in order to stabilize the cytoskeletal filaments. Artificial axoplasm solutions containing molecules with amine groups, such as solutions containing amino acids, are not suitable since these molecules react with aldehydes and interfere with the fixation. Phosphate is a good choice for the buffer since it ranks low in the Lyotropic Series and therefore does not destabilize protein interactions (see Section 11.1). After glutaraldehyde fixation the axoplasm is post-fixed with 1% osmium tetroxide in SPME solution, stained *en bloc* with 0.5% (aq) uranyl acetate, and finally dehydrated through a graded ethanol series into propylene oxide, and embedded in Epon (Poly/Bed 812, Polysciences, Inc.). Thin sections are stained with aqueous uranyl acetate and lead citrate. Since axoplasm poses no permeability barrier, many of the steps in the preparation, including the fixation, could probably be shortened. Some other fixation protocols that have been used for squid axoplasm and for the giant axon are described by Metuzals (1969), Hodge and Adelman (1980) and Metuzals *et al.* (1983).

13.2. Preservation of microfilaments

The above conventional fixation, dehydration and embedding protocol preserves microtubules and neurofilaments in axoplasm, but actin microfilaments are more susceptible to destruction during fixation and dehydration, and a modified procedure is necessary to ensure good preservation. This is important because our recent studies indicate that microfilaments are a major component of the inner and cortical cytoskeletons of axons (see Section 19).

The two treatments that are critical for microfilament preservation are post-fixation with osmium tetroxide, and dehydration. Conventional osmication and dehydration protocols can fragment and destroy microfilaments, even after aldehyde fixation. McDonald (1984) obtained improved preservation of microfilaments by post-fixation with osmium ferricyanide rather than osmium tetroxide. In addition, the destructive effects of dehydration were minimized by further fixation with tannic acid and uranyl acetate, and by dehydration in acetone rather than ethanol.

Fath and Lasek (1988) used this protocol for the visualization of actin microfilaments in squid axoplasm. Axoplasm was fixed by immersion for 30 minutes in SPME solution (see Section 13.1) containing 1% glutaraldehyde. In some cases, axoplasm was extruded directly into the fixative. In other cases the axoplasm was first extruded into Buffer X (see Section 10.2) containing 10 μM phalloidin (to stabilize the microfilaments), and extracted with gentle stirring for 4 hours at ambient temperature prior to immersion in the fixative. Extraction of the soluble proteins in this way improved the visibility of the microfilaments by removing some of the dense matrix that surrounds them in axoplasm (see Section 19). After primary fixation, the axoplasm was post-fixed according to the osmium ferricyanide procedure of McDonald (1984). Briefly, the fixed axoplasm was first washed for 5 minutes in 0.1 M sodium phosphate buffer (pH 7.2), and then reacted for 5 minutes with a solution of 0.8% potassium ferricyanide in 0.1 M sodium phosphate buffer. The axoplasm was then post-fixed for 15 minutes in a mixture of 0.5% osmium tetroxide and 0.8% potassium ferricyanide in 0.1 M sodium phosphate (on ice, in the dark), followed by 15 minutes in 0.5% tannic acid in 0.1 M sodium phosphate, and rinsed in deionized water. Finally, the axoplasm was stained *en bloc* with aqueous uranyl acetate, dehydrated through graded acetones, and embedded in plastic for sectioning.

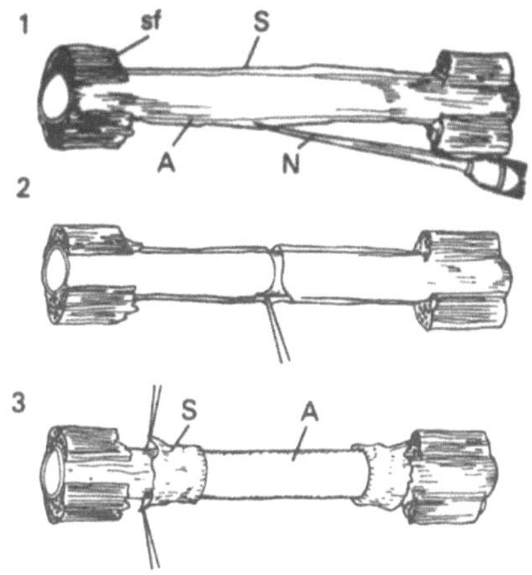

Figure 6. Removal of the Schwann cell sheath from the giant axon. 1. Small nerve fibers (*sf*) are removed from a portion of the giant nerve fiber. The sheath (*S*), axolemma (*A*) and needle (*N*) are labelled. 2. The sheath is cut circumferentially with the needle. 3. The sheath (*S*) is pulled back from the opening on either side to expose the surface of the axon (*A*). Reproduced with permission from Metuzals *et al.* (1981b).

13.3. Fixative penetration

Good fixation also depends on good penetration of the fixative into the axoplasm. Best fixation is achieved if the fixative is in direct contact with the axoplasm. The simplest approach is to extrude axoplasm as described in Section 6.1, and then to fix by immersion. However, the compressive forces during extrusion in the conventional manner tend to cause some disruption of the longitudinal organization of the cytoskeletal polymers. Some improvement is achieved by extruding the axoplasm onto glass, and then immersing the axoplasm on the glass into fixative. The adhesive effect of the glass allows the axoplasm to be extruded under tension which helps to restore the longitudinal alignment of the polymers that is present in the axon (see Section 6.1). Compression of the cytoskeleton can also be avoided by isolating axoplasm by the slitting method described in Section 6.2, but for optimum preservation of the organization of the cytoskeleton, the axoplasm must be fixed *in situ.* However, fixation by immersion of the whole giant axon in fixative does not produce good preservation of the axoplasm. Apparently, the collagenous sheath becomes a barrier to the diffusion of the fixative during fixation.

13.4. Removing the axon sheath

Improved preservation of axoplasm *in situ* has been obtained by first desheathing the axon according to a method developed by Ichiji Tasaki in 1975 and described by Metuzals *et al.* (1981b); see Fig. 6. The close association of the glial cells with the axon means that the sheath cannot be removed from the axon by conventional dissection procedures. To desheath the axon, the giant nerve fiber is first fine-cleaned to remove associated small nerve fibers and then immersed for 1 hour, without loss of excitability, in seawater containing 1 mg/ml trypsin. The trypsinized fiber is then subjected to mild fixation for 15 minutes in a 0.5% solution of glutaraldehyde in seawater. After fixation the fiber is placed in a hypertonic solution of 0.4 M sucrose in seawater, which causes the axon to shrink away from the sheath. The elevated sheath, consisting of the Schwann cell layer, basement membrane and outer collagenous layer (see Section 3), can then be transected with a needle by pinching the sheath between the needle and the base of the dissecting dish. A circumferential incision of the sheath is then made in the same manner by inserting the tip of the needle into this opening. The sheath is then everted, exposing about 1 – 2 cm of desheathed axon. Desheathed axons can then be fixed by direct immersion in seawater containing aldehyde fixative and processed for transmission electron microscopy in the normal manner, or the axon can be prepared for scanning electron microscopy of the surface of the axon (see Section 14.2).

Improved fixation of axons may also be facilitated by pretreatment of the fine-cleaned fiber with collagenase to remove the collagenous layer of the sheath (Lasek, 1989), though this procedure renders the fiber extremely fragile.

13.5. Cannulation

Hodge and Adelman (1980) obtained much improved fixation and preservation of organization of the axoplasm by internal irrigation of axons with fixative following cannulation (see Section 6.3). One could also slit a fine-cleaned axon along its length as described in Section 6.2 and then immediately flood with fixative. Both of these techniques allow the fixative direct access to the axoplasm *in situ*, without the need for separation of the axoplasm from the sheath.

13.6. Negative staining

The cytoskeletal elements in axoplasm can also be visualized by negative staining. We have used two preparations. In one, the axoplasm is gently homogenized at a concentration of about 10 mg/ml wet weight (200 μg/ml protein) in a suitable solution, and then a small aliquot is applied to a carbon-coated grid and stained with 1% aqueous uranyl acetate for electron microscopy (see Huxley, 1963, for procedure). If solutions containing high salt concentrations are used, such as Buffer P and Buffer X, then it is necessary to rinse the grids with a low-salt solution in order to obtain good staining. If required, the proteins can be fixed with glutaraldehyde before applying them to the grid. Phosphate buffer is not recommended since it forms a precipitate with uranyl acetate.

In another approach, a carbon-coated grid can be gently pressed directly against the surface of freshly extruded axoplasm, allowing proteins to stick to the carbon support (Wais-Steider *et al.*, 1988). The proteins are then stained with uranyl acetate as described above. This procedure provides excellent negatively-stained images of the cytoskeletal polymers in axoplasm, without the need for dilution or homogenization, and essentially in their native state.

14. The organization of the cytoskeleton

14.1. Longitudinal organization

Axons are cylindrical in shape, and their three-dimensional structure can be usefully described by two aspects, one radial and the other longitudinal. The cylindrical shape of the axon, in its longitudinal dimension, is formed and largely maintained by the long polymers of the axonal cytoskeleton (neurofilaments and microtubules), which are orientated along the long axis of the axon. The high degree of alignment of these cytoskeletal elements is the most striking feature of the organization of the axonal cytoskeleton. This structural property of axoplasm was first inferred over 50 years ago by Bear *et al.* (1937a), based on their observations on the birefringence of axoplasm in polarized light.

One good way to appreciate the longitudinal alignment of the cytoskeletal elements in the giant axon is to examine an intact axon by differential interference contrast microscopy. Electron microscopy provides far more structural detail, but light microscopy gives a much better impression of the long-range order. Using this technique, Metuzals and Izzard (1969) showed that axoplasm *in situ* was composed of highly linear elements that were orientated along the long axis of the axon. These linear elements apparently represent individual bundles of microtubules that are now known to be refractile enough to be detected by this type of light microscopy (see Section 18.2). Electron microscopy confirms that the microtubules in axoplasm *in situ* are very straight and that they are aligned in parallel to each other along the long axis of the axon. The neurofilaments, in contrast, trace a slightly more sinuous path through the axon. In extruded axoplasm, the longitudinal organization of the axoplasm is essentially preserved, but the straightness of the long polymers is lost due to the compressive forces that are employed in the extrusion process (see Section 6.1).

14.2. Helicity

Axons often have a natural internal helicity that is superimposed on the longitudinal alignment of the cytoskeletal polymers in the axoplasm (for discussion, see Lasek *et al.*, 1983). This helicity is particularly evident in the axoplasm of squid and marine fanworm giant axons. For example, Metuzals *et al.* (1981b) demonstrated helicity in the cytoskeleton in the squid giant axon by scanning electron microscopy of the surface of the axon after removal of the sheath (see Section 13.4 for description of procedure). The surface of the desheathed axon had a ridge-and-groove topology, which traced a right-handed helical path along the axon. The angle of the helix was about 10° to the long axis, which corresponds to a pitch of about 18 mm for an axon with a diameter of 0.5 mm. The ridges and grooves may have been an artifact due to the osmotic shrinkage employed in the desheathing procedure, but nevertheless, they did reveal an underlying helical twist to the cytoskeletal polymers in the periphery of the axon. This helicity has also been observed directly by differential interference contrast microscopy of axoplasm *in situ* (Metuzals and Izzard, 1969).

A more striking example of helical organization in axoplasm is the giant axon of the marine fanworm, *Myxicola infundibulum*, which is also unusual in that it contains a very high proportion of neurofilaments and relatively few microtubules. The helicity is so pronounced in these axons that three orders of helical configuration can be identified (Gilbert, 1972; 1975a); a *ripple helix* that has a pitch of about 13 μm, a *segmental helix* that corresponds to the helicity described above for the squid giant axon, and a macroscopic *gel helix*, which forms the axon into large coils during the substantial length changes that occur when the worm contracts. The ripple helix may also be weakly expressed in squid axoplasm (Gilbert, 1972).

Gilbert was able to isolate axoplasm from the fanworm and he showed that the neurofilament-rich axonal cytoskeleton retains the helical properties of the axon. These observations indicate that helicity may be an intrinsic property of the axonal cytoskeleton. This proposal is supported by the fact that, in squid, giant axons from the right and left sides of the animal both have the same helical handedness (Metuzals *et al.*, 1981b). The abundance of neurofilaments in the squid and fanworm giant axons, and in other axons with helical cytoskeletons, suggests that neurofilaments contribute to the intrinsic helicity of the axonal cytoskeleton (Gilbert, 1972; 1975b; Metuzals *et al.*, 1981a; Lasek *et al.*, 1983). For example, Gilbert showed that fanworm axoplasm behaves like a weakly coiled spring. This property is apparently due to the neurofilaments, which trace a helical path that forms the *ripple helix* of the axon (Gilbert, 1975a). If the axoplasm is stretched, the neurofilaments straighten, and if the stretch is released, they then tend to recoil.

In addition, the molecular structure of neurofilaments also supports the proposal that these fibrous polymers contribute to the helical organization of the cytoskeleton. Like other intermediate filaments, neurofilaments are composed of rod-shaped subunits that are arranged in a rope-like helical pattern along the axis of the neurofilament (Krishnan *et al.*, 1979; Aebi *et al.*, 1983), and this helical arrangement of the fibrous subunits within the neurofilaments may generate a slight torque within these structures so that they tend to twist when force is applied to them by the cytoskeletal translocation mechanisms of the slow axonal transport system. In this way, the helical characteristics of the neurofilaments may largely determine the higher order helical properties of the axonal cytoskeleton, in axons with large numbers of neurofilaments (Gilbert, 1975a).

The cytoskeletal helicity is a very weak property of axoplasm in squid, since observations on axoplasm isolated by the axon slitting procedure (see Section 6.2) show no sign of helical recoil after separation from the sheath, and the axoplasm retains the original proportions of the axon (see Fig. 5). In contrast to the squid giant axon, the helical properties of axoplasm in the giant axon of the fanworm *Myxicola infundibulum* are more pronounced, and they may have an important role in the mechanics of the axonal cytoskeleton when these axons are compressed and stretched during contraction and elongation of the animal. This difference between squid and the fanworm axoplasm may reflect the large number of microtubules in squid giant axons compared to the fanworm; microtubules are more rigid than neurofilaments and they tend to trace a straighter path and are therefore less extensible than axoplasmic neurofilaments (see Section 18).

In addition, the fanworm axoplasm *in situ* is actually subjected to repeated stretch and compression by the motility of the worm, and the continual working of the cytoskeleton in this way probably exaggerates the intrinsic potential of neurofilaments to generate the higher order helical twist that is observed in these and other annelid axons. In contrast with the extensible tissues surrounding the nerve cords of annelids, the giant axons of the squid are embedded in relatively inextensible tissues that are supported by a rigid chitinous pen (which serves as the backbone of the animal). Thus, in the squid giant axon, working of the cytoskeleton by external forces probably contributes little, if at all, to the helicity of the cytoskeleton.

14.3. Radial organization

In the radial dimensions of axons, two regions of the cytoplasm can be defined; a central or inner region, and a peripheral or cortical region (see Lasek, 1984 and 1988, for further discussion). The cytoskeleton in these regions is distinct in both composition and organization. The large size of the squid giant axon enables separation of these two regions of the axoplasm and direct analysis of their biochemistry and ultrastructure.

15. The cortical cytoskeleton

15.1. Axoplasm is attached to the plasma membrane

The axoplasmic cytoskeletal elements in the cortex of the axon are connected to the plasma membrane. In squid it is possible to demonstrate this by direct mechanical manipulation. For example, Metuzals and Tasaki (1978) have used polarizing microscopy to observe the cannulation of giant axons for internal perfusion. In this process, a glass cannula is inserted into the cut end of an axon and then advanced axially along the inside of the axon, using gentle suction to remove a central core of axoplasm (see Section 6.3). Metuzals and Tasaki noted that in some cases strong suction produced a birefringent distortion of the axoplasm that extended to the inner surface of the plasma membrane, where the axolemma became indented towards the tip of the cannula. Apparently suction produces a strain on fibrous elements in the cortical axoplasm that are attached to the inner surface of the plasma membrane, and this causes an orientation of these elements towards the tip of the cannula, producing the birefringence. Sato *et al.* (1984) reported similar evidence for attachment based on their observations of the movement of magnetic spheres in axoplasm under a magnetic field. Movement of the spheres from the periphery of the axon towards the center was seen to cause a local inward distortion of the axolemma. If sufficient force was applied the spheres would overcome this resistance and move more freely.

Observations on the extrusion process also indicate an association of the peripheral cytoskeleton with the plasma membrane. When axoplasm is extruded from the giant axon it shears at an annular plane beneath the axolemma. This plane of shear defines a cortical layer of axoplasm within the axon, the *membrane-associated cortical cytoskeleton*, that has some distinct properties from the inner cytoplasm. We have used video-enhanced differential interference contrast microscopy to examine the extrusion process (Brady and Lasek, 1989). A fine-cleaned giant axon was placed on a glass slide in the specimen plane of the microscope and pressure applied at one end so as to extrude the axoplasm from the cut end (see Section 6.1). By slowly increasing the pressure it was possible to reach a point at which the extrusion of the axoplasm could be controlled. With just sufficient pressure it was possible to observe the axoplasm moving through the axon. Our observations indicate that the axoplasm shears at a relatively sharp boundary that is located roughly 10 μm beneath the plasma membrane.

This is in agreement with the observations of Baker *et al.* (1962), using electron microscopy and action potential velocity measurements on extruded nerve fibers, and of Kobayashi *et al.* (1986), using electron microscopy. Thus, for an axon 500 µm in diameter, extrusion removes about 92% of the total axoplasm.

The plane of shear produced by the extrusion process may not correspond directly to a pre-existing natural partition between cortical and inner axoplasm. For example, electron microscopic studies of the organization of the axoplasm within intact axons indicate that the cortical cytoskeletal elements that are directly adherent to the plasma membrane are arranged within a thin 0.1 – 1.0 µm layer immediately beneath the membrane (see Section 15.4). Perfusion studies, which are described in Section 15.2, indicate that the remainder of the axoplasm within the axon is not tightly attached to the plasma membrane since it can readily be removed by cannulation and the flow of perfusion solution. Although the axoplasmic cytoskeletal polymers that are deeper than 1 – 2 µm from the plasma membrane are not attached to the plasma membrane, they may experience considerable resistance to movement under extrusion pressure (or suction pressure in the case of cannulation) because of frictional interactions with the adjacent stationary polymers of the membrane-associated cortical cytoskeleton. This viscous resistance to sliding under pressure may contribute to the tendency of the axoplasm to shear in a plane that is farther from the plasma membrane than that which separates it from the membrane-associated cytoskeleton. This subject of the interaction between axonal cytoskeletal elements within the axon and their resistance to longitudinal sliding is discussed further in Section 21.

15.2. Internal perfusion of axons

Internal perfusion of the squid giant axon is a useful technique for studying the membrane-associated cortical cytoskeleton. Methods for perfusing the giant axon are described by Adelman and Gilbert (1990). In brief, the procedure involves removing a central core of axoplasm from the giant axon, thereby allowing the flow of solutions through the inside of the axon (see Section 6.3). The perfused axon can then be preserved by internal perfusion with fixative and processed for transmission electron microscopy (see Section 13), or it can be cut open to expose the inner surface of the axolemma for scanning electron microscopy.

By altering the composition of the perfusion solution, the perfusion technique can be used to chemically dissect the cortical axoplasm (Metuzals and Tasaki, 1978; Baumgold *et al.*, 1981). Perfusion with an isotonic solution containing potassium fluoride (360 mM potassium fluoride, 40 mM potassium phosphate, 4% glycerol, pH 7.3) flushes away most of the inner axoplasm, but preserves a meshwork of subaxolemmal filaments associated with the inner surface of the membrane. In contrast, perfusion with more chaotropic salts, such as potassium iodide, which solubilize cytoskeletal protein filaments (see Section 11.1), causes extraction of the subaxolemmal meshwork. This membrane-associated filamentous meshwork can also be removed by proteolysis. For example, media containing pronase or millimolar calcium, which activates the endogenous calcium-dependent protease (see Sections 11.2 and 17.4), also extract the cortical cytoplasm (Metuzals and Tasaki, 1978). Scanning electron microscopy after perfusion with chaotropic salts or proteases shows that much of the inner surface of the axolemma is devoid of filaments, and extensive blebbing of the membrane is observed in these regions. These observations indicate that the cortical cytoskeleton is composed of protein filaments that are associated with the plasma membrane of the axon through non-covalent interactions that can be overcome by altering the ionic environment. The blebbing of the membrane in regions devoid of these filaments suggests that the cortical cytoskeleton may be important in maintaining membrane stability.

15.3. Protein composition of the cortical cytoskeleton

Since the association of the cortical cytoskeleton with the axolemma is preserved by potassium fluoride but is extracted by potassium iodide, perfusion with these salts can be used to specifically analyze the composition of the cortical cytoplasm. Baumgold *et al.* (1981) and Kobayashi *et al.* (1986) used silver staining of polyacrylamide gels to analyze the protein composition of fractions after perfusion with potassium iodide. Baumgold *et al.* used the squid *Loligo pealei* and Kobayashi *et al.* used the squid *Doryteuthis bleekeri*. Because of the small amounts of protein that elute it is necessary to dialyze the salts from the fractions and then concentrate them by lyophilization before analysis by SDS PAGE. These studies indicated that the cortical cytoskeleton contained tubulin and actin and a variety of other proteins, including a number of high molecular weight.

The separation afforded by mechanical extrusion allows more extensive biochemical analysis of the composition of the cortical and inner axoplasm. In order to analyze the cortical axoplasm remaining in the sheath after extrusion, Kobayashi *et al.* (1986) slit open the extruded sheaths of *Doryteuthis bleekeri* axons and extracted them, without homogenization, in a solution containing 0.6 M NaCl, 10 mM $CaCl_2$ and 10 mM sodium 2-(N-morpholino)ethane sulfonate (MES) buffer, pH 6.8, as described by Sakai and Matsumoto (1978). In this way the proteins in the cortical axoplasm were extracted but the proteins in the sheath cells (fibroblasts and glia) were not.

Two high molecular weight polypeptides were identified that are present in high concentration in the cortical axoplasm. One protein of M_r 260,000, a putative microtubule-binding protein named axolinin, was shown to co-sediment with microtubules but not with microfilaments, and to produce bundling of microtubules *in vitro* (Murofushi *et al.*, 1983). The other, a putative microfilament-binding protein of M_r 255,000, was shown to co-sediment with microfilaments but not with microtubules and to increase the low-shear viscosity of actin filament suspensions *in vitro*, which suggests that it can act as a cross-linker. Both proteins have been purified from axonal sheaths and from a bundle of small nerve fibers called the fin nerve. The proteins have distinct solubility properties and the M_r 255,000 protein showed no reactivity with antisera raised to axolinin. Both proteins are concentrated in the cortex, but are also present in the inner axoplasm; axolinin and the M_r 255,000 protein constitute 1.5 and 1.7%, respectively, of the total axoplasmic protein in extruded axoplasm, and 8.9 and 5.2%, respectively, of the axonal sheath extract. Assuming that the cortical axoplasm represents 8% of the total axoplasm (see Section 15.1), we calculate that about 32% of the axolinin and 20% of the M_r 255,000 protein in the axon are present in the cortical region.

Another high-molecular weight cytoskeletal protein that is present in the cortex of the giant axon is brain spectrin (fodrin), which migrates as a doublet of M_r 235,000 and 240,000 on SDS-polyacrylamide gels. The identity of these two polypeptides has been confirmed on nitrocellulose blots using a polyclonal antiserum to mammalian brain spectrin (Fath and Lasek, 1989). The apparent molecular weights (by SDS PAGE) of

Table V. The distribution of some proteins in cortical and inner axoplasm.

	Inner cytoskeleton	*Cortical cytoskeleton*
Total axoplasm[1]	92%	8%
Actin[2]	92%	8%
Axolinin[3]	68%	32%
M_r 255,000 protein[3]	80%	20%
Spectrin[4]	80%	20%

Source of data:
[1] Based on calculation for a 500 μm diameter axon with a 10 μm-thick region of cortical axoplasm.
[2] Fath and Lasek (1988).
[3] Calculated from Kobayashi *et al.* (1986).
[4] Fath and Lasek, (1989)

the two polypeptides are identical to those for spectrin from avian and mammalian brain. Like axolinin and the M_r 255,000 protein, spectrin may also cross-link cytoskeletal proteins within the membrane-associated cytoskeleton.

Our studies also indicate that significant amounts of spectrin are present in extruded axoplasm. To quantify the distribution of spectrin, we have compared the amounts of spectrin in the inner and cortical axoplasm by densitometry of SDS-polyacrylamide gels (Fath and Lasek, 1989). Spectrin has been considered to be principally localized subjacent to the plasma membrane (Levine and Willard, 1981). Our analyses indicate that 80% of the spectrin in the giant axon is recovered in the extruded axoplasm and 20% in the remaining 8% of the axoplasm in the sheath. Thus spectrin is 2 – 3 fold more concentrated in the cortex, but most of the protein is present in the inner cytoplasm.

Table V compares the distribution of these three high-molecular weight proteins in the cortical and inner axoplasm of the giant axon. Axolinin, the M_r 255,000 protein and spectrin are all present in higher concentrations in the cortex. However, because the cortical axoplasm is only a small fraction of the total axoplasmic volume, most of each of these proteins is present in the inner cytoplasm and is not associated with the plasma membrane. Thus these proteins apparently contribute to the inner cytoskeleton as well as to the cortex. For example, in addition to cross-linking the membrane cytoskeleton, these proteins may also cross-link filaments in the inner cytoskeleton (See Goodman and Zagon, 1986, for review).

15.4. Architecture of the cortical cytoskeleton

Tsukita *et al.* (1986) have studied the morphology of the subaxolemmal cytoskeleton in giant axons from the squid *Doryteuthis bleekeri* after internal perfusion with solutions containing potassium fluoride, which stabilizes the polymers and their associations. For thin section electron microscopy, axons were fixed by internal perfusion with aldehyde fixative containing tannic acid. For freeze-etch replica electron microscopy, perfused axons were fixed with glutaraldehyde and then cut open along their length and rapidly frozen by touching the inner surface against a copper block cooled by liquid helium.

These methods have revealed that the cortical cytoskeleton has a distinct architecture from that of the inner cytoplasm. A dense filamentous meshwork, a few micrometers in thickness, lines the inner surface of the axolemma. The major cytoskeletal polymers in this zone are microfilaments and microtubules; neurofilaments, which are the major cytoskeletal element in the inner axoplasm, are relatively few in number. The high concentration of microtubules in the cortex of the giant axon has also been noted by Sakai *et al.* (1985). They found that the mean microtubule density within 1 μm of the axolemma was about 80 per μm^2 (cross-sectional area), which is about 3 – 4 fold greater than the density in the inner axoplasm.

Within the cortex, the cytoskeletal polymers are organized into discrete microfilament- and microtubule-associated membrane domains. In the microtubule domains, microtubules run in parallel beneath the axolemma along the long axis of the axon, and sometimes appear to be connected to the plasma membrane by fine filamentous strands. Locally the microtubules are often displaced inwards from the axolemma by punctate clusters of microfilaments that lie directly beneath the membrane. These microfilament domains occupy about 15% of the surface of the axolemma, and also appear to be connected to the plasma membrane by fine filamentous strands. In addition, microfilaments are seen outside these clusters in association with a matrix surrounding the microtubules, and this is also a feature of the microtubules in the inner cytoplasm (see Sections 18 and 19). A schematic illustration of the inner and cortical axoplasm is shown in Fig. 7.

The fine filamentous matrix that surrounds the microtubules and microfilaments may be composed of proteins such as axolinin, the M_r 255,000 protein and spectrin, which are all present at higher concentrations in the cortex of the axoplasm. To examine this possibility, Tsukita *et al.* (1986) used immunogold electron microscopy in thin section to localize the microtubule-associated protein, axolinin, in the cortical axoplasm. Their results indicate that axolinin is concentrated in the microtubule-associated membrane domains within the cortex. This distribution is consistent with a role for this protein in the organization of the cytoskeleton in regions where microtubules are the principal cytoskeletal polymers.

15.5. The cortical cytoskeleton and membrane excitability

The intimate association of the cortical cytoskeleton with the plasma membrane is probably important in providing mechanical stability to the lipid bilayer. However, in addition there are indications that the membrane-associated cytoskeleton may also be important in the maintenance of the electrical excitability of the axon. The evidence for this rests primarily on the observation that many agents which extract the cortical cytoskeletal elements also alter the ionic permeability of the membrane. For example, proteases and chaotropic salts, which are known to remove the cortical cytoskeleton (see Section 15.2), also abolish the excitability of the axon (Tasaki and Takenaka, 1964; Tasaki *et al.*, 1965; Metuzals and Tasaki, 1978). Much of this evidence is equivocal, however, because many of the treatments that abolish excitability may also have direct effects on channel proteins within the membrane. For example, the effects of intracellular perfusion with proteases on the sodium conductance of the giant axon

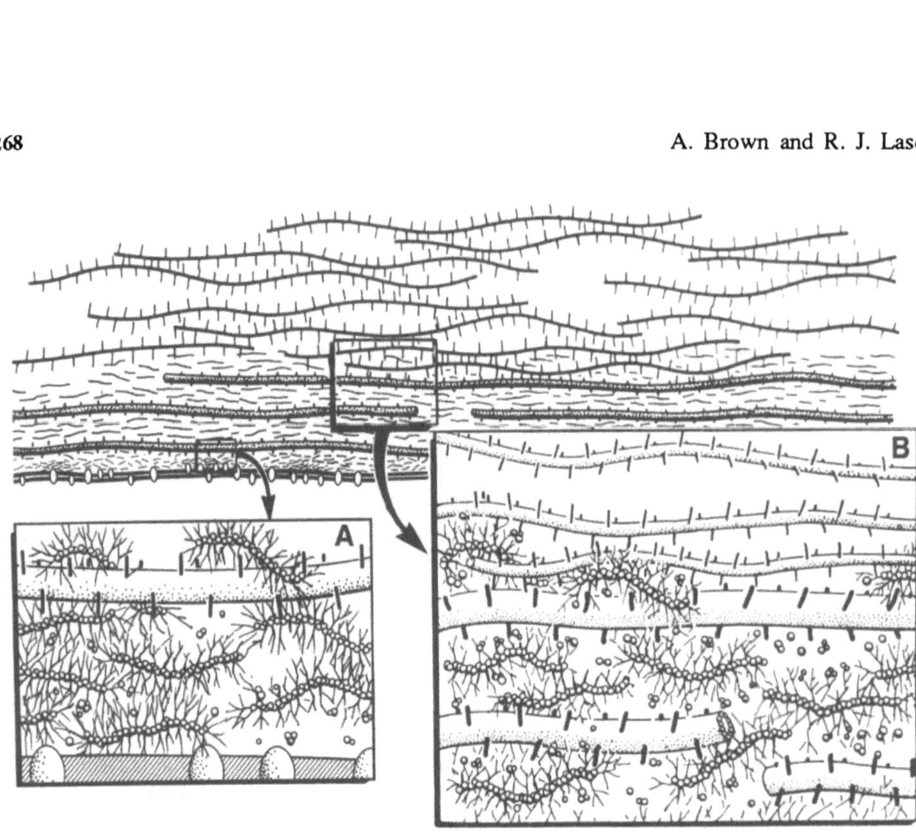

Figure 7. A schematic illustration of the inner and cortical regions of the axonal cytoskeleton. The number and relative diameter of the neurofilaments, microtubules and microfilaments are shown as they are in axoplasm. The polymers are orientated along the long axis of the axon. Note that the microfilaments are much shorter than the microtubules and neurofilaments, but that they are much more numerous. The lateral projections from the polymers represent side-arms and associated proteins. The small globular particles represent soluble proteins of the cytomatrix, such as metabolic enzymes. A. Inset showing a microfilament-associated domain directly apposed to the inner surface of the axolemma. The microfilaments are attached to integral membrane proteins via radial cross-links. In other regions microtubules can also be directly apposed to the membrane (see text). B. Inset showing organization characteristic of the inner cytoskeleton. Microfilaments cluster in the microtubule domains, which are present throughout the inner axoplasm and are interspersed by neurofilaments.

have widely been attributed to direct proteolytic cleavage of the sodium channel (Armstrong *et al.*, 1973; Sevcik and Narahashi, 1975; Rojas and Rudy, 1976).

However, studies using more specific treatments also suggest an involvement of the cortical cytoskeleton in the maintenance of membrane excitability. For example, the effect of chaotropic ions on the sodium conductance of the axon can be mimicked by a variety of agents that depolymerase microtubules (Matsumoto and Sakai, 1979a; Matsumoto *et al.*, 1984). These agents include drugs such as colchicine, podophyllotoxin and vinblastine, and submillimolar concentrations of calcium ions. Furthermore, studies reported by Matsumoto and Sakai (1979b) and Matsumoto *et al.* (1979) indicate that excitability can be restored to these inexcitable axons by subsequent perfusion with media containing microtubules and various cofactors. These results have been interpreted to indicate a specific requirement for cortical microtubules in the maintenance of the electrical excitability of the squid axon. For example, it is possible that microtubules in the cortical cytoskeleton may interact directly or indirectly with channel proteins in the membrane, and that this interaction may be necessary for the proper functioning of these channels. Indeed, there is considerable precedent for the interaction of the membrane cytoskeleton with integral membrane proteins in other cells (for example, see Goodman and Zagon, 1986), and it seems plausible that this interaction might have specific functional consequences for these proteins. The reader who is interested in these studies is referred to a review by Sakai *et al.* (1985) for more information.

16. The inner cytoskeleton

The inner cytoskeleton forms the bulk of the axoplasm in the axon and it has a different protein composition than the cortex; though all the axonal proteins that have been identified in squid axoplasm appear to be present in both the inner and cortical regions, their relative concentrations can differ greatly. For example, Table V shows that certain cytoskeletal binding proteins are more concentrated in the cortical axoplasm. In contrast, other proteins such as neurofilaments appear to be partially excluded. This difference in composition is reflected by a different cytoskeletal organization.

In the inner axoplasm, neurofilaments are the major cytoskeletal polymer and together with microtubules and microfilaments, these polymers largely determine the architecture of the inner cytoskeleton. The neurofilaments and microtubules are both very long (greater than tens of micrometers), while the microfilaments are much shorter (less than 1 μm). The long polymers (microtubules and neurofilaments) are both aligned parallel to the long axis of the axon. Estimates based on analysis of the amounts of cytoskeletal protein as monomer and polymer in axoplasm (see Section 20) indicate that the ratio of neurofilaments to microtubules may be as low as 2:1 (Fath and Lasek, 1988), which represents a much larger proportion of microtubules than in large vertebrate axons. For example, the ratio of neurofilaments to microtubules in the internodal regions of the large somatic motor axons of the chick oculomotor nerve is about 75:1 (Price and Lasek, 1989). However, in other respects the organization of the cytoskeleton in these axons appears to be very similar.

Neurofilaments, microtubules and microfilaments are present throughout the inner cytoplasm, but their distribution is not homogeneous. Microtubules often cluster together and are surrounded by a dense filamentous matrix, and this is most apparent in longitudinal sections (for example, see Miller and Lasek. 1985). Often membranous organelles and endoplasmic reticulum are associated with these domains. This filamentous matrix is composed of actin microfilaments and other cytoplasmic proteins. The microfilaments in the inner axoplasm are concentrated in these domains forming longitudinal columns through the axoplasm (Fath and Lasek, 1988). A schematic illustration of the inner cytoskeleton is shown in Fig. 7.

17. Neurofilaments in axoplasm

Neurofilaments are the major cytoskeletal elements of the inner axoplasm of the squid giant axon. Because of their abundance in the giant axon and the relatively large amount of pure axoplasm that can be obtained from the squid giant axon, studies of squid neurofilaments have been important in understanding neurofilament structure and physiology. The neurofilament proteins in axoplasm are essentially fully polymerized (see Section 20) and studies *in vitro* indicate that they are very stable to changes in the ionic composition of the medium. To achieve full disassembly of squid neurofilaments requires high concentrations of chaotropic salt and high pH (see Section 11.1). For example, complete disassembly can be achieved by dialysis of axoplasm against solutions containing 1 M potassium chloride or 0.5 M potassium thiocyanate, at pH 8.

Electron microscopy of thin sections of intact squid axoplasm reveals that the neurofilaments are aligned in parallel along the long axis of the axon, though they trace a somewhat more wavy path than the microtubules. Neurofilaments can also be visualized in homogenized axoplasm by electron microscopy of negatively-stained preparations (see Section 13.6). By this procedure the neurofilaments measure about 8.5 nm in diameter and are smooth in appearance (Lasek *et al.*, 1979; Wais-Steider *et al.*, 1983). In homogenates the filaments are many micrometers in length, but they are likely to be much longer in intact axoplasm. In marked contrast to microtubules, the neurofilaments in homogenized axoplasm appear to be very flexible polymers that often entangle.

17.1. Purification of squid neurofilaments

Neurofilaments have been purified from squid axoplasm by a variety of procedures, including precipitation with cytochrome-c (Gilbert, 1976), centrifugation on discontinuous sucrose gradients (Lasek *et al.*, 1979), Millipore™ filtration (Roslansky *et al.*, 1980) and cycles of disassembly and reassembly of the neurofilament subunits (Steinert *et al.*, 1982; Brown, 1989). Enriched preparations of neurofilaments can also be readily obtained from axoplasm by a simple washing procedure (Wais-Steider *et al.*, 1983). Freshly extruded axoplasm is quickly weighed on a piece of Parafilm™ and then placed in about 30 volumes of *Glu-500* washing solution (500 mM glutamate, 10 mM magnesium chloride, 2 mM EGTA, 2 mM dithiothreitol, 0.01% azide, pH 7.2) at 4 °C. The axoplasm in this solution is kept on ice and the solution changed frequently over the course of a day in order to extract the soluble proteins. During this time the axoplasm swells greatly but retains its mechanical integrity. These enriched preparations may be used directly for investigations, or as a starting material for further purification of neurofilaments.

Neurofilaments can also be purified from squid brain optic lobes by cycles of disassembly and reassembly (Zackroff and Goldman, 1980; Steinert *et al.*, 1982). Two cycles yielded about 0.5 mg of protein from 30 optic lobes. The major polypeptides in this preparation appear to correspond to those in neurofilament preparations from axoplasm (see Section 17.2), though some of this protein may also originate from intermediate filaments of glial origin.

17.2. Polypeptide composition of squid neurofilaments

Unlike microfilaments and microtubules, which are highly conserved, neurofilaments from different phyla differ both in the number and size of their component polypeptides. For example, the neurofilaments of warm blooded vertebrates (birds and mammals) are composed of three polypeptides (the *neurofilament triplet*) with an apparent M_r by SDS PAGE of about 200,000, 150,000 and 70,000. In contrast, neurofilaments from the giant axon of the annelid fanworm *Myxicola infundibulum* are composed of a single parent polypeptide of apparent M_r by SDS PAGE of 172,000 (Eagles *et al.*, 1981).

Squid neurofilaments have a distinct composition from those of both annelids and vertebrates. Purification of neurofilaments from squid axoplasm indicates that they are composed of two polypeptides of M_r 60,000 (P60)and 200,000 (P200), and a high molecular weight component (Band 1). Estimates of the apparent M_r of Band 1 have varied from 300,000 (Gallant *et al.*, 1986) to 800,000 (Eagles *et al.*, 1980). This variation is probably due to difficulties in the calibration of M_r in this region of the gel, rather than to an actual variation in molecular mass.

What is the relationship between the three squid neurofilament proteins? Peptide mapping of the neurofilament proteins from *Loligo forbesi* using proteolytic enzymes indicates that P60 and P200 are distinct proteins, but that P200 and Band 1 show great similarity (Eagles *et al.*, 1980). These similarities and differences have been confirmed by staining Western blots of SDS-polyacrylamide gels (on which the neurofilament proteins have been resolved) with monoclonal antibodies. Western blotting is a procedure for the electrophoretic transfer of proteins from SDS-polyacrylamide gels on to nitrocellulose sheets, thereby affording better accessibility of the proteins to antibodies (Towbin *et al.*, 1979). In this way, it has been shown that aNFP, which is an antibody that was raised against squid axoplasm, binds to P200 and Band 1, but not to P60 (Cohen *et al.*, 1987). Dephosphorylation of these polypeptides (see Section 17.3) abolishes the binding of this antibody, which suggests that it recognizes a common phosphorylated epitope on the two high-M_r proteins. In contrast, Anti-IFA, an antibody that recognizes a conserved epitope amongst all intermediate filament proteins (Pruss *et al.*, 1981), binds to P200 and P60, but not to Band 1 (Brown and Eagles, 1986; Gallant *et al.*, 1986). It is likely that Band 1 is a cross-linked oligomer of P200 in which the Anti-IFA epitopes are masked.

As with the neurofilaments of the fanworm *Myxicola infundibulum* and with those of vertebrates, neurofilaments in squid axoplasm are substrates for two axoplasmic enzymic activities; phosphorylation and calcium-dependent proteolysis. These activities were first identified in squid axoplasm but they have also been identified in other neurons, and they are thought to play an important part in the physiology and metabolism of neurofilaments in axons (for example, see Roots, 1983; Nixon *et al.*, 1987). Though there is still much to be learned about the physiological role of these enzymes, their mode of action has helped to define structural similarities between neurofilaments of different phyla, in spite of phylogenetic differences in the polypeptide composition of these cytoskeletal elements.

17.3. Phosphorylation of squid neurofilaments

Neurofilaments in axoplasm are actively phosphorylated by a specific endogenous Mg^{++}-dependent, Ca^{++}- and cyclic-nucleotide-independent protein kinase (Pant *et al.*, 1978, 1979b, 1986; Eagles *et al.*, 1980). ^{32}P-radiolabelled neurofilaments can be prepared by incubating freshly extruded axoplasm with ATP labeled with P^{32} in the γ position (γ^{32}-P-ATP) or with $[^{32}P]P_i$. In the latter case the inorganic radioactive phosphate is incorporated into ATP by endogenous mitochondrial oxidative phosphorylation. Phosphorylation can be achieved in intact axoplasm or in homogenized axoplasm. For labelling intact axoplasm, the isotope can be injected directly into the extruded axoplasm or added in a small volume of artificial axoplasm solution. It is also possible to achieve labelling by incubating intact axons in [^{32}P]P_i but larger amounts of isotope are necessary to achieve equivalent incorporation (Brown, 1989).

In these ^{32}P-phosphorylated preparations, P200 and Band 1 are the major labelled polypeptides and there is no detectable phosphorylation of P60. Indeed, extensive phosphorylation of the high molecular mass polypeptides is a characteristic feature of neurofilaments from all sources. An indication of the extent of phosphorylation of the squid polypeptides has been provided by dephosphorylation with exogenous alkaline phosphatase (Brown and Eagles, 1986; Cohen *et al.*, 1987). Treatment with this enzyme causes a substantial increase in the mobility of both P200 and Band 1 by SDS PAGE, but no detectable change for P60. For example, the apparent M_r by SDS PAGE of the dephosphorylated P200 is about 160,000 (Brown, 1989). Thus the apparent M_r values by SDS PAGE are overestimates. In fact, this is also a characteristic property of the high molecular mass polypeptides of neurofilaments from other sources. For mammalian neurofilaments this has been shown to be due in part to the high phosphate content of these proteins (Julien and Mushynski, 1982; Carden *et al.*, 1985).

Studies on the phosphorylation of neurofilaments in other organisms have shown that neurofilaments in the cell bodies of neurons are not heavily phosphorylated, and that phosphorylation of neurofilaments is associated with their entry into the axon (for example, see Oblinger, 1988; Shaw *et al.*, 1986). This has lead to speculation that phosphorylation causes structural changes in the peripheral domains of the neurofilament proteins that are associated with the space-filling properties of neurofilaments in axons. In squid it is possible to excise the stellate ganglion that contains the cell bodies of the giant axon and compare its protein composition to axoplasm. Using the intermediate-filament antibody Anti-IFA, Gallant *et al.* (1986) have shown that the major reactive proteins in the ganglion are a polypeptide of M_r 60,000, that presumably corresponds to P60, and a polypeptide of M_r 65,000. In addition there was a minor a polypeptide of M_r 180,000, and another of M_r 74,000 that probably corresponds to a protein of similar M_r that copurifies with neurofilaments from axoplasm (Brown and Eagles, 1986; and see Section 20.2). Neither P200 (M_r 220,000 on their gels) nor Band 1 were detectable in the ganglion.

It is likely that the immunoreactive M_r180,000 protein in the ganglion corresponds to a non-phosphorylated precursor to the P200 polypeptide in axoplasm. On the other hand, the identity of the M_r 65,000 is not clear. One possibility is that it is a precursor of the P60 polypeptide, though alternatively it may be the intermediate filament protein of the glial cells that also populate the ganglion. In addition, it is interesting to speculate that the absence of Band 1 in the ganglion may also reflect a post-translational modification associated with entry into the axon, but in this case there is a covalent cross-linking of two or more P200 polypeptides or their non-phosphorylated precursors (see Section 17.2).

17.4. Calcium-dependent proteolysis

Neurofilaments in axoplasm are also substrates for an endogenous calcium-dependent proteinase (Gilbert *et al.*, 1975; Pant *et al.*, 1979a and 1982; Pant and Gainer, 1980). The protease is only active at calcium concentrations greater than 0.5 mM and it preferentially cleaves the neurofilament proteins and spectrin in axoplasm, though it may also cleave other cytoskeletal proteins. Based on its calcium-sensitivity, the enzyme is probably equivalent to the protein calpain II that has been identified in a wide variety of vertebrate tissues (Murachi, 1983). Exposure of axoplasm to millimolar concentrations of calcium, such as those present in seawater, results in rapid dissolution of the axoplasm due to fragmentation and destruction of the cytoskeletal polymers (see Section 11.2).

17.5. The structure of the squid neurofilament

Brown and Eagles (1986) investigated the calcium-dependent proteolysis of [^{32}P]phosphorylated neurofilaments in axoplasm of the squid *Loligo forbesi* in order to localize the major sites of phosphorylation and probe the arrangement of the polypeptides in the filament. In the presence of millimolar concentrations of calcium, the protease preferentially cleaves the P200 and Band 1 neurofilament proteins, liberating the phosphorylated domains from the filament. Intact filaments remain after limited proteolytic cleavage has removed essentially all of the phosphorylated domains. These studies indicate that proteolysis defines two distinct filament domains; a core domain that forms the backbone of the filament and a peripheral domain that is highly phosphorylated and is not essential for filament integrity.

The organization of the neurofilament polypeptides was probed further using the monoclonal antibody Anti-IFA, which recognizes both P200 and P60 on Western blots of SDS-polyacrylamide gels. This antibody is known to bind to a conserved epitope present in the filament-forming core domain of all intermediate filament proteins (Pruss *et al.*, 1981; Geisler *et al.*, 1983). Axoplasm was subjected to limited calcium-dependent proteolysis and the neurofilaments were then separated from soluble proteolytic fragments by centrifugation. Anti-IFA was shown to bind to the proteolytic fragments in the filament fraction, but not to the phosphorylated fragments that are liberated from the filament.

These results indicate that, despite their distinct polypeptide composition, neurofilaments from squid conform to the same structural principles as those found for other phyla (for example, see Geisler *et al.*, 1983). Specifically, all neurofilament proteins share a common alpha-helical rod-shaped domain that is analogous to the core domains of other intermediate filament proteins and that forms the filament core. In addition, the neurofilament polypeptides are unique in possessing large highly charged carboxy-terminal extensions of the rod domain that lie peripheral to the filament core and that are not required for filament formation or the maintenance of filament integrity. Variations in the length of this peripheral domain account for the large differences in polypeptide size for neurofilaments from different phyla. Such variations also explain the relationship between the multiple neurofilament polypeptides that comprise neurofilaments in animals such as squid and mammals.

The proposed structure of the squid neurofilament is summarized in the diagram in Fig. 8. The neurofilament is depicted as having a cylindrical core with a laterally projecting peripheral domain. Single P200 and P60 polypeptides are shown that are meant to be representative of the location of these polypeptides throughout the neurofilament. Limited proteolysis defines two major domains in P200, one in the filament core that binds Anti-IFA (P100p), and the other located peripheral to the filament core that is phosphorylated (P110s). P60 resides at least predominantly within

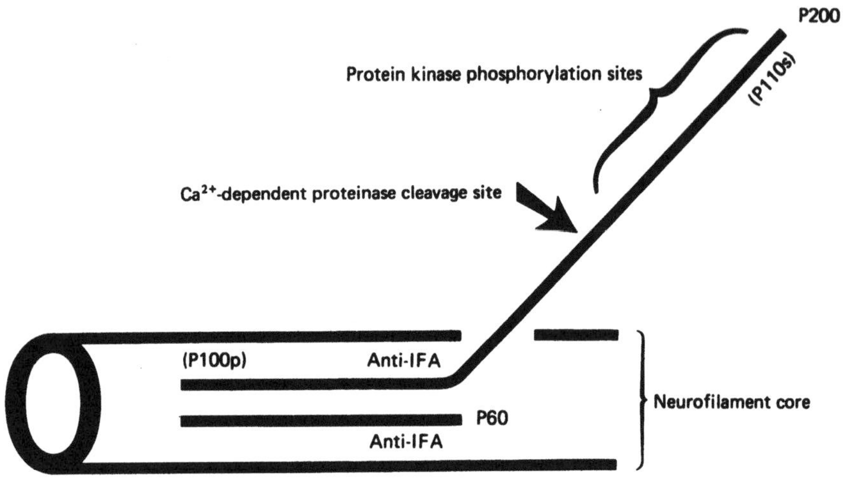

Figure 8. Model for the squid neurofilament. The proposed sites of phosphorylation, calcium-dependent proteolytic cleavage and anti-IFA binding are shown (see Section 17.5 of text for explanation). From Brown and Eagles (1986).

the filament core and is relatively resistant to limited proteolysis. The disposition of Band 1 appears to be predominantly peripheral to the filament core, but its precise nature and location is unclear.

18. Microtubules in axoplasm

Axoplasm from the squid giant axon contains many microtubules. In contrast, axoplasm from the giant axon of the marine fanworm *Myxicola infundibulum* contains relatively few microtubules and many more neurofilaments. This difference may explain, in part, the much larger amount of organelle traffic that is seen in squid axoplasm.

One way to visualize microtubules in squid axoplasm is by electron microscopy of negatively-stained preparations. Measurements of negatively-stained microtubules indicate a minimum diameter of 26 nm, though larger apparent diameters may result from flattening of the tubular structure on the support film during the preparation (Wais-Steider *et al.*, 1988). Observations on the microtubules in these preparations suggest that these polymers are quite different from neurofilaments in their mechanical properties and that they have the potential to confer quite different properties on the cytoskeleton. Specifically, the tubular form of microtubules makes them more rigid than neurofilaments. Consequently, the microtubules in homogenized axoplasm are short and straight, since they tend to snap rather than bend when they are sheared.

In negatively-stained preparations of diluted and homogenized axoplasm there are fewer microtubules are than there are in intact axoplasm, and this is because a substantial proportion of the microtubules are metastable and depolymerize in dilute solution. These metastable microtubules can be stabilized to dilution by using solutions containing 50% glycerol (Wais-Steider *et al.*, 1988) or 10 μM taxol. In contrast to the metastable population, some microtubules in axoplasm are very stable and resist depolymerization by a variety of conditions including cold, submillimolar calcium concentrations, and some anti-mitotic drugs (Brady, 1989). The stability and assembly dynamics of microtubules and the other cytoskeletal polymers is discussed in Section 20.

Electron microscopy of thin sections of intact axoplasm shows that the microtubules cluster forming longitudinal bundles that are aligned parallel to the long axis of the axon. Surrounding these microtubules is a fine filamentous matrix that is denser than the surrounding cytoplasm, and is composed of actin microfilaments and other proteins (see Section 19). These microtubules domains are most apparent in longitudinal sections, and they are often associated with vesicles and smooth endoplasmic reticulum.

18.1. Purification of microtubules

Several investigators have purified microtubules from squid tissues for biochemical studies. For example, Sakai and Matsumoto (1978) purified microtubules from squid axoplasm by one cycle of temperature-dependent disassembly and reassembly and studied their properties *in vitro*. However, it should be noted that some of the microtubules in axoplasm are stable (see Section 20) and may not purify by this protocol. Thus purification by this procedure will select for soluble tubulin and metastable microtubules. For studies that require larger quantities of protein (a single axon contains only about 55 μg of tubulin), tubulin can also be purified from the optic lobes of squid brain (see Vale, Schnapp, Reese, and Sheetz, 1985). Squid optic lobes, which form the major mass of the squid brain, provide a good source of tissue for protein purification. The lobes lie immediately behind the squid eyes and can be dissected out in a few minutes. About 0.5 g wet weight of lobe tissue can be obtained from a single squid.

18.2. Video-enhanced contrast light microscopy of axoplasm

The study of microtubules in squid axoplasm has been greatly facilitated by the development of video-enhanced differential interference contrast light microscopy (Allen and Allen, 1983). This technique allows the detection of single microtubules and membranous organelles in living cytoplasm and it has allowed rapid progress in understanding the mechanisms of organelle translocation in axoplasm (Weiss, Meyer, and Langford, 1990). Neurofilaments and microfilaments are below the limits of detection by this technique and can only be visualized by electron microscopy.

Observation of intact axons by differential interference contrast microscopy gives a striking impression of the longitudinal alignment of the microtubules in the axoplasm, but the density of the cytoplasm does not allow the clear resolution of the individual polymers. Single microtubules can only be delineated at the periphery of extruded axoplasm or in dispersed preparations. The lengths of microtubules in squid axoplasm have not been accurately measured, but our observations of many microtubules at the periphery of extruded axoplasm suggest that they are in excess of 50 μm in length. In homogenized or gently dispersed preparations, the microtubules are much shorter, which indicates that they are very susceptible to shearing.

Weiss *et al.* (1988) have used video-enhanced differential interference contrast microscopy to examine the length changes of microtubules in extruded axoplasm. Their observations indicate that length changes are a rare event. This is consistent with the existence of stable microtubules in axoplasm (see Section 20) and suggests that most of the microtubules in axoplasm do not display the *dynamic instability* characteristic of some other cell types, such as actively crawling epithelial cells, in which many microtubules rapidly grow and shrink in length (Mitchison and Kirschner, 1984). This observation underlines one of the many specific differences between the cytoskeletal elements of axons and those of cells, such as fibroblasts, that are capable of mitosis.

18.3. Microtubule-associated proteins in axoplasm

A number of proteins that bind to microtubules have been isolated from squid axoplasm. One of these proteins is kinesin, which is an oligomeric protein with a native relative molecular mass of about 600,000 and is composed of M_r 110 – 120,000 and M_r 60 – 70,000 polypeptides (Vale, Reese, and Sheetz, 1985). Kinesin has been purified from squid axoplasm on the basis of its affinity for microtubules in the presence of 5'-adenylyl imidodiphosphate (AMP-PNP), which is a non-hydrolyzable analog of ATP (Lasek and Brady, 1985; Vale, Reese, and Sheetz, 1985). The kinesin preparation is sufficient to cause unidirectional ATP-dependent movement of latex beads along microtubules, but additional proteins appear to be required for vesicle translocation in axoplasm. The direction of translocation is towards the plus-ends of the microtubules (Vale, Schnapp, Mitchison, *et al.*, 1985). Studies on microtubules in other axons indicate that microtubules are orientated with their plus-ends away from the cell body (Burton and Paige, 1981; Heidemann *et al.*, 1981). This suggests that kinesin is a component of the cross-bridges that mediate anterograde vesicle transport along microtubules in squid axoplasm (Miller and Lasek, 1985; Vale, 1987).

Subsequent to its identification in squid axoplasm, kinesin has also been identified in many other cells (see Hollenbeck, 1988, for review). Studies on this protein from these sources have indicated that kinesin is an adenosine-5'-triphosphatase (ATPase) and that its ability to hydrolyze ATP is stimulated in the presence of microtubules (i.e., see Kuznetsov and Gelfand, 1986). In the future it is likely that much of the biochemical characterization of kinesin will be done on protein purified from sources other than the squid nervous system because the amount of tissue available from squid is limited. However, squid axoplasm will continue to be important in elucidating the mechanisms and dynamics of axoplasmic transport since there are many features of axoplasm that cannot be reconstituted in cell-free assays, and there are likely to be many more components of the translocating mechanisms that have yet to be identified (Brady *et al.*, 1985).

Just as for the axons of vertebrates, squid axoplasm also contains a number of high-molecular weight microtubule-associated proteins. For example, Sakai and Matsumoto (1978) identified five high molecular weight polypeptides on SDS-polyacrylamide gels that all sedimented with microtubules from extruded axoplasm (*Doryteuthis bleekeri*) after one cycle of temperature-dependent disassembly and reassembly. One of these proteins is axolinin, which has a M_r of 260,000 and is present in large amounts in the cortical axoplasm (see Section 15.3). Another of these proteins is the M_r 292,000 protein that has been identified in axoplasm of the squid *Loligo pealei* by Gilbert and Sloboda (1986). This protein cross-reacts with antibodies to porcine brain MAP-2 and copurifies with microtubules from squid brain optic lobes.

The M_r 292,000 protein was identified as an ATP-binding protein using 8-azido-ATP, and Pratt (1986) showed that it associated in an ATP-dependent manner with a sedimentable fraction containing microtubules and vesicles, but it is not clear whether or not it is an ATPase. Pratt (1986) also identified a number of other high-molecular weight proteins that sediment with fractions containing microtubules and vesicles from squid axoplasm, but these proteins were not characterized further.

Because of the current interest in microtubule-based organelle translocation, rapid progress is being made on the identification and characterization of microtubule-associated proteins in squid axoplasm. For further discussion of the role of microtubule-associated proteins in organelle transport, consult Weiss, Meyer, and Langford (1990).

19. Microfilaments in axoplasm

19.1. Microfilaments are numerous in the inner cytoskeleton

Actin microfilaments are the least studied of the cytoskeletal polymers in axons, though much is known about their structure and function in some other cell types, such as muscle cells, fibroblasts, and red blood cells. Actin comprises about 6% of the protein in the squid giant axon. Analysis of the axoplasm and sheath after extrusion indicates that about 92% of the total actin in the giant nerve fiber is extruded; only 8% remains in the sheath, and a portion of this actin is located in extra-axonal cells. Thus greater than 92% of the actin in the giant axon is present within the inner axoplasm (Fath and Lasek, 1988).

Fath and Lasek (1988) used quantitative biochemistry together with light and electron microscopy to investigate the distribution and organization of the microfilaments in extruded axoplasm. Electron microscopy of negatively-stained preparations of extruded axoplasm revealed the presence of many 4 – 6 nm filaments that were decorated by myosin subfragment-1 (S-1) with the arrowhead pattern that is characteristic of actin microfilaments. For negative staining, axoplasm from one axon (about 3 – 4 µl) was extruded directly into 200 – 500 µl Buffer X and gently dispersed in a small-volume teflon/glass homogenizer. Negative staining was performed as described in Section 13.6. For S-1 decoration, a drop of dispersed axoplasm was placed on a carbon-coated grid and then replaced with a drop of 0.4 mg/ml S-1 in standard salts solution (50 mM KCl, 5 mM $MgCl_2$, 6 mM sodium phosphate, pH 7.2). The S-1 was prepared by chymotryptic digestion of rabbit muscle myosin. In order to measure the microfilament lengths, the neurofilaments and microtubules, which tend to obscure the microfilaments, were solubilized using 0.6 M potassium iodide in which actin filaments are stable. Measurements made in this way showed that axoplasmic microfilaments are short polymers (mean length of 0.5 µm), in marked contrast to the microtubules and neurofilaments which are tens of micrometers in length.

The observation that microfilaments are substantial components of the inner cytoskeleton in the axon is a further illustration of the usefulness of the extrusion paradigm for quantitative determinations of the distribution of axoplasmic elements in the cortical and inner axoplasm. Studies on vertebrate axons using fluorescent staining methods have suggested that axonal microfilaments are principally located in the cortex. This idea became entrenched in the literature (for example, see review by Pachter *et al.*, 1984), and because microfilaments are essential elements in actomyosin-mediated motility, models of axonal motility were constrained by the notion that most and possibly all of the microfilaments and certain associated proteins, such as spectrin (see Section 15.3), were located subjacent to the plasma membrane. Based on this notion, one of the most actively supported models of slow axonal transport explicitly assumed that slow transport operated through force generating mechanisms located subjacent to the plasma membrane (for example, see Levine and Willard, 1981). Although there were some clues that the cortex was not the only the part of the axon where microfilaments were plentiful (Morris and Lasek, 1982; Heriot *et al.*, 1985), it was not until quantitative studies using the extrusion model were developed that the abundance of microfilaments and their associated proteins in the inner axoplasm was established.

19.2. Two classes of microfilaments in the inner cytoskeleton

The state of actin as monomer and polymer has been investigated by Morris and Lasek (1984). Based on studies using pharmacological agents and a kinetic equilibration paradigm that they developed, they proposed that axoplasmic actin exists in three forms: soluble (monomeric), metastable (dilution-labile) microfilaments and stable microfilaments. Their results suggested that 56% of the actin in extruded axoplasm was polymeric; 26% stable and 30% metastable. The soluble actin concentration in axoplasm is about 0.37 mg/ml, which is greater than the critical concentration *in vitro*, and this suggests that factors in axoplasm, such as actin binding proteins, may sequester monomeric actin in a form that does not readily polymerize. The stability and assembly dynamics of actin microfilaments and the other cytoskeletal polymers in axoplasm are discussed in more detail in Section 20.

Electron microscopy has indicated that the stable and metastable microfilaments differ in their mean length as well as in their stability. Fath and Lasek (1988) measured microfilament length in the presence and absence of phalloidin, which is a pharmacological agent that stabilizes actin filaments, and they showed that the stable microfilaments in axoplasm have a mean length of 0.79 μm, and that the metastable microfilaments are shorter, with a mean length of 0.41 μm. Thus both morphological and kinetic criteria define two distinct populations of actin filaments in axoplasm. Based on these observations, they proposed that these two populations of microfilaments that differ in their stability may play distinct roles in the dynamic architecture of the inner cytoskeleton in the axon.

19.3. Organization of microfilaments in the inner cytoskeleton

Since phalloidin binds to polymeric actin, but not to monomer, it can be used to localize microfilaments in cells. By staining axoplasm with rhodamine-phalloidin, Fath and Lasek (1988) showed that the microfilaments are not distributed uniformly in the inner axoplasm, but are organized in columns that are orientated along the long axis of the axon. Each column is about 1 μm wide and extends for many micrometers through the axoplasm. Microfilaments can also be visualized in intact axoplasm by electron microscopy of thin sections. Axoplasm was stained with S-1 to unambiguously identify the microfilaments, and then fixed using the potassium ferricyanide procedure for optimum preservation (see Section 13.2). Using this procedure it appears that the microfilaments are concentrated in the vicinity of the microtubules bundles and are often aligned along their long axis. This indicates that the columns of actin may correspond to the microtubule bundles in axoplasm.

A number of experiments on vesicle transport in axoplasm support the localization of actin microfilaments in the vicinity of microtubules. For example, Brady *et al.* (1984) showed that DNase I and gelsolin both inhibit anterograde and retrograde vesicle transport in extruded axoplasm. Both of these proteins can cause the fragmentation of actin filaments *in vitro*. Since vesicle transport along microtubules can be reconstituted *in vitro*, apparently in the absence of actin, it is not likely that microfilaments are involved in force generation for vesicle transport. Rather, these experiments suggest that DNase I and gelsolin inhibit vesicle transport in axoplasm by disrupting the cytoskeletal architecture of the microtubule domains. Another actin-binding protein that interferes with organelle movement in axoplasm is synapsin I, which has been shown to inhibit vesicle transport in extruded axoplasm, and in a phosphorylation-dependent manner (McGuinness *et al.*, 1989). It is thought that the dephosphorylated form of synapsin I can bind vesicles to the microfilament matrix that surrounds microtubules, thereby interfering with their translocation through the axoplasm. These experiments suggest that microfilaments in the inner axoplasm play an important role in the functional architecture of those microtubule domains that are associated with the transport of membranous organelles.

19.4. Myosin from squid brain

With the current interest in microtubule-based organelle transport in axoplasm, little attention has been given to the role of myosin in the squid giant axon. This is particularly surprising given the large number of microfilaments in axoplasm and the fact that neurons were among the first non-muscle cells in which actin and myosin were identified (Berl *et al.*, 1973). See and Metuzals (1976) have purified and characterized myosin from squid brain optic lobes. Analysis of the purified myosin by SDS PAGE revealed a single band that co-migrated with the heavy chain of rabbit muscle myosin, and several minor bands that may have corresponded to myosin light chains. The purified squid brain myosin demonstrated ATPase activity that was stimulated by F-actin from rabbit muscle. In addition, electron microscopy of negatively-stained preparations showed that the squid myosin decorated rabbit muscle F-actin with the characteristic arrowhead pattern, and polymerized to form bipolar myosin filaments at low ionic strength. These observations indicate that if a serious commitment to the study of myosin-mediated mechanisms in the giant axon were made, it could advance our understanding of the biochemistry and organization of intra-axonal dynamics along a path that has been almost completely neglected.

20. Studying the stability of the cytoskeleton

20.1. Assembly dynamics of cytoskeletal polymers in axons

In addition to the opportunities for biochemical and ultrastructural investigations on axoplasm, the squid giant axon offers a special and perhaps unique opportunity for

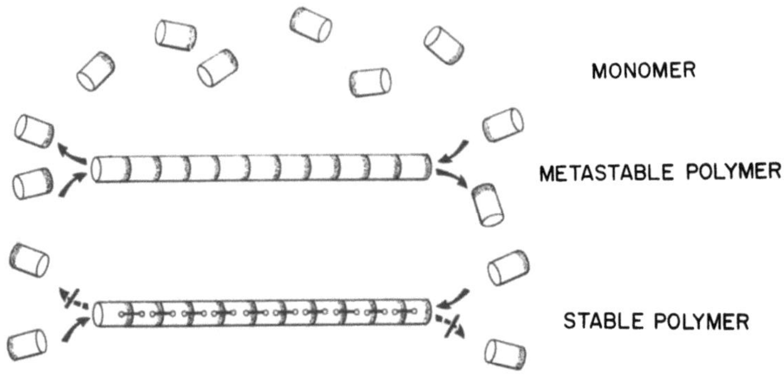

Figure 9. Diagram illustrating the three states of cytoskeletal protein in axoplasm. Monomer subunits exchange freely with the ends of the metastable polymer, and the steady-state concentration of metastable polymer is dependent on the monomer concentration. In contrast, monomer can not dissociate from the ends of stable polymers. Reproduced from Lasek (1988).

analyzing the stability and dynamics of the axoplasmic cytoskeletal elements. The axon has three distinct cytoskeletal polymer systems: the neurofilament system, the microtubule system, and the microfilament system. These systems consist of their self-assembling protein subunits (neurofilament proteins, tubulin, and actin) and their respective polymers (neurofilaments, microtubules, and microfilaments). In the axon, these systems operate at a steady state that is apparently fairly constant along the length of the axon between the cell body and axon terminal.

The cytoskeletal polymers themselves can be sub-classified with regard to their assembly dynamics into two distinct populations; metastable polymers and stable polymers (see Fig. 9). Metastable polymers (also termed dynamic or equilibrium polymers) continually exchange protein subunits with the diffusible monomer pool, and their length is a function of the amount of monomer available for polymerization. Metastable polymers are very sensitive to decreases in the concentration of available free monomer, and if the monomer concentration decreases sufficiently, they will disassemble entirely, so that no metastable polymers remain. In contrast with metastable polymers, stable polymers are insensitive to decreases in the concentration of available monomeric subunits.

If extruded axoplasm is immersed in an artificial axoplasm solution, the monomeric proteins are free to diffuse out of the axoplasm, and the concentration of monomer in the axoplasm decreases accordingly. This decreases the availability of monomer for assembly onto the ends of the metastable polymers. Thus the subunits that are continually dissociating from the metastable polymers are not replaced, and the polymers decrease in length. If the volume of the solution is large relative to the volume of the axoplasm then net disassembly will continue until all of the metastable polymers have been extinguished.

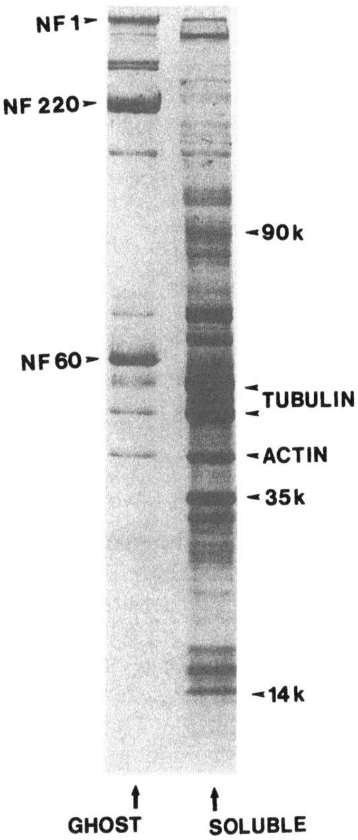

Figure 10. Stable and soluble proteins in axoplasm. Axoplasm was extracted in Buffer P for 2 – 3 hours as described in the text and then analyzed by SDS PAGE and stained with Coomassie Blue. The polypeptide composition of the extracted axoplasm (GHOST) and of the extraction medium (SOLUBLE) are shown. The three neurofilament proteins are labelled on the left, and a number of other proteins in axoplasm are labelled on the right. NF-1, NF220 and NF60 correspond to Band 1, P200 and P60 in the nomenclature used in Section 17. Reproduced from Morris and Lasek (1982).

Morris and Lasek (1982, 1984) used this experimental approach to quantitatively analyze the elution of proteins from isolated axoplasm. For these studies a solution (Buffer P) that closely simulates the solution conditions inside the axon was developed to preserve the types of protein associations that occur in axoplasm *in situ* (see Section 10.1). Because of the large diameter of the axoplasm, the egress of diffusible proteins occurs over a time course of minutes and can therefore be measured readily. In this Section we describe these studies and show how analysis of the extent and kinetics of elution can be used to infer the amount of cytoskeletal protein as monomer, metastable polymer, and stable polymer in axons. These methods could also be applied to the study of the association of other cytoplasmic proteins of the cytoskeleton, and this is an opportunity that has yet to be exploited.

20.2. The axoplasmic ghost

When axoplasm is extruded into a solution that simulates the solution conditions of the axon, the loss of soluble proteins (including free cytoskeletal subunits) by diffusion occurs rapidly, and the loss due to the disassembly of dilution-labile cytoskeletal polymers occurs more slowly. After a few hours the axoplasm approaches equilibrium with the surrounding medium, and what remains is comprised of the stable

Table VI. The protein composition of axoplasm and the axoplasmic ghost.

	Axoplasm	*Ghost*
Tubulin	22%	12%
Neurofilament	13%	54%
Actin	6%	5%
35kD protein	6%	n.d.
90kD protein	1%	n.d.
Other proteins	52%	29%

The values are percentages of the total protein in axoplasm and in the axoplasmic ghost. Not detected (n.d.). Data from Morris and Lasek (1982, 1984).

cytoskeletal polymers of axoplasm, their associated proteins, and the axoplasmic membranous elements: mitochondria, vesicles, and endoplasmic reticulum. Though many of the axoplasmic proteins have been lost, the extracted axoplasm, or *axoplasmic ghost*, retains its original cylindrical shape and mechanical integrity.

In order to determine the proportions of stable polymer for the cytoskeletal proteins in axoplasm, Morris and Lasek have quantitatively analyzed the protein composition of the extracted axoplasm and extraction medium (Buffer P). The amount of each protein was determined by densitometry of SDS-polyacrylamide gels using bovine serum albumin as a standard. The volume of the extraction medium determines the extent of dilution of the axoplasmic proteins. Thus, in order to achieve full disassembly of metastable polymers, the solution volume must be large relative to the axoplasm volume. In a typical experiment, axoplasm from one axon (about 5 μl) was extruded into 1 or 2 ml of Buffer P, which represents a 200 to 400-fold dilution.

Fig. 10 shows the polypeptide composition of the extracted and stable proteins after the axoplasm had equilibrated with the Buffer P, and Table VI shows the relative proportions of a number of the major proteins in axoplasm and in the axoplasmic ghost. About 80% of the axoplasmic proteins are soluble in Buffer P and essentially all of these soluble proteins are extracted within 2 hours. The remaining 20% of the protein constitutes the axoplasmic ghost. All three neurofilament polypeptides remain associated with this stable axonal cytoskeleton, which indicates that the neurofilaments are stable to dilution. In contrast, only 17% of the tubulin and 26% of the actin are present as stable polymer (see Section 20.3), and many other proteins in axoplasm, such as the M_r 35,000 protein, are completely lost from the axoplasm. Because of the stability of the neurofilaments, the neurofilament proteins are the major component of the axoplasmic ghost, and these polypeptides constitute 54% of the total ghost protein. Together, neurofilament protein, tubulin and actin, constitute over 70% of the ghost protein. Electron microscopy of the axoplasmic ghost indicates that neurofilaments are the major cytoskeletal structures; microtubules are much fewer than in unextracted axoplasm and much of the filamentous matrix that surrounds them also appears to have been extracted. These observations support the conclusion that the cytoskeletal polymers are indeed the morphological correlate of the stable cytoskeletal proteins.

A number of other cytomatrix proteins remain in the axoplasmic ghost, and this may indicate that they have a high affinity for the stable cytoskeletal polymers. For example, very little spectrin (see Section 15.3) is extracted in Buffer P. In fact, we have noted that spectrin often copurifies with neurofilaments (for example, see Brown and Eagles, 1986), and this suggests that it may bind tightly to these cytoskeletal elements. Another protein that remains in the axoplasmic ghost is a protein with an approximate M_r of 70,000, which we have also consistently observed in squid neurofilament preparations. This protein binds the antibody Anti-IFA (Brown and Eagles, 1986) and therefore may be related to internexin, an intermediate filament-associated protein that is present in intermediate filament preparations from a wide variety of sources (Pruss *et al.*, 1981; Napolitano *et al.*, 1985; Pachter and Liem, 1985).

20.3. Monomer-polymer equilibria in axoplasm

We have shown above that a large proportion of the actin and tubulin in axoplasm is extracted in dilute solution. This extracted protein includes both monomer and metastable polymer. Thus the amount of monomer and metastable polymer in axoplasm cannot be determined directly. To determine the concentration of actin and tubulin as monomer and metastable polymer in axoplasm, Morris and Lasek (1984) took two approaches. In both approaches, densitometry of SDS-polyacrylamide gels was used to quantify the amount of protein in the same way as for the studies described in Section 20.2.

One method, the *kinetic equilibration paradigm (KEP)*, employs the basic principles of diffusion to distinguish freely diffusible monomer from proteins that are present in the form of polymer. In this approach axoplasm is extruded into a large volume (0.5 – 1 ml) of artificial axoplasm solution (Buffer P) and the rate at which proteins elute from the cylinder of axoplasm is measured by analyzing the amount of protein in the bathing medium at different times. The KEP analysis can be applied to a single piece of axoplasm (the *single axon KEP*) by transferring the axoplasm through a series of baths, or alternatively separate pieces of axoplasm can be used for each time-point determination. The latter method (the *ratio KEP*) is preferable because repeated transfer between solutions makes the axoplasm more fragile and small fragments of axoplasm can break off and remain in the bath. For the ratio KEP analysis, axoplasm is extruded into 1 ml of Buffer P, extracted for a measured length of time, and then removed from the bath with forceps. The axoplasm and bath are analyzed by SDS PAGE and the amount of protein is quantified by densitometry of the polyacrylamide gel. For each extraction time the proportion of protein extracted is calculated and the results for a range of times are used to plot the elution kinetics.

Morris and Lasek (1984) found that the rate of elution of actin and tubulin from the axoplasm was not monophasic, but that it could be modelled as the sum of two exponential components. One component elutes at the rates predicted by diffusion theory for actin and tubulin monomer. 95% of this component elutes from the axoplasm within the first 10 minutes. The other component elutes more slowly and corresponds to the elution of monomer from the pool of metastable polymers. This rate of elution is limited by the rates of disassembly for these polymers, which is slower than the rate of diffusion from the axoplasm. 95% of the actin and tubulin in this component elutes within two hours. Quantification of the amount of protein in these two rate components yields the total amounts of protein as monomer and metastable polymer in the axoplasm.

Table VII. Cytoskeletal proteins as monomer and polymer in axoplasm.

	Actin	*Tubulin*	*Neurofilament*
Stable polymer	26%	17%	>95%
Metastable polymer	30%	55%	–
Monomer	44%	28%	–

Values given as percentage of the total amount of that protein present in axoplasm. Data from Morris and Lasek (1984).

The other method is *pharmacological* and employs taxol and phalloidin to stabilize the microtubules and microfilaments, respectively. Taxol and phalloidin are stored at room temperature as a 10 mM stock solution in dimethylsulfoxide (DMSO) and diluted in Buffer P to give a final drug concentration of 10 µM in 0.1% DMSO. Axoplasm is extruded into 1 ml Buffer P either with or without the stabilizing drug and then extracted to equilibrium (2 or 3 hours). Analysis of the extracted axoplasm and bathing medium by SDS PAGE and densitometry allows comparison of the amount of actin or tubulin that elutes in the presence and absence of the appropriate drug. The difference between the amount of that protein in extracted axoplasm in the presence and absence of the stabilizing drug represents the amount of metastable polymer in the axon.

Morris and Lasek (1984) found that the results of the kinetic and pharmacological methods showed good agreement, and the mean values are shown in Table VII. These results indicate that 56% of the actin in axoplasm and 72% of the tubulin in axoplasm are present as polymer. Since axoplasm contains 25 mg/ml protein (see Section 9), we can calculate the actual monomer and polymer concentrations from the known proportions of actin and tubulin in axoplasm (shown in Table VI). These calculations indicate that the monomer concentration in axoplasm is about 0.6 mg/ml for actin and about 1.6 mg/ml for tubulin. These concentrations are greater than the critical concentrations for assembly of these proteins *in vitro* and this suggests that factors are present in the axoplasm that inhibit the polymerization of actin and tubulin, or that elevate the critical concentration for their assembly. No detectable neurofilament protein elutes in these experiments and this supports the proposal that essentially all of the neurofilament protein in axoplasm is stably polymerized.

21. Mechanical studies on axoplasm

21.1. Axoplasm has mechanical integrity

The ability to isolate intact axoplasm from the squid giant axon provides one of the rare opportunities in nature for the biologist to see and handle pure cytoplasm. Biologists who have had this opportunity tend to be struck by the mechanical integrity of the axoplasm when they are reminded that it is composed mostly of water and contains only 2.5% protein by weight (see Section 9). This mechanical integrity is due to the long polymers of the cytoskeleton, and treatments that disrupt the cytoskeletal polymers, such as proteolytic degradation or chaotropic ions, cause the axoplasm to liquify (see Section 11). The cytoskeletal polymers therefore confer solid-like properties on the axoplasm so that it maintains its cylindrical form after isolation, even when it is handled, albeit gingerly, with forceps.

21.2. Axoplasm is anisotropic

The axoplasmic cytoskeleton appears homogeneous to the naked eye, but light and electron microscopy indicate that the cytoskeletal polymers are highly aligned along the long axis of the axon (see Section 14.1). This organization of the cytoskeleton makes axoplasm highly anisotropic. For example, the longitudinal alignment of the polymers confers distinct mechanical properties on the radial and longitudinal dimensions of axoplasm.

A graphic illustration of this structural and mechanical anisotropy of axoplasm was described by Chambers and Kao (1952) based on their observations on the microinjection of substances into axons. They showed that bubbles of air or oil injected into the axoplasm assumed an ovoid shape with their long axes orientated parallel to the long axis of the axon. Injection of millimolar concentrations of calcium into the axons, which is now known to activate the calcium-dependent protease (see Sections 11.2 and 17.4), caused the axoplasm to liquify, and the bubbles then assumed a spherical shape. These observations indicate that axoplasm has a mechanical anisotropy. The oil and air tend to form bubbles in an aqueous environment because of the surface tension forces of the water. Apparently the axoplasm offers less resistance to these surface tension forces in its longitudinal dimension than in its radial dimension. With these observations, Chambers and Kao were able to infer a *linear ultrastructure* in axoplasm at a time before electron microscopy had revealed the existence of cytoskeletal polymers. We now know that this linear ultrastructure is due to the cytoskeletal polymers that are orientated along the long axis of the axon. These observations therefore indicate that axoplasm has different mechanical properties in the dimension parallel to the cytoskeletal polymers than perpendicular to the polymers.

21.3. The macroscopic mechanical properties of axoplasm

Two approaches have been taken for the quantitative analysis of the mechanical properties of axoplasm from the squid giant axon. In one approach, Rubinson and Baker (1979) measured the resistance of axoplasm to flow through small-bore cellulose acetate capillary tubing. They showed that the axoplasm behaved in a viscous manner and was characterized by a yield-strength, which is a force below which the axoplasm did not flow. However, it is likely that the methods used in these measurements, which involved high rates of shear, resulted in considerable disruption of the organization of the cytoskeleton. Thus it is likely that many factors contributed to the resistance to flow, including frictional interactions with the capillary wall and shearing of entangled polymers.

Using a different approach, Sato *et al.* (1984) implanted single magnetic spheres (100 μm in diameter) into axons and then measured their movement in a magnetic field. Their analyses indicated that axoplasm had viscoelastic properties. In other

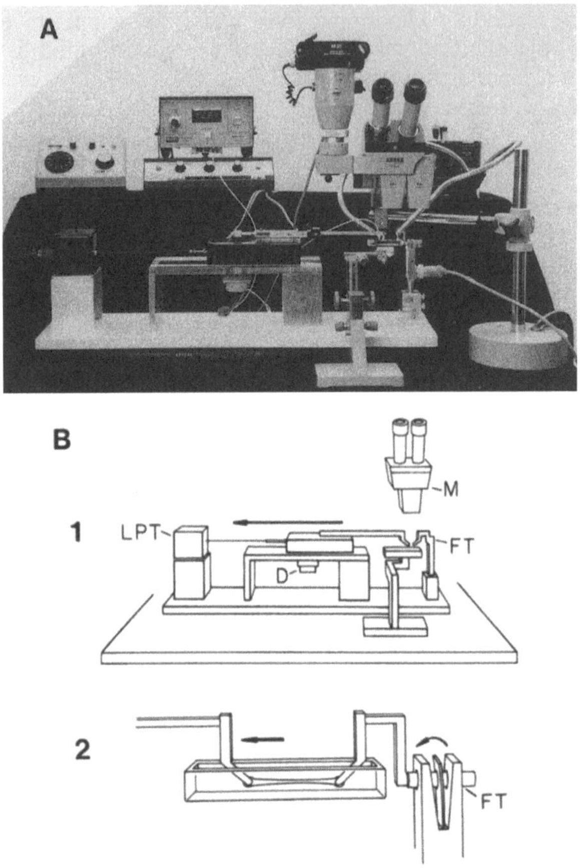

Figure 11. The apparatus for stretching axoplasm. A. Photograph of the apparatus. B. Schematic diagram illustrating the essential features of the apparatus. In the upper drawing (1) the apparatus is shown with the chamber lowered to reveal the two angled clips. To stretch axoplasm, the axoplasm is immersed in liquid in the chamber and then mounted between the clips. The lower drawing (2) shows a close-up view of a piece of axoplasm being stretched in the chamber. The right clip is attached to a the force transducer (*FT*). The left clip is attached via a metal rod to the drive carriage that is driven by a stepper motor (*D*). The movement of the left clip is measured by a linear position transducer (*LPT*) that is attached to the drive carriage by a retractable sprung cable. The straight arrows indicate the direction of movement during stretch. The resistance of the axoplasm to stretch is measured by the extent of displacement (curved arrow) of the vertical tongue of the force transducer, to which the right clip is attached. The axoplasm is observed and photographed from above with a binocular microscope (*M*).

words, the magnetic spheres experienced both viscous and elastic resistances to their movement through the axoplasm. Since these studies used intact axons and very low rates of shear, it is likely that they involved less disruption of the cytoskeletal organization than the studies of Rubinson and Baker (1979). However, the large size of the magnetic spheres caused visible disruption of the axoplasm and entanglement of the cytoskeletal polymers, and it is likely that polymer shearing was also a contributing factor to the viscous resistance that was measured in these studies. Nevertheless, by changing the orientation of the axon relative to the poles of the magnet, it was possible to see differences in the resistance to movement in the radial and longitudinal dimensions. These analyses confirmed observations of Chambers and Kao (1952) that the axoplasm is mechanically anisotropic. Specifically, they showed that there was a greater elastic resistance to movement in the radial dimension of the axon (perpendicular to the long axis of the cytoskeletal polymers) than in the longitudinal dimension of the axon (parallel to the cytoskeletal polymers).

The anisotropy of axoplasm, and specifically its mechanical properties in the longitudinal dimension, are likely to be important in many aspects of the mechanics of axonal motility. For example, studies on squid axoplasm have shown that there is a continuous and bustling traffic of membranous vesicles along the microtubules in the axon, and these constitute the fast component of axoplasmic transport (Weiss, Meyer, and Langford, 1990). In order for the vesicles to move through the axoplasm, they have to overcome the viscous and elastic resistances that they encounter as they move past adjacent cytoplasmic structures along the long axis of the axon (Lasek and Miller, 1986).

In order to more effectively study the mechanical properties of axoplasm in its longitudinal dimension, we have developed a novel approach in which axoplasm is stretched along its long axis at slow rates (George and Lasek, 1989; Brown and Lasek, 1989). By measuring the resistance of the axoplasm to stretch, the mechanical properties of the axoplasm can be investigated. This method does not involve disruption of the anisotropic organization of the axoplasm, and the measurements therefore reflect the mechanical properties along the long axis of the cytoskeletal polymers. In this Section we describe the apparatus and experimental design, and indicate how these studies may aid our understanding of the dynamics of the axonal cytoskeleton.

21.4. The stretch apparatus

Three essential requirements for the stretch measurements are a pulling device, a force transducer, and a means to amplify and record the transducer output. Fig. 11 shows a photograph and a schematic diagram of the stretch apparatus. A segment of axoplasm is suspended between two metal clips (Schwartz temporary clip, Roboz Surgical Instrument Company) in a clear plexiglass (perspex) chamber. The dimensions of the chamber are 60 mm long, 12 mm wide and 10 mm deep. The plexiglass is 2 mm thick. Inside the chamber is a block of plexiglass cut at an angle of 45 degrees onto which a thin mirror is mounted. This provides an additional image of the axoplasm from the side when the axoplasm is viewed from above. The mirror is close enough to the axoplasm that the two images appear in the same field of view, though at slightly different depths of focus. We use an aluminum *first-surface* mirror (12 × 49 × 0.8 mm) which was purchased from Edmund Scientific and cut to size using a glass knife. The mirror can be glued to the plexiglass mount or, less permanently, it can be secured with an inert silicone lubricant (Dow Corning 111, Dow Corning Corporation). The chamber is mounted on a micromanipulator to allow adjustment of its position relative to the clips. Axoplasm is transparent but can be seen in dark-field because the long cytoskeletal polymers scatter light. We illuminate the axoplasm from the back of the chamber using fiber optic light guides and view the chamber from above using a low power binocular microscope. The base of the chamber is made of black plexiglass to provide a black background for the dark-field image. The internal volume of the chamber with the mirror block in place is about 4.5 ml, but 2 ml of liquid is sufficient to cover the axoplasm.

The force is measured using a Model FT03 force displacement transducer (Grass Instrument Company) without the isotonic springs. The transducer gives a voltage output that is proportional to the displacement of the transducer tongue. The transducer is mounted vertically to the steel base of the stretch apparatus. The position of the mount and the height of the transducer are adjustable to allow alignment with the pulling device. The clip is mounted to the transducer in such a way that the axis of the axoplasm is in line with the shaft of the transducer so that no torque is applied to the transducer tongue. To prevent damage to the transducer when the axoplasm is being attached to the clips, we designed a small clamp consisting of two set screws that attaches to the housing of the transducer and allows the tongue to be temporarily secured. The maximum sensitivity of the FT03 transducer is about 1 mg force. Transducers of much higher sensitivity can be obtained (for example, from the Cambridge Instrument Company), but we still use the FT03 because the metal clip is easily attached to the transducer tongue, and its construction is rugged enough to withstand the handling when the axoplasm is mounted in the clips. The FT03 is also comparatively low cost.

The pulling device is a converted Sage syringe pump (Model 352, Orion Instrument Company). In the version of the stretch apparatus shown in Fig. 11, the stepper motor and drive carriage have been removed from the pump control and mounted on the raised platform of the stretch apparatus. The stepper motor is connected to the motor control panel via a cable. This allows mechanical isolation of the stretch apparatus from the operating unit. The Sage syringe pump controller that we use allows a broad range of pull rates from 2 μm/sec to greater than 1 mm/sec, and different ranges of rates can be obtained by substituting motors with different gear ratios. A switch on the control panel allows the direction of the motor to be reversed. Stepper motors are preferable to DC motors for this application because they have a long life, are variable over a wider range of rates, and are highly reproducible independent of the load. A gear wheel on the shaft of the motor engages with a toothed bar in the drive carriage, converting the rotational motion of the motor into a linear movement of the drive carriage along the raised platform. The clip that holds the left end of the axoplasm is attached to the drive carriage via a metal rod. The position and movement of the puller rod are monitored with a linear position transducer (LPT, Unimeasure, Inc.) that is attached to the drive carriage via a retractable sprung cable. The LPT gives a voltage output that is proportional to the extension of the cable.

The design and placement of the apparatus must allow operation of the pump controller, microscope and camera without transmitting vibration to the transducer. This is achieved by placing the camera and pump controls and the microscope base on a separate adjacent table or shelf. The microscope is mounted on a stand that swings over the apparatus to allow free access to the chamber. More recently, we have also enclosed the apparatus in a transparent plexiglass box in order to protect the transducer from air currents and to minimize temperature changes over time. Recently, we have modified the stretch apparatus by removing the rack-and-pinion drive carriage and replacing it with a worm-drive (Stoelting Company, 5 mm pitch thread), which allows a tighter coupling of the motor to the puller-rod, with less lateral play. In this version of the stretch apparatus, the motor shaft is attached directly to the worm-drive via a close-fitting chuck. The puller rod that holds the clip at the left end of the axoplasm now attaches to the translating portion of the worm-drive. For our experiments at the Marine Biological Laboratory in Woods Hole, we have recorded the voltage output of the FT03 force transducer and linear position transducer on a two-channel strip chart recorder (Grass Instrument Company Model 79 Polygraph equipped with a Model 7P1 low level DC pre-amplifier and Model 7DA DC pen driver amplifier on each channel).

It is important for these studies to isolate axoplasm as a continuous unbroken length and to minimize any distortion of the organization of the cytoskeletal polymers.

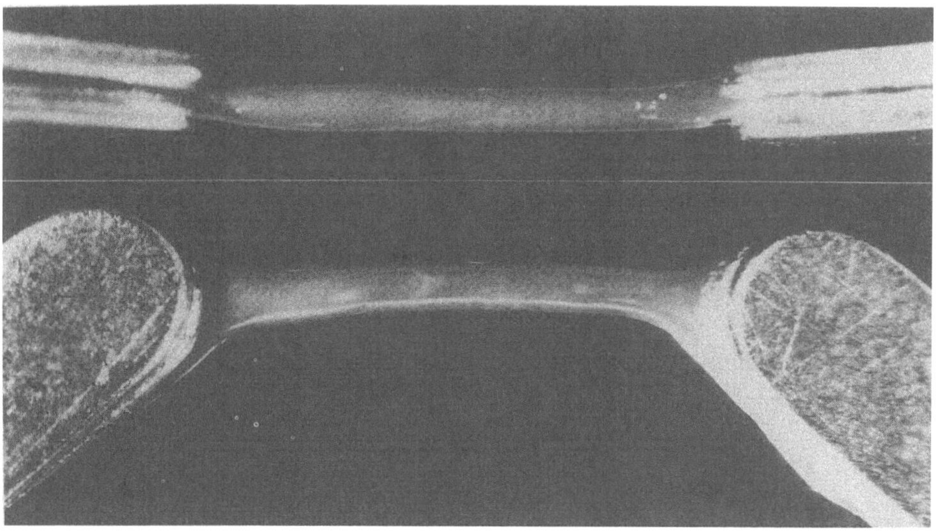

Figure 12. Photograph of a piece of axoplasm suspended in liquid between the clips of the stretch apparatus. The length of the axoplasm between the clips is about 4.5 mm and its diameter is about 0.4 mm. The upper image is of the axoplasm viewed from above and the lower image is a reflection of the same axoplasm from the side. The two images can be seen and photographed in the same field of view of the microscope.

In our stretch experiments so far we have used axoplasm isolated by the extrusion method, but the slitting method may be preferable since it allows axoplasm to be isolated without applying tension or compression to the cytoskeleton (see Section 6.2).

To stretch axoplasm, the isolated axoplasm is immediately transferred into 2 ml of liquid in the chamber. In our experiments we have stretched the axoplasm in paraffin oil so as to maintain a constant volume and prevent the extraction of diffusible proteins that occurs in aqueous solution (see Section 20). A minimum axoplasm length of about 10 mm is required to successfully mount the axoplasm in the clips, but longer lengths are preferable. The axoplasm is first mounted in the right clip, which is attached to the force transducer. To mount the axoplasm, the clip is held open with a pair of hemostatic forceps and the axoplasm is raised in between the clip blades using fine forceps. The clip blades are then allowed to close onto the axoplasm. The inside of the clip blades can be slightly roughened with abrasive paper to improve the grip. Then the left clip, which is attached to the puller rod, is advanced to a distance of 4 mm from the right clip and the left end of the axoplasm is also attached. Angled clips are used so that the tips are not obscured when viewed from above. The axoplasm can be photographed during the experiment through a trinocular attachment to the microscope. A photograph of a piece of axoplasm between the clips is shown in Fig. 12.

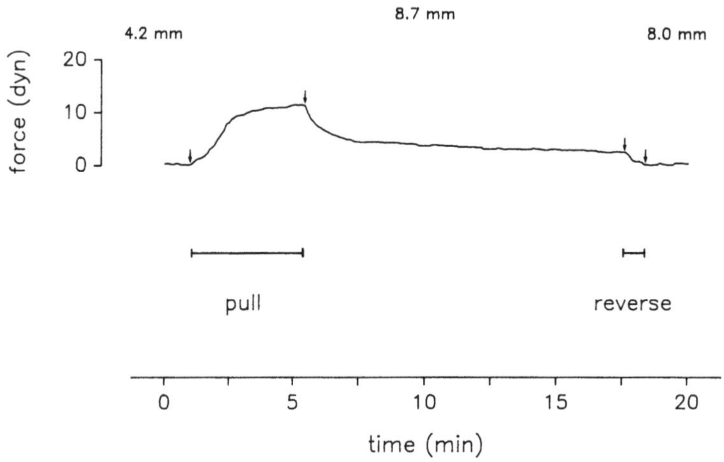

Figure 13. A typical stretch profile for axoplasm. This segment of axoplasm was extruded onto parafilm and then stretched in paraffin oil at a rate of 17.2 µm/sec. The stretch profile represents the voltage output from the force transducer that is attached to one end of the axoplasm. The movement of the puller, which is attached to the other end of the axoplasm, is indicated below the trace. The small arrows on the stretch profile indicate the points at which the puller motor was started and stopped. The units on the ordinate are force in dynes and the units on the abscissa are time in minutes. The axoplasm was stretched for about 4 minutes (*pull*). When the force reached a plateau, the motor was switched off and the force was allowed to decay at constant extension. After about 12 minutes, the puller was reversed (*reverse*) at the same rate in order to remove the remaining elastic tension in the axoplasm. The lengths are shown for the axoplasm before the stretch (4.2 mm), at the maximum extent of the stretch (8.7 mm), and at its new resting length after the reversing the puller (8.0 mm).

21.5. Stretch analysis

Fig. 13 shows a typical stretch profile for axoplasm. The output from the force transducer and the movement of the puller are shown. The puller was off at the beginning of this trace, and this resting force is taken to be zero. About 40 seconds into the trace, the puller was switched on, and the axoplasm was stretched at a constant rate of 17.2 µm/sec. The force rose slowly at first and then more rapidly. As the stretch continued, the rate of rise of force then began to decrease until it reached a constant value (or plateau) of about 12 dynes (1 dyne = 10^{-5} Newtons) of force. When the force reached this plateau value, the length of the axoplasm was 8.7 mm, which was greater than twice the original length. In one batch of thirteen stretch experiments, we observed plateau forces ranging from 5 to 25 dynes.

As the axoplasm is stretched it becomes longer and narrower in diameter. Because the axoplasm is in oil, which is immiscible with water, the overall volume of the axoplasm remains constant. If we continue to stretch the axoplasm past the point at which plateau is reached, the force very gradually decreases as the axoplasm becomes thinner and thinner. Our observations of the axoplasm through the microscope during stretching indicate that, in practice, certain regions thin more than others. This indicates that the axoplasm is not uniform along its length. This could be caused by slight variations in the diameter of the axoplasm along its length and also by local weakening of the axoplasm caused by local compression or tension during its isolation.

Measurements of the axoplasm using periodic carbon-dust markings indicate that all regions of the axoplasm elongate, but that the narrow regions undergo the greatest elongation and the greatest thinning. Since the force produced by the stretch is maintained along the whole length of the axoplasm, the narrower regions receive proportionally more force per unit cross-sectional area than thicker regions, and this probably explains why they undergo the greatest thinning. Eventually the thin regions become so narrow that the axoplasm breaks. This may arise if insufficient polymer overlap remains to maintain the continuity of the axoplasm, or if the force borne by the small number of polymers in the thin region exceeds their breaking strength, causing them to snap.

In the stretch analysis in Fig. 13, the axoplasm was stretched until a constant plateau force had been reached, and then the motor was switched off and the axoplasm was held at its new extended length of 8.7 mm. As soon as we stopped stretching the axoplasm, the force rapidly decayed. The decay is most rapid initially and becomes slower with time. This time-dependent decay in force at constant length has a characteristic and very reproducible shape that can be modelled as a complex exponential function. If the axoplasm is left for long enough (many hours), the force can eventually decay completely, whereupon the axoplasm is seen to sag between the clamps.

In practice we generally allow the axoplasm to relax for long enough to clearly establish the kinetics of the decay in force, and then we advance the motor in reverse at the same rate until the remaining tension in the axoplasm has been removed. Baseline drift during the experiment sometimes makes it difficult to judge the point of zero force, but we have found that a reliable indicator is to look for the point at which the axoplasm begins to sag between the clamps. At this point the axoplasm is no longer under tension and its length represents a resting length. In fact, we have found that the axoplasm routinely increased about two-fold in resting length during such stretch-hold-reverse cycles. We have analyzed many pieces of axoplasm in this way over two consecutive summers at Wood's Hole and we have found that the shape of this profile is highly reproducible. In a few cases, we have been able to repeat this stretch-hold-reverse cycle up to three times before the axoplasm finally thins to breaking point. Each time the plateau force is lower, but the general form of the stretch profile is the same. The decrement in the plateau force is apparently due to the thinning of the axoplasm in successive stretches.

21.6. Interpretation of the stretch profile

The general form of the stretch profile that we have described is immediately recognizable to mechanical engineers and materials scientists as characteristic behavior of a viscoelastic substance. One of the best studied viscoelastic tissues in biology is connective tissue (for example, see Silver, 1987). One example is tendon, which is a highly anisotropic tissue that is composed of bundles of collagen polymers which are aligned longitudinally along the tension-bearing axis (Kastelic and Baer, 1980). The mechanical behavior of tendon under stretch has been extensively studied. The mechanical response to stretch can be analyzed into a time-independent (*instantaneous*) elastic response and a time-dependent viscous response.

The rising phase of the response of tendon to stretch has a similar general form as that seen for axoplasm (see Fig. 13), and is characterized by four regions (Baer *et al.*, 1986). Analyses of tendon during stretch have allowed determination of the structural basis for this mechanical behavior. The initial part of the curve is called the toe region and it corresponds to an elastic straightening of the collagen fibrils. If the tendon is stretched further, the force rises markedly and enters a linear region which corresponds to a reversible extension of the collagen fibrils themselves. The slope of this region corresponds to the modulus of elasticity for the collagen fibrils. With further stretch, the force enters a yield region in which the force tends towards a plateau. In this region, the force is high enough to cause slippage of the collagen fibrils relative to each other, and this results in an irreversible extension of the tendon. It the force on the tendon is now released, it is found to have a new and longer resting length. By correlating the organizational changes in the tendon with the mechanical behavior, these studies indicate that the structural correlate of elasticity in the tendon is straightening and extension of the collagen fibrils, and the structural correlate of viscous flow in tendon is the slippage of the collagen fibers relative to each other.

However, this mechanical behavior differs from that of axoplasm in a number of important respects. Firstly, the forces required to deform axoplasm are very low compared to those for tendon collagen. For example, a typical piece of axoplasm plateaus at about 20 mg force, while to deform tendon to the same extent would require kilograms of force for the equivalent cross-sectional area of tissue (Baer *et al.*, 1986; Silver, 1987). A second major difference is that tendon is highly elastic and the rising phase of the force is characterized by a linear elastic region that represents a reversible elongation of the collagen fibrils. In contrast, we have found that no phase of the extension of axoplasm is totally elastic. For example, if axoplasm is held at even small extensions of its original length, a time-dependent decay in force is seen similar to that in Fig. 13. Furthermore, we noted earlier that if axoplasm is stretched to plateau and then held at this length, the force can eventually decay to zero. This indicates that the resistance to flow in axoplasm is low and chiefly limited by its time-dependence. Thus, if axoplasm is stretched and then held at this new length for sufficient time, substantial increases in the resting length of the axoplasm are observed. For example, the two-fold increase in resting length of the axoplasm in Fig. 13 (after about 12 minutes of relaxation) corresponded to greater than 90% of the maximum extent of stretch.

21.7. A mechanical model

These observations indicate that the mechanical behavior of axoplasm in its longitudinal dimension can be considered to be a combination of elasticity and time-dependent viscous flow. To emphasize the dominant property of viscous flow, we refer to axoplasm in its longitudinal dimension as *elastoviscous* rather than viscoelastic. The behavior of axoplasm can be explained by the following mechanical model in which both elastic and viscous forces contribute to the stretch profile: as the axoplasm is stretched, instantaneous extension of elastic components results in an increase in resistive force and this is accompanied simultaneously by a slower time-dependent relaxation due to viscous flow.

In our experiments, we have found that the mechanical behavior of axoplasm under tension depends greatly on the rate of stretch deformation. The above mechanical model, in which instantaneous elastic extension and time-dependent viscous flow both contribute simultaneously to the stretch profile, can explain these observations. Specifically, since the viscous component of the deformation is time-dependent, rapid stretches are essentially elastic and the axoplasm elongates until the force is sufficient to cause it to break, but if the axoplasm is stretched slowly it can flow to attain a new resting length. This behavior is characteristic of viscoelastic materials in general. A common example is putty or plasticine, which extends to great lengths if stretched slowly but snaps if stretched rapidly. We found that the appropriate rate of stretch for axoplasm must be determined empirically. If axoplasm is stretched very slowly (rates less than 1 µm/sec), substantial increases in resting length can be achieved, but the force is too low to measure. On the other hand, if axoplasm is stretched too rapidly then the force rises rapidly and the axoplasm breaks. In our investigations on axoplasm, we have chosen rates of stretch that allow us to measure both the viscous and elastic components. We have found that stretch rates of about 20 µm/sec produce a measurable force without causing the axoplasm to break.

In the plateau region, the force generated by the extension of the elastic components is sufficient to balance the rate of stretch with the rate of viscous flow, and thus the plateau represents a purely viscous deformation with the elastic elements maintained at a constant extension. When stretching is stopped in the plateau phase, and the axoplasm held at its new length, viscous flow continues as the elastic components gradually return to their original resting length. If we wait long enough, the force can eventually decay completely, whereupon the axoplasm sags between the clamps. At this point, the tension generated by the stretch has been completely dissipated by viscous flow, and the axoplasm now has a new resting length. Alternatively, the remaining elastic tension can be dissipated more rapidly by reversing the motor, but the longer the force is allowed to decay, the greater the extent of viscous flow and thus the longer the final resting length.

21.8. The structural basis for elasticity and flow

What is the ultrastructural basis for elasticity and flow in axoplasm in these experiments? Based on the analogy with the mechanics of viscoelastic biological materials such as tendon, and on our knowledge of the architecture of axoplasm, we propose that the structural correlate of flow in axoplasm under longitudinal tension is the sliding of axonal polymers relative to each other, and that the viscous resistance to flow represents the frictional interactions of the polymers with other axoplasmic elements as they slide through the axoplasm (George and Lasek, 1986). By similar analogy, the weak elasticity in axoplasm may arise from a spring-like straightening or elongation of the cytoskeletal polymers during the stretch. Since the force required to stretch axoplasm is many orders of magnitude less than that required to stretch collagen fibrils (see Section 21.6), the contribution of stretching the polymers themselves to the elasticity is likely to be negligible and the major contribution will come from the *configurational elasticity* of the polymers. For example, the neurofilaments and microtubules are oriented helically along the axon (see Section 14.2), and stretching axoplasm will tend to straighten the polymers, causing an elastic resistance. This elastic effect is particularly dramatic for axoplasm from the giant axon of the marine fanworm *Myxicola infundibulum*, which has a pronounced helical organization. In addition, the polymers in extruded axoplasm are not perfectly straight, and elasticity may also arise from the tendency of the polymers to return to their slightly wavy configuration.

21.9. Polymer sliding in axons

Our measurements using the stretch apparatus indicate that axoplasm readily flows when stretched along the long axis of the cytoskeletal polymers at low rates. This suggests that the lateral interactions between the cytoskeletal polymers in axoplasm may be relatively weak. Support for this has come from observations on the transport of membranous vesicles in axoplasm (fast axoplasmic transport) which indicate that the vesicles experience a similar resistance to movement in axoplasm as they do to movement *in vitro* along isolated microtubules (Lasek and Miller, 1985).

These considerations are of interest with respect to the mechanism of slow axoplasmic transport in axoplasm. Studies on axons in other organisms indicate that slow transport represents the movement of cytoskeletal and cytomatrix proteins from their site of synthesis in the neuronal cell body along the axon towards the terminals (Lasek *et al.*, 1984). In one model of this process, Lasek (1986) proposed that the cytoskeletal polymers are the vehicle for slow axoplasmic transport, and that the mechanism of movement is the sliding translocation of the polymers along the long axis of the axon. In this model, transported proteins move as components of, or in association with, these cytoskeletal elements.

For cytoskeletal polymers to slide in axons they must be relatively free to move so that the forces applied to each polymer can translocate them through the axon. If the plateau force in our stretch experiments represents the force required to induce polymer sliding in axoplasm (see Section 21.8), then we can estimate the force required to slide an individual polymer in axoplasm. We will consider only the long cytoskeletal polymers (neurofilaments and microtubules) because it is the overlap of these long elements that is mainly responsible for the longitudinal continuity and structural integrity of axoplasm, and thus for the mechanical properties of axoplasm in the longitudinal dimension. In one batch of thirteen separate stretch experiments, the mean plateau force was about 20 dynes, which corresponded to about 50 dynes/mm^2 cross-sectional area per mm length of axoplasm. To calculate the force required to slide a single polymer in axoplasm, we need to know exactly how many sliding interactions there are when axoplasm is stretched. At present we cannot determine this directly, but we can make an approximation. We (Brown and Lasek, 1989) know from electron microscopy that the density of neurofilaments and microtubules in axoplasm is about 50/μm^2 (cross-sectional area). Thus there are about 5×10^7 neurofilaments and microtubules per mm^2 in a cross-section of axoplasm. If we assume that all these polymers in the axoplasm slide relative to each other, then we can calculate that the force required to slide a 1 mm length of polymer in axoplasm would be about 1×10^{-6} dynes, which is equal to 1×10^{-11} Newtons, or 10 pN (pico Newtons).

The force generated by a single dynein cross-bridge during microtubule sliding in cilia is about 1 pN (Kamimura and Takahashi, 1981), and the force generated by a single myosin head during the sliding of actin and myosin filaments in muscle contraction has been estimated to be about 5 pN (Oplatka, 1972). Thus, based on the above assumptions, the force necessary to slide cytoskeletal polymers in axoplasm is low enough to be generated by a small number of myosin or dynein-like cross-bridges per millimeter length of polymer (George and Lasek, 1986). Such calculations are very approximate but they are important in order to illustrate the magnitude of the plateau force in biological terms. They show that the forces required to cause polymer sliding in axoplasm are within the range that could be generated by known biological force-generating proteins. Further studies may allow us to define the assumptions more precisely and obtain more rigorous estimates. By studying the resistance to polymer sliding in axoplasm under different conditions, the stretch studies may allow us to identify factors that determine the rate of polymer sliding in the axon and thereby affect the rate of slow axoplasmic transport *in vivo*.

22. Summary

Ultimately we are most interested in understanding the axons of mammals and especially those of man, but often it is difficult or impossible to do the kinds of experiments that we would like to on these organisms. The squid giant axon is especially useful for studies on the cytoskeleton of axons for three reasons. First and foremost is its large size, especially in the radial dimension. This allows biologists to conduct experiments that would be difficult or impossible with smaller cells. Second is the large amount of structural and biochemical information that is now available for the cytoskeleton of the giant axon, which rivals that of any other kind of axon. Thirdly, information obtained for the squid giant axon can lead directly to understanding other kinds of axons, notably those of man.

Seven hundred million years of evolutionary change separate squid from man, and during this time some of the proteins in axons have apparently diverged considerably. Perhaps the most notable differences are for the neurofilament proteins, which differ greatly in both their size and number. In fact, we expect that both man and squid have some new proteins that were not present in the axons of their common ancestral precursor, though no cytoskeletal proteins have yet been identified that are not shared by both organisms. With this cautionary note about the differences between the neurons of squid and humans clearly in mind, we find that at the level of the structure, composition, and organization of axonal cytoskeletons, squid and humans are more alike than they are different, and we can be fairly confident that information obtained about the organization and dynamics of cytoskeletal elements in the squid giant axon will continue to be very useful for understanding the biology of other axons, including our own, whether they are large or small.

DEDICATION. This article is dedicated to the memory of Peter F. Baker F.R.S. (deceased March 1987) for his outstanding contributions to experimental physiology and to the use of the squid as an experimental animal. A.B. is grateful to have worked in his laboratory at the Marine Biological Laboratory in Plymouth, England, for two squid seasons shortly before his death.

References

Abercrombie, R. F., Gammeltoft, K., Jackson, J., and Young, L., 1986, An intracellular calcium-binding site on neurofilament proteins of *Myxicola* giant axon. *J. Gen. Physiol.* **88**:9a.

Adelman, W. J., Jr. and Gilbert, D. L., 1990, Electrophysiology and biophysics of the squid axon, *this volume*.

Adelman, W. J. Jr., Moses, J., and Rice, R. V., 1977, An anatomical basis for the resistance in series with the excitable membrane of the squid giant axon. *J. Neurocytol.* **6**:621–646.

Aebi, U., Fowler, W. E., Rew, P., and Sun, T.-T., 1983, The fibrillar structure of keratin unravelled, *J. Cell Biol.* **97**:1131–1143.

Alema, S., Calissano, P., Rusca, G., and Giuditta, A., 1973, Identification of a calcium-binding brain-specific protein in the axoplasm of squid giant axons, *J. Neurochem.* **20**:681–689.
Allen, R. D. and Allen, N. S., 1983, Video-enhanced microscopy with a computer frame memory, *J. Microscopy* **129**:3–17.
Andrews, A. T., 1981, *Electrophoresis*, Oxford University Press, New York.
Armstrong, C. M., Bezanilla, F., and Rojas, E., 1973, Destruction of sodium conductance inactivation in squid axons perfused with pronase, *J. Gen. Physiol.* **62**:375–391.
Arnold, J. M., Gilbert, D. L., Daw, N. W., Summers, W. C., Manalis, R. S., and Lasek, R. J., 1974, *A Guide to the Laboratory Use of the Squid Loligo pealei*, Marine Biological Laboratory, Woods Hole, Massachusetts.
Baer, E., Cassidy, J. J., and Hiltner, A., 1986, Hierarchical structure of collagen and its relationship to the physical properties of tendon, in: *Collagen: Biochemistry, Biotechnology and Molecular Biology* (M. Nimni, ed.), CRC Press, Inc..
Baker, P. F. and Crawford, A. C., 1972, Mobility and transport of magnesium in squid axons, *J. Physiol.* **227**:855–874.
Baker, P. F. and Schlaepfer, W. W., 1978, Uptake and binding of calcium by axoplasm isolated from giant axons of *Loligo* and *Myxicola*, *J. Physiol.* **276**:103–125.
Baker, P. F., Hodgkin, A. L., and Shaw, T. I., 1962, Replacement of the axoplasm of giant nerve fibers with artificial solutions, *J. Physiol.* **180**:424–438.
Baumgold, J., Terakawa, S., Iwasa, K., and Gainer, H., 1981, Membrane-associated cytoskeletal proteins in squid giant axons, *J. Neurochem.* **36**:759–764.
Bear, R. S., Schmitt, F. O., and Young, J. Z., 1937a, The ultrastructure of nerve axoplasm, *Proc. Roy. Soc. Ser. B* **123**:505–519.
Bear, R. S., Schmitt, F. O., and Young, J. Z., 1937b, Investigations on the protein constituents of nerve axoplasm, *Proc. Roy. Soc. Ser. B* **123**:520–529.
Berl, S., Puszkin, S., and Nicklas, W. J., 1973, Actomyosin-like protein in brain, *Science* **179**:441–446.
Bicher, H. and Ohki, S., 1972, Intracellular pH electrode experiments on the squid giant axon, *Biochim. Biophys. Acta* **255**:900–904.
Blaustein, M. P., Russell, J. M., and De Weer, P., 1974, Calcium efflux from internally dialysed squid axons, *J. Supramol. Struct.* **2**:558–581.
Boron, W. F. and DeWeer, P., 1976, Intracellular pH transients in squid giant axons caused by CO_2, NH_3, and metabolic inhibitors, *J. Gen. Physiol.* **67**:91–112.
Brady, S. T., 1989, *personal commun.*
Brady, S. T. and Lasek, R. J., 1989, *unpub. observ.*
Brady, S. T., Lasek, R. J., Allen, R. D., Yin, H. L., and Stossel, T. P., 1984, Gelsolin inhibition of fast axonal transport indicates a requirement for actin microfilaments, *Nature* **310**:56–58.
Brady, S. T., Lasek, R. J., and Allen, R. D., 1985, Video microscopy of fast axonal transport in extruded axoplasm: a new model for study of molecular mechanisms, *Cell Motil. Cytoskel.* **5**:81–101.
Brinley, F. J. and Scarpa, A., 1975, Ionized magnesium concentration in axoplasm of dialysed squid axons, *FEBS Letts.* **50**:82–85.
Brown, A., 1989, *unpub. observ.*
Brown, A. and Eagles, P. A. M., 1986, Squid neurofilaments: phosphorylation and calcium-dependent proteolysis *in situ*, *Biochem. J.* **239**:191–197.
Brown, A. and Lasek, R. J., 1989, *in prep.*
Burton, R. R., and Paige, J. L., 1981, Polarity of axoplasmic microtubules in the olfactory nerve of the frog, *Proc. Nat. Acad. Sci. USA* **78**:3269–3273.
Caldwell, P. C., 1958, Studies on the internal pH of large muscle and nerve fibers, *J. Physiol.* **142**:22–62.
Caldwell, P. C., 1960, The phosphorus metabolism of squid axons and its relation to the active transport of sodium, *J. Physiol.* **152**:561–590.
Carden, M. J., Schlaepfer, W. W., and Lee, V. M.-Y., 1985, The structure, biochemical properties, and immunogenicity of neurofilament peripheral regions are determined by their phosphorylation state, *J. Biol. Chem.* **260**:9805–9817.

Chambers, R. and Kao, C.-Y., 1952, The effect of electrolytes on the physical state of the nerve axon of the squid and of stentor, a protozoan, *Exp. Cell Res.* **3**:564–573.

Cohen, R. S., Pant, H. C., House, S., and Gainer, H., 1987, Biochemical and immunocytochemical characterization and distribution of phosphorylated and nonphosphorylated subunits of neurofilaments in squid giant axon and stellate ganglion, *J. Neurosci.* **7**:2056–2074.

Deffner, G. G. J., 1961, The dialyzable free constituents of squid blood; a comparison with nerve axoplasm, *Biochim. Biophys. Acta* **47**:378.

De Weer, P., 1976, Axoplasmic free magnesium levels and magnesium extrusion from squid giant axons, *J. Gen. Physiol.* **68**:159–178.

DiPolo, R., Requena, J., Brinley, F. J., Mullins, L. J., Scarpa, A., and Tiffert, T., 1976, Ionized calcium concentration in squid giant axons, *J. Gen. Physiol.* **67**:433–467.

DiPolo, R., Rojas, H., Vergara, J., Lopez, R., and Caputo, C., 1983, Measurement of intracellular ionized calcium in squid axons using calcium-selective electrodes, *Biochim. Biophys. Acta* **728**:311–318.

Eagles, P. A. M., Gilbert, D. S., and Maggs, A., 1980, Neurofilament structure and enzymic modification, *Biochem. Soc. Trans.* **8**:484–487.

Eagles, P. A. M., Gilbert, D. S., and Maggs, A., 1981, The polypeptide composition of axoplasm and of neurofilaments from the marine fanworm *Myxicola infundibulum*, *Biochem. J.* **199**:89–100.

Fath, K. R. and Lasek, R. J., 1988, Two classes of actin microfilaments are associated with the inner cytoskeleton of axons, *J. Cell Biol.* **107**:613–621.

Fath, K. R. and Lasek, R. J., 1989, *unpub. observ.*

Friefelder, D. M., 1985, *Principles of Physical Chemistry with Applications to the Biological Sciences*, 2nd ed., Jones and Bartlett Publishers, Inc., Boston, MA, pp. 1–809.

Gainer, H., Tasaki, I., and Lasek, R. J., 1977, Evidence for the glia-neuron protein transfer hypothesis from intracellular perfusion studies of squid giant axons, *J. Cell Biol.* **74**:524–530.

Gainer, H., Gallant, P. E., Gould. R. M., and Pant, H. C., 1984, Biochemistry and metabolism of the squid giant axon, in: *Current Topics in Membranes and Transport*, vol. **22** (P. F. Baker, ed.), pp. 57–90, Academic Press, Inc. (London) Ltd.

Gallant, P. E., Pant, H. C., Pruss, R. M., and Gainer, H., 1986, Calcium-activated proteolysis of neurofilament proteins in the squid giant neuron, *J. Neurochem.* **46**:1573–1581.

Geisler, N., Kaufmann, G., Fischer, S., Plessmann, U., and Weber, K., 1983, Neurofilament architecture combines structural principles of intermediate filaments with carboxy-terminal extensions of increasing size between triplet proteins, *EMBO J.* **2**:1295–1302.

George, E. B. and Lasek, R. J., 1986, Rheology of extruded axoplasm: evidence for polymer sliding, *Biol. Bull.* **171**:469.

George, E. B. and Lasek, R. J., 1989, *in prep.*

Gilbert, D. S., 1972, Helical structure of *Myxicola* axoplasm, *Nature New Biol.* **237**:195–198.

Gilbert, D. S., 1975a, Axoplasm architecture and physical properties as seen in the *Myxicola* giant axon, *J. Physiol.* **253**:257–301.

Gilbert, D. S., 1975b, Axoplasm chemical composition in *Myxicola* and solubility properties of its structural proteins, *J. Physiol.* **253**:303–319.

Gilbert, D. S., 1976, Neurofilament rings from giant axons, *J. Physiol.* **266**:81–83P.

Gilbert, D. S., Newby, B. J., and Anderton, B. H., 1975, Neurofilament disguise, destruction and discipline, *Nature* **256**:586–589.

Gilbert, S. P. and Sloboda, R. D., 1986, Identification of a MAP-2-like ATP-binding protein associated with axoplasmic vesicles that translocate on isolated microtubules, *J. Cell Biol.* **103**:947–956.

Glauert, A. M., 1975, *Fixation, Dehydration and Embedding of Biological Specimens*, Elsevier North-Holland Biomedical Press, pp. 1–207.

Goodman, S. R. and Zagon, I. S., 1986, The neural cell spectrin skeleton: a review, *Am. J. Physiol.* **250**:C347–C360.

Gould, R. M. and Alberghina, M., 1990, Lipid metabolism in the squid nervous system, *this volume.*

Head, J. F. and Kaminer, B., 1980, Calmodulin from the axoplasm of the squid, *Biol. Bull.* **159**:485.

Heidemann, S. R., Landers, J. M., and Hamborg, M. A., 1981, Polarity orientation of axonal microtubules, *J. Cell Biol.* **91**:661–665.

Henkart, M., Reese, T. S., and Brinley, F. J., 1978, Endoplasmic reticulum sequesters calcium in the squid giant axon, *Science* **202**:1300–1303.

Heriot, K., Gambetti, P., and Lasek, R. J., 1985, Proteins transported in slow components a and b of axonal transport are distributed differently in the transverse plane of the axon, *J. Cell Biol.* **100**:1167–1172.

Hodge, A. J. and Adelman, W. J. Jr., 1980, The neuroplasmic network in *Loligo* and *Hermissenda* neurons, *J. Ultrastruct. Res.* **70**:220–241.

Hodgkin, A. L. and Katz, B., 1949, The effect of calcium on the axoplasm of giant nerve fibers, *J. Exp. Biol.* **26**:292–294.

Hodgkin, A. L., 1958, Ionic movements and electrical activity in giant nerve fibers, *Proc. Roy. Soc. Ser. B.* **148**:1–37.

Hollenbeck, P. J., 1988, Kinesin: its properties and possible functions, *Protoplasma* **145**:145–152.

Hoskin, F. C. G. and Rosenberg, P., 1965, Penetration of sugars, steroids, amino acids and other organic compounds into the interior of the squid giant axon, *J. Gen. Physiol.* **49**:47.

Huneeus, F. C. and Davison, P. F., 1970, Fibrillar proteins from squid axons. I. Neurofilament protein, *J. Mol. Biol.* **52**:415.

Huxley, H., 1963, Electron microscope studies on the structure of native and synthetic protein filaments from striated muscle, *J. Mol. Biol.* **7**:281–308.

Julien, J.-P. and Mushynski, W. E., 1982, Multiple phosphorylation sites in mammalian neurofilament polypeptides, *J. Biol. Chem.* **257**:10467–10470.

Kamimura, S. and Takahashi, K., 1981, Direct measurement of the force of microtubule sliding in flagella, *Nature* **293**:566–568.

Kastelic, J. and Baer, E., 1980, Deformation in tendon collagen, in: *Soc. Exp. Biol. Symp.* **34**, *The mechanical properties of biological materials,* (J. F. V. Vincent and J. D. Currey, eds.), Soc. Exp. Biol., Great Britain.

Kobayashi, T., Tsukita, S., Tsukita, S., Yamamoto, Y., and Matsumoto, G., 1986, Subaxolemmal cytoskeleton in squid giant axon. I. Biochemical analysis of microtubules, microfilaments, and their associated high-molecular weight proteins, *J. Cell Biol.* **102**:1699–1709.

Koechlin, B. A., 1955, On the chemical composition of squid giant nerve fibers with particular reference to its ion pattern, *J. Biophys. Biochem. Cytol.* **1**:511–529.

Krishnan, N. and Singer, M., 1974, Localization of cations in the peripheral nerve fiber by the pyroantimonate method, *Exp. Neurol.* **42**:191–205.

Krishnan, N., Kaiserman-Abramof, I. R., and Lasek, R. J., 1979, Helical substructure of neurofilaments isolated from *Myxicola* and squid giant axons, *J. Cell Biol.* **82**:323–335.

Kuznetsov, S. A. and Gelfand, V. I., 1986, Bovine brain kinesin is a microtubule-activated ATPase, *Proc. Nat. Acad. Sci. USA* **83**:8530–8534.

Laemmli, U. K., 1970, Cleavage of structural proteins during assembly of the head of bacteriophage T4, *Nature* **227**:680–685.

Larrabee, M. G. and Brinley, F. J., 1968, Incorporation of labelled phosphate into phospholipids in squid giant axons, *J. Neurochem.* **15**:533–546.

Lasek, R. J., 1984, The structure of axoplasm, in: *Current Topics in Membranes and Transport,* vol. **22** (P. F. Baker, ed.), pp. 39–53, Academic Press, Inc. (London) Ltd..

Lasek, R. J., 1986, Polymer sliding in axons, *J. Cell Sci. [Suppl.]* **5**:161–179.

Lasek, R. J., 1988, Studying the intrinsic determinants of neuronal form and function, in: *Intrinsic Determinants of Neuronal Form and Function* (Lasek, R.J. and M. M. Black, eds.), pp. 1–58, Alan R. Liss, Inc., New York.

Lasek, R. J., 1989, *unpub. observ.*

Lasek, R. J. and Brady, S. T., 1985, Attachment of transported vesicles to microtubules in axoplasm is facilitated by AMP-PNP, *Nature* **316**:645–647.

Lasek, R. J. and Miller, R. H., 1986, How can vesicles move freely through the filamentous matrix that surrounds the microtubules in axons? in: *Microtubules and Microtubule Inhibitors* (M. DeBrabander, and J. DeMay, eds.), pp. 197–204, Elsevier, Amsterdam.

Lasek, R. J., Krishnan, N., and Kaisermann-Abramof, I. R., 1979, Identification of the subunit proteins of 10 nm neurofilaments isolated from axoplasm of squid and *Myxicola* giant axons, *J. Cell Biol.* **82**:336–346.

Lasek, R. J., Metuzals, J., and Kaiserman-Abramof, I. R., 1983, Cytoskeletons reconstituted *in vitro* indicate that neurofilaments contribute to the helical structure of axons. In *Developing and Regenerating Nervous Systems*, pp. 1–18, Alan R. Liss, Inc., NY.

Lasek, R. J., Garner, J. A., and Brady, S. T., 1984, Axonal transport of the cytoplasmic matrix, *J. Cell Biol.* **99**:212s–221s.

Levine, J. and Willard, M., 1981, Fodrin: axonally transported polypeptides associated with the internal periphery of many cells, *J. Cell Biol.* **90**:631–643.

Martin, R., 1965, On the structure and embryonic development of the giant fiber system of the squid *Loligo vulgaris*, *Z. Zellforsch. Mikrosk. Anat.* **67**:77–85.

Martin, R., 1969, The structural organization of the intracerebral giant fiber system of cephalopods. The chiasma of the first order giant axons, *Z. Zellforsch. Mikrosk. Anat.* **97**:50–68.

Matsumoto, G. and Sakai, H., 1979a, Microtubules inside the plasma membrane of squid giant axons and their possible function, *J. Membrane Biol.* **50**:1–14.

Matsumoto, G. and Sakai, H., 1979b, Restoration of membrane excitability of squid giant axons by reagents activating tyrosine-tubulin ligase, *J. Membrane Biol.* **50**:15–22.

Matsumoto, G., Kobayashi, T. and Sakai, H., 1979, Restoration of excitability of squid giant axon by tubulin-tyrosine ligase and microtubule proteins, *J. Biochem.* **86**:1155–1158.

Matsumoto, G., Ichikawa, M., Tasaki, A., Murofushi, H., and Sakai, H., 1984, Axonal microtubules necessary for generation of sodium current in squid giant axons: I. Pharmacological study on sodium current and restoration of sodium current by microtubule proteins and 260 K protein, *J. Membrane Biol.* **77**:77–91.

McColl, J. D. and Rossiter, R. J., 1951, Lipids of the nervous system of the squid *Loligo pealei*, *J. Exp. Biol.* **28**:116–124.

McDonald, K., 1984, Osmium ferricyanide fixation improves microfilament preservation and membrane visualization in a variety of animal cell types, *J. Ultrastruct. Res.* **86**:107–118.

McGuinness, T., Brady, S. T., Gruner, J., Sugimori, M., Llinas, R., and Greengard, P., 1989, Phosphorylation-dependent inhibition by synapsin I of organelle movement in squid axoplasm, *J. Neurosci., in press.*

Metuzals, J., 1969, Configuration of a filamentous network in the axoplasm of the squid (*Loligo pealei L.*) giant nerve fiber, *J. Cell Biol.* **43**:480–505.

Metuzals, J. and Izzard, C. S., 1969, Spatial patterns of threadlike elements in the axoplasm of the giant nerve fiber of the squid (*Loligo pealei L.*) as disclosed by differential interference microscopy and by electron microscopy, *J. Cell Biol.* **43**:456–479.

Metuzals, J. and Tasaki, I., 1978, Subaxolemmal filamentous network in the giant nerve fiber of the squid (*Loligo pealei*) and its possible role in excitability, *J. Cell Biol.* **78**:597–621.

Metuzals, J., Montpetit, V. and Clapin, D. F., 1981a, Organization of the neurofilamentous network, *Cell Tiss. Res.* **214**:455–482.

Metuzals, J., Tasaki, I., Terakawa, S., and Clapin, D. F., 1981b, Removal of the Schwann cell sheath from the giant nerve fiber of the squid: an electron microscopic study of the axolemma and associated structures, *Cell Tiss. Res.* **221**:1–15.

Metuzals, J., Hodge, A. J., Lasek, R. J., and Kaiserman-Abramof, I. R., 1983, Neurofilamentous network and filamentous matrix preserved and isolated by different techniques from squid giant axon, *Cell Tiss. Res.* **228**:415–432.

Miller, R. H. and Lasek, R. J., 1985, Cross-bridges mediate anterograde and retrograde vesicle transport along microtubules in squid axoplasm, *J. Cell Biol.* **101**:2181–2193.

Mitchison, T. and Kirschner, M., 1984, Dynamic instability of microtubule growth, *Nature* **312**:237–242.

Morris, J. R. and Lasek, R. J., 1982, Stable polymers of the axonal cytoskeleton: the axoplasmic ghost, *J. Cell Biol.* **92**:192–198.

Morris, J. R. and Lasek, R. J., 1984, Monomer-polymer equilibria in the axon: direct measurement of tubulin and actin as polymer and monomer in axoplasm, *J. Cell Biol.* **98**:2064–2076.

Murachi, T., 1983, Calpain and calpastatin, *Trends Biochem. Sci.* **8**:167–169.

Murofushi, H., Minami, Y., Matsumoto, G., and Sakai, H., 1983, Bundling of microtubules *in vitro* by a high molecular weight protein prepared from the squid axon, *J. Biochem.* **93**:639–650.

Napolitano, E. W., Pachter, J. S., Chin, S. S. M., and Liem, R. K. H., 1985, Beta-internexin, a ubiquitous intermediate filament-associated protein, *J. Biol. Chem.* **101**:1323–1331.

Nixon, R. A., Lewis, S. E., and Marotta, C. A., 1987, Post-translational modification of neurofilament proteins by phosphate during axoplasmic transport in retinal ganglion cell neurons, *J. Neurosci.* **7**:1145–1158.

Oblinger, M. M., 1988, Characterization of post-translational processing of mammalian high molecular weight neurofilament protein *in vivo*, *J. Neurosci.* **7**:2510–2521.

Oplatka, A., 1972, On the mechanochemistry of muscle contraction, *J. Theor. Biol.* **34**:379–403.

Pachter, J. and Liem, R. K. H., 1985, Alpha-internexin, a 66 kD intermediate filament binding protein from mammalian CNS tissues, *J. Biol. Chem.* **101**:1316–1322.

Pachter, J., Liem, R. K. H., and Shelanski, M., 1984, The neuronal cytoskeleton, *Adv. Cell. Neurobiol.* **5**:113–142.

Pant, H. C. and Gainer, H., 1980, Properties of a calcium-activated protease in squid axoplasm which selectively degrades neurofilament proteins, *J. Neurobiol.* **11**:1–12.

Pant, H. C., Shecket, G., Gainer, H., and Lasek, R. J., 1978, Neurofilament protein is phosphorylated in the squid giant axon, *J. Cell Biol.* **78**:R23–27.

Pant, H. C., Terakawa, S., and Gainer, H., 1979a, A calcium-activated protease in squid axoplasm, *J. Neurochem.* **32**:99–102.

Pant, H. C., Yoshioka, T., Tasaki, I., and Gainer, H., 1979b, Divalent cation dependent phosphorylation of proteins in the squid giant axon, *Brain Res.* **162**:303–314.

Pant, H. C., Gallant, P. E., Gould, R. M., and Gainer, H., 1982, Distribution of calcium-activated protease activity and endogenous substrates in the squid nervous system, *J. Neurosci.* **2**:1578–1587.

Pant, H. C., Gallant, P. E., and Gainer, H., 1986, Characterization of a cyclic-nucleotide- and calcium-independent neurofilament protein kinase activity in axoplasm from the squid giant axon, *J. Biol. Chem.* **261**:2968–2977.

Pratt, M. M., 1986, Stable complexes of axoplasmic vesicles and microtubules: protein composition and ATPase activity, *J. Cell Biol.* **103**:957–968.

Price, R. L. and Lasek, R. J., 1989, *unpub. observ.*

Pruss, R. M., Mirsky, R., Raff, M. C., Thorpe, R., Dowding, A. J., and Anderton, B. H., 1981, All classes of intermediate filaments share a common antigenic determinant defined by a monoclonal antibody, *Cell* **27**:419–428.

Robertson, J. D., 1949, Ionic regulation in some marine invertebrates, *J. Exp. Biol.* **26**:182–200.

Rojas, E. and Rudy, B., 1976, Destruction of the sodium conductance inactivation by a specific protease in perfused nerve fibers from *Loligo*, *J. Physiol.* **262**:501–531.

Roots, B. I., 1983, Neurofilament accumulation induced in synapses by leupeptin, *Science* **221**:971–972.

Rosenberg, P., 1981, The squid giant axon: methods and applications, in: *Methods in Neurobiology*, vol. **1** (R.Lahue, ed.), pp. 1–133, Plenum Press, New York.

Rosenberg, P. and Hoskin, F. C. G., 1963, Demonstration of increased permeability as a factor in the effect of acetylcholine on the electrical activity of venom-treated axons, *J. Gen. Physiol.* **46**:1065.

Roslansky, P. F., Cornell-Bell, A., Rice, R. V., and Adelman Jr., W. J., 1980, Polypeptide composition of squid neurofilaments, *Proc. Nat. Acad. Sci. USA* **77**:404–408.

Rubinson, K. A. and Baker, P. F., 1979, The flow properties of axoplasm in a defined chemical environment: influence of anions and calcium, *Proc. Roy. Soc. Lond. Ser. B* **205**:323–345.

Sakai, H. and Matsumoto, G., 1978, Tubulin and other proteins from squid giant axon, *J. Biochem.* **83**:1413–1422.

Sakai, H., Matsumoto, G., and Murofushi, H., 1985, Role of microtubules and axolinin in membrane excitation of the squid giant axon, *Adv. Biophys.* **19**:43–89.

Sato, M., Wong, T. Z., Brown, D. T., and Allen, R. D., 1984, Rheological properties of living cytoplasm: a preliminary investigation of squid axoplasm (*Loligo pealei*), *Cell Mot. Cytoskel.* **4**:7–23.

See, Y-P. and Metuzals, J., 1976, Purification and characterization of squid brain myosin, *J. Biol. Chem.* **251**:7682–7689.

Sevcik, C. and Narahashi, T., 1975, Effects of proteolytic enzymes on ionic conductances of squid axon membranes, *J. Membrane Biol.* **24**:329–339.

Shaw, G., Osborn, M., and Weber, K., 1986, Reactivity of a panel of neurofilament antibodies on phosphorylated and dephosphorylated neurofilaments, *Eur. J. Cell Biol.* **42**:1–9.

Silver, F. H., 1987, *Biological Materials: Structure, Mechanical Properties, and Modelling of Soft Tissues*, New York Univ. Press, New York, pp. 1–228.

Spyropoulos, C. S., 1960, Cytoplasmic pH of nerve fibers, *J. Neurochem.* **5**:185–194.

Steinert, P. M., Zackroff, R., Aynardi-Whitman, M., and Goldman, R. D., 1982, Isolation and characterization of intermediate filaments, *Methods in Cell Biology, part A* **24**:399–419.

Studier, F. W., 1973, Analysis of bacteriophage T7 early RNAs and proteins on slab gels, *J. Mol. Biol.* **79**:237–248.

Tasaki, I. and Takenaka, T., 1964, Effects of various potassium salts and proteases upon excitability of intracellularly perfused squid giant axons, *Proc. Nat. Acad. Sci. U.S.A.* **52**:804–810.

Tasaki, I., Singer, I., and Takenaka, T., 1965, Effects of external and internal environment on excitability of squid giant axons: a macromolecular approach, *J. Gen. Physiol.* **48**:1095–1123.

Towbin, H., Staehelin, T., and Gordon, T., 1979, Electrophoretic transfer of proteins from polyacrylamide gels to nitrocellulose sheets; procedure and some applications, *Proc. Nat. Acad. Sci.* **76**:4350–4353.

Tsukita, S., Tsukita, T., Kobayashi, T., and Matsumoto, G., 1986, Subaxolemmal cytoskeleton in squid giant axon. II. Morphological identification of microtubule- and microfilament-associated domains of axolemma, *J. Cell Biol.* **102**:1710–1725.

Vale, R. D., 1987, Intracellular transport using microtubule-based motors, *Ann. Rev. Cell Biol.* **3**:347–378.

Vale, R. D., Reese, T. S., and Sheetz, M. P., 1985, Identification of a novel force-generating protein, kinesin, involved in microtubule-based motility, *Cell* **42**:39–50.

Vale, R. D., Schnapp, B. J., Mitchison, T., Steuer, E., Reese, T. S., and Sheetz, M. P., 1985, Different axoplasmic proteins generate movement in opposite directions along microtubules *in vitro*, *Cell* **43**:623–632.

Vale, R. D., Schnapp, B. J., Reese, T. S., and Sheetz, M. P., 1985, Organelle, bead, and microtubule translocations promoted by soluble factors from the squid giant axon, *Cell* **40**:559–569.

Villegas, G. M. and Villegas, R., 1984, Squid axon ultrastructure. in: *Current Topics in Membranes and Transport*, vol. **22** (P. F. Baker, ed.), pp. 3–37, Academic Press, Inc. (London) Ltd..

von Hippel, P. H. and Wong, K-Y., 1964, Neutral salts: the generality of their effects on the stability of macromolecular conformations, *Science* **145**:577–510.

Wais-Steider, C., Eagles, P. A. M., Gilbert, D. S., and Hopkins, J., 1983, Structural similarities and differences amongst neurofilaments, *J. Mol. Biol.* **165**:393–400.

Wais-Steider, C., White, N. S., Gilbert, D. S., and Eagles, P. A. M., 1988, X-ray diffraction patterns from microtubules and neurofilaments in axoplasm, *J. Mol. Biol.* **197**:205–218.

Weiss, D. G., Langford, G., Seitz-Tutter, D., and Keller, F., 1988, Dynamic instability and motile events of native microtubules from squid axoplasm, *Cell Motil. Cytoskel.*, **10**:1–11.

Weiss, D. G., Meyer, M. A., and Langford, G. M., 1990, Studying axoplasmic transport by video microscopy and using the squid giant axon as a model system, *this volume*.

Young, J. Z., 1936, The giant nerve fibers and epistellar body of cephalopods, *Quarterly J. Microsc. Sci.* **78**:367–386.
Young, J. Z., 1939, Fused neurons and synaptic contacts in the giant nerve fibers of cephalopods, *Phil. Trans. Roy. Soc. Ser. B* **229**:465–503.
Zackroff, R. V. and Goldman, R. D., 1980, *In vitro* reassembly of squid brain intermediate filaments (neurofilaments): purification by assembly-disassembly, *Science* **208**:1152–1155.
Zwahlen, M. J., Sandri, C., and Greef, N. G., 1988, Transglial pathway of diffusion in the Schwann sheath of the squid giant axon, *J. Neurocytol.* **17**:145–159.

Chapter 15

Studying Axoplasmic Transport by Video Microscopy and Using the Squid Giant Axon as a Model System

DIETER G. WEISS, MONICA A. MEYER, and
GEORGE M. LANGFORD

1. Present status of the results obtained by the use of intact axons and membrane-free or cell-free preparations

Microtubules are directly involved in many basic processes of eukaryotic cell function including division, growth, and locomotion of cells, and organelle transport. As a consequence, the study of microtubule function has become a major field of cell biology. The squid giant axon has recently been introduced as the ideal system to study microtubules and microtubule-associated motility such as organelle transport. At the same time this preparation made axoplasmic transport one of the most important paradigms for the study of all microtubule-based motility.

It was Robert D. Allen who made this progress possible by developing the Allen video-enhanced contrast differential interference contrast microscope (AVEC-DIC) (Allen, Allen, and Travis, 1981; Allen, Travis, *et al.*, 1981; Allen, 1985) to directly observe *living* microtubules and small organelles such as axoplasmic vesicles in intact cells (Travis *et al.*, 1983; Hayden and Allen, 1984). Allen, in collaboration with Scott Brady and Raymond Lasek, introduced the extruded squid axoplasm preparation which turned out to be an excellent model for the field of microtubule-associated motility (Allen *et al.*, 1982; Brady *et al.*, 1982). Hence, microtubule-associated events can be studied either in the intact squid giant axon, or *in vitro* using extruded axoplasm.

1.1. Studies on intact axons

The studies with intact squid giant axon and other neurons employing fluorescence microscopy and electron microscopy have yielded an enormous body of information

---

D. G. WEISS • Marine Biological Laboratory, Woods Hole, MA 02543 and Institut für Zoologie, Technische Universtät München, D-8046 Garching, Fed. Rep. Germany   M. A. MEYER • Marine Biological Laboratory, Woods Hole, MA 02543   G. M. LANGFORD • Marine Biological Laboratory, Woods Hole, MA 02543 and Department of Physiology, School of Medicine, University of North Carolina, Chapel Hill, NC 27599

on the three-dimensional arrangement of microtubules and their involvement in axoplasmic transport (Kreutzberg and Gross, 1977; Hodge and Adelman, 1980, 1983; Smith, 1980; Tsukita and Ishikawa, 1980; Schnapp and Reese, 1982; Metuzals *et al.*, 1983; Fahim *et al.*, 1985; Miller and Lasek, 1985; Tsukita *et al.*, 1986; Arai and Matsumoto, 1988). However, observation of microtubules and microtubule-associated events in living cells, while the process under study was taking place, was impossible with the exception of a few cases, because most of these high resolution techniques require fixed material. The movement of the various classes of organelles including axoplasmic vesicles could however be studied in detail in the intact squid giant axon using video microscopy (Allen *et al.*, 1982).

1.2. Studies on native microtubules and organelles

The intact squid giant axon is a useful model system for studying axoplasmic transport, but the extruded axoplasm is a far better preparation. It closes the gap between the use of the intact axon and the purified brain microtubule preparations and opens new means for investigation.

The key feature making the giant axon superior to other systems is the fact that its axoplasm can be extruded with its structural and physiological properties remaining almost unchanged. This means that the microtubules' native microenvironment is maintained with respect to their associated and interacting soluble proteins as well as regarding the composition of the surrounding medium.

Another property of great advantage is that axoplasm contains mainly prelysosomal organelles, while mature, functioning lysosomes are absent (Holtzman and Novikoff, 1965; Holtzman *et al.*, 1973; Smith, 1980). This allows one to observe these preparations for several hours without degradation. In this respect the axoplasm differs greatly from all other types of cell-free preparations, such as brain homogenates or permeabilized cells. Contrary to the intact cell, this system also allows for the free accessibility of drugs, nucleotides, inhibitors, and other biochemical probes (Brady *et al.*, 1982, 1985). Furthermore the optical properties in slightly dissociated axoplasm are such that light microscopy, especially the newly developed technique of video microscopy, can be applied to its limits.

When squid axoplasm is extruded into a well-defined artificial axoplasmic medium (Morris and Lasek, 1982, 1984; Brady *et al.*, 1982, 1985; Gainer *et al.*, 1984; Baker, P. F., 1984), individual, native microtubules and organelles, including the axoplasmic or synaptic vesicles, separate from the bulk phase and can be observed directly by video microscopy (Brady *et al.*, 1982; Allen *et al.*, 1983, 1985; Martz *et al.*, 1984).

Based on these pioneering studies it was found that organelles move bidirectionally on one microtubule (Allen *et al.*, 1985; Schnapp *et al.*, 1985) and that microtubules themselves move actively on a glass surface (Allen and Weiss 1985; Allen *et al.*, 1985; Vale, Schnapp, Reese, and Sheetz, 1985a). The system was found to be undisturbed to the degree that motility of organelles has very similar characteristics as in intact cells (Allen *et al.*, 1985; Brady *et al.*, 1985; Weiss *et al.*, 1987).

The motility and assembly dynamics of individual native microtubules can be analyzed especially clearly in the vicinity of the bulk extruded axoplasm. The microtubule assembly and disassembly dynamics (dynamic instability) was studied with video microscopy in native axoplasmic microtubules in the presence of the natural microtubule-associated proteins (MAPs) (Seitz-Tutter *et al.*, 1988; Weiss *et al.*, 1988). For these studies, the axoplasm was kept in a small volume of artificial axoplasm solution which is not a microtubule-stabilizing buffer (such as e.g., the one introduced by Shelanski *et al.*, 1973) but one which very closely mimics the physiological conditions in the axon and allows for the presence of all endogenous MAPs (Morris and Lasek, 1982, 1984; Langford *et al.*, 1986).

Generally speaking, native microtubules observed in this semi-artificial microenvironment are considered an intermediate system to study, because the biochemical complexity of the entire cell is essentially maintained, while the three-dimensional arrangement of the cytoskeletal structures is, at least to a certain degree, maintained. The ultrastructure of native microtubules is well characterized (Hodge and Adelman, 1980, 1983; Miller and Lasek, 1985; Langford *et al.*, 1987; Wais-Steider *et al.*, 1987). We should, however, always remember that even in this system, structures may have been partially disrupted which might alter proper functioning. The properties of microtubule-based motility in this microenvironment are very similar to the properties of axoplasmic transport as known from a great number of studies on a variety of intact nerves reported in journals (Breuer *et al.*, 1975; Forman *et al.*, 1977; Smith, 1982, 1988; Weiss, 1986b; Weiss *et al.*, 1987) and in books (Weiss, 1982; Iqbal, 1986).

Another method used to prepare axoplasm for observation by video microscopy is to gently homogenize the extruded material in a special, handmade glass homogenizer (see Section 4.5 for details).

It has been shown that most of the motile properties of the microtubules are retained after mild homogenization (Allen *et al.*, 1985). This method yields many short microtubules, so that gliding can be observed much more frequently. Excessive homogenization can lead to complete inhibition of microtubule motility and therefore this paradigm has not become a popular method of preparation.

1.3. Studies on purified components and reconstituted preparations

For *in vitro* studies, well-established procedures to reassemble purified microtubule proteins from brain are used (Weisenberg, 1972; Shelanski *et al.*, 1973, Borisy *et al.*, 1975). The usefulness of highly purified microtubule systems to study microtubule function is limited by the fact that the validity of such systems, when used, has to be proven for each putative microtubule function under study, because an unknown number of cellular components required for proper function may have been extracted or destroyed during the purification process leading to a situation where the observed behavior may differ from the physiological one. However, this approach offered the unique opportunity that proteins involved in microtubule-based movements could be specified and isolated (Brady, 1985; Scholey *et al.*, 1985; Vale, Reese, and Sheetz, 1985; Pryer *et al.*, 1986; Cohn *et al.*, 1987; Paschal and Vallee, 1987; Do *et al.*, 1988; Euteneuer *et al.*, 1988). This entire field of research originated from the above-mentioned studies on the squid giant axon and later on squid optic lobes.

In parallel to purification of the enzymes involved, experiments have also been performed with purified and fluorescence-labeled synaptic vesicles. These were injected into squid giant axons and vesicle movement was observed with fluorescence microscopy (Gilbert and Sloboda, 1984; Schroer *et al.*, 1985). In a related experiment, vesicles purified from squid axoplasm were observed to translocate along flagellar microtubules (Gilbert *et al.*, 1985).

The ability of microtubules to glide over glass surfaces in the presence of adenosine-5'-triphosphate (ATP) was first observed in our native microtubule preparation

(Allen *et al.*, 1985; Weiss and Allen, 1985) and served later as the most convenient *motility assay* for putative translocator proteins in systems containing purified microtubules and the proteins to be tested.

By means of this assay and based on the findings of Lasek and Brady (1985) and Brady (1985) that organelles will bind to squid axonal microtubules in the presence of the non-hydrolyzable ATP-analog 5'-adenylyl imidodiphosphate (AMP-PNP), a novel protein was identified from squid brain which turned out to be the first of a new class of force generating enzymes (Brady, 1985; Vale, Reese, and Sheetz, 1985). This protein, subunit Mr 116 kiloDaltons (kD) and 65 kD, which is soluble in the presence of ATP but binds to microtubules in its absence was named *kinesin* because it supports the gliding movements of microtubules and the transport of carboxylated latex beads (Vale, Schnapp, Reese, and Sheetz, 1985b; Vale, Schnapp, Mitchison *et al.*, 1985). Kinesin-like proteins which have meanwhile been described from a fair number of sources are composed of two identical polypeptides of 116 – 130 kD, which contain the ATPase at the microtubule-binding site, and two identical subunits of about 65 kD (Kuznetsov and Gelfand, 1986; Bloom and Brady, 1988; Kuznetsov *et al.*, 1988). The results concerning the ATPase activity of kinesin are very contradictory but at least all authors agree that there is a several fold increase of the otherwise very low ATPase activity upon addition of microtubules (Kuznetsov and Gelfand, 1986; Scholey *et al.*, 1986; Penningroth *et al.*, 1987). The gliding velocity in kinesin-containing preparations is identical to that found in the native system and also to that of bead translocation (0.3 – 0.6 μm/s) (Allen *et al.*, 1985; Vale, Schnapp, Reese, and Sheetz, 1985a, 1985b; Vale, Schnapp, Mitchison *et al.*, 1985; Pryer *et al.*, 1986; Cohn *et al.*, 1987; Saxton *et al.*, 1988). The ultrastructure of pig brain kinesin has been described by Amos (1987).

Recently, Vallee and coworkers, also using microtubule gliding as assay system, have been able to identify one of the formerly known high molecular weight MAPs as a cytoplasmic dynein, which causes gliding of microtubules in the opposite direction as kinesin. Dyneins generally move microtubules with the (+) end frontal and at much higher velocities (1 – 9 μm/s) (Paschal and Vallee, 1987; Paschal *et al.*, 1987; Shpetner *et al.*, 1988; Vallee *et al.*, 1988; Vallee, 1989).

Although both proteins are discussed as the force-generating enzymes (motors) for anterograde (kinesin) and retrograde (dynein) axonal transport, the movement of vesicles and other organelles which is most abundant in native axoplasmic microtubules, could not yet or could only barely be induced in a reconstituted system (Vale, Schnapp, Reese, and Sheetz, 1985b). Vesicles have, however, been shown to bind to purified microtubules in the presence of kinesin when ATP is absent, and are released, when ATP is added, but without inducing any active movement (Schroer *et al.*, 1988). Crude microsomal fractions translocate along microtubules and form membranous networks (Sheetz, 1987; Dabora and Sheetz, 1988a, 1988b; Vale and Hotani, 1988).

The presently very active field of studying force-generating ATPases emerged from the use of squid axoplasmic microtubules (Hollenbeck and Brady, 1985; Vale, Reese, and Sheetz, 1985), but since biochemical work requires larger amounts of tissue, experiments were soon performed mainly on squid optic lobes, and more recently on mammalian brain tissues (Kuznetsov and Gelfand, 1986; Lye *et al.*, 1987; Paschal *et al.*, 1987; Neighbors *et al.*, 1988; Vallee *et al.*, 1989). The gene encoding the heavy chain of *Drosophila* kinesin has been cloned (Yang *et al.*, 1988).

2. Gross dissection of the squid giant axon

The gross dissection should be carried out on a squid board according to procedures outlined by Gilbert (1974), Rosenberg (1981), and Adelman and Gilbert (1990). We remove the giant axon from the squid mantle with some muscle and connective tissue still attached. Particular care should be taken to prevent stretching (Koike, 1987) and damage to the axonal sheath which may cause Ca^{2+} influx. This would result in destruction of the axoplasm, mainly by activation of Ca^{2+}-dependent proteases (Pant *et al.*, 1978; Gainer *et al.*, 1984; Gallant *et al.*, 1986). Damaged areas are conspicuous because of their constricted and opaque appearance. One should discard *such* axons. Thin sewing thread should be used to tie off the axon just distal to the stellate ganglion and proximal to the first main branch point to prevent Ca^{2+} influx from the ends. It should be noted that axons do not keep as well in Ca^{2+}-free sea water for prolonged periods of time, therefore regular sea water should be used for these steps (see details in Section 3).

3. Fine dissection of the squid giant axon

Transfer the axon to a petri dish (100 × 15 mm) with dental wax pressed firmly along the bottom inner rim, filled with Millipore™-filtered sea water. Tweezers aid in anchoring the thread in the wax on opposite sides of the dish. The axon should be taut, but under no circumstances stretched.

Further dissection requires two pairs of very fine watchmaker's tweezers (e.g., Dumont No.5), 1 pair of fine dissection scissors as used in eye surgery, the cut tip (2 cm) of a 30 gauge hypodermic needle inserted in a needle holder, and a spool of thread. The petri dish with axons should be kept on ice as much as possible.

Under a dissecting microscope, clean off all remaining muscle and connective tissue, without injury of any kind to the axonal sheath. Connective tissue and attached muscle can be removed by gently pulling fibers laterally or obliquely away from the axon with tweezers while cutting with scissors exactly between the tissue and the axon. Alternatively, use tweezers to hold muscle and connective tissue fibers to the side, while teasing them from the axon with the 30 G needle. In effect, the tip of the needle functions as a scalpel blade and is best used in long light strokes. Long, clearly separated strips of excess tissue are removed by cutting at both ends. Small branches can be seen coming off the giant axon on both sides. These should not be cut too close to their point of origin.

Long axons can be tied into smaller segments. This may be done to eliminate damaged sections (only if recognized immediately), or to minimize potential stresses on the axon during cleaning of excess connective tissue and muscle. Unused portions of intact and well-tied axons can be kept at 0 – 4 °C in the uncleaned state floating in Millipore™-filtered sea water for up to 24 hours. Only the portion of the axon being prepared for immediate use should be cleaned.

4. Preparation of the axon for light microscopy

Axoplasmic transport of lysosomal-like organelles and mitochondria can be observed using conventional light microscopy (Berlinrood *et al.*, 1972; Breuer *et al.*, 1975, Forman *et al.*, 1977, Smith, 1980, Smith, 1982; Burdwood, 1965). Axoplasmic vesicles, i.e., synaptic vesicle precursors, are 4 times smaller and microtubules 10 times smaller than the limit of resolution in classical light microscopy and can, therefore, only be observed by video microscopy (AVEC-DIC) as developed by Allen (Allen, Allen, and Travis, 1981; Allen, Travis, *et al.*, 1981; Schnapp, 1986; Weiss, 1986a; Weiss *et al.*, 1989) and by Inoué (1981, 1986). Fluorescence microscopy and specific labels (antibodies, dyes) are used when the identity of the objects which are visualized by the above techniques are to be revealed (e.g., Fath and Lasek, 1988).

4.1. Slides and holders

The specimen's region of interest should be close to the coverslip surface, where the best image is obtained. If highest magnification video microscopy is intended, it is possible that the optics can only be set for Köhler illumination at this surface and a few tens of micrometers below (upright microscope), since high magnification objectives are usually designed for optical imaging of objects at a distance of 170 µm from the front lens element. Unless you work with dry objectives (oil immersion is however preferred) the use of No. 0 coverslips (80 – 120 µm thick) instead of No. 1 (approx. 170 µm thick) is recommended (e.g., Gold Seal from Clay Adams Co., Parsippany, NJ). Similarly, thinner slides may be useful (0.8 – 0.9 instead of 1 mm).

Aqueous samples must be prevented from drying out by completely sealing the coverslip with nail polish or, if *live* specimen such as microtubules, extruded cytoplasm or cultured cells are observed, with valap. This consists of equal parts by weight of vaseline, lanolin and paraffin (51 – 53 °C) and liquefies at approximately 65 °C. It is applied around the edge of the coverslip with a cotton tip applicator. If the specimen is in suspension no more than 5 – 10 µl aliquots should be used with regular size coverslips in order to produce very thin specimen (approx. 10 µm) for best image quality.

If working with an inverted microscope, the slide should be placed on the stage with the top coverslip pointing downward. With most microscope stages this will interfere with the valap sealant and flat positioning of the slide will not be possible. It is instead recommended that a brass or stainless steel frame measuring 0.8 – 1.0 mm in thickness and equal in size to a regular slide be used. The central portion is cut out so that a U-shaped holder is obtained (Fig. 1). Now a larger coverslip (24 × 50 or 60 mm) is attached on two sides with adhesive tape on the upper side so that it covers the opening. If necessary, this should be the No. O coverslip since its top surface is the one to be observed. After the sample has been applied, the top coverslip of regular size and thickness is added and sealed.

If the preparation is to be perfused with another solution after initial mounting, 'spacers' can be added by placing one half strip of adhesive tape along the long sides of the slide or larger coverslip. A perfusion chamber can be created on the slide by leaving the central portion of the two lateral edges of the coverslip unsealed (Brady *et al.*, 1985). Solutions applied dropwise at one edge will be drawn through the preparation when the pointed tip of a thinly cut, wedge-shaped strip of filter paper is laid at

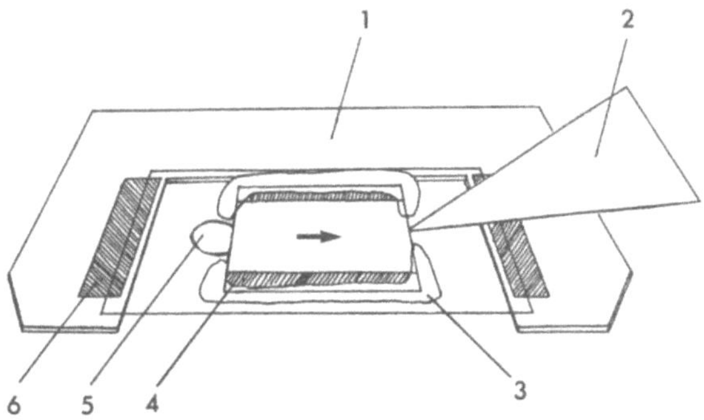

Figure 1. Slide preparation suitable for superfusion of the live axon or native axoplasm. *1*, metal frame; *2*, filter paper wick; *3*, valap sealant; *4*, adhesive tape spacer; *5*, drop of medium to replace the original medium; *6*, adhesive tape.

the opposite open edge (Fig. 1). Use qualitative #5 or #6 (fine crystalline) filter paper, as used in microbiology, to ensure very slow fluid uptake.

For the observation of intact axons, spacers are also used (one layer of scotch tape = 24 μm; No. 0 coverslips = approximately 100 μm). If thicker specimens are to be observed, only differential interference contrast or anaxial illumination techniques (Allen *et al.*, 1969; Kachar, 1985) are recommended. At the highest magnifications only the first 10 or 20 μm next to the objective front lens will yield good serially sectioned images, while the quality degrades quickly on deeper focusing into the tissue.

4.2. Solutions

The intact axons are observed in Millipore™-filtered seawater. The extruded and purified preparations require the use of artificial axoplasm solutions as originally developed by Morris and Lasek (1982, 1984). It was found that a dilution to about half the concentration increases the motile activity (Allen *et al.*, 1985; Vale, Schnapp, Reese, and Sheetz, 1985a). The increase of ATP from the endogenous 1 – 3 mM (Deffner, 1961; Gainer *et al.*, 1984) up to 20 mM was found to allow for a prolonged period during which motility can be observed (5 or more hours as compared to less than one hour) at normal speed) (Allen *et al.*, 1985). Typical examples for artificial axoplasm solutions are given in Table I.

4.3. Intact axons

The intact axon, cleaned from surrounding tissue as good as possible is a well suited system for the observation of intracellular organelle movement. Allen *et al.*, (1982) were the first to observe axoplasmic transport of submicroscopic 40 – 70 nm size vesicles using this preparation. There are very few other neurons which have the optical properties to allow for the observation of vesicles (Lynn *et al.*, 1986, Gulden *et al.*, 1988, Smith, 1988), although many other systems have been studied since then.

Table I. Composition of the most commonly used artificial axoplasm solutions*

Component	*Buffer 1*	*Buffer 2*	*Buffer 3*	*Buffer 4*
K^+-Aspartate	350	175	200	160
Taurine	130	65	37	52
Betaine	70	35	20	28
Glycine	50	25	15	20
HEPES‡	10	10	8	16
$MgCl_2$	12.9	6.45	3.7	5.2 – 20
EGTA**	10†	5	2.8	4
$CaCl_2$	3†	1.5	0.86	1.2
D-Glucose	1	0.5	0.3	0
ATP (K^+ salt)	1	2	0 – 20	0 – 20
pH (KOH)	7.2	7.2	7.2	7.2

Units are in mM, with the exception of pH.

Buffer 1: *Buffer X* (Brady *et al.*, 1985).
Buffer 2: *Half-strength Buffer X* (Vale, Schnapp, Reese, and Sheetz, 1985a).
Buffer 3: *Axoplasmic Dissociation Buffer* (Allen *et al.*, 1985).
Buffer 4: *Axoplasmic Buffer* presently used in our laboratory.

* All these artificial axoplasmic buffers are derived from analysis of axoplasm composition resulting in formulation of *Buffer P* which very closely mimics the axoplasm contents (Morris and Lasek, 1982; see also Gainer *et al.*, 1984 for review of the chemical composition of axoplasm).
† variable
‡ HEPES (N-2-hydroxyethlypiperazine-N'-2-ethanesulfonic acid)
** EGTA (ethylene glycol-bis(2-amino ethyl ether)N,N,N',N'-tetraacetic acid)

4.4. Extrusion of axoplasm

Assemble the following items for extrusion: 1. Glass slide and coverslip or coverslip mounted on brass or stainless steel slide frame and second coverslip (for use with inverted microscopes) (Fig. 1). 2. Filter paper 3. Razor blade 4. Axoplasmic buffer on ice. 5. Micropipettor 10 – 50 µl. 6. Extrusion needle: 20 G needle with Teflon or polyethylene tubing (e.g., Clay Adams # 7440, inner diameter 1.4, outer diameter 1.9 mm) which fits snugly over the needle, extending 1 cm past needle tip.

Only the portion of the axon to be used for a given preparation should be removed from the Millipore™ filtered sea water, rinsed for 1 – 2 min. in Ca^{2+}-free sea water, and gently touched to the filter paper for blotting. This should be done very briefly; then cut open the wider (ganglion) end close to the thread with a razor blade. Holding the remaining thread with tweezers, transfer the axon to a coverslip or to the prepared slide. The extrusion needle is best used with the bevel directed upward,

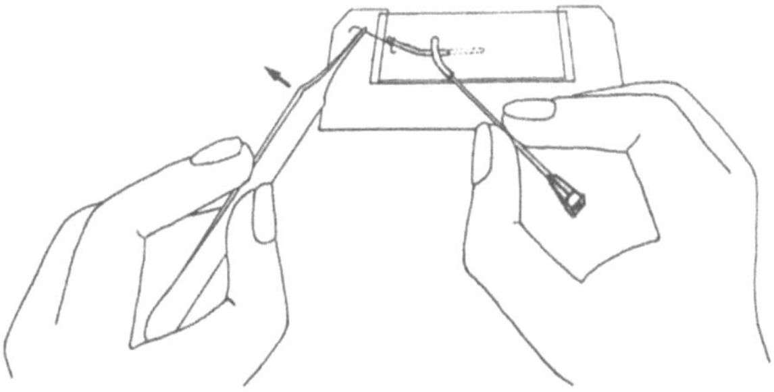

Figure 2. Schematic drawing of the extrusion process showing how to hold axon and instruments.

applying pressure just past the needle tip. It may help to glide the extrusion needle over the surface of the axon, as if to smooth it down, prior to extrusion.

Holding the axon by the thread, put gentle pressure near the threaded end with the extrusion needle, while pulling slowly on the axon until the axoplasm has been extruded in its entirety (much like a tube of toothpaste) (Fig. 2). If the axon is pulled too hard or too quickly the surrounding sheath may break before extrusion is complete. It is possible to finish the extrusion by picking up the end of the broken sheath with tweezers, and again gently pulling while applying pressure with the extrusion needle. Immediately surround the axoplasm with the aliquot of buffer previously drawn into the micropipette tip. This may now be either covered with a second coverslip and sealed with Valap for microscopic visualization, or further treated as described below.

4.5. Homogenization

If highly dissociated axoplasm is desired, extrude the axoplasm as above onto a separate large coverslip and surround with buffer. A long Pasteur pipette is used to transfer both axoplasm and 20 – 40 µl of buffer to a miniature homemade glass homogenizer. This is made from a glass tube (25 mm long, outer diameter 4 mm, inner diameter 2 mm) by carefully melting one end. The pestle is made from a 1.2 mm diameter glass rod with the tip melted until a 2 mm sphere develops. The pestle should give a good lateral fit. If required, the inner surface of the bottom can be made more hemispherical by using polishing paste for grinding with the pestle. After 4 – 10 strokes in this homogenizer, 5 – 10 µl of the homogenate are transferred with a long Pasteur pipette to the slide or coverslip. It should be noted that such a minute volume may, depending on the relative humidity, evaporate within a few seconds. It is, therefore, essential to close the preparation as quickly as possible to prevent drying or concentration of the buffer. Seal the preparation with Valap as previously described.

5. Light microscopy

Observation of axoplasmic transport of vesicles and other organelles and particularly the visualization of microtubules requires both a good knowledge of light microscopic techniques and of the principles of video microscopy. These are extensively described by Inoué (1986), by Schnapp (1986), and by Weiss *et al.* (1989). An overview on the development of video microscopy and its application to the squid

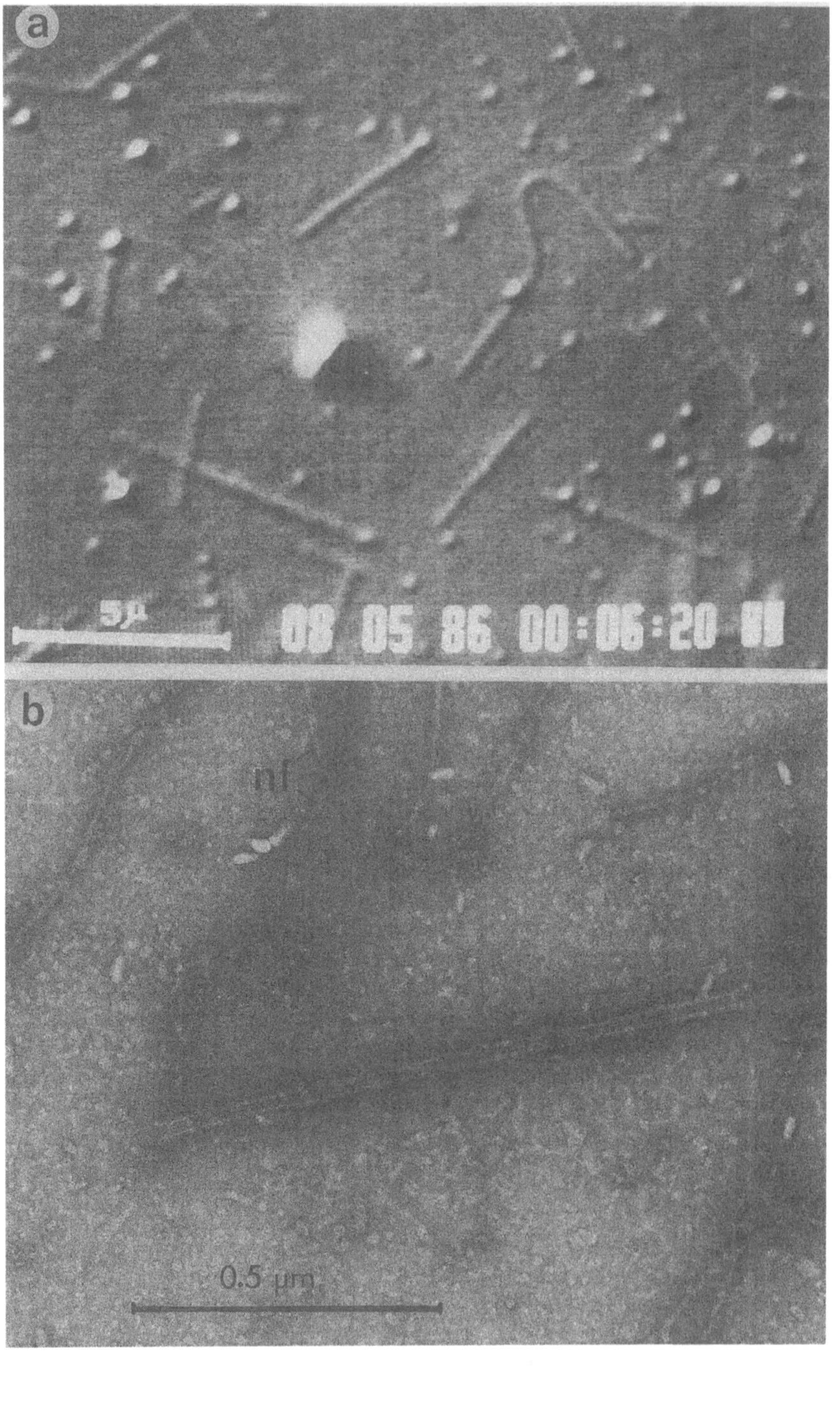
a
5μ
08 05 86 00:06:20
b
nf
0.5 μm

system is given by Allen (1987), while some of the results are reviewed by Weiss (1986a) and Shotton (1988).

The motile behavior in squid axoplasm is observed best by Allen video- enhanced contrast differential interference contrast (AVEC-DIC) microscopy (Allen, Allen, and Travis, 1981; Allen, Travis, *et al.*, 1981; Allen, 1985; Weiss, 1986a). We use the following setup with good results:

> An internally corrected 100 × planapochromatic oil immersion objective of the inverted Zeiss Axiomat, Zeiss IM, or Axiovert microscope, equipped with a 50 W mercury arc lamp is used, with additional magnification changers yielding a magnification on the television (TV) monitor (screen width 25 cm) of 10.000 ×. Real time analog contrast enhancement and digital image processing is performed with an image processor (Allen *et al.*, 1985). The procedure and the equipment required were described in detail by Weiss *et al.*, (1989). Basically the following steps are required: (a) AVEC analog contrast enhancement of the full aperture image at an instrumental compensator setting of 1/9 lambda, with gain and offset adjusted from about one-third to one-half of their respective ranges; (b) digital subtraction of the fixed pattern of mottle in the optic path; (c) manipulation of the gray scale by stretching of the pixel (video dot) brightness histogram (digital contrast enhancement); (d) in some cases, reduction of pixel noise by a real-time digital rolling average (recursive filtering) operation over two frames.

Video-enhanced microscope images can be recorded simultaneously in real time by a normal speed video cassette recorder, using broadcast quality tape such as U-matic or Super VHS (video home system), and by a time lapse recorder for display at an acceleration of approximately 10 times or by an optical disk recorder. The preparations can be studied in parallel by electron microscopy as described later in this chapter (Allen *et al.*, 1985; Schnapp *et al.*, 1985; Langford *et al.*, 1987).

For quantitative motion analysis we transfer the X-Y coordinates of particles tracked by an automatic X-Y tracker (Hamamatsu C1055, Photonic Microscopy Inc.) to an IBM-compatible microcomputer for further data analysis (Weiss *et al.*, 1986, 1988). If the contrast of the moving organelles is very low, we perform single frame analysis where the X-Y tracker is used in an interactive way to track the paths of the particles. This latter technique was found to be the only one applicable when the coordinates of microtubule ends are recorded (Seitz-Tutter *et al.*, 1988; Weiss *et al.*, 1988) (Fig. 3).

A software package that includes the generally used features needed for the analysis of time series as first developed for movie film analysis by Koles *et al.* (1982a, 1982b) was written for the analysis of video sequences of intracellular particle motion (PARTI-MOVI written in Microsoft® Pascal by Franz Keller) (Weiss *et al.*, 1986, 1988). The data sets were limited to 512 pairs of coordinates. Due to the resolution of the video system of 512 × 512 pixels, during the determination of positions, noise is inevitably introduced. Therefore, a digital low-pass filter (Butterworth type) is applied. A fast Fourier transform (FFT) analysis (Cooley and

Figure 3. a. Dispersed native axoplasm as seen in AVEC-DIC microscopy. Magnification: 6,000 ×. b. A similar preparation as in a. but prepared for negative contrast electron microscopy. Note the presence of neurofilaments (*nf*) which are not visualized by light microscopy. Magnification: 75,000 ×.

Tukey, 1965; Chatfield, 1978) is used to obtain frequential information about the translational function. Our system allows for the detection of moving organelles as well as for rapid acquisition of quantitative data (on-line in some applications) at high resolution. The temporal resolution is only limited by the video frequency (25 or 30 frames/sec). In most cases we found it useful to utilize only 10 frames/sec. Other motion analysis programs and techniques were developed by several authors (Hodge and Adelman, 1984; Cohn *et al.*, 1987; Gelles and Schnapp, 1988; Schnapp *et al.*, 1988; Smith, 1988) (Fig. 3b).

6. Preparation of the axoplasm for electron microscopy

6.1. Negative contrast electron microscopy

Negative staining is one of the highest resolution techniques in transmission electron microscopy. It yields high contrast images with resolution down to 2 – 4 nm. It is the method of choice for looking at the fine structure of microtubules and was the technique originally used by Amos and Klug (1974) to determine the tubulin subunit lattice in microtubules. We have used it extensively to visualize MAPs that decorate the outer surface of microtubules and to characterize the membranous organelles in squid axoplasm as well as the native microtubules (Langford *et al.*, 1987).

Samples are prepared for negative contrast electron microscopy by the method described by Langford (1983). A drop of the homogenized axoplasm or a similar preparation is placed on a 400 mesh copper grid that has been freshly coated with Formvar (0.25% solution in dichlorethane) and thinly sputtered with carbon. The grids are rinsed and stained simultaneously with 8 – 10 drops of 1% aqueous uranyl acetate. The excess stain is removed by touching the edge of the grid to filter paper. The specimens are observed and photographed in the electron microscope at magnifications between 5,000 and 50,000 ×.

To obtain high quality negative staining, one should use protein solutions or suspensions of 10 – 100 μg/ml. Soluble proteins which coat the grid surface improve the quality of staining by facilitating the uniform spreading of stain.

Negatively stained grids do not always stain uniformly and the quality of staining may vary. Therefore, it is important to search for areas on the grid that show the best staining with the least amount of specimen damage.

The primary drawbacks to negative staining are the flattening of large organelles caused by the severe surface tension forces that occur when the stained grids are dried. In addition the apparent diameter of negatively stained microtubules is greater than 25 nm due to their tendency to flatten somewhat during drying. The extent of flattening depends upon the composition of the solution in which the microtubules are suspended. Taxol-stabilized microtubules show a greater tendency to flatten than taxol-free microtubules, for example. Preparations in sucrose and glycerol exhibit a poorer quality of staining. Flattening can be reduced by fixing the microtubules with glutaraldehyde prior to staining. However, glutaraldehyde will produce configurational changes of filamentous proteins. MAPs, for example, coil into globular shape when fixed with glutaraldehyde (Langford, 1983).

Another drawback to negative staining is the tendency of some proteins to become positively stained by uranyl acetate so that they do not show contrast against the dark film of stain on the grid. Hence some proteins are not visible by negative staining. Uranyl acetate is the preferred stain for microtubules since phosphotungstic acid destroys them. In addition the fine structural details of microtubules that have been negatively stained by uranyl acetate are better understood.

6.2. Other forms of electron microscopy

Studies on the rapid freezing of unfixed axoplasm have been used by Schnapp *et al.* (1985) for observation of native microtubules and vesicles. The principal advantage of this technique is the preservation of macromolecular structure without the destructive effect of chemical fixation and stains. The technique is, however, limited in resolution compared to negative staining.

Other types of preservative procedures for squid axoplasm can be used for transmission electron microscopy (EM). Refer to Hodge and Adelman (1980, 1983) for procedures to fix axoplasm for thin sectioning. Miller and Lasek (1985) have also developed methods for fixing extruded axoplasm for thin sectioning.

7. Conclusion

Studies of the squid giant axon have contributed greatly to the advancement of our understanding of cytoskeletal organization and cytoplasmic dynamics in axons. These studies were possible because extruded squid axoplasm can be obtained easily, is accessible to treatment with pharmacological compounds, and is relatively free of proteolysis. Furthermore, video microscopy, which permits the direct visualization of the structures involved in organelle transport, was applied to this model system.

As a result the direct involvement of microtubules in axonal organelle transport was shown beyond any doubt, the dynamics, instability, and motility of cytoplasmic microtubules themselves were demonstrated, and a whole new class of cytoplasmic force-generating ATPases was discovered. Today, the frontier in the field of intracellular motility, which originated from studying squid axons, has already moved on to many other systems. It is to be expected that squid axoplasm will provide several more surprises to those of us trying to unravel cytoplasmic structure and function, because it is probably the most convenient animal model system currently in use for the study of the cytoplasm.

ACKNOWLEDGMENTS. This chapter is dedicated to the memory of Robert D. Allen (1927 – 1986) who not only provided the optical techniques which made the progress in the field of axonal transport possible, but also applied the techniques to squid giant axon thus contributing most of the basic findings. And it also needs to be pointed out that the progress of this field crucially depended on the scientific environment provided by the Woods Hole community of scientists and the Marine Biological Laboratory where most of the results were obtained.

We personally and gratefully acknowledge the help, advice and support from the many friends and colleagues at the Marine Biological Laboratory at Woods Hole, especially Robert D. Allen, Scott Brady, Anthony Brown, Harold Gainer, Susan Gilbert, Robert Gould, Antonio Guiditta, Raymond Lasek, Harish Pant, Dieter Seitz-Tutter, Marcia Simpson, and others, who over the years generously shared their experience, discussed their procedures with us, or critically read this manuscript. Our experimental research on which this chapter is based was supported by research grants from NSF (DCB 851 7983), NATO (30/0874/86) and Deutsche Forschungsgemeinschaft (We 790/11). Special thanks are due to Dieter Seitz-Tutter for providing the video micrograph and to Beatrix Rubin for preparing the drawings.

References

Adelman, W. J., Jr. and Gilbert, D. L., 1990, Electrophysiology and biophysics of the squid axon, *this volume*.

Allen, R. D., 1985, New observations on cell architecture and dynamics by video-enhanced contrast optical microscopy, *Ann. Rev. Biophys. Biophys Chem.* **14**:256–290.

Allen, R. D., 1987, The microtubule as an intracellular engine. *Sci. Amer.* **255**:42–49.

Allen, R. D., Allen, N. S., and Travis, J. L., 1981a, Video-enhanced contrast, differential interference contrast (AVEC-DIC) microscopy: A new method capable of analyzing microtubule-related motility in the reticulopodial network of Allogromia laticollaris, *Cell Motil.* **1**:291–302.

Allen, R. D., Brown, D. T., Gilbert, S. P., and Fujiwake, H., 1983, Transport of vesicles along filaments dissociated from squid axoplasm, *Biol. Bull.* **165**:523.

Allen, R. D., David, G. B., and Nomarski, G., 1969, The Zeiss-Nomarski differential interference equipment for transmitted-light microscopy, *Zeitschr. wiss. Mikrosk. mikro. Tech.* **69**:193–221.

Allen, R. D., Metuzals, J., Tasaki, I., Brady, S. T., and Gilbert, S. P., 1982, Fast axonal transport in squid giant axon, *Science* **218**:1127–1129.

Allen, R. D., Travis, J. L., Allen, N. S., and Yilmaz, H., 1981b, Video-enhanced contrast polarization (AVEC-POL) microscopy: A new method applied to the detection of birefringence in the motile reticulopodial network of *Allogromia laticollaris*, *Cell Motil.* **1**:275–289.

Allen, R. D. and Weiss, D. G., 1985, An experimental analysis of the mechanisms of fast axonal transport in the squid giant axon, in: *Cell Motility: Mechanism and Regulation, 10 Yamada Conference, Sept. 1984*, (H. Ishikawa, S. Hatano, and H. Sato, eds.), pp. 327–333, University of Tokyo Press, Tokyo.

Allen, R. D., Weiss, D. G., Hayden, J. H., Brown, D. T., Fujiwake, H., and Simpson, M., 1985, Gliding movement of and bidirectional transport along native microtubules from squid axoplasm: Evidence for an active role of microtubules in cytoplasmic transport, *J. Cell Biol.* **100**:1736–1752.

Amos, L. A., 1987, Kinesin from pig brain studied by electron microscopy, *J. Cell Sci.* **87**:105–111.

Amos, L. A. and Klug, A., 1974, Arrangement of subunits in flagellar microtubules, *J. Cell Sci.* **14**:523–550.

Arai, T. and Matsumoto, G., 1988, Subcellular localization of functionally differentiated microtubules in squid neurons: regional distribution of microtubule-associated proteins and β-tubulin isotypes, *J. Neurochem.* **51**:1825–1838.

Baker, P. F., (ed.), 1984, *The Squid Axon, Current Topics in Membranes and Transport. Vol. 22*, Academic Press, New York.

Berlinrood, M., McGee-Russel, S. M., and Allen R. D., 1972, Patterns of particle movement in nerve fibres *in vitro*: An analysis of photokymography and microscopy, *J. Cell Sci.* **11**:875–886.

Bloom, G. S., Wagner, M. C., Pfister, K. K., and Brady, S. T., 1988, Native structure and physical properties of bovine brain kinesin and identification of the ATP-binding subunit polypeptide, *Biochemistry* **27**:3409–3416.

Borisy, G. G., Marcum, Olmsted, J. M., Murphy, D. B., and Johnson, K. A., 1975, Purification of tubulin and associated high molecular weight proteins from porcine brain and characterization of microtubule assembly *in vitro*, *Ann. N.Y. Acad. Sci.* **253**:107–132.

Brady, S. T., 1985, A novel brain ATPase with properties expected for the fast axonal transport motor, *Nature* **317**:73–75.
Brady, S. T., Lasek, R. J., and Allen, R. D., 1982, Fast axonal transport in extruded axoplasm from squid giant axon, *Science* **218**:1129–1131.
Brady, S. T., Lasek, R. J., and Allen, R. D., 1985, Video microscopy of fast axonal transport in extruded axoplasm: A new model for study of molecular mechanisms, *Cell Motil.* **5**:81–101.
Breuer, A. C., Christian, C. N., Henkart, M., and Nelson, P. G., 1975, Computer analysis of organelle translocation in primary neuronal cultures and continuous cell lines, *J. Cell Biol.* **65**:562–576.
Burdwood, W., 1965, Bidirectional particle movement in neurons, *J. Cell Biol.* **27**:115a.
Chatfield, C., 1978, *The Analysis of Time Series*, Chapman & Hall, London.
Cohn, S. A., Ingold, A. L., and Scholey, J. M., 1987, Correlation between the ATPase and microtubule translocating activities of sea urchin kinesin, *Nature* **328**:160–1163.
Cooley, J. W. and Tukey, J. W., 1965, An algorithm for the machine calculation of complex Fourier Series, *Math. Comput.* **19**:297–301.
Dabora, S. C. and Sheetz, M. P., 1988a, The microtubule-dependent formation of a tubulo-vesicular network with characteristics of the ER from cultured cell extracts, *Cell* **54**:27–35.
Dabora, S. L. and Sheetz, M. P., 1988b, Cultured cell extracts support organelle movement on microtubules *in vitro*, *Cell Motil. Cytoskel.* **10**:482–495.
Deffner, G. G. J., 1961, The dialyzable free organic constituents of squid blood; A comparison with nerve axoplasm, *Biochim. Biophys. Acta* **47**:378–388.
Do, C. V., Sears, E. B., Gilbert, S. P., and Sloboda, R. D., 1988, Vesikin, a vesicle associated ATPase from squid axoplasm and optic lobe, has characteristics in common with vertebrate MAP1 and MAP2, *Cell Motil. Cytoskel.* **10**:246–254.
Euteneuer, U., Koonce, M. P., Pfister, K. K., and Schliwa, M., 1988, An ATPase with properties expected for the organelle motor of the giant amoeba, *Reticulomyxa*, *Nature* **332**:176–178.
Fahim, M. A., Lasek, R. S., Brady, S. T., and Hodge, A. S., 1985, AVEC-DIC and electron microscopic analyses of axonally transported particles in cold-blocked squid giant axons, *J. Neurocytol.* **14**:689–704.
Fath, K. R. and Lasek, R. J., 1988, Two classes of actin microfilaments are associated with the inner cytoskeleton of axons, *J. Cell Biol.* **107**:613–621.
Forman, D. S., Padjen, A. L., and Siggins, G. R., 1977, Axonal transport of organelles visualized by light microscopy: cinemicrographic and computer analysis, *Brain Res.* **136**:197–213.
Gainer, H., Gallant, P. E., Gould, R., and Pant, H., 1984, Biochemistry and metabolism of the squid giant axon, *Curr. Topics Membr. Transp.* **22**:57–90.
Gallant, P. E., Pant, H. C., Pruss, R. M., and Gainer, H., 1986, Calcium-activated proteolysis of neurofilament proteins in the squid giant neuron, *J. Neurochem.* **46**:1573–1581.
Gelles, J., Schnapp, B. J., and Sheetz, M. P., 1988, Tracking kinesin-driven movements with nanometre-scale precision, *Nature* **331**:450–453.
Gilbert, D. L., 1974, Physiological uses of the squid with special emphasis in the use of the giant axon, in: *A Guide to the Laboratory Use of the Squid Loligo pealei*, (Arnold, J., *et al.*, eds.), pp. 45–55, Marine Biological Laboratory, Woods Hole, MA.
Gilbert, S. P., Allen, R. D., and Sloboda, R. D., 1985, Translocation of vesicles from squid axoplasm on flagellar microtubules, *Nature* **315**:245–248.
Gilbert, S. P. and Sloboda, R. D., 1984, Bidirectional transport of fluorescently labeled vesicles introduced into extruded axoplasm of squid *Loligo pealei*, *J. Cell Biol.* **99**:445–452.
Gulden, J., Weiss, D. G., and Clasen, B., 1988, The velocity fluctuations of organelles transported in crustacean and human axons are random, *Eur. J. Cell Biol. (Suppl. 22)* **46**:24.
Hayden, J. H. and Allen, R. D., 1984, Detection of single microtubules in living cells: particle transport can occur in both directions along the same microtubule, *J. Cell Biol.* **99**:1785–1793.

Hodge, A. J. and Adelman, W. J., Jr., 1980, The neuroplasmic network in *Loligo* and *Hermissenda* neurons. *J. Ultrastruct. Res.* **70**:220–241.
Hodge, A. J. and Adelman, W. J., Jr., 1983, The neuroplasmic lattice. Structural characteristics in vertebrate and invertebrate axons, in: *Structure and Function in Excitable Cells*, (D. C. Chang, I. Tasaki, I., W. J. Adelman, Jr., and H. R. Leuchtag, eds.), pp. 75–111, Plenum Press, New York.
Hodge, A. J. and Adelman, W. J., Jr., 1984, Direct visualization of particle velocity distribution by pseudostereoscopic viewing of time-lapsed sequential images: Application to fast axonal transport, *Cell Motil.* **4**:231–239.
Hollenbeck, P. J. and Bray, D., 1985, A novel 330 kD protein and its possible role in axonal transport, in: *Microtubules and Microtubule Inhibitors* (M. De Brabander, J. De Mey, eds.), pp. 205–211, Elsevier, Amsterdam.
Holtzman, E. and Novikoff, A. B., 1965, Lysosomes in the rat sciatic nerve following crush, *J. Cell Biol.* **27**:651–669.
Holtzman, E., Teichberg, S., Abrahams, S. J., Citkowitz, E., Crain, S. M., Kawai, N., and Peterson, E. R., 1973, Notes on synaptic vesicles and related structures, endoplasmic reticulum, lysosomes and peroxisomes in nervous tissue and the adrenal medulla, *J. Histochem. Cytochem.* **21**:349–385.
Inoué, S., 1981, Video image processing greatly enhances contrast, quality, and speed in polarization-based microscopy, *J. Cell Biol.* **89**:346–356
Inoué, S., 1986, *Video Microscopy*, Plenum Press, New York.
Iqbal, Z. (ed.), 1986, *Axoplasmic Transport*, CRC Press, Boca Raton, FL.
Kachar, B., 1985, Asymmetric illumination contrast: a method of image formation for video light microscopy, *Science* **227**:766–768.
Keller, F., 1986, Computergestzte Analyse von Organellenbewegungen in Zellen dargestellt an den Axopodien von Raphidiophrys ambigua und Echinosphaerium nucleofilum, *Masters Thesis*, Faculty of Biology, University of Munich, Munich.
Koike, H., 1987, The disturbance of the fast axonal transport of protein by passive stretching of an axon in *Aplysia*, *J. Physiol.* **390**:489–500.
Koles, Z. J., McLeod, K. D., and Smith, R. S., 1982a, The determination of the instantaneous velocity of axonally transported organelles from filmed records of their motion, *Can. J. Physiol. Pharmacol.* **60**:670–679.
Koles, Z. J., McLeod, K. D., and Smith, R. S., 1982b, A study of the motion of organelles which undergo retrograde and anterograde rapid axonal transport in *Xenopus*, *J. Physiol.* **328**:469–484.
Kreutzberg, G. W. and Gross, G. W., 1977, General morphology and axonal ultrastructure of the olfactory nerve of the pike, *Esox lucius*, *Cell Tissue Res.* **181**:443–457.
Kuznetsov, S. A. and Gelfand, V. I., 1986, Bovine brain kinesin is a microtubule-activated ATPase, *Proc Natl. Acad. Sci.* **83**:8530–8534.
Kuznetsov, S. A., Vaisberg, E. A., Shanina, N. A., Magretowa, N. N., Chernyak, V. Y., and Gelfand, V. I., 1988, The quaternary structure of bovine brain kinesin, *EMBO J.* **7**:353–356.
Langford, G. M., 1983, Length and appearance of projections on neuronal microtubules *in vitro* after negative staining: evidence against a crosslinking function for MAPs, *J. Ultrastruct. Res.* **85**:1–10.
Langford, G. M., Allen, R. D., and Weiss, D. G., 1987, Substructure of sidearms on squid axoplasmic vesicles and microtubules visualized by negative contrast electron microscopy, *Cell Motil. Cytoskel.* **7**:20–30.
Langford, G. M., Williams, E., and Peterkin, D., 1986, Microtubule-associated proteins (MAPs) of dogfish brain and squid optic ganglia, *Ann. N.Y. Acad. Sci.* **466**:440–443.
Lasek, R. J. and Brady, S. T., 1985, Attachment of transported vesicles to microtubules in axoplasm is facilitated by AMP-PNP, *Nature* **316**:645–647.
Lye, R. J., Porter, M. E., Scholey, J. M., and McIntosh, J. R., 1987, Identification of a microtubule-based cytoplasmic motor in the nematode *C. elegans*, *Cell* **51**:309–318.

Lynn, M. P., Atkinson, M. B., and Breuer, A. C., 1986, Influence of translocation track on the motion of intra-axonally transported organelles in human nerve, *Cell Motil. Cytoskel.* **6**:339–346.

Martz, D., Lasek, R. J., Brady, S. T., and Allen, R. D., 1984, Mitochondrial motility in axons: membranous organelles may interact with the force generating system through multiple surface binding sites, *Cell Motil.* **4**:89–101.

Metuzals, J., Hodge, A. J., Lasek, R. J., and Kaiserman-Abramof, I. R., 1983, Neurofilamentous network and filamentous matrix preserved and isolated by different techniques from squid giant axon, *Cell Tissue Res.* **228**:415–432.

Miller, R. H. and Lasek, R. J., 1985, Crossbridges mediate anterograde and retrograde vesicle transport along microtubules in squid axoplasm, *J. Cell Biol.* **101**:2181–2193.

Morris, J. R. and Lasek, R. J., 1982, Stable polymers of the axonal cytoskeleton: The axoplasmic ghost, *J. Cell Biol.* **92**:192–198.

Morris, J. R. and Lasek, R. J., 1984, Monomer-polymer equilibria in the axon: Direct measurement of tubulin and actin as polymer and monomer in axoplasm, *J. Cell Biol.* **98**:2064–2076.

Neighbors, B. W., Williams, R. C., Jr., and McIntosh, J. R., 1988, Localization of kinesin in cultured cells, *J. Cell Biol.* **106**:1193–1204.

Pant, H. C., Terakawa, S., and Gainer, H., 1978, A calcium activated protease in squid axoplasm, *J. Neurochem.* **32**:99–102.

Paschal, B. M., Shpetner, H. S., and Vallee, R. B., 1987, MAP 1C is a microtubule-activated ATPase which translocates microtubules *in vitro* and has dynein-like properties, *J. Cell Biol.* **105**:1273–1282.

Paschal, B. M. and Vallee, R. B., 1987, Retrograde transport by the microtubule-associated protein MAP 1C, *Nature* **330**:181–183.

Penningroth, S. M., Rose, P. M., and Peterson, D. D., 1987, Evidence that the 116 kDa component of kinesin binds and hydrolyzes ATP, *FEBS Lett.* **222**:204–210.

Pryer, N. K., Wadsworth, P., and Salmon, E. D., 1986, Polarized microtubule gliding and particle saltation produced by soluble factors from sea urchin eggs and embryos, *Cell Motil. Cytoskel.* **6**:537–548.

Rosenberg, P., 1981, The squid giant axon: methods and applications, in: *Methods in Neurobiology, Vol I*, (R. Lahue, ed.), pp. 1–133, Plenum Press, New York.

Saxton, W. M., Porter, M. E., Cohn, S. A., Scholey, J. M., Raff, E. C., and McIntosh, J. R., 1988, *Drosophila* kinesin: Characterization of microtubule motility and ATPase, *Proc. Natl. Acad. Sci.* **85**:1109–1113.

Schnapp, B. J., 1986, Viewing single microtubules by video light microscopy, *Methods in Enzymology* **134**:561–573.

Schnapp, B. J., Gelles, J., and Sheetz, M. P., 1988, Nanometer-scale measurements using video light microscopy, *Cell Motil. Cytoskel.* **10**:47–53.

Schnapp, B. J. and Reese, T. S., 1982, Cytoplasmic structure in rapid-frozen axons, *J. Cell Biol.* **94**:667–679.

Schnapp, B. J., Vale, R. D., Sheetz, M. P., and Reese, T. S., 1985, Single microtubules from squid axoplasm support bidirectional movements of organelles, *Cell* **40**:455–462.

Scholey, J. M., Neighbors, B., McIntosh, J. R., and Salmon, E. D., 1986, Isolation of microtubules and a dynein-like Mg-ATPase from unfertilized sea urchin eggs, *J. Biol. Chem.* **259**:6516–6525.

Scholey, J. M., Porter, M. E., Grissom, P. M., and McIntosh, J. R., 1985, Identification of kinesin in sea urchin eggs and evidence for its localization in the mitotic spindle, *Nature* **318**:483–486.

Schroer, T. A., Brady, S. T., and Kelly, R. B., 1985, Fast axonal transport of foreign synaptic vesicles in squid axoplasm. *J. Cell Biol.* **101**:568–572.

Schroer, T. A., Schnapp, B. J., Reese, T. S., and Sheetz, M. P., 1988, The role of kinesin and other soluble factors in organelle movement along microtubules, *J. Cell Biol.* **107**:1785–1792.

Seitz-Tutter, D., Langford, G. M., and Weiss, D. G., 1988, Dynamic instability of native microtubules from squid axon is rare and independent of gliding and vesicle transport, *Exp. Cell Res.* **178**:504–511.

Sheetz, M. P., 1987, What are the functions of kinesin? *BioEssays* **7**:165–168.

Shelanski, M. L., Gaskin, F., and Cantor, C. R., 1973, Microtubule assembly in the absence of added nucleotides, *Proc. Natl. Acad. Sci.* **70**:765–768.

Shotton, D. M., 1988, Review: Video-enhanced light microscopy and its applications in cell biology, *J. Cell Sci.* **89**:129–150.

Shpetner, H. S., Paschal, B. M., and Vallee, R. B., 1988, Characterization of the microtubule-activated ATPase of brain cytoplasmic dynein (MAP 1C), *J. Cell Biol.* **107**:1001–1009.

Smith, R. S., 1980, The short term accumulation of axonally transported organelles in the region of localized lesions of single myelinated axons, *J. Neurocytol.* **9**:39–65.

Smith, R. S., 1982, Axonal transport of optically detectable particulate organelles, in: *Axoplasmic Transport,* (D. G. Weiss, ed.), pp. 181–192, Springer-Verlag, Berlin.

Smith, R. S., 1988, Studies on the mechanism of the reversal of rapid organelle transport in myelinated axons of *Xenopus laevis, Cell Motil. Cytoskel.* **10**:296–308.

Travis, J. L., Kenealy, J. F. X., and Allen, R. D., 1983, Studies on the motility of the Foraminifera. II. The dynamic microtubular cytoskeleton of the reticulopodial network of Allogromia reticularis, *J. Cell Biol.* **97**:1668–1676.

Tsukita, S. and Ishikawa, H., 1980, The movement of membranous organelles in axons. Electron microscopic identification of anterogradely and retrogradely transported organelles, *J. Cell Biol.* **84**:513–530.

Tsukita, S., Tsukita, S., Kobayashi, T., and Matsumoto, G., 1986, Subaxolemmal cytoskeleton in squid giant axon. II. Morphological identification of microtubule- and microfilament-associated domains of axolemma, *J. Cell Biol.* **102**:1710–1725.

Vale, R. D. and Hotani, H., 1988, Formation of membrane networks *in vitro* by kinesin-driven microtubule movement, *J. Cell Biol.* **107**:2233–2241.

Vale, R. D., Reese, T. S., and Sheetz, M. P., 1985, Identification of a novel, force-generating protein, kinesin, involved in microtubule-based motility, *Cell* **42**:39–50.

Vale, R. D., Schnapp, B. J., Mitchison, T., Steuer, E., Reese, T. S., and Sheetz, M. P., 1985, Different axoplasmic proteins generate movement in opposite directions along microtubules *in vitro, Cell* **43**:623–632.

Vale, R. D., Schnapp, B. J., Reese, T. S., and Sheetz, M. P., 1985a, Movement of organelles along filaments dissociated from axoplasm of the squid giant axon, *Cell* **40**:449–454.

Vale, R. D., Schnapp, B. J., Reese, T. S., and Sheetz, M. P., 1985b, Organelle, bead and microtubule translocations promoted by soluble factors from the squid giant axon, *Cell* **40**:559–569.

Vallee, R. B., Shpetner, H. S., and Paschal, B. M., 1989, The role of dynein in retrograde axonal transport, *TINS* **12**:66–70.

Vallee, R. B., Wall, J. S., Paschal, B. M., and Shpetner, H. S., 1988, Microtubule-associated protein 1C from brain is a two-headed cytosolic dynein, *Nature* **332**:561–563.

Wais-Steider, C., White, N. S., Gilbert, D. S., and Eagles, P. A. M., 1987, X-ray diffraction pattern from microtubules and neurofilaments in axoplasm, *J. Mol. Biol.* **197**:205–218.

Weisenberg, R. C., 1972, Microtubule formation *in vitro* in solutions containing low calcium concentrations, *Science* **177**:1104–1105.

Weiss, D. G. (ed.), 1982, *Axoplasmic Transport,* Springer, Berlin.

Weiss, D. G., 1986a, Visualization of the living cytoskeleton by video-enhanced microscopy and digital image processing, *J. Cell Sci.* (Suppl.) **5**:1–15.

Weiss, D. G., 1986b, The mechanism of axoplasmic transport, in: *Axoplasmic Transport,* (Z. Iqbal, ed.), pp. 275–307, CRC Press, Boca Raton, FL.

Weiss, D. G. and Allen, R. D., 1985, The organization of force generation in microtubule-based motility, in: *Microtubules and Microtubule Inhibitors* (M. De Brabander, and J. De Mey, eds.), pp. 232–240, Elsevier, Amsterdam.

Weiss, D. G., Keller, F., Gulden, J., and Maile, W., 1986, Towards a new classification of intracellular particle movements based on quantitative analyses, *Cell Motil. Cytoskel.* **6**:128–135.

Weiss, D. G., Langford, G. M., Seitz-Tutter, D., and Keller, F., 1988, Dynamic instability and motile events of native microtubules from squid axoplasm, *Cell Motil. Cytoskel.* **10**:285–295.

Weiss, D. G., Maile, W., and Wick, R. A., 1989, Video microscopy, in: *Light Microscopy in Biology. A Practical Approach, Chapter 8*, pp. 221–278, IRL Press, Oxford.

Weiss, D. G., Seitz-Tutter, D., Langford, G. M., and Allen, R. D., 1987, The native microtubule as the engine for bidirectional organelle movement, in: *Axonal Transport* (R. S. Smith, and M. A. Bisby, eds.), pp. 91–111, Alan R. Liss, New York.

Yang, J. T., Saxton, W. M., and Goldstein, L. S. B., 1988, Isolation and characterization of the gene encoding the heavy chain of Drosophila kinesin, *Proc. Natl. Acad. Sci.* **85**:1864–1868.

Chapter 16

Lipid Metabolism In The Squid Nervous System

ROBERT M. GOULD and MARIO ALBERGHINA

1. Introduction

As biochemists, we choose the squid giant axon as an experimental model because its axoplasm is abundant and easy to isolate. From single *Loligo pealei* giant axons, yields generally range from 3 to 10 μl. This specialized cytoplasm has been found to possess active and wide-ranging lipid metabolism (see Section 2). Further studies on squid axon lipid metabolism are warranted because axons of mammalian nerves also metabolize lipids (Gould, 1976), and this metabolism appears to be sensitive to the physiological state of the axon. That is, it is modified during nerve degeneration and regeneration, and when nerve conduction slows in experimental diabetic neuropathy (Gould, 1989). The squid provides an unique model in which to conduct biochemical studies (following Sections) which could shed light on the potential role(s) of lipid metabolism in activities unique to axons and their terminal specializations.

To learn about squid axon lipid metabolism we must either apply approaches that are currently available or develop new ones. We shall first examine the approaches that have been used to study axonal lipid metabolism in squid and mammalian nerves and then consider their advantages and limitations. In various Sections, we will also consider approaches that have not yet been used.

Pure squid axoplasm, isolated by extrusion (see Section 3.2), catalyzes the incorporation of labeled precursors into lipid. For example, ^{32}P-orthophosphate, ^{3}H-myo-inositol and ^{3}H-glycerol are actively incorporated into phosphatidylinositol and polyphosphoinositides (phosphate) (Larrabee and Brinley, 1968; Gould, Pant, *et al.*, 1983). Activities of phosphatidylinositol synthase, the enzyme responsible for *de novo* phosphatidylinositol synthesis, and other enzymes of phosphoinositide metabolism can be assayed in pure squid axoplasm (Gould, Spivack, *et al.*, 1983).

Autoradiographic approaches were used to localize phosphatidylinositol synthesis to mammalian axons (Gould, 1976; Gould, Holshek, Silverman, *et al.*, 1987). In fact, myelinated axons in mouse sciatic nerve were shown to be highly active in catalyzing the incorporation of tritiated myo-inositol into phosphatidylinositol. Furthermore, with an alternative enzyme assay-based approach, we demonstrated the axonal location phosphatidylinositol synthase in rat sciatic nerves by its association with fast *axonal* transport (Kumara-Siri and Gould, 1980).

Unfortunately, the relatively high lipid metabolic activity of Schwann cells has made it impractical to use these autoradiographic and enzyme-based approaches to

R. M. GOULD • Marine Biological Laboratory, Woods Hole, MA 02543 and New York State Institute for Basic Research in Developmental Disabilities, 1050 Forest Hill Rd., Staten Island, NY 10314. M. ALBERGHINA • Marine Biological Laboratory, Woods Hole, MA 02543 and Institute of Biochemistry, Faculty of Medicine, University of Catania, Viale Andrea Doria 6, 95125 Catania, Italy.

study other facets of mammalian axon-based lipid metabolism. Furthermore, approaches with isolated axonal membranes from mammalian sources are currently impractical. Most *starting* nervous tissues contain very low portions of axonal membranes. Therefore, isolated axonal membrane fractions would most certainly be contaminated with other neuronal and glial cell membranes. Since it is likely that contaminating membranes participate in neural lipid metabolism, it would be impossible to use such fractions to characterize *axon-based* lipid metabolism.

Another possibility is to isolate fractions from nerves with high axonal membrane contents, i.e., the garfish olfactory nerves and squid retinal fibers. Studies to purify axolemmal membranes from these nerves (Chacko *et al.*, 1974; Condrescu *et al.*, 1984), have in part succeeded, since sodium channels and sodium, potassium adenosine-5'-triphosphatases (ATPases) appear to be reliable markers for these membranes. However, a nagging problem in designing fractionation procedures to purify axon and glial membranes is the lack of specific biochemical markers for glial cell membranes. Therefore, if one is to study the structural and metabolic properties of membranes present in axoplasm, the squid giant axon is a unique source material. Furthermore, studies with this tissue should provide a basis for developing strategies to isolate and characterize axonal membranes from other sources (see Section 6).

In addition to the squid giant axon, other tissues from the squid nervous system are of potential interest to the lipid biochemist. Among these tissues are the *optic lobe*, a rich source of acetylcholine-containing nerve endings (Cohen *et al.*, 1990), the *retinal fibers*, rich in axolemmal membrane (see Section 3.4.1) and the *retina*, rich in photoreceptor membranes (Saibil, 1990). We will briefly consider how strategies applied to the giant axon could be used for studying lipid metabolism with these systems. We feel that studies with these evolutionary simple squid neural preparations may provide insight into the development of roles for lipid metabolism in the modulation of neural activities.

2. Historical perspective

In this Section, we summarize studies on lipid composition, metabolism and enzyme activities of squid giant axon and other squid neural tissues. Additional information can be found in earlier reviews (Rosenberg, 1981; Gainer *et al.*, 1984).

2.1. Lipid composition

From ultrastructural and biochemical observations it is clear that the specialized cytoplasm of axons is richly represented by cytoskeletal structures and poorly represented by membrane organelles. Therefore, it was not surprising that McColl and Rossiter (1951), the first to analyze the lipid compositions of squid neural tissues, found that axoplasm contained relatively little lipid compared with the central ganglia (mainly optic lobes) and pallial nerves (connectives running between the central and stellate ganglia). With the relatively crude methods used at that time to separate lipid classes, McColl and Rossiter showed that the lipid composition of axoplasm and the central ganglia (neuron-rich sources), contained lower proportions of cholesterol and a *crude* sphingomyelin fraction, which includes ether lipids and plasmalogens, than lipids from glial cell-rich sources: the pallial nerve and giant axon sheath (see Section 3.2). Their observations were in accord with subsequent studies of Sheltawy and Dawson (1966), who showed that sphingomyelin, ethanolamine plasmalogen and cholesterol were in relative abundance in myelinated (glial cell-dominated) versus non-myelinated (axon-rich) nerves (Table I). In a similar vein, myelin-typical glycolipids, cerebrosides and sulfatides (Nonaka and Kishimoto, 1979), were not detected in myelin-lacking squid neural tissues (McColl and Rossiter, 1951; Yamaguchi *et al.*, 1987; Tanaka *et al.*, 1987).

In studies of the lipolytic properties of toxins that block giant axon conduction, Rosenberg and his colleagues (e.g., Rosenberg and Condrea, 1968; Condrea and Rosenberg, 1968; Rosenberg and Khairallah, 1974) determined the phospholipid compositions of the squid giant axon and extruded axoplasm (Table I). Like McColl and Rossiter (1951), they found that the phospholipid content of axoplasm (0.09 μg lipid phosphate/mg wet weight of tissue) was far lower than that of the whole giant axon (0.23 μg/mg), an entity containing about 80% axoplasm by weight (Bear and Schmitt, 1939). Not surprisingly, the sheath (glial cell-rich tissue remaining after the extrusion of axoplasm) was relatively lipid-rich (>0.5 μg/mg).

Total lipid, phospholipid and cholesterol contents of the various squid neural tissues and fractions are presented in Table I. For comparison, lipid compositions of representative neural tissues from other animals are presented in Table II. Total lipid content of squid neural tissues; i.e., optic lobe, stellate ganglion and fin nerve, are notably lower (13 to 19%) than the lipid contents of other neural tissues, particularly myelin-containing samples (Table II). Phospholipids (58 to 67%) and cholesterol (22 to 39%) are the dominant lipids in squid neural tissues. Only traces of glycolipid, mainly glucocerebroside, are present (Yamaguchi *et al.*, 1987).

In squid neural tissues and fractions (Table I), as well as in neural tissues and fractions from other sources (Table II), phosphatidylcholine, phosphatidylethanolamine and cholesterol are the major lipids. Among phospholipids, phosphatidylcholine levels predominate in all squid tissues and fractions with optic lobe as the exception. Like squid optic lobe, peripheral nerves of garfish and rabbit have higher contents of ethanolamine versus choline phosphoglycerides. In squid, phosphatidylcholine levels range between 31% (optic lobe) and 56% (axoplasm) of total phospholipid. Membranes with low phosphatidylcholine levels generally contain relatively high amounts of sphingomyelin (Tables I and II, also Barenholz and Gatt, 1982). In squid, phosphatidylethanolamine levels ranged between 27% (stellar nerve) and 52% (optic lobe) of total phospholipid. The portion of phosphatidylethanolamine in the plasmalogen form (3%, only analyzed in optic lobe) was lower than in other invertebrate (Okamura *et al.*, 1985; Okamura *et al.*, 1986) and vertebrate (Table II; Ansell and Spanner, 1982) neural tissues. The low content of plasmalogen in squid ethanolamine lipids may provide a clue for understanding its function.

After choline and ethanolamine phosphoglycerides, sphingomyelin, phosphatidylserine and phosphatidylinositol are the next most abundant membrane phospholipids. Among these lipids, sphingomyelin content was most variable. As noted above, membranous preparations with low sphingomyelin content tended to be *neuron*-dominant; i.e., axoplasm (0%), optic lobe (2%), stellate ganglia (2%), retinal fiber axolemma (4%), rat brain (7%) and human gray matter (10%), whereas *glial cell*-dominant membranes had high sphingomyelin levels. Contrasting this trend is the generally high sphingomyelin content of axolemmal membranes; i.e., squid stellar nerve (10%), lobster leg nerve (13%) and crab leg nerve (18%). In many biological membranes, particularly plasma membranes, there is a positive correlation between levels of cholesterol and sphingomyelin (Patton, 1970; Barenholz and Gatt, 1982). In

Table I. Lipid compositions of squid neural tissues

	LIPIDS								
TISSUES	*Lipid**	*PL***	*PC†*	*PE*	*PS*	*PI*	*SPM*	*CL*	*Chol*
Stellar nerve[A]	nd	nd	52	27	7	2	12	nd	nd
Giant axon[A]	nd	nd	45	31	8	5	11	nd	nd
Axoplasm[A]	nd	nd	56	31	7	7	0	nd	nd
Sheath[A]	nd	nd	40	28	11	5	16	nd	nd
Optic lobe[B]	19	62	31	52	7‡	6‡	2	2	33
Stellate ganglion[B]	16	59	48	35	8‡	7‡	2	2	30
Fin nerve[B]	13	59	49	34	4‡	4‡	8	0	39
Fractions:									
Stellar n. axolemma[C]	70	58	46	34	10	nd	10	nd	28
Stellar n. gliolemma[C]	51	66	44	33	14	nd	9	nd	25
Retinal n. axolemma[D]	45	67	43	41	10	nd	4	nd	22

Comparisons of total lipid, phospholipids and cholesterol levels among squid neural tissues and fractions.
* Total lipid values are given as percentage of tissue or fraction dry weight.
** Phospholipid (PL) values are based on percentages of total lipid.
† Values for individual phospholipids, including phosphatidylcholine (PC), phosphatidylethanolamine (PE, may include plasmalogen), phosphatidylserine (PS), phosphatidylinositol (PI), sphingomyelin (SPM) and cardiolipin (CL) are expressed as percentage of total phospholipid. Cholesterol (Chol) values are given as percent of total lipid. As relatively minor components, values for glycolipids and free fatty acids are not included. The levels can be obtained from the original publications.
‡ PS and PI values are calculated assuming (as stated in the ref. [B]) that these lipids are present in nearly equal amounts. The symbol *nd* indicates that the values were not given in the study.
The references are:
[A] Condrea and Rosenberg, 1968
[B] Yamaguchi *et al.*, 1987
[C] Camejo *et al.*, 1969
[D] Zambrano *et al.*, 1971

squid and other neural tissue membranes listed here, this relationship appears to hold. Perhaps the high sphingomyelin and cholesterol contents of glial cell-rich preparations reflects the high proportion of plasma membrane in these cells.

Phosphatidylserine and phosphatidylinositol levels were relatively constant among squid tissues and membranes, with phosphatidylinositol levels being lower than levels of the other common phospholipids. Perhaps this constancy reflects a universal role of these lipids in maintaining membrane integrity.

Significant amounts of ceramide 2-aminoethylphosphonate were detected in squid optic lobe (Yamaguchi *et al.*, 1987). Incorporation of tritiated glucose into this lipid by the giant fiber lobe (Tanaka *et al.*, 1987) is indicative of its wider distribution in squid neural tissue. In line with this notion is the detection of ceramide 2-aminoethylphosphonate in other invertebrates (e.g., Bolognani *et al.*, 1981; Hori and

Table II. Lipid compositions of other neural tissues

	LIPIDS								
TISSUES	*Lipid**	*PL†*	*PC‡*	*PE*	*PS*	*PI*	*SPM*	*Other*	*Chol*
Garfish olfactory n.[A]	20	73	41	32(56)	12	4	9		27
Garfish trigeml. n.[A]	48	64	32	41	12	1	13		27
Lobster leg n.[B]	nd	70	31	24(75)	7	1	16	20	30
Lobster claw n.[B]	nd	67	35	16(57)	10	2	20	18	33
Crab leg n.[B]	nd	62	34	23(67)	7	3	12	21	38
Crab claw n.[B]		63	32	24(64)	8	2	14	20	37
Rabbit sciatic n.[B]	nd	49	16	37(82)	14	1	24	9	38
Rat brain[C]	37	55	40	36(59)	13	4	7		23
Human gray matter[C]	33	68	39	34(39)	13	4	10		22
Human white matter[C]	54	44	29	34(74)	18	2	17		28
Fractions:									
Garfish olf. n. axolemma[D]	66	74	42	34	12	4	8		26
Lobster leg n. axolemma[D]	76	77	25	43	12	3	13		23
Crab leg n. axolemma[D]	68	71	31	36	14	1	18		29
Rat brain axolemma[D]	51	62	45	39	9	5	3		21
Rat brain myelin[C]	70	39	29	42	18	3	8		27

Comparisons of the quantities of total lipid, phospholipids and cholesterol in invertebrate and vertebrate neural tissues and fractions.
* Total lipid values are given as percentage of tissue or fraction dry weight.
† Phospholipid (PL) values are the percentage of total lipid.
‡ Values for individual phospholipids are expressed as percent of total phospholipid. Values in parentheses (after PE) values represent the percentage of PE in the plamalogen form. Levels of cardiolipin, phosphatidate, alkylether phospholipids, phosphatidylglycerol and polyphosphoinositides have been obtained for some samples, reference [B], though they are not included here.
References are:
[A] Chacko *et al.*, 1972
[B] Sheltawy and Dawson, 1966
[C] Norton, 1981
[D] DeVries, 1981

Nozawa, 1982) and in the whole body of squid *Loligo edulis* (Hori *et al.*, 1967). The possible failure to detect this lipid in earlier studies might have been due to its co-migration with more abundant phosphatidylcholine in various chromatographic systems.

Values for minor-occurring phospholipids; i.e., polyphosphoinositides, phosphatidylglycerol, ceramide 2-aminoethylphosphonate, phosphatidate and cardiolipin are not included in the Tables, since they were often not reported. Cardiolipin (2%), a recognized lipid component of mitochondrial inner membranes (Hostetler, 1982), is present in optic lobe and stellate ganglion (2% of total phospholipid), though apparently not in fin nerve (Yamaguchi *et al.*, 1987). It will be of interest to see if fin nerve mitochondria lack cardiolipin.

With improved technology, lipid compositional analyses are becoming both easier to perform and more sensitive (Jungalwala, 1985; Christie, 1987). High pressure liquid chromatograph (high performance liquid chromatgraphy) (HPLC), which uses column chromatography at high pressure, is costly and limited to one sample at a time. However, with HPLC in the same run both the lipid composition by class, e.g., phosphatidylcholine, phosphatidylinositol, and the acyl side chain distributions among the classes, e.g., the portion of phophatidylcholine that is 1-palmitoyl-2-arachidonyl phosphatidilcholine, etc. can be determined. Thin layer chromatography (TLC), which uses silica gel on support plastic or glass is relatively simple and cheap and multiple samples can be processed at one time. More comprehensive compositional information will be valuable in designing metabolic and enzyme studies (following Sections) to pinpoint the lipids that play key roles in neural-specific activities.

2.2. Lipid metabolism

Two results that have shaped and will continue to shape experiments on lipid metabolism by the squid giant axon are that: (1) Axoplasm displays a multitude of lipid metabolizing enzymes. (2) The activities of these enzymes in axoplasm are low compared with activities in other preparations. Hence, in studies with the giant axon, the Schwann cells and not the axon dominate lipid metabolism.

Larrabee and Brinley (1968), the first to study lipid metabolism in the squid giant axon, already uncovered these facets of giant axon lipid metabolism. They found that the intact giant axon incorporates ^{32}P-orthophosphate into several phospholipids, including phosphatidylethanolamine, phosphatidylinositol and phosphatidic acid. These lipids accumulated in both the Schwann cell-rich sheath and in axoplasm, extruded after the incubation. Their observation, that radioactive lipids accumulate in the axon, was perhaps the first indication that an axonal lipid metabolism takes place independently of axonal transport. In a follow-up study (Gould, Pant, *et al.*, 1983), we showed that tritiated choline, inositol, serine, ethanolamine and glycerol behaved similarly to ^{32}P-phosphate. When giant axons were incubated with these precursors, phospholipid products accumulated mainly (80 to 95% of the lipid radioactivity) in the Schwann cell-rich sheath.

Larrabee and Brinley (1968) showed that extruded axoplasm incorporated ^{32}P-phosphate into a variety of phospholipids, though incorporation was much lower than in the intact giant axon or stellate ganglia. We (Gould, Pant, *et al.*, 1983) found that many other lipid precursors were incorporated by separated axoplasm, indicating that it contained a variety of enzyme systems involved in lipid metabolism. Subsequently, we used axoplasm as an enzyme source to characterize lipid enzymes of axoplasm (see Section 2.3).

Because labeled lipids accumulate more actively in the sheath versus the axoplasm, two sources of the lipids that accumulate in axoplasm should be considered: (1) synthetic organelles in the axoplasm, and/or (2) Schwann cells surrounding the giant axon. Several lines of evidence favor the second source, since it is known that Schwann cells transfer *select* newly-synthesized proteins to the giant axon (Lasek *et al.*, 1977: Gainer *et al.*, 1977; Tytell and Lasek, 1984; Tytell, 1987). Furthermore, as some of these proteins are membranous (i.e., they incorporate glucosamine, Tytell, 1987), associated membrane lipids would be expected to accompany them. Second, ^{32}P-labeled phospholipids that accumulate in axoplasm closely resemble sheath lipids (Larrabee and Brinley, 1968; Gould, Pant, *et al.*, 1983) and not the lipids formed by extruded axoplasm. Following incubation of the giant axon with ^{32}P-phosphate, radioactive lipids of both sheath and axoplasm follow the pattern: phosphatidylethanolamine > phosphatidylcholine > phosphatidylserine > phosphatidylinositol. When extruded axoplasm is incubated with ^{32}P-phosphate, a completely different labeling pattern is obtained: phosphatidylinositol 4-phosphate >> phosphatidylinositol 4,5-bisphosphate > phosphatidate = phosphatidylinositol. It is still not known why axoplasm shows this preference toward minor-occurring lipids. Perhaps it is an indication that the lipid metabolizing system in axoplasm is both unique and simple. ^{32}P-orthophosphate incorporation into phosphatidylinositol-4-phosphate and phosphatidic acid occurs by single step kinase reactions. Probably because they did not use acidic extraction, Larrabee and Brinley (1968) fail to find the dominant ^{32}P-orthophosphate incorporation into polyphosphoinositides, particularly phosphatidylinositol 4-phosphate.

Since phosphatidylethanolamine and phosphatidylinositol are labeled with ^{32}P-orthophosphate, the synthesis of these lipids must have involved its prior incorporation via adenosine-5'-triphosphate (ATP) into lipid intermediates, cytidine 5'-diphosphate-ethanolamine (CDP-ethanolamine) and CDP-diacylglycerol, respectively. Therefore, the metabolic data obtained by Larrabee and Brinley (1968) not only demonstrated that terminal step in the synthesis of these phospholipids occurs in axoplasm, but that the *de novo* pathways are present.

Another approach to study axon-based phospholipid metabolism in the intact giant axon is to inject labeled precursor directly into axoplasm (Larrabee and Brinley, 1968; Gould, Jackson, *et al.*, 1983). Although there is some indication that lipid synthesis takes place, incorporation was always extremely low and variable (see Section 4.1.2).

For an unknown reason, Larrabee and Brinley (1968) were unable to detect ^{32}P-phosphate label in phosphatidylcholine, either after incubation of giant axons or stellate ganglia. Brunetti *et al.* (1979) used ^{3}H-choline to show phosphatidylcholine synthesis in the squid giant axon and giant fiber lobe. Since they also detected labeled phosphorylcholine and CDP-choline intermediates, they suggested that the *de novo* pathway was at least partly responsible for phosphatidylcholine synthesis in this preparation. They also demonstrated that choline was incorporated into sphingomyelin, indicating that the giant axon contained requisite enzymes for this synthetic pathway. However, because they did not study the properties of extruded axoplasm, choline lipid metabolism was not localized to this compartment.

We (Gould *et al.*, 1982; Gould, Pant, *et al.*, 1983; Tanaka *et al.*, 1987) expanded the Larrabee and Brinley (1968) studies to a host of lipid precursors and showed that extruded axoplasm, albeit at low rates, had activities for synthesizing all its own phospholipids. Studies have not yet been conducted which demonstrate that axoplasm synthesizes cholesterol or specific glycolipids, though it catalyzes the incorporation of tritiated glucosamine into lipid (Gould, Pant, *el al*, 1983).

2.3. Lipid enzymes

To characterize the active phosphatidylinositol (see Section 1) and polyphosphoinositide (see Section 2.2) metabolism in squid axoplasm, we assayed key metabolic enzymes (Gould, Spivack, *et al.*, 1983). Enzymes are classified according to the Enzyme Commission (EC) number (Barman, 1969, 1974). There are four sets of numbers in the EC number separated by periods. The first number refers to the class name, e.g., 2 refers to oxidoreductases. The second number refers to the subclass name, e.g., enzymes acting on the CH–NH group of donors. The third number refers

Table III. Phosphatidylinositol Synthase Activity in Various Neural Tissues and Fractions

A. Squid neural tissues		
Tissue	*Activity*	
Axoplasm	88 ± 10	
Giant fiber lobe	531 ± 61	
Optic Lobe	696 ± 59	
Retinal fiber	563 ± 225	
Fin nerve	207 ± 28	
B. Squid axoplasm subfractions		
Subfraction	*Activity*	
Axoplasm	12 ± 3	
	Control	*KI-treated*
S2	0.3 ± 0.1	–
F1	18.2 ± 4.8 (1.5)	16.4 ± 10.1 (1.4)
F2	4.6 ± 3.2 (0.4)	3.9 ± 3.1 (0.3)
F3 (vesicles)	1.6 ± 1.3 (0.1)	67.8 ± 27.3 (5.6)
F4	0.9 ± 0.6 (0.1)	4.2 ± 4.5 (0.4)
C. Homogenates and axolemmal fractions		
Origin	*Activity*	
Squid retinal fibers	633 ± 232	
R. f. axolemma	301 ± 43	
Garfish olfactory nerve	78 ± 18	
O. n. axolemma	318 ± 27	

Phosphatidylinositol synthase was assayed as described (see Section 5.2.1). *A*. Specific activities of the different tissues (nmol/h/mg protein) are based on results of four separate experiments with fresh material. *B*. The axoplasm was subfractionated as described (see Section 3.2.1) and assays were performed in triplicate on three independent fractionations. It is not known why the activities in these studies (carried out at the end of the summer of 1988) were lower than in earlier work. *C*. Squid retinal fiber axolemma were fractionated as described (see Section 3.4.1) and garfish olfactory nerve fractionated in a similar manner by Dr. John Elam according to the procedure of Cancalon and Beidler (1975).

to the sub-subclass name, e.g., enzymes with nictinamide adenine dinucleotide (NAD) or nicotinamide adenine dinucleotide 3'-phosphate (NADP) as acceptors. The fourth number refers to the specific enzyme. For example, EC 1.5.1.11 refers to octopine dehydrogenase. The systematic name for EC 1.5.1.11 is octopine: NAD oxidoreductase, which refers to the enzyme reaction which is octopine + NAD + H_2O = arginine + pyruvate + reduced NAD.

Phosphatidylinositol synthase (CDP-1,2- diacylglycerol:myo-inositol 3-phosphatidyltransferase, EC 2.7.8.11), diacylglycerol kinase (ATP:1,2-diacylglycerol phosphotransferase, EC 2.7.1.-), phosphatidylinositol kinase (EC 2.7.1.67) and phosphatidylinositol transfer protein (also catalyzes phosphatidylcholine transfer) are all present in axoplasm. Under optimal assay conditions, activities were generally lower than those in rat brain and sciatic nerve. Phosphatidylinositol 4-P kinase activity was not detected under the assay conditions that were used. More recently, calcium-dependent base exchange activity for serine incorporation into phosphatidylserine (Holbrook and Gould, 1988), phospholipase A_2 (both calcium-dependent and independent forms) and acyltransferase activities have been detected and characterized in squid axoplasm (Alberghina and Gould, 1988).

In Table III, recent results comparing the activity of phosphatidylinositol synthase in axoplasm with activities in other squid neural tissues are presented. With all preparations exogenous CDP-diglyceride (didecanoyl) was added, since it stimulated (>10 fold) the rate of myo-inositol incorporation. Other reaction components: buffer, magnesium, bovine serum albumin, potassium chloride and ethylene glycol-bis(2-amino ethyl ether)N,N,N',N'-tetraacetic acid (EGTA), were included at levels optimal for squid axoplasm. On a protein basis, the activity in axoplasm fell between 12.6% (optic lobe) and 42.5% (fin nerve) of the activity in other *enzyme* preparations. The other enzymes that we tested, phospholipase A_2, acyltransferase and serine-base exchange were also present at lower levels in axoplasm than in other squid neural tissue preparations.

Subcellular fractionation (Gould, Spivack, *et al.*, 1983) (see Section 3.2.1) to locate axonal organelles involved lipid metabolism. In the present study (Table III), the overall activity was low. Here, we found that phosphatidylinositol synthase activity was selectively enriched in the same subfraction (F2-KI) that is contained in axonal transport organelles. This fraction was used in reconstituted axonal transport studies (Schroer *et al.*, 1988). Similarly, using this fractionation scheme, we found that acyltransferase and, to a lesser extent, phospholipase A_2 also were concentrated in this membrane fraction. Further studies are needed to define this fraction in terms of classic subcellular marker enzymes. We also analyzed axolemmal fractions prepared from squid retinal fibers and garfish olfactory nerve for phosphatidylinositol synthase, acyltransferase and phospholipase A_2 activities. Both axolemma fractions contained these activities. The activity for phosphatidylinositol synthase was consistently lower in the squid retinal fiber axolemmal fraction than in whole nerve homogenates (Table III). In contrast enzyme activity in the axolemmal fraction from the garfish olfactory nerve was higher than that in the total nerve homogenate.

Whether or not these fractionation approaches will define unique characteristics of the lipid enzymes present in axonal membranes, remains to be determined.

3. Tissue preparations

This Section describes how to prepare squid neural tissues for use in lipid metabolism studies (see Section 4).

3.1. The giant axon

The two largest (hindmost) giant axons are used in virtually all studies. In the squid, *Loligo pealei*, these axons measure up to eight centimeters in length (to the major bifurcation) and 0.8 millimeters in diameter (Rosenberg, 1981; Villegas and Villegas, 1984). To facilitate extrusion (see Section 3.2) and to reduce competition for precursor in incubation studies (see Section 4.1.1), adherent fibers and the stellate ganglion are usually removed.

For studies with intact giant axons, it is sound practice to test electrical excitability both before and after incubation. Rosenberg (1981) describes equipment and procedures needed for these measurements. The ability of giant axons to conduct nerve impulses reflects both the intactness of their axolemmal membrane and the purity of their axoplasm.

During each summer our laboratory (2 to 3 people) uses, on average, 5 to 6 large squid (10 to 12 giant axons) each day. Since squid deteriorate rapidly in captivity, we prefer dissecting them the day of their capture. Squid kept overnight in tanks of running seawater are used if they appear to be healthy.

Giant axons are dissected in two steps: (1) removal from the mantle and (2) separation of adherent small fibers. Each step is performed at a separate dissecting microscope. With experience, dissections take roughly 10 to 20 minutes each. The dissection of the squid giant axon has been described by Gilbert (1974) and Rosenberg (1981) and in this volume (Adelman and Gilbert, 1990; Brown and Lasek, 1990). In practice, researchers usually learn the dissection from colleagues. For our purposes we recommend removing the fin nerves before the giant axon is exposed. Once the internal organs and pen are taken out, the squid mantle is turned over. The fins and underlying skin are grasped and together pulled forward away from the mantle. The fin nerves usually slide out through holes in the mantle. Their removal makes it easier to dissect the giant axon. Once freed of connective tissue and muscle (distal end), ends of the axon are tied with cotton thread, and the giant axon removed from the mantle without excessive stretching. A Castroviejo microscissor (RS-5658) from Roboz Surgical Inst. Co, Inc. is used for this dissection procedure.

For removing adherent small fibers, the threads attached to the axon are stuck to dental wax at opposite sides of glass petri dish (100 mm in diameter) that is filled with ice-cold calcium-deficient seawater (obtained from the Chemistry Room at the Marine Biological Laboratory, Woods Hole, MA, see Section 7). The preparation is viewed with darkfield illumination at 12 to 16 × magnification. Small fibers are first teased from the proximal end of the giant axon with the tip of fine (Dumont #5 watchmaker) forceps. They are then grasped and pulled backward away from the giant axon. Care must be taken to cut the small branches that derive from the giant axon at a millimeter or more from the trunk. A Vannas ultra microscissor (Roboz No. RS-5610) is useful for this part of the dissection. Pulling out these branches will damage the giant axon and cause it to lose excitability. Morphological evidence of damage is the appearance of white spots and/or constrictions along the axon. The axon becomes soft and narrow at these spots, because calcium leaks into the axon and both precipitates proteins and activates endogenous proteases which destroy the neurofilaments (Brown and Lasek, 1990). Liquification of axoplasm can also occur if the dissecting dish becomes contaminated. To prevent this, the dish is frequently cleaned in soapy water and rinsed thoroughly with distilled water.

3.2. Extruding axoplasm

The general procedure is described by Brown and Lasek (1990). In brief, a giant axon (see Section 3.1) is blotted on filter paper or Kimwipe™ to remove adherent seawater. After the proximal end is cut, the axon is laid on a piece of Parafilm™ that has been placed on a large rubber stopper (≥ 100 mm in diameter). By rolling a capillary tube or piece of fine tubing over the axon in a distal to proximal direction, axoplasm extrudes from the cut end. To prevent contamination of the axoplasm with tissue debris and adherent seawater, the thread tied to the distal axon is drawn away from the extruding axoplasm as the roller nears the cut end. The axoplasm nearest the distal tie should not be extruded.

The extruded axoplasm is sucked into a capillary tube (25 µl) and the length (La) in mm recorded. Axoplasm is then expelled into suitable medium (see below). The total volume (Va) of axoplasm collected from several axons (generally 8 to 12) is calculated from:

$$\text{Va (in } \mu\text{l)} = \text{La} \times 25\ \mu\text{l/Lc} \quad (1)$$

where Lc is the length (in mm) of the 25 µl capillary tube.

For precursor incubation studies (see Section 4.2), we disperse extruded axoplasm in buffer X (see Section 7). For enzyme studies (see Section 5.2), it is dispersed in a 10 mM tris (hydroxymethyl) aminomethane-HCl (Tris-HCl), 0.25 M sucrose solution (TS), prepared at the optimum pH for the enzyme assay. The viscous axoplasm (approximately 50 µl) is homogenized in 300 µl of medium with a conical ground glass homogenizer (1 ml, from Micrometrics, Sarasota, FL). Extruded axoplasm and other tissue preparations are used for enzyme studies are often stored frozen. When used within a few days, the material is kept in a –20 °C freezer. For longer storage, samples are kept at –80 °C. For each *new* assay, the effect of storage on enzyme activity should be checked. A company, Calamari Ink Specimen Preparations, P.O. Box 661, Woods Hole, MA 02543, (508) 540–6614, sells quick frozen squid giant axons and other squid parts. This material (only recently available) could be used for studies to characterize enzymes in squid tissues at one's home institution.

3.2.1. Axoplasmic subfractions

Axoplasm consists of mitochondria, various other membrane organelles, and cytoskeletal polymers in a cytosolic matrix. Membrane fractions (mitochondria and *crude* microsomes) can be separated by differential sedimentation. However, cytoskeletal polymers associate with and contaminate these membranes. Including high salt, potassium chloride or preferably potassium iodide, in the fractionation buffer will help dissociate cytoskeletal proteins from the membrane fractions (Baumgold *et al.*, 1981). We used differential centrifugation in potassium chloride containing sucrose solutions to localize lipid enzymes in axoplasm (see Section 2.3). We adapted the fractionation procedure of Schroer *et al.* (1988) in more recent studies (Alberghina and Gould, 1988). This procedure includes a potassium iodide treatment to solubilize cytoskeletal polymers from the membranes and a discontinuous sucrose gradient to subfractionate crude microsomes. Schroer *et al.* (1988) showed that the membrane organelles of one particular fraction are transported in a reconstituted axonal transport system.

For this fractionation, freshly extruded axoplasm (50 to 75 µl) is collected in 210 µl of fractionation medium (FM) consisting of: half strength buffer X (see Section 7) with added ATP (2 mM), dithiothreitol (1 mM), leupeptin (10 µg/ml) and tosyl L-arginine methyl ester (TAME, 10 µg/ml). Axoplasm is dispersed by trituration, about 50 times with a P-200 pipet tip. The suspension is centrifuged at 12000 × g for 4 minutes in a 1.5 ml microcentrifuge tube. The clear upper layer (150 to 200 µl) is transferred to another 1.5 ml tube and diluted with an equal volume of FM (designated

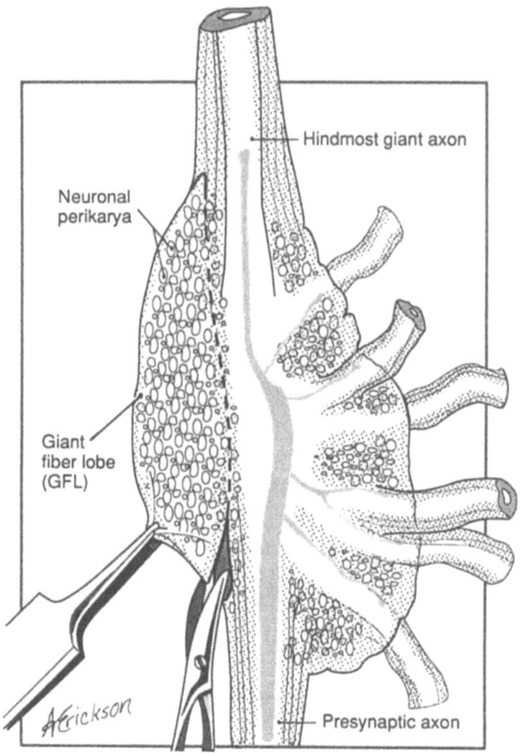

Figure 1. Artists view of the right stellate ganglia (based on Fig. 1A of Llinas, 1984). The hindmost giant axon exits the preparation at the top of the illustration much as it would in a squid mantle opened for dissection. The gray image branching from the presynaptic axon depicts the giant synapse arrangement as visualized by filling this structure with dye. The uppermost giant synapse is straight and over a millimeter long. This preparation is recommended for autoradiographic studies (see Section 5.3.2).

C-S). The cloudy lower layer (50 to 100 µl) is resuspended by trituration, 10 times in 150 µl of FM containing 0.6 M potassium iodide (KI). This solution is centrifuged at 12000 × g for 4 minutes. The supernatant (KI-S, 200 to 250 µl) is diluted with an equal volume of FM. C-S and KI-S are each layered over two identical 600 µl discontinuous sucrose (prepared in FM) gradients consisting of (from bottom to top): 60 µl of 45% sucrose; 140 µl of 15% sucrose; 100 µl of 12% sucrose and 100 µl of 10% sucrose. The tubes are centrifuged (4 °C) in an SW 50.1 rotor at 114,000 × g for 90 minutes. The top layers (C-S2 and KI-S2) of each gradient are collected. The bottom of each tube is then punctured and the material collected in four fractions (F1, first 2 drops; F2, next 4 drops; F3, next 4 drops; F4, remaining 2 – 3 drops). Fractions from identical (C or KI) tubes are pooled and organelles in F2 (C and KI, 15% sucrose) concentrated. The F2 fractions are diluted with an equal volume (140 µl) of FM and centrifuged (4 °C) in an SW 50.1 rotor at 190,000 × g for 35 minutes. The supernatants are removed and the (invisible) pellets are each suspended by trituration (15 times) in 10 µl of FM. This fraction, containing organelles that are transport-competent (Schroer *et al.*, 1988), is particularly rich in phosphatidylinositol synthase and acyltransferase (see Section 2.3).

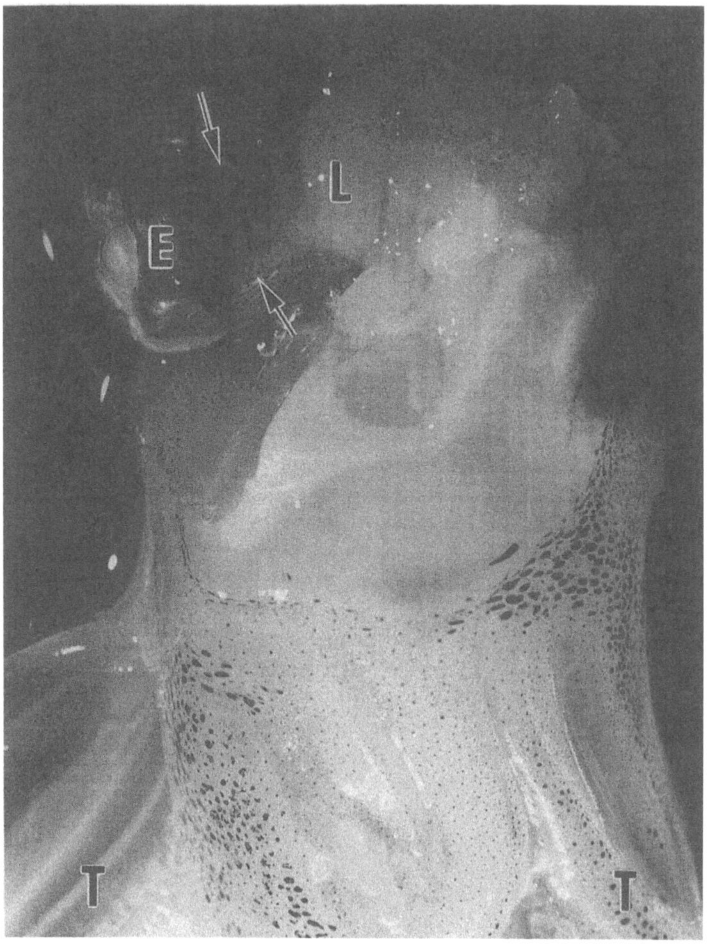

Figure 2. A squid head photographed above with exposed right eye (*E*) and optic lobe (*L*). The retinal fibers (between arrows) connect these structures. The tentacles (*T*) are displayed out toward the bottom of the photo. Pigment granules in the skin dot this structure.

3.3. Giant fiber lobe (GFL)

The GFL and parent stellate ganglia are used in precursor incubation and enzyme assay experiments (Gould *et al.*, 1982; Tanaka *et al.*, 1987). Since GFLs contain many of the neuronal perikarya responsible for forming and maintaining the giant axons (Villegas and Villegas, 1984), they are uniquely suited for comparative biochemical studies with axoplasm. The GFL is a crescent-shaped structure lying on the ventromedial side of the stellate ganglion (Fig. 1). To remove it from the ganglion, the overlying stub of pallial nerve (left when the pen is removed) and associated connective tissue are first cut away. A cut is then made at the base of the ganglia (Fig. 1, scissors), and the slit enlarged (Fig. 1, dotted line) until the GFL can be taken. For precursor incubation studies, the GFL is rinsed in fresh seawater, blotted dry and incubated directly with precursor (see Section 4.3). For enzyme studies (see Section 5.2), GFLs are rinsed in calcium-deficient seawater (see Section 7) and collected in ice-cold Tris-HCl (10 mM), sucrose (0.25 M) solution (TS) used with enzyme assays (see Section 5.2). They are transferred to a measured volume (10 to 12 GFL/300 µl) of fresh TS and homogenized by hand in a 1 ml ground glass homogenizer. The homogenate is centrifuged at 12000 × g for 4 minutes to pellet connective tissue and cell debris. If not used immediately, the turbid supernatant is aliquoted and stored (see Section 3.2).

3.4. Retinal fibers

Retinal fibers are dissected from severed squid heads (Fig. 2). The tentacles and beak are removed and the head divided in half. Each eye and optic lobe is freed as a unit from cartilage and connective tissue. The retinal fibers (Fig. 2, between arrows) are exposed when the optic lobe is lifted from the eye. Adherent material, with the exception of the retinal fiber band, is removed from the back surface of the eye. The connections between the fibers and optic lobe are severed and the retinal fiber band (left lying on the back of the eye) is grasped with fine forceps and removed. The fibers are rinsed in calcium-deficient seawater and collected in ice-cold TS. Pooled fibers are homogenized (10 strokes) in 5 to 10 volumes of ice-cold TS with a motor-driven (5 or 10 ml) Potter-Elvehjem homogenizer. The insoluble debris is removed by centrifugation at 12000 × g for 4 minutes and the turbid supernatant used immediately for enzyme studies (see Section 5.2) or aliquoted and stored (see Section 3.2).

3.4.1. Retinal fiber axolemma

The retinal fiber axolemmal fraction is prepared as described by Condrescu *et al.* (1984) with minor modification. Pooled fibers (0.5 to 1.5 g wet weight) are homogenized (10 strokes) in 5 to 10 volumes of a fractionation medium (FM-2) consisting of: 0.32 M sucrose, 1 mM ethylene tetraacetic acid (EDTA), 1 µM phenylmethylsulfonyl fluoride (PMSF) and 10 mM Tris-HCl, pH 7.4, in a motor-driven 10 ml Potter-Elvehjem homogenizer. The homogenates are centrifuged (4 °C) at 2000 × g for 8 minutes in a Sorvall SS-34 rotor. The supernatants are removed and the pellet rehomogenized in one-half volume of FM-2 and recentrifuged. This washing step is repeated. The combined supernatants (S-1) are centrifuged (4 °C) in a Beckman 40 Ti rotor at 12000 × g for 25 minutes. The supernatant (S-2) is collected and centrifuged (4 °C) in a Beckman SW 41 rotor at 100,000 × g for 45 minutes to pellet *crude* microsomes. The pellet is resuspended in 10 ml of FM-2, and 5 ml are layered in each of two tubes, over 8 ml of 1.12 M sucrose in FM-2. The tubes are centrifuged (4 °C) at 35000 × g for 60 minutes. The interfaces are collected, pooled and diluted with five volumes of 10 mM Tris-HCl, pH 7.4, and sedimented (4 °C) in an SW 41 rotor at 100,000 × g for 45 minutes. Pellets are suspended in TS, aliquoted and stored (see Section 3.2).

3.5. Optic lobe

Optic lobes are generated during the retinal fiber dissection (see Section 3.4). For enzyme studies, they are dispersed in 5 to 10 volumes of ice-cold TS with a motor-driven (10 or 20 ml) Potter-Elvehjem homogenizer. The homogenate is transferred to 1.5 ml microcentrifuge tubes and spun at 12000 × g for 4 minutes to pellet insoluble material. The supernatants are aliquoted and stored (see Section 3.2). For precursor incorporation studies, optic lobes are sectioned into 1 to 2 mm cubes with a sharp razor blade and incubated in seawater.

At the beginning of each summer, we extract total lipids from optic lobes (see Section 5.1) to use as a carrier for separating samples with low endogenous lipid contents.

3.6. Fin nerves

Fin nerves, obtained during the initial giant axon dissection (see Section 3.1), are cut into 1 to 2 mm segments with a sharp razor blade and then homogenized with a 10 ml Potter-Elvehjem homogenizer. Small samples are homogenized by hand. Following homogenization, large pieces are removed with forceps and the remaining suspension centrifuged in microcentrifuge tubes at 12000 × g for 4 minutes. Turbid supernatants are aliquoted and stored (see Section 3.2).

Because of the relatively large size of the axons and its homogeneous nature, intact fin nerves can be used for biochemical studies to determine the effects of electrical stimulation and depolarization on precursor incorporation and lipid product metabolism, and in autoradiographic studies to locate sites of lipid synthesis (see Section 5.3.1). For these studies, the nerves are incubated in seawater containing the labeled precursor.

3.7. Photoreceptor membranes

The following procedure is adapted from one used by Dr. Phyllis Robinson, Marine Biological Laboratory and Brandeis University (see also Saibil, 1990):

For biochemical and physiological studies, the squid are often dark-adapted for 1 to 2 hours before decapitation and the retinal dissection is carried out under infrared illumination. For enzyme studies, dark adaptation and dissection under dim light are probably unnecessary. The eyes are removed from the head and hemisected. The posterior cups, consisting largely of retinal tissue, are rinsed in cold, oxygenated calcium-deficient seawater (see Section 7) and frozen in liquid nitrogen. After thawing, these are placed in approximately ten volumes of isolation medium: 400 mM potassium chloride; 2 mM manganese chloride; 7.5 mM N-2-hydroxyethylpiperazine-N'-2-ethanesulfonic acid (HEPES), pH 7.8; 10 μM aprotinin; 0.1 mM phenylmethylsulfonyl fluoride; 5 mM dithiothreitol. The photoreceptor membranes separate as the cups are gently shaken on a rotary shaker. These *crude* photoreceptor membranes are collected by centrifugation at 12000 × g for 4 minutes. The photoreceptor membranes and remaining retinal cup are separately homogenized with glass homogenizers and then centrifuged at 12000 × g for 4 minutes to remove cellular debris. The supernatants are aliquoted and stored (see Section 3.2).

3.8. Giant synapse and pallial nerve

The methods for preparing this giant synapse is described by Stanley (1990). We will discuss the use of this preparation in autoradiographic studies to locate sites of lipid synthesis and turnover (see Section 5.3.2).

4. Studying lipid metabolism

Information in this Section is based largely on personal experience gained from working with the squid giant axon. Other squid nervous system preparations are discussed, though with less familiarity, because we feel that studies of lipid metabolism in these phylogenetically simple neural systems may provide novel information on its putative role(s) in coupling physiological stimulation with cellular response.

4.1. The giant axon

As an experimental model the squid giant axon is uniquely suited for studying: (1) the scope of *axonal* lipid metabolism and its regulation, and (2) the possible axonal contributions of this metabolism to the specialized physiological processes of axon and synaptic membranes. An attraction in using the giant axon derives from the increasing body of neurophysiological and cell biological information on: nerve impulse propagation (Adelman and Gilbert, 1990), axoplasmic transport (Brown and Lasek, 1990; Weiss *et al.*, 1990), calcium metabolism (Baker and DiPolo, 1984) and synaptic transmission (Llinas, 1984; Smith and Augustine, 1988; Stanley, 1990).

In principle, the methods applied to studying *axonal* lipid metabolism in giant axon and axoplasmic preparations (see Sections 4.1 and 4.2) can readily be used with other squid nervous tissues (see Section 4.3).

4.1.1. Incubation of giant axons

When intact giant axons are incubated with radioactive precursors (see Section 5.1), labeled lipids rapidly accumulate in both the giant axon and surrounding Schwann cells (see Section 2.2). We describe experimental details for the incubation of giant axons with labeled precursors and the characterization of lipid products.

4.1.1.1. Procedures

Disposable incubation chambers (40 × 5 × 2 mm^3), are constructed with: microscope slides, dental wax, vaseline and disposable petri dishes (Fig. 3). The short strands of dental wax are pressed along each narrow end of the slide. The rectangular chamber is completed with bridging strands of vaseline or silicon grease extruded from a 10 or 20 ml syringe onto the slide. Such a chamber holds up to 0.5 ml of solution. It should be prefilled with seawater to test for leaks. One to four axons at 100 μl/axon are incubated per chamber. Normal Millipore™-filtered (0.45 microns) seawater or artificial seawater (obtained in the Chemistry Room, Marine Biological Laboratory,

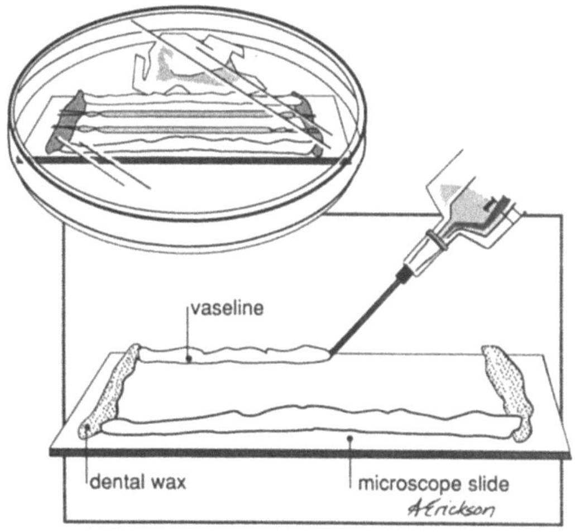

Figure 3. The lower figure illustrates a microscope slide that has been made into an incubation chamber with strands of dental wax and vaseline. In the upper figure, the chamber containing two giant axons, is enclosed in a petri dish with wetted tissues.

Woods Hole, MA, see Section 7) is used. Giant axons (see Section 3.1), tied at each end, are blotted on filter paper or Kimwipe™ to remove excess seawater and then placed in the chamber. The threads are pressed into the dental wax to hold down and stabilize the giant axons during the incubation. We find that 1 millicurie/ml (mCi/ml) of radioactive precursor is usually sufficient (depending on the precursor) to label *axonal* lipids at a reasonable rate (see Sections 2.2 and 4.1.1.2). During the incubation, each chamber is kept in a disposable petri dish (100 mm diameter) with wetted tissues to prevent evaporation (Fig. 3). Incubations are at room temperature, 20 °C to 22 °C. Higher temperatures, particularly above 27 °C, must be avoided as Schwann cells swell and vesiculate (Tytell, 1989) and the giant axons lose excitability. After incubation, giant axons are removed from the chamber, rinsed in several changes of cold seawater and blotted. The distribution of radioactive lipids is usually determined on separated extruded axoplasm and sheath samples (see Section 3.2).

Alternatively, intact segments of giant axon, which include the cortical axoplasm and axolemma, are separated from surrounding Schwann cells and analyzed. Metuzals *et al.* (1981) developed this separation procedure (see also Brown and Lasek, 1990). Briefly, the contacts between the axolemmal and the enveloping Schwann cell membranes are broken during a one hour incubation with trypsin (1 mg/ml). We add trypsin directly to the incubation medium (seawater plus radioactive precursor) an hour before the axons are removed. The axons are then lightly fixed with 0.5% glutaraldehyde in seawater at 4 °C for 15 minutes to stiffen the Schwann cell layer, and next transferred to cold hyperosmotic seawater made with 0.4 M sucrose. In this solution the giant axon shrinks from its Schwann cell layer, which can then be removed. A hole is made in the Schwann cell layer by pressing it with a sharp needle against the dissecting dish. Two needles are inserted into the break and the cut made circumferential. The Schwann cell layer is separated and peeled away from a section (approximately 2 cm) of giant axon. Both the Schwann cell layer and freed giant axon segment are extracted and their radioactive lipids analyzed (see Section 5.1).

4.1.1.2. Experimental considerations

The usefulness of the intact giant axon has already been demonstrated (see Section 2.2). It incorporates most precursors into lipid in both time- and concentration-dependent fashion. Radioactive lipids usually accumulate fast enough, i.e., $>10^3$ dpm/hr (disintegrations per minute/hr), for detailed characterization (see Section 5.1). Because lipids accumulate in both axoplasm and the sheath, the products in these distinct compartments can be compared. Furthermore, since giant axons can be separated from Schwann cells (above), the radioactive lipids in these compartments can alternatively be compared. Finally, the *intactness* of the preparation makes it possible to determine the effects of physiological perturbation, e.g., high frequency electrical stimulation, on lipid metabolism.

In designing these studies one should realize that lipid accumulation in the axon compartment is strongly influenced by the surrounding Schwann cells. Firstly, with all the precursors we have tested, lipid metabolism is dominated by the Schwann cells. This domination is predictable from the known paucity of membrane (lipids) in axoplasm (see Section 2.1). Secondly, it appears that lipids formed by the surrounding Schwann cells are actively transferred to the giant axon (see Section 2.2).

4.1.2. Injection of giant axons

To bypass the Schwann cell barrier and, thereby, limit Schwann cell influence on *axonal* lipid metabolism, precursors can be injected directly into intact giant axons. Unfortunately, the procedure is technically demanding and the injected precursors are poorly incorporated into lipid (see Section 2.2). We, therefore, recommend that one use extruded axoplasm (see Section 4.2) to characterize lipid metabolism inherent to axoplasm. It may be possible to find conditions to study axon-based lipid metabolism with the injected axon (see Section 4.1.2.2).

4.1.2.1. Procedures

Since injection procedures are technically difficult, they are best carried out as a collaborative venture with someone already experienced with the method. Besides injection, radioactive precursors can be introduced into axoplasm by perfusion. This alternative is technically even more difficult than injection, and, furthermore, may involve proteolytic treatment (Rosenberg, 1981; Adelman and Gilbert, 1990), which could inactivate axonal lipid enzymes and consequently reduce precursor incorporation. For those interested, the method is clearly described by Rosenberg (1981) and Adelman and Gilbert (1990).

The injection procedure that we use was developed by Dr. Ichii Tasaki (Tasaki *et al.*, 1961), who participated in our study (Gould, Jackson, *et al.*, 1983). Fresh giant axons (see Section 3.2) are freed of most, though not all, adherent fibers. Remaining small fibers provide structural stability. The giant axon is placed in a grooved lucite chamber that is wired for stimulation and recording (Tasaki *et al.*, 1961; Rosenberg, 1981; Adelman and Gilbert, 1990). The axon is positioned across parallel grooves in the chamber and over small platforms placed a few millimeters from each groove. The outer surfaces of the grooves and inner surfaces of the platforms are coated with wax to prevent leakage of seawater from the chamber and the consequential drying of the preparation. Once situated in the chamber on the proximal side to the right for right-handed injection, the axon is stimulated at low frequency (0.5 to 2 Hz) to monitor its *intactness* during the injection. The chamber is illuminated from below and the injection watched continually through a dissecting microscope. A glass micropipette, 5 to 8 cm long and 50 to 120 μm outer diameter, is filled (5 to 10 μl) with concentrated i.e., 50 to 100 μcuries/μl (μCi/μl), precursor dissolved in buffer X (see Section 7) or other suitable internal medium. Chlorophenol red, 0.025%, is included as a visual indicator. To facilitate the injection, both the giant axon and micropipette are prealigned, i.e., the micropipette is moved forward above the axon to make sure that it holds a parallel path. If not, the position of the axon should be altered. Once the pipette and axon are aligned, the proximal end of the axon is hemisected at a point just above the right-hand platform. Sharp Vannas ultra microscissors (Roboz Instrument Co.) should be reserved for this procedure. The micropipette is advanced into the opening and on through 2 to 3 cm of axoplasm. Care must be taken to prevent the micropipette tip from nicking the axolemmal membrane, for if the membrane is damaged and excitability lost, it will be necessary to start with a fresh axon. The precursor-containing solution (1 to 3 μl) is injected into the axoplasm with air pressure, applied via an attached 10 ml syringe, just before and during the slow withdrawal of the micropipette from the giant axon. Once the pipette tip is out, the axon is tied shut and incubated (see Section 4.1.1). If desired, injected axons can be electrically stimulated during the incubation.

4.1.2.2. Experimental considerations

The injection approach is designed to restrict the participation of the Schwann cells, since lipid precursors, deposited directly into axoplasm, may remain in the axoplasm for long periods of time (Tasaki *et al.*, 1961; Gould and Jackson, 1989). As the preparation remains intact and electrically excitable, effects of electrical stimulation and depolarization on lipid metabolism can be studied.

Unfortunately, extremely low amounts (usually >0.1%) of injected precursor (includes studies with ^{32}P-phosphate, γ-^{32}P-labeled ATP, ^{3}H-choline, ^{3}H-myo-inositol, ^{3}H-serine and ^{3}H-acetate) are incorporated into lipid even during incubations lasting several hours (see Section 2.2). Furthermore, in these studies we were unable to show that lipid formation increased systematically with time or precursor concentration. It might be possible to enhance precursor incorporation into lipid by this preparation, by including in addition to precursor, appropriate substrates and/or cofactors (see Section 7).

4.1.3. Axonal transport of lipid metabolizing enzymes

The availability of sensitive assays for lipid enzyme (see Section 5.2), make it feasible to study the association of these enzymes with axonal-transport organelles. Axonal transport is blocked at a point along the axon and the time-dependent changes in enzyme activity measured in narrow sections of axoplasm just proximal and distal to the blockage where transport organelles accumulate.

Chilling a narrow portion of the giant axon causes axoplasmic organelles to accumulate in a time-dependent fashion in defined regions just proximal (anterograde transport) and distal (retrograde transport) to the *cold block* (Fahim *et al.*, 1985; Miller and Lasek, 1985). At times when axonal transport organelles are accumulating, the regions of axoplasm are harvested and assayed for enzymes of lipid metabolism. A

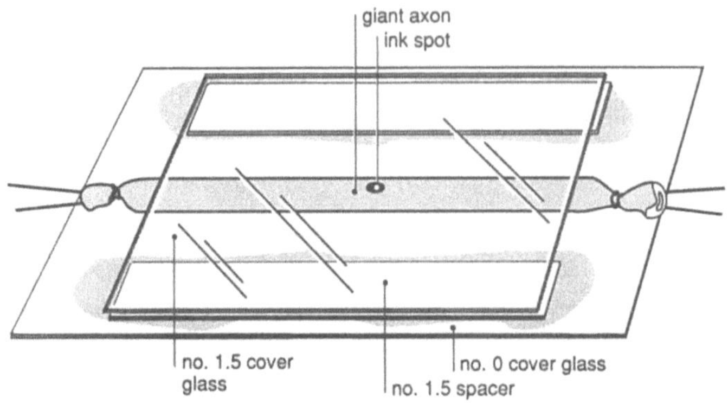

Figure 4. Illustration of a tied off giant axon marked with squid ink and placed in an incubation chamber for *cold-block*. Thin coverglass and spacers are used to flatten the giant axon and keep the entire cross-section close to the block and suitably cooled.

time-dependent increase in enzyme activity in proximal (and distal) regions would indicate axonal transport. Autoradiographic approaches could also be conducted to locate sites of lipid synthesis in regions where axonal transport organelles are accumulating (see Section 5.3.1).

4.1.3.1. Procedure

To set up a *cold-block*, cleaned giant axons (see Section 3.2) are first marked at their centers with a small spot of squid ink (Fig. 4). They are then incubated in special chambers, which are readily constructed on 0-thickness coverglass (22 × 60 mm). Slivers (24 × 4 mm) of 1.5-thick coverglass are attached to the edges of the large coverslip with silicon grease as spacers. Once the giant axon is positioned in the chambers, it is covered with a 1.5-thick coverglass (24 × 24 mm). The giant axon is bathed in seawater, which will perfuse around it by capillary action from an open end of the chamber. The chambers are then positioned on the *cold-block* device (Fig. 5).

To construct such a device, an aluminum block (9.5 × 7.5 × 5 cm^3, heater block insert, #5891-C75, Thomas Scientific, Swedesboro, NJ), a length of stainless steel pipe (15 × 0.5 cm, o.d.) and silicon caulking are used (Fig. 5). A groove (0.8 × 0.8 cm) is machined into the face of the block and is lined with the caulking for insulation. The pipe is positioned in the groove so that its top surface is flush with the block and the caulking allowed to harden. During incubations the temperature of the pipe is held between −5 °C and 0 °C with circulating ethanol-water pumped through it with a constant temperature circulator bath. To hold the temperature of the aluminum block at 18 °C to 20 °C, it may be necessary to place the devise in a water bath. Sharp temperature gradients form about 2 mm on either side of the pipe's center. After incubations lasting 1 to 3 hours, the giant axon is removed, blotted free of seawater, and the central *marked* section of axon (0.3 mm) excised. Axoplasm is carefully extruded from the adjacent 4 mm proximal (P-1) and distal (D-1) regions. Additional axoplasm is extruded from the remaining axon (P-2 and D-2). Each axoplasm sample is dispersed in a small volume (10 to 20 μl) of TS (see Section 5.2., at the pH used for enzyme assay). For less sensitive assays, axoplasm pooled from several *cold-blocked* axons might be needed.

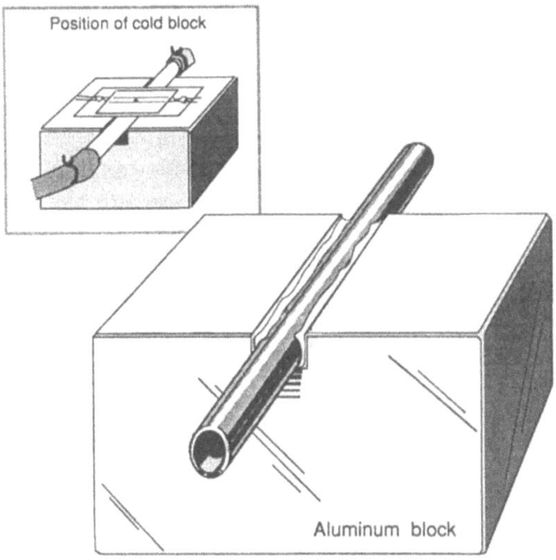

Figure 5. Picture of a *cold-block* devise, constructed from a machined aluminum block, fitted with a stainless steel pipe and separated by some caulking material (see also Fig. 1, Fahim *et al.*, 1985). The device is connected to a circulator bath and one or more slides placed on it (upper portion). The central region of the axon is chilled, while the rest of the axon stays at the temperature of the block. Organelles moving at rapid rates in the warm-regions of axon accumulate when they reach the colder regions. This device can also be used to cold block extruded axoplasm (see Section 5.2.1).

4.1.3.2. Experimental considerations

Several enzyme assays, i.e., phosphatidylinositol synthase, diacylglycerol kinase and acyltransferase, are sensitive enough to measure activity accurately in 4 mm lengths of axoplasm (see Section 5.2). It is relatively straightforward to use these assays to determine the possible time-dependent (over several hours) increases enzyme activity in regions of organelle accumulation, i.e., P-1 versus P-2 and D-1 versus D-2. The activities of transported lipid enzymes are expected to increase relative to bulk protein, which is largely cytoskeletal (see Section 2.1), and would not redistribute during the short times used for *cold block*.

Vogel *et al.* (1988) have successfully used this paradigm to demonstrate that G-proteins rapidly accumulate proximal, though not distal, to a *cold blocked* giant axon. We have also conducted preliminary studies demonstrating that phosphatidylinositol synthase actively accumulates in both anterograde and retrograde directions. Furthermore, the high activity of this enzyme in a subcellular fraction rich in axonal transport vesicles (see Section 2.3) is consistent with its association in these vesicles. We are currently using autoradiographic approaches (see Section 5.3.1) to compliment and strengthen these findings.

It is interesting that the G proteins and phosphatidylinositol synthase appears to behave differently, i.e., one accumulates in both directions, whereas the other accumulates in only the anterograde direction. Such information will be useful in gaining understanding of the mechanisms involved in the processing of different axonally transported proteins for their return to the cell soma.

An axonal transport study of this sort is a major undertaking. Small amounts of samples from each giant axon must be reproducibly extruded and the subsequent handling procedures must be carefully managed. Furthermore, since some *cold-blocked* giant axons inevitably develop white spots during the incubation, extra axons need to be committed to each experiment.

4.2. Extruded axoplasm

Axon-specific lipid metabolism is readily studied with extruded axoplasm (see Section 3.2) and its subfractions (see Section 3.2.1). Both from precursor incorporation (see Section 2.2) and from enzyme assay studies (see Section 2.3), we now realize that axonal lipid metabolism in this specialized cytoplasm is both wide-ranging and relatively weak (see Section 2.2). Future biochemical studies should greatly increase our understanding of the scope and regulation of this metabolism. In addition, quantitative electron microscopic autoradiographic approaches with extruded axoplasm will complement biochemical studies designed to locate sites of lipid metabolic machinery in axoplasm (see Section 5.3.1).

4.2.1. Procedures

Radioactive precursors of high specific activity (see Section 5.1), incubated with extruded axoplasm in a suitable internal medium, such as buffer X (Brown and Lasek, 1990), are incorporated into lipid in time- and concentration-dependent fashion (see Section 2.2). Depending on the precursor, the nature of the lipid products will vary. To characterize the lipids, relatively quick and simple extraction and separation methods are used (see Section 5.1). The incubation medium can easily be modified to study the effects of potential regulators, e.g., ions, other substrates and lipid products, on the incorporation rate and on the distribution of label among lipid products. The properties of axoplasm can be compared with axoplasmic subfractions (see Section 4.2.3) and with homogenates and subfractions of the giant fiber lobe as well as other squid preparations.

A variety of sensitive assays can be used to demonstrate and characterize enzymes of lipid metabolism in axoplasm (see Section 5.2). To select among *lipid* enzymes to study, one should consider results of precursor incubation studies as these will pinpoint the more actively metabolized lipids.

Pooled axoplasm, dispersed in a suitable buffer, is used as *enzyme* source. Properties of the enzyme activity, present in extruded axoplasm, can be compared with the activity in other squid preparations (following Sections). Since many activities appear to be quite stable to freezing, assays can be performed on samples that have been aliquoted and stored.

Due to Schwann cell and axolemma barriers, it is appealing to use extruded axoplasm for autoradiographic localization studies. The design of such studies is outlined in Section 5.3.1.

4.2.2. Experimental considerations

Axoplasm is relatively simple to obtain. It can be collected and used after storage for enzyme studies. Unlike axoplasm that is injected with precursor (see Section 4.1.2), it will readily incorporate radioactive precursors into lipid. However, in these studies, incorporation rates are generally lower than in homogenates from other squid tissues. Also, rates of incorporation may vary during the summer. Low incorporation rates may, at least in part, be overcome by increasing the amounts of axoplasm, amounts or specific activity of radioactive precursor, and/or time of incubation. It may also be possible to augment the incorporation by fortifying the incubation medium with appropriate substrates and cofactors (see Section 7). To limit variability, we suggest that large experiments be conducted with single *pooled* batches of axoplasm.

4.2.3. Axoplasmic subfractions

Axoplasm appears to have a far more limited set of organelles than the neuronal soma. It contains *mitochondria*, a stationary *endoplasmic reticular network* (Ellisman and Lindsey, 1983; Terasaki *et al.*, 1986; Vale and Hotani, 1988) and various *rapidly-transported organelles*, including clear and dense core vesicles and *lysosomes*. It lacks *nuclear membranes*, *rough endoplasmic reticulum*, and *Golgi apparatus*. Subcellular fractionation methods should allow the separation and subsequent biochemical characterization of axoplasmic organelles, i.e., their protein, lipid and enzyme makeup. The assignment of *lipid* enzymes to specific subcellular organelles may provide clues to their function (see Section 6).

4.2.3.1. Procedures

Squid axoplasm has been subfractionated by differential and gradient centrifugation procedures (see Section 2.3). Some lipid enzymes associate exclusively with the organelles (crude microsomes) while others partition into both microsomal and cytosolic fractions (see Section 2.3). Separated fractions can be used in precursor incubation studies (never tried) or for enzyme assay (see Section 2.3). If they are to be used in precursor incubation studies, the medium will most likely have to be fortified with appropriate substrates and cofactors to stimulate incorporation.

4.2.3.2. Experimental considerations

The major constituents of axoplasm, cytoskeletal proteins, would, at best, play a regulatory role in lipid metabolism. Axoplasmic membranes probably contain most *lipid* enzymes, though some are also present in cytosol (see Section 2.3). Studies that determine the subcellular location of *lipid* enzymes may serve several purposes: i.e., (1) Provide a biochemical basis for classifying axoplasmic organelles. (2) Provide information on the possible roles axoplasmic organelles may play. (3) Provide a means to determine if membrane association is involved in enzyme regulation. Protein kinase C and CDP-choline synthase (cytidine 5'-triphosphate:choline phosphate cytidylyltransferase, EC 2.7.7.15) are two enzymes whose regulation is influenced by association with membranes (Nishizuka, 1986; Pelech and Vance, 1989).

It takes 2 to 3 hours to prepare the relatively large amounts of starting axoplasm (50 to 100 μl from 8 to 12 giant axons) and an additional 3 to 4 hours to fractionate this material (see Section 3.2.1). The current procedure of collecting fractions as drops from the tiny gradient fractionation tubes is not very precise. Therefore, the imprecision of the current method and the small quantities of membrane fractions may affect the reproducibility among experiments.

4.3. Giant fiber lobe, retinal fibers, retinal fiber axolemma, optic lobe, optic lobe synaptosomes, fin nerve, retina and photoreceptor membranes

All squid neural tissues and fractions that have been tested actively incorporate precursor (see Section 2.2) into lipid and appear to contain high levels of *lipid* enzymes (see Section 2.3). Comparative experiments on the giant axon and extruded axoplasm use the giant fiber lobe, since it is a relatively homogeneous preparation containing the neuronal perikarya from which the giant axons originate. Cells from the giant fiber lobe have been cultured (Rice *et al.*, 1990) and may also prove to be a useful model, since they develop processes and express sodium and potassium channels in their plasma membranes (Brismar and Gilly, 1987; Parsons and Chou, 1989). Major aims of lipid studies with all of these preparations might be to: (1) Establish differences in the properties of axoplasm directed lipid metabolism from other tissues. (2) Demonstrate that lipid metabolism responds to physiologically relevant perturbation.

4.3.1. Procedures

Intact preparations are used soon after they are dissected. Incubations with radiolabeled precursors are in formulated seawater or natural Millipore™-filtered seawater. We use formulated seawater, since its composition is constant and it is stocked at the Chemistry Room of the Marine Biological Laboratory, Woods Hole, MA (see Section 7). Unlike blood, seawater lacks proteins or amino acids (Brown and Lasek, 1990). It might be worthwhile to supplement seawater with these and other potential affecters and cofactors (see Section 7) to increase lipid metabolism. In studies with cells in culture, the incubations should be carried out in the growth medium, if levels of unlabeled precursor are inconsequential. If the medium contains high levels of precursor, a minimal essential salt medium should be used for the short incubations with labeled precursor. The complete medium, possibly with added unlabeled precursor would provide an ideal *chase* environment.

For studies with tissue homogenates or fractions, we recommend buffer X, or half strength buffer X, as these media have been shown to provide a supportive *internal* environment for axoplasmic transport studies (Brown and Lasek, 1990; Weiss *et al.*, 1990). In designing studies with these preparations, it should be noted that lipid precursors are, with the exception of myo-inositol, incorporated far better (4 to 80 fold on a protein basis) by intact versus homogenized stellate ganglia and optic lobe synaptosomes (Gould *et al.*, 1982).

A basic strategy that can be used in planning studies to characterize lipid metabolism in these tissue preparations is as follows: First, monitor the time-dependent incorporation of the chosen precursor into lipids by the preparation(s) (see Section 5.1 for details). Second, once these non-stimulated incorporation studies are completed, the study should be repeated with the system stimulated physiologically. Third, after labeling the lipids in a pre-incubation (1 to 2 hours), the preparation is placed into a *chase* medium and the time-dependent *turnover* of labeled lipids monitored. Fourth, once turnover rates have been determined, the effects of physiological stimulation on lipid turnover rates could be studied. Fifth, if a response of lipid metabolism to stimulation is found, assays of the potentially relevant enzymes can be used to uncover the mechanism underlying the enzyme response. This path from basic precursor incorporation studies to enzyme regulation may be useful in demonstrating the way in which physiological stimulation is translated by the preparation. However, before starting the effort the appropriate precursor(s) and preparation(s) must be chosen (see Section 5.1).

4.3.2. Experimental considerations

It is best in designing studies to relate lipid metabolism to physiological stimulation to choose a relatively homogenous preparation, and one that is physiologically responsive. For example, both fin nerves and retinal fiber are relatively homogenous though only the former is relatively easy to electrically stimulate. The retina is rich in photoreceptor membranes that respond to light signals (Saibil, 1990) and optic lobes are highly cholinergic and secrete transmitter upon depolarization (Dowdall and Whittaker, 1973; Cohen *et al.*, 1990). With these preparations, once basic parameters are worked out for a non-excitable situation, the effect of excitation on lipid metabolism should be rather straightforward.

However, studies of this nature are time-consuming and success is not guaranteed. For many investigators, the time to do experiments on squid is limited. The season when the squid are collected and the method of collecting the squid contribute to the general health and age of the squid, which can possibly cause experimental variation.

5. Useful lipid techniques

In this Section techniques and approaches that we use for studying squid neural tissue lipid metabolism are described.

5.1. Precursor selection and lipid product identification

Many radioactive lipid precursors are commercially available. We have used tritiated choline, ethanolamine, glycerol, methyl-labeled methionine, myo-inositol and serine, and ^{32}P-orthophosphate and ATP in our initial studies to determine the scope of lipid metabolism in the squid giant axon and in extruded axoplasm (see Section 2.2). Of these precursors, tritiated choline, myo-inositol and serine were the most selective for *single* phospholipid classes; i.e., phosphatidylcholine, phosphatidylinositol and phosphatidylserine, respectively. This selectivity could be of potential value in future studies that examine the effects of functional perturbation on phospholipid metabolism, since any change in precursor incorporation (as measured by lipid extractability; see Section 5.1), would indicate that the metabolism of the specific phospholipid (class) was responding to the perturbation. Furthermore, follow-up autoradiographic studies could be used to locate the morphological site(s) of the metabolic response (see Section 5.3).

Other lipid precursors, glucose, acetate, fatty acids, and including many listed above are incorporated into multiple lipid products. On the one hand, the complex incorporation patterns of these precursors require more care in the analysis. On the other hand, their widespread distribution insures a far greater probability that a response in lipid metabolism to a physiological perturbation will be detected.

In order to identify metabolic products of precursor incorporation (see Section 4.1) or enzyme assay (see Section 5.2), extraction procedures of Folch *et al.*, (1957) or Bligh and Dyer (1956) are usually followed (Kates, 1986). Complete extraction of acidic lipids, especially polyphosphoinositides, requires acidified solvents (Hauser and Eichberg, 1973). We generally extract lipids from tissues and enzyme incubations by a modified Folch procedure.

Tissue samples of 20 mg wet weight or less are homogenized by hand in a 1 ml conical ground glass homogenizer with 0.4 ml of chloroform:methanol (C:M), 1:1, vol:vol. For larger samples the amount of extraction solvent and other solutions is increased proportionally. The homogenate is transferred to a microcentrifuge tube (1.5 ml) and the homogenizer rinsed twice with additional (0.4 ml) C:M, 1:1, vol:vol. Tubes containing the pooled extracts are sonicated (15 to 30 seconds in a bath sonicator) and the insoluble material pelleted by centrifugation at 12000 × g for 3 minutes. The extracts are transferred to culture tubes (13 × 100 cm) and the pellet resuspended by sonication in 0.8 ml of C:M, 1:1, vol:vol., and the sequence repeated. The insoluble pellet is next extracted with 1 ml of C:M, 2:1, vol:vol. One ml of chloroform is added to the combined extracts to give a total lipid extract in 4 ml (≥20 volumes) of C:M, 2:1, vol:vol. The insoluble residue is extracted with 0.8 ml of distilled water and this extract is added to the pooled organic solvent extract. The phases are mixed on a vortex mixer (15 to 30 seconds) and then separated by centrifugation at 2000 revolutions/min (rpm) for 3 minutes. The upper phase is transferred to a separate tube and the lower lipid-rich phase is washed (usually twice) with 1.6 ml of theoretical upper phase (C:M:water, 3:48:47 by vol). The washed lower phase contains the labeled lipids separated from precursors and aqueous metabolites. The samples are warmed in a water bath and solvent removed under a stream of nitrogen. The lipids are redissolved in 1 ml of chloroform and a 0.1 ml portion taken for scintillation counting (measurement of total lipid radioactivity). The remaining lipid extract is generally used in determining the label distribution among different lipid classes (see below). Similarly, the label distributions among aqueous phase and proteinaceous residue components can be determined (see below).

For enzyme incubations, with homogenized tissues, an aliquot of the reaction mix is extracted directly with C:M, 2:1, vol:vol., 2 or 4 ml. This extraction solvent volume should be sufficient (≥20 times the sample volume) to solubilize all aqueous material into a single organic phase. Afterwards, an aqueous phase is produced with the addition of 0.2 volumes of water or acid (see below). Upon vortex mixing and centrifugation, the lipid will partition into the lower, chloroform-rich layer, while the water soluble precursors and metabolites will partition into the upper layer. Further details are given for separate enzyme assays (see Section 5.2).

With precursors, i.e., myo-inositol, phosphate and serine, that are selectively incorporated into acidic lipids, HCl is added to the extraction solvents to a final concentration of 0.5%. The extract is partitioned with one-fifth volume of 1 N HCl and 0.01 N HCl is used in the *theoretical* upper phase washes. These additions improve the extraction and subsequent recovery of acidic lipids. With general precursors (e.g., glucose, glycerol and fatty acids), that are incorporated into both neutral and highly acidic lipids, it is perhaps safest to divide the sample and use both normal and acidic extraction protocols.

Lipids are routinely separated on silica gel 60 plastic-backed TLC plates (#5506, Merck, Alltech Associates, Deerfield, IL). The dried lipids are dissolved in 20 µl of chloroform or C:M, 2:1, vol:vol, and spotted (1 cm from the bottom) as bands, 0.8 to 1 cm long, on plates that are 10 cm high and cut to accommodate samples and lipid standards. The plates are briefly activated with a stream of hot air from a heavy duty warm air dryer (#09-201-5, Fisher Scientific) and immediately placed in the developing solvent (see Section 7). Solvent #1 (C:M:acetic acid:water, 65:45:1:4 by vol) separates all major phospholipids in a single run. The major lipids appear as yellow spots when the plates are exposed (in closed jars) to iodine vapor. The distance of migration of the lipid relative to the solvent front is measured and defined as the Rf values (see Section 7 for Rf values). Phosphatidylcholine and phosphatidylethanolamine are the major spots seen on the TLC separations of tissue lipids.

We have modified this separation procedure for some precursors. (1) Choline lipids, which migrate more slowly, are better separated when the plate is redeveloped a second time. It is only necessary to run this second development (in the same solvent after reactivation) half-way up the plate. Standards of lysolecithin, sphingomyelin and phosphatidylcholine are included in external lanes. (2) Ethanolamine plasmalogens migrate with other ethanolamine phosphoglycerides during development. We first separate all the lipids with solvent #1. Phosphatidylinositol and phosphatidylethanolamine standards are included in the external lanes. Diradyl (two fatty acyl groups) and lyso (one fatty acid) ethanolamine phospholipid are separated in this initial run. The outer lanes are removed and exposed to iodine to visualize the standards. The lanes are repositioned on the plate, which is cut just above the phosphatidylinositol spot and below the phosphatidylethanolamine spot. The upper portion of the plate, containing diradyl ethanolamine phosphoglyceride is exposed in a closed jar to HCl vapors, for 10 minutes at 20 °C, to hydrolyze the plasmalogens (Horrocks and Sun, 1972). Then this portion of the plate is dried under hot air to remove the acid and reactivate the plate, and rechromatographed in solvent #1. Plasmalogens, converted to lysophosphatidylethanolamine, migrate slower (about half the distance) than other (alkylacyl and diacyl) ethanolamine phosphoglycerides. The relative distribution of ethanolamine label among lysolipid, plasmalogen, and other ethanolamine phosphoglycerides (and other TLC separated radioactive lipids) is determined by scraping the gel containing individual lipids from the plates, putting the scrapings into vials and counting them (see below). The radioactivity in alkylacyl and diacyl species, included with the fastest migrating ethanolamine phosphoglycerides, can be determined by treating the gel scrapings (containing these lipids) with methylamine (Clarke and Dawson, 1982) to hydrolyze the ester-linked fatty acids. The resultant glycerophosphorylethanolamine (derived from diacyl species) is partitioned from lysoethanolamine phosphoglyceride (derived from alkylacyl species) and radioactivity in the two phases counted.

Polyphosphoinositides remain at the origin in solvent #1 and most other chromatographic systems. With precursors (^{3}H-myo-inositol and ^{32}P-phosphate) that are actively incorporated into these lipids, we use the chromatographic system devised by Shaikh and Palmer (1976). Remember to use acid in the extraction solvents. Thin layer plates (silica gel 60) are impregnated with 0.25% potassium oxalate in methanol:water, 1:1 vol:vol, for 10 to 20 minutes and then allowed to dry. Unlabeled phosphatidylinositol 4-phosphate and phosphatidylinositol 4,5-bisphosphate (each 5 to 10 μg) are added to the samples before spotting. The plates are developed with solvent #2 (C:M:acetone:acetic acid:water, 40:15:13:12:8 by vol, see Section 7 for Rf values).

Neutral lipids are separated with developing solvent #3 (hexane:diethylether:acetic acid, 70:30:1.5 by vol, see Section 7 for Rf values). With certain precursors, i.e., glycerol, glucose, acetate and fatty acids, it is desirable to separate phospholipids, glycolipids and neutral lipids on one plate. A 20 cm high plate is used. It is first pre-run in developing solvent #1 to displace contaminating material in the silica gel to the top of the plate. Otherwise, the contaminants cause a spreading of neutral lipids. The lipid samples and phosphatidylethanolamine standard (external lanes) are spotted, the plate heat-activated and developed in solvent #1 to a height of 12 cm. Neutral lipids migrate with the solvent front. The external lanes are removed and the phosphatidylethanolamine spot visualize. The plate is cut above this spot and the upper portion of the plate is dried, reactivated and run in developing solvent #3 to separate the neutral lipids. Other suitable one and two dimensional TLC separation systems are described by Marinetti (1976) and Kates (1986).

The distribution of radioactivity among lipids is determined by scraping the silica gel containing each separate lipid (visualized with iodine) into a mini-vial (after the iodine has evaporated), solubilizing the lipid with methanol (0.6 ml) and counting the sample with a suitable scintillation cocktail. Sometimes it is necessary to first locate the labeled lipids by fluorography (tritium) or autoradiography (^{14}C and ^{32}P).

Our experience in separating water-soluble metabolites is limited to tritiated choline and myo-inositol. Choline metabolites, present in the aqueous extract, are separated on PEI-cellulose F plastic backed sheets (#5504, Alltech) with developing solvent #4 (butanol:acetic acid:water, 5:2:3 by vol, see Section 7 for Rf values). Most likely, ethanolamine and its metabolites, phosphoethanolamine and CDP- ethanolamine, migrate like the corresponding choline derivatives in this system. Aqueous products are located with fluorography and the radioactivity in the labeled spots counted after scraping the plate (see above).

Water soluble products of phosphoinositide metabolism, i.e., inositol phosphates, are separated on SEP-PAK minicolumns (Waters Associates) by the procedure of Wreggett and Irving (1987).

5.2. Enzymes of lipid metabolism

Faced with relatively small volumes of axoplasm (see Section 3.2), enzyme assays should be simple, sensitive and highly reproducible. We recommend that assay conditions be tested with stored squid tissue homogenates or neural tissues from other species before the summer research period begins.

Enzymes of lipid metabolism are rarely assayed under physiological conditions, largely because they are membrane-bound and their substrates are amphipathic. Since exogenous lipid substrates interact poorly with membrane-bound enzymes in an aqueous environment, detergents are often used to enhance enzyme-substrate interaction. In some cases, *artificial* aqueous substrates are used instead of detergent. For example, phosphatidylinositol synthase uses a short chain CDP-diglyceride (didecanoyl, diC-10) far more effectively in the absence of detergent (see Section 5.2.1) than longer chain CDP-diglyceride in the presence of detergent. Another way to increase lipid substrate-enzyme interactions is with phospholipid transfer proteins. Phosphatidylcholine transfer protein has been used to stimulate the interaction of substrate with the enzyme synthesizing sphingomyelin (Voelker and Kennedy, 1982; van den Hill *et al.*, 1985). In many instances, however, the relatively nonphysiological conditions used for enzyme assays, including: *artificial* substrate, detergent, pH, monovalent and/or divalent ion concentration, do not mimic the conditions under which the enzyme is functioning in the native membrane. Hence, one should realize that the assay conditions may impair the ability of the enzyme to respond to physiologically-relevant agents that are introduced into the assay system.

5.2.1. Phosphatidylinositol synthase (CDP-diacylglycerol: myo-inositol transferase) (EC 2.7.8.11)

The enzyme assay is based on the incorporation of radioactive [2-^{3}H]-myo- inositol into lipid. Assays are conducted in disposable tubes (12 × 75 mm) into which 22.8 nmol (3 μl) CDP-didecanoyl-*sn*-diglyceride (from a 6.25 mg/ml stock in chloroform, Serdary Research Labs, London, Ontario, Canada) is first added. After the solvent has evaporated, a mixture (25 μl final volume) containing: 100 mM Tris-HCl, pH 8.5; 10 mM $MgCl_2$; 0.3 M KCl; 5 mM EGTA; 5 mM β-mercaptoethanol; 0.3 mg bovine albumin (A-6003, Sigma Chem. Co.); 1 mM myo-inositol; and 0.5 μCi [2-^{3}H]-myo-inositol (specific activity 15 Ci/mmol, ART-261, American Radiolabeled Chemicals, St. Louis) is incubated with *enzyme* (10 to 50 μg of protein) for 5 or 10 minutes at 20 °C. Radioactive lipid products are separated by acidic extraction (see Section 5.1). Usually the entire washed lower phase is counted, since radioactive myo-inositol is mainly (>95%) incorporated into phosphatidylinositol. Alternatively, an aliquot from the reaction mixture is spotted on filter discs and the aqueous radioactivity extracted with trichloroacetic acid (TCA) (Gould, Spivack, *et al.*, 1983).

5.2.2. 1,2-Diacylglycerol kinase (ATP:diacylglycerol phosphotransferase) (EC 2.7.1.-)

1,2-Diacylglycerol kinase activity is measured by the procedure of Lapetina and Hawthorne (1971) with minor modification. The assay measures the incorporation of ^{32}P-phosphate from γ-labeled ATP into phosphatidic acid. Incubations are in disposable tubes (12 × 75 mm) into which 20 nmol (3 μl in chloroform) of 1,2-dioleoyl-*sn*-glycerol (D-0138, Sigma) or 1,2-didecanoyl-*sn*-glycerol (B-272, Serdary Research Laboratory, London, Ontario, Canada) is first added. After the solvent has evaporated a mixture (60 μl final volume) containing: 50 mM imidazole buffer, pH 7.4; 10 mM $MgCl_2$; 3 mM deoxycholate; 10 mM KF; 5 mM ATP; approximately 1 μCi of ^{32}P-γ-labeled ATP (NEG-002, DuPont Co., NEN Research Products, Boston, MA.); and 0.3 mg bovine albumin is incubated with *enzyme* (25 to 100 μg of protein) for 10 minutes at 20 °C. Total lipids are separated by acid extraction (see Section 5.1) and individual lipids then separated on oxalate-impregnated TLC plates with solvent #2, and the labeled spots scraped and counted (see Section 5.1).

5.2.3. Phosphatidylinositol kinase (EC 2.7.1.67)

Phosphatidylinositol kinase activity is measured by a modification of the procedure of Kai *et al.* (1966). The assay measures the transfer of phosphate from γ-labeled ATP into phosphatidylinositol 4-phosphate. Phosphatidylinositol (60 μmol in 3 μl of chloroform, P-0639, Sigma) is added to the disposable culture tube (12 × 75 mm) and the solvent allowed to evaporate. A mixture (60 μl final volume) of: 25 mM Tris-HCl, pH 8.0; 375 μg of bovine albumin; 1 mM EGTA; 10 mM $MgCl_2$; 5 mM ATP; 0.5 μCi of γ-labeled ATP (DuPont-NEN, Section 5.2.2), and detergent (1–4% Triton X-100, or 0.5–2% Cutscum) is incubated with the *enzyme* (25 to 100 μg of protein) for 5 or 10 minutes at 20 °C. The reaction is stopped with acidic C:M and extracted (see Section 5.1). Radioactivity in phosphatidylinositol 4-phosphate and other lipids are measured after separation in solvent #2 (see Section 5.1).

5.2.4. Phospholipase A_2 (EC 3.1.1.4)

Phospholipase A_2 activity is measured by the procedure of Alberghina *et al.*, (1988). The assay measures the liberation of the oleoyl moiety from exogenous ^{14}C-labeled phosphatidylcholine substrate. Reactions are in disposable culture tubes (12 × 75 mm) into which 66 nmol (in 21 μl of chloroform) of L-3-phosphatidylcholine, 1-palmitoyl-2-[1-^{14}C]-oleoyl (specific activity, 3374 dpm/nmol, CFA 656, Amersham Corp., Arlington Hts., IL) is first added. After the solvent has evaporated under a stream of nitrogen gas, a mixture (in 0.3 ml final volume) of: 0.1 M Tris-HCl, pH 7.4; and 0.5 mg sodium taurocholate is added. To check for calcium-activation, three conditions are tested with each sample: no addition, 5 mM $CaCl_2$, and 2 mM EGTA. After brief sonication (10 seconds in an ultrasonic cleaner, e.g., Model 250, EM/C), *enzyme* (50 to 400 μg of protein) is added to start the reaction. Incubations are for two hours at 22 °C with constant shaking. Control tubes, lacking enzyme, are included in each experiment. The reactions are stopped with 0.8 ml of cold C:M 1:1, vol:vol, and then 0.15 ml of 0.1 M KCl is added. The phases are mixed on a vortex mixer (30 seconds) and then separated by centrifugation for 3 minutes at 2000 rpm. The lower phase is transferred to a second tube and the upper phase washed with an additional 200 μl of chloroform. The chloroform wash (lower layer) is removed and combined with the original organic solvent extract. These pooled extracts are applied to small columns, made with Pasteur pipettes containing 5 cm of heat-activated silicic acid. Liberated fatty acids are selectively eluted with chloroform (6 ml) into mini-vials. Phosphatidylcholine (substrate) and lysophosphatidylcholine remain on the column. The solvent is evaporated in a hood with a stream of warm air (Oster air blower, Fisher Scientific) and the radioactive fatty acid measured by liquid scintillation counting.

5.2.5. Acyl-CoA: 1-acyl-*sn*-glycero-3-phosphocholine acyltransferase (EC 2.3.1.23)

Acyl-CoA: 1-acyl-*sn*-glycero-3-phosphocholine acyltransferase activity is measured according to the procedure of Baker and Thompson (1973) with slight modification. The assay measures the transfer of the oleoyl moiety from [1-^{14}C]- oleoyl-CoA into phosphatidylcholine. The incubation mixture (in 1 ml final volume) contains: 70 mM Tris-HCl, pH 7.4; 2 mM $MgCl_2$; 100 μM lysophosphatidylcholine from egg yolk (4-6003, Supelco., Inc., Bellefonte, PA, emulsified at 1 mg/ml in 1 mM Tris-HCl buffer, pH 7.4, by sonication); 50 μM oleoyl-CoA, and 0.07 μCi [1-^{14}C]-oleoyl-CoA, specific activity 52 mCi/nmol, (CFA. 634, Amersham Corp. Arlington Hts., IL.). After a brief preincubation, *enzyme* (100 to 200 μg protein) is added and the tubes incubated at 22 °C for 10 minutes with continuous shaking. Control incubations, lacking enzyme, are included in each experiment. Four ml of cold C:M, 2:1, vol:vol, containing carrier lipid (approximately 2 to 5 μg lipid phosphorus, from squid optic lobe, Section 3.5) is added to stop the reaction. Phases are mixed on a vortex mixer (30 seconds) and separated by centrifugation at 2000 rpm for 3 minutes. The lower organic phase is transferred to a second tube and the solvent removed under nitrogen. The lipid, dissolved in 20 μl of chloroform, is spotted on TLC plates. Following separation in solvent #1 (see Sections 5.1 and 7), the phosphatidylcholine spot (contains >95% of lipid radioactivity), visualized with iodine vapor, is scraped from the plate, and counted (see Section 5.1).

5.2.6. Serine base-exchange

Serine base-exchange activity is measured by a slight modification of the method of Kanfer (1972). The assay measures the incorporation of [^{3}H]-serine into lipid. Incubations are in disposable tubes (13 × 100 mm) with a mixture (200 μl final volume) containing: 100 mM HEPES, pH 7.6; 2.5 mM $CaCl_2$; 25 μM L-serine; and 5 μCi L-[G-^{3}H]-serine (specific activity 5-20 μCi, no. 20050, ICN Biomedicals, Irvine, CA.). *Enzyme* (50 to 100 μg of protein) is added to start the reaction, and the tubes are incubated at 20 to 22 °C for 30 to 180 minutes with continuous shaking. Control incubations, lacking enzyme, are included in each experiment. The reaction is stopped with 4 ml of C:M 2:1, vol:vol, and extracted (see Section 5.1). Lipids are separated on TLC plates in solvent #1 (see Section 7). Phosphatidylserine and phosphatidyl-ethanolamine spots (contain >95% of lipid radioactivity) are scraped and counted (see Section 5.1).

5.2.7. Phospholipid transfer proteins

The assays of phosphatidylcholine and phosphatidylinositol transfer proteins, described in an earlier publication (Gould, Spivack, *et al.*, 1983), are based on the exchange of radioactive phospholipids from labeled organelle or liposome donors to unlabeled acceptors. The donors and acceptors are separated by centrifugation, and a non-exchangeable carrier, e.g., [^{14}C]-cholesterol oleate, is used to measure non-specific sticking of donors to acceptors. These assays are rather tedious, and in our experiments were conducted with *enzyme* that had been frozen, stored and assayed at the University of Alberta.

Assays based on the transfer of fluorescent phospholipids between donor and acceptor liposomes have recently been developed (Somerharju *et al.*, 1981; Somerharju *et al.*, 1987; Van Paridon *et al.*, 1987). The principal of this newer assay is based on the fact that fluorescent phospholipids transferred from liposomes in which they are at high concentration quenched to liposomes lacking fluorescent lipids results in increased fluorescence. Such assays would be more sensitive and the transfer process could be monitored continuously.

5.2.8. Octopine dehydrogenase (EC 1.5.1.11); procedure from Dr. Michael Dowdall, (1989) Univ. of Nottingham

The assay of octopamine dehydrogenase is included because of its potential usefulness as a cytosolic marker for squid axoplasm and other squid neural tissues (see below). The need for a suitable cytoplasmic marker enzyme stems from recent interest in the mechanism of regulation of membrane associated enzymes, such as diacylglycerol kinase (see Section 5.2.2), calcium-independent phospholipase A_2 (see Section 5.2.4), protein kinase C and CDP-choline synthase, that are stimulated by a translocation from the cytosol to a membrane compartment (see Section 6).

The classic cytoplasmic marker of mammalian tissues, lactate dehydrogenase, cannot be used in squid neural tissues. Its activity in squid optic lobe, and probably other squid tissues as well, is only 5% that measured in mammalian brain (Dowdall and Whittaker, 1973; Dowdall and Downie, 1988).

Octopine dehydrogenase catalyzes the reversible reductive-condensation of L-arginine and pyruvic acid, with reduced nicotinamide adenine dinucleotide (NADH) as cofactor, to form octopine (van Thoai *et al.*, 1969). Reaction mixtures (1 ml final volume) containing: 4 mM L-arginine; 4 mM sodium pyruvate; 0.09 mM NADH; 0.1 M phosphate buffer, pH 6.6 are incubated at 20 °C in a quartz cuvette. Reactions are continuously monitored at 340 mμ with a spectrophotometer. Under these conditions optic lobe homogenates had activities of 24.6 ± 3.5 μmol/min/g wet weight. Retinal fiber homogenates had comparable activity. However, because axoplasm and fin nerve had considerably lower octopine dehydrogenase activities, it can not be assured that this assay will be helpful in fractionation studies with axoplasm or fin nerve as starting material.

5.3. Localizing radioactive lipids in squid neural tissues by quantitative EM autoradiography

Only two published studies have applied quantitative autoradiographic approaches to locate macromolecules in the squid giant axon. Lasek and coworkers (1977) used autoradiography to demonstrate the transfer of newly synthesized proteins from the Schwann cells to the giant axon; Rawlins and Villegas (1978) used this approach to locate acetylcholine receptor sites in the Schwann cell layer. The large size of the giant axon, the giant synapse, and even the smaller axons associated with the giant axon make them highly suitable preparations to use in localizing and quantifying the distribution of newly-formed (and further metabolized) lipids, proteins, glycoproteins and nucleic acids. We first discuss the basic design of an autoradiographic study to locate sites of phospholipid synthesis in squid neural tissue. In the two subsections, we consider specific research projects. *Techniques in Autoradiography* by Rogers (1979) is a book we highly recommend for learning the basic approach and techniques involved. More directed information useful for applying EM autoradiography to lipid metabolism is found in our recent studies (Gould, Connell, *et al.*, 1987; Gould, Holshek, Silverman, *et al.*, 1987).

To quantify the distribution of newly-formed lipid in the giant axon and other tissues (see below), four steps are required: (1) labeling the sample, (2) preserving lipids during dehydration and embedment, (3) preparing the autoradiographs and (4) analyzing the grain distributions.

Squid tissues are most conveniently labeled during incubation in seawater with tritiated precursor (see Section 4.1.1). Keeping the dissection simple and the labeling period short should help maintain viability and good ultrastructure. Because of the large size of some structures, i.e., the giant axon and giant synapse, it may seem desirable to introduce precursors directly into them by microinjection. However, from experience with the giant axon (see Section 4.1.2.2), injected precursors are poorly incorporated into lipid. Without more active incorporation, it would not be possible to use microinjection in autoradiographic studies for: (1) The dominant water-soluble precursors and metabolites will be difficult to completely remove, and (2) there would be too little labeled lipid to detect. Perhaps, fortification of the injectate (see Section 4.1.2.2) may help. It is possible that the injected giant synapse may display higher lipid synthetic capacity than the giant axon.

Once tissues are labeled, the metabolism is stopped and ultrastructure preserved with a fixation step. We use the formulation of Pumplin and Reese (1978): 2.5% glutaraldehyde in 0.1 M cacodylate, pH 7.6, 0.8 M sucrose as fixative. Other fixatives are given by Brown and Lasek (1990). Fixative is added directly to the preparation. With amine containing precursors, a chase period with unlabeled precursor (5 to 10 mM) is conducted prior to the addition of fixative to reduce non-specific sticking of labeled precursors and consequential high background. After fixation the tissue is soaked in bi-daily changes of fixation buffer (FB, 0.1 M cacodylate, pH 7.6, 0.8 M sucrose, 10 mM unlabeled precursor). Pieces of fixed tissue can be removed and extracted (see Section 5.1) to ensure that water soluble radioactivity is mostly removed. When >95% of the radioactivity is in lipid, the tissue processing should be continued.

The tissue is postfixed with 1% osmium tetroxide for 1 to 2 hours in FB, rinsed in 0.1 M cacodylate, pH 5.5 (3 changes for 5 minutes each) and *en bloc* stained with lead aspartate (Walton, 1979). To *en bloc* stain the tissue, it is incubated with lead aspartate (66 mg of lead nitrate in 10 ml of stock aspartic acid (1 gm/250 ml, pH 5.5) for 1 hour at 55 °C, and then rinsed with 3 changes of 0.1 M cacodylate buffer, pH 5.5, and with water. Stained tissues are dehydrated with a graded series of cold acetones and embedded in Spurr resin (Gould, Holshek, Silverman, *et al.*, 1987; Gould and Armstrong, 1989).

The preparation of autoradiographs is described by Rogers (1979). We check the morphological preservation and labeling in light microscope autoradiographs before cutting samples for electron microscopy. An adequate exposure to emulsion at the electron microscopy level takes about 2 to 4 times as long as an exposure for light microscope autoradiographs. Usually 2 to 3 slides containing electron microscope sections are sufficient to insure a correct exposure for quantitative analysis.

The initial analysis should be simple and comparative in nature (see examples below). With the increased availability of image analysis equipment, it may be possible to gain sufficient quantitative information from light microscopic images. Once results from a simple analysis are digested, more sophisticated analyses can be used (Blackett and Parry, 1977; Salpeter *et al.*, 1978; Williams, 1982; Miller *et al.*, 1985).

5.3.1. Lipid metabolism in squid axoplasm

Two questions that might be answered with quantitative electron microscopic autoradiography are: (1) What portions of lipid metabolism occur in the giant axon and what portions occur in the surrounding Schwann cells? (2) Which membrane organelles participate in axonal lipid synthesis and metabolism? Autoradiography approaches can be used to address these same questions with smaller fibers (see Fig. 6) that surround the giant axon and with fin nerves, since these nerves have relatively large axons (>10 μm in diameter). Many of the axons are even larger than the large myelinated axons of rodent peripheral nerve (Gould *et al.*, 1988).

Labeling giant axons and fin nerves for autoradiographic analysis should be in precisely the same manner as described for biochemical studies (see Sections 4.1.1 and 4.3). However, for autoradiographic studies, we suggest that short (2 to 3 cm) tied off segments of giant axon from small squid (5 to 8 cm mantle length) with their neighboring small fibers left attached (see Section 3.1) be used. Incubations, with tritiated precursor of 1 to 2 hours should be sufficient to label lipids (and proteins) for autoradiographic analysis (Gould and Tytell, 1989). We recommend that four identical preparations be labeled in each study so that the data from the analyses can be tested by statistical methods.

After the nerves are incubated with radioactive precursor, then *chases* of 5 minutes each in seawater containing 10 mM unlabeled precursor are used to lower levels of

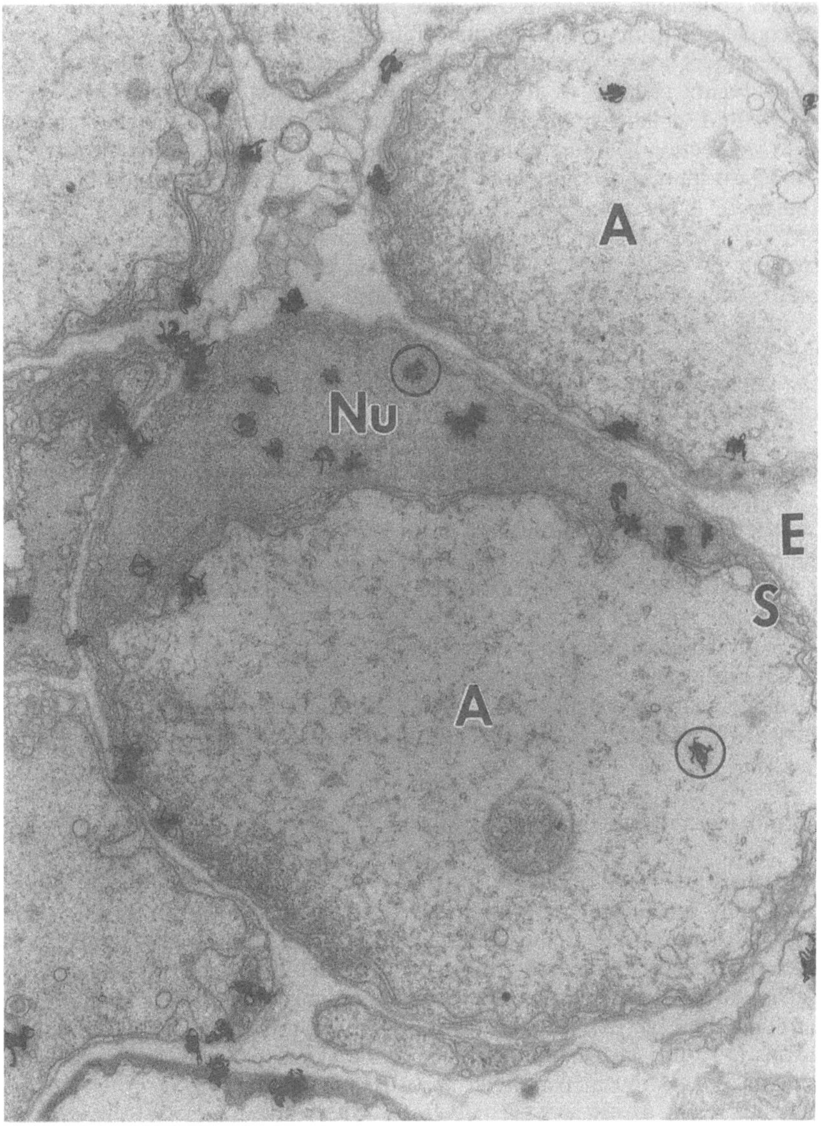

Figure 6. Autoradiograph of *small* fibers that surround the giant axon. These fibers, along with the giant axon, were labeled for 45 minutes with L-^{3}H-leucine and then *chased* for 15 minutes in seawater containing 10 mM *cold* leucine. The tissue was processed as described (see Section 5.3) and the micrograph photographed at 6850 ×. For quantification the photograph is magnified 2.7 ×. This micrograph is printed at 11,650 × magnification. Circles around two of the grains represent 50% probability circles (Williams, 1973). The grains would be assigned to axon and S/Nu compartments, respectively. Symbols: *A*, axon; *S*, Schwann cell cytoplasm; *Nu*, Schwann cell nucleus; *E*, extracellular space.

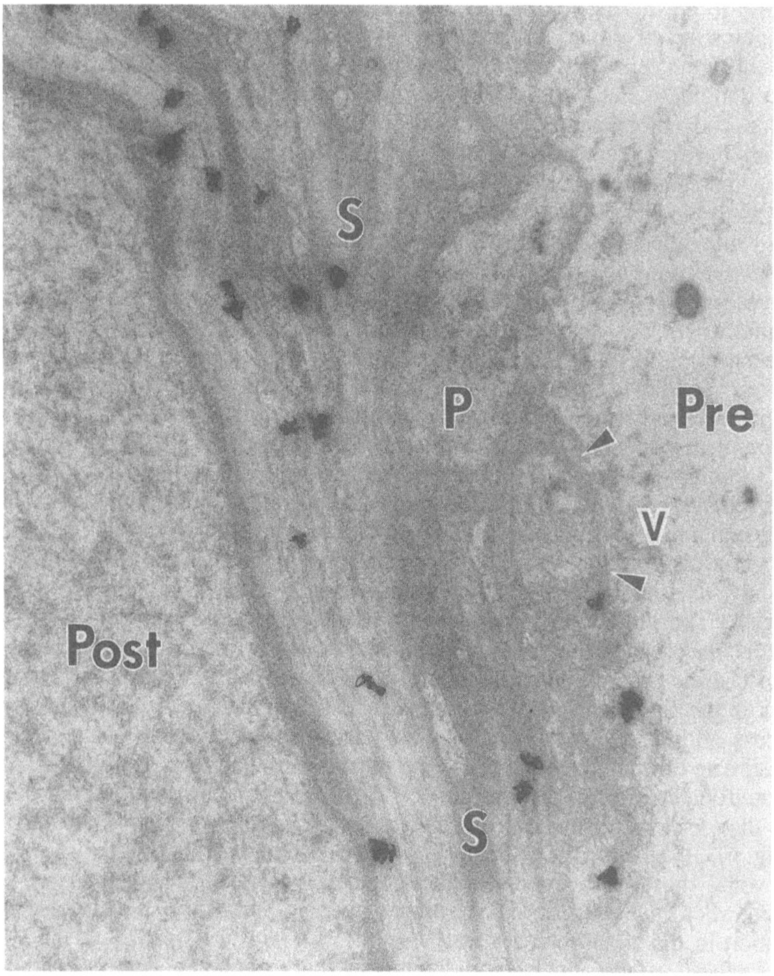

Figure 7. Portion of a giant synapse labeled for 3 hours with ^{3}H-choline. The tissue was processed to selectively retain newly formed lipid (see Section 5.3). The position of postsynaptic (post) and presynaptic (pre) axons are indicated. One of two nearby postsynaptic processes (*P*) penetrate the Schwann cell layer (*S*) and abut the presynaptic axon at active zones (between arrowheads). These zones contain concentrations of synaptic vesicles (*V*). The micrograph is taken at 3050 × and magnified to 5750 ×. Most of the newly formed lipid, mainly phosphatidylcholine, is associated with Schwann cell processes.

radioactive precursors and aqueous metabolites. This step is particularly important with amino-containing precursors, as serine or methionine, since endogenous labeled precursor can be covalently bonded to the tissue during the fixation step.

Following the chase, the nerves, in slightly stretched position, are fixed (see above) in the incubation chamber. After fixation (≥ two hours), the nerves are washed in changes of buffered sucrose in FB for several days to allow further removal of non-lipid (protein) radioactivity. End segments can be removed and lipid extracted (see

Section 5.1) to ensure that the nerve contains mostly (>95%) radioactive lipid. When tritiated amino acids, sugar, or a purine (pyrimidine) base is used to label protein, glycoprotein or nucleic acid, respectively, most (>95%) of the radioactivity in the end segments should pellet with TCA insoluble material. Following poststaining, dehydration and embedment (see Section 5.3), light microscopic autoradiographs are prepared to check labeling and morphology of the nerves. These can be used for analysis in the same way that is described for electron microscopic autoradiographs.

To analyze electron microscopic autoradiographs, the ultrastructure is first subdivided into select *single* and *compound* items (Williams, 1973). In this example, we wish to determine the proportion of labeled macromolecules in Schwann cells and axons, (see Fig. 6). Subdivisions would include of *single items*: axon (A), Schwann cell cytoplasm (S), Schwann cell nucleus (Nu), extracellular space (E), and other (O), i.e., anything else. If other morphological entities were observed with regularity, e.g., small axons bundles (B), they could either be included as a separate item or micrographs selected which minimize their representation. Because many grains will cover more than one item, *compound items*, e.g., A/S, S/E, A/S/E, A/S/Nu are also analyzed. Grains and circles are also assigned to these compartments.

One relatively simple way to determine and quantify axonal lipid metabolism in these preparations is to compare lipid (choline and myo-inositol) with that of protein (leucine) grain distributions (see Gould, Holshek, Silverman, *et al.*, 1987). To conduct such a study, it is assumed that amino acid incorporation into protein does not occur in axoplasm. The initial comparison should be made between one lipid and a protein precursor incubated under the same experimental condition, e.g., 45 minute labeling and 15 minute chase. Once the data is collected (n=4) statistical comparisons are made to determine if the lipid label in axon compartments (A, A/S) exceeds that of protein. If an excess portion of lipid grains was associated with the axon, this portion would give an estimate of lipid synthesis involving the axon compartment. Preliminary studies with tritiated leucine (also Lasek *et al.*, 1977), demonstrate the high degree of protein synthesis that occurs in Schwann cells of fibers surrounding the giant axon and smaller axons (Fig. 6). If an axonal preference for lipid was demonstrated with a short labeling time, it would most likely represent local axonal synthesis, since the other possible mechanism, Schwann cell-to-axon transfer, would probably also cause protein accumulation in the axon compartment.

We suggest an additional autoradiographic study, with the same preparations, to determine if the axoplasmic organelles involved in lipid metabolism are axonally transported. The difficulty of assigning silver grains to specific axoplasmic organelles is the smallness and wide distribution of axonal organelles and difficulty in classification based on their ultrastructural characteristics. They comprise a rather heterogeneous population which includes: clear and dense-core vesicles and tubulo-vesicular profiles. Another complication is the presence of phospholipid transfer proteins in axoplasm (see Section 2.3). These proteins would most likely maintain a dynamic population of labeled phospholipids (of unknown size) in cytosolic, rather than membranous, axoplasmic areas.

The *cold block* paradigm offers the chance to accumulate populations of transport organelles in specific locations in the giant axon (and smaller axons). The autoradiographic studies should be able to pinpoint sites of incorporation of precursors; e.g., tritiated choline, myo-inositol, serine, ethanolamine and glycerol into lipid to these organelles. Autoradiographic and biochemical studies with mouse and rat sciatic nerve (see Section 1) have demonstrated that axoplasmic organelles accumulating at a block (ligature) contain phosphatidylinositol synthase. Preliminary enzyme studies (see

Section 4.1.3) indicate that this activity accumulates, i.e., is transported in both anterograde and retrograde directions, in *cold-blocked* squid axoplasm. Complementing such a biochemical study, an autoradiographic study could also be conducted to determine whether anterograde and retrograde transport organelles were incorporating precursors into lipid. The analysis would be with regions in which transport particles had accumulated and regions further away where particles would not have accumulated. In the regions of particle accumulation, the axon (A) compartment would be subdivided into organelle-rich (Ao) and organelle-lacking (A1) regions. In addition, since axonal transport organelles accumulate in longitudinal strands (Fahim *et al.*, 1985; Miller and Lasek, 1985), analysis could be conducted on sections of nerve cut longitudinally, for in these sections the size of the *clustered* transport particles, i.e., grain source, would be significantly increased.

Furthermore, the *cold block* experiments could be performed in an additional way. Axoplasm, extruded onto a cover glass, could be incubated over a *cold block* allowing transport organelles to accumulate (Miller and Lasek, 1985; Brady *et al.*, 1985). The labeled precursor in buffer X (see Section 7) is added directly to the axoplasm through the open ends of the chamber. During the incubation, organelles in axoplasm incorporate precursor into lipid and also accumulate around the cold block. Following a brief *chase* with buffer X containing unlabeled precursor to reduce the level of water-soluble precursor, the preparation is exposed to fixative by perfusion through the chamber. The following embedding steps would be carried out directly on the coverslip (Miller and Lasek, 1985):

In this study, there is no sheath and, hence, no Schwann cell contribution. Since during the incubations the axoplasm will swell (Brown and Lasek, 1990), and transport particles begin to leave accumulation sites, short (1 to 2 hour) incubation periods are recommended. Also precursor must be present at sufficient levels to label lipids at a level for detection and quantitative analysis. Because all the labeled lipid originates in axoplasm, it will be interesting to compare axon grain distributions (Ao versus A1) with those obtained from intact axon incubations. Differences may indicate that Schwann cells metabolically influence organelles undergoing axoplasmic transport.

5.3.2 Inositol lipid metabolism by the squid giant synapse

The squid giant synapse is a uniquely large synapse which makes numerous contacts with the postsynaptic giant axon (Stanley, 1990). The large size and homogeneity with a small number of ultrastructural components are properties which are well suited parameters for an autoradiographic study. With the recognized presence of inositol lipid metabolism in axons (see Section 1), and the known function of these lipids as second messengers (e.g., Saibil, 1990), we feel it would be of interest to study active inositol lipid synthesis and metabolism in the presynaptic terminal, in order to find out if this metabolism responds to stimulated synaptic transmission. In fact, we proposed that a major role of *lipid* enzymes in the giant axon is played at their terminal neuromuscular junctions. Hence, study of lipid incorporation at the presynaptic terminal has direct relevance (see Section 6).

Labeled precursor are introduced in one of three ways: (1) direct microinjection into the presynaptic terminal, (2) incubation with precursor in the seawater bathing the stellate ganglia-pallial nerve preparation (see Section 3.8), or (3) perfusion through the arterial supply. We used the second approach in a preliminary study (Gould, Holshek, and Pumplin, 1987), because we did not have a facility for microinjection, and because of our concern that injected precursor, as was the case with the giant axon (see Section 4.1.2), might be poorly incorporated into lipid. Also, we were concerned with contamination problems in the case of arterial perfusion. In this study, we found that most of the newly-formed lipid (choline and inositol labeled) resides in regions other than the presynaptic terminal (Fig. 7). In future studies, we would use microinjection to specifically label presynaptic terminals. To demonstrate active lipid synthesis, labeled preparations would be extracted for lipid (see Section 5.1) to demonstrate that sufficient incorporation (>10,000 dpm/terminal) had occurred for autoradiographic analysis.

In our preliminary study, we incubated preparations with either tritiated choline or myo-inositol (at 50 μCi/ml) in oxygenated seawater for 180 minutes. We then placed the preparations in *chase* medium for 10 minutes. Some preparations were presynaptically stimulated during the chase with the supposition that this stimulation might selectively decrease the amounts of labeled lipid in the vesicle containing active zones. Electrical stimulation has been shown to deplete vesicles in the active zone (Pumplin and Reese, 1978).

Micrographs from areas such as the one in Fig. 7 would be taken for analysis. Such high magnifications would be needed to visualize the vesicles in the active zones. Single item categories might include: *Po*, postsynaptic axon; *Pj*, projections of the postsynaptic axon; *Pr*, presynaptic cytoplasm: *Pv*, vesicle-rich active zone; *Pl*, presynaptic membrane areas lacking active zones; and *S*, Schwann cell processes in the presynaptic terminal regions. Appropriate *compound items* would also be used. Two types of study could be conducted: (1) a comparative study (see Section 5.3.1) to demonstrate that phospholipids versus proteins accumulated in the vesicle-rich active zones, and (2) a comparative study to find out if presynaptic stimulation, and the related membrane cycling, would selectively affect lipid radioactivity in the vesicle-rich regions. These studies are important in light of the need for direct proof that active lipid metabolism occurs in presynaptic terminals as a basis for speculations on its participation in local secretion and membrane cycling events.

6. Future directions

In writing this chapter, we tried to make the reader aware of the limitations, as well as the advantages, of the squid giant axon as an experimental model. More than twenty years have passed since *axonal* lipid metabolism was first reported in squid axons (see Section 2.2). Since that time, although advances have been made in our understanding of the properties of this metabolism, we are no closer to understanding its role in vital axonal functions. One reason might be that the squid giant axon is not an appropriate model to use for such studies, particularly those involving the excitable axolemmal membrane.

Its large size, the property that makes it so invaluable in studies that characterize axonal lipid metabolism, places *bulk axoplasm* too far from the excitable membrane to sense, and hence, respond to electrical activity. Furthermore, the outer ten micron rim of axoplasm, that probably contains higher contents of organelles involved in excitation and related calcium metabolism (Henkart *et al.*, 1978), remains associated with the sheath during extrusion (Brown and Lasek, 1990).

Without a role in excitable membrane function, we must conjure up other roles for the lipid-metabolizing organelles that are present in bulk axoplasm. One promising role is as a supplier of components needed for synaptic transmission. The importance of this role may be realized when one considers the myriad of neuromuscular junctions that each giant axon must activate in causing mantle contraction and escape for the squid. In fact, it is appealing to place active lipid metabolism in the specialized synaptic regions, for it is in these regions, and not in the axon, that the membrane cycling and remodeling events needed to support transmitter release occur.

Keeping this role as a starting point, we must reckon with the fact that these lipid enzymes, common to axons and synaptic terminals, probably display widely different behavior in these functionally different neuronal domains. A biologically important question that arises is: How are these enzymes regulated? The activities must remain relatively silent during the long passage through the giant axons and then develop a different character to participate in membrane turnover events occurring in the terminal.

In line with these ideas we can begin to suggest potential modulators of each *lipid* enzyme's activity as it moves from axon to synaptic terminal. Calcium plays an active stimulatory role in transmitter release (e.g., Blaustein, 1988; Smith and Augustine, 1988). It is also a recognized modulator of many *lipid* enzymes. In fact, the *lipid* enzymes present in squid axoplasm (see Section 5.2), i.e., phosphatidylinositol synthase, lysophosphatidylcholine acyltransferase, phospholipase A_2, and serine base exchange enzymes, are all modulated by calcium. Proteases also appear to play an unique role in membrane cycling events at the nerve terminal (e.g., Sahenk and Lasek, 1988). They are needed for the anterograde-retrograde conversion of membranous organelles. Perhaps future studies on the regulation of axonal *lipid* enzymes should consider these potentially active modulators. Furthermore, since these modulators have known effects on the cytoskeleton (Cohen *et al.*, 1990; Lasek, 1988), studies to find out if cytoskeletal polymers and monomers affect *lipid* enzyme activities may also be of interest.

Axoplasm contains a unique subset of membrane organelles derived from the neuronal perikarya. Subcellular fractionation procedures have been designed (see Section 3.2.1) and have begun to be used (see Section 4.2.3; Tytell, 1987) to characterize there organelles. Information gained from such studies may play a role in developing fractionation procedures to isolate axonal membranes from more complex nervous tissues, since these would contain far higher total amounts of axonal membranes.

Immunological and molecular biological approaches have so far not been tried in the field of axonal lipid metabolism. This has been due, at least in part, to few researchers using this tissue and the small quantities of membranes present in each giant axon. We hope that clones of *lipid* enzymes from other sources (e.g., Kodaki and Yamashita, 1987; Thornton *et al.*, 1988; Yuan *et al.*, 1988; Han *et al.*, 1988; Wion *et al.*, 1988; Kramer *et al.*, 1989) can be used to isolate clones from squid stellate ganglia and optic lobe libraries. These clones will make it possible to determine protein sequence and consequently generate antibodies to lipid enzymes to use in immunological and immunocytochemical localization studies. Furthermore, these might make it possible to look for differences in the structure of enzymes that are present in axons and terminal regions.

In conclusion, the squid nervous system, particularly the squid giant axon, is a model which should continue to provide us with important and novel information on the properties of axonal (lipid) metabolism needed in understanding the etiology of some nervous system disorders.

7. Appendix

7.1 Solutions and media

1. Artificial seawater (compound, millimolar concentration, weight per liter): NaCl, 423, 24.72; KCl, 9, 0.67; $CaCl_2 \cdot 2H_2O$, 9.27, 1.36; $MgCl_2 \cdot 6H_2O$, 22.94, 4.66; $MgSO_4 \cdot 7H_2O$, 25.5, 6.29; $NaHCO_3$, 2.15, 0.18.

2. Calcium free seawater: NaCl, 436.71, 25.53; KCl, 9, 0.67; $MgCl_2 \cdot 6H_2O$, 22.94, 4.66; $MgSO_4 \cdot 7H_2O$, 25.5, 6.29; $NaHCO_3$, 2.15, 0.18.

3. Buffer X (weight per 25 ml): Aspartate (K^+ salt), 350, 1.498; taurine, 130, 0.407; betaine, 70, 0.175; glycine, 50, 0.094; $MgCl_2 \cdot 6H_2O$, 12.9, 0.0657; EGTA (disodium salt), 10, 0.0951; $CaCl_2 \cdot 2H_2O$, 3, 0.11; HEPES, 20, 0.119; ATP, 1, 0.00138; dextrose, 1, 0.0451.

4. Fortifying components (taken from Gibco Laboratories Catalogue and Reference Fortifying Guide, 1987): Minimal essential medium (MEM) components related to lipid metabolism (mg/l): choline chloride, 1; myo-inositol, 2; and cytidine, 10. We would also suggest adding: serine, 10; and ethanolamine, 10; and include glucose at 5 to 10 mM.

7.2. Chromatography solvents

A Rf value is the migration distance relative to the solvent front. A lipid migrating with and Rf of 0.5, would migrate half the distance of the solvent front on a TLC sheet. This value is affected by humidity; therefore, the plates are heated with an air blower to remove moisture just before placing them in the tank.

7.2.1. Solvent 1

Chloroform:methanol:acetic acid:water, 65:45:1:4, by vol.

7.2.1.1. Rf values for solvent 1

Rf's for lipid standards: lysolecithin, 0.04; sphingomyelin, 0.07; lysophosphatidylserine, 0.08; phosphatidylcholine, 0.12; lysophosphatidylethanolamine, 0.22; phosphatidylserine, 0.23; phosphatidylinositol, 0.28; phosphatidate, 0.43; phosphatidylethanolamine, 0.47; phosphatidylglycerol, 0.50; sulfatides, 0.64 (hydroxy-form) and 0.67; galactocerebroside, 0.74 (hydroxy-form) and 0.79. Neutral lipids migrate with the solvent front.

7.2.2. Solvent 2

Chloroform:methanol:acetone:acetic acid:water, 40:15:13:12:8 by vol on plates impregnated with 0.25% potassium oxalate.

7.2.2.1. Rf values for solvent 2

Rf values for lipid standards: phosphatidylinositol 4,5-bisphosphate, 0.14; phosphatidylinositol 4-phosphate, 0.18; phosphatidylinositol, 0.27; phosphatidylcholine, 0.33; phosphatidylserine, 0.34; phosphatidylethanolamine, 0.47; sphingomyelin, 0.47; phosphatidate, 0.60.

7.2.3. Solvent 3

Hexane:diethylether:acetic acid, 70:30:1.5 by vol.

7.2.3.1. Rf values for solvent 3

Rf values for lipid standards: phospholipids, 0.0; monoacylglycerols, 0.0; diacylglycerols, 0.04; cholesterol, 0.055; fatty acids, 0.155; triacylglycerols, 0.23; fatty acid methylesters, 0.63; cholesterol esters, 0.93.

7.2.4. Solvent 4

Butanol:acetic acid:water, 5:2:3, by vol.

7.2.4.1. Rf values for solvent 4

Rf values for aqueous standards: cytidine 5'-diphospho (CDP)-choline, 0.15; glycerophosphorylcholine, 0.37; phosphocholine, 0.43; choline, 0.61; acetylcholine, 0.75.

ACKNOWLEDGEMENTS. As research scientists who work feverishly each summer to obtain interesting and publishable results on our *squid* projects, and then return to our home institutions to catch up on other research projects and miscellaneous diversions, we have greatly enjoyed this opportunity to write more fully our thoughts about squid research, and to relate to fellow investigators our opinions of the advantages and shortcomings in using it as a model for studying lipid metabolism in nerve processes and terminals.

We would especially like to thank the editors, William Adelman, Daniel Gilbert and John Arnold for encouraging us to contribute to their book and for their patience in waiting for us to complete this chapter. One of us (R.M.G) is especially grateful to Dr. Rex Dawson for suggesting the squid giant axon as an experimental system and to Drs. Ray Lasek, Jesse Baumgold, Harold Gainer, Harish Pant and Ichii Tasaki for teaching him about the preparation. We would like to thank Dr. Yasuo Kishimoto as both a collaborator and critic of this chapter, and Jesse Baumgold and Mike Tytell, for their critical reading of a drafts of this chapter. We would like to thank Eileen Feehan for typing the manuscript, Ann Erickson for the lucid illustrations and Mary Ellen Nascimento for her help with the photography.

We would like to acknowledge the NIH (Grant NS-13980) for supporting our research program on the squid nervous system.

References

Adelman, W. J., Jr. and Gilbert, D. L., 1990, Electrophysiology and biophysics of the squid axon, *this volume*.

Alberghina, M., Buonacera, P., Agodi, A., and Giuffrida Stella, A. M., 1988, Occurrence of phospholipase A_1-A_2 and lysophosphatidylcholine acyltransferase activities in axolemma-enriched fractions of brain stem, optic pathway, and cranio-spinal nerves of the rabbit, *J. Neurosci. Res.* **19**:79–87.

Alberghina, M. and Gould, R. M., 1988, Association of phospholipid metabolizing enzymes with axolemmal membranes (squid retinal fibers and garfish olfactory nerve) and axoplasmic vesicles from squid axoplasm, *Biol. Bull.* **175**:313.

Ansell, G. B. and Spanner, S., 1982, Phosphatidylserine, phosphatidylethanolamine and phosphatidylcholine, in: *Phospholipids*, (J. N. Hawthorne, and G. B. Ansell, eds.), pp. 1–41, Elsevier Biomedical Press, Amsterdam.

Baker, P. F. and Dipolo, R., 1984, Axonal calcium and magnesium homeostasis, *Curr. Top. in Membr. Transp.* **22**:195–247.

Baker, R. R. and Thompson, W., 1973, Selective acylation of 1-acylglycerophosphorylinositol by rat brain microsomes: Comparison with 1-acylgycerophosphorylcholine, *J. Biol. Chem.* **248**:7060–7065.

Barenholz, Y. and Gatt, S., 1982, Sphingomyelin: metabolism, chemical synthesis, chemical and physical properties, in: *Phospholipids*, (J. N. Hawthorne, and G. B. Ansell, eds.), pp. 129–177, Elsevier Biomedical Press, Amsterdam.

Barman, T. E., 1969, *Enzyme Handbook. Vol. I & II*, Springer-Verlag, New York.

Barman, T. E., 1974, *Enzyme Handbook. Supplement I*, Springer-Verlag, New York.

Baumgold, J., Terakawa, S., Iwasa, K., and Gainer, H., 1981, Membrane-associated cytoskeletal proteins in squid giant axons, *J. Neurochem.* **36**:759–764.

Bear, R. S. and Schmitt, F. O., 1939, Electrolytes in the axoplasm of the great nerve fibers of the squid, *J. Cell Comp. Physiol.* **14**:205–215.

Blackett, N. M. and Parry, D. M., 1977, A simplified method of hypothetical grain analysis of electron microscopy autoradiographs, *J. Histochem. Cytochem.* **25**:206–214.

Blaustein, M. P., 1988, Calcium transport and buffering in neurons, *TINS* **11**, 438–443.

Bligh, E. G. and Dyer, W. J., 1959, A rapid method of total lipid extraction and purification, *Can. J. Biochem. Physiol.* **37**:911–917.

Bolognani, L., Masserini, M., Bodini, P. A., and Bolognani, A. M., 1981, Lipid composition in ganglia of mollusca, *J. Neurochem.* **36**:821–825.

Brady, S. T., Lasek, R. J., and Allen, R. D., 1985, Video microscopy of fast axonal transport in extruded axoplasm: a new model for study of molecular mechanisms, *Cell Motil.* **5**:81–101.

Brismar, T. and Gilly, W. F., 1987, Synthesis of sodium channels in the cell bodies of squid giant axons, *Proc. Natl. Acad. Sci. U.S.A.* **84**:1459–1463.

Brown, A. M. and Lasek, R. J., 1990, The cytoskeleton of the squid giant axon, *this volume*.

Brunetti, M., Giuditta, A., and Porcellati. G., 1979, The synthesis of choline phosphoglycerides in the giant fibre system of the squid, *J. Neurochem.* **32**:319–324.

Camejo, G., Villegas, G. M., Barnola, F. V., and Villegas, R., 1969, Characterization of two different membrane fractions isolated from the first stellar nerves of the squid *Dosidicus gigas*, *Biochim. Biophys. Acta* **193**:247–259.

Cancalon, P. and Beidler, L. M., 1975, Distribution along the axon and into various subcellular fractions of molecules labeled with [^{3}H]leucine and rapidly transported in the garfish olfactory nerve, *Brain Research* **89**:225–244.

Chacko, G. K., Goldman, D. E., Malhotra, H. C., and Dewey, M. M., 1974, Isolation and characterization of plasma membrane fractions from garfish *Lepisosteus osseus* olfactory nerve, *J. Cell Biol.* **62**:831–843.

Chacko, G. K., Goldman, D. E., and Pennock, B. E., 1972, Composition and characterization of the lipids of garfish (*Lepisosteus osseus*) olfactory nerve, a tissue rich in axonal membrane, *Biochim. Biophys. Acta* **280**:1–16.

Christie, W. W., 1987, *High Performance Liquid Chromatograph and Lipids: A Practical Guide*, Pergamon Press, New York, pp. 1–387.
Clarke, N. G. and Dawson, R. M. C., 1980, Alkaline O→N-transacylation: A new method for the quantitative deacylation of phospholipids, *Biochem. J.* **195**:301–306.
Cohen, R. S., Pant, H., and Gainer, H., 1990, Squid optic lobe synaptosomes: Structure and function of isolated synapses, *this volume.*
Condrea, E. and Rosenberg, P., 1968, Demonstration of phospholipid splitting as the factor responsible for increased permeability and block of axonal conduction induced by snake venoms. II. Study on squid giant axon, *Biochim. Biophys. Acta* **150**:271–284.
Condrescu, M., Osses, L., and DiPolo, R., 1984, Partial purification and characterization of the (Ca^{2+} + Mg^{2+})-ATPase from squid optic nerve plasma membrane, *Biochim. Biophys. Acta* **769**:261–269.
DeVries, G. H., 1981, Isolation of axolemma-enriched fractions from mammalian CNS, *Research Methods in Neurochem.* **5**:3–38.
Dowdall, M. J., 1989, *personal commun.*
Dowdall, M. J. and Whittaker, V. P., 1973, Comparative studies in synaptosome formation: the preparation of synaptosomes from head ganglia of the squid, *Loligo pealei, J. Neurochem.* **20**:921–935.
Dowdall, M. J. and Downie, D. L., 1988, Octopine dehydrogenase activity in *Loligo pealei*; An active cytoplasmic enzyme marker for subcellular studies, *Biol Bull.* **75**:315.
Ellisman, M. H. and Lindsey, J. D., 1983, The axoplasmic reticulum within myelinated axons is not transported rapidly, *J. Neurocytol.* **12**:393–411.
Fahim, M. A., Lasek, R. J., Brady, S. T., and Hodge, A. J., 1985, AVEC-DIC and electron microscopic analyses of axonally transported particles in cold-blocked squid giant axons, *J. Neurocytol.* **14**:689–704.
Folch, J., Lees, M., and Stanley, G. H. S., 1957, A simple method for the isolation and purification of total lipids from animal tissues, *J. Biol. Chem.* **226**:497–509.
Gainer, H., Gallant, P. E., Gould, R., and Pant, H., 1984, Biochemistry and metabolism of the squid giant axon, *Curr. Top. Membr. Transp.* **22**:57–90.
Gainer, H., Tasaki, I., and Lasek, R. J., 1977, Evidence for the glia neuron protein transfer hypothesis from intracellular perfusion studies of squid giant axons, *J. Cell Biol.* **74**:524–530.
Gilbert, D. L., 1974, Physiological uses of the squid with special emphasis on the use of the giant axon, in: *A Guide to Laboratory Use of the Squid Loligo pealei,* (J. M. Arnold *et al.*, eds.), pp.45–54, Marine Biological Laboratory, Woods Hole, MA.
Gould, R. M., 1977, Inositol lipid synthesis localized in axons and unmyelinated fibers of peripheral nerve, *J. Neurochem.* **117**:169–174.
Gould, R. M., 1989, *unpublished.*
Gould, R. M. and Armstrong, R., 1989, Use of lead aspartate block staining in quantitative EM autoradiography of phospholipids: Application to myelinating peripheral nerve, *J. Histochem. Cytochem.* **37**: *in press.*
Gould, R. M., Connell, F., and Spivack, W., 1987, Phospholipid metabolism in mouse sciatic nerve in vivo, *J. Neurochem.* **48**:853–859.
Gould, R. M., Holshek, J., and Pumplin, D. W., 1987, Incorporation of tritiated inositol and choline into phospholipids in the squid stellate ganglia with special reference to the giant synapse, *Biol. Bull.* **173**:443.
Gould, R. M., Holshek, J., Silverman, W., and Spivak, W. D., 1987, Localization of phospholipid synthesis to Schwann cells and axons, *J. Neurochem.* **48**:1121–1131.
Gould, R. M. and Jackson, M., 1989, *unpublished.*
Gould, R. M., Jackson, M., and Tasaki, I., 1983, Phospholipid synthesis in the injected squid giant axon, *Biol. Bull.* **165**:526.
Gould, R. M., Pant, H. Gainer, H., and Tytell, M., 1982, Studies of phospholipid synthesis in mammalian and invertebrate axons and terminals, in: *Phospholipids in the Nervous System, Vol. 1: Metabolism,* (L. A. Horrocks, G. B. Ansell, and G. Porcellati, eds.) pp. 221–229, Raven Press, New York.

Gould, R. M., Pant, H. Gainer, H., and Tytell, M., 1983, Phospholipid synthesis in the squid giant axon: Incorporation of precursors, *J. Neurochem.* **40**:1293–1299.
Gould, R. M., Spivack, W., Cataneo, R., Holshek, J., and Konat, G., 1988, Lipid and myelination, in: *Multidisciplinary approach to myelin disease*, (G. S. Crescenzi, ed.), pp. 87–102, Plenum Press, New York.
Gould, R. M., Spivack, W. D., Robertson, D., and Poznansky, M. J., 1983, Phospholipid synthesis in the squid giant axon: Enzymes of phosphatidylinositol metabolism, *J. Neurochem.* **40**:1300–1306.
Gould, R. M. and Tytell, M., 1989, *unpublished.*
Han, J. H., Stratowa, C., and Rutter, W. J., 1987, Isolation of full-length putative rat lysophospholipase cDNA using improved methods for mRNA isolation and cDNA cloning, *Biochemistry* **26**:1617–1625.
Henkart, M. P., Reese, T. S., and Brinley, F. J., Jr., 1978, Endoplasmic reticulum sequesters calcium in the squid giant axon, *Science* **202**:1300–1303.
Holbrook, P. G. and Gould, R. M., 1988, Calcium-dependent incorporation of serine into phosphatidylserine in the squid giant axon: Physiological role in excitable membranes?, *Biol. Bull.* **175**:305.
Hori, T., Itasaka, O., Sugita, M., and Arakawa, I. 1967, Distribution of ceramide 2-aminoethylphosphonate and ceramide aminoethylphosphate (sphingoethanolamine) in some aquatic animals, *J. Biochem.* **62**:67–70.
Hori, T. and Nozawa, Y., 1982, Phosphonolipids. in: *Phospholipids*, (J. N. Hawthorne, and G. B. Ansell, eds.), pp. 95–125, Elsevier Biomedical Press, Amsterdam.
Hostetler, K. Y., 1982, Polyglycerophospholipids:phosphatidylglycerol, diphosphatidylglycerol and bis(monoacylglycero)phosphate. in: *Phospholipids*, (J. N. Hawthorne, and G. B. Ansell, eds.), pp. 215–254, Elsevier Biomedical Press, Amsterdam.
Jungalwala, F. B., 1985, Recent developments in techniques for phospholipid analysis, in: *Phospholipids in Nervous Tissues*, (J. Eichberg, ed.), pp.1–40, John Wiley, New York.
Kanfer, J. N., 1972, Base-exchange reactions of the phospholipids in rat brain particles, *J. Lipid Res.* **13**:468–476.
Kodaki, T. and Yamashita, 1987, Yeast phosphatidylethanolamine methylamine pathway, *J. Biol. Chem.* **262**:15428–15435.
Kramer, R. M., Hession, C., Johansen, B., Hayes, G., McGray, P., Chow, E. P., Tizard, R., and Pepinsky, R. B., 1989, Structure and properties of a human non-pancreatic phospholipase A_2, *J. Biol. Chem.* **264**:5718–5775.
Kumara-Siri, M. H. and Gould R. M., 1980, Enzymes of phospholipid synthesis: Axonal versus Schwann cell distribution, *Brain Research* **186**:315–330.
Larrabee, M. J. and Brinley, F. J., Jr., 1968, Incorporation of labelled phosphate into phospholipids in squid giant axons, *J. Neurochem.* **115**:533–545.
Lasek, R. J., 1984, The structure of axoplasm, *Curr. Top. Membr. Transp.* **22**:39–53.
Lasek, R. J., 1988, Studying the intrinsic determinants of neuronal form and function, in: *Intrinsic Determinants of Neuronal Form and Function*, (R. J. Lasek, and M. M. Black, eds.), pp. 3–58, Alan R. Liss, New York.
Lasek, R. J., Gainer, H., and Przybylski, R. J., 1977, Transfer of newly synthesized proteins from Schwann cells to the squid giant axon, *Proc. Natl. Acad. Sci. U.S.A.* **71**:1188–1192.
Llinas, R. R., 1984, The squid giant synapse, *Curr. Top. Membr. Transp.* **22**:519–546.
Marinetti, G. V., 1976, *Lipid Chromatographic Analysis, 2nd. ed., Vols. 1, 2, and 3*, Dekker, New York.
McColl, J. D. and Rossiter, R. J., 1951, Lipids of the nervous system of the squid *Loligo Pealii*, *J. Exp. Biol.* **28**:116–124.
Metuzals, J., Tasaki, I., Terakawa, S., and Clapin, D. F., 1981, Removal of the Schwann cell from the giant nerve fiber of the squid: An electron-microscope study of the axolemma and associated axoplasmic structures, *Cell Tissue Res.* **221**:1–15.
Miller, R. H. and Lasek, R. J., 1985, Cross-bridges mediate anterograde and retrograde vesicle transport along microtubules in squid axoplasm, *J. Cell Biol.* **101**:2181–2193.
Nishizuka, Y., 1986, Studies and perspectives of protein kinase C, *Science* **233**:305–312.

Nonaka, G. and Kishimoto, Y., 1979, Levels of cerebrosides, sulfatides, and galactosyl diglycerides in different regions of rat brain: change during maturation and distribution in subcellular fractionation of gray and white matter of sheep brain, *Biochim. Biophys. Acta* **572**:432–441.

Norton, W. T., 1981, Formation structure, and biochemistry of myelin, in: *Basic Neurochemistry*, (G. J. Siegel, R. W. Albers, B. W. Agranoff, and R. Katzman, eds.), pp.63–88, Little, Brown and Co., Boston.

Okamura, N., Stoskopf, M. Hendricks, F., and Kishimoto, Y., 1985, Phylogenetic dichotomy of nerve glycosphingolipids, *Proc. Natl. Acad. Sci. USA*. **82**:6779–6782.

Okamura, N., Yamaguchi, H., Stoskopf, M., Kishimoto, Y., and Saida, T., 1986, Isolation and characterization of multilayered sheath membrane rich in glucocerebroside from shrimp ventral nerve, *J. Neurochem.* **47**:1111–1116.

Parsons, T. D. and Chow, R. H., 1989, Neuritic outgrowth in primary cell culture of neurons from the squid, *Loligo pealei, Neurosci Lett.* **97**:23–28.

Patton, S. 1970, Correlative relationship of cholesterol and sphingomyelin in cell membranes, *J. Theor. Biol.* **29**:489–491.

Pelech, S. L. and Vance, D. E., 1989, Signal transduction via phosphatidylcholine cycles, *TIBS* **14**:28–30.

Pumplin, D. W. and Reese, T. S., 1978, Membrane ultrastructure of the giant synapse of the squid *Loligo pealei, Neuroscience* **3**:685–696.

Rawlins, F. A. and Villegas, J., 1978, Autoradiographic localization of acetylcholine receptors in the Schwann cell membrane of the squid nerve fiber, *J. Cell Biol.* **77**:371–376.

Rice, R. V., Mueller, R. A., and Adelman, W. J. Jr., 1990, Tissue culture of squid neurons, glia, and muscle cells, *this volume*.

Rogers, A. W., 1979, *Techniques in Autoradiography, 3rd ed.*, Elsevier/North-Holland Biomedical Press, Amsterdam.

Rosenberg, P., 1981, The squid giant axon: Methods and applications, in: *Methods in Neurobiology, Vol . 1*, (R. Lahue, ed.), pp. 1–133, Plenum Press, New York.

Rosenberg, P. and Condrea, E., 1968, Maintenance of axonal conduction and membrane permeability in presence of extensive phospholipid splitting, *Biochem. Pharmacol.*, **17**:2033–2044.

Rosenberg, P. and Khairallah, E., 1974, Effect of phospholipases A and C on free amino acid content of the squid axon, *J. Neurochem.* **23**:55–64.

Sahenk, Z. and Lasek, R. J., 1988, Inhibition of proteolysis blocks anterograde-retrograde conversion of axonally transported vesicles, *Brain Research* **460**:199–203.

Saibil, H., 1990, Structure and function of the squid eye, *this volume*.

Salpeter, M. M., McHenry, F. A., and Salpeter, E. E., 1978, Resolution in electron microscopic autoradiography. IV. Application to analysis of autoradiographs, *J. Cell Biol.* **76**:127–145.

Schroer, T. A., Schnapp, B. J., Reese, T. S., and Sheetz, M. P., 1988, The role of kinesin and other soluble factors in organelle movement along microtubules, *J. Cell Biol.* **107**:1785–1792.

Shaikh, N. A. and Palmer, F. B., 1976, Deposition of lipids in the developing central and peripheral nervous systems of the chicken, *J. Neurochem.* **26**:597–603.

Sheltawy, A. and Dawson, R. M. C., 1966, The polyphosphoinositides and other lipids of peripheral nerve, *Biochem. J.* **100**:12–18.

Smith, S. J. and Augustine, G. J., 1988, Calcium ions, active zones and synaptic transmitter release, *TINS* **11**:458:464.

Somerharju, P., Brockerhoff, H., and Wirtz, K. W. A., 1981, A new fluorimetric method to measure protein-catalyzed phospholipid transfer using 1-acyl-2-parinaroylphosphatidylcholine, *Biochim. Biophys. Acta* **649**:521–528.

Somerharju, P. J., van Loon, D., and Wirtz, K. W. A., 1987, Determination of the acyl chain specificity of the bovine liver phosphatidylcholine transfer protein. Application of pyrene-labeled phosphatidylcholine species, *Biochemistry* **26**:7193–7199.

Stanley, E., 1990, The preparation of the squid giant synapse for electrophysiological investigation, *this volume*.

Tanaka, T., Yamaguchi, H., Kishimoto Y., and Gould, R. M., 1987, Lipid metabolism in various regions of squid giant nerve fibre, *Biochim. Biophys. Acta.* **922**:85–94.

Tasaki, I., Teorell, T., and Spyropoulos, C. S., 1961, Movement of radioactive tracers across squid axon membrane, *Am. J. Physiol.* **200**:11–22.

Terasaki, M., Chen, L. B., and Fujiwara, K., 1986, Microtubules and the endoplasmic reticulum are highly interdependent structures, *J. Cell Biol.* **103**:101–108.

Thornton, J., Howard, S. P., and Buckley, J. T., 1988, Molecular cloning of a phospholipid-cholesterol acyltransferase from *Aeromonas hydrophila*. Sequence homologies with lecithin-cholesterol acyltransferase and other lipases, *Biochim. Biophys. Acta* **959**:153–159.

Tytell, M., 1987, Characterization of glial proteins transferred into the squid giant axon, in: *Glial-Neuronal Communication in Development and Regeneration, NATO ASI Series H, Vol. 2*, (H. H. Althaus, and W. Seifert, eds.), pp. 24–261, Springer Verlag, Berlin.

Tytell, M., 1989, *personal commun.*

Tytell, M., and Lasek, R. J., 1984, Glial polypeptides transferred into the squid giant axon, *Brain Research* **324**:223–232.

Vale, R. D. and Hotani, H., 1988, Formation of membrane networks in vitro by kinesin-driven microtubule movement, *J. Cell Biol.* **107**:2233–2241.

van den Hill, A., van Heusden, G. P. H., and Wirtz, K. W. A., 1985, The synthesis of sphingomyelin in the Morris hepatoma 7777 and 5123D is restricted to the plasma membrane, *Biochim. Biophys. Acta* **833**:354–357.

van Paridon, P. A., Gadella, T. W. J., Somerharju, P. J., and Wirtz, K. W. A., 1987, Properties of the binding sites for the sn-2-acyl chains on the phosphatidylinositol transfer protein from bovine brain, *Biochemistry* **27**:6208–6214.

van Thoai, N., Huc, C., Pho, D. B., and Olomucki, A., 1969, Octopine dehydrogenase: Purification and catalytic properties, *Biochim. Biophys. Acta* **191**:46–57.

Villegas, G. M. and Villegas, R., 1984, Squid axon ultrastructure, *Curr. Top. in Membr. Transp.* **22**:3–34.

Vogel, S. S., Hess, S. D., and Reese, T. S., 1988, G proteins in the squid *Loligo pealei*: Characterization, quantification and examination of possible role in axonal transport, *Biol. Bull.* **175**:318.

Voelker, D. R. and Kennedy, E. P., 1982, Cellular and enzymatic synthesis of sphingomyelin, *Biochemistry* **21**:2753–2759.

Weiss, D. G., Meyer, M. A., and Langford, G. M., 1990, Studying axoplasmic transport by video microscopy and using the squid giant axon as a model system, *this volume*.

Williams, M. A., 1973, Electron microscopic autoradiography: its application to protein biosynthesis, in, *Techniques in Protein Biosynthesis, Vol. 3*, (P. M. Campbell, and J. R. Sargent, eds.), pp. 219–272, Academic Press, New York.

Williams, M. A., 1977 Quantitative methods in biology, in: *Practical Methods in Electron Microscopy, Vol. 6*, (A. Glauert, eds.), pp. 85–169, North Holland Biomedical Press, Amsterdam.

Wion, K. L., Kirchgessner, T. G., Luisis, A. J., Scholtz, M. C., and Lawn, R. M., 1987, Human lipoprotein lipase complementary DNA sequence, *Science* **235**:1638–1642.

Wreggett, K. A. and Irvine, R. F., 1987, A rapid separation method for inositol phosphates and their isomers, *Biochem. J.* **245**:655–660.

Yamaguchi, H., Tanaka, T., Ichioka, T., Stoskopf, M., Kishimoto, Y., and Gould, R., 1987, Characterization and comparison of lipids in different squid nervous tissues, *Biochim. Biophys. Acta* **922**:78–84.

Yuan, Z., Liu, W., and Hammes, G. G., 1988, Molecular cloning and sequencing of DNA complementary to chicken fatty acid synthase mRNA, *Proc. Natl. Acad. Sci. U.S.A.* **85**:6328–6331.

Zambrano, F., Cellino, M., and Canessa–Fischer, M., 1971, The molecular organization of nerve membranes, IV. The lipid composition of plasma membranes from squid retinal axons, *J. Membrane Biol.* **6**:289–303.

Part V

SENSORY SYSTEMS

Chapter 17

Structure and Function of the Squid Eye

HELEN R. SAIBIL

1. Introduction

This account is not intended as a comprehensive review of work on the remarkable eye of the squid. There is an excellent review on the physiology of cephalopod vision by Messenger (1981), and a brief account of research methods by Daw (1974). The purpose of this chapter is to briefly introduce the eye and its capabilities and then focus on areas where work on squid has contributed or will potentially contribute to biochemistry, cell and molecular biology. I hope to justify this particular choice of topics by demonstrating that the squid eye, while difficult for experimental biology, provides a valuable and unique preparation for studies in the cellular and molecular areas.

1.1. The cephalopod camera eye

Cephalopods have invertebrate-type photoreceptors in what is optically a vertebrate-type eye. Their large, spherical eye is a famous example of convergent evolution with the fish eye. They have the same optics and comparable speed, sensitivity and resolution. The evolutionary convergence may be attributable to the fact that this is the optimal design for an eye made of biological materials. A remarkable genetic parallel also exists between the structural proteins of cephalopod and vertebrate lenses. In both cases, the crystallins have been recruited from pre-existing proteins with enzymatic or other functions, including several different enzymes in the vertebrates and glutathione S transferase in cephalopods (Tomarev and Zinovieva, 1988). By far the largest eyes in the animal kingdom are found in squid: eye diameters of 25 - 40 cm have been reported for giant squid (Roper and Boss, 1982; citations by Messenger, 1981 and Land, 1981). Fig. 1 shows a drawing of the cephalopod eye and optic lobe. The eye sits in a cartilaginous orbit, and the corneal area is continuous with the skin. The sensory axons exit the back of the eye and pass through a dorso-ventral chiasm, terminating in an ipsilateral optic lobe. Vision is the main sensory modality for these

H. R. SAIBIL • Department of Zoology, Oxford University, Oxford OX1 3PS, England. Present address: Department of Crystallography, University of London Birkbeck College, Malet Street, London WC1E 7HX, England.

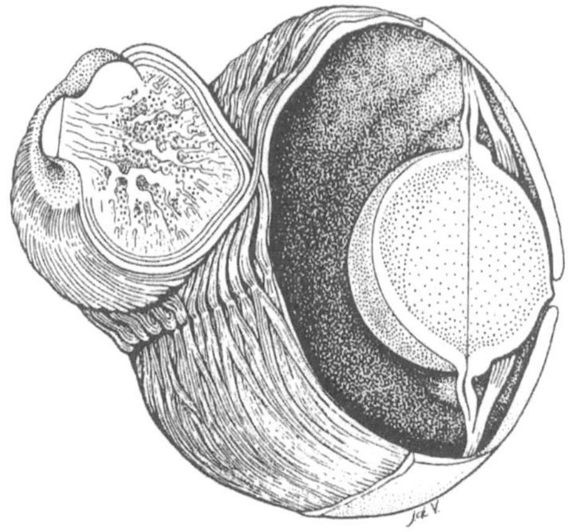

Figure 1. The cephalopod eye and optic lobe. This drawing is of an *Octopus* eye, but squid is very similar. From Young (1962b), by permission of the Royal Society.

animals. The eyes can comprise up to 50% of the body weight in a small cephalopod, and the optic lobes have 50 – 80% of the brain neurons.

2. Visual information produced by the squid eye

2.1. Spatial resolution and sensitivity

The excellent discussion by Land (1981) is summarized here. The large spherical lens is corrected for spherical aberration, by a refractive index gradient of n=1.33 in the periphery to n = 1.52 in the center (Sivak, 1982). The receptors are narrow and densely packed at 70,000/mm^2, giving them an angular separation of 1.3 min of arc, which is comparable to the human retina. For maximal resolution, each receptor should act as an optically independent unit. The waveguide properties calculated by Land give an acceptance cone of semi-angle 13°, but the fully open lens delivers a 21.4° cone, which would intersect 300 receptors. High resolution can only be obtained by extending some of the screening pigment granules to the tips of the cells. This isolates the receptors by blocking oblique rays, excluding over 60% of the incident light. Conversely, to maximize sensitivity, there is a penalty of lower spatial resolution. Light and dark adaptation are effected by migration of screening pigment granules (see Section 4.2.1).

2.2. Polarization sensitivity

2.2.1 Photoreceptor membrane and photopigment orientation

A detailed analysis of polarization sensitivity in vision has been published by Waterman (1981). In invertebrates with rhabdomeric photoreceptors, polarization discrimination arises from the arrangement of linear 11-*cis* retinal chromophores in the plane of the cylindrical photoreceptor membrane. In Fig. 2 a diagram of chromophore

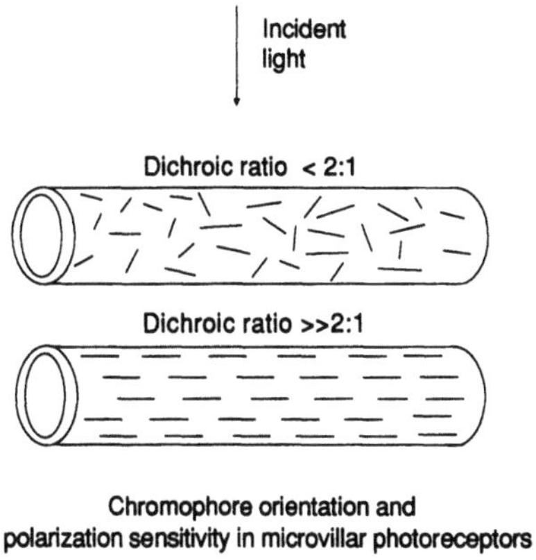

Figure 2. Alignment of rhodopsin chromophores in photoreceptor microvilli.

polarization orientation in microvilli is shown. The high polarization sensitivity in cephalopods supports the aligned model (Tomita, 1968). In addition, the whole retinal mosaic is aligned with the horizontal and vertical axes of the animal.

2.2.2. Contrast enhancement

What function does this polarization discrimination serve? An attractive speculation published by Land (1981), but attributed by him to an unidentified source, proposes that it serves to break the mirror camouflage of silvery fish (light intensity reflected off the sides of fish matches the background illumination, but the mirror reflection is polarized). Saidel *et al.* (1983) have shown that polarization information is encoded as intensity variation in optic nerve responses, and suggest, less specifically, that it provides an enhancement of contrast by eliminating reflected highlights.

3. The squid eye as an experimental preparation

3.1. Disadvantages

3.1.1. Delicate

Most species of squid are difficult to obtain in good condition and keep alive in captivity. Behavior experiments have almost all been done on *Octopus* and *Sepia*, which are much easier to keep because of their more sedentary life styles. Squid are difficult to catch in a big tank (especially in the dark!) because they move very fast and are always swimming around. They very quickly learn to avoid the net after the first individual in a tank is caught.

3.1.2. Electrophysiology difficult

Squid retina is difficult to keep alive and needs perfusion with 100% O_2. Although several studies of cephalopod retinal physiology have been done, recent work on visual transduction physiology in invertebrates has concentrated on species with large and robust photoreceptor cells, particularly the ventral photoreceptor of *Limulus.*

3.1.3. Spectral overlap

For *Loligo* rhodopsin λ_{max} = 493 nm, and for its photostable product at physiological pH, acid metarhodopsin λ_{max} = 500 nm. In *Octopus* and *Eledone,* rhodopsin λ_{max} = 470 nm and acid metarhodopsin $\lambda_{max} \geq$ 500 nm. The better spectral separation in the octopod species allows the photochemical reaction to be driven in either the rhodopsin-to-metarhodopsin direction or in the reverse direction using appropriate wavelength filters. The ability to reverse the photoexcitation reaction is useful for photochemical studies and for lateral and rotational diffusion studies on rhodopsin.

3.2. Advantages

3.2.1. Large numbers and large eyes

Because of their fish-like shoaling behavior, squid are caught in relatively large numbers. A 40 cm long *Loligo forbesi* has about four times as much rhodopsin as a cow. This has obvious advantages for protein chemistry studies.

3.2.2. Simple retinal neuroanatomy

The retina contains only two types of neural element: photoreceptors and the terminal arborizations of centrifugal neurons whose cell bodies are in the optic lobe. The relatively complex optic nerve responses are thus attributable only to the receptor cells, to their lateral interactions, and in the intact eye/optic lobe preparation, to the centrifugal fibers terminating in the synaptic plexus of the retina. The large nerve fibers in the retina and optic lobe facilitate studies with Golgi staining. A population of supporting cells has cell bodies near the top of the inner segment layer, just below the pigment granules.

3.2.3. Pure photoreceptor membrane preparation

Alone among invertebrates, squid rhabdoms can be readily detached virtually free of other retinal layers. Other cephalopods are similar but the preparations are not as clean as in squid. The outer segment layer is formed of a paracrystalline array of microvilli with only fine processes of the supporting cells and little receptor cell cytoplasm.

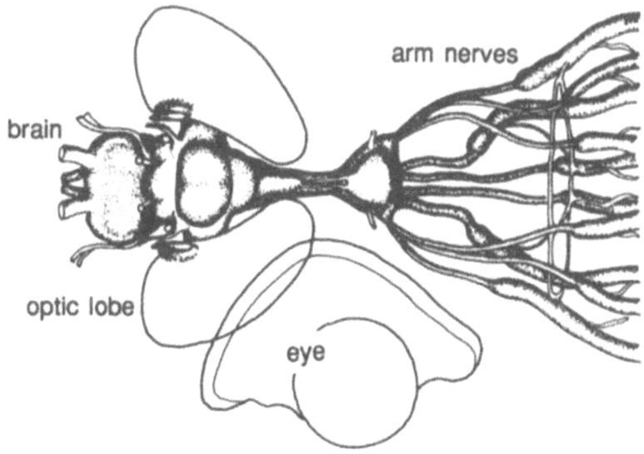

Figure 3. Eyes and optic lobes in relation to the squid brain and arm nerves, reconstructed from sections of a small *Loligo vulgaris*. From Young (1976), by permission of the Royal Society.

3.2.4. Photostable metarhodopsin

Unlike vertebrate rhodopsin, cephalopod rhodopsin has a photostable metarhodopsin. Many experiments can be done in the light, with rhodopsin traceable by its pink color.

3.3. Protocol for dissection of the eye

Pinch up the corneal skin in front of the eye to make a cut. Widen the cut into a circular opening and cut away the cartilage to expose the yellowish, bean-shaped optic lobe behind the eyeball. Take care not to damage the optic lobe which has the retinula cell axons running on its surface. In a large animal, cut the fiber tract running out the back of the optic lobe and gently lift the eye out of its orbit. In small species the two eyes cannot be separated without damaging at least one of the optic lobes which are in close apposition. For biochemical work, the eye can be cut out of its orbit without the optic lobe, as long as the retina is used or frozen immediately. The diagram in Fig. 3 shows the disposition of the eyes and optic lobes in relation to the brain and arm nerves.

Note about working in the dark: Many of the procedures described in this chapter need to be done with minimal light exposure of the photopigments, and conditions for dim red or infrared (IR) illumination are given. For those who are not experienced with this type of work, it is of course necessary to learn the manipulations for each procedure while working in the light.

4. The retina

4.1. Cellular organization

The cellular organization of the cephalopod retina was comprehensively described by J. Z. Young (1962a). The squid retina is practically unicellular, and the receptors

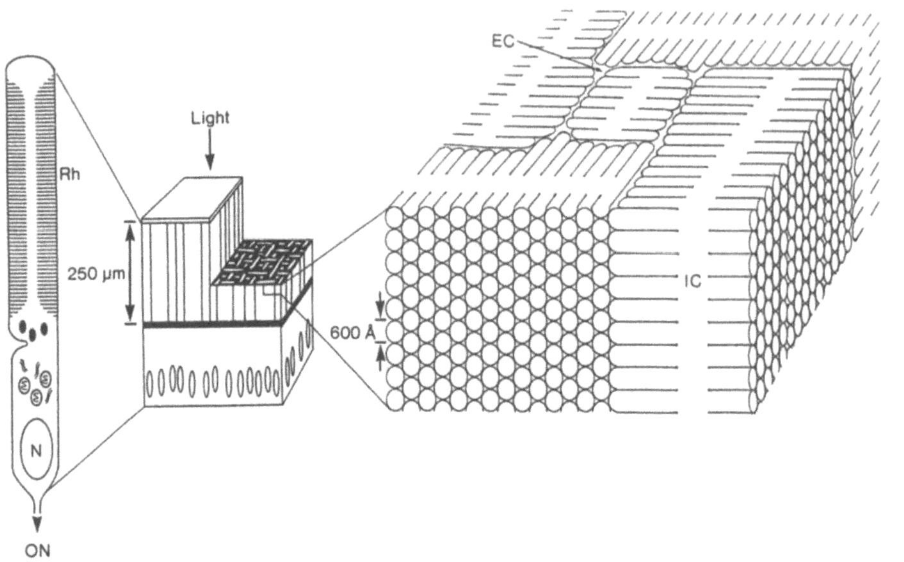

Figure 4. Arrangement of retinula cells in the retina and of microvilli in the rhabdoms, showing the ribbon-like retinula cells and the orthogonal orientation of microvilli in neighboring cells. *Rh*, rhabdom; *ON*, optic nerve; *EC*, extracellular compartment; *IC*, intracellular compartment. From Saibil and Hewat (1987), reproduced from *The Journal of Cell Biology* by copyright permission of the Rockefeller University Press.

depolarize in response to light. Its photoreceptors are formed of membrane cylinders, or microvilli, in the rhabdom structure characteristic of higher invertebrates. The photoreceptive *outer segments* (by analogy with vertebrate rod outer segments although they point towards the inside of the eye) face the light. At the base of the inner segments, the visual cells narrow into the sensory axons which terminate in the optic lobe.

4.1.1. Fixation methods for the visual cells

Fig. 4 shows the arrangement of retinula cells in the retina, and the organization of microvilli in the rhabdoms, as revealed by electron microscopic studies (Zonana, 1961; Cohen, 1973a; Saibil, 1982). The retinula cells are divided into the photoreceptive outer segments (also called distal or rhabdomeric segments) and inner segments, at the level of the pigment granule layer. In the proximal zone, both retinula cells and glia contain pigment granules. The glial cell bodies sit above flat-topped capillaries, and they send only fine processes through the photoreceptor layer to the amorphous limiting membrane at the distal surface of the retina. The retinula cell somas are filled with lamellar bodies, and mitochondria are lined up along the membranes (see Fig. 5). Further down in the retina are the nuclei and plexiform layer with receptor collaterals and centrifugal fiber synapses, and the optic fiber collecting layer is at the base of the cells.

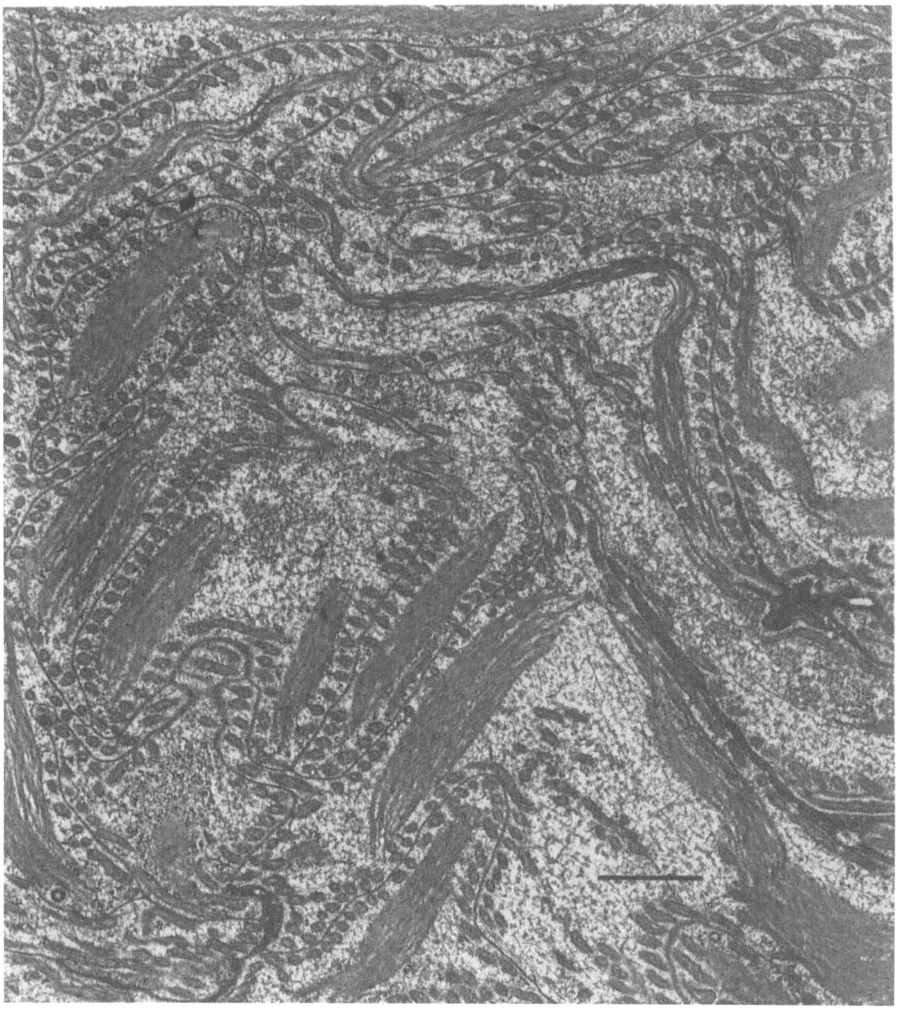

Figure 5. (EM) section through the somal region of the inner segments, showing the abundant lamellar bodies and numerous mitochondria lined up along the membranes. Micrograph by the author, using the fixation protocol of section V.1.a. Calibration bar, 2 μm.

The receptor mosaic forms a continuous sheet of closely packed, hexagonal arrays of microvilli, as shown in Fig. 6. This makes squid retina a difficult tissue to fix, as the dense membrane array greatly impedes the entry of glutaraldehyde. Perhaps because of the slow fixation, leakiness of the cells to ions during fixation is also a problem. Fixation at sea water or intraocular fluid (Packard, 1972) ionic strength results in considerable swelling of the photoreceptor layer and distortion of the receptor mosaic. Another problem is a calcium-activated protease, mainly attacking the cytoskeleton, as reported for other invertebrate photoreceptors (de Couet *et al.*, 1984) and also in squid axoplasm.

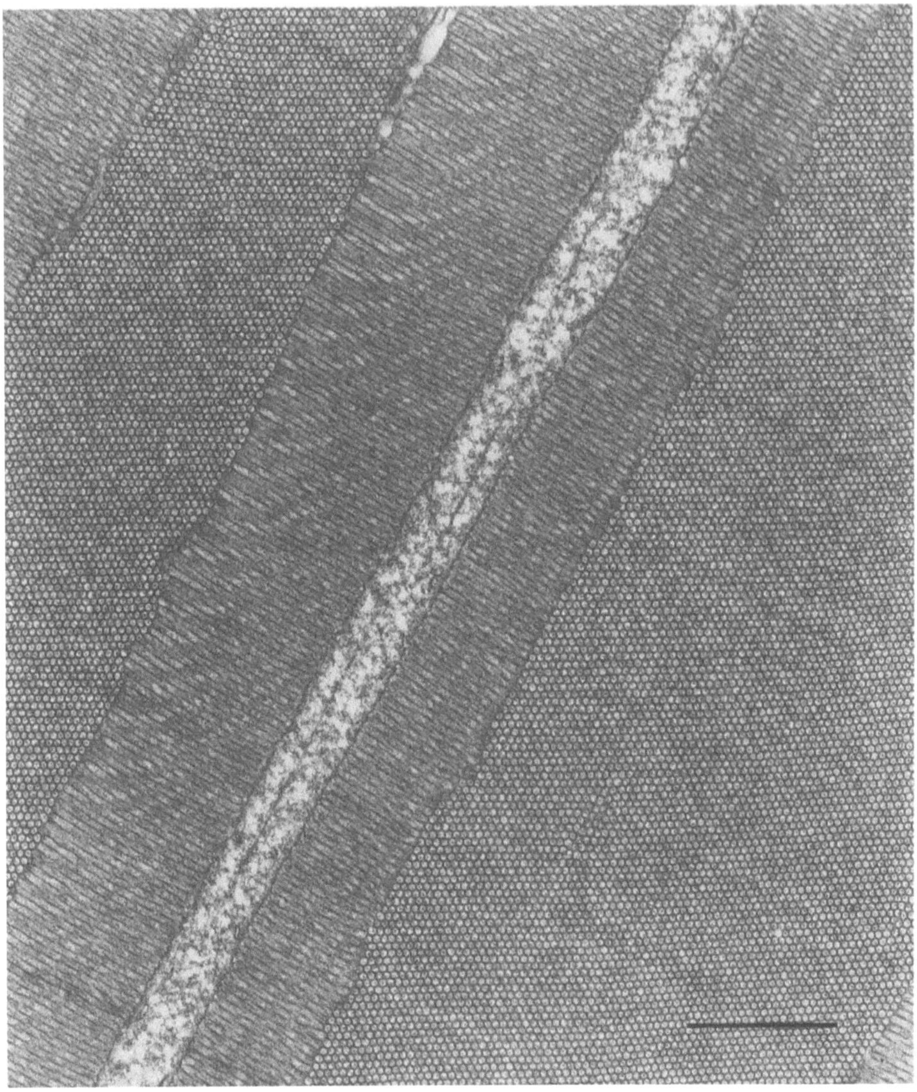

Figure 6. EM section through the photoreceptor layer, illustrating the hexagonal packing and alternating orientation of the microvilli. The cytoplasm is filled fairly uniformly with filamentous and granular material. Micrograph by the author, as in Fig. 5. Calibration bar, 1 μm.

Most of these problems can be solved. Cohen's methods (1973a) use perfusion of fixative into the systemic heart, which is particularly important for studies of the neural layers and optic lobe. The strategy developed by Saibil (1982; Saibil and Hewat, 1987) for high resolution studies of the photoreceptor membranes, but which also gives good fixation of the inner segments, is to use small animals and expose the

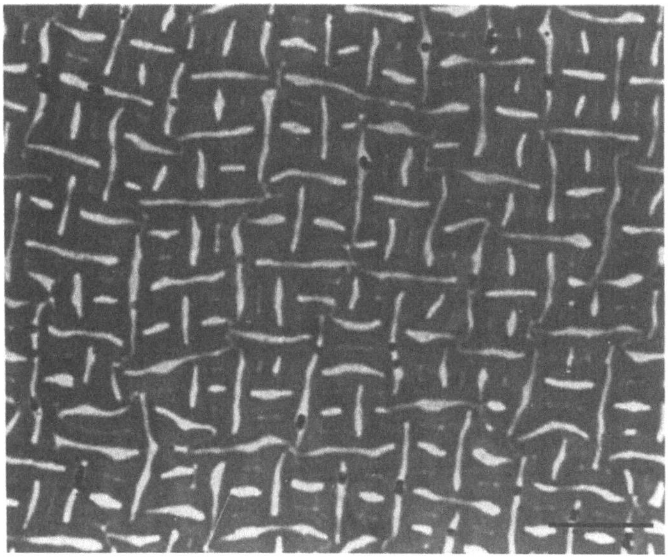

Figure 7. A tangential view of the receptor mosaic in the light microscope. Areas containing microvilli are dark and cytoplasm is light. Micrograph by Dr. J. Patterson and E. J. Fisher (1989), using the same fixation protocol as above. Calibration bar, 10 μm.

vitreal surface of the retina immediately to fixative. The best fixation was found with the following protocol:

4.1.1.1. Protocol

Working in dim red or IR illumination, decapitate a dark-adapted squid (ideally ≤ 5 – 10 cm in length), remove beak and tentacles with a v-shaped cut; dissect away the skin around the eyes, and hemisect the eyes. Immediately place the tissue in room-temperature fixative:

5% sucrose (w/v), 5% dextran 70,000 (W/V), 475 mM NaCl, 20 mM $MgSO_4$, 10 mM ethylene glycol-bis(2-amino ethyl ether)N,N,N',N'-tetraacetic acid (EGTA), 25 mM 3-(N-morpholino)propanesulfonic acid (MOPS) pH 7.4, 2% glutaraldehyde.

Fix for 1 h in the dark with O_2 bubbling through the fixative (this may accelerate the action of glutaraldehyde, and agitation of the solution speeds diffusion into the tissue, but it also requires the use of a fume hood). Cut 0.5 mm strips of retina and wash in cold fixative without glutaraldehyde. Post fix in 0.5% OsO_4 in 5% dextran, 10% sucrose, one-half the original salt and buffer concentrations for 30 min at 0° C.

4.1.1.2. Quality of fixation

To judge the quality of primary fixation, cut a longitudinal razor-blade slice and check that the retina is not swollen or buckled, and observe the birefringence of the photoreceptor layer. This can be done with a reasonable dissecting microscope by taping on crossed polaroids. With good fixation, the tissue is ordered and the photoreceptors are highly birefringent. A tangential light microscope view of the receptor mosaic fixed in this way is shown in Fig. 7. From measurements of cell profiles, Fisher and Patterson (1988) found evidence for three preferred cell lengths.

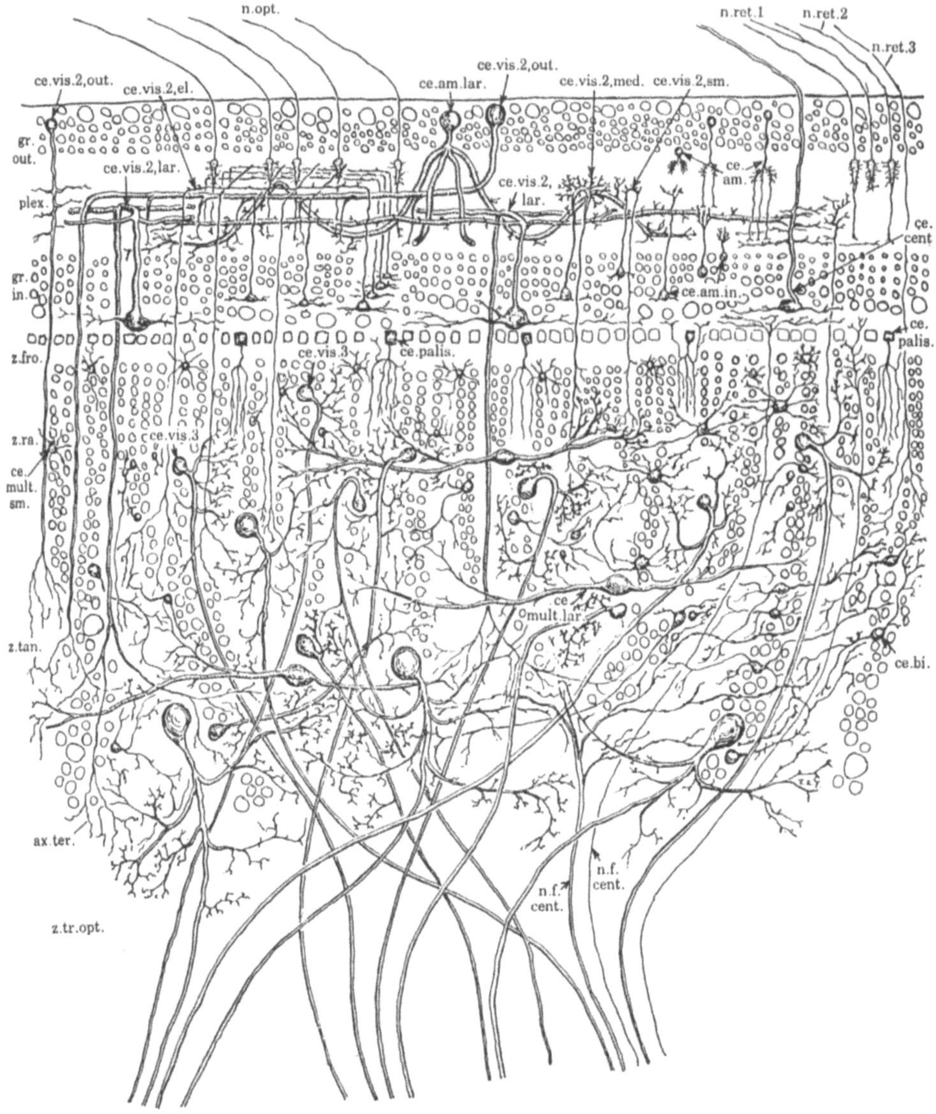

Figure 8. Diagram of cell types found in the optic lobe of *Loligo* as seen in a transverse section. From Young (1974), by permission of the Royal Society. Abbreviations: *ax.ter.* = axon terminal; *ce.am.* = amacrine cell; *ce.am.in.* = inner amacrine cell; *ce.am.lar.* = large amacrine cell; *ce.bi.* = bipolar cell; *ce.cent.* = cell with centrifugal nerve fibers; *ce.mult.lar.* = large multipolar cell; *ce.palis.* = cell of palisade layer; *ce.vis.2* = second-order visual cell; *ce.vis.2,el* = second-order visual cell with elongated receptive field; *ce.vis.2,lar.* = large second-order visual cell; *ce.vis.2,med.* = medium second-order visual cell; *ce.vis.2,out.* = second-order visual cell of outer granule cell layer; *ce.vis.2,sm.* = small second-order visual cell; *ce.vis.3* = third-order visual cell; *gr.in.* = inner granule cell layer; *gr.out.* = outer granule cell layer; *n.f.cent.* = centrifugal nerve fiber; *n.opt.* = optic nerve; *n.ret.1–3* = retinal nerve fiber, types 1 – 3; *plex.* = plexiform zone; *z.fro.* = frontier zone of optic lobe; *z.ra.* = zone of radial columns of medulla; *z.tan.* = zone of tangential bundles; *z.tr.opt.* = zone of optic tract bundles in optic lobe.

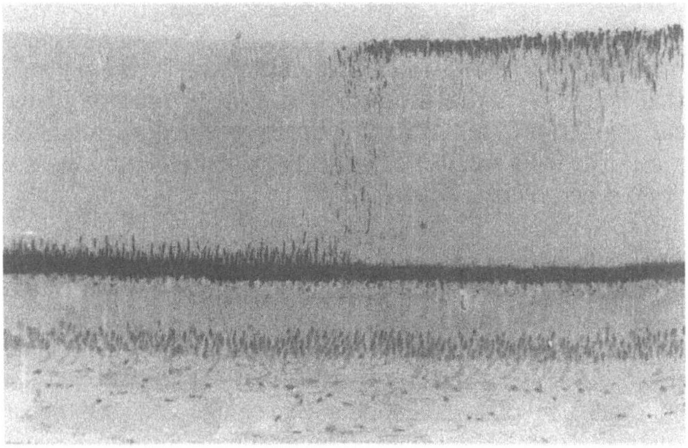

Figure 9. Pigment granule migration at the edge of an illuminated region of the squid retina. From Daw and Pearlman (1974), reproduced from *The Journal of General Physiology* by copyright permission of the Rockefeller University Press.

4.1.2. Neuroanatomy

Cohen (1973b) perfused retinas with aldehyde/OsO_4 or OsO_4 fixatives to investigate the synaptic plexus. The receptors interact through collateral fibers, although the identification of gap junctions or synapses was uncertain. Centrifugal fibers synapse onto the receptors. The centrifugal fibers originate from cells in the inner granular layer of the optic lobe (Young, 1974), and they appear to use dopamine as a transmitter (Lam *et al.*, 1974; Tansey, 1980; Silver *et al.*, 1983).

J. Z. Young (1974) carried out a detailed study of cellular interactions in the squid optic lobe, using Golgi-Kopsch staining, and Patterson (1985) developed an improved Golgi-rapid method for the cephalopod retina. The carrot-shaped photoreceptor nerve endings terminate in three different layers of the optic lobe (Young, 1974). Their synapses often include finger-like protrusions from one cell into another (Cohen, 1973c). A diagram of the squid optic lobe is shown in Fig. 8. The fibers in squid are larger than those in *Octopus*, and Golgi staining is more successful, so that a more detailed analysis is possible in the squid. The optic lobe contains many analogies to the non-photoreceptor layers of the vertebrate retina. The optic nerve fibers enter the optic lobe and pass through the outer granule cell layer to the outer plexiform layer, where most of them terminate. The outer part of the optic lobe contains four well-defined layers, and the medulla, the largest part of the lobe, contains five less well defined zones. Young proposed the following interpretation: the outer plexiform layer contains the receptive fields of second order visual cells, with specific sizes and orientations. Coded information is then passed to the medulla, where appropriate responses are selected and transmitted to other brain centers. The plexiform zone contains amacrine cells of various sizes with spreading tangential processes, while the medulla is arranged in radial columns, perhaps analogous to columns in the vertebrate cerebral cortex.

4.2. Intracellular transport

Because the retinula cells are very long (500 μM) and there is no pigment epithelium in this non-inverted retina, intracellular transport is a major function of the receptors. The outer segment cytoplasm is full of longitudinally oriented microtubules and microfilaments (see Fig. 6). The cells are also reported to exhibit photomechanical movements (Young, 1963; Tasaki and Nakaye, 1984).

4.2.1. Pigment granule migration

Light/dark adaptation in the cephalopod retina is mediated by pigment granule movement (Young 1963), shown in Fig. 9. Daw and Pearlman (1974) showed that the migration was dependent on rhodopsin photoexcitation. Screening produces higher resolution by isolating cells optically (see Section 2.1). The motile mechanism has not been investigated, but in other systems membrane-bound pigment granules are transported by microtubules and this mechanism seems likely to be operating here. A light-activated motor, perhaps like the mechanism for myoid retinomotor movements in fish, could make use of internal messengers such as calcium or cyclic nucleotides to couple filament activation to light stimulation.

4.2.2. Retinoid transport and membrane turnover

On the basis of ^{3}H retinoid incorporation into the *Octopus* retina, Breneman *et al.* (1986) have suggested that retinoids are stored in retinochrome in the inner segments in the dark and transferred to rhodopsin in the outer segments after illumination. Retinochrome is also reported to be at the base of the outer segments (Hara and Hara, 1976; Robles *et al.*, 1987), where it could supply 11-*cis* retinal to rhodopsin directly. A probable mechanism for retinoid transport, which is likely to be very similar in the squid, is via the soluble retinoid binding protein discovered by Ozaki *et al.* (1987) (see Section 5.2.3). Opsin is synthesized in the inner segments prior to insertion in rhabdomeric membrane (Robles *et al.*, 1984, 1987). Signs of membrane breakdown are evident in the most distal layer of the outer segments, where the microvillar packing is disrupted (Saibil & Hewat, 1987; Tsukita *et al.*, 1988). The cytoplasm in this layer is filled with membrane material (see Fig. 1 in Tsukita *et al.*, 1988) and it appears darker than the rest of the outer segment cytoplasm in phase contrast micrographs of the retina (see Fig. 1a in Saibil, 1982). Robles *et al.* (1987) report opsin-containing multivesicular bodies in *Octopus* inner segments and propose that these are involved in rhabdom turnover.

4.3. Electrophysiology

4.3.1. Electroretinogram (ERG)

The ERG is a gross, extracellular recording which is useful for monitoring visual response characteristics such as spectral sensitivity, adaptation, flicker fusion frequency and polarization sensitivity. It can be recorded from the intact eye or from slices of retina. Lange and Hartline (1974) developed an isolated eye-optic lobe preparation which gave stable recordings of retinal potentials for up to 10 hours. The eye and its optic lobe were rapidly excised and placed in chilled oxygenated (~10° C, 100% O_2) sea water. The cartilage layer at the back of the eye was carefully removed to allow adequate oxygenation of the tissue. The ERG was recorded by applying a large suction electrode to the back of the eye.

4.3.2. Photoreceptor potentials and ionic mechanisms

Using 100 μm slices of isolated retina manipulated under IR illumination, Hagins *et al.* (1962; Hagins, 1965) showed that the site of illumination of the photoreceptor outer segments is the source of the photoreceptor potential. Later, Pinto and Brown (1977) recorded intracellular potentials from larger slices of isolated retina. They found that the preparation was very sensitive to hypoxia and perfused with artificial sea water (ASW), bubbled with 100% O_2, at ≥5 ml/min. The ASW composition to resemble *Loligo* hemolymph contained:

423 mM NaCl, 10 mM KCl, 25 mM $MgCl_2$, 25 mM $MgSO_4$, 10 mM CaCl2, 10 mM N-2-hydroxyethlypiperazine-N'-2-ethanesulfonic acid (HEPES) pH 7.8, 5.6 mM dextrose.

They found that the receptor potential is largely generated by increased Na^+ influx, and that light stimulation is followed by an increase in cytoplasmic free Ca^{++}. Duncan and Pynsent (1979) studied intracellular and extracellular voltage responses in the *Sepiola* retina, in order to analyze the intracellular response waveforms without distortion by the large extracellular potentials. The response waveforms are very similar to those of turtle cones, apart from a change in sign.

4.3.3. Optic nerve responses

The isolated eye-optic lobe preparation of Lange and Hartline (1974) was successfully used to record nerve action potentials from single units in optic nerve bundles on the back of the eye, with suction electrodes. Relatively complex visual processing takes place in the retina, despite its neuroanatomical simplicity. Hartline and Lange (1974; 1984) obtained *on*, *sustained*, *off* and *on-off* responses, which were all dependent on the state of adaptation of the retina. The processing which generates these complex responses must lie in the lateral interactions between the photoreceptors in the synaptic plexus of the retina, as they can be observed when the optic nerves are cut, suppressing the centrifugal impulses. Afferent and efferent components of action potential activity in *Octopus* optic nerves were studied in a similar preparation by Patterson and Silver (1983).

5. Molecular organization of squid photoreceptors

5.1. Isolation of the microvillar membranes

5.1.1. Preparation of the retinas

Healthy animals, protected from bright light after capture, should be dark adapted for several hours in the largest tank compatible with catching them in dim red (>620 nm, Kodak 1A/2) light or under IR illumination with a IR viewer:

Decapitate and quickly dissect out the eyes individually in a large animal, or the eye-optic lobe in a small one. Resect the lenses and briefly rinse the eyecups in low calcium high salt buffer such as 0.4 M NaCl, 10 mM EGTA, 5 mM $MgCl_2$, 20 mM MOPS buffered to a pH of 7.4. Freeze the tissue rapidly by plunging into liquid N_2-cooled isopentane (pentane and hexane can also be used). The eyecups can be stored indefinitely in liquid N_2. They deteriorate surprisingly rapidly at −20° C, especially if Ca^{++} is present. Minus 70° C is adequate for several months.

5.1.2. Protocol for isolation of the membranes

A diagram of this procedure is shown in Fig. 10. If the dark adapted squid retina is thawed into Ca^{++}-free solution, it splits very cleanly with the outer segment layer released as large pink sheets. On the other hand, if the retina is thawed in high Ca^{++} solution, there is severe proteolysis of the preparation, and the outer segments come off in small bits with lots of black pigment. (The split-off outer segments also include much more black pigment in preparations from *Sepia* and *Octopus*, but their protein compositions seem very similar to that of squid photoreceptors).

5.1.2.1. Method of Saibil (adapted from Saibil and Hewat, 1987 and Baer and Saibil, 1988)

Degas or flush solutions with argon or nitrogen.

Shake the eyecups briskly in 0.4 M NaCl, 10 mM EGTA, 5 mM $MgCl_2$, 20 mM MOPS at a pH of 7.4 to detach the photoreceptor layer.

Collect the photoreceptor layer fragments and spin briefly (1 min in a cooled microcentrifuge with 1.5 ml tubes).

Homogenize the pellet in 40% sucrose with 10 passes of a moderately tight-fitting homogenizer, taking account of the pellet volume to give a final sucrose concentration of 36%.

Perform a sucrose flotation by layering buffer over the homogenate and spin for 5 min.

Most of the screening pigment granules present should come down in a black pellet.

Wash the pink interface material with 1 min spins in the required buffer.

Washing in low salt, ethylenediamine-tetraacetic acid (EDTA) releases ommin from the remaining pigment granules, and a small amount of actin.

Refreezing this preparation causes reductions in enzyme activities.

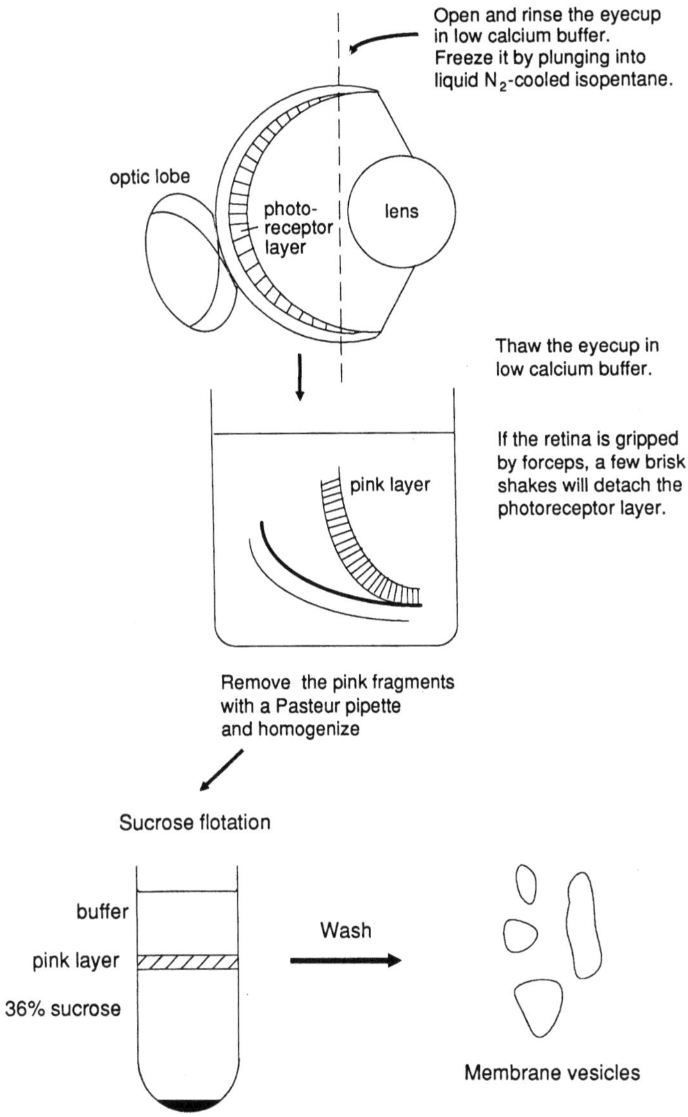

Figure 10. Diagram of photoreceptor membrane isolation.

This method can be scaled up with no change by using multiple microcentrifuge tubes, or with longer spins in larger tubes in a medium speed centrifuge. The yield of this membrane isolation is about 0.1 – 0.2 mg rhodopsin/*Alloteuthis* eye 3 – 6 mm in diameter, or 4 mg/*Loligo* eye 4 cm in diameter. A sodium dodecyl sulfate (SDS) gel of the purified membranes and the detergent soluble and insoluble fractions is shown in Fig. 11 (lanes M, S, P).

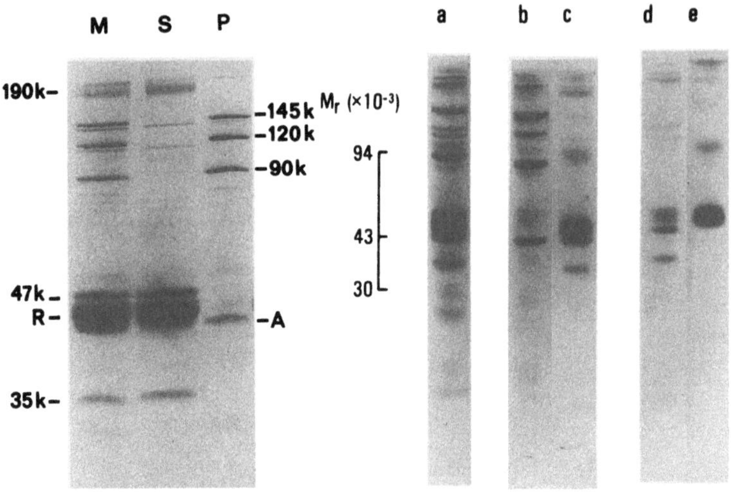

Figure 11. SDS gels of photoreceptor membrane preparation and rhodopsin. M, S, P, 10% polyacrylamide gels from Saibil and Hewat (1987), reproduced from *The Journal of Cell Biology* by copyright permission of the Rockefeller University Press; *a–e*, 7% polyacrylamide gels from Vandenberg and Montal (1984c), reproduced from *Biochimica et Biophysica Acta* by permission of Elsevier Publications. *M*, proteins of *Alloteuthis* photoreceptor microvilli, isolated by the method of Saibil; *S*, fraction soluble in octyl glucoside; *P*, fraction insoluble in octyl glucoside. *a*, proteins of *Loligo* photoreceptor microvilli, isolated by the method of Vandenberg; b, octyl glucoside insoluble fraction; *c*, octyl glucoside soluble fraction; *d*, proteins adhering to DEAE cellulose; *e*, rhodopsin and its oligomers, purified by the method of Vandenberg.

5.1.2.2. Method of Vandenberg (Vandenberg, 1982; Vandenberg and Montal, 1984c)

Dissect eyecups from dark-adapted *Loligo*, using Kodak filter #2 and store at −80° C. Use deoxygenated solutions.

Thaw 50 – 70 eyecups and shake in 680 mM NaCl.

Spin the resulting suspension at 35,000 × g for 30 min.

Homogenize the pellet with 10 slow passes of a motor-driven teflon-glass in 120 ml 40% (w/v) sucrose, 300 mM NaCl, 100 mM Na phosphate, 2 mM EDTA at a pH of 6.8.

Overlay 20 ml aliquots with 13 ml 33% and 6 ml 10% (w/v) sucrose in the same buffer.

Spin the step gradient at 23,000 rpm for 90 min in a Beckman SW 27 rotor.

Collect the pink membranes at the 10/33% interface.

Pool and wash 3 times with 820 mM NaCl at 35,000 × g for 20 min.

Wash once with 35 ml 1 mM EDTA pH 7 at 35,000 × g for 40 min.

Wash 4 times with 35 ml 820 mM NaCl at 35,000 × g for 20 min.

Wash once with 35 ml 1 mM EDTA pH 7 at 35,000 × g for 35 min.

Resuspend in 10 mM imidazole, 1 mM EDTA at a pH of 7.

Divide into aliquots, pellet and freeze in liquid N_2 for storage at −80° C.

A SDS gel of this preparation and detergent soluble and insoluble fractions is shown in Fig. 11 (lanes a – c).

5.1.3. Lipid and protein composition of the preparation

There are various estimates of the lipid/protein ratio. However, they are difficult to interpret because a substantial part of the protein is not in the membrane but is associated with the membrane-bound cytoskeleton. Akino and Tsuda (1979) and Paulsen *et al.* (1983) found about equal amounts of protein and lipid, with exceptionally high unsaturation of phospholipids, and 15 – 30% of the lipid was reported to be cholesterol. Anderson *et al.* (1978) found 75% protein, but their preparation involved homogenization of entire retina and may have contained protein contaminants from other parts of the retina.

The identified proteins in this preparation are rhodopsin (Hagins, 1973), actin (Saibil, 1982) and the β-subunit of guanosine 5'-triphosphate (GTP)-binding protein (Tsuda *et al.*, 1986). Vandenberg and Montal (1984c) suggest that rhodopsin aggregation, a frequent problem with this preparation, is caused by oxidation.

5.1.4. Measurement of rhodopsin concentration

Dilute a small volume of membrane suspension into 2% octyl glucoside or sucrose monolaurate at pH 10 on ice and incubate for 5 – 20 min. Pellet the insoluble material with a 5 min spin in the microcentrifuge, or higher speed if ultraviolet (UV) spectra are required. Scan the extract between 700 and 350 nm without exposing it to room light. Bleach the sample (this can be slow in sucrose monolaurate) and rescan. The rhodopsin concentration is calculated from the optical density difference at λ_{max} (493 nm for *Loligo*, extinction coefficient 40,600 cm^2/mole; Hubbard and St George, 1958).

Squid metarhodopsin is unstable in most detergents, with stability decreasing from digitonin to sucrose monolaurate to octyl glucoside. It is stabilized by high salt (0.5 M NaCl) in sucrose monolaurate (Nashima *et al.*, 1978). The concentration measurement can be done with cold 0.5% Triton-X-100 if all steps are done very quickly, but this will lead to an underestimatation of the alkaline metarhodopsin concentration.

5.2. Purification of retinal binding proteins

5.2.1. Rhodopsin

The absorption spectra of rhodopsin and acid and alkaline metarhodopsins are shown in Fig. 12. The sequence of *Octopus* rhodopsin has been determined by cDNA cloning and is homologous to the vertebrate and *Drosophila* pigments (Ovchinnikov *et al.*, 1988). Only one rhodopsin is thought to be present in the squid retina, with the notable exception of the bioluminescent, deep-sea species, *Watasenia scintillans*. In the retina of this species, three visual pigments have been discovered, rhodopsin (λ_{max} = 484 nm), 3-dehydroretinal-opsin (λ_{max} = 500 nm), and 4-hydroxyretinal-opsin (λ_{max} = 470 nm) (Kito *et al.*, 1986; Matsui *et al.*, 1988a). While some insect opsins contain 3-hydroxyretinal, this is the first report of a 4-hydroxyretinal-based visual pigment. In a ventral region of the retina where the photoreceptor layer is twice as thick as in the rest of the retina, the 3-dehydroretinal (vitamin A2) pigment is present in a proximal pinkish layer, and a distal yellowish layer contains the 4-hydroxyretinal (termed vitamin

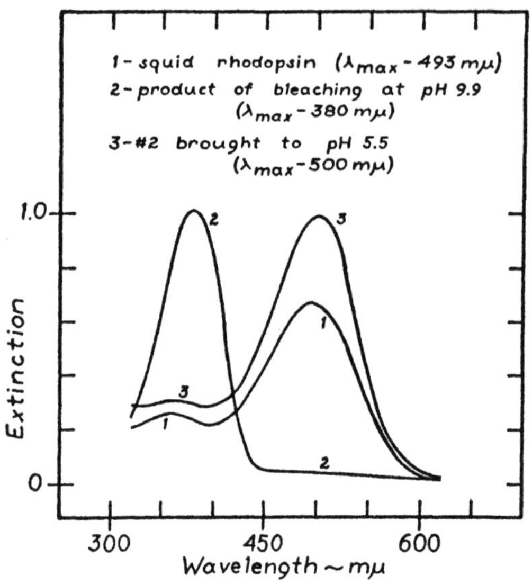

Figure 12. Absorption spectrum of rhodopsin. From Hubbard and St. George (1958), reproduced from *The Journal of General Physiology* by copyright permission of the Rockefeller University Press. *1*, rhodopsin; *2*, alkaline metarhodopsin; *3*, acid metarhodopsin.

A4) pigment. This arrangement is proposed to serve for detection of luminescence of other members of this species in the very low light level environment of the deep sea (Matsui *et al.*, 1988b).

5.2.1.1. Method of Nashima, Mitsudo and Kito (Nashima *et al.*, 1979)

Todarodes pacificus or *Watasenia scintillans* are captured at night and the eyes stored at −20° C.

Shake off the outer segments in 4% (w/v) NaCl.

Rhodopsin purification in 2% sucrose monolaurate (SML) is as follows:

Extract the membranes with 0.4 ml/eye of SML in 50 mM tris (hydroxymethyl) aminomethane (Tris) pH 7.

Equilibrate a diethylamino ethyl (DEAE) cellulose anion exchange column 10 × 1.2 cm with 0.2% SML, Tris.

Apply 20 – 25 ml of crude extract to the column and elute with 0.2% SML, Tris.

Apply the eluate to a Concanavalin A sepharose column 10 × 1.2 cm in 100 mM Tris pH 7, 1 M NaCl, 1 mM $CaCl_2$, 1 mM $MnCl_2$.

Wash the Concanavalin A column with ≥100 ml of SML, Tris.

Elute the rhodopsin with SML, Tris containing 50 mM α-methyl glucoside.

5.2.1.2. Method of Vandenberg (Vandenberg, 1982; Vandenberg and Montal, 1984c)

Extract membranes containing 0.15 – 0.2 μmol rhodopsin extract into 6 ml solubilization buffer composed of:

50 mM octyl glucoside, 10 mM imidazole, 1 mM EDTA pH 7 containing 0.5 mg/ml diphytanoyl phosphatidylcholine (a fluid but not oxidation sensitive phospholipid).

Spin out the insoluble material 35,000 × g 40 min.

Prepare DEAE cellulose for batch chromatography as follows:

- Wash in distilled H_2O until the supernatant is clear.
- Wash 1 h in 0.5 M HCl.
- Rinse with H_2O.
- Wash 1 h in 0.5 M NaOH.
- Rinse with H_2O until the rinse is neutral.
- Equilibrate with solubilization buffer.

Incubate the rhodopsin extract with the prepared DEAE. Rhodopsin does not bind at low ionic strength, while other proteins do.

Remove the rhodopsin-containing solution by sedimenting the DEAE cellulose at low speed, re-extracting with 2 ml solubilization buffer to remove the remaining rhodopsin.

Yield: 70 – 90% of rhodopsin solubilized.

An SDS gel of this purification is shown in Fig. 11, lanes d and e.

5.2.2. Retinochrome

A second major photopigment in the cephalopod retina with a retinaldehyde chromophore was discovered by Hara and Hara in 1965 (see Hara and Hara, 1976) and called retinochrome. It is found in the lamellar bodies, or myeloid bodies, of the inner segments (Fig. 5), and it can also be found in outer segment preparations (see Section 4.2.2). This unusual photopigment protein, 24 kiloDalton (kD), contains all *trans* retinal in the dark, which is photoisomerized to 11-*cis* by light to form metaretinochrome (Ozaki *et al.*, 1983). It thus undergoes the reverse photoreaction to rhodopsin. It can donate 11-*cis* retinal directly to squid or bovine opsin if the retinochrome and opsin-containing membranes are in close contact, but not if they are separated by a dialysis membrane (Seki *et al.*, 1982). Its λ_{max} (495 – 476) is sensitive to salt and pH (Ozaki *et al.*, 1982). The chromophore site is thus more reactive than that of rhodopsin. In this respect it resembles a vertebrate cone pigment. Its spectrum is shown in Fig. 13.

5.2.2.1. Purification method of Hara and Hara (Hara and Hara, 1982)

Todarodes pacificus are collected at night and the enucleated eyes are immediately frozen. Working at 4° C in dim red light:

Remove the anterior half of the eye to prepare the eyecup;

Shake off the outer segments in 67 mM phosphate buffer pH 6.5;

Spread the eyecup on filter paper and peel out the inner half of retina with forceps;

Cut it in fine pieces, homogenize in a glass-teflon homogenizer in a small volume of buffer, and pellet;

Suspend the pellet in 43% (w/v) sucrose in phosphate buffer and spin 12,000 rpm 15 min;

Save the supernatant and repeat the sucrose step on the pellet;

Pool the 2 supernatants, dilute with an equal volume of buffer and spin 12,000 rpm 20 min. All subsequent washes should be done at this speed and time.

The brownish orange-red pellet contains most of the lamellar bodies.

Wash in 0.5% Na_2CO_3 (to remove contaminating soluble pigment); do several washes in H_2O.

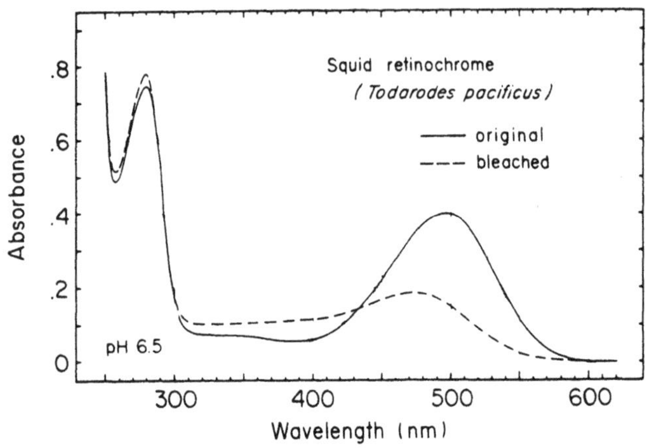

Figure 13. Absorption spectrum of retinochrome. From Hara *et al.* (1981), by permission of Pergamon Press. (———), retinochrome; (- - - -), metaretinochrome.

Harden with phosphate buffer.

Incubate in Weber-Edsall solution (0.6 M KCl, 0.04 M $NaHCO_3$, 0.01 M Na_2CO_3) 30 min at 27° C. This is to extract soluble or weakly bound proteins.

Spin and wash 3 times with H_2O, and lyophilize.

Extract the dried material 3 times with petroleum ether to remove lipids.

Remove the solvent completely and wash once with H_2O.

Detergent extraction: mix with 2% aqueous digitonin, shake gently in the cold 20 min.

Yield: equal amounts of rhodopsin.

A very similar method has been used by Sperling and Hubbard (1975) to purify retinochrome from the Atlantic squid *Loligo pealei*, which was found to contain only one third to one quarter the amount of retinochrome relative to rhodopsin.

5.2.3. Retinal binding protein

A soluble, 51 kilodalton retinal binding protein has been isolated and characterized by Ozaki *et al.* (1987). They have devised a purification procedure using anion exchange and size-exclusion chromatography. They have shown that it can take up 11-*cis* retinal from metaretinochrome and donate it to opsin, suggesting that it acts as a shuttle in the regeneration of rhodopsin in the squid retina. It is an abundant protein, amounting to about 10% of rhodopsin. It binds both retinal and retinol, and has some properties resembling those of vertebrate retinoid binding proteins.

5.3. Microvillus structure

5.3.1. X-ray diffraction

The enormous array of photoreceptor microvilli, with a paracrystalline lattice of 600 Å unit cells correlated over macroscopic distances, makes the squid retina a unique tissue for structural studies. The first low angle X-ray diffraction patterns of squid retina were obtained by Worthington (1976), demonstrating that the hexagonal array of membranes was sufficiently ordered to diffract X-rays. Later studies with more highly oriented and ordered samples (Saibil, 1982; Saibil and Hewat, 1987) allowed recording of the full hexagonal lattice pattern. High quality diffraction has so far only been obtained from glutaraldehyde fixed retinas, as unfixed material is too labile over the 16 – 24 hr it takes to record the pattern. Aldehyde-fixed material, while still hydrated, is subject to certain fixation artifacts, particularly aggregation of cytoskeletal filaments and possibly shrinkage of the extracellular space between membranes. The main utility of X-ray diffraction in this system has been to analyze the membrane structure. The very large unit cell in this material requires a well focussed low-angle camera (point focus < 200 μm), and the specimen should be cooled. Lattice diffraction can be recorded to $(90\ \text{Å})^{-1}$ and oriented, diffuse intensity out to $(35\ \text{Å})^{-1}$ (see Section 5.3.2). The very small intermembrane spacing seen with X-rays is also obtained by freeze substitution electron microscopy (EM) (see Section 5.3.3).

5.3.2. Electron microscope image analysis

The purpose of EM image analysis of a repeating structure is to average together many theoretically identical units. The repeating signal (true structure) will be reinforced with averaging, while random deviations (noise) will tend to cancel out. Conventionally this has been done by spatial filtering of the optical or computed diffraction pattern. But if the lattice is disordered, which is often the case in biological specimens, the information in the average is degraded. A more effective method in this case is real space correlation averaging (refs. in Saibil and Hewat, 1987). This method takes account of displacements of unit cells from their ideal lattice positions, and averages them together in their actual positions, discarding more seriously distorted ones from the average. This method is particularly effective for invertebrate photoreceptor membranes. The microvillar lattice has the peculiar property of a degree of long-range order coexisting with local variation in microvillus diameter. Since the repeating unit is a membrane structure, it is locally deformable, but as the membranes are firmly linked together, local distortions are counterbalanced by opposing distortions in neighboring regions (for example an expanded microvillus will be surrounded by compressed ones) and the long-range lattice orientation is preserved. A diagram of the membrane junctions revealed by diffraction and correlation averaging methods is shown in Fig. 14.

5.3.3. Rapid freezing, freeze etching and freeze substitution EM

This avoids the problems of aldehyde fixation and gives a more faithful view of the subcellular structure (Tsukita *et al.*, 1988). The disadvantage of this method is the shallow depth of tissue (≤30 μm) in which it is possible to have sufficiently rapid freezing. In the squid retina, the surface layer accessible to rapid freezing without damaging dissection of the unfixed tissue is unfortunately the most disordered part of the microvillar lattice, and image averaging has not been done on freeze-substituted squid photoreceptors. However, Tsukita *et al.* have obtained clear images of the cytoskeletal filaments. They have taken advantage of rapid freezing to compare dark and light-adapted retinas, and found that the cytoskeleton structure is disrupted in light-stimulated microvilli. This may take place on a time scale compatible with an

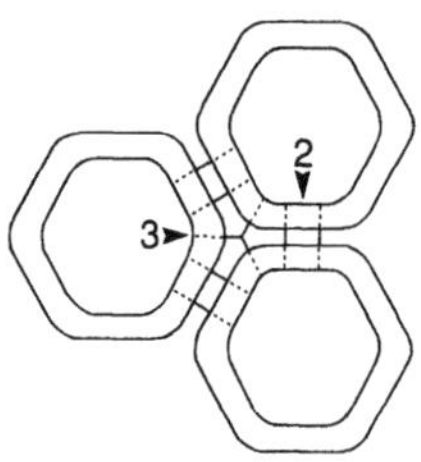

Figure 14. Diagram of the two-fold and three-fold membrane junctions in squid photoreceptor microvilli derived from X-ray diffraction and electron microscope image analysis. From Saibil and Hewat (1987), reproduced from *The Journal of Cell Biology* by copyright permission of the Rockefeller University Press.

involvement of the cytoskeleton in phototransduction, possibly as a consequence of light-induced changes in cytoplasmic messenger concentrations.

5.4. Assays for rhodopsin-activated signalling enzymes

Rhodopsin is a member of a large family of membrane receptors which mediate the responses of cells to external signals by controlling the concentrations of intracellular messengers such as cyclic nucleotides and phosphoinositides. The signal is relayed across the membrane when the activated receptor catalyzes guanine nucleotide exchange on GTP-binding proteins. Photoreceptor membranes are very favorable for studying the biochemical mechanisms of signal transduction, as they contain high concentrations of receptor and signalling enzymes, and they are highly specialized for transduction. In vertebrate rod and squid photoreceptor outer segments, rhodopsin is the unique receptor and comprises the major fraction of the membrane protein. In the invertebrate system, there is evidence that one pathway involves GTP-binding protein activation of phosphatidylinositol bisphosphate (PIP_2) phospholipase C (Payne, 1986; Tsuda, 1987).

5.4.1. GTP-binding proteins

If the microvillar membranes are prepared under dim red or IR illumination, and care is taken to maintain reducing conditions (deoxygenated buffers with ≥1 mM dithiothreitol), light stimulates guanosine 5' triphosphatase (GTPase) and rhodopsin kinase activities. A low K_m GTPase activity is stimulated 3 – 4 fold by light (Calhoon *et al.*, 1980; Vandenberg and Montal, 1984a; Saibil and Michel-Villaz, 1984). High hydrolysis in the dark can be suppressed by adding adenosine-5'-triphosphate (ATP) or ATP regenerating system. Light also stimulates guanine nucleotide binding (Vandenberg and Montal, 1984a). The two bacterial toxins which have been used to identify α-subunits of GTP-binding protein, cholera and pertussis toxins, both catalyze light-dependent labelling of proteins in cephalopod photoreceptors. Vandenberg and Montal (1984a) found a 45 kD cholera toxin substrate if they performed the labelling on membranes from fresh retinas, and reported its aggregation to a 120 kD band upon freezing. In *Octopus* membranes, Tsuda *et al.* (1986) found a 40 kD pertussis toxin substrate. These two distinct GTP-binding protein α-subunits can be observed in squid photoreceptor membranes (Baverstock *et al.*, 1989).

5.4.2. Phosphoinositides

Vandenberg and Montal (1984b) showed that, in addition to rhodopsin phosphorylation, radiolabelled ATP was rapidly incorporated into phosphatidylinositol phosphate (PIP), PIP_2 and phosphatidic acid in squid membranes, and that PIP and PIP_2 decreased in the light. By incubating eyecups in ^{3}H-inositol, Szuts *et al.* (1986) found a rapid light-stimulated increase in IP_3 release, with only the 1,4,5 isomer detected by high-performance liquid chromatography (HPLC). Brown *et al.* (1987) incorporated ^{3}H inositol by either incubating eyecups or by intraocular injection, and also found light-induced IP_3 release and reduction of PIP_2. These effects were not blocked by treatment with pertussis toxin. Using an *in vitro* approach, Baer and Saibil (1988) added exogenously labelled PIP_2 to purified membranes and found light stimulated hydrolysis which required GTP, implicating a GTP-binding protein. The corresponding physiological work has mostly been done on *Limulus* and fly photoreceptors.

5.4.3. Cyclic guanosine monophosphate (cGMP)

Weaker evidence that cGMP acts a messenger in invertebrate phototransduction has been obtained using the squid retina and isolated photoreceptor membranes. Two groups have found a light-stimulated increase in cGMP, which is found in the photoreceptor membranes and occurs rapidly in the intact retina (Saibil, 1984; Johnson *et al.*, 1986). Johnson *et al.* found some cGMP-sensitive regions in *Limulus* photoreceptors. These *hot spots* responded to cGMP injection in a manner that mimicked the response to light.

6. Conclusions

6.1. Signal transduction

Squid photoreceptors are an excellent system for biochemical studies of receptor-activated GTP-binding proteins and PIP_2 phospholipase C. They are likely to exhibit interactions between parallel pathways, as the sole receptor activates at least 2 GTP-binding proteins. An advantage is the rapid and precise activation by light in this highly specialized transducing system. However, there is a lack of physiological studies on the squid retina, for the reasons mentioned before (see Section 3.1.2).

6.2. Membrane-cytoskeleton and membrane-membrane interactions

These photoreceptor microvilli comprise an extensive and highly ordered array of membrane-cytoskeleton connections. The proportion of membrane contacts with other membranes or with the cytoskeleton is exceptionally high, and the protein composition is simple. A disadvantage is the limited accessibility of the tissue for rapid freezing methods.

ACKNOWLEDGEMENTS: I am grateful to the staff of the Marine Biological Association, Plymouth, U. K. and to Dr S. von Boletzky of the Laboratoire Arago, Banyuls-sur-mer, France, for the excellent supply of cephalopods, essential for work in this field. I thank Professor T. Hara, Mr. K. Langmack, Dr. J. Lisman and Professor J. Z. Young for their helpful comments on the manuscript, and Dr. J. Patterson for providing the micrograph in Figure 7.

References

Akino, T. and Tsuda, M., 1979, Characteristics of phospholipids in microvillar membranes of *Octopus* photoreceptor cells, *Biochim. Biophys. Acta* **556**:61–71.

Anderson, R. E., Benolken, R. M., Kelleher, P. A., Maude, M. B., and Wiegand, R. D., 1978, Chemistry of photoreceptor membrane preparations from squid retinas, *Biochim. Biophys. Acta* **510**:316–326.

Baer, K. M. and Saibil, H. R., 1988, Light- and GTP-activated hydrolysis of phosphatidylinositol bisphosphate in squid photoreceptor membranes, *J. Biol. Chem.* **263**:17–20.

Baverstock, J., Fyles, J., and Saibil, H., 1989, in: *Receptors, Membrane Transport and Signal Tranduction. NATO ASI Series Volume H29*, (A. E. Evangelopoulos, J. P. Changeux, L. Packer, T. Sotiroudis, and K. Wirtzoc, eds.), pp. 76–84, Springer-Verlag, Berlin.

Breneman, J. W., Robles, L. J., and Bok, D., 1986, Light-activated retinoid transport in cephalopod photoreceptors. *Exp. Eye Res.* **42**:645–658.

Brown, J. E., Watkins, D. C., and Malbon, C. C., 1987, Light-induced changes in the content of inositol phosphates in squid (*Loligo pealei*) retina, *Biochem. J.* **247**:293–297.

Calhoon, R., Tsuda, M., and Ebrey, T. G., 1980, A light-activated GTPase from Octopus photoreceptors, *Biochem. Biophys. Research Commun.* **94**:1452–1457.

Cohen, A. I., 1973a, An Ultrastructural Analysis of the Photoreceptors of the Squid and Their Synaptic Connections. I. Photoreceptive and non-synaptic regions of the retina, *J. Comp. Neur.* **147**:351–378.

Cohen, A. I., 1973b, An ultrastructural analysis of the photoreceptors of the squid and their synaptic connections. II. Intraretinal synapses and plexus, *J. Comp. Neurol.* **147**:379–398.

Cohen, A. I., 1973c, An ultrastructural analysis of the photoreceptors of the squid and their synaptic connections. III. Photoreceptor terminations in the optic lobes, *J. Comp. Neurol.* **147**:399–426.

de Couet, H. G., Stowe, S., and Blest, A. D., 1984, Membrane-associated actin in the rhabdomeral microvilli of crayfish photoreceptors, *J. Cell Biol.* **98**:834–846.

Daw, N. W., 1974, Squid visual system, in: *A guide to laboratory use of the squid Loligo pealei*, (J. M. Arnold, W. C. Summers, D. L. Gilbert, R. S. Manalis, N. W. Daw, and R. J. Lasek, eds.), pp. 66–68, Marine Biological Laboratory, Woods Hole, Mass.

Daw, N. W. and Pearlman, A. L., 1974, Pigment migration and adaptation in the eye of the squid, *Loligo pealei*, *J. Gen. Physiol.* **63**:22–36.

Duncan, G. and Pynsent, P. B., 1979, An analysis of the wave forms of photoreceptor potentials in the retina of the cephalopod *Sepiola atlantica*, *J. Physiol.* **288**:171–188.

Fisher, E. J. and Patterson, J. A., 1988, Photoreceptor sizes in the retina of the squid, *Alloteuthis subulata*, *J. Physiol.* **398**:97P.

Hagins, F. M., 1973, Purification and partial characterization of the protein component of squid rhodopsin, *J. Biol. Chem.* **248**:3298–3304.

Hagins, W. A., 1965, Electrical signs of information flow in photoreceptors, *Cold Spring Harbor Symp. Quant. Biol.* **30**:403–415.

Hagins, W. A., Zonana, H. V., and Adams, R. G., 1962, Local membrane current in the outer segments of squid photoreceptors, *Nature* **194**:844–847.

Hara, R., Hara, T., Tokunaga, F., and Yoshizawa, T., 1981, Photochemistry of retinochrome, *Photochem. Photobiol.* **33**:883–891.

Hara, T. and Hara, R., 1982, Cephalopod retinochrome. in: *Methods in Enzymology* **81 (Part H)**, (L. Packer, ed.): 190–197, Academic Press, New York.
Hara, T. and Hara, R., 1976, Distribution of rhodopsin and retinochrome in the squid retina, *J. Gen. Physiol.* **67**:791–805.
Hartline, P. H. and Lange, G. D., 1974, Optic nerve responses to visual stimuli in squid, *J. Comp. Physiol.* **93**:37–54.
Hartline, P. H. and Lange, G. D., 1984, Visual systems of cephalopods, in: *Comparative Physiology of Sensory Systems*, (L. Bolis, R. D. Keynes, and S. H. P. Maddrell, eds.), pp. 335–355, Cambridge Univ. Press, Cambridge.
Hubbard, R. and St. George, R. C. C., 1958, The rhodopsin system of the squid, *J. Gen. Physiol.* **41**:501–528.
Johnson, E., Robinson, P., and Lisman, J., 1986, Involvement of cGMP in the excitation of invertebrate photoreceptors, *Nature* **324**:468–470.
Kito, Y., Seki, T., Suzuki, T., and Uchiyama, I., 1986, 3-dehydroretinal in the eye of a bioluminescent squid, *Watasenia scintillans*, *Vision Res.* **26**:275–279.
Lam, D. M. K., Wiesel, T. N., and Kaneko, A., 1974, Neurotransmitter synthesis in cephalopod retina, *Brain Res.* **82**:365–368.
Land, M. F., 1981, Optics and vision in invertebrates, in: *Handbook of Sensory Physiology*, Vol. **VII/6B**, (H. Autrum, ed.), pp. 471–592, Springer, Berlin.
Lange, G. D. and Hartline, P. H., 1974, Retinal responses in squid and octopus, *J. Comp. Physiol.* **93**:19–36.
Matsui, S., Seidou, M., Uchiyama, I., Sekiya, N., Hiraki, K., Yoshihara, K., and Kito, Y., 1988a, 4-Hydroxyretinal, a new visual pigment chromophore found in the bioluminescent squid, *Watasenia scintillans*, *Biochim. Biophys. Acta* **966**:370–374.
Matsui, S., Seidou, M., Horiuchi, S., Uchiyama, I., and Kito, Y., 1988b, Adaptation of a deep-sea cephalopod to the photic environment. Evidence for three visual pigments, *J. Gen. Physiol.* **92**:55–66.
Messenger, J. B., 1981, Comparative physiology of vision in molluscs. in: *Handbook of Sensory Physiology*, Vol. **VII/6C**, (H. Autrum, ed.), pp. 93–200, Springer, Berlin.
Nashima, K., Mitsudo, M., and Kito, Y., 1978, Studies on cephalopod rhodopsin. Fatty acid esters of sucrose as effective detergents, *Biochim. Biophys. Acta* **536**:78–87.
Nashima, K., Mitsudo, M., and Kito, Y., 1979, Molecular weight and structural studies on cephalopod rhodopsin, *Biochim. Biophys. Acta* **579**:155–168.
Ovchinnikov, Yu. A., Abdulaev, N. G., Zolotarev, A. S., Artamonov, I. D., Bespalov, I. A., Dergachev, A. E. and Tsuda, M., 1988, Octopus rhodopsin. Amino acid sequence deduced from cDNA, *FEBS Lett.* **232**:69–72.
Ozaki, K., Hara, R., and Hara, T., 1982, Dependency of absorption characteristics of retinochrome on pH and salts, *Exp. Eye Res.* **34**:499–508.
Ozaki, K., Hara, R., Hara, T., and Kakitani, T., 1983, Squid retinochrome. Configurational changes of the retinal chromophore, *Biophys. J.* **44**:127–137.
Ozaki, K., Terakita, A., Hara, R., and Hara, T., 1987, Isolation and characterization of a retinal-binding protein from the squid retina, *Vision Res.* **27**:1057–1069.
Packard, A., 1972, Cephalopods and fish: the limits of convergence, *Biol. Rev.* **47**:241–307.
Patterson, J. A., 1985, A modified, prefixed, Golgi rapid technique for the cephalopod retina, *J. Neurosci. Meth.* **12**:219–225.
Patterson, J. A. and Fisher, E. J., 1989, *personal commun.*
Patterson, J. A. and Silver, S. C., 1983, Afferent and efferent components of *Octopus* retina, *J. Comp. Physiol.* **151**:381–387.
Paulsen, R., Zinkler, D., and Delmelle, M., 1983, Architecture and dynamics of microvillar photoreceptor membranes of a cephalopod, *Exp. Eye Res.* **36**:47–56.
Payne, R., 1986, Phototransduction by microvillar photoreceptors of invertebrates: mediation of a visual cascade by inositol trisphosphate, *Photobiochem. Photobiophys.* **13**:373–397.
Pinto, L. H. and Brown, J. E., 1977, Intracellular recordings from photoreceptors of the squid (*Loligo pealii*), *J. Comp. Physiol.* **122**:241–250.

Robles, L. J., Cabebe, C. S., Aguilo, J. A., Anyakora, P. A., and Bok, D., 1984, Autoradiographic and biochemical analysis of photoreceptor membrane renewal in *Octopus* retina, *J. Neurocytol.* **13**:145–164.

Robles, L. J., Watanabe, A., Kremer, N. E., Wong, F., and Bok, D., 1987, Immunocytochemical localization of photopigments in cephalopod retinae, *J. Neurocytol.* **16**:403–415.

Roper, C. F. E. and Boss, K. J., 1982, The giant squid, *Sci. Am.* **246**(4):82–89.

Saibil, H. R., 1982, An ordered membrane-cytoskeleton network in squid photoreceptor microvilli, *J. Mol. Biol.* **158**:435–456.

Saibil, H., 1984, A light-stimulated increase of cyclic GMP in squid photoreceptors, *FEBS Letters* **168**:213–216.

Saibil, H. and Hewat, E., 1987, Ordered transmembrane and extracellular structure in squid photoreceptor microvilli, *J. Cell Biol.* **105**:19–28.

Saibil, H. and Michel-Villaz, M., 1984, Cross-reactions between rhodopsin and GTP-binding protein in bovine and squid photoreceptors, *Proc. Natl. Acad. Sci. USA* **81**:5111–5115.

Saidel, W. M., Lettvin, J. Y., and MacNichol, E. F., 1983, Processing of polarized light by squid photoreceptors, *Nature* **304**:534–536.

Seki, T., Hara, R., and Hara, T., 1982, Reconstitution of squid and cattle rhodopsin by the use of metaretinochrome in their respective membranes, *Exp. Eye Res.* **34**:609–621.

Silver, S. C., Patterson, J. A., and Mobbs, P. G., 1983, Biogenic amines in cephalopod retina, *Brain Res.* **273**:366–368.

Sivak, J. G., 1982, Optical properties of a cephalopod eye (the shortfinned squid, *Illex illecebrosus*), *J. Comp. Physiol.* **147**:323–327.

Sperling, L. and Hubbard, R., 1975, Squid retinochrome, *J. Gen. Physiol.* **65**:235–251.

Szuts, E., Wood, S., Reid, M., and Fein, A., 1986, Light stimulates the rapid formation of inositol trisphosphate in squid retinas, *Biochem. J.* **240**:929–932.

Tansey, E. M., 1980, Aminergic fluorescence in the cephalopod brain, *Phil. Trans. Roy. Soc. B* **291**:127–145.

Tasaki, I. and Nakaye, T., 1984, Rapid mechanical responses of the dark-adapted squid retina to light pulses, *Science* **223**:411–413.

Tomarev, S. I. and Zinovieva, R. D., 1988, Squid major lens polypeptides are homologous to glutathione S-transferase subunits, *Nature* **336**:86–88.

Tomita, T., 1968, The electrical response of single photoreceptors, *Proc. Inst. Elec. Electron. Eng.* **56**:1015–1023.

Tsuda, M., 1987, Photoreception and phototransduction in invertebrate photoreceptors, *Photochem. Photobiol.* **45**:915–931.

Tsuda, M., Tsuda, T., Terayama, Y., Fukada, Y., Akino, Y., Yamanaka, G., Stryer, L., Katada, T., Ui, M., and Ebrey, T., 1986, Kinship of cephalopod photoreceptor G-protein with vertebrate transducin, *FEBS Letters* **198**:5–10.

Tsukita, S., Tsukita, S., and Matsumoto, G., 1988, Light-induced structural changes of cytoskeleton in squid photoreceptor microvilli detected by rapid-freeze method, *J. Cell Biol.* **106**:1151–1160.

Vandenberg, C. A., 1982, *Light-regulated biochemical events in squid photoreceptors*, Ph.D. Thesis, University of California at San Diego, La Jolla, California.

Vandenberg, C. A. and Montal, M., 1984a, Light-regulated biochemical events in invertebrate photoreceptors. 1. Light-activated guanosine triphosphatase, guanine nucleotide binding, and cholera toxin catalyzed labelling of squid photoreceptor membranes, *Biochem.* **23**:2339–2347.

Vandenberg, C. A. and Montal, M., 1984b, Light-regulated biochemical events in invertebrate photoreceptors. 2. Light regulated phosphorylation of rhodopsin and phosphoinositides in squid photoreceptor membranes, *Biochem.* **23**:2347–2352.

Vandenberg, C. A. and Montal, M., 1984c, Purification of squid rhodopsin and reassembly into lipid bilayers, *Biochim. Biophys. Acta* **771**:74–81.

Waterman, T. H., 1981, Polarization sensitivity, in: *Handbook of Sensory Physiology*, Vol. **VII/6B**, (H. Autrum, ed.), pp. 281–496, Springer, Berlin.

Worthington, C. R., Wang, S. K., Folzer, C. M., 1976, Low angle X-ray diffraction patterns of squid retina, *Nature* **262**:626–628.
Young, J. Z., 1962a, The retina of cephalopods and its degeneration after optic nerve section, *Phil. Trans. Roy. Soc. Series B* **245**:1–18.
Young, J. Z., 1962b, The optic lobes of *Octopus vulgaris*, *Phil. Trans. Roy. Soc. B* **245**:1–58.
Young, J. Z., 1963, Light and dark adaption in the eyes of some cephalopods, *Proc. Zool. Soc. Lond.* **140**(Part 2):255–272.
Young, J. Z., 1974, The central nervous system of *Loligo*. I. The optic lobe, *Phil. Trans. Roy. Soc. Series B* **267**, 263–302.
Young, J. Z., 1976, The nervous system of *Loligo*. II. Suboesophageal centres, *Phil. Trans. Roy. Soc. B* **274**:101–167.
Zonana, H. V., 1961, Fine structure of the squid retina, *Johns Hopkins Hosp. Bull.* **109**:185–205.

Chapter 18

Development of the Squid's Visual System

I. A. MEINERTZHAGEN

1. Introduction

Like other cephalopods, squid have magnificent visual systems, the development of which has been too little studied in recent times. The structure of the eye (Fig. 1) is described in another chapter (Saibil, 1990), while the adult organization of the optic lobe is well described in *Loligo* (Kopsch, 1899; Young, 1974). This chapter will review the development of both, and the methods appropriate to that task, in an account drawn entirely from the literature; works on coleoids other than teuthoids are cited only when evidence is lacking in squid and to identify areas in which interesting findings would be forthcoming. The visual system will be interpreted conservatively so as not to include the extraocular photoreceptors in the paraolfactory vesicles, found in squid (Mauro, 1977; Young, 1978); the development of another system of cephalopod extraocular photoreceptors found in octopods, the epistellar body, has already been described (Sacarrão, 1956). The prospects for developmental studies on squid have recently improved with the prolonged culture of healthy animals in captivity (Hanlon, 1990). Squid mating behavior is described for *Loligo pealei* by Arnold (1962, 1990); embryos are obtained by methods described by Arnold (1974, 1990) and Hanlon (1990). Developmental stages are provided for *Loligo vulgaris* by Naef (1928), as amended by Meister (1972), and by Arnold (1965a, 1974, 1990) who provides a comprehensive illustrated timetable for *Loligo pealei*; a table of the events of eye morphogenesis at closely spaced intervals (Marthy, 1973) cross refers to these.

2. Development of the eye

Eye development has long been described in species of *Loligo* (Grenacher, 1874; Lankester, 1875; Bobretsky, 1875, 1876; summarized in Lankester, 1883). The essentially definitive works of Faussek (1893, 1900) are superseded only by the modern re-examination of Meister (1972); all such studies have utilized conventional light microscopic methods. By contrast with these descriptive studies, experimental work on the development of these advanced single-lens eyes, reviewed in Arnold and Williams-Arnold (1976), still leaves unanswered many major questions (see Section 2.5).

I. A. MEINERTZHAGEN • Life Sciences Centre, Dalhousie University, Halifax, Nova Scotia, Canada B3H 4J1.

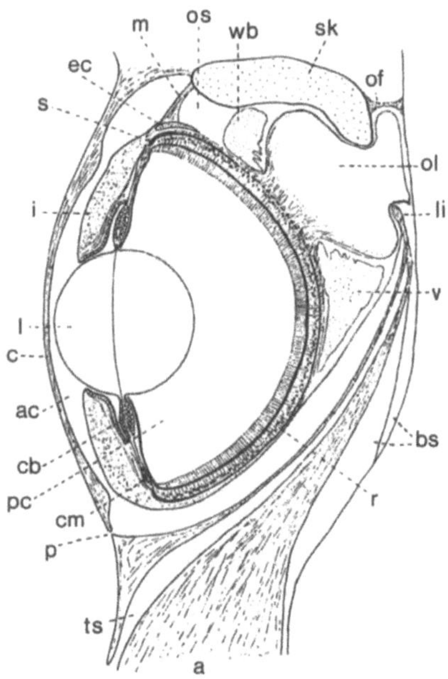

Figure 1. The eye of *Loligo pealei* in horizontal section. *a*, muscular base of the arms; *ac*, anterior chamber of the eye; *bs*, buccal sinuses (outer and inner); *c*, cornea; *cb*, ciliary body; *cm*, ciliary muscle; *ec*, accessory eye cartilage; *i*, iris; *l*, lens; *li*, lateral skull ligament; *m*, extraocular muscle, total nine per eye, in orbital fascia; *of*, orbital fascia; *ol*, optic lobe; *os*, orbital sinus; *p*, aquiferous pore; *pc*, posterior chamber of the eye; *r*, retina; *s*, sclera; *sk*, skull; *ts*, tentacular sac; *wb*, white body (a hemopoietic tissue). Adapted from Williams (1909).

Routine methods of wax histology for *light microscopy* are difficult to use on early embryos. The large amount of yolk makes the sections brittle, especially following dehydration and clearing in xylol; to avoid this, go from Bouin's fixative to three one-hour changes of dry Dioxane, thence to molten Paraplast® for three one hour impregnations (Arnold, 1974). For earlier stages, and in any case, embedment in epoxy after osmium fixation yields superior section uniformity and resolution. Use of hydrophilic acrylic resins (e.g., L. R. White, London Resin Company) combines the resolution of semi-thin sections with access to conventional histological stains.

The development of selected ocular tissues has also been documented, partially from electron microscopy (EM) observations. For descriptive studies of ocular morphogenesis the use of *scanning electron microscopy* (SEM) is also advocated. Interior views of the eye vesicle and its contents are readily obtained (Yamamoto, 1985) from specimens cleaved after fixation by routine EM methods (see Section 2.6). Arnold (1984) uses the following protocol: fix in 2% glutaraldehyde in Millipore™-filtered sea-water buffered with a few drops of 900 mOsm collidine-HCl (pH 7.4) for 1 to 1.5 hours; wash in collidine buffered sea-water (pH 7.4) for 0.5 to 1 hour; post-fix in 1% OsO_4 in 900 mOsm sea-water for 0.5 hours; finally dehydrate in a graded acetone series, critical-point dry through CO_2 and sputter coat.

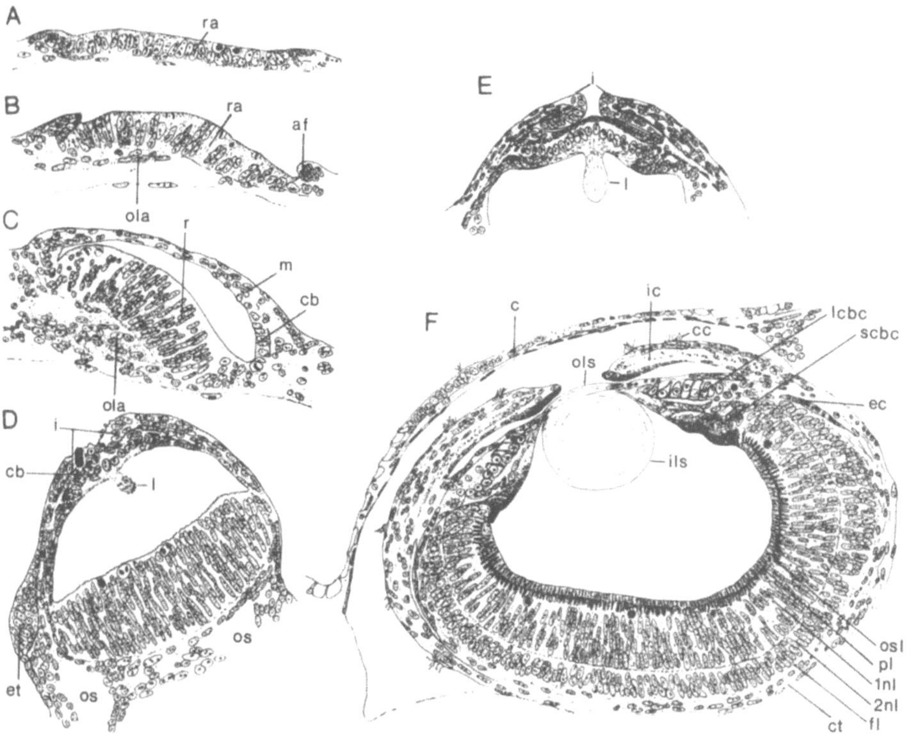

Figure 2. The development of the eye primordium, seen in section in *Loligo vulgaris*. A, stage VII; B, stage VIII; C, stage IX; D, stage XI; E, stage XII; F, stage XVIII; D and E, successive stages in the formation of the iris. Abbreviations: *af*, annular fold; *c*, cornea; *cb*, ciliary body; *cc*, ciliated cells; *ct*, connective tissue; *ec*, accessory eye cartilage; *et*, ectodermal thickening; *fl*, fiber layer; *i*, iris; *ic*, iridocytes; *ils*, inner lens segment; *l*, lens; *lcbc*, large ciliary body cells; *m*, mesoderm; *1nl*, first nuclear layer of photoreceptors; *2nl*, second nuclear layer of photoreceptors; *ola*, optic lobe anlage; *ols*, outer lens segment; *os*, orbital sinus; *osl*, outer-segment layer; *pl*, pigment layer; *r*, retina; *ra*, retina anlage; and *scbc*, small ciliary body cells. Adapted from Meister (1972).

2.1. Formation of an eye vesicle

The eye develops from a thickened ectodermal monolayer (stage VI: Naef, 1928), which forms an oval mitotic placode on the dorsal surface of the head lobe and becomes multilayered (Fig. 2A). The placode curls upwards at its perimeter by the outgrowth of an annular ectodermal fold, which draws itself over the placode like a veil, with the circular leading edge acting as a purse-string; the process is cytochalasin sensitive (Arnold and Arnold, 1978). The placode thus internalizes rather than invaginates (Fig. 2B). The first cells of the future optic lobe are apparent at this time (Meister, 1972). The ectodermal folds gradually enclose the primordium of the retina, so that the first conspicuous developmental stage is an optic cup with a thickened base. Subsequently the lips of the cup close to a narrow constriction to produce an eye vesicle (Fig. 2C). A multilayered epithelium lining the floor of the vesicle will give rise to the retina (see Section 2.6). The ectoderm at the point of this closure fuses, thereby generating a vesicle which will become the posterior eye chamber and which is sealed off by three layers of the primary eye fold. These are: an outer ectodermal layer; an inner ectodermal layer in which large cells destined to produce the lens are visible for the first time; and, separating these two, a layer of mesoderm (Fig. 2D). The two ectodermal layers differentiate simultaneously, the inner one producing the inner lens segment (see Section 2.2) and the outer one the iris (see Section 2.3) and outer lens segment. Finally, the cornea forms to enclose these (see Section 2.4).

2.2. Development of the lens

The lens forms as an outer and inner segment, as described by Arnold (1967), who resolves conflicting earlier ideas (Metschnikoff, 1867; Hollerbach, 1901, Sacarrão, 1954); Faussek's (1900) account is in error. The lens starts as a small population of large, central *Reisenzellen* that form a cone protruding from the inner ectoderm layer into the eye chamber (Figs. 2E; 3A, 3B). This will form the inner lens segment, to which smaller, lateral lentigen cells (Arnold, 1967) later add and, upon differentiating (Meister, 1972), contribute to the forming lens. Cells in the same layer (Fig. 3A), the *corpus epithelial* of earlier authors, later form the ciliary body. The lens initially forms from a cluster of lentigen processes, one per cell, which enlarges by new processes adding (Fig. 3B). Microtubules are found within the processes (Arnold, 1966a), and microvilli project from the process' tips, perhaps implicated in their extension. The lentigen cell bodies then begin to retract and migrate outwards forming a ring, leaving the lens suspended in the posterior eye chamber. Further resolution of cellular morphogenesis could be usefully procured from SEM methods (Willekens *et al.*, 1984). Lentigen process outgrowth is colchicine sensitive, but the accumulation of lens material is not, albeit this now occurs in the cell bodies following colchicine block (Arnold, 1966b). Lens material is transported from the cells into the lens. The much smaller outer lens segment forms subsequent to the inner, by analogous events, from a secondary group of lentigen cells.

Antisera may be raised to the structural crystallins in this lens material using conventional immunological methods (Brahma, 1978). Immunoreactivity first appears at Naef stage X and intensifies thereafter; only the lens and lentigenic cells show the immunoreactivity and they do so simultaneously and evenly (Brahma, 1978). The proteins are now known to share sequence homologies with mammalian glutathione S-transferase (Tomarev and Zinovieva, 1988). However, their ancestry must be remote.

2.3. Development of the iris

The iris forms above the inner lens segment, directly from the outer ectodermal layer of the primary eye fold (Meister, 1972) (Fig. 2E), and not from a secondary periocular eye fold as earlier authors claimed. From here develops a secondary infolding situated above the lens, which extends towards the eye's periphery (Figs. 2E, F). In *Sepia* the iris incorporates mesoderm from the primary eye fold to produce iris

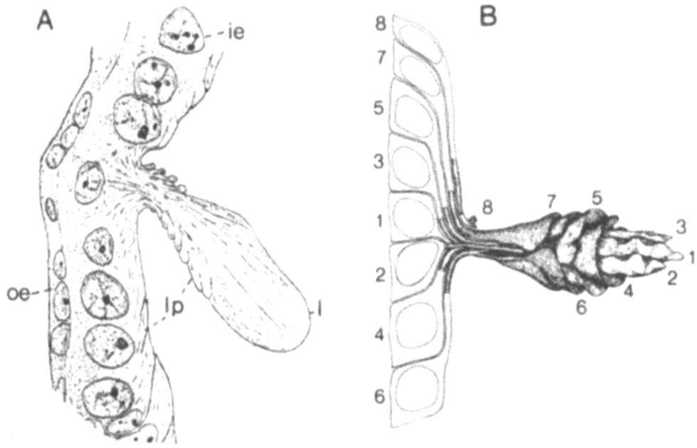

Figure 3. A. Section through the ciliary body and lens of *Loligo pealei*. *oe*, outer ectodermal layer; *ie*, inner ectodermal layer; *l*, lens; *lp*, processes of lentigen cells (from Williams, 1909). B. The arrangement of lentigen cells and their processes in the formation of the lens. The processes surround each other axially, like an onion bulb. Cells are numbered in a radial direction *1–8*; their processes are numbered correspondingly (from Arnold, 1967).

musculature (Lemaire and Richard, 1979). The mesoderm then disappears, leaving the suspensory ligament of the lens.

2.4. Development of the cornea

A quite new skin fold lastly forms the cornea from the edge of the forward growing tentacles, and thus has a very different ectodermal origin from the rest of the eye (Naef, 1928). In *Loligo* and also some octopods, the primary lid is totally closed, or patent only with a small pore to the anterior eye chamber. It arises behind the eyeball and progresses forward, closing over the eye by muscular contraction of the leading edge (Arnold, 1984). Closure is insensitive to cytochalasin B or colchicine, implying that muscular contraction is the morphogenetic event. Colchicine (10^{-3} M) is applied as a solution dissolved directly in sea-water; cytochalasin is first dissolved in dimethylsulfoxide (DMSO) and diluted to 2 μg/ml in filtered sea-water (Arnold, 1984).

2.5. Experimental embryology of the eye

This is an area of contended interpretation, with experimental technique, especially in operating upon and subsequent rearing of experimental embryos the likely reason for most differences between workers. Arnold (1974, 1990) provides a complete manual of methods for the procurement, de-chorionation and subsequent rearing of operated animals. Recent improvements in microscopical methods (with the advent of video microscopy methods: Inoué, 1986) are one area of technological update. In early studies, particularly, even the purity of seawater, now readily accomplished with Millipore™ filters and sterile technique, cannot be assumed. Ranzi's early observations (Ranzi, 1928a; 1928b; 1930; 1931a; 1931b; 1932), partially on *Loligo*, provide a marketplace of ideas that indicate the regulative development of the eye vesicle and the independence of lens development from it; they are a basis for critical modern re-examinations. Partial eye vesicles round off to form smaller than normal but otherwise complete eyes; a single optic vesicle, when bisected so that each half invaginates separately, produces two eyes (Ranzi, 1931a), but this has not been reproduced in stage

IX embryos by Marthy (1973). Chemically induced developmental abnormalities of the optic placode, when the latter fails to invaginate, produce differentiated retinal cellular elements independent of normal organogenesis (Ranzi, 1932). The lens too may develop independently of the rest of the eye, when this fails to form a completely closed invagination, for example.

In his experiments, Ranzi operated upon and reared embryos entirely while they were still within the chorion, and the progress of more recent experimental embryological studies on the eyes required the advent of culture methods for squid embryos, developed first by Arnold (1965b, as amended 1974). A simple *culture medium for maintaining operated embryos* under sterile conditions contains 10% horse serum in sterile sea-water with a penicillin-streptomycin antibiotic mixture added (Arnold, 1974). See Section 3.2 for the composition of a more complex medium.

The results have been used to support both inductive (Arnold, 1965b, 1971; Arnold and Williams-Arnold, 1976) and regulative (Marthy, 1973, 1978, 1979) phenomena in the development of the eye vesicle. According to Arnold (1965b) the yolk syncytium contains a morphogenetic map with an inductive influence upon the overlying outer layers of cells, the embryo proper. Thus, after removing the eye primordium, leaving the underlying denuded yolk syncytium intact, he obtained a regenerate eye vesicle from surrounding cells which moved in to close over the wound (Arnold, 1965b). In refining the experimental analysis of these results, Marthy (1973) performed total, sub-total, or partial excisions of eye primordia. Partial excisions resulted in complete regenerates of eye vesicles; the same was also true when the primordium alone was excised (sub-total excisions), leaving the denuded yolk syncytium, but only in young embryos. All regenerates, whether from partial or sub-total excisions showed stage dependence in the completeness of their regeneration. Young embryos regenerate more completely than old; the completeness of regeneration reflecting the sequence of eye development, with the iris fold being one of the last steps in morphogenesis and the character still susceptible to deletion in the oldest embryos of an age series. Finally, removal of the eye primordium and surrounding ectoderm resulted in slow wound closure and no regenerate eye formation (Marthy, 1973).

Complementary explant experiments have also been conducted. Explants of the eye primordium from the outer ectodermal cell layer, cultured alone, survive but fail to differentiate an eye vesicle; explants which included both the primordium and underlying yolk syncytium do differentiate into a vesicle (Arnold, 1965b). This difference Marthy (1973) tentatively attributed to nutritional factors derived from the yolk droplets still adhering to the yolk syncytium. Excised intact eye-cup primordia (Marthy, 1970; 1973) grafted upon a denuded yolk syncytium differentiate eye, again in a stage-dependent fashion, with eye vesicles forming only in implants from late embryos. Marthy's view is of the eye primordium as a morphogenetic field, with the capacity for autonomous differentiation into an eye vesicle and capable, under experimental conditions, of regulation. Predictably this regulative ability decreases with age, with formation of the iris fold susceptible to partial deletions of the primordium, even though the eye vesicle itself still forms. Marthy (1973, 1978, 1979) thus concludes from his results that the yolk syncytium has no causal influence over the eye field, and proposes that Arnold's results were due to incomplete removal of the eye placode, the remnants of which regulate to produce a complete vesicle.

Arnold and Williams-Arnold (1976), on the other hand, attribute Marthy's failure to replicate all of Arnold's (1965b) results in regenerating an eye vesicle from denuded yolk syncytium to differences in experimental technique which caused delayed wound healing in Marthy's operated embryos. The major evidence for an inductive role of the yolk syncytium in eye development comes, however, from explants of dissociated outer layer cells which, when reaggregated and placed ectopically over the denuded yolk syncytium, develop into complete vesicles, thus assuming a fate which depends on their position over that layer (Arnold, 1965b). That result is furthermore backed by the effects of exposing young embryos to cytochalasin B (Arnold and Williams-Arnold, 1974). Organogenesis is affected in such animals. Some of the effects are long after, and in a pattern correlated with, the time of embryonic exposure to the drug; they are apparently exerted through a pattern of developmental information set up in the egg cortex and expressed through the nuclei of the yolk syncytium.

These studies illustrate the precision required of both surgery and experimental design, as well as the difficulties in interpreting divergent results. Further analysis of both existing and future findings in this area needs to be based upon both histological analyses of experimentally operated eyes and upon the pattern of cellular proliferation and migration under control and experimental conditions. The patterns of thymidine incorporation or of bromodeoxyuridine immunocytochemical staining (Gratzner, 1982) are one of the most obvious of many outstanding studies needed on cephalopod eyes; their most obvious focus is the cell layers of the retina, conveniently accessible to local thymidine or bromodeoxyuridine injection via the posterior chamber.

2.6. Retinal differentiation

The retina differentiates late in eye development, along with further development of the other eye tissues.

Meister (1972) provides an accessible chronology of differentiation in *Loligo vulgaris*, from light microscopy. Hatchling squid already have the adult retinal layers distinguishable. These layers are, progressing from the eye's interior to the sclera: a limiting membrane; the receptor outer segment layer, basal to which is a pigment layer; a basement membrane, distal to which lies a layer of supporting cell nuclei; and, proximal, the nuclei of the receptor cell proximal segments; a plexiform layer containing the synaptic processes of receptor and efferent axons; and, the layer of receptor axons themselves with iridophores: finally, the connective tissue of the argentea (Fig. 2F) (Cohen, 1973a). The pattern of differentiation generally follows this same sequence (Meister, 1972). The outer segment layer first separates distally from a nuclear layer at stage XII (Fig. 2E). Deposition of ommin pigment granules follows thereafter (stage XIII), followed by the segregation of nuclei into two layers (stage XV and later), then, finally, elongation of the rhabdomes and condensation of the pigment in the proximal receptor layer (at stage XVIII, Fig. 2F) (Meister, 1972). EM observations are lacking in squid embryos, but available in the cuttlefish *Sepiella* (Yamamoto, 1985; Yamamoto *et al.*, 1985). Several fixatives described in the literature have given satisfactory preservation. These include those on cuttlefish embryos (Yamamoto, 1985) as well as adult squid retinae (Cohen, 1973a, 1973b, 1973c).

2.6.1. EM fixation for squid retina

2.6.1.1. Method of Cohen (1973a)

Anesthetize squid with urethane, catheterize the systemic heart with a polyethylene catheter and perfuse the vascular system by either syringe or gravity feed (reservoir height of 75 cm), with a primary fixative of the following composition: 2.5% glutaraldehyde in 0.2 M phosphate buffer (pH 7.4) containing 0.14 M NaCl. For gravity perfused squid, perfuse 1 liter of fixative for 1 hour, additionally injecting 0.5 ml fixative intraocularly. For syringe perfusion, use 30 ml fixative for 30 min, and inject 0.5 ml fixative intraocularly. Dissect out and slice the retina and/or optic lobe and further fix the slices for 1 hour. Wash in 0.2 M phosphate buffer with 0.3 M NaCl, store cold overnight and post-fix in 2% OsO_4 in 0.2 M phosphate (or 1.25% $NaHCO_3$) containing 0.14 M NaCl (pH 7.4). For syringe perfusion, perfuse with 1% OsO_4 as the primary fixative in either 0.2 M phosphate buffer and 0.14 M NaCl, or in veronal acetate and balanced salts, with sucrose added to attain 900 mOsm.

2.6.1.2. Method of Yamamoto (1985)

Remove embryo from egg capsule into fixative containing 2.5% glutaraldehyde in 0.1 M cacodylate buffer (pH 7.4) and 0.4 M sucrose; excise eye vesicles and cut along their major axes. Transfer eye pieces to fresh fixative; after 2 hours at room temperature, transfer to sucrose-cacodylate buffer, wash, and post-fix in 1% OsO_4 in sucrose-cacodylate mixture.

2.6.2. Electrophysiological differentiation

When compared with the anatomical development of the retina, electrophysiological differentiation probably holds no surprises, but data are lacking in squid. In cuttlefish, electroretinogram (ERG) development parallels photoreceptor rhabdomeric microvillar elongation (Yamamoto, 1985) and outer segment elongation (Yoshida *et al.*, 1976), preceded by the appearance of early receptor potentials (ERPs), the latter increasing with microvillar P-face particles (Yamamoto *et al.*, 1985). Early qualitative behavioral responses of the embryo to light begin much later than these (Yoshida *et al.*, 1976).

2.6.2.1. Recording ERG and optic nerve activity in squid

Following is a brief summary of the extensive methods of Lange and Hartline (1974) on adult *Loligo opalescens*. Decapitate a fresh squid; rapidly excise an eye and attached optic lobe; transfer to a small chamber of chilled oxygenated sea-water and carefully dissect a thin layer of cartilage from the back of the eye to allow oxygenation. The 5 × 5 × 5 cm chamber is lined with plastic, containing sea-water equilibrated with 100% oxygen; it is cooled on its base with a Peltier device, with one side a clear plexiglass window, into which is set a hemispheric viewing port. The eye, mounted on a holder, is set with its lens at the center of the hemisphere, around which a radially directed fiber optic device coupled to the stimulating light source may be moved on a perimeter device. Glass micropipette suction electrodes drawn to a fire polished tip of 0.05 to 0.5 mm are used to record impulse activity from optic nerve fibers, in response to stimulation, and larger bore versions, applied to the back of the eye, record ERG activity. These methods, and related ones used on *Alloteuthis* (Patterson, 1984), are in principle suitable for juvenile squid using appropriately scaled down equipment. A physiological saline for squid retina (Patterson, 1984) is 475 mM NaCl, 5 mM KCl, 10 mM $CaCl_2$, 50 mM $MgCl_2$, 10 mM glucose, 10 mM myo-inositol, 20 mM tris (hydroxymethyl) aminomethane-HCl (Tris-HCl) (TRIZMA) preset pH crystals at pH 7.0.

Yamamoto *et al.* (1985) have made ERG measurements in developing cephalopod retinae. They used pre-hatchling cuttlefish embryos, starting at stages roughly comparable with Naef stage XXVI in *Loligo*, for which their protocol should also be suitable. Recordings are undertaken in darkness at 20 °C, with embryos dark-adapted for 1 hour before removal from the egg capsule. The eye is excised and placed lens-up on black filter paper wet with sea-water; underneath is the indifferent electrode, a chlorided silver plate. Recordings are made with a sea-water filled capillary microelectrode, having a tip diameter of 5 to 10 μm, inserted through a side of the lens vesicle into the posterior chamber and manipulated to maximize the ERG amplitude. Stimulating light is delivered from a light guide via a shutter with a 3 min minimum interstimulus interval. Saturating responses of 10 mV were obtained in *Sepiella* at the hatchling stage, using a light intensity of 30,000 lux (Yamamoto, 1985).

2.6.3. Golgi impregnation of the retina

Comparison between the states of electrophysiological development and cell morphology would be possible following impregnation by the Golgi-rapid method for cephalopod retina. The following method, which has been used successfully on the retinae of *Sepia* and *Octopus* (Patterson, 1985), should also work well on squid. It derives from methods used on insect tissues: see also the Golgi methods for optic lobe studies (Section 3.4).

Anesthetize squid in 2% v/v ethanolic sea-water until flaccid; excise eyes and attached optic lobes rapidly; dissect away the iris and puncture the ciliary body at points around the lens to penetrate the vitreous, which is then displaced by the gentle injection of fixative containing 2% glutaraldehyde, 2% paraformaldehyde, 7% sucrose, and 0.2 M piperazine-N,N'-bis(2-ethanesulfonic acid) (PIPES), adjusted to pH 7.4 prior to use. Transect the optic nerve to remove the optic lobe and immerse the eyes in fixative for 2 to 4 hours. Dissect away the anterior part of the eye and store the eye-cup in fixative, for a period up to a year.

Impregnate the retina as follows: Dice eye-cup into pieces 3 × 3 to 5 × 15 mm and wash several times for 2 hours in 7% sucrose in a buffer prepared from 0.2 M boric acid (90 ml) and 0.05 M borax (10 ml) stock solutions; the solution initially has a pH of 7.4, but will change with time. Change 3 times for a total duration of 3 hours. The 2.5% dichromate preferentially impregnates photoreceptors and glia; increasing the concentration to 3.0 – 3.5% enhances the frequency of impregnated centrifugal terminals. Place 1 to 3 pieces into sachets made by rolling and stapling the ends of 50 mm long segments of 10 mm wide dialysis tubing (Visking Tubing, Scientific Instrument Centre, London, U.K. or similar). Place 2 to 5 sachets into a wide-mouthed jar of dichromate solution of the same composition as before, but containing 1 ml of 2% OsO_4 per 20 ml. Seal and store 3 to 10 days (optimally 7 days), in the dark at room temperature. Remove, blot, and place in 0.75% $AgNO_3$ in borax-boric acid buffer (pH 7.4) in a wide-mouth jar containing 30 ml; change 3 times during 10 min, seal, and leave 1 to 4 hours in darkness at room temperature. Remove the retinal tissue, place in distilled water, and brush off any deposits. Section freehand or with a sledge microtome, either after freezing in a drop of Tissue Tex II OCT compound or embedding in epoxy, following brief dehydration. Mount in Depex (Gurr) or similar.

3. Development of the optic lobe

3.1. Morphogenetic dependence of the optic lobe upon eye development

The dependence of the optic lobe's development upon maturation in the eye has been claimed from observations on embryos of *Loligo*, treated with 0.98% LiCl in seawater for 12 to 48 hours, which develop abnormally. Some of these embryos grow asymmetrical eyes, in which suppressed eye development is invariably accompanied by a small corresponding optic lobe (Ranzi, 1928a). A similar conclusion is reached by Arnold (1965b). Since a reduced optic lobe may also form behind a normal eye, albeit rarely, the evidence is compatible with first, a dependence of the optic lobe upon the eye, but, second, does not rule out the possible additional direct action of LiCl upon embryonic development of the optic neuropiles. Moreover, Ranzi neither defined accurately the different stages of embryos nor attempted to investigate the developmental effects at different ages. Dependence of the optic lobe upon eye development is most readily explained as the consequence of lost or reduced retinal innervation. Distinction between the hypotrophy due to a reduced volumetric contribution of receptor terminals and hypoplasia due to loss of interneurons following that reduction has, however, yet to be made. Marthy's (1973) findings in some cases fail to replicate those of Ranzi. He records the formation of a normal optic lobe viewed in whole mounts opposite either a small, single eye vesicle or no eye vesicle at all, following extirpation of the central part of the eye primordium (at, respectively, stage VII – VIII or stage VIII and later). This discrepancy is not discussed although the formation of a small optic lobe following entire removal of an eye rudiment (at stage VI) was considered the possible outcome of excessive extirpation of ectoderm, whereas, conversely, the formation of a normal optic lobe following total removal of an eye rudiment (at stage VII or later) was considered the possible consequence of more skillful extirpation.

Morphometric measurements and cell counts are needed to define the exact contribution of hypoplasia and hypotrophy to the reduction of specific regions of the optic lobe neuropile following reduction of retinal innervation. An influence of a reduced optic lobe upon an otherwise apparently normal eye (Ranzi, 1928a) via the centrifugal innervation (Cajal, 1917) remains a theoretical possibility. Ranzi's later observation (1930), that extirpation of the eye and part of the adjacent optic lobe in stage XX and later *Sepia* embryos leaves the remaining portion of the optic lobe normally developing *in situ*, is most likely explained by the relative morphogenetic insensitivity to retinal input of the deep regions of the optic lobe medulla, which receive many other inputs from widespread central nervous system (CNS) sources (Young, 1974 in *Loligo*).

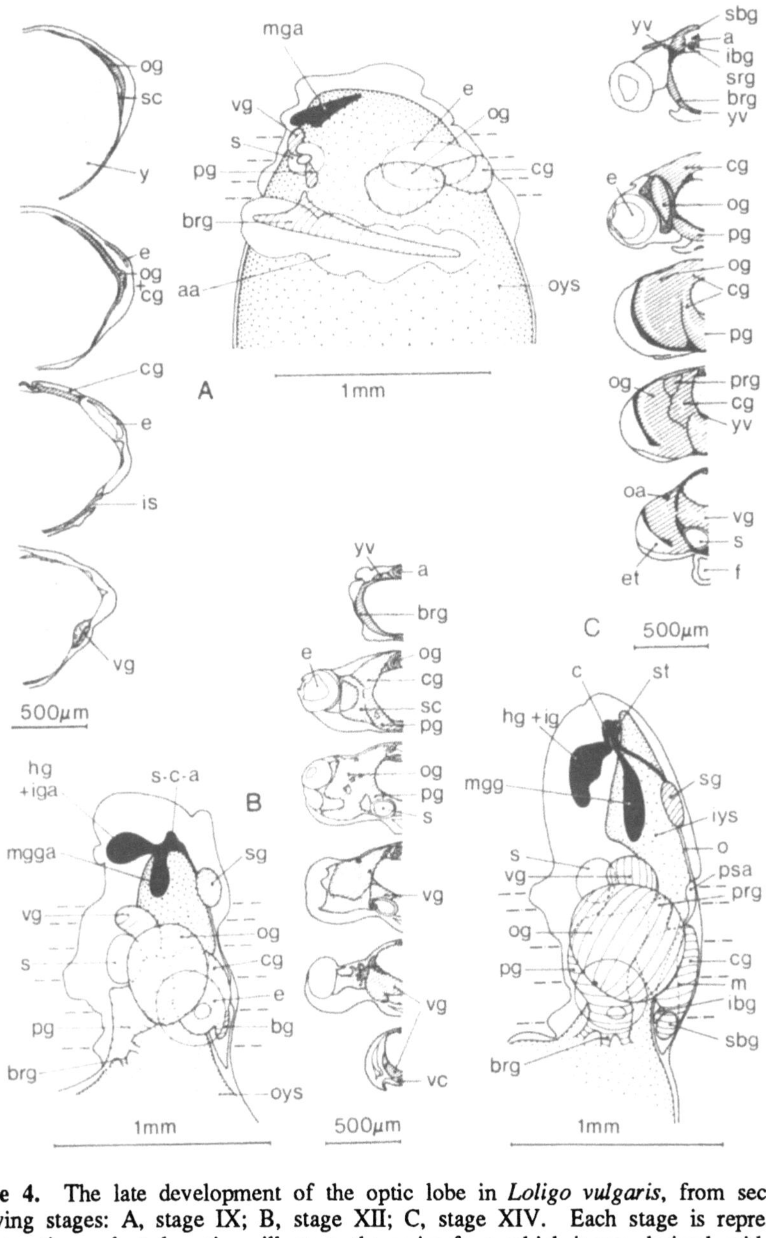

Figure 4. The late development of the optic lobe in *Loligo vulgaris*, from sections at the following stages: A, stage IX; B, stage XII; C, stage XIV. Each stage is represented by a reconstruction; selected sections illustrate the series from which it was derived, with their levels indicated by interrupted lines. Abbreviations: *a*, arm; *aa*, anlage of the arm; *bg*, buccal ganglion; *brg*, brachial ganglion; *c*, cecum; *cg*, cerebral ganglion; *e*, eye; *et*, ectoblast thickening; *f*, funnel; *hg*, hindgut; *ibg*, inferior buccal ganglion; *ig*, ink gland; *iga*, anlage of the ink gland; *is*, invaginating statocyst; *iys*, inner yolk sac; *m*, mouth; *mga*, anlage of the midgut; *mgg*, midgut gland; *mgga*, anlage of the midgut gland; *o*, oesophagus; *oa*, ophthalmic artery; *og*, optic ganglion; *oys*, outer yolk sac; *pbg*, posterior buccal ganglion; *pg*, pedal ganglion; *prg*, peduncular ganglion; *psa*, posterior salivary ganglion; *s*, statocyst; *sbg*, superior buccal ganglion; *sc*, sinus

3.2. Developmental origins of the optic lobe

Each optic lobe arises from the lateral lobe of a primordium that also produces the cerebral ganglion frontally. The primordium of each side is thus paired, as Faussek (1900) first noticed; Meister's (1972) description of its development is the most complete for *Loligo*, (see also Fioroni, 1978) (Fig. 4). The primordium of the optic lobe is an ectodermal thickening that is first conspicuous close to the eye primordium in stage VII (Naef, 1928) embryos. It remains connected with the primordium of the ipsilateral cerebral ganglion and later develops connections with the pedal ganglion and abutts the brachial ganglion. Just after the eye vesicle closes, nerve fibers are first visible in the eye and the optic lobe which originally lay frontal to and beneath the vesicle (Fig. 4B). This is likely to be the first detectable retinal innervation of the optic lobe, which thus appears precociously in the timetable of retinal differentiation. It will require EM observation to confirm the timing of the growth of these fibers. Later, the eyes move frontally trailing the optic lobes behind them (Fig. 4C). Further development sees the gradual emergence of the complex architecture of the mature optic neuropiles.

Tissue culture methods have been a relatively unexplored avenue of enquiry in squid developmental neurobiology since Fedorow's early observations (Fedorow, 1933). Marthy and Aroles (1987) provide the composition of a medium that supports the long-term culture of optic lobes in squid from Naef stage XIII – XIV embryos (i.e., just after the postulated time of initial retinal innervation). Individual cells resembling neurons survive an average of 5 to 7 days, revealing patterns of neurite outgrowth that match those of mature interneurons *in vivo*.

3.2.1. Culture conditions

The composition of a tissue culture medium (Marthy and Aroles, 1987) are: 3.0 ml liquid tissue culture (TC) medium (Gibco); 1.5 ml fetal calf serum (Gibco); 7.0 ml perivitelline fluid collected from squid egg strings close to hatching; 1.0 ml L-glutamine (1.0 g/l); 1.0 ml glycine (0.5 g/l); 1.0 ml NaCl (0.67 g/10 ml); and 50 to 100 international units (IU) penicillin per ml medium. Sterilize by Millipore™ filtration (0.25 µm). Culture conditions are: salinity of 35‰; pH 7.2 (adjusted with NaOH); and temperature 15 to 18°C. Embryos are passed through 4 to 5 changes of Millipore™ (0.45 µm) filtered sea-water and 20 to 30 explant fragments are left undisturbed under normal lighting for the first 4 to 5 days in 3 ml of medium in suitable flasks (e.g., Falcon 3012, Corning 2500) with a 25 cm² surface. Medium is renewed at 2 to 3 week intervals.

cephalicus (orbital sinus); *s-c-a*, anlage of the stomach and cecum; *sg*, stellate ganglion; *srg*, subradular ganglion; *st*, stomach; *vc*, vena cephalica; *vg*, visceral ganglion; *y*, yolk; *yv*, yolk vein. (From Fioroni, 1978, after Meister, 1972).

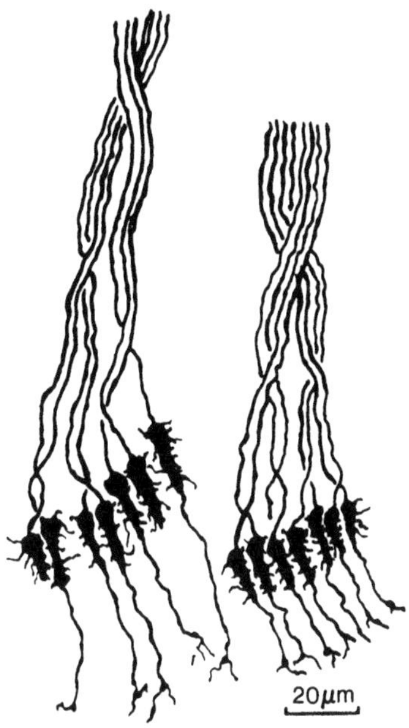

Figure 5. The local chiasmata of optic nerve axons about to enter the cortex of the optic lobe in *Loligo vulgaris* (From Young, 1974).

3.3. Fiber tracts

No embryological investigation of the development of the fiber pathways within the optic lobe has yet been reported. Growth of the chiasmal projection between the retina and the cortex of the optic lobe would reward investigation in *Loligo*, in which the rearrangement of axon pathways is illustrated most clearly (Fig. 5). To date, silver staining methods have been used (Young, 1974), but silver degeneration (Nauta) methods are also effective on cephalopod tissue (Messenger, 1976), while the development or utilization of axonal antibody probes would further facilitate the task of staining young fibers; operations on eye vesicles (e.g., following splitting of the eye primordium: Marthy, 1973) offer the opportunity to perturb the growth of the chiasma experimentally. To demonstrate the normal pattern of connections established within fiber tracts during neurogenesis, or to assay the outcome of experimental perturbations induced during the growth of those tracts, silver stains for degenerating fibers offer a useful prospective method. A general procedure follows.

3.3.1. Nauta degeneration method

The Nauta degeneration method for cephalopods described by Messenger (1976) is preferred because it can be used on serial paraffin sections, avoiding more difficult frozen sections. Following nerve section or damage, the squid is allowed to survive 0.5 to 12 days, optimally 1.5 to 2.5 days, at 20 to 25 °C. Decapitate and rapidly fix the brain or optic lobe in cold 10% neutral formalin in sea-water; wash in tap and distilled water; dehydrate; clear in cedarwood oil; impregnate 2 hours in a 37 °C bath of cedarwood oil, toluene and wax (1:1:1); and embed in wax. Section, bring to water, and immerse 6 hours in 100 parts 50% ethanol/1 part 0.880 ammonia. Wash 3 times in distilled water and transfer through the following: a fresh mixture of 1.5% $AgNO_3$ (85 parts) and pyridine (15 parts) in the dark at room temperature, a fresh ammoniacal silver bath of 4.5% $AgNO_3$ (40 ml), ethanol (20 ml), 0.880 ammonia (14.4 ml), and 2.5% NaOH (3.0 ml), for 3 min at 35 °C (use only for 6 slides); fresh reducer of 10% ethanol (450 ml), 1% citric acid (17.5 ml), formaldehyde (17.5 ml) for 1 min; distilled water wash; and then fixer bath of 1% sodium thiosulfate for 2 min. The method stains degenerating terminals and axons; the stain intensity may be modified by varying the reducer concentration of citric acid.

3.4. Interneurons

The optic lobe interneurons of adult squid are catalogued in the account of Young (1974). The only previous work was that of Kopsch (1899), who introduced his variant of the Golgi method with this species, as supplemented by Cajal (1917) in a study primarily on *Sepia*. That species exhibits post-embryonic growth of receptor terminals in the neuropile layer of the optic lobe cortex (Cajal, 1917), which could also exist in squid.

3.4.1. Silver impregnation and staining methods for cephalopod tissue

The methods given in Sections 3.4.1.1 and 3.4.1.2 are two of the most generally suitable given by Stephens (in Young, 1971).

3.4.1.1. Cajal's block silver method

Fix optic lobe or other brain material in 10% neutral formalin in sea-water, in the dark at a cool temperature, from 1 day to several months. Wash, in running tap-water for 9 hours, then in distilled water for more than 2 hours with 2 to 3 changes. The material is then placed in turn in 70% alcohol for 1 hour, 90% alcohol for 7 to 24 hours, 2 to 5% $AgNO_3$ for 15 to 48 hours in the dark at 37°C, and distilled water for 30 to 60 min. Then the material is reduced in a solution of 1 g pyrogallol, 10 ml formalin, and 90 ml distilled water for at least 7 hours at room temperature. Next it is washed thoroughly in distilled water. After that it is dehydrated in a graded alcohol series and then put into equal parts of absolute alcohol and methyl salicylate for at least 1 hour. Finally, the material is cleared in methyl salicylate containing 1% celloidin. Material is then embedded in wax and sectioned. The method provides consistent, stable preparations for nerve fibers. Times may be reduced for small pieces of tissue from immature animals.

3.4.1.2. Golgi-Kopsch method

Fix fresh optic lobe for 1 to 7 days in 3.5% potassium dichromate in 4 parts of sea-water and 1 part 40% formaldehyde. Transfer to 3.5% dichromate in distilled water for at least 14 days. Rinse in several changes of 0.75% $AgNO_3$ until no further precipitate forms and place in 0.75% $AgNO_3$ at room temperature in the dark for 3 days. Dehydrate and embed in celloidin according to the original method, or in epoxy, and section on a sledge microtome. The method reveals individual neurons, in favorable preparations impregnated in their entirety against a clear background.

3.4.1.3. Golgi-rapid method

This method is described by Kemali (1977). Dissect out tissue under oxygenated sea-water and immerse for 12 hours in 3% potassium dichromate in either 43 ml distilled or sea-water, mixed with 7 ml of 2% OsO_4 in sodium barbital. Then transfer to 0.75% $AgNO_3$ for 12 hours, after 3 washes in the osmium/barbital solution for tissue fixed in the dichromate/sea-water mixture, to prevent silver precipitation. Embed and section as before, or in wax as Kemali suggests.

3.4.2. Other methods to stain pathways

Horeseradish peroxidase (HRP) (Saidel, 1981) and cobalt (Saidel, 1979) backfilling methods, which have been used in octopus, should also be appropriate for squid.

3.4.2.1. HRP labeling method

Make an incision on the medial side of an eye in an anesthetized squid to expose the white membrane, and cut through this to penetrate the orbital sinus surrounding the optic lobe and its enveloping white body. Various methods of introducing HRP (Sigma, type VI) into the optic lobes have been reported in octopus (Saidel, 1982). The method which produces the most discrete labelling is to dip a fine insect pin (00 or 000, or finer 0.15, or 0.1 mm Minutien pins for small squid, from Fine Science Tools or similar company) into a gelled HRP solution and penetrate the optic lobe for several seconds, allowing the enzyme to diffuse into the surrounding space. Alternatively, pressure inject 30% HRP through a sawn-off micropipette. Following injection, blot the optic lobe surface and irrigate the sinus with sea-water to flush out the HRP; close the wound with a stitch, if needed, and allow to recover. Survival times of 2 to 8 days are suggested at 12 to 15 °C.

To process the tissue, perfuse the arterial system with 0.1 M phosphate buffer at a pH of 7.4, and fix in 2% glutaraldehyde in the same buffer for up to 2 hours. Then remove the optic lobe into fresh fixative with 30% sucrose for 4 to 8 hours, followed by fresh buffer and sucrose for 4 to 12 hours. Frozen sections are cut following gelatin embedment. These times need to be reduced for small immature stages. Sections are then reacted with chromogen to reveal the distribution of HRP labelled processes.

3.4.2.2. Cobalt backfills

Saidel (1979) has also used another method for the octopus, which should likewise be applicable to the squid. The optic lobes are dissected out as in Section 3.4.2.1. The cut retinal bundles are dipped into small cups fabricated from fine Intramedic tubing. The cup is first filled with distilled water to dilate the cut ends, and is then replaced with 0.3 to 1.0 M $CoCl_2$ for 4 to 20 hours at 10 to 12°C. Diffusion times need to be adjusted according to temperature and axon dimensions. Rinse in sea-water for 5 min, then in 10% NH_4S in sea-water for 45 to 60 min; fix in 10% formalin or a standard alcohol, acetic acid, and formol fixative. Intensification is possible using Timm's procedure, for which various more recent alternatives (Tyrer *et al.*, 1980) may repay experiment.

3.5. Synaptic organization

The synaptic organization of the photoreceptor terminals is described by Cohen (1973c), and in the optic lobe is known in outline from the analysis of Haghighat *et al.* (1984), who provide criteria and procedures for fixation suitable for an initial study of the fine-structural features of synaptogenesis. A more general review is also available (Ducros, 1979).

4. Growth of the visual system

Growth of the visual system, especially of the optic lobes, has been estimated by various volumetric means in different cephalopod species, but mainly in octopus and somewhat in cuttlefish, with few studies on squid species. Most works have utilized direct approaches of the sort now routinely accomplished with simple computer based morphometric systems. Reference to a major text on stereological methods (e.g., Weibel, 1979) is the best conceptual preparation for studies of this type and often the source of technical short-cuts to their numerical analysis. In addition to morphometric methods, cell number has also been estimated from total desoxynucleic acid (DNA) content, according to methods given in Packard and Albergoni (1970), and by the rapid method of nuclear counts derived from fresh brain homogenates (Giuditta *et al.*, 1971) summarized below. These methods should be directly applicable to squid. Cell counts are preferable to volumetric or wet-weight determinations; and age (or, better, stage, but not body weight) is the preferred abscissa, if allometric studies are actually to be compared with other developmental relationships. The cell numbers increase dramatically, as demonstrated in octopus, but the source of new cells is totally unknown. Where possible the cell populations need to be separately identified, since relative size increases of particular neuron populations provide more valuable information as to their developmental relationships than do changes in their sum. Only Packard (1969) compares the number of photoreceptors with optic lobe cells, but again for octopus. During growth from about 0.5 to 500 g body weight, the ratio of optic lobe cells to photoreceptors decreases from about 65 to 17:1, implying a lower convergence ratio in the adult than in juvenile forms. Similar data would be valuable for squid, as would the relative contributions of neuronal and glial growth to these totals.

4.1. Nuclear counts

This method for deriving nuclear counts is described by Giuditta *et al.* (1971). Dissect brain into as discrete a region as possible, place on ice, blot, and record wet weight. Then homogenize in 0.32 M sucrose containing 1 mM Tris-HCl at pH 7.4 and 1 mM $MgCl_2$. Homogenize samples in 20 times their volume up to a maximum of 2 ml, under carefully controlled conditions in a glass homogenizer with a Teflon pestle rotating at 800 to 900 rpm for 30 sec. According to the authors, more severe treatment disrupts nuclei, thereby lowering the counts of released intact nuclei. Nuclei are then counted under phase contrast optics in a Bürker chamber, counting 10 randomly distributed squares, and repeating counts either with more squares or supplemental aliquots to obtain acceptable measures of scatter. In octopus, the nuclei are clearly distinguishable by these methods from cellular debris. However, aliquots of the small optic lobe nuclei may need to be diluted to reduce their numbers below a convenient counting value, which is approximately 50. The method needs to be calibrated against control counts from histological sections, e.g., on preparations stained by the Feulgen method. This is numerically less daunting for juvenile squid than for adult squid.

4.2. Eye growth

Although eye growth has not been studied closely in squid, an early study shows that the growth of the eye's diameter, to which lens diameter is also proportional, increases in *Loligo vulgaris* with the power 0.73 of the length of the dorsal mantle surface (Teissier, 1933). No assessment is available of the changes to the physiological optics (Sivak, 1982) caused by these increases in the diameter of the lens, or the diameter and thickness of the retina, nor of anticipated differences in the spacing of receptors. Comparison of these changes with those in octopus (Packard, 1969), and their consequences for vision in young squid are currently being explored (Shears, 1989).

4.3. Brain growth

In *Loligo vulgaris*, the growth of the brain's diameter varies with mantle length raised to the power 0.70, thus roughly in parallel with the growth in eye diameter (Teissier, 1933). What this means for cell numbers in the retina (a square function of retinal diameter) and optic lobes (presumably more nearly a cube function of brain diameter) is unclear. Some idea of the general strategies of growth in the optic lobes may be found from comparative data (Frösch, 1971). The optic lobes are large in adult squid, as they are for all cephalopods, between one and two times as large as the rest of the brain, for different species (Wirz, 1959). As a proportion of the entire brain they are relatively smaller in the freshly hatched larva, however, and their subsequent growth thus relatively greater, than is the case in other cephalopods (Frösch, 1971). The effect of this growth seems to be to regulate Wirz's (1959) index of cerebralization, the rough quotient of the vertical system and the optic lobes, which stays approximately constant from the freshly hatched to adult stages in *Loligo* and nine other cephalopod species (Frösch, 1971). Adult eye size varies enormously too. It increases in deep-sea forms, the giant squid *Architeuthis* having the largest of all known eyes (37 cm in diameter: Beer, 1897; 37 - 40 cm: Akimushkin, 1965); and proportionately so as well in certain forms with small body size. In addition to these volumetric changes, changes in the qualitative patterns of protein concentration also occur in the optic lobe, but also in octopus (Giuditta *et al.*, 1975). Finally, experiential factors influence optic lobe development also, at least in octopus reared in different aquaria chosen to provide varied visual experience (Giuditta and Prozzo, 1974). The habits of squid render them far less suited to this sort of experimental treatment, even though their visual capabilities make the outcome of such experiments just as likely to be similar.

ACKNOWLEDGEMENTS. I wish to thank Dr. John M. Arnold, one of the editors, and John Shears (Department of Biology, Dalhousie University), who both kindly commented upon a draft of the manuscript. The author is supported by NSERC grant A-0065.

References

Akimushkin, I. I., 1965, *Cephalopods of the Seas of the USSR*, Program for Scientific Translations, Jerusalem, Israel.

Arnold, J. M., 1962, Mating behavior and social structure in *Loligo pealii*, *Biol. Bull.* **123**:53–57.

Arnold, J. M., 1965a, Normal embryonic stages of the squid, *Loligo pealii* (Lesueur), *Biol. Bull.* **128**:24–32.

Arnold, J. M., 1965b, The inductive role of the yolk epithelium in the development of the squid, *Loligo pealii* (Lesueur), *Biol. Bull.* **129**:72–78.

Arnold, J. M., 1966a, On the occurrence of microtubules in the developing lens of the squid *Loligo pealii*, *J. Ultrastruct. Res.* **14**:534–539.

Arnold, J. M., 1966b, Squid lens development in compounds that affect microtubules, *Biol. Bull.* **131**:383.

Arnold, J. M., 1967, Fine structure of the development of the cephalopod lens, *J. Ultrastruct. Res.* **17**:527–543.

Arnold, J. M., 1971, Cephalopods, in: *Experimental Embryology of Marine and Fresh-water Invertebrates*, (G. Reverberi, ed.), pp. 265–311, North-Holland Pub. Co., Amsterdam.

Arnold, J. M., 1974, Embryonic development, in: *A Guide to Laboratory Use of the Squid Loligo pealei*, (Arnold, J. M. *et al.*, eds.), pp. 24–44, Marine Biological Laboratory, Woods Hole, Massachusetts.

Arnold, J. M., 1984, Closure of the squid cornea: a muscular basis for embryonic tissue movement, *J. Exp. Zool.* **232**:187–195.

Arnold, J. M., 1990, Embryonic development of the squid, *this volume*.

Arnold, J. M. and Williams-Arnold, L. D., 1974, Cortical-nuclear interactions in cephalopod development: cytochalasin B effects on the informational pattern in the cell surface, *J. Embryol. Exp. Morphol.* **31**:1–25.

Arnold, J. M. and Williams-Arnold, L. D., 1976, The egg cortex problem as seen through the squid eye, *Am. Zool.* **16**:421–446.

Arnold, J. M. and Williams-Arnold, L. D., 1978, Preliminary studies on the closure of the optic vesicle of *Loligo pealei* embryos, *Biol. Bull.* **155**:426.

Beer, T., 1897, Die Akkomodation des Kephalopodenauges, *Pflüg. Arch. ges. Physiol.* **67**:541–586.

Bobretsky, N. W., 1875–1876, Über die Embryologie der Cephalopoden, *Schrift Kiev Naturf. Ges.* **4**:

Brahma, S. K., 1978, Ontogeny of lens crystallins in marine cephalopods, *J. Embryol. Exp. Morph.* **46**:111–118.

Cajal, S. Ramón y., 1917, Contribución al conocimiento de la retina y centros ópticos de los cefalópodos, *Trab. Lab. Invest. Biol. Univ. Madr.* **15**:1–82.

Cohen, A. I., 1973a, An ultrastructural analysis of the photoreceptors of the squid and their synaptic connections. I. Photoreceptive and non-synaptic regions of the retina, *J. Comp. Neurol.* **147**:351–378.

Cohen, A. I., 1973b, An ultrastructural analysis of the photoreceptors of the squid and their synaptic connections. II. Intraretinal synapses and plexus, *J. Comp. Neurol.* **147**:379–398.

Cohen, A. I., 1973c, An ultrastructural analysis of the photoreceptors of the squid and their synaptic connections. III. Photoreceptor terminations in the optic lobes, *J. Comp. Neurol.* **147**:399–426.

Ducros, C., 1979, Synapses of cephalopods, *Int. Rev. Cytol.* **56**:1–22.

Faussek, V., 1893, Über den sogenannten "Weissen Körper" sowie über die embryonale Entwicklung derselben, der Cerebralganglien und des Knorpels bei Cephalopoden, *Mém. Acad. Imp. Sci. St. Pétersbourg, VII Sér.* **41**:1–27.

Faussek, V., 1900, Untersuchungen über die Entwicklung der Cephalopoden, *Mitt. Zool. Stat. Neapel.* **14**:83–237.

Fedorow, B. G., 1933, Über die in-vitro-Kultur des Nervengewebes der Cephalopoden, *Biol. Zentralbl.* **53**:41–49.

Fioroni, P., 1978, *Cephalopoda, Tintenfische (Morphogenese der Tiere. Reihe 1, Leiferung 2).* Gustav Fischer, Jena.

Frösch, D., 1971, Quantitative Untersuchungen am Zentralnervensystem der Schlüpfstadien von zehn mediterranen Cephalopodenarten, *Rev. Suisse Zool.* **78**:1069–1122.

Giuditta, A., Cimarra, P., and Prozzo, N., 1975, Protein patterns in different lobes and during development of Octopus brain, *J. Neurochem.* **24**:1131–1133.

Giuditta, A., Libonati, M., Packard, A., and Prozzo, N., 1971, Nuclear counts in the brain lobes of *Octopus vulgaris* as a function of body size, *Brain Res.* **25**:55–62.

Giuditta, A., Moore, B. W., and Prozzo, N., 1977, A brain-specific protein from *Octopus vulgaris*, LAM, *J. Neurochem.* **29**:235–244.

Giuditta, A. and Prozzo, N, 1974, Postembryonic growth of the optic lobe of *Octopus vulgaris*, Lam., *J. Comp. Neurol.* **157**:109–116.

Gratzner, H. G., 1982, Monoclonal antibody to 5-bromo- and 5-iododeoxyuridine: A new reagent for detection of DNA Replication, *Science* **218**:474–475.

Grenacher, H., 1874, Zur Entwicklungsgeschichte der Cephalopoden. Zugleich ein Beitrag zur Morphologie der höheren Mollusken, *Z. Wissen. Zool.* **24**:419–498.

Haghighat, N., Cohen, R. S., and Pappas, G. D., 1984, Fine structure of squid (*Loligo pealei*) optic lobe synapses, *Neuroscience* **13**:527–546.

Hanlon, R., 1990, Rearing and maintenance of squid, *this volume.*

Hollerbach, C., 1901, *Linse und Corpus epithelial im Cephalopodenauge und ihre Entwicklung*, Dissertation, Bern (Cited in Fioroni, 1978).

Inoué, S., 1986, *Video Microscopy*, Plenum Press, New York.

Kemali, M., 1977, A modification of the rapid Golgi method, devised for vertebrates, applied successfully to the octopus and the squid nervous system, *Biol. Cell.* **29**:219–220.

Kopsch, F., 1899, Mitteilungen über das Ganglion opticum der Cephalopoden, *Int. Monats. Anat. Physiol.* **16**:3–24.

Lange, G. D. and Hartline, P. H., 1974, Retinal responses in squid and octopus, *J. Comp. Physiol.* **93**:19–36.

Lankester, E. R., 1875, Observations on the development of the Cephalopoda, *Quart. J. Micros. Sci.* **15**:37–47.

Lankester, E. R., 1883, Mollusca, in: *Encyclopaedia Britannica (9th Ed.)*, Volume XVI, pp. 632–695, Charles Scribner's Sons, New York.

Lemaire, J. and Richard, A., 1979, Organogenèse de l'oeil du Céphalopode *Sepia officinalis* L., *Bull. Soc. Zool. France* **103**:373–377.

Marthy, H.-J., 1970, Beobachtungen beim Transplantieren von Organanlagen am Embryo von *Loligo vulgaris* (Cephalopoda, Decapoda), *Experientia* **26**:160–161.

Marthy, H.-J., 1973, An experimental study of eye development in the cephalopod *Loligo vulgaris*: determination and regulation during formation of the primary optic vesicle, *J. Embryol. Exp. Morphol.* **29**:347–361.
Marthy, H.-J., 1978–1979, Embryologie expérimentale chez les Céphalopodes, *Vie Milieu, sér. AB* **28–29**:121–142.
Marthy, H.-J. and Aroles, L., 1987, *In vitro* culture of embryonic organ and tissue fragments of the squid *Loligo vulgaris* with special reference to the establishment of a long term culture of ganglion-derived nerve cells, *Zool. Jahrbüch. Physiol.* **91**:189–202.
Mauro, A., 1977, Extra-ocular photoreceptors in cephalopods, *Symp. Zool. Soc. Lond. (The Biology of Cephalopods)* **38**:287–308.
Meister, G., 1972, Organogenese von *Loligo vulgaris* LAM. (Mollusca, Cephalopoda, Teuthoidea, Myopsida, Loliginidae), *Zool. Jahrbüch. Anat.* **89**:247–300.
Messenger, J. B., 1976, A paraffin Nauta method for staining degeneration in cephalopod C.N.S., *J. Physiol. (Lond.)* **259**:18P–20P.
Metschnikoff, E., 1867, Le développement des Sépioles, *Ann. Mag. Nat. Hist.* **20**:449–453.
Naef, A., 1928, De Cephalopoden. *Fauna e Flora del Golfo di Napoli, Mongraphia 25, Teil 1, Band 2 Embryologie*, pp. 1–357.
Packard, A., 1969, Visual acuity and eye growth in *Octopus vulgaris* (Lamarck), *Monit. Zool. Ital. (N.S.)* **3**:19–32.
Packard, A. and Albergoni, V., 1970, Relative growth, nucleic acid content and cell numbers of the brain in *Octopus vulgaris* (Lamarck), *J. Exp. Biol.* **52**:539–552.
Patterson, J. A., 1984, An *in vitro* preparation for the study of visual physiology in the squid, *Alloteuthis subulata*, *J. Physiol. (Lond.)* **360**:18P.
Patterson, J. A., 1985, A modified, prefixed, Golgi-rapid technique for the cephalopod retina, *J. Neurosci. Methods* **12**:219–225.
Ranzi, S., 1928a, Correlazioni tra organi di senso e centri nervosi in via di sviluppo, *Roux' Arch. Entwicklungsmech. Organ.* **114**:364–370.
Ranzi, S. C., 1928b, Contributo al problema della ciclopia. La Ciclopia nei cefalopodi e qualche notizia sui ciclopi nei vertebrati, *Att. Pont. Accad. Sci. Nuov. Lincei* **82**:74–84.
Ranzi, S., 1930, Ulteriori ricerche sulle correlazioni tra organi di senso e centri nervosi in via di sviluppo, *Boll. Zool. (Napoli)* **1**:131–135.
Ranzi, S., 1931a, Resultati di ricerche di embriologia sperimentale sui Cefalopodi, *Arch. Zool. Ital.* **16**:403–408.
Ranzi, S., 1931b, Sviluppo di parti isolate di embrioni di Cefalopodi (Analisi sperimentale dell'embriogenesi), *Pubbl. Staz. Zool. Napoli* **11**:104–146.
Ranzi, S., 1932, Sull'indipendenza dell'istogenesi dell'organogenesi, *Att. Pont. Accad. Sci. Nuov. Lincei.* **85**:27–28.
Sacarrão, G. da F., 1954, Quelques aspects sur l'origine et le développement du type d'oeil des Céphalopodes, *Rev. Fac. Cienc. Lisboa Ser. C* **4**:123–158.
Sacarrão, G. da F., 1956, Contribution à l'étude du développement embryonnaire du ganglion stellaire et de la glande épistellaire endocrine des Céphalopodes, *Arq. Mus. Bocage* **27**:137–152.
Saibil, H., 1990, Structure and function of the squid eye, *this volume.*
Saidel, W. M., 1979, Relationship between photoreceptor terminations and centrifugal neurons in the optic lobe of octopus, *Cell Tiss. Res.* **204**:463–472.
Saidel, W. M., 1982, Connections of the octopus optic lobe, *J. Comp. Neurol.* **206**:346–358.
Shears, J. C., 1989, *personal comm.*
Sivak, J. G., 1982, Optical properties of a cephalopod eye (the short finned squid, *Illex illecebrosus*), *J. Comp. Physiol.* **147**:323–327.
Teissier, G., 1933, Sur la croissance du système nerveux central chez les Céphalopodes, *C. R. Soc. Biol. Paris* **112**:777–779.
Tomarev, S. I. and Zinovieva, R. D., 1988, Squid major lens polypeptides are homologous to glutathione S-transferases subunits, *Nature* **336**:86–88.

Tyrer, N. M., Shaw, M. K., and Altman, J. S., 1980, Intensification of cobalt-filled neurons in sections (light and electron microscopy), in: *Neuroanatomical Techniques. Insect Nervous System*, (N. J. Strausfeld and T. A. Miller, eds.), pp. 429–446, Springer-Verlag, New York.

Weibel, E. R., 1979, *Stereological Methods, Vol. 1, Practical Methods for Biological Morphometry*, Academic Press, Orlando, Florida.

Willekens, B., Vrensen, G., Jacob, T., and Duncan, G., 1984, The ultrastructure of the lens of the cephalopod *Sepiola*, *Tissue and Cell* **16**:941–950.

Williams, L. W., 1909, *The Anatomy of the Common Squid Loligo pealii, Lesueur*, E. J. Brill, Leiden.

Wirz, K., 1959, Étude biométrique du système nerveux des Céphalopodes, *Bull. Biol. France Belg.* **93**:78–117.

Yamamoto, M., 1985, Ontogeny of the visual system in the cuttlefish, *Sepiella japonica*. I. Morphological differentiation of the visual cell, *J. Comp. Physiol.* **232**:347–361.

Yamamoto, M., Takasu, N., and Uragami, I., 1985, Ontogeny of the visual system in the cuttlefish, *Sepiella japonica*. II. Intramembrane particles, histofluorescence, and electrical responses in the developing retina, *J. Comp. Neurol.* **232**:362–371.

Yoshida, M., Ohtsu, K., and Nakaye, T., 1976, Development of the cuttle fish retina. in: *Structure of the Eye III*, (E. Yamada, and S. Mishima, eds.), Maruzen, Japanese Journal of Ophthalmology, Tokyo, pp. 215–221.

Young, J. Z., 1971, *The Anatomy of the Nervous System of Octopus vulgaris*, Oxford Univ. (Clarendon) Press, London.

Young, J. Z., 1974, The central nervous system of *Loligo*. I. The optic lobe, *Phil. Trans. Roy. Soc. Lond. B* **267**:263–301.

Young, R. E., 1978, Vertical distribution and photosensitive vesicles of pelagic cephalopods from Hawaiian waters, *Fish. Bull. U. S.* **76**:583–615.

Chapter 19

The Statocysts of Squid

BERND U. BUDELMANN

1. Introduction

Cephalopods, especially squids, are the most swift-moving of all the aquatic invertebrates. Any such swift-moving animal requires sensory devices for proper spatial control. No wonder then that cephalopods – during the course of evolution and probably in competition with fish (Packard, 1972) – developed sophisticated sense organs, the *statocysts*, that provide the animals with information about their attitudes and movements in space. Among all aquatic invertebrates, these sense organs reach the highest level of complexity in octopod and decapod cephalopods, with receptor systems for the detection of both linear (gravity) and angular accelerations (Budelmann, 1980, 1988).

Cephalopod statocysts have attracted scientists for more than 100 years. Research of the last 30 years has shown that many parallels exist in morphological as well as physiological details between the cephalopod statocysts and the vertebrate vestibular apparatus (e.g., Williamson and Budelmann, 1985; Budelmann, 1988); these parallels include the control system for compensatory eye movements (Budelmann and Young, 1984).

Cephalopod statocysts show a wide range of levels of differentiation in the different taxonomic groups and these differences have been correlated with the various forms of locomotion employed by these animals (Maddock and Young, 1984; Budelmann, 1988). Despite these differences, however, cephalopod statocysts can be categorized in three basic morphological types (Budelmann, 1988): (1) The *nautilus type* (in *Nautilus*); this is the most simple one of the cephalopod statocysts, with just a gravity receptor system and no further differentiation (similar to the gastropod and bivalve statocysts). (2) The *octopod type* (in octopods, e.g., *Octopus*); this is a sphere-like sac, containing one gravity receptor system and an angular acceleration receptor system that is divided into nine subsections. (3) The *decapod type* (in the sepioids, e.g., the cuttlefish *Sepia*, and in the teuthoids, e.g., the squid *Loligo*); this is a cavity of irregular shape, containing a gravity receptor system that is subdivided into three separate organs and an angular acceleration receptor system that is divided into four subsections. The only intermediate type of statocyst has been described in *Vampyroteuthis*, which is a phylogenetic relic somewhere near the ancestor of octopods and decapods (Stephens and Young, 1976).

B. U. BUDELMANN • Marine Biomedical Institute and Department of Otolaryngology, University of Texas Medical Branch, Galveston, Texas 77550.

In deference to the subject of this book, the following chapter will focus on the *decapod type* of statocyst of the teuthoid squids. Within this group, however, details of the statocyst organization may vary depending upon the species. Furthermore, this chapter will include, when appropriate, details of the sepioid statocysts as well, and will occasionally refer to the *octopod type* of statocyst, since most details on structure and function are known for it.

2. Short summary of research on cephalopod statocysts

In 1867, Owsjannikow and Kowalevsky gave the first morphological description of cephalopod statocysts (in fact, this was the first detailed description of any invertebrate statocyst). Some years later, in behavioral experiments, Delage (1886), Fröhlich (1904) and Muskens (1904) showed that in octopods the statocysts are necessary for proper equilibrium orientation and proper compensatory eye movements. Hamlyn-Harris (1903) extended the morphological description of the statocysts of *Octopus* and *Sepia* and included statocysts of several other cephalopod species as well. By 1924, Ishikawa used characteristics of the statocysts of 59 species of 31 genera for a phylogenetic study, an attempt that has recently been taken up again by Young (1984) for the statocysts of cranchiid squids.

Kolmer (1926), and particularly Klein (1932) who used silver stains of the *Sepia* statocyst, raised the point that some of the statocyst receptor cells might be secondary sensory cells (at that time a daring suggestion, contradicting the dogma that secondary sensory cells are present exclusively in vertebrate sense organs). Much later, their suggestion has been proved with the techniques of electron microscopy and cobalt tracing (Budelmann, 1977; Budelmann and Thies, 1977; Colmers, 1977), and in a comprehensive morphological description of the *Octopus* angular acceleration receptor system Budelmann *et al.* (1987) have shown that both primary and secondary sensory cells occur side by side in the same epithelium.

In 1960, the renowned British neuroanatomist J. Z. Young gave a thorough light microscopical description of the *Octopus* statocyst. At the same time, Boycott (1960) in behavioral experiments confirmed the importance of the *Octopus* statocyst for equilibrium orientation. These papers, no doubt, renewed scientists' interest in cephalopod statocysts and initiated further research in many directions. In behavioral experiments (post-rotatory head movements and nystagmus), Dijkgraaf (1961, 1963a) and later Collewijn (1970) and Messenger (1970) confirmed the presence of an angular acceleration receptor system in the *Octopus* and *Sepia* statocyst, and Maturana and Sperling (1963) successfully recorded in *Octopus* the first action potentials from this system.

Barber (1966) and Vinnikov *et al.* (1967) extended the light microscopical studies of the statocysts to the level of electron microscopy and this type of research has been thoroughly continued (Budelmann *et al.*, 1973; Colmers, 1977, 1981; Budelmann, 1979); the most comprehensive description became that of the *Octopus* angular acceleration receptor system (Budelmann *et al.*, 1987).

On the physiological side, Budelmann (1970, 1975) analyzed compensatory eye movements in *Octopus* and *Sepia* and discovered a new, possibly cephalopod specific, principle of function of a gravity receptor system. In 1973, Budelmann and Wolff were able to demonstrate unequivocally that the *Octopus* angular acceleration receptor system is not only sensitive to angular accelerations, but also to gravity (a phenomenon that has later been confirmed also for the receptor systems in the vertebrate semicircular canals).

Electrophysiological experiments were continued by Williamson and Budelmann (1985) and Williamson (1985, 1986). Analyzing the physiological characteristics of the afferent and efferent fiber system of the *Octopus* angular acceleration receptor system, they discovered this system to be of dual angular acceleration sensitivity, with dynamic response characteristics that parallel those of the vertebrate semicircular canal system. In the squid angular acceleration receptor system, only limited data exists on intracellular recordings from the receptor cells (Williamson, 1988a, 1989a). Using another approach, Maddock and Young (1984) correlated morphological data of the shapes and dimensions of the cephalopod angular acceleration receptor systems with the animals' swimming (buoyant vs. non-buoyant) behavior.

The putative transmitters acting in the statocyst sensory epithelia are dopamine, noradrenaline, and acetylcholine; all occur in the efferent fiber system (Budelmann and Bonn, 1982; Auerbach and Budelmann, 1986; Williamson, 1989b). The transmitter(s) in the afferent fiber system is not yet known.

The central pathways of the nerves innervating the statocyst sensory epithelia have been analyzed by Young (1971, 1976) and later, in more detail and including the efferent pathways, by Colmers (1982) and Budelmann and Young (1984) for the *Octopus* systems. The latter study includes a comprehensive description of the statocyst-oculomotor system and shows striking parallels in organization with the vertebrate vestibulo-oculomotor system.

To date, little is known on the structure and function of the nautilus type of statocyst (Griffin, 1900; Young, 1965; Hartline *et al.*, 1979). Also, surprisingly few papers of those on the decapod type of statocyst are specifically devoted to those of teuthoid squids and, with the exception of some data from intracellular recordings (Williamson, 1988a, 1989a), are purely anatomical at the level of light microscopy (Stephens and Young, 1978, 1982; Young, 1984).

It is still a point of discussion whether parts of the cephalopod statocyst are involved in some kind of hearing (Hubbard, 1960; Dijkgraaf, 1963b; see Hanlon and Budelmann, 1987), but evidence already exists that crista units are sensitive to vibrations (Maturana and Sperling, 1963; Williamson, 1988b).

3. Dissection of the squid statocyst

Whichever squid species will be used, best preparations can be obtained from a small, fresh (unfixed, unfrozen) specimen; in smaller specimen (and species), the tissue and cartilage is more transparent and thus the preparation is easier. Formalin-fixed (10%) material usually gives better results compared to 70% alcohol or frozen material (although, if the formalin is not neutral, it dissolves the statoliths and statoconia). Frozen material is easier to dissect than fixed material; the disadvantage is that the preservation is generally poor and the statoliths and statoconial layers are detached from the sensory epithelia.

If the animal needs to be sacrificed, an overdose of anesthetic (see Messenger *et al.*, 1985; see also below) or decapitation is appropriate. Since the statocyst cavities extend to the very posterior edge of the cranial cartilage, care is needed to avoid damage to the statocysts during decapitation.

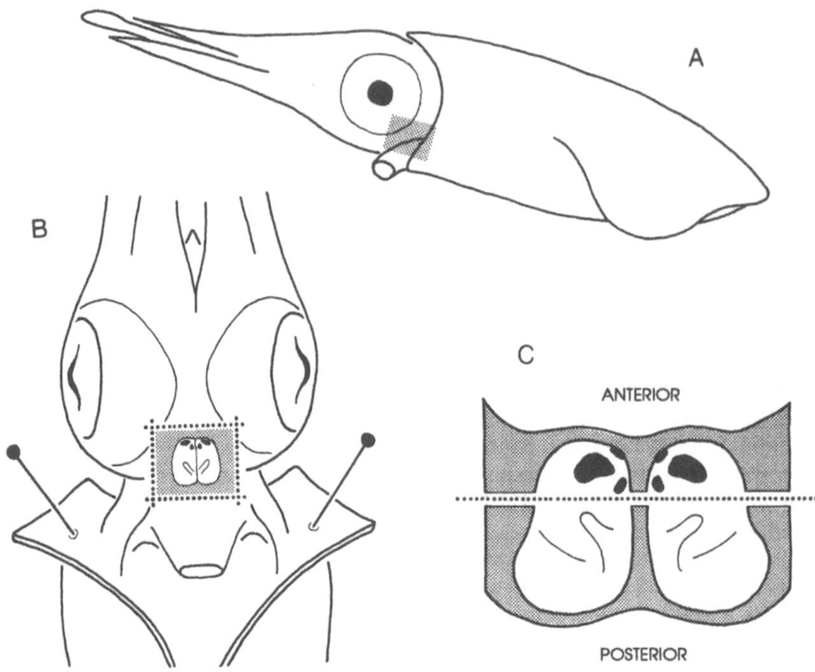

Figure 1. Diagram of a squid to show the position of the statocysts. (*A*) Lateral view; dotted area indicates the location of the statocysts. (*B*) Ventral view of the head and neck region, with the mantle cut open and the funnel bent posteriorly. The dotted lines indicate the cuts to be made to remove the *statocyst block* (*C*) from the head. (*C*) Statocysts seen from ventral. The dotted line posterior to the statoliths and statoconial layers indicates a transverse vertical cut to open the statocyst cavities (as seen in figure 2).

The two statocysts are cavities of irregular shape, located in the posterior/ventral bulge of the cranial cartilage. To dissect the statocysts, two approaches are possible, one from dorsal, the other from ventral. In any case, the tissue should be kept under water during the preparation. A dark background (e.g., black-colored wax, or a piece of black plastic) and good illumination (preferably fiber optics) greatly enhance the contrast between the different types of tissue, especially fiber tracts, vessels and nerves, and thus facilitates the dissection.

Preparation from ventral (see Fig. 1). This approach is easier and requires less experience. Fix the specimen in a dish, ventral side up. If the body is still attached to the head, open the mantle cavity ventrally with a cut half way to posterior. This allows one to bend the funnel posteriorly. In unfixed material, small white patches are visible through the transparent tissue: two crystal-like large patches of irregular shape and two small patches close to the midline. These are the statoliths and statoconial layers, respectively, of the gravity receptor systems. The large anterior crystal-like statoliths are attached to the anterior wall of the statocysts and thus indicate the most anterior border of the statocyst cavities. After removal of all soft tissue covering the cartilage above the statocysts, the statoliths, statoconial layers, and the outline of the

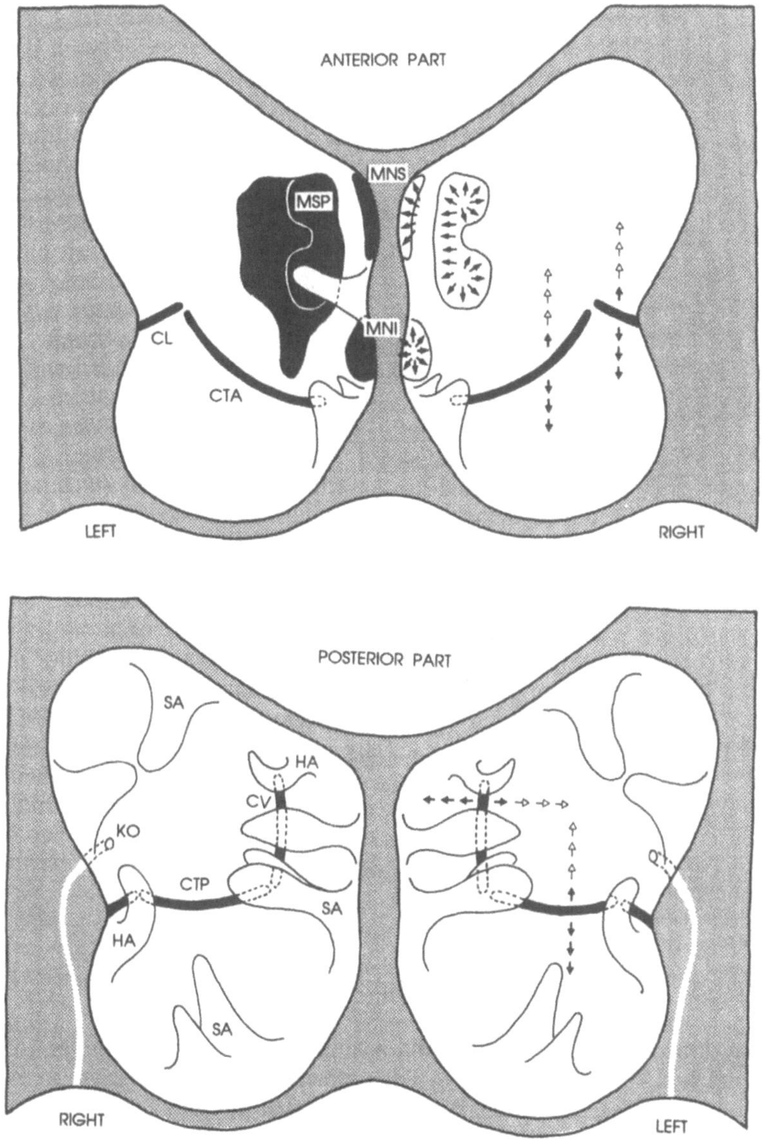

Figure 2. Diagram of the inside of the statocyst cavities of a squid after a transverse vertical cut. The upper half of the diagram shows the anterior parts of the statocysts looking forward; the lower half of the diagram shows the posterior parts of the statocysts looking rearward. The diagrams on the left show the three gravity receptor systems (*MSP*, macula statica princeps; *MNS*, macula neglecta superior; *MNI*, macula neglecta inferior) covered by the statoliths and statoconial layers, and the course of the anterior transverse (*CTA*), longitudinal (*CL*), posterior transverse (*CTP*) and vertical crista (*CV*) sections; the cupulae are not illustrated. The arrows in the diagrams on the right indicate the pattern of polarization of the hair cells of the three maculae and four crista sections. Each arrow stands for a group (maculae) or row (cristae; see figure 4) of hair cells polarized in the same direction. *KO*, opening of Kölliker's canal; *HA*, hamuli; *SA*, straight anticrista.

irregular statocyst cavities become clearly visible, with many almost transparent knobs (*anticristae*) inside the cavities being most obvious.

To continue the preparation, a block of cartilage, containing the two statocysts, should be cut out of the head (see Fig. 1). The cuts should be vertical, deep through the cartilage and preferably parallel to the main axes (longitudinal and transverse) of the animal; all cuts should be some distance from the statocyst cavities to avoid their damage. Since the cavities reach to the posterior end of the cranial cartilage, the posterior cut will just separate soft tissue from the cartilage. After these cuts, remove the *statocyst block* out of the head and clean it from all surrounding soft tissue. Care must be taken when removing the brain tissue, since the dorsal and anterior statocyst walls are very thin. Also, some of the cartilage surrounding the statocysts might be cut away if desired. At this stage, before opening the cavities, one should inspect the transparent statocysts from all sides to get a three-dimensional understanding of the complex form of the cavities.

The statocyst cavities can be cut open in any desired plane, but one should avoid cutting through any of the statolith/statoconia covered sensory epithelia. If this is the first statocyst preparation, one should use a vertical transverse cut (preferably with a razor blade splinter), directly posterior to the posteriormost patches of statoconia (Fig. 1). The statocysts can now be opened for further inspection of the inside (Fig. 2).

Preparation from dorsal. Fix the specimen in a dish, dorsal side up, and remove the skin above the head and both eyes, then remove the eye balls and optic lobes. With a horizontal cut remove the dorsal part of the cartilage, including the supra-oesophageal half of the brain (this can be done by just one cut, or by several until one reaches the level of the oesophagus which runs right through the middle of the brain in anterior/posterior direction). Then carefully remove the remaining sub-oesophageal brain tissue until the two statocysts become visible; avoid damaging the thin cartilage layer between the brain and statocyst cavities. When the anterior, lateral and posterior outlines of the statocysts are visible, a block of cartilage containing the two statocysts can be cut out of the head as described above. See previous paragraph for further treatment of the statocysts.

3.1. Statocyst operations on living squid

In any kind of squid, operations on statocysts of living animals are difficult to perform and it is beyond the scope of this chapter to describe the surgical techniques in detail. A few notes, however, may serve as a first assistance.

The only possible approach to the statocyst cavities is from ventral. After deep anesthesia (in seawater: 0.5 – 1.0% urethane w/v, 2% ethanol v/v, 3 – 4% magnesium chloride w/v, or cooling to 3 – 5°C; see Messenger *et al.*, 1985), the animal should be placed in a preparation dish (preferably under water) with its neck over a small ridge to slightly raise the posterior part of the head containing the statocysts. The initial cut into the skin and muscle tissue should be far laterally of the statocyst in question, about halfway between the midline of the animal and the lateral side of the eye. This is necessary to avoid cutting into the large colorless blood sinus that ventrally overlies both statocyst cavities. It is absolutely essential that the approach towards the statocysts be made underneath this blood sinus, without damaging it. With a very fine blade splinter a window can be cut into the ventral bulge (*ventral sac*) of the statocyst cavity. This allows one to detach the statolith, statoconial layers and cupulae from their sensory epithelia with a strong jet of saline water from a syringe, thus abolishing the function of the statocyst receptor systems. Any rough treatment of the statocyst cavity (e.g., with a wad of cotton wool) should be avoided, since this treatment can easily cause uncontrolled damage to those parts of the brain that are separated from the statocyst cavity by only a very thin layer of cartilage.

Sophisticated operations on the statocysts, such as partial ablation of some of the sensory structures (see Budelmann, 1975) are possible. However, much experience and specially designed surgical tools and equipment are necessary.

4. Structure and function of the squid statocyst

The squid statocysts, like the vertebrate vestibular apparatus, are structurally and functionally subdivided into two main receptor systems, one for the detection of gravity and other types of linear acceleration (*macula/statolith* and *macula/statoconia systems*, respectively), and the other for the detection of angular accelerations (*crista/cupula system*). In both these systems, the sensory epithelia (*maculae, cristae*) are composed of many receptor cells. These transform the mechanical energy of the acceleration stimuli, or more precisely the movements of the accessory structures (*statoliths, statoconia, cupulae*) relative to the sensory epithelia, into electrical signals that finally reach the brain for further processing.

4.1. The receptor cells

All statocyst receptor cells are hair cells. Unlike vertebrate vestibular hair cells, each carries many kinocilia (between 50 and 150), but no stereovilli (or stereocilia). Each kinocilium is about 6 μm in length and is characterized by the internal $9 \times 2 + 2$ tubuli content. All kinocilia of a single hair cell are arranged such that they form an elongated, flap-like ciliary group. To these groups, accessory statolith, statoconia, or cupula structures are attached. Any displacement of these structures, due to acceleration stimuli, causes small deflections (shear) of the ciliary groups and thus stimulates the hair cells (Budelmann *et al.*, 1973; Budelmann, 1979).

Based on ultrastructural criteria, such as the orientation of the $9 \times 2 + 2$ tubuli content within the cilia, or the orientation of the cilia's basal feet, each cilium is morphologically polarized in one direction. All the cilia of a hair cell are polarized in the same direction. This results in a characteristic pattern of polarization of all hair cells of the sensory epithelium (see Fig. 2). Any movement of the accessory structures relative to the epithelium, and thus to the hair cells, excites (depolarizes) or inhibits (hyperpolarizes) each hair cell to a different extent because of the specific polarization direction each hair cell has within the epithelium and thus relative to the motion (Budelmann, 1988). On the way to the brain, the resulting complex pattern of electrical signals produced by the various hair cells is most likely altered by first-order afferent neurons and efferent signals that come from the brain (Williamson, 1985, 1986; Williamson and Budelmann, 1985). Whatever the final pattern of signals might be, it yields information about the animals position and movement in space.

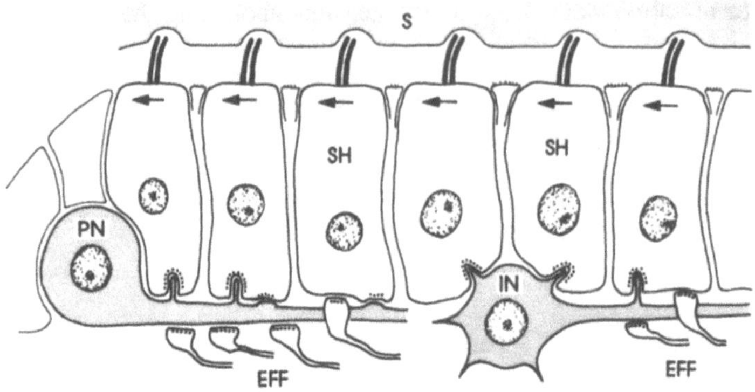

Figure 3. Cellular and synaptic organization in the periphery of a macula of the gravity receptor system. Although based on data of *Octopus*, the diagram represents the basic organization of the squid maculae as well (see text). Secondary sensory hair cells (*SH*) are in afferent synaptic contact with two types of first-order afferent neurons (*PN*, perimacular neuron; *IN*, intramacular neuron). *Arrows* in the distal end of the hair cells indicate the hair cell's polarization; *EFF*, efferent synaptic endings; *S*, mucous layer of statolith or statoconial layer (based on Colmers, 1981; from Budelmann, 1988; reprinted with the permission of Springer-Verlag).

4.2. The gravity receptor system

Orientation. The gravity receptor system is the most distinct structure in the anterior part of the statocyst. It is subdivided into three separate organs (Fig. 2). The largest is the *macula statica princeps* (MSP) at the anterior wall of the cavity. It is oriented vertically in the transverse plane and carries a large compact statolith of irregular shape, with a long ventral spine. The form of the statolith is often used as a taxonomic characteristic for both living and fossil squid species (Clarke, 1978). Careful removal of the statolith shows that the underlying hair cell epithelium, covered with remnants of mucus, is much smaller than the outline of the statolith; the epithelium looks like a *3*, with its open part laterally. The other two organs are much smaller and are covered by thin layers of statoconia, the areas of which exactly match those of the underlying hair cell epithelia. Dorsally, at the medial wall, is the *macula neglecta superior* (MNS); it looks like a semicircle and is oriented also vertically, but in the sagittal plane. Ventrally, in the anterio-medial corner of the cavity, is the pear-shaped *macula neglecta inferior* (MNI). All three maculae are arranged roughly at right angles to each other (Budelmann *et al.*, 1973; Budelmann, 1979; Stephens and Young, 1982).

Innervation. All three maculae are innervated by just one macula nerve. Underneath the MSP it splits into a dorsal and a ventral branch when passing through the cartilage. The dorsal branch innervates the dorsal half of the MSP and the MNS, the ventral branch innervates the ventral half of the MSP and the MNI (Stephens and Young, 1982).

Morphology. Each macula consists of between 1000 and 2000 hair cells. Their pattern of polarization is in an almost 360 degrees radial (MSP, MNI) and 180 degrees fan-like (MNS) outward direction towards the macula periphery (Fig. 2) (Budelmann *et al.*, 1973; Budelmann, 1979). The ultrastructural organization of the maculae is complex. Although the squid (and cuttlefish) maculae have not yet been analyzed in detail, preliminary data show a similar organization as for the *Octopus* macula: secondary sensory hair cells are in afferent synaptic contact with various first-order afferent neurons, and both hair cells and neuron receive many efferent endings (Fig. 3) (Colmers, 1977, 1981; Budelmann, 1989).

Physiology. The gravity receptor system provides the animal with information about its attitude in space and thus elicits compensatory reactions, such as head, eye and funnel movements that are necessary for proper equilibrium orientation. In *Sepia*, when all gravity receptor organs in both statocysts are destroyed, the animal is completely disoriented during swimming: it zigzags, corkscrews and somersaults irregularly. This disorientation is less obvious when the statolith and statoconial layers are removed just unilaterally; then, the animal keeps its head inclined towards the unoperated side (see Fig. 7). When the sensory epithelia are additionally destroyed on the operated side (e.g., by anesthetization with Novocain), the effect becomes more obvious: the animal permanently rolls around its longitudinal axis towards the unoperated side (Budelmann, 1975).

Careful measurements of compensatory eye and head movements (see below, Section 6) indicate that the gravity receptor systems of the cuttlefish function similarly to that of the *Octopus* system. There, information about position in space is encoded in the orientation of the excitation pattern of all hair cells relative to macula-fixed coordinates (it is not encoded in the strength of stimulation of the hair cells, as in the gravity receptor systems of vertebrates and crustacea). This is obvious from experiments with animals (*Octopus* and *Sepia*) in a centrifuge where an increase in the magnitude of gravity, and thus in the strength of stimulation of the hair cells, had no effect on behavioral reactions (Budelmann, 1970, 1975). However in *Sepia*, and therefore presumably also in squids, other experiments indicate that in some of the gravity receptor systems a change in the magnitude of stimulation of the receptor cells might be effective as well (Budelmann, 1975).

4.2.1. Statolith growth rings and aging

The large compact statolith is composed of calcium carbonate, with its aragonite crystals embedded in a protein matrix (Radtke, 1983; Lowenstam *et al.*, 1984). The statolith grows with age/size of the squid, as does the underlying hair cell epithelium (Stephens and Young, 1982). In fresh or non-formalin fixed statolith material a regular lamination becomes visible after grinding, decalcification, or even when viewed whole in glycerine. Clarke (1966) first noted these laminations in squid statoliths and suggested that they might be useful for age determination. To date, it is well-documented that the laminations, as in fish otoliths (Panella, 1971), represent growth rings which may be formed at a rate of one ring per day (Spratt, 1978; Kristensen, 1980; Lipinski, 1986; Morris, 1988).

4.3. The angular acceleration receptor system

Orientation. The angular acceleration receptor system, at first sight, is less obvious than the gravity receptor system. It is noticeable as an opaque white line that winds as a ridge inside the cyst cavity, along the anterior, lateral and posterior walls. It is subdivided into four sections that run approximately in the three main planes of the animal: transverse, longitudinal and vertical (Fig. 2; Budelmann 1977, 1988). In

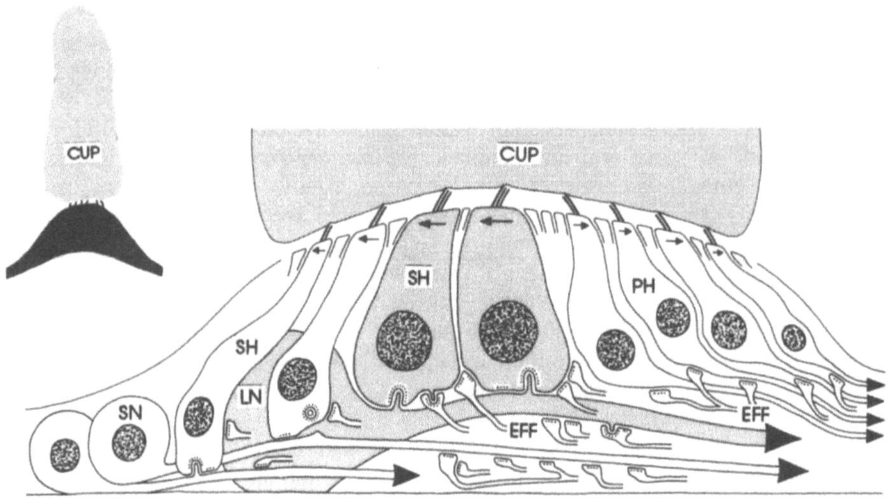

Figure 4. Cellular and synaptic organization of the crista of the angular acceleration receptor system. Although based on data of *Octopus*, the diagram represents the basic organization of the squid crista as well (see text). The cross section shows primary (*PH*) and secondary (*SH*) sensory hair cells, which are in afferent synaptic contact with large (*LN*) and small (*SN*) first-order afferent neurons, and many efferent synaptic endings (*EFF*). *Arrows* in the distal ends of the hair cells indicate the direction of the cell's polarization. Inset in the upper left shows a cross section of the crista, with the cupula (*CUP*) attached to it, at lower magnification (from Budelmann *et al.*, 1987; reprinted with the permission of The Royal Society).

many squids, the beginning and end of most of the sections are roofed over by cartilaginous lobes (*anticristae*; see Section 4.3.1.); sometimes these lobes leave only the middle portion of the crista section exposed (Stephens and Young, 1982; Young, 1984).

To follow the course of the crista ridges more precisely, the *crista transversalis anterior* (CTA) runs along the anterior cyst wall, ventral to the gravity receptor systems, from medial to lateral in a transverse, slightly dorsal direction. After a turn of almost 90°, the crista continues as *crista longitudinalis* (CL) at the lateral wall running from anterior to posterior. At the posterior wall it turns again and continues first in a transverse direction as *crista transversalis posterior* (CTP) from lateral to medial and, after another turn, as *crista verticalis* (CV) in a vertical upward direction.

To each of the four crista sections a cupula is attached. These are very delicate structures that are often detached from the crista ridge during the preparation. Because of their transparency, they are almost invisible. The cupulae are sail-like structures that are attached to the cilia of the hair cells along the whole length of the crista section. They protrude freely into the cyst cavity and act like swinging doors that move at right angles to the course of their crista section. The topographical arrangement of the four crista sections indicate that the transverse crista/cupula sections (CTA, CTP) respond to angular acceleration stimuli around the animal's transverse axis (i.e., during pitch), the longitudinal crista/cupula section (CL) during stimuli around the longitudinal axis (i.e., during roll), and the vertical crista/cupula section (CV) during stimuli around the vertical axis (i.e., during yaw).

Innervation. Two crista nerves have been described. They enter the statocyst cavity separately through the anterior statocyst wall. The smaller medial nerve, *n. crista minor*, innervates the anterior transverse crista section (CTA), and the larger, more lateral nerve, *n. crista major*, innervates the remaining crista sections (CL, CTP, CV). However, a small bundle of the n. crista major (of presumably only efferent fibers) branches off this nerve to innervate also the anterior transverse crista section (Stephens and Young, 1982).

Morphology. Each of the crista sections consists of between 600 and 800 hair cells, with four rows of regularly arranged large hair cells and three to four rows of less regularly arranged smaller hair cells, totalling about 3000 hair cells in the whole crista system in *Loligo* (Budelmann *et al.*, 1973). The pattern of polarization of these cells is at right angles to the course of the crista ridge, in each section with oppositely polarized hair cells (Figs. 2 and 4; Budelmann, 1977).

The ultrastructural, neuronal and synaptic organization of the crista epithelium reaches an unusual degree of complexity for such a receptor system, even when compared with vertebrate standards. Although the decapod crista system has not yet been analyzed in detail, preliminary data already indicate a similar organization as in the *Octopus* crista system. There, each section is composed of primary sensory hair cells (with an axon leaving the cell), and two types of secondary sensory cells (without an axon) that are in afferent synaptic contact with two (in squids probably three) types of first-order afferent neurons; all these elements receive many efferent endings (Fig. 4; Budelmann, 1989).

Physiology. The main function of the angular acceleration receptor system is to provide the animal with information about rotational movements, i.e., during angular accelerations around any of its body axes. This information is used to initiate, when appropriate, compensatory reactions, such as funnel, head and eye movements, including nystagmus (compare Section 6). Although in squids no electrophysiological recordings have yet been performed during natural stimulation, the morphological organization of their crista system and preliminary electrophysiological data already indicate a function similar to that of the *Octopus* crista system (Williamson, 1985, 1986; Williamson and Budelmann, 1985). Angular acceleration stimuli will first – because of inertia – result in a movement of the fluid inside the cyst cavity relative to the crista ridge. The cupulae, following this movement to a certain extent, thereby cause the cilia to deflect (shear) in one or the other direction (depending on the direction of angular acceleration), and thus stimulate the hair cells.

In the squid *Alloteuthis*, the secondary sensory hair cells have a resting membrane potential of about – 50 mV (Williamson, 1988a) and some of the cells are electrically coupled along the length of the crista section (Williamson, 1989a).

In *Octopus*, two morphologically different crista subsystems have been described that both code angular velocity (as do the ampullary organs in the vertebrate semicircular canals), but differ in their sensitivities by a factor of 10 (Williamson and Budelmann, 1985). Since in the decapod statocyst all crista/cupula sections are morphologically more or less alike, no such distinct differences in sensitivity are to be expected between the four crista/cupula sections. However, the various cartilaginous protrusions (*anticristae*; Section 4.3.1.) certainly influence the endolymph movement. Consequently, when differences occur in the arrangement of anticristae between the four crista section domains, they may account for some differences in sensitivity of the crista section. For behavioral reactions elicited by the angular acceleration receptor system, see Section 6.

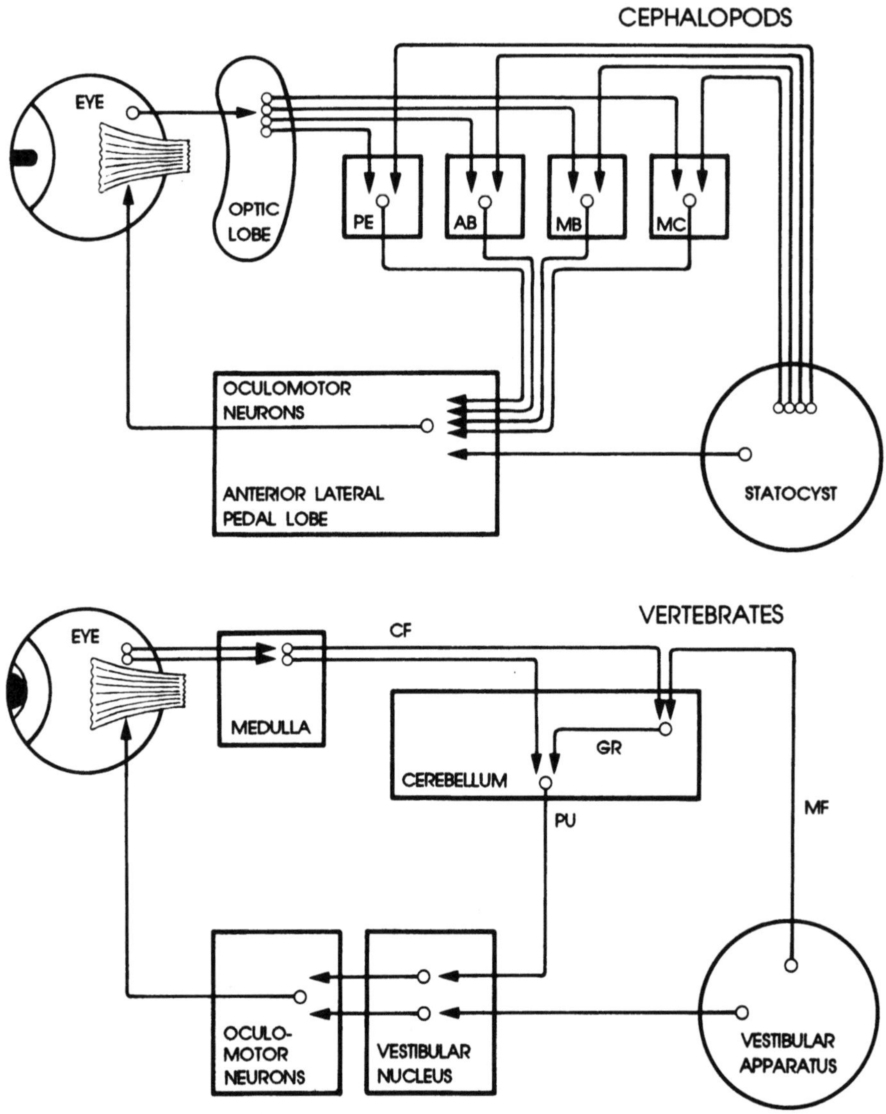

Figure 5. Simplified diagrams to show the similarities in brain pathways of the oculomotor control system in cephalopods and vertebrates. A direct and an indirect pathway connect the equilibrium receptor organs (statocyst, vestibular apparatus) with the motoneurons of the eyes in both the statocyst-oculomotor system of cephalopods and the vestibulo-oculomotor system of vertebrates. *AB*, anterior basal lobe; *CF*, climbing fibers; *GR*, granule cells; *MB*, median basal lobe; *MC*, magnocellular lobe; *MF*, mossy fibers; *PE*, pedal lobe; *PU*, Purkinje cells (from Budelmann and Young, 1984; reprinted with the permission of The Royal Society).

4.3.1. Anticristae

Other distinct and rather unusual structures in squid statocysts are the many cartilaginous lobes, *anticristae*, that protrude into the cyst cavity (Fig. 2). They vary largely in number, shape and distribution from species to species, and even within species individual differences may occur. The total number of anticristae per statocyst varies from 0 to 43 (e.g., there are 13 in *Loligo vulgaris*). In newly-hatched animals, anticristae are only poorly developed, or even absent; their size increases with the age of the animal (Ishikawa, 1924; Stephens and Young, 1982; Young, 1984). There is a distinction into *hamuli* and *straight anticristae*. Hamuli are those at the beginning and end of the crista ridge and at its three intersections (where the ridge turns almost 90° into another direction); hamuli often bend far over the crista ridge. All the other anticristae are more or less unbent; thus, they are termed *straight anticristae*. A few (sensory?) ciliated cells and neurons may occur at the base of and below the hamuli (Dilly *et al.*, 1975; Stephens and Young, 1982).

The anticristae certainly influence the fluid (endolymph) flow inside the cyst cavity during angular accelerations. Interestingly, in neutrally buoyant (and thus presumably slow-moving) squids the statocyst cavities are relatively large and the number and size of the anticristae are reduced. This provides the large inertial mass of fluid necessary to sense the small accelerations (i.e., small cupula movements) that occur during slow movements. In contrast, in fast-moving squids, such as *Loligo*, the statocysts are relatively small, as is the mass of fluid since the anticristae are well-developed; in some species parts of the cyst cavity may even be restricted to almost canals. Thus, their systems can be expected to be less sensitive to angular accelerations (Stephens and Young, 1982; Maddock and Young, 1984).

4.4. Ciliated cells

In addition to the hair cells of the statocyst sensory epithelia, there are other cells scattered all over the statocyst walls that carry cilia of 15 – 20 μm in length (Dilly *et al.*, 1975; Stephens and Young, 1982). These ciliated cells are especially abundant up to 250 μm ventrally (CTA, CL, CTP) and medially (CV) from the crista sections. In *Sepia officinalis*, their average number is 600 – 700 per crista section (Budelmann, 1977, 1989).

The function of the ciliated cells is unknown. They are presumably not sensory, since they are actively beating towards the crista ridge. This can easily be seen in thin cross sections of a crista, made with a razor blade and viewed in a phase contrast microscope. The beating causes a small endolymph movement directed towards the crista ridge (Budelmann, 1989). It is difficult to think of any biological significance of this minute endolymph movement, as compared to the much larger movements that occur during angular accelerations. But the abundance of ciliated cells in close vicinity to the crista ridge is certainly not accidental.

4.5. Kölliker's canal

During development, squid statocysts are formed from ectodermal invaginations. Remnants of this invagination persist as *Kölliker's canal*. It can easily be seen as an opaque line in the lateral wall of each statocyst, with its opening to the cyst cavity above the longitudinal crista section and in front of a large anticrista lobe (Fig. 2). The canal ends ventrally, outside the cranial cartilage, probably among the connective tissue (Stephens and Young, 1982).

The canal is lined with cells with very long cilia that almost completely fill the lumen of the canal. The cilia beat actively (thus they can be easily viewed with a stereomicroscope). They produce a current flow that is directed from the statocyst cavity into the canal. This may indicate that the canal is somehow involved in the regulation of the volume and chemical composition of the endolymph fluid inside the statocyst cavity (Stephens and Young, 1982).

5. Central projections of the statocyst sensory epithelia

The gravity and angular acceleration receptor systems are connected with various parts of the squid brain (compare Fig. 5): in the sub-oesophageal mass with the lateral pedal, pedal and ventral magnocellular lobes, and in the supra-oesophageal mass with the peduncle, and anterior and median basal lobes (Young, 1976). With the exception of the pedal lobes, all of these receive visual input and are connected with the anterior lateral pedal lobe, which is the oculomotor center. Thus, the oculomotor neurons in the anterior lateral pedal lobe get their information via two pathways (just as in the *Octopus* system; Budelmann and Young, 1984): via a direct pathway, and via an indirect pathway in which the visual information about movement is integrated with that coming from the statocysts. Since this organization is similar to that known for the vertebrate vestibulo-oculomotor system, it demonstrates a striking convergent evolution between a complex vertebrate and invertebrate sensorimotor system (Fig. 5; compare Budelmann and Young, 1984).

6. Compensatory eye and head movements

In cephalopods, the best investigated behavioral reactions that are controlled by the statocyst receptor systems are compensatory eye and head movements. They serve to stabilize the visual image on the retina during head and body movements. Three such compensatory movements have been analyzed in detail: counterrolling of the eyes, post-rotatory nystagmus, and compensatory head movements.

6.1. Counterrolling of the eyes

This compensatory reaction can easily be watched in those cephalopods that have non-circular pupils, such as *Octopus* or *Sepia*: when these animals move around a transverse body axis, their pupils remain nearly horizontal regardless of the animal's position in space. Counterrolling is a fast compensatory reflex; in *Octopus*, one gets the impression the animal moves around its fixed eyes. Of course, there are morphological restraints for such a movement. Careful measurements (Fig. 6) of these eye movements showed that *Sepia* is able to counterrotate its eyes at least 45° in either direction when moved around a transverse body axis (this is an enormous amount by

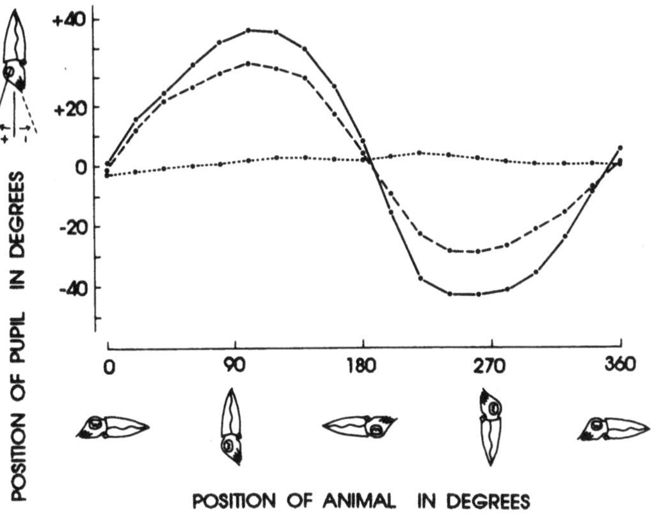

Figure 6. Compensatory counterrolling of the eyes of *Sepia* as a function of body position relative to gravity during rotation around an animal's trans-transverse body axis. The curves are results from normal animals (solid line), animals with the gravity receptor systems destroyed unilaterally (dashed line) and bilaterally (dotted line) (from Budelmann, 1975; reprinted with the permission of Gustav Fischer Verlag).

any standards). In a step-by-step elimination of the six gravity receptor systems in *Sepia*, the reaction becomes more and more reduced, and after a total destruction of all the systems the reaction is gone completely (Fig. 6; Budelmann, 1975). This clearly indicates that the compensatory counterrolling of the eyes is controlled by just the gravity receptor systems of the statocysts.

6.2. Post-rotatory nystagmus

This type of eye movement is well-known to be elicited by the angular acceleration receptor systems. It occurs when an animal is rotated in the horizontal plane at constant speed and without visual cues, and is then abruptly stopped. Because of inertia, the endolymph inside the statocyst cavity continues to rotate to some extent and thereby stimulates the vertical crista/cupula section. This, in turn elicits post-rotational eye flicks, *nystagmus*, with a slow and a fast phase. The reaction is almost completely suppressed when the animal is able to fix visual cues, and it is completely abolished by bilateral destruction of the statocyst sensory epithelia (Dijkgraaf, 1963a; Collewijn, 1970; Messenger, 1970).

6.3. Compensatory head movements

This reflex can be observed when an animal is moved around its longitudinal body axis. In such a situation, it countermoves its head relative to the body (Fig. 7). Careful measurements of these movements showed that Sepia is able to countermove its head almost 30° in either direction. After unilateral removal of the statolith and statoconial layers, the amplitude of this reflex is clearly diminished, with the head kept inclined to the unoperated side (Fig. 7). After bilateral removal of the statoliths and statoconial layers, the head reflex is gone completely (Fig. 7). This clearly indicates

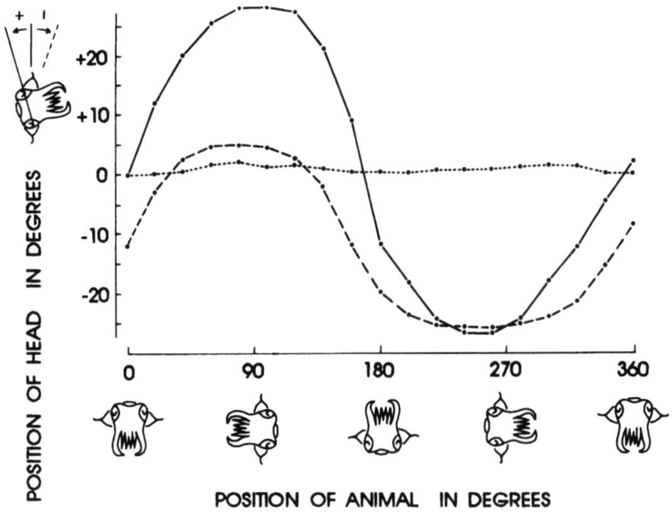

Figure 7. Compensatory head movements of *Sepia* as a function of body position relative to gravity during rotation around the animal's longitudinal body axis. The curves are results from normal animals (solid line), animals with the gravity receptor systems destroyed unilaterally (dashed line) and bilaterally (dotted line) (from Budelmann, 1975; reprinted with the permission of Gustav Fischer Verlag).

that also the compensatory head movements are controlled by the statocysts' gravity receptor systems (Budelmann, 1975).

7. Conclusion

The statocysts of squids, and of cephalopods in general, are fascinating objects to study. Research of the past roughly 20 years has shown that cephalopods have statocysts and a control system for compensatory eye movements that rival the sophistication of their vertebrate counterparts, the vestibular end organs and vestibulo-oculomotor system. Since comparative information on similarities and differences of equivalent vertebrate and invertebrate systems can substantially contribute to the understanding of basic underlying principles of morphology, physiology, and even pathology, and since in cephalopods the various parts of the system are easily accessible for operations, the cephalopod systems may serve in future experimental research as useful invertebrate models of the vertebrate system.

ACKNOWLEDGMENT. The author thanks Dr A. Novicki for improvements to the English text.

References

Auerbach, B. and Budelmann, B. U., 1986, Evidence for acetylcholine as neurotransmitter in the statocyst of *Octopus vulgaris*, *Cell Tissue Res.* **243**:429–436.

Barber, V. C., 1966, The fine structure of the statocysts of *Octopus vulgaris*, *Z. Zellforsch.* **70**:91–107.

Boycott, B. B., 1960, The functioning of the statocysts of *Octopus vulgaris*, *Proc. Roy. Soc. B* **152**:78–87.

Budelmann, B. U., 1970, Die Arbeitsweise der Statolithenorgane von *Octopus vulgaris*, *Z. Vergl. Physiol.* **70**:278–312.
Budelmann, B. U., 1975, Gravity receptor function in cephalopods with particular reference to *Sepia officinalis*, *Fortschr. Zool.* **23**:84–98.
Budelmann, B. U., 1977, Structure and function of the angular acceleration receptor systems in the statocysts of cephalopods, *Symp. Zool. Soc. Lond.* **38**:309–324.
Budelmann, B. U., 1979, Hair cell polarization in the gravity receptor systems of the statocysts of the cephalopods *Sepia officinalis* and *Loligo vulgaris*, *Brain Res.* **160**:261–270.
Budelmann, B. U., 1980, Equilibrium and orientation in cephalopods, *Oceanus* **23**:34–43.
Budelmann, B. U., 1988, Morphological diversity of equilibrium receptor systems in aquatic invertebrates, in: *Sensory Biology of Aquatic Animals* (J. Atema, R. R. Fay, A. N. Popper, and W. N. Tavolga, eds.), pp. 757–782, Springer-Verlag, New York.
Budelmann, B. U., 1989, *unpub. observations.*
Budelmann, B. U. and Bonn, U., 1982, Histochemical evidence for catecholamines as neurotransmitters in the statocyst of *Octopus vulgaris*, *Cell Tissue Res.* **227**:475–483.
Budelmann, B. U. and Thies, G., 1977, Secondary sensory cells in the gravity receptor system of the statocyst of *Octopus vulgaris*, *Cell Tissue Res.* **182**:93–98.
Budelmann, B. U. and Young, J. Z., 1984, The statocyst-oculomotor system of *Octopus vulgaris*: Eye muscles, eye muscle nerves, statocyst nerves, and the oculomotor centre in the central nervous system, *Philos. Trans. R. Soc. Lond. B* **306**:159–189.
Budelmann, B. U. and Wolff, H. G., 1973, Gravity response from angular acceleration receptors in *Octopus vulgaris*, *J. Comp. Physiol.* **85**:283–290.
Budelmann, B. U., Barber, V. C., and West, S., 1973, Scanning electron microscopical studies of the arrangements and numbers of hair cells in the statocysts of *Octopus vulgaris*, *Sepia officinalis*, and *Loligo vulgaris*, *Brain Res.* **56**:25–41.
Budelmann, B. U., Sachse, M., and Staudigl, M., 1987, The angular acceleration receptor system of the statocyst of *Octopus vulgaris*: Morphometry, ultrastructure, and neuronal and synaptic organization, *Philos. Trans. R. Soc. Lond. B* **315**:305–343.
Clarke, M. R., 1966, A review of the systematics and ecology of oceanic squids, *Adv. Marine Biol.* **4**:91–300.
Clarke, M. R., 1978, The cephalopod statolith – an introduction to its form, *J. Mar. Biol. Assoc. U.K.* **58**:701–712.
Collewijn, H., 1970, Oculomotor reactions in the cuttlefish, *Sepia officinalis*, *J. Exp. Biol.* **52**:369–384.
Colmers, W. F., 1977, Neuronal and synaptic organization in the gravity receptor system of the statocyst of *Octopus vulgaris*, *Cell Tissue Res.* **185**:491–503.
Colmers, W. F., 1981, Afferent synaptic connections between hair cells and the somata of intramacular neurons in the gravity receptor system of the statocyst of *Octopus vulgaris*, *J. Comp. Neurol.* **197**:385–394.
Colmers, W. F., 1982, The central afferent and efferent organization of the gravity receptor system of the statocyst of *Octopus vulgaris*, *Neuroscience* **7**:461–476.
Delage, Y., 1886, Sur une fonction nouvelle des otocystes chez les invertébrés, *C.R. Acad. Sci. (Paris)* **103**:798–801.
Dijkgraaf, S., 1961, The statocyst of *Octopus vulgaris* as a rotation receptor, *Pubbl. Staz. Zool. Napoli* **32**:64–87.
Dijkgraaf, S., 1963a, Nystagmus and related phenomena in *Sepia officinalis*, *Experientia* **19**:29–30.
Dijkgraaf, S., 1963b, Versuche über Schallwahrnehmung bei Tintenfischen, *Experientia* **50**:50.
Dilly, P. N., Stephens, P. R., and Young, J. Z., 1975, Receptors in the statocyst of squids, *J. Physiol.* **249**:59–61P.
Fröhlich, A., 1904, Studien über Statocysten, I. Versuche an Cephalopoden, *Pflügers Arch. Ges. Physiol.* **102**:415–472.
Griffin, L. E., 1900, The anatomy of *Nautilus pompilius*, *Mem. Acad. Sci. Wash.* **8**:103–230.
Hamlyn-Harris, R., 1903, Die Statocysten der Cephalopoden, *Zool. Jahrb., Abt. Anat. Ontog.* **18**:327–358.

Hanlon, R. T. and Budelmann, B. U., 1987, Why cephalopods are probably not 'deaf.' *Am. Naturalist* **129**:312–317.
Hartline, P. H., Hurley, A. C., and Lange, G. D., 1979, Eye stabilization by statocyst mediated oculomotor reflex in *Nautilus*, *J. Comp. Physiol.* **132**:117–126.
Hubbard, S. J., 1960, Hearing and the octopus statocyst, *J. Exp. Biol.* **37**:845–853.
Ishikawa, M., 1924, On the phylogenetic position of the cephalopod genera of Japan based on the structure of statocysts, *J. Coll. Agr. Tokyo Imp. Univ.* **7**:165–210.
Klein, K., 1932, Die Nervenendigungen in der Statocyste von *Sepia*, *Z. Zellforsch. Mikrosk. Anat.* **14**:481–516.
Kolmer, W., 1926, Statoreceptoren, in: *Handbuch der normalen und pathologischen Physiologie 11* (A. Bethe, G. v. Bergmann, G. Embden, A. Ellinger, eds.), pp. 767–790, Springer-Verlag, Berlin.
Kristensen, T. K., 1980, Periodical growth rings in cephalopod statoliths, *Dana* **1**:39–51.
Lipinski, M., 1986, Methods for the validation of squid age from statoliths, *J. Mar. Biol. Assoc. U.K.* **66**:505–526.
Lowenstam, H. A., Traub, W., and Weiner, S., 1984, *Nautilus* hard parts: A study of the mineral and organic constituents, *Paleobiology* **10**:268–279.
Maddock, L. and Young, J. Z., 1984, Some dimensions of the angular acceleration receptor systems of cephalopods, *J. Mar. Biol. Assoc. U.K.* **64**:55–79.
Maturana, H. M. and Sperling, S., 1963, Unidirectional response to angular acceleration recorded from the middle cristal nerve in the statocyst of *Octopus vulgaris*, *Nature, Lond.* **197**:815–816.
Messenger, J. B., 1970, Optomotor responses and nystagmus in intact, blinded and statocystless cuttlefish (*Sepia officinalis* L.), *J. Exp. Biol.* **53**:789–796.
Messenger, J. B., Nixon, M., and Ryan, K. P., 1985, Magnesium chloride as an anaesthetic for cephalopods, *Comp. Biochem. Physiol.* **82C**:203–205.
Morris, C. C., 1988, Statolith growth lines and statocyst function in the Cephalopoda, Ph.D. Dissertation, University of Cambridge, U.K., 140 pp.
Muskens, L. J. J., 1904, Über eine eigentümliche kompensatorische Augenbewegung der Octopoden mit Bemerkungen über deren Zwangsbewegungen, *Arch. (Anat.) Physiol. (Lpz.)* **28**:49–56.
Owsjannikow, P. H. and Kowalewsky, A., 1867, Über das Centralnervensystem und das Gehörorgan der Cephalopoden, *Acad. Imp. Sci. St. Petersbourg* **11**:1–36.
Packard, A., 1972, Cephalopods and fish: The limits of convergence, *Biol. Rev.* **47**:241–307.
Panella, G., 1971, Fish otoliths: Daily growth layers and periodical patterns, *Science* **173**:1124–1127.
Radtke, R. L., 1983, Chemical and structural characteristics of statolith form from the short-finned squid *Illex illecebrosus*, *Mar. Biol.* **76**:47–54.
Spratt, J. D., 1978, Age and growth of the market squid, *Loligo opalescens* Berry, in Monterey Bay, *Fish. Bull.* **169**:35–44.
Stephens, P. R. and Young, J. Z., 1976, The statocyst of *Vampyroteuthis infernalis* (Mollusca: Cephalopoda), *J. Zool. Lond.* **180**:565–588.
Stephens, P. R. and Young, J. Z., 1978, Semicircular canals in squids, *Nature* **271**:444–445.
Stephens, P. R. and Young, J. Z., 1982, The statocyst of the squid *Loligo*, *J. Zool. Lond.* **197**:241–266.
Vinnikov, Y. A., Gasenko, O. G., Bronstein, A. A., Tsirulis, T. P., Ivanov, V. P., and Pyatkina, G. A., 1967, Structural, cytochemical and functional organization of statocysts of cephalopoda, *Symposium on Neurobiology of Invertebrates*, pp. 29–48, Akadémiai Kiado, Budapest.
Williamson, R., 1985, Efferent influences on the afferent activity from the octopus angular acceleration receptor system, *J. Exp. Biol.* **119**:251–264.
Williamson, R., 1986, Efferent activity in the *Octopus* statocyst nerves, *J. Comp. Physiol. A* **158**:125–132.
Williamson, R., 1988a, Intracellular recordings from hair cells in the statocysts of the squid, *Alloteuthis subulata*, *Pflügers Arch.*, Suppl. 1, **411**:R196.

Williamson, R., 1988b, Vibration sensitivity in the statocyst of the northern octopus, *Eledone cirrosa*, *J. Exp. Biol.* **134**:451–454.
Williamson, R., 1989a, Electrical coupling between secondary hair cells in the statocyst of the squid *Alloteuthis subulata*, *Brain Res.* **486**:67–72.
Williamson, R., 1989b, Electrophysiological evidence for cholinergic and catecholaminergic efferent transmitters in the statocyst of *Octopus*, *Comp. Biochem. Physiol.* **93C**:23–27.
Williamson, R., and Budelmann, B. U., 1985, The response of the *Octopus* angular acceleration receptor system to sinusoidal stimulation, *J. Comp. Physiol. A* **156**:403–412.
Young, J. Z., 1960, The statocysts of *Octopus vulgaris*, *Proc. R. Soc. Lond. B* **152**:3–29.
Young, J. Z., 1965, The central nervous system of *Nautilus*, *Philos. Trans. R. Soc. Lond. B* **249**:1–25.
Young, J. Z., 1971, *The Anatomy of the Nervous System of Octopus vulgaris*, Clarendon Press, Oxford.
Young, J. Z., 1976, The nervous system of *Loligo*, II. Suboesophageal centres, *Philos. Trans. R. Soc. Lond. B* **274**:101–167.
Young, J. Z., 1984, The statocysts of cranchiid squids (Cephalopoda), *J. Zool. Lond.* **203**:1–21.

Part VI

INTEGRATED SYSTEMS

Chapter 20

Gas Transport in the Blood

CHARLOTTE P. MANGUM

1. Introduction

The transport of respiratory gases to and from gas exchange sites is clearly the central function of the blood in squid. In view of the squid's inability to live without oxygen for even a brief period (Dykens and Mangum, 1978), one would expect the squid gas transport system to be highly adapted to maintain an adequate supply. The cephalopod oxygen transport system has been characterized as a rather poor one, primarily on the grounds that the blood oxygen carrying capacity, in this case a simple function of concentration of the carrier molecule hemocyanin (Hc), is lower than in most fishes (e.g., O'Dor and Webber, 1986). This view (often overtly tongue-in-cheek) is a welcome departure from the attitude that a physiological system, regardless of its properties, must be the best of all possible ones under the particular set of circumstances. But it also ignores the considerable genetic constraints placed upon the evolution of one taxon from another. The Order Teuthoidea evolved from more primitive cephalopods that almost certainly already transported oxygen bound to a molecule that resembles present day teuthoid Hc. Moreover, the Class Cephalopoda did not evolve from the same kinds of animals that gave rise to the fishes, which may well have transmitted the genetic information for only hemoglobins. It is likely that cephalopods evolved from animals that relied on a distinctive molluscan type of Hc for their metabolic oxygen supply (Mangum *et al.*, 1987). Hc is not carried inside a blood cell but is dissolved in the blood. Hc is a copper containing respiratory pigment and the copper gives the oxygenated form (oxyHc)(HcO_2) a blue color. The deoxygenated form (deoxyHc) is colorless.

This chapter examines first the physiological and molecular properties of squid Hcs and then the performance of the gas transport system as a whole. Since less is known about the teuthoid system than some other cephalopod systems, it will be necessary to discuss in detail the other groups as well. I shall attempt to show that the squid gas transport system is, in fact, a highly evolved one. And I shall at least emphasize questions that, once answered, should further elucidate the reasons why the teuthoid system is built the way it is.

C. P. MANGUM • Dept. of Biology, College of William and Mary, Williamsburg, Virginia 23185.

2. Respiratory properties of cephalopod bloods

2.1. Squid

In 1926, the first full report of the oxygen equilibrium properties of a cephalopod Hc (along with others) emerged from the laboratory of A. C. Redfield. The paper is remarkable for many reasons, including the astonishing (and, in the eyes of a pair of British contemporaries, incredible) finding of the reversed Bohr shift of gastropod and chelicerate Hcs. It is also remarkable in that, in the present author's experience (e.g., Mangum and Lykkeboe, 1979), the data for gastropod and chelicerate Hcs are correct *numerically*, not merely in principle. As I shall show below, the data for *Loligo pealei* also agree reasonably well with quite recent ones. Redfield's precocious accuracy is fortunate because this Hc has been re-examined only recently, at least under physiological conditions, and data for only a handful of other squids are available.

At the time of these investigations, the oxygen equilibrium properties of human blood had been explored in some detail and the first investigations of other hemoglobins were just beginning to appear. Early reports of unexpectedly high oxygen affinities of the extracellular hemoglobins of the annelid lugworm *Arenicola* and the gastropod snail *Planorbis* had led to the erroneous conclusion that these proteins cannot deoxygenate under normoxic conditions and thus that they must function as oxygen stores for hypoxia not normoxic oxygen carriers. Early studies on the Hcs had focused on the oxygen carrying capacity of Hc-containing bloods, which was deemed low relative to the blood of not only man but also *Arenicola*. Only in very recent years has the exceptional character of the high hemoglobin concentration in *Arenicola* become clear. Moreover, Redfield and Ingalls (1933) mention a value for prebranchial oxygen content in *Loligo pealei* that should correspond to more than 200 mg Hc/ml, which means that the molecules occupy about 20% of the available space in the blood.

In any event, it was Redfield's stated goal to shift the focus away from concentration to more meaningful respiratory properties couched in the most modern physicochemical terms of the day, so as to provide a rigorous comparison of the Hcs with the hemoglobins, to examine the differences between the Hcs in the four taxa in which they were known at the time and to understand the physiological basis of such differences.

The 1926 contribution by Redfield, Coolidge, and Hurd (1926) can only be described as a watershed in respiratory physiology. First they confirmed the earlier findings of the oxygen carrying capacity of cephalopod blood, averaging 6 or more times that of seawater (Table I). They also made the following observations: 1) Squid Hc has a very *low* oxygen affinity at 21–24 °C in the virtual absence of carbon dioxide (which probably means a pH of about 7.6), in striking contrast to the annelid and gastropod hemoglobins known at the time. Quite correctly they pointed out that the real question was not whether the squid oxygen carrier can deoxygenate at the tissues but rather whether it can become highly oxygenated at the gill. 2) The oxygen binding as a function of the oxygen pressure is commonly expressed graphically by plotting the log of the ratio of the degree of oxygen saturation to the degree of oxygen unsaturation against the log of the partial pressure of oxygen (PO_2). This plot is referred to as the Hill plot. The Hill coefficient, *n*, is the slope of this plot at 50% oxygenation, which corresponds to the maximum slope. The steeper the slope, the greater is the degree of cooperativity, i.e., during oxygenation the molecule undergoes a transition from a lower to a higher affinity conformation. Squid Hc exhibits cooperativity, as indicated by a

Table I. Quantitative Indices of Gas Transport in Cephalopods[†,‡]

	Loligo pealei	*Octopus dofleini*	*Octopus vulgaris*	*Sepia officinalis*	*Nautilus pompilius*
Body wt (kg)	?	6–18	0.6–2.9	1.5–1.6	0.4–0.6
Exp. temp. (°C)	23	8–11	20.5–22	17–19	17–18
HcO_2 capacity (mM)	1.87	1.72	1.52	1.53; 1.44	1.03
PiO_2 (torr)	>150	127	145; 50	133–150; 30-67	159
PaO_2 (torr)	**120**	69	79; 25	101; 28	99
$HcaO_2$ (%)	93	94	98; 71	96; 32	**100**
CaO_2 (mM)	1.91	1.47	1.50; 1.00	1.60; 1.07	**1.25**
HcO_2 transport (%)	94	94	95; 98	95; 79	88
PvO_2 (torr)	**48**	9.7	30; 12	25; 15	20
$HcvO_2$ (%)	5	21	14; **ca. 3**	71; 31	65
CvO_2 (mM)	0.17	0.18	0.20; <0.10	0.52; 0.47	**0.31**
$PaCO_2$ (torr)	**2.2**	3.1	**1.9; 2.7**	**2.7; 2.8**	?
$CaCO_2$ (mM)	1.79	1.89	1.90; 3.40	3.40; 4.50	?
pHa	**ca. 7.34**	7.13	7.48; 7.61	7.58; 7.69	**ca. 7.45**
$PvCO_2$ (torr)	**6.0**	4.5	**3.5; 4.3**	**4.7; 1.7**	?
$CvCO_2$ (mM)	3.72	3.41	2.70; 4.20	6.20; 3.00	?
pHv	**ca. 7.22**	7.08	7.38; 7.48	7.60; 7.72	**ca. 7.40**
$ExtrO_2$ (%)	**91**	**80**	**87; 95**	**80; 83**	**75**
$ExtrCO_2$ (%)	**52**	**45**	**30; 19**	**45; -50**	?
EbO_2 (%)	**67**	**51**	**43; 34**	**65; 42**	**57**
Sources	Redfield & Goodkind (1929)	Johansen (1965); Johansen & Lenfant (1966)	Houlihan *et al.* (1982)	Johansen *et al.* (1982)	Johansen *et al.* (1978)

Values separated by semicolons show measurements for normoxic and hypoxic conditions.
† Values in **bold** are calculated rather than directly measured.
‡ Symbols and Derived Quantities:

PiO_2 = inhalant PO_2
PaO_2, PvO_2 = post-, prebranchial PO_2
$PaCO_2$, $PvCO_2$ = post-, prebranchial PCO_2
$HcaO_2$, $HcvO_2$ = post-, prebranchial % HcO_2
CaO_2, CvO_2 = post-, prebranchial O_2 concentration
$CaCO_2$, $CvCO_2$ = post-, prebranchial CO_2 concentration
pHa, pHv = post-, prebranchial pH

$$ExtrO_2 \text{ (}O_2\text{ extraction from blood) (\%)} = \frac{(CaO_2 - PvO_2)}{CaO_2} 100$$

$$ExtrCO_2 \text{ (}CO_2\text{ extraction from the blood) (\%)} = \frac{(CvCO_2 - CaCO_2)}{CvCO_2} 100$$

$$EbO_2 \text{ (efficiency of }O_2\text{ uptake at the gill) (\%)} = \frac{(PaO_2 - PvO_2)}{(PiO_2 - PvO_2)} 100$$

large value of *n*, which can as big as 4.2. This property may facilitate oxygen transport. 3) Cooperativity is not a fixed property but increases at high PCO_2. 4) The decrease in oxygen affinity at low pH is known as the Bohr shift. The Bohr coefficient, commonly used as a measure of this Bohr shift, is defined as $\Delta \log P_{50}/\Delta pH$, where P_{50} is the oxygen pressure in torr when the Hc is 50% oxygenated. Note that the more negative the Bohr coefficient, the greater is the Bohr shift. Unlike some of the other Hcs examined by Redfield *et al.* (1926), squid Hc was shown to possess a normal Bohr shift. This shift was greater than that of any other oxygen carrier known at the time. The data of Redfield and Ingalls (1933) allows one to calculate a Bohr coefficient of about −1.04 at 25 °C and −1.37 at 15 °C. To make this calculation, I have assigned pH values to their PCO_2 levels using unpublished data for the relationship between the two. These values are also consistent with recent ones (−1.15 at 15 °C) for the same species collected by Pörtner (1990). 5) If the pH dependence is so great that an oxygen carrier fails to become fully oxygenated at low pH, this phenomenon is known as the Root shift (Root, 1931). Redfield *et al.* observed that, at only moderately high (6%) carbon dioxide levels, squid Hc is still partially deoxygenated in pure oxygen. This is the first report of the Root shift which should be called the Redfield shift. 6) Squid Hc has buffering properties. 7) DeoxyHc (deoxygenated Hc) has a greater carbon dioxide affinity than oxyHc (oxygenated Hc). This is known as the Haldane effect. 8) Subsequently Redfield and Ingalls (1933) explored the lability of P_{50} and cooperativity, showing in an extension (6.96 to 9.03) well beyond the physiological pH range, that cooperativity and P_{50} vary inversely with pH. Unlike human hemoglobin as well as other Hcs there is no sign of a reversed Bohr shift anywhere. 9) Cooperativity and P_{50} vary directly with temperature, diminishing at low temperature as the molecule assumes a more closed structure. 10) Redfield and Ingalls (1933) also showed that the oxygen affinity of squid blood diluted with seawater, almost certainly that to which the animals were acclimated, changed only as expected due to a small attendant change in pH; there was no clear change in cooperativity. They did realize the importance of this observation. Unlike the heme proteins, the concentration of Hcs does not affect their oxygen binding properties. 11) Finally, cooperativity and oxygen affinity of blood diluted by the same factor with phosphate buffer diminish sharply. Redfield and Ingalls (1933) could not have recognized their *de facto* discovery of the only known allosteric effectors of cephalopod HcO_2 binding, inorganic ions. While the seawater would have changed inorganic ion levels in the blood of these osmotic conformers very little, the six fold dilution with phosphate buffer would have reduced monovalent ions to a small fraction of the physiological level and divalent cations to almost nil. Recent data (Mangum, 1989) suggest that the Hc of *Lolliguncula brevis*, a rare example of a euryhaline squid, is uniquely insensitive to inorganic ions, which vary considerably within the salinity range (Mangum, 1981), and only moderately sensitive to pH, which may also vary with salinity.

I am aware of only a few other published oxygen binding data for squid Hcs. The P_{50} values for *Loligo pealei* Hc reported by DePhillips *et al.* (1969) are much greater than Redfield's, almost certainly due to the absence of inorganic ions in their purified preparations. Surprisingly, their values for cooperativity (Fig. 1) agree quite well with Redfield's; the Bohr factor in their data also seems to be the same. Pörtner's (1990; Fig. 1) very recent data also agree quite closely with Redfield's; however, the difference in experimental temperature would be expected to widen the gap somewhat. While Pörtner's samples had been frozen and Redfield's had not, freezing does not appear to influence oxygen binding of other cephalopod Hcs (see below).

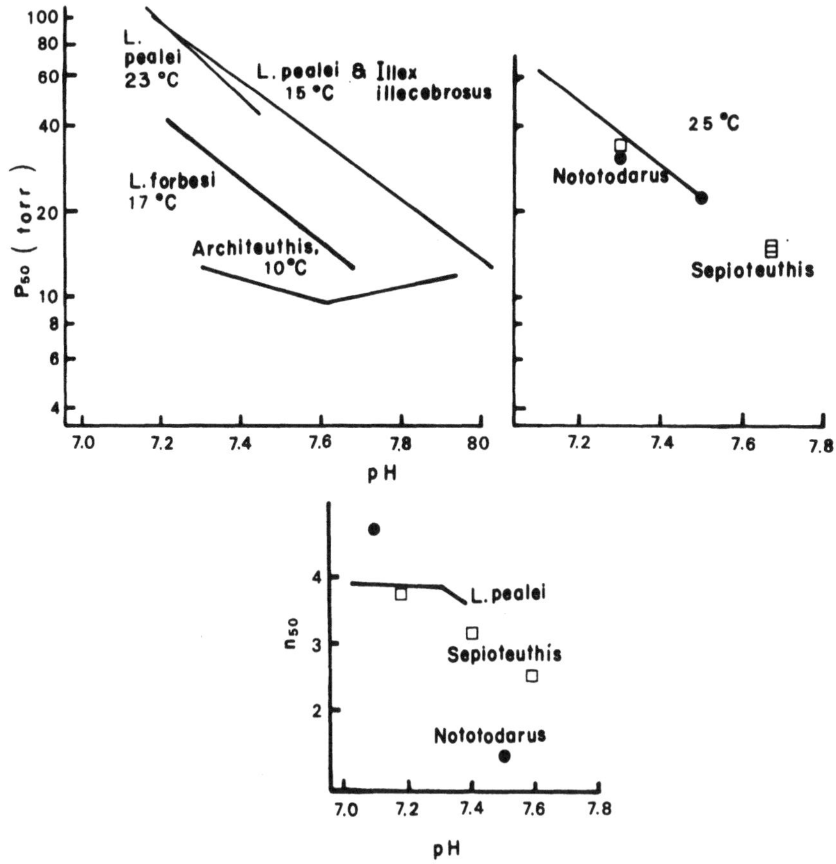

Figure 1. Oxygen binding of squid hemocyanins. Solid lines reproduce data for *Loligo pealei* at 23 °C from Redfield and Ingalls (1933) and at 15 °C from Pörtner (1990), for *Architeuthis monachus* from Brix (1983), for *Loligo forbesi* from Brix *et al.* (1981), for *Illex illecebrosus* from Pörtner (1990) and for *Nototodarus sloani philippinensis* and *Sepiotheuthis lessoniana* from Lykkeboe and Johansen (1982). Only one line is shown for Pörtner's data because the two Hcs are virtually indistinguishable. All of the samples were buffered with carbon dioxide and the data collected by tonometry. Pörtner's and Lykkeboe and Johansen's samples had been frozen. Symbols show unpublished data for freshly collected, never frozen whole blood diluted by 10% with 0.05 M mono[tris(hydroxymethyl)-aminomethane] maleate (Tris maleate) buffered seawater, obtained by the present author using the cell respiration procedure. Circles, *Nototodarus sloani philippinensis*; squares, *Sepiotheuthis lessoniana*.

Loligo forbesi Hc has a higher oxygen affinity than *Loligo pealei* or *Illex illecebrosus* Hcs (Brix *et al.*, 1981; Pörtner, 1990; Fig. 1), though its pH dependence (−1.15) is about the same. It also appears to have a lower carbon dioxide combining capacity (Fig. 2).

Oxygen binding data for two tropical squids reported by Lykkeboe and Johansen (1982), who used previously frozen samples, are shown in Fig. 1 together with my own

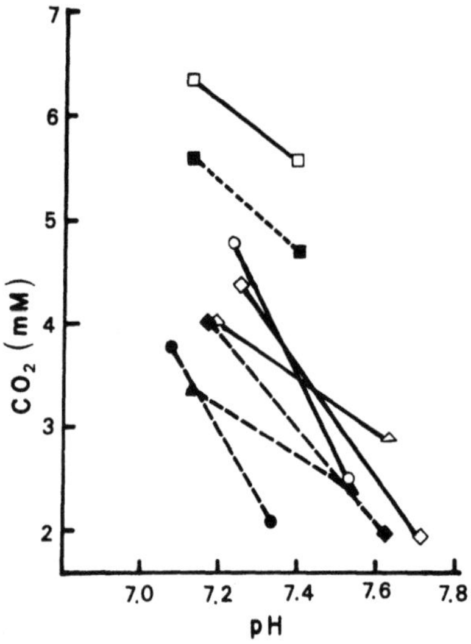

Figure 2. Carbon dioxide binding by squid Hcs. Closed symbols and solid lines show HcO_2 and open symbols and dashed lines show deoxyHc. Squares, *Loligo pealei* (Redfield *et al.*, 1926; 23 °C). Other sources and temperatures as in Fig. 1. Diamonds, *Loligo forbesi*; circles, *Sepiotheuthis lessoniana*; triangles, *Nototodarus sloani philippinensis*.

data obtained at the collection site immediately after sampling and therefore never frozen. The P_{50} values are essentially identical indicating no effect of freezing, which has also been found for octopod Hcs (Miller, 1989). In addition, Lykkeboe and Johansen (1982) used exogenous carbon dioxide to control pH whereas I used Tris maleate buffered saline, indicating little or no difference between the two.

The two tropical species have surprisingly similar HcO_2 affinities considering their taxonomic separation into different suborders. Comparison of the tropical squid Hcs with those of their cold temperate and boreal relatives is difficult due to the differences in experimental temperature. At the moment it is safest to conclude that, unlike crustacean Hcs (Mangum, 1982), latitudinal adaptation among squid Hcs is not clear. The Bohr shifts, −1.09 for *Sepiotheuthis lessoniana* and −1.16 for *Nototodarus sloani philippinensis* Hc are similar to that of *Loligo pealei* Hc. In other respects, however, the respiratory properties of the two tropical squid bloods must be quite different from one another. *Sepiotheuthis lessoniana* has a much higher HcO_2 carrying capacity (1.93 mM) than *Nototodarus sloani philippinensis* (0.74 mM) (Lykkeboe and Johansen, 1982). The pH dependence of Hc cooperativity differs in the two tropical species and in *Loligo pealei* (Fig. 1). At high pH the cooperativity of *Nototodarus sloani philippinensis* Hc becomes surprisingly small, for reasons that are totally obscure since its respiratory physiology is unknown. I have recently found little or no cooperativity of *Loligo brevis* Hc, in this case using previously frozen samples. A systematic examination of the effects of freezing on a wide range of cephalopod Hcs would be useful. Finally, the carbon dioxide equilibrium curves appear to be quite different, with the binding capacity much lower in the tropical species than *Loligo pealei* and the slope much greater in *Sepiotheuthis lessoniana* than the other two (Fig. 2). However, recent data of Pörtner (1989) on *Loligo pealei* suggest less difference than indicated in Fig. 2.

Most surprising results were reported for oxygen binding of the Hc of the giant boreal squid *Architeuthis monachus* (Brix, 1983). Like most other cephalopod Hcs, this molecule has a low oxygen affinity and it is highly cooperative (Fig. 1). The temperature dependence or the apparent heat of oxygenation or enthalpy (ΔH) of the oxygen reaction appears to be greater (about −18 kcal/mole in the interval 10–15 °C, pH 7.3) than that of *Loligo pealei* Hc (about −8 in the interval 15–25 °C, at the same pH; calculated from Redfield and Ingalls, 1933). However, the thermal sensitivity of most Hcs, like other oxygen carriers, generally varies inversely with temperature (Miller and Van Holde, 1981; Mauro and Mangum, 1982a, 1982b; Burnett *et al.*, 1988) although a few exceptions have been reported (Sanders and Childress, 1988). The truly unique feature of the data for *Architeuthis monachus* is the reversed Bohr shift at high pH (Fig. 1), which might be expected to lie within the physiological range in this boreal species. As indicated above, Redfield and Ingalls (1933) found no indication of a reversed Bohr shift of *Loligo pealei* Hc despite a careful search in a pH range greatly in excess of the physiological one. Nor did any of the other investigators cited here. Since he did not emphasize it in his report, Brix (1983) may regard the finding as uncertain. He was also careful to note that the animal had been dead for two days when he took his sample. In any event it will be of considerable interest to enlarge the data base on the respiratory properties of squid Hcs, including a re-examination of this one when possible.

Using thawed samples, Brix *et al.* (1989) obtained oxygen binding data for two squids, *Loligo vulgaris* and *Todarodes sagittatus*. *Todarodes sagittatus* Hc, described in detail, shows a typically large Bohr shift (−0.9 to −1.2) and typical temperature dependence (ΔH = −9 kcal/mole). Perhaps of most interest, if the points were shown in Fig. 1, they would be very similar to those for *Loligo pealei* and *Illex illecebrosus* at the very same temperature (Pörtner, 1989). Like crustacean Hcs and unlike some of the heme proteins (Mangum, 1983b, 1985), squid Hcs may prove to be fairly conservative with respiratory properties of the majority falling within a relatively narrow range.

2.2. Other cephalopod Hcs

2.2.1. Gas transport

Because the squid sample is so small, respiratory properties of other cephalopod Hcs are summarized here. In general, all other known cephalopod Hcs have low oxygen affinities at their physiological temperatures and in the physiological pH range (Fig. 3). Latitudinal adaptation is not always present. At 20 °C, HcO_2 affinity is much lower in the Mediterranean *Octopus vulgaris* than in the cold temperate zone *Octopus dofleini* (Houlihan *et al.*, 1982; Miller, 1985). In the tropical *Octopus macropus*, however, HcO_2 affinity is somewhat higher than in *Octopus dofleini*.

All other cephalopod Hcs are cooperative, usually highly so, and cooperativity is strikingly pH dependent. In all but *Nautilus pompilius*, HcO_2 affinity is highly pH dependent (Fig. 3); it is thus tempting to suggest that this feature, which certainly enhances oxygen transport, arose within the cephalopods. While it is the pH dependence of P_{50} that continues to receive the most attention from physiologists, the pH dependence of cooperativity is also a very important feature (see Section 3.2).

Nautilus pompilius Hc is also insensitive to NaCl (Bonaventura *et al.*, 1981) while *Octopus vulgaris* (Tallandini and Salvato, 1981) and *Octopus dofleini* (Miller and Van Holde, 1985) Hcs respond to NaCl, as well as to $MgCl_2$. Inorganic ions and H+ have opposite effects on these as well as other Hcs: NaCl and $MgCl_2$ raise oxygen affinity while H^+ greatly lower it. The inorganic ion insensitivity of *Nautilus pompilius* Hc may well be related to its very small pH dependence. As indicated above, inorganic ions

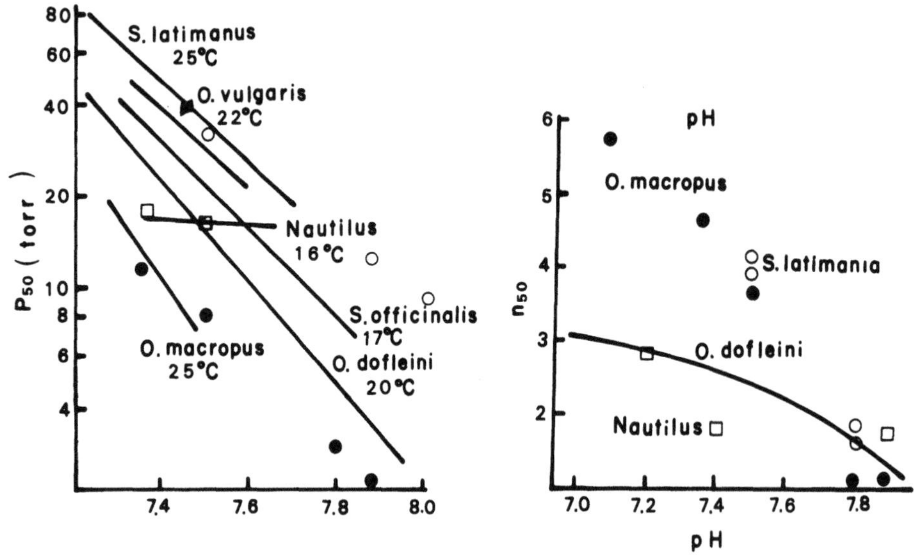

Figure 3. Oxygen binding by other cephalopod Hcs. Solid lines reproduce data from the literature, as cited in Fig. 1 and Table I. Symbols show data collected by the author, as in Fig. 1.

are not likely to be important physiological effectors of HcO_2 binding in most cephalopods.

Only two attempts to identify other effectors have been made, both somewhat confusing. Wolf and Decleir (1982) reported that both adenosine-5'-triphosphate (ATP) and guanosine 5'-triphosphate (GTP) lower the oxygen affinity of purified *Sepia officinalis* Hc, apparently in the absence of inorganic ions other than the HCl used to adjust pH. Under these conditions the Hc is virtually uncooperative and its oxygen affinity is higher than that of whole blood (Brix *et al.*, 1981). Its aggregation state must be open to question. Moreover, the physiological relevance to an extracellular oxygen carrier is unclear; no attempts to repeat the experiment have been reported. Miller and Mangum (1988) found that *Octopus dofleini* Hc purified by sedimentation, which does not alter oxygen binding, and redissolved in its native plasma exhibits a smaller response to carbon dioxide than purified Hc in a physiological saline. At most pH values there is also a lower oxygen affinity for Hc in plasma. While the change may be due to carbon dioxide binding of the divalent cations (Miller and Mangum, 1988), it could conceivably involve an unidentified effector in native blood.

Gastropod bloods do not appear to contain unidentified effectors. Native respiratory properties can be restored to stripped Hcs by adding inorganic ions in the correct ratios (Mangum and Lykkeboe, 1979) and hypoxic exposure induces no change (Mangum, 1983). I am willing to hazard a guess that cephalopod Hcs will also prove to be insensitive to organic effectors but the available information is simply inadequate to warrant a definitive conclusion.

2.2.2. pH dependence: a special case

Gunnar Lykkeboe and co-workers (Lykkeboe *et al.*, 1980; Brix *et al.*, 1981; Lykkeboe and Johansen, 1982) have recently drawn attention to a seemingly paradoxical property of sepioid blood that mandates reconsideration of one of Redfield's favorite conclusions, that the function of the large Bohr shift is primarily to enhance oxygen unloading to the tissues. The argument is somewhat complex even for respiratory physiologists and therefore it is stated below in intentionally simplified terms.

A Bohr shift involves an oxygen-linked binding of H^+ to the carrier. That reaction is expressed quantitatively as the Haldane coefficient, the change in Hc bound H^+ relative to the change in Hc bound oxygen. A Haldane shift is an oxygen linked binding of carbon dioxide to the carrier. In a fully aerobic animal it is the carbon dioxide that generates the H^+. Therefore coefficients describing the interdependent Bohr and Haldane effects should be the same. Moreover, unless the respiratory quotient (the ratio of the amount of carbon dioxide produced to the amount of oxygen consumed) is very different from 1, a Bohr shift, indicated by a Bohr coefficient more negative than -1, will not be beneficial to the animal, because the carbon dioxide produced will not generate enough H^+ to take advantage of this shift. Why would a Hc be so pH dependent that Bohr coefficients are commonly smaller than -1?

Measurements of the effect of carbon dioxide on blood pH provide some answers. While the Haldane effect on *Sepia officinalis* Hc is clear at high PCO_2, it disappears at 0.7 torr and therefore equimolar solutions of oxy- and deoxyHc should have the same pH. But in fact the oxyHc solution is still more acid than the deoxyHc solution, just as it is when the Haldane effect is present. In other words, at the same total CO_2 level, more H+ are in free solution and less are bound when oxygen is bound. So total CO2 must consist of more of the un-ionized species when oxygen is bound and more ionized carbon dioxide when oxygen is not bound. In addition, when carbon dioxide is washed out, the pH of the oxyHc solution increases more than the pH of the deoxyHc solution. So there must be some carbon dioxide in the oxyHc solution, not present in the deoxyHc solution, that does not increase H^+ activity. DeoxyHc behaves differently from oxyHc because this carbon dioxide fraction binds only to oxyHc and it becomes molecular carbon dioxide, doing what it should do to pH, when oxyHc becomes deoxyHc.

That component of total CO_2 is different from the fraction that is a physiologically meaningful Haldane effector and a meaningful Haldane coefficient should not include it. When it is subtracted, the functional Haldane coefficient approximates unity. This implies that no appreciable change in pH occurs when the blood passes through the microcirculation in the tissues. Indeed, if the affinity of deoxyHc for H^+ is so great, small increases in pH would be expected, which in fact are sometimes found in blood returning from the tissues (see Section 4.4).

Is a pH decrease and therefore Bohr proton-induced unloading even possible? Yes, but only when the oxygen-linked carbon dioxide fraction is great enough to generate a functional Haldane coefficient well below 1. Often this situation does occur in *Sepia officinalis* but not *Loligo forbesi*. Pörtner (1989) has concluded that oxygen linked carbon dioxide binding by *Loligo pealei* Hc also does not occur; Miller and Mangum (1988) reached the same conclusion for octopod Hc.

Brix *et al.* (1981) and Lykkeboe and Johansen (1982) argued that the primary physiological significance of the big Bohr shift of sepioid Hc may be branchial and temporal rather than tissue and spatial. In other words the pH difference between the two sites of gas exchange is less important than the difference at the gills of normoxic and hypoxic animals. It is now known that sepioids and octopods respond to hypoxia by hyperventilating (see Section 4.4), bringing about a respiratory alkalosis that elevates blood oxygen affinity and thus facilitates oxygen uptake from hypoxic water. While this suggestion is strongly supported by recent findings, it does not eliminate a comparable function as the blood passes the tissues. As in gastropods (Mangum and Polites, 1980), the venous reserve in cephalopods may be very small (see Section 4). The reversed Bohr shift of the gastropod Hcs and the pH rise due to the excessive normal Bohr shift of some cephalopod Hcs may actually serve to minimize deoxygenation by raising HcO_2 affinity, thus preventing further depletion of the oxygen reserve, in gastropods for the target organs further along in the serially arranged cardiovascular system or in cephalopods for high levels of locomotion. Once again, the same reasoning that seems so obvious for a mammalian system may not be general. Finally, Pörtner (1989) has suggested that a consideration of the carbon dioxide arising from direct oxygen uptake and consumption in the vast amount of squid tissue that lies within the limiting diffusion distance of the ambient water may substantially alter an interpretation assuming that all carbon dioxide in the blood arises from the metabolism of blood transported oxygen (see section 3.1).

The pH dependence of cooperativity is also an important respiratory property. Within the physiological pH range the Hill coefficient n varies sharply and inversely with pH (Miller, 1985). At pH 7.61, the highest value found in *Octopus vulgaris* (Houlihan *et al.*, 1982), n for *Octopus dofleini* Hc is a little above 2, while at 7.38, the lowest value, n rises to almost 3. The potential consequences of this change are shown in Fig. 4. At pH 7.61 and 22 °C P_{50} of *Octopus vulgaris* Hc is about 20 torr. If its cooperativity is the same as that of *Octopus dofleini* Hc and if postbranchial blood PO_2 were between 30 and 40 torr, which must occur in less extreme hypoxia than the level examined by Houlihan *et al.* (1982), its Hc would be 10% less oxygenated than if n were almost 3. Similarly, at pH 7.38, P_{50} is about 50 torr. If prebranchial blood PO_2 were 20–30 torr, which must also occur during moderate hypoxia, its Hc would be about 10% less oxygenated at the tissues at native cooperativity than if n were only a little greater than 2. The pH sensitivity of cooperativity, then, should impair oxygen uptake at the gill of hypoxic animals and enhance utilization of the venous reserve.

3. Molecular properties of cephalopod Hcs

The molecular basis of some of the respiratory properties described above is becoming clear, at least in part. In addition, an understanding of the structure of these molecules should eventually clarify their evolutionary relationships. Finally, progress is being rapidly made in elucidating aspects of biosynthesis. A somewhat oversimplified discussion of pertinent molecular features is given below in hopes of promoting awareness on the part of investigators working at different levels.

3.1. Biosynthesis

Strong morphological evidence that the cephalopod Hcs are synthesized in the branchial glands was presented by Dilly and Messenger (1972), who found Hc-like molecules in intracellular vacuoles. The definitive evidence was obtained by isolating the mRNAs for Hc from the branchial glands of the squid *Loligo vulgaris* (Vanderbeke

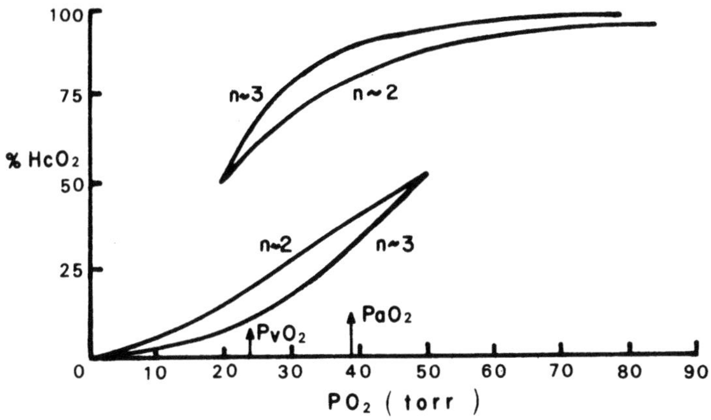

Figure 4. Potential consequences of the pH dependence of cooperativity of cephalopod hemocyanins, using *Octopus* as an example. See text for explanation.

et al., 1981). Moreover, injection of messengers from *Loligo forbesi* and *Sepia officinalis* into oocytes of the toad *Xenopus laevis* induces the synthesis of material that behaves immunologically and electrophoretically like the Hc monomers of these species (Vanderbeke *et al.*, 1982). Thus the membrane bound polyribosomes and the mRNAs are quite large and the basic product of primary synthesis from amino acids is the monomer (Préaux *et al.*, 1986). This is of interest because of its huge size and complexity (see Section 3.2).

3.2. Molecular structure

Probably due to often higher Hc concentrations, the blue color of the blood and the respiratory function of the molecule responsible for it was recognized in molluscs before arthropods. According to Lontie (1976), the term *hemocyanin* was first applied to the cephalopod molecule, by Leon Fredericq in 1878. A sample of his crystals, the first ones prepared, still exists in a display case in the University of Liège. The chronology may introduce some difficulties for biochemical nomenclature. For the molluscan Hcs clearly have in common with the arthropod Hcs only the binuclear copper composition of the active site; in all other respects, they are so different that their independent origin is a virtual certainty.

Arthropod Hcs are built, like many macromolecules, as multiples of loosely linked polypeptide chains, about 60–80 kD (kiloDaltons), each with a single active site (for review see Van Holde and Miller, 1982). In all cases at least two and usually more than two different chains are found. The heterogeneity is believed to be important in assembling the native polymer (Markl, 1986), a physiologically important process (Snyder and Mangum, 1982; Mangum, 1985). All known molluscan Hcs are built from truly gigantic (300–450 kD) polypeptide chains, each monomer having 7–8 covalently linked binuclear Cu sites. The ratio of active site to polypeptide of a molluscan Hc is about 2/3 that of an arthropod Hc, so a molluscan Hc is not just a group of arthropod monomers stuck together in a different way. Primary sequence data are beginning to emerge for both Hcs. Most probably, little or no structural similarity will be seen at this level as well.

A difference between single and multiple active site monomers is also found among the heme proteins (reviewed by Mangum, 1985). Some investigators designate those built from subunits with a single oxygen binding domain as hemoglobins and those composed of subunits with multiple active sites as erythrocruorins, which is unfortunate because it violates the original definition of the terms and also, in this case, implies a fundamental dichotomy where none may exist (Keilin and Hartree, 1951; Terwilliger and Terwilliger, 1984).

However, the dichotomy between the arthropod and molluscan Hcs is fundamental. Should the hemoglobin-erythrocruorin distinction set the stage, it would be the arthropod molecules designated hemocyanins while the molluscan ones would be termed cyanocruorins, despite the historical precedent. In the present contribution the term hemocyanin is used to describe both, with the implicit recognition that the category is not phylogenetically natural.

Sepioid monomers contain 8 O_2 binding domains (Gielens *et al.*, 1983) while octopod monomers have only 7 (K. I. Miller, in Van Holde, 1983; Gielens *et al.*, 1986; Lamy *et al.*, 1986). The Cu content of the cephalopod (and other molluscan) chains appears to be fairly constant, about one pair per 50,000 polypeptide by mass, while the sedimentation coefficients vary quite a lot, from 3.3 to 4.2 kD (Van Holde and Miller, 1984). The value (330 kD) for nautilus Hc (Bonaventura *et al.*, 1981) suggests 7 O_2 binding domains. The value for *Loligo pealei* Hc (Van Holde and Cohen, 1965), the earliest in the modern literature, is 385 kD which suggests 8. The question is of considerable evolutionary and physiological importance, since the larger number clearly reflects a greater oxygen binding capacity per osmotically active particle.

The cephalopod (and other molluscan) Hcs also differ in subunit heterogeneity. While the few octopod, sepioid and teuthoid Hcs examined appear to consist of only one type of subunit (Van Holde and Miller, 1982; Brenowitz *et al.*, 1986), *Nautilus pompilius* consists of three electrophoretically separable chains, differing in oxygen affinity by a factor of more than 2 in the physiological pH range (Bonaventura *et al.*, 1982). Moreover, the various active domains within a chain have oxygen affinities differing by a factor as little as about 2 (*Nautilus pompilius*) and as great as about 10 (*Octopus dofleini*). The difference in the octopod molecule, which lacks subunit heterogeneity, must reflect heterogeneity within the monomer, not between different monomers. This point is of physiological interest because the rearrangement of the proportions of functionally distinct subunits is a mechanism of adaptation of HcO_2 transport in the arthropods (Mangum and Rainer, 1988; deFur *et al.*, 1988). Only if heterogeneity is present could the mechanism be available in the molluscs. Many, if not all, molluscan Hcs may simply be intrinsically unadaptable.

Miller and Van Holde (1989) have recently shown that at least one of the 7 O_2 binding domains of octopus Hc has a lower oxygen affinity and lower Mg^{+2} and H^+ sensitivity than the native polymer. What is perhaps of greatest interest is that this fragment of the monomer is sensitive to Mg^{+2} at all; it is not cooperative.

Finally, the cephalopod Hcs differ somewhat in their association-dissociation behavior. The removal of divalent cations is necessary for dissociation of Hcs, as is raising pH above a threshold value. While this value is very high (8.5) for *Nautilus pompilius* Hc, it is only 7.6 for *Loligo pealei* and 7.0 for *Octopus dofleini* (Van Holde and Cohen, 1965; Bonaventura *et al.*, 1981; Van Holde and Miller, 1985). Although pH values in the range 7–8 are clearly physiological, the inorganic ion conditions for dissociation are not. All known cephalopods, including squids, are osmotic conformers to the ambient medium. Few, however, are even moderately euryhaline and the interesting exception, *Loligo brevis*, is a strong regulator of Ca^{+2} if not NaCl and Mg^{+2} (Mangum, 1981). Strong divalent cation regulation would seem to be a requirement for euryhalinity in coleoids, for respiratory rather than osmotic reasons. Nonetheless it will be interesting to learn the conditions necessary for dissociation of the Hc of this euryhaline species.

The assembled molecule found in the blood of squids as well as all other cephalopods known is a decamer, a multiple of 10 of the huge polypeptide chains. Larger multiples are typical of gastropod and bivalve Hcs; the smaller particles occasionally observed are believed to be artifacts. Again, this property is of considerable importance *per se*, especially in cephalopods. If the number of O_2 binding sites is held constant, their presence in a small multiple enhances the number of osmotically active macromolecules in the blood and thus its colloid osmotic pressure (Snyder and Mangum, 1982; Mangum, 1986). In most animals blood colloid osmotic pressure is important primarily in water balance with the ambient medium but in cephalopods, with two extracellular fluid compartments, it must play a role in tissue hydration as well.

Negatively stained squid Hcs appear in electron micrographs as hollow cylinders about 35 nm wide and 17 nm high (see van Bruggen *et al.*, 1981) for a complete description). In side view the cylinders have three rows of substructures, the latter believed to represent rings of the individual O_2 binding domains. When viewed from either end the cylinders appear as closed tubes. There is also a so-called collar region, an inner ring of a smaller diameter than the outer three. In this view there is also a fivefold and, in the case of *Loligo pealei*, often a tenfold rotational symmetry around the axis of the cylinder. The shape of this molecule is also of potential physiological importance. A compact shape of an extracellular macromolecule minimizes blood viscosity and thus the cardiovascular workload. For example, the dissociated monomers of gastropod Hc have a higher viscosity than the native polymers (Snyder and Mangum, 1982). One would expect the less compact shape of cephalopod (35 × 17 nm) than typical gastropod (35 × 38 nm) polymers to confer greater viscosity. Unfortunately, no measurements are available.

Some information is also available about the consequences of molecular integrity for oxygen binding. Bonaventura *et al.* (1977) showed that fragments of *Octopus dofleini* Hc obtained by proteolysis with subtilisin, which should represent the individual functional domains, differ in oxygen affinity from one another, as described above. Of pertinence here, the average oxygen affinity (weighted by % composition of the 4 fragments cleanly separated on an ion exchange gel column) is 15 torr at pH 8 and 20 °C, in the presence of ethylenediamine tetraacetic acid (EDTA); in contrast, the value for the native polymer at the same pH and temperature (but also in a physiological saline) is about 3.1 torr (Miller and Van Holde, 1986). Fragments from an unidentified octopus show a decrease in oxygen affinity in response to NaCl, which is surprising because the oxygen affinity of other cephalopod Hcs increases in response to the addition of inorganic ions (see Section 2.1). The fragments also have a large Bohr shift in the physiological pH range.

A 50 kD component of *Octopus vulgaris* Hc obtained by trypsin digestion of the monomers also exhibits a normal Bohr shift (Ricchelli *et al.*, 1986). But the Bohr shift is very small in the physiological pH range. Consequently, the putative single domain fragments may have either a much higher (at pH 7) or lower (at pH 8) oxygen affinity than the native polymer. In addition this component has lost its cooperativity in full. Regrettably, in these experiments it is necessary to employ different experimental conditions of inorganic ions to either maintain the integrity of the native polymer or to keep the digestion products from reassociating. Therefore it is not possible to attribute the changes to size or association alone and not to the actions of allosteric effectors such as Ca^{+2}.

Octopus dofleini Hc dissociates in full to monomers or to a mixture of monomers and dimers, depending on the pH (Van Holde and Miller, 1985). Regardless, the native molecule is more cooperative and has a higher oxygen affinity and greater pH dependence than dissociation products (Miller and Van Holde, 1986). It is perhaps more interesting that the greatest difference in this case is in P_{50}. The monomers and dimers are almost as cooperative as the native polymer and their Bohr shift is still extremely large; the Bohr coefficient equals −1 as opposed to −1.7 for the native polymer at 20 °C. Again, the different ionic conditions required to keep the polymers intact and the dissociation products from reassembling preclude a simple interpretation of these results in physiological terms. One can conclude, however, that native size is not necessary for a substantial fraction of two critical respiratory properties, cooperativity and the pH dependence of oxygen affinity.

4. Performance of the oxygen transport system

A large body of information on *in vivo* respiratory variables has enabled us to understand in considerable detail why the brachyuran crustacean system, for example, is designed as it is and what its limitations are (e.g., Mangum, 1980, 1983, 1985; McMahon, 1985, 1986). We have little comparable knowledge of squid physiology and information is available for only a few other cephalopods.

4.1. Oxygen transport in *Loligo pealei*

Even though the respiratory properties of *Loligo pealei* Hc were not as exotic (in mammalian terms) as others he had investigated, it was this oxygen transport system that Redfield chose to study *in vivo*. Like Parsons and Parsons (1924), who had merely wanted to draw blood from animals regarded as *troublesome*, Redfield and Goodkind (1929) nailed their experimental animals to a board, surgically exposed the systemic heart and vigorously ventilated the gills and mantle with running seawater, simulating the ambient PO_2 in the water column. The artificial ventilation probably did simulate the naturally normoxic conditions though not the rhythmicity of water flow. While they did not determine Hc oxygenation, it is clear from oxygen content data that the samples taken from the branchial vessels were highly oxygenated (Table I). Using the equation given in Redfield and Goodkind's Table II, the arterial pH equals the first ionization constant of carbonic acid plus 1.21 and assuming that this first ionization constant is 5.975 at 23 °C (Howell, 1989), Howell and Gilbert calculated that at an implied temperature of 23 °C that the arterial pH was only 7.185. Henderson (1928) also gave a value of the arterial pH of 7.251, which was obtained from Redfield. As a result, Howell and Gilbert (1976) suggested that Redfield's animals were highly acidotic, which would not be very surprising given the treatment accorded them. Pörtner (1989), however, also calculated pH values from Redfield and Goodkind's (1929) results and reached quite different conclusions. His values, given in Table I, are in fact fairly close to the measured ones at that temperature, e.g., 7.37 for postbranchial blood pH (Mangum and Shick, 1972) and 7.42 for prebranchial blood (Howell and Gilbert, 1976).

Redfield and Goodkind (1929) presented convincing data that fulfill one of Redfield's original goals, namely to show that a low oxygen affinity molecule can become highly oxygenated at the gill. Moreover, they concluded that the large Bohr shift brought about by the influx of carbon dioxide from the tissues into the blood is responsible for one third of the total amount of oxygen supplied to the tissues. Finally, they showed that death occurs when HcO_2 decreases to a point where oxygenation at the gill is very small. Curiously, they also concluded that squids made hypercapnic, which presumably would lower their blood oxygen affinity, would enable these organisms to survive a lower water PO_2 than normocapnic animals since more oxygen would be delivered to the tissues. In view of the already small venous reserve, it is difficult to understand how this could be possible. Their Fig. 2 appears to indicate quite the reverse.

Pörtner (1989) has reminded me that, like the mammalian (and also the gastropod) one, the cephalopod cardiovascular system is a dual circulation, in this case consisting of cephalic and pallial components. Using cannulated animals he found that the venous reserve remaining after cephalic circulation is not severely depleted (*ca.* 40% of the postbranchial value), and suggests that the much lower values repeatedly found in the literature result from virtually complete exploitation of the oxygen supply by the powerful mantle musculature, which is responsible for the awesome locomotor capabilities of the animals. Finally, Pörtner suggests that such exploitation may be possible only because the combustion of oxygen that reaches the mantle mitochondria without passing first into the blood generates carbon dioxide that does find its way into the blood, thus lowering pH and consequently oxygen affinity. In support of this hypothesis he cites the finding of aerobic muscle fibers in the superficial regions of the mantle, as opposed to the anaerobic fibers found only slightly beneath.

This is certainly an interesting idea that warrants further consideration. While the rate of oxygen consumption by isolated strips of superficial mantle tissue is not high relative to that of other tissues (Mangum and Towle; 1982), the integument is hardly the inert barrier often found in highly terrestrial animals.

4.2. Oxygen transport in octopus

More than 35 years were to elapse until the next investigation of the performance of a cephalopod oxygen transport system, which was undertaken on *Octopus dofleini* by Kjell Johansen and Claude Lenfant (Johansen, 1965; Lenfant and Johansen, 1965; Johansen and Lenfant, 1966). Using the same individuals, they made numerous paired observations of respiratory parameters ranging from branchial ventilation and oxygen uptake to blood gases and pH to *in vitro* HcO_2 binding. The investigation represents a major advance in understanding HcO_2 transport because of the large number of independently and directly measured *in vivo* variables and because it introduced the technique of taking anaerobic blood samples from cannulated, free-ranging animals. In this case the cephalic aorta and the vena cava cephalica were both cannulated without obstructing blood flow, thus allowing the simultaneous sampling of post- and prebranchial blood. While they noted that the animals were sampled only after full recovery from the anesthetic, the actual period was not specified.

Like Redfield and Goodkind (1929), they found that the Hc is highly oxygenated at the gill and that the venous reserve is very small (Table I). Thus oxygen extraction from the blood is very great, as is its efficiency. The size of the venous reserve is even more remarkable in view of the very low experimental temperature. While latitudinal adaptations of the Hcs can occur in other Hc-containing groups, they are not great enough to result in identical oxygenation states at the sites of gas exchange; specifically, venous reserves tend to be very large in cold water species with their high oxygen affinity bloods (Mangum, 1980; Mauro and Mangum, 1982b). At present there is no reason to suppose that latitudinal adaptations in cephalopods are more perfect.

Despite the efficiency of oxygen uptake into branchial blood, Johansen and Lenfant (1966) emphasized the significant PO_2 difference (averaging 22 torr) between exhalant water and postbranchial blood which, they suggested, may be due to the shunting of a fraction of the blood through non-respiratory portions of the gills, although some of the numerous alternative explanations are mentioned. This question has not been investigated subsequently.

Johansen and Lenfant (1966) also reported high levels of carbon dioxide, which they regarded as uncertain and possibly incorrect because they had obtained their values from a graphic display and extrapolated the numbers from values for standards well above the observed range. At the present writing, their PCO_2 values do seem somewhat high relative to those found in other aquatic animals at low temperatures. But the parameter that is most conspicuously deviant is blood pH. At 8–11 °C the values are well below neutrality (by about –0.2; this would be equivalent to a blood pH at 37 °C of about 6.7). Indeed, the independently determined acid-base variables reproduced in Table I are not internally consistent. If the pairs for pH and total CO_2 are correct, $PaCO_2$ would not be very different from the value in Table I but $PvCO_2$ should be 7.7 torr. If the pairs for PCO_2 and total CO_2 are correct, pHa should be 7.21 and pHv 7.30, still less than the pH at neutrality (pN). Thus it seems likely that these early blood pH values are the ones in serious error. It also seems likely that the animals were, in fact, acidotic though not in the absolute sense (relative to pN). Johansen and Lenfant (1966) concluded that the large changes in pH enhance both oxygen uptake at the gill and delivery to the tissues.

An important addition to our knowledge of gas transport in octopi was recently made by Houlihan *et al.* (1982), using the smaller *Octopus vulgaris* at a higher temperature. Given the different experimental conditions, the general concurrence is notable. Again, *Octopus vulgaris* blood is highly oxygenated at the gill and the venous reserve is small. PCO_2 is somewhat lower than the earlier values but the pH values are the ones that seem to differ the most, and also seem to agree with more recent measurements on the other species (e.g., Table I).

Houlihan *et al.* (1982) also examined the effects of hypoxia. They found that blood PO_2 declines and total CO_2 rises linearly as water PO_2 falls. The decrease in blood PO_2 is offset in part by a concomitant rise in blood pH, which implicates pronounced hyperventilation; the pH difference between the gills and the tissues is maintained. The net result is a decline in oxygenation at both gas exchange sites and a rather small decline in the volume of oxygen supplied to the tissues.

Houlihan *et al.* (1982) made an important though discouraging point that can only undermine our confidence in much of the available information on gas transport in cephalopods: while *Octopus vulgaris* normally regulates its total oxygen uptake almost perfectly in the PO_2 range 70–159 torr, their cannulated animals (used in these experiments shortly after recovery from anesthesia) regulated poorly. While being nailed to a board is almost certainly more stressful than being cannulated, one must conclude that this formidable and, indeed, awe-inspiring procedure does not produce animals which are fully competent in a respiratory sense. More remains to be learned about the performance of the cephalopod oxygen transport system.

4.3. Oxygen transport in *Nautilus pompilius*

On the first Philippines Expedition of R/V *Alpha Helix* Johansen *et al.* (1978) made a number of observations on gas transport in the chambered nautilus. The available information for this species is not as extensive as that for the others and more of the parameters have been calculated rather than directly measured (Table I). Blood pH appears to be rather low, given the experimental temperature, as in most of the other species, and the post-prebranchial difference is small. The most conspicuous and perhaps primitive features of this Hc are its lower concentration in the blood, resulting in a lower oxygen carrying capacity (Table I), and its small Bohr shift (Fig. 1). Changes in blood pH in any dimension can have little importance (see Johansen *et al.*, 1978). Once again, postbranchial blood PO_2 is high and Hc oxygenation is essentially complete. Most notably, the venous reserve is large, as in most animals except cephalopods. Even if cannulation proves to cause acidosis, this phenomenon is likely to be normal. As pointed out by Johansen *et al.* (1978), little is known about physico-chemical variability of the *Nautilus* habitat and not a great deal more is known about its behavior; therefore the function of this reserve is unclear. The low oxygen carrying capacity must diminish the volume supplied to the tissues. However, its metabolic oxygen requirement is also small relative to other cephalopods (Johansen *et al.*, 1978).

It should be pointed out that, while Redmond *et al.* (1978) concluded surprisingly and apparently erroneously (Wells and Wells, 1985) that *Nautilus pompilius* is unable to regulate oxygen uptake in declining water PO_2, the observations were made at a higher temperature and on different animals than those studied by Johansen *et al.* (1978).

4.4. Gas transport in *Sepia officinalis*

The final organismic investigation of oxygen transport in cephalopods was made by Johansen *et al.* (1982) on *Sepia officinalis*. These cannulated animals were held for 24 hr prior to sampling. The authors noted that the cannulation lowered the gas exchange area, although they regarded the impairment as small. Detailed observations on total respiratory regulation were not made. Once again it is clear that in normoxic water the blood is highly oxygenated at the gill (Table I). In this case, however, the venous reserve is somewhat larger than in the other coleoids. Blood pH values are also more in line with expected values at that temperature. Prebranchial blood pH often exceeded postbranchial blood pH, albeit on the average not significantly (*viz.* the 95% confidence interval around the mean difference calculated from their Table I overlaps zero). Exposure to hypoxia (85 min) brings about sharp declines in blood PO_2, offset by a rise in pH, much greater in post- than prebranchial blood. The net result is a reduction in oxygenation state at the gill but no clear change at the tissues. The volume supplied to the tissues decreases by one third, while the fraction transported by the carrier also decreases, but still remains quite high (Table I).

5. Methods of investigating gas transport

Methods of anesthetization and catheterization of squid blood vessels are described by O'Dor, Pörtner, and Shadwick (1990). Methods of measuring blood gases and pH have recently been described by Cameron (1978) and Bridges (1981). The coverage of those subjects here is intended to be supplemental, to detail some of my personal experiences and to emphasize features unique to molluscan bloods.

5.1. Measurement of blood gases

All electrometric procedures require equilibration between an electrode and a sample. I strongly recommend the use of a chart recorder to ascertain the end point.

Accurate values for total HcO_2 capacity can be obtained from Hc concentration measured by absorbance. Nickerson and Van Holde (1971) reported the specific extinction (E, 1 cm, 1%) of *Loligo pealei* Hc at 345 nm as 2.79. The calculation is valid because the optical density at this wavelength is a property of the active site; thus a non-functional precursor or derivative will not absorb in that region with the same spectrum. However, it is valid *only* if the molecule is dissociated to its monomeric sub-units; otherwise the absorbance will include the scattering of light by the large Hc particles. Some light scattering by the 450×10^3 monomer may still occur, though it should not distort the results nearly as seriously as the intact decamer. Dilution of squid blood by about 1:20 with 0.05 M Tris buffer (pH 8.9) containing 0.05 M EDTA dissociates the native polymer and produces an optical density in the most sensitive range.

Total oxygen concentration of post- and prebranchial blood is often calculated from oxygen equilibrium curves, if determinants such as PO_2, pH and inorganic ions are known at the experimental temperature. Alternatively, it can be measured (Cameron, 1978; Bridges, 1981). Another clinical instrument still in wide use employs an inert scrubber gas to elicit and carry the oxygen to a galvanic electrode (Valeri *et al.*, 1972). The sensitivity of the procedure is limited by the range of the electrode, which was designed for mammalian blood. It can be improved by enlarging the sample size and then correcting the result. I have successfully increased it by a factor of five.

Values for PCO_2, total CO_2, and pH (see below) are used to calculate HCO_3^- and (in animals with higher blood pH than squids) CO_3^{+2}, according to the Henderson-Hasselbalch equation in the form:

$$pH = pK' + \log \frac{HCO_3^- + CO_3^{+2}}{\alpha PCO_2} \qquad (1)$$

where pK' is the constant describing the equilibrium between the components of the CO_2 system in aqueous media
and α is the absorption coefficient or solubility of molecular carbon dioxide in aqueous media.

While Cameron (1986) correctly noted that pK' and also α are known only for human and a few other bloods, these constants are also well known for natural waters at a variety of temperatures and solute contents. In our experience (Mangum *et al.*, 1985) there is little or no difference between results obtained using experimentally determined values for crab blood and those from the recent oceanographic literature (Stumm and Morgan, 1981). We have not, however, attempted a comparison at low temperature.

Cameron (1978) and deFur *et al.* (1980) concluded that the components of the carbon dioxide system are not in equilibrium in bloods lacking carbonic anhydrase in the circulation, because the uncatalyzed reactions are so slow. DeFur *et al.* (1980) recommended calculating PCO_2 from total CO_2 and pH, although they suggested that discrepancies arise from the base excess associated with skeletal stores of $CaCO_3$, which is not relevant to squid bloods. Moreover, the status of carbonic anhydrase in squids has not, to my knowledge, been investigated. The problem of carbon dioxide equilibrium warrants further consideration.

The reader may wish to consult the large oceanographic literature on the subject of acid-base determination, which is directly relevant to squid and other bloods: e.g., Pytkowicz *et al.* (1966), Bates and Macaskill (1975), Khoo *et al.* (1977).

Numerous low values for blood pH, which are unlikely to be correct, continue to be reported in the non-respiratory literature. The problem arises from the little appreciated difference between low ionic strength (0.1) calibration buffers and high ionic strength ($\geq$ 0.6 in most squids) bloods. When a reference electrode is transferred from low to high ionic strength, the change in ionic mobility at the salt bridge gives rise to an asymmetry potential at the glass electrode that takes a long time to decay and may require several repetitions of the unstirred sample. Many investigators interpret the subsequent rise to the equilibrated pH value as *drift* and erroneously accept the initial low value. The problem can be circumvented by first equilibrating the electrode to a *throw-away* sample of similar ionic strength as the unknown. These electrodes are highly stable and do not require frequent recalibration.

When the pH of a sample that need not be handled anaerobically is to be determined, e.g., one on which oxygen binding measurements are being made, very accurate pH measurements can be made with Ross-type electrodes, the response time of which is so fast that equilibration is unlikely to be interpreted as *drift*.

Calibrations of and measurements with pH electrodes must be made *at* the experimental temperature. Sensitivity of the electrodes is an important consideration. Another important consideration is the considerable thermal sensitivity of blood pH. Many values in the older literature are undoubtedly incorrect because these values did not take into account this thermal sensitivity. Attention to accuracy of pH determination is especially critical in the investigation of squid oxygen transport because of the very great pH dependence of both cooperativity and oxygen affinity of squid Hc.

5.2. HcO_2 equilibrium measurements

The determination of oxygen equilibria of molluscan bloods and Hcs is relatively easy (although a few special precautions such as attention to pH and inorganic ion conditions are especially critical in squids). The reasons include: 1) Optical homogeneity, once blood cells are removed. Absorbance measurements accurately reflect oxygenation state (Richey *et al.*, 1983), as also indicated by the virtual identity of oxygen equilibrium data obtained by optical and non-optical methods (Mangum and Lykkeboe, 1979; Mangum and Burnett, 1986). 2) Molecular stability. Oxidation is rare and, at least at atmospheric pressure, precipitation occurs only when samples of pure blood cannot be obtained; precipitation may be mitigated by draining impure blood into a cocktail of protease inhibitors (Miller, in Mangum *et al.*, 1986). 3) Independence of mass action principles. Concentration affects neither the oxygen binding constants nor the action of effectors on them. 4) Independence of organic effectors, so far as known. Oxygen binding of stripped gastropod blood can be restored to the native state in full by adding the appropriate inorganic ions (Mangum and Lykkeboe, 1979). Moreover, data for blood from normoxic and hypoxic animals agree fully, thus providing no basis for a hypothesis of unidentified organic effectors (Mangum, 1983).

The importance of inorganic ions, however, cannot be emphasized enough. Cephalopod Hc responds to NaCl as well as divalent cations (see Sections 2, 3), perhaps specifically and allosterically. Inorganic composition of the blood must be known and controlled carefully if physiologically meaningful data are to be obtained.

Purification of molluscan Hc is usually accomplished by gel filtration, although the molecules are so big that sedimentation in an ultracentrifuge is also used.

5.2.1. Tonometry

The modern version of the classical tonometric procedure was described in detail by Riggs (1951) and Riggs and Wolbach (1956). This method should be quite accurate for squid Hcs. When the PO_2 of samples equilibrated as described was measured with a Radiometer PO_2 electrode, I found that the measurements agreed quite well with the calculated values.

The accuracy of the calculated PO_2 is improved if the value for the vapor pressure at the temperature and solute content of the sample is used. Accuracy is also improved if the gas injected is humidified in full rather than its relative humidity estimated by a sling psychrometer.

In my laboratory we use gas mixing pumps to prepare the equilibration atmosphere from highly pure commercial gases, which are scrubbed further before entering the pump, and humidified in wash bottles afterwards. These pumps also drive the resulting mixture, which is far more precise than can be measured by oxygen electrodes, through the tonometers at atmospheric pressure. Thus the sample is never exposed to changes in hydrostatic pressure, which can cause denaturation.

Other modifications of classical tonometric measurements have been made to accommodate smaller samples. The most widely used is the diffusion chamber described by Sick and Gersonde (1969), which originally employed continuously changing ratios of nitrogen gas and dioxygen and thus resulted in a plot from which an infinite number of points could be obtained. To investigate carbon dioxide effects we modified it to a stepwise procedure to ensure equilibration of the carbon dioxide system (Lykkeboe, in Mangum and Lykkeboe, 1979); it is the stepwise procedure that is in the widest use today. While we used a colorimeter with fixed wavelength sensitivities, which did not coincide with the active site of molluscan Hcs, there is no reason why a continuous wavelength spectrophotometer could not be substituted.

Other modifications of tonometric procedures have been made to eliminate the light scattering caused by the presence of red blood cells, which is not relevant here. Two quite elegant procedures, also developed for hemoglobins but surely adaptable to 345 nm with a suitable photometer, have been developed by Gill (1981) and Imai (1981). In both cases PO_2 is measured polarographically with an electrode seated in the sample chamber and oxygenation is registered by the attendant changes in absorbance. Either cell could be used in continuous or stepwise modes, with a little modification.

The fundamentally different and non-optical cell respiration method registers PO_2 and oxygenation continuously and simultaneously (Mangum and Lykkeboe, 1979;

Mangum *et al.*, 1983). A simple description of the method prepared for undergraduate class use is available from the present author on request. Colman and Longmuir (1963) introduced a procedure for examining oxygenation as well as deoxygenation, which was found to be faulty by Hughes *et al.* (1976). This phase of the method is in fact unnecessary.

The potential pitfalls of the cell respiration method arise from: 1) Nonlinearity of the response below P_{100}, due to either the intrinsic properties of the electrode, in which case it can usually be corrected by cleaning with abrasive and/or changing the membrane, or to intrinsic properties of the particles used to lower PO_2. For reasons totally obscure to me, commercial yeast reconstituted from dry preparations often give a nonlinear response, at least in saline, while cells from yeast cakes do not. 2) Far more important, failure to accurately identify P_0 or P_{100}. This occurs most often when the oxygen carrier has the combination of a low oxygen affinity and little cooperativity.

Some of the advantages of the method are irrelevant to squid bloods, such as 1) its suitability for optically heterogeneous as well as homogeneous preparations, 2) ease of use with high concentrations of the oxygen carrier and 3) absence of a gas-liquid interface and thus rarity of precipitation, which 4) in fact does not appreciably alter the result unless it is very great. On the whole this is a forgiving method. More relevant advantages include: 1) Accuracy, especially for estimating cooperativity. 2) Availability of an infinite number of points including high and low oxygenation states, which are often needed for data analysis according to the Monod-Wyman-Changeux model. 3) A continuous increase in carbon dioxide in approximately physiological additions (unless measurements in the absence of carbon dioxide are desired). Finally, 4) Availability of additional information, such as oxygen capacity (Mangum and Burnett, 1986).

The disadvantages are primarily two: 1) The measurement cannot be carried out in the absence of carbon dioxide and thus it is poorly suited to the determination of carbon dioxide effects. 2) Data analysis is tedious and time consuming. While I am presently attempting to circumvent the latter disadvantage by incorporating a modern data acquisition system, the extent to which the manual procedure can be omitted is not yet clear. Finally, while I have never actually attempted to work at hyperbaric oxygen tensions, it would probably require building a calibration chamber.

6. Summary and conclusions

Squid Hcs occur in the blood in high concentrations, which confer a moderately high oxygen carrying capacity relative to many more primitive molluscs. As in other cephalopods, squid Hcs transport all but a negligible fraction of the oxygen consumed from blood. While the molecule has a low oxygen affinity at physiological pH, it also has a very large Bohr shift. Many years ago Redfield and his co-workers reported that almost all of the oxygen transported to the tissues was delivered, leaving little remaining as a venous reserve for bursts of motor activity. They concluded that the very great oxygen extraction of the blood by the tissues was due to the extreme normal Bohr shift. Modern experiments using free-ranging animals with cannulated blood vessels indicate that, in many other cephalopods under normoxic conditions, the venous reserve is utilized to an extent rarely found outside of the molluscs (but also found in prosobranch gastropods). Although cannulated animals may not fully represent the natural condition, under at least some experimental conditions the performance of their HcO_2 transport systems resembles that of others.

The very great proton affinity of the molecule sometimes results in increases in pH and, presumably, HcO_2 affinity at the tissues. This seemingly maladaptive consequence may serve to protect a routinely active animal from a too severe depletion of its venous oxygen reserve. In other circumstances carbon dioxide production from the consumption of oxygen in the mantle which is not supplied by the blood may lower blood pH to the point where the venous reserve is dangerously low. The rather large and also highly pH dependent cooperativity of HcO_2 may enhance oxygen delivery, especially during moderate hypoxia, though it must impair oxygen uptake.

Cephalopod Hcs resemble those of other molluscs in that they are cylindrical molecules built of gigantic polypeptide chains, each with several oxygen binding domains. Although important properties such as pH dependence, cooperativity, and the low oxygen affinity reach their maxima only when the native polymer is fully assembled, they may reside in substantial part in the monomeric subunits. Assembly to the native polymer has some respiratory significance. Its physical consequences, in terms of water balance and blood viscosity, are not known.

Regrettably, all but one modern investigation of the *in vivo* performance of the cephalopod HcO_2 transport system have focused on groups other than squids and, at this writing, Pörtner's findings are not in print. It is fairly clear that the primitive cephalopod system, exemplified by *Nautilus*, resembles that of other groups in ways that coleoid systems may not. Thus a considerable amount of evolutionary refinement of HcO_2 transport appears to have occurred within the Class Cephalopoda.

Towards the end of his active scientific career Redfield (1968) wrote an interesting history of his all too brief participation in respiratory physiology and noted *a distinct lull in work on hemocyanin* (p. 3), which he hoped was at an end. At present it is sadly correct to say that, while considerable progress in understanding other cephalopod Hcs has been made since then, understanding oxygen transport in squids is just beginning.

ACKNOWLEDGMENTS. The author is supported by the Physiological Processes Program of the National Science Foundation (DCM 88-16172). I thank the many colleagues who sent me their preprints and reprints. I am especially grateful to H. O. Pörtner, who not only shared with me his unpublished findings but also found a number of errors in the original typescript.

References

Bates, R. G. and Macaskill J. B., 1975, Acid-base measurements in seawater, in: *Analytical Methods in Oceanography*, (T. R. P. Gibb, ed.), pp. 110–123, American Chemical Society, Washington D.C.

Bonaventura, C., Bonaventura, J., Miller, K. I., and Van Holde, K. E., 1981, Hemocyanin of the chambered nautilus: structure-function relationships, *Arch. Biochem. Biophys.* **211**:589–598.

Bonaventura, J., Bonaventura, C., and Sullivan, B., 1977, Properties of oxygen-binding domains isolated from subtilisin digests of six molluscan hemocyanins, in: *Structure and Function of Haemocyanin*, (J. V. Bannister, ed.), pp. 206–216, Springer-Verlag, Berlin.

Bridges, C. R., 1981, PO_2 and oxygen content measurement on blood samples; using polarographic oxygen sensors, in: *Polarographic Oxygen Sensors*, (E. Gnaiger and H. Forstner, eds.), pp. 219–223, Springer-Verlag, Berlin.

Brix, O., 1983, Giant squids may die when exposed to warm water currents, *Nature* **303**:422–423.
Brix, O., Bardgard, A., Cau, A., Colosimo, A., Condo, S., and Giardina, B., 1989, Oxygen-binding properties of cephalopod blood with special reference to environmental temperatures and ecological distribution, *J. Exp. Zool. in press.*
Brix, O., Lykkeboe, G., and Johansen, K., 1981, The significance of the linkage between the Bohr and Haldane effects in cephalopod bloods, *Resp. Physiol.* **44**:177–186.
Burnett, L. E., Scholnik, D. A., and Mangum, C. P., 1988, Temperature sensitivity of molluscan and arthropod hemocyanins, *Biol. Bull.* **174**:153–162.
Cameron, J. N., 1986, *Principles of Physiological Measurement*, Academic Press, Orlando, Florida.
Colman, C. H. and Longmuir, I. S., 1963, A new method for registration of oxyhemoglobin dissociation curves, *J. Appl. Physiol.* **18**:420–423.
deFur, P. L., Mangum, C. P., Reiber, C. L., and Reese J. E. F., 1988, Respiratory responses of the blue crab *Callinectes sapidus* to longterm hypoxia, (in review).
DePhillips, H. A., Nickerson, K. W., Johnson, M., and Van Holde, K. E., 1969, Physical studies of hemocyanins. IV. Oxygen-linked dissociation of *Loligo pealei* hemocyanin, *Biochem.* **8**:3665–3672.
Dykens, J. A. and Mangum, C. P., 1979, The design of cardiac muscle and the mode of metabolism in molluscs, *Comp. Biochem. Physiol.* **62A**: 549–554.
Gielens, C., Benoy, C., Préaux, G., and Lontie, R., 1986, Presence of only seven functional units in the polypeptide chain of the haemocyanin of the cephalopod *Octopus vulgaris*, in: *Invertebrate Oxygen Carriers*, (B. Linzen, ed.), pp. 221–222, Springer-Verlag, Berlin.
Gielens, C., Bosman, F., Préaux, G., and Lontie, R., 1983, Structural studies by limited proteolysis of the haemocyanin of *Sepia officinalis*, *Life Chem. Rept.* **Suppl. 1**: 121–128.
Gill, S. J., 1981, Measurement of oxygen binding by means of a thin-layer optical cell, *Methods in Enzymology* **76**:427–437 (E. Antonini. L. Rossi-Bernardi, and E. Chiancone, eds.), Academic Press, New York.
Henderson, L. J., 1928, *Blood. A Study in General Physiology*, Yale Univ. Press, New Haven, Connecticut.
Houlihan, D. F., Innes, A. J., Wells, M. J., and Wells, J., 1982, Oxygen consumption and blood gases of *Octopus vulgaris* in hypoxic conditions, *J. Comp. Physiol.* **148**:35–40.
Howell, B. J., 1989, *personal commun.*
Howell, B. J. and Gilbert, D. L., 1976, pH-temperature dependence of the hemolymph of the squid, *Loligo pealei*, *Comp. Biochem. Physiol.* **55A**:287–289.
Hughes, G. M., O'Neill, J. G., and van Aardt, W. J., 1976, An electrolytic method for determining oxygen dissociation curves using small blood samples: the effect of temperature on trout and human blood, *J. Exp. Biol.* **65**:21–38.
Imai, K., 1981, Measurement of accurate oxygen equilibrium curves by an automatic oxygenation apparatus, *Methods in Enzymology* **76**:438–439, (E. Antonini, L. Rossi-Bernardi and E. Chiancone, eds.), Academic Press, New York.
Johansen, K., 1965, Cardiac output in the large cephalopod *Octopus dofleini*, *J. Exp. Biol.* **42**: 475–480.
Johansen, K., Brix, O., and Lykkeboe, G., 1982, Blood gas transport in the cephalopod, *Sepia officinalis*, *J. Exp. Biol.* **99**:331–338.
Johansen, K. and Lenfant, C., 1966, Gas exchange in the cephalopod, *Octopus dofleini*, *Amer. J. Physiol.* **210**:910–918.
Johansen, K., Redmond, J. R., and Bourne, G. B., 1978, Respiratory exchange and transport of oxygen in *Nautilus pompilius*, *J. Exp. Zool.* **205**:27–36.
Keilin, D. and Hartree, E. F., 1951, Relationship between haemoglobin and erythrocruorin, *Nature* **168**:266–269.
Khoo, K. H., Ramette, R. W., Culberson, C. H., and Bates, R. G., 1977, Determination of hydrogen ion concentrations in seawater from 5 to 40 °C: standard potentials at salinities from 20 to 45‰, *Anal. Chem.* **49**:29–34.

Lamy, J., Lamy, J. N., Leclerc, M., Compin, S., Miller, K. I., and Van Holde, K. E., 1986, Preliminary results on the structure of *Octopus dofleini* hemocyanin, in: *Invertebrate Oxygen Carriers*, (B. Linzen, ed.), pp. 231–235, Springer-Verlag, Berlin.

Lenfant, C. and Johansen, K. 1965, Gas transport by hemocyanin-containing blood of the cephalopod *Octopus dofleini*, *Amer. J. Physiol.* **209**:991–998.

Lontie, R., 1977, On the active site of molluscan haemocyanin and of tyrosinases. Opening address, in: *Structure and Function of Haemocyanin*, (J.V. Bannister, ed.), pp. 150–155, Springer-Verlag, Berlin.

Lykkeboe, G., Brix, O., and Johansen, K., 1980, Oxygen-linked CO_2 binding independent of pH in cephalopod blood, *Nature* **287**:330–331.

Lykkeboe, G. and Johansen, K., 1982, A cephalopod approach to rethinking about the importance of the Bohr and Haldane effects, *Pacif. Sci.* **36**:305–314.

Mangum, C. P., 1980, Respiratory function of the hemocyanins, *Amer. Zool.* **20**:19–38.

Mangum, C. P., 1981, The influence of inorganic ions and pH on HcO_2 transport systems, in: *Invertebrate Oxygen-binding Proteins*, (J. Lamy and J. Lamy, eds.), pp. 811–822, Marcel Dekker, New York.

Mangum, C. P., 1982, On the relationship between P_{50} and the mode of gas exchange in tropical crustaceans, *Pacif. Sci.* **36**:403–410.

Mangum, C. P., 1983a, Adaptability and inadaptability among HcO_2 transport systems: an apparent paradox, *Life Chem. Rept.* **Suppl. 1**:335–352.

Mangum, C. P., 1983b, Oxygen transport in the blood, in: *Biology of the Crustacea* (L. H. Mantel, ed.), pp. 373–429, Academic Press, New York.

Mangum, C. P., 1985, Oxygen transport in the invertebrates, *Amer. J. Physiol.* **248**: R505–R514.

Mangum, C. P., 1986, Osmoregulation in marine and estuarine animals: its influence on respiratory gas exchange and transport, *Boll. Zool.* **53**:1–7.

Mangum, C. P., 1989 *unpublished data.*

Mangum, C. P. and Burnett, L. E., 1986, The CO_2 sensitivity of the hemocyanins and its relationship to Cl sensitivity, *Biol. Bull.* **171**:248–263.

Mangum, C. P. and Lykkeboe, G., 1979, The influence of inorganic ions and pH on the oxygenation properties of the blood in the gastropod mollusc *Busycon canaliculatum*, *J. Exp. Zool.* **207**:417–430.

Mangum, C. P., McMahon, B. R., deFur, P. L., and Wheatly, M. G., 1985, Gas exchange, acid-base balance, and the oxygen supply to the tissues during a molt of the blue crab *Callinectes sapidus*, *J. Crust. Biol.* **5**:188–206.

Mangum, C. P., Miller, K. I., Scott, J. L., Van Holde, K. E., and Morse, M. P., 1987, Bivalve hemocyanin: structural, functional and phylogenetic relationships, *Biol. Bull.* **173**: 205–221.

Mangum, C. P. and Polites, G., 1980, Oxygen uptake and transport in the prosobranch mollusc *Busycon canaliculatum*. I. Gas exchange and the response to hypoxia, *Biol. Bull.* **158**:77–90.

Mangum, C. P. and Rainer, J. S., 1988, The relationship between subunit composition and oxygen binding of blue crab hemocyanin, *Biol. Bull.* **174**:77–82.

Mangum, C. P. and Shick, J. M., 1972, The pH of body fluids of marine invertebrates, *Comp. Biochem. Physiol.* **42A**: 693–698.

Mangum, C. P. and Towle, D. W., 1982, The nautilus siphuncle as an ion pump, *Pacif. Sci.* **36**:273–282.

Mangum, C. P., Terwilliger, R. C., Terwilliger, N. B., and Hall, R., 1983, Oxygen binding of intact coelomic cells and extracted hemoglobin of the echiuran *Urechis caupo*, *Comp. Biochem. Physiol.* **76A**:253–257.

Markl, J., 1986, Evolution and function of structurally diverse subunits in the respiratory protein hemocyanin from arthropods, *Biol. Bull.* **171**:90–115.

Mauro, N. A. and Mangum, C. P., 1982a, The role of the blood in the temperature dependence of oxidative metabolism in decapod crustaceans. I. Intraspecific responses to seasonal differences in temperature, *J. Exp. Zool.* **219**:179–188.

Mauro, N. A. and Mangum, C. P., 1982b,. The role of the blood in the temperature dependence of oxidative metabolism in decapod crustaceans. II. Interspecific adaptations to latitudinal change, *J. Exp. Zool.* **219**:189–196.

McMahon, B. R., 1984, Functions and functioning of crustacean hemocyanin, in: *Respiratory Pigments in Animals,* (J. Lamy, J-P. Truchot, and R. Gilles, eds.), pp. 35–58, Springer-Verlag, Berlin.
McMahon, B. R., 1986, Oxygen binding by hemocyanin: compensation during activity and environmental change, in: *Invertebrate Oxygen Carriers,* (B. Linzen, ed.), pp. 299–320, Springer-Verlag, Berlin.
Miller, K. I., 1985, Oxygen equilibria of *Octopus dofleini* hemocyanin, *Biochem.* **24**: 4582–4586.
Miller, K. I., 1989, *personal commun.*
Miller, K. I. and Mangum, C. P., 1988, An investigation of the nature of Bohr, Root and Haldane effects in *Octopus dofleini* hemocyanin, *J. Comp. Physiol. B*:547–552.
Miller, K. I. and Van Holde, K. E., 1981, The effect of environmental variables on the structure and function of hemocyanin from *Callianassa californiensis*. I. Oxygen binding, *J. Comp. Physiol.* **143**:253–260.
Miller, K. I. and Van Holde, K. E., 1986, Oxygen-linked dissociation and oxygen binding by subunits of *Octopus dofleini* hemocyanin, in: *Invertebrate Oxygen Carriers* (B. Linzen, ed.), pp. 417–420, Springer-Verlag, Berlin.
Miller, K. I. and Van Holde, K. E., 1989, *personal commun.*
Nickerson, K. W. and Van Holde, K. E., 1971, A comparison of molluscan and arthropod hemocyanin. I. Circular dichroism and absorption spectra, *Comp. Biochem. Physiol.* **64A**:433–436.
O'Dor, R., Pörtner, H. O., and Shadwick, R. E., 1990, Squid as elite athletes: Locomotory, respiratory, and circulatory integration, *this volume.*
O'Dor, R. K. and Webber, D. M., 1986, The constraints on cephalopods: why squid aren't fish, *Canad. J. Zool.* **64**:1591–1605.
Parsons, T. R. and Parsons, W., 1924, Observations on the transport of carbon dioxide in the blood of some marine invertebrates, *J. Gen. Physiol.* **6**:153–166.
Pörtner, H. O., 1989, *personal commun.*
Pörtner, H. O., 1990, A graphical presentation of the effects of pH *in vitro* and *in vivo* on oxygen binding by cephalopod haemocyanin, *in preparation.*
Préaux, G., Vandamme, A., de Bethune, B., Jacobs, M-P., and Lontie, R., 1986, Hemocyanin-mRNA-rich fractions of cephalopodan decabrachia and of crustacea, their *in vivo* and *in vitro* translation, in: *Invertebrate Oxygen Carriers* (B. Linzen, ed.), pp. 485–488, Springer-Verlag, Berlin.
Pytkowicz, R. M., Kester, D. R., and Burgener, B. C., 1966, Reproducibility of pH measurements in seawater, *Limnol. Oceanogr.* **11**:417–419.
Redfield, A. C., 1966, Opening address, in: *Physiology and Biochemistry of Haemocyanins,* (F. Ghiretti, ed.), pp. 1–4, Academic Press, London.
Redfield, A. C., Coolidge, T., and Hurd, A. L., 1926, The transport of oxygen and carbon dioxide by some bloods containing hemocyanin, *J. Biol. Chem.* **69**:475–509.
Redfield, A. C. and Goodkind, R., 1929, The significance of the Bohr effect in the respiration and asphyxiation of the squid, *Loligo pealei*, *J. Exp. Biol.* **6**:340–349.
Redfield, A. C. and Ingalls, E. N., 1933, The oxygen dissociation curves of some bloods containing hemocyanin, *J. Cell. Comp. Physiol.* **3**:169–202.
Redmond, J. R., Bourne, G. B., and Johansen, K., 1978, Oxygen uptake by *Nautilus pompilius*, *J. Exp. Zool.* **205**:45–50.
Ricchelli, F., Filippi, B., Gobbo, S., Simoni, E., Tallandini, L., and Zatta, P., 1986, Functional and structural properties of the 50,000 D subunit of *Octopus vulgaris* hemocyanin, in: Invertebrate Oxygen Carriers, (B. Linzen, ed.), pp. 235–239, Springer-Verlag, Berlin.
Richey, B., Decker, H., and Gill, S. J., 1983, A direct test of the linearity between optical density change and oxygen binding in hemocyanins, *Life Chem. Rept.* **Suppl. 1**:309–312.
Riggs, A., 1951, The metamorphosis of hemoglobin in the bullfrog, *J. Gen Physiol.* **35**: 23–40.
Riggs, A. and Wohlbach, R. A., 1956, Sulfhydryl groups and the structure of hemoglobin, *J. Gen. Physiol.* **39**:585–605.
Root, R. W., 1931, The respiratory function of the blood of marine fishes, *Biol. Bull.* **61**: 427–456.

Sanders, N. K., Arp, A. J., and Childress, J. J., 1988, Oxygen binding characteristics of the hemocyanins of two deep-sea hydrothermal vent crustaceans, *Resp. Physiol.* **71**:57–68.
Sick, H. and Gersonde, K., 1969, Method for continuous registration of O_2-binding curves of hemoproteins by means of a diffusion chamber, *Anal. Biochem.* **32**:362–376.
Snyder, G. K. and Mangum, C. P., 1982, The relationship between the size and shape of an extracellular oxygen carrier and the capacity for oxygen transport, in: *Physiology and Biochemistry of Horseshoe Crabs*, (J. Bonaventura, C. Bonaventura, and S. Tesh, eds.), pp. 173–188, Allan Liss, New York.
Stumm, W. and Morgan, J. J., 1981, *Aquatic chemistry*, John Wiley & Sons, New York.
Tallandini, L. and Salvato, B., 1981, Allosteric modulations in the oxygen binding of *Octopus vulgaris* hemocyanin, in: *Invertebrate Oxygen-binding Proteins*, (J. Lamy and J. Lamy, eds.), pp. 727–738, Marcel Dekker, New York.
Terwilliger, R. C. and Terwilliger, N. B., 1983, Oxygen binding domains in invertebrate hemoglobins, *Life Chem. Rept.* **Suppl. 1**:227–238.
Valeri, C. R., Zaroulis, C. G., and Marchoni, L., 1972, A simple method for measuring oxygen content in blood, *J. Lab. Clin. Med.* **79**: 1035–1040.
van Bruggen, E. F. J., Schutter, W. G., van Breeman, J. F. L., Bijholt, M. N. C., and Wichertjes, T., 1981, Arthropodan and molluscan haemocyanins, in: *Electron Microscopy of Proteins*, (J. R. Harris, ed.), pp. 1–38, Academic Press, London.
Van Holde, K. E., 1983, Some unresolved problems concerning hemocyanins, *Life Chem. Rept.* **Suppl. 1**:403–412.
Van Holde, K. E. and Cohen, L. B., 1965, Physical studies of hemocyanins. I. Characterization and subunit structure of *Loligo pealei* hemocyanin, *Biochem.* **3**:1803–1808.
Van Holde, K. E. and Miller, K. I., 1982, Haemocyanins, *Quart. Rev. Biophys.* **15**:1–129.
Van Holde, K. E. and Miller, K. I., 1984, Cephalopod hemocyanins: structure and function, in: *Respiratory Pigments in Animals* (J. Lamy, J-P. Truchot and R. Gilles, eds.), pp. 87–96, Springer-Verlag, Berlin.
Van Holde, K. E. and Miller, K. I., 1985, Association-dissociation equilibria of *Octopus* hemocyanin, *Biochem.* **24**:4577–4582.
Vanderbeke, E., Cleuter, Y., Marbaix, G., Préaux, G., and Lontie, R., 1981, Isolation and translation of a haemocyanin-mRNA-containing fraction from *Loligo vulgaris*, *Arch Int. Physiol. Biochem.* **89**:B135–B136.
Vanderbeke, E., Cleuter, Y., Marbaix, G., Préaux, G., and Lontie, R., 1982, Synthesis of haemocyanin in *Xenopus laevis* oocytes microinjected with the polyadenylated RNA fraction isolated from the gills of *Sepia officinalis* and *Loligo vulgaris*, *Biochem. Int.* **5**:23–29.
Wells, M. J. and Wells, J., 1985, Ventilation volume and oxygen uptake by *Nautilus*, *J. Exp. Biol.* **118**:297–312.
Wolf, G. and Decleir, W., 1981, A study of hemocyanin in *Sepia officinalis*: functional properties of the adult molecule, in: *Invertebrate Oxygen-binding Proteins* (J. Lamy and J. Lamy, eds.), pp. 749–754, Marcel Dekker, New York.

Chapter 21

An Organophosphorus Detoxifying Enzyme Unique to Squid

FRANCIS C. G. HOSKIN

1. Introduction

Three tissues or components of the squid – nerve, hepatopancreas and saliva – contain high levels of an enzyme that hydrolyzes the P-X bond of a group of compounds popularly termed *nerve gases*, where -X is fluoride or cyanide. The history of this particular enzyme has little to do with the squid. That came later.

Mazur (1946) first reported the hydrolysis of a powerful, essentially irreversible cholinesterase inhibitor, diisopropyl phosphorofluoridate (DFP), by an enzyme in rabbit and human tissues. For want of a natural substrate, Mazur declined to assign a name. Others were less cautious, and DFPase was followed by Sarinase (isopropyl methylphosphonofluoridate = Sarin), Somanase (2,3,3-trimethylpropyl methylphosphonofluoridate = Soman), Tabunase (ethyl N,N-dimethylphosphoramidocyanidate = Tabun) (see Figure 1) and so on. As a result of three recent meetings (see Reiner *et al.*, 1989) the name *organophosphorus acid anhydrase* shortened to OPA anhydrase has been tentatively adopted. The term will be used here, even for work entitled by the earlier names, and even when referring to one or another of the several OPA anhydrases.

Mounter (1963) reviewed the enzymatic degradation of organophosphorus anticholinesterase compounds, with special attention to the OPA anhydrases, the plural usage justified then and since by his phrase, "...several hydrolytic enzymes of overlapping specificities". In summary, the OPA anhydrases are ubiquitous with relatively greater amounts in liver and kidney of the vertebrates, and in some bacteria. Usually the activity is stimulated several-fold by Mn^{2+} and sometimes by other divalent cations in the 10^{-3} M range. In hindsight, the ambiguities were probably caused by the use of tissue homogenates and sonicates, or at most only several-fold purifications, initially by cold ethanol fractionation. We now know that this, in turn, was due to ammonium sulfate inactivation and, at that time, a lack of good column procedures.

From about 1945 until his death in 1983, Professor David Nachmansohn vigorously championed a theory of axonal conduction that involved the various components of the generally accepted cholinergic synaptic transmission system (Nachmansohn, 1959). Opposition to such a view was equally vigorous (Katz, 1960). While the theory is now mainly of historical interest, one consequence of predictions

F. C. G. HOSKIN • Biology Department, Illinois Institute of Technology, Chicago, IL 60616.

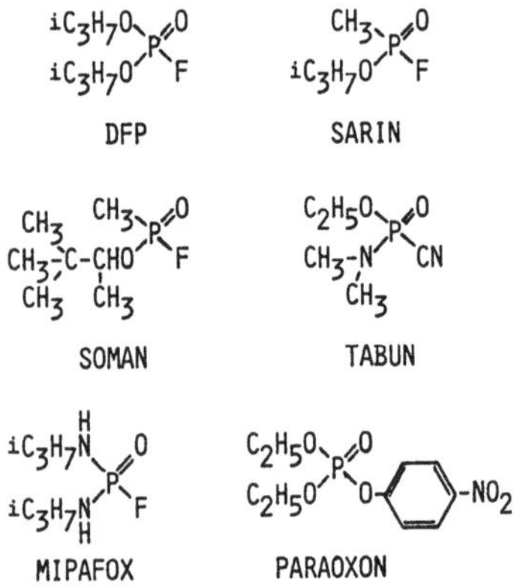

Figure 1. Organophosphorus compounds referred to in text.

of what powerful lipid-soluble acetylcholinesterase (AChE) inhibitors should do to axonal conduction was the discovery of an OPA anhydrase in the squid giant axon (Hoskin *et al.*, 1966), and then in squid ganglia and hepatopancreas from which it is purified (Hoskin and Long, 1972; Garden *et al.*, 1975), in cuttlefish and octopus nerve (Hoskin and Brande, 1973), and in squid saliva (Hoskin and Prusch, 1983). It is found in low levels, if at all, in other tissues of the squid, and not in members of other phyla, or even in other classes of this same phylum. The properties of this enzyme that we term *squid type* OPA anhydrase, in contrast to those of the ubiquitous *Mazur type* OPA anhydrases (found even in other tissues of the squid) is a major theme of this chapter.

Because virtually any degradative change in organophosphorus AChE inhibitors drastically reduces their inhibitory potency, and because AChE inhibition is directly related to toxicity, there is a practical interest in the enzymatic hydrolysis of these compounds. The potential applications of a nerve gas detoxifying enzyme immobilized on a resin (Hoskin and Roush, 1982) or on a fabric (Hoskin *et al.*, 1989) or even on an electrode (Rajan *et al.*, 1985) are obvious. There is, however, a limiting factor of both practical and fundamental significance. Many of the most potent organophosphorus AChE inhibitors exist as stereoisomers. AChE or the OPA anhydrases or both may, indeed do, react selectively with these isomers. It may seem trivial to note that if the most inhibitory and thus the most toxic isomers were also the most readily hydrolyzed, then they would not be the most toxic. Some, perhaps most, of the OPA anhydrases appear to be stereospecific in the *wrong* direction, i.e., they hydrolyze the less toxic isomers. The squid type OPA anhydrase appears, in contrast, to hold some practical promise in this context. In this chapter, the term *stereoselectivity* will denote partial rather than absolute stereospecificity.

The stereoselectivity of the OPA anhydrases, the narrow distribution of the squid enzyme, especially its presence in squid nerve, and recently emerging evidence of a divalent cation requirement for the squid type OPA anhydrase (Hoskin *et al.*, 1988) give the illusion of an approach to the question of a natural substrate and a physiological role. The most that can be done at this point is to describe the methods of enzyme extraction and determination in the hope that other biologists may be encouraged to examine the question.

2. Methods

Tissues or fractions of one or more members of virtually every phylum of the animal kingdom have been shown to contain measurable levels of OPA anhydrase, typified by the hydrolysis of DFP or Soman, or in the earlier work Tabun, or analogous compounds (see Figure 1). A broad selection of prokaryotes (Mounter, 1963) including an interesting obligate thermophile (Chettur *et al.*, 1988), the protozoan *Tetrahymena thermophila* (Landis *et al.*, 1985), and several fungi (Hoskin *et al.*, 1987) show OPA anhydrase activity, although the pH optima of the fungal sources raises some question whether these are phosphatases. For obvious reasons the squid will be used to illustrate the methods of OPA anhydrase determination. In a subsequent section the evidence will be reviewed showing that squid type OPA anhydrase is a distinctly different enzyme from the ubiquitous, Mazur type in our terminology, OPA anhydrases.

Other chapters describe the dissection of the squid giant axon and the extrusion of axoplasm. Collection of axoplasm on a pre-weighed coverslip, re-weighing, and dropping directly into a stirred reaction medium allows rapid dispersion. A Cahn electrobalance can be used right beside the dissecting table. Optic ganglia are obtained by cutting through the head mid-sagittally and scooping the pea-shaped ganglion out of its chitinous housing directly behind the eyeball. A little chipping and trimming with sturdy scissors and forceps is sometimes necessary. The pale gray-green hepatopancreas within the mantle will be obvious. The ink-sac should be removed early, as well as some of the white connective tissue. Used and otherwise discarded dental probes (various shapes, angles, and edges; usually obtainable from one's own dentist) make excellent *scoopers*. For enzyme purification kilogram quantities of optic ganglia and hepatopancreas are easily obtained.

Our idea in examining squid saliva was that the toxin known to be in the saliva might be a natural analogue of the *nerve gases*, DFP for example, and that the squid might carry its own protection, squid type OPA anhydrase for example. Our findings are clear, but the original idea appears unfounded (Hoskin and Prusch, 1983; see final paragraph for references to the tetrodotoxin of the blue-ringed octopus, not of the squid). It is noteworthy that the OPA anhydrase found in squid saliva is, clearly, extracellular. For those wishing to pursue this point, or even the not yet flatly refuted idea of toxin and protection, the drawings of L. W. Williams (1909), most easily obtainable at the library of the Marine Biological Laboratory, Woods Hole, are essential for location and gross dissection of the posterior salivary gland. A small incision is made in the side of the excised gland (magnification as for the axon) and saliva is collected by cannulation with a 5 μl Drummond micropipette.

The hydrolysis of a mole of a compound shown in Figure 1 results in the production of a mole of a substituted phosphoric or phosphonic acid, pK_a (constant describing the equilibrium between the dissociated and undissociated forms of an acid) ≈ 2, and hydrofluoric acid, $pK_a \approx 3$, or for Tabun, hydrogen cyanide, $pK_a \approx 9$; this latter is a significant point. Diethyl p-nitrophenyl phosphate (Paraoxon) is not hydrolyzed by squid type OPA anhydrase (Hoskin, 1969). For other substrates there may be other products, but these will serve to illustrate OPA anhydrase, the determination of which depends on acid production, fluoride ion production, and loss of AChE inhibitory potency.

Acid production can be measured by the older manometric method (Umbreit *et al.*, 1957) which measures the release of CO_2 from a bicarbonate buffer. Exact conditions and actual data are presented by Hoskin *et al.* (1966) and Hoskin (1971). This cumbersome method has one advantage: it is a closed system. At the end of a determination with one of the very hazardous substrates such as Sarin or Soman, a side-arm of the reaction vessel is opened, an excess of strong alkali is added, and the vessel is closed. After a few hours the contents can be safely disposed of with careful washing.

The other acid production method is titrimetric, in which very small volumes of carefully standardized alkali are delivered on demand via a pH electrode-electronic-mechanical feedback circuit to maintain a constant pH. The delivery of alkali as a function of time is recorded. Exact conditions for a typical determination are given by Hoskin and Long (1972). It will be evident from Figures 4 – 6 of Hoskin and Long (1972) that conditions can be varied. User handbooks are essential, and equipment manufacturers will invariably provide instruction. Both of these acid production methods have shortcomings. The 1:1, Tabun:acid ratio can be corrected for, but reduces the sensitivity. In the manometric method the pH is not exactly constant; in the pH-stat method, the volume changes. However, the resulting errors are small. The more troublesome aspect is that bulky equipment must be crowded into a fume hood.

The fluoride electrode (Frant and Ross, 1966) provides a compact, sensitive, and rapid method of OPA anhydrase determination when the substrates are P-F compounds. Two of these, DFP and Soman, have been most useful for characterizing squid type OPA anhydrase in contrast to the Mazur type enzymes, as will be documented in the next section. Exact conditions and primary data have been published from this laboratory over more than a decade (Hoskin, 1976; Hoskin and Roush, 1982; Hoskin, 1985; Chettur *et al.*, 1988). A 400 mM KCl – 40 mM NaCl salt solution is used to provide a kind of general purpose intracellular medium; 20 mM buffer is required, and this concentration of bis(hydroxethyl)imino-tris(hydroxymethyl)-methane (BIS-TRIS), 1,3-bis(tris(hydroxymethyl)methylamino)-propane (BIS-TRIS-propane), N-2-hydroxy-ethlypiperazine-N'-2-ethanesulfonic acid (HEPES), or piperazine-N,N'-bis(2-ethane-sulfonic acid) (PIPES) (Good *et al.*, 1966)(see any chemical catalogue for *Good buffers*) does not adversely affect the reaction. Other buffers can be used, the desired pH and the pK_a's being the consideration. Orion Research equipment is reliable, e.g., Orion 901 meter, 951 printer, and 96-09 combination electrode; manuals are adequate without further instruction. For standardizing the electrode, the buffer to be used should be made, e.g., 2×10^{-5} M and 2×10^{-4} M in F^- (use NaF dried over P_2O_5) and stored in plastic bottles with a small stirring bar. For the actual OPA anhydrase assay which lasts at most about 30 min, a 10 ml glass beaker is suitable (see Figure 1 of Hoskin and Roush, 1982) and is preferable to plastic which may actually retain Soman, for example, if that is the substrate.

OPA anhydrase determination by loss of AChE inhibitory potency is slow and provides a limited number of points, but it has one outstanding advantage. The combination of this method with one of the others, preferably the fluoride ion method, will reveal stereospecificity (or selectivity) where this is possible. The interested reader is referred to Chettur *et al.* (1988), and to the references therein for the work of others. The method is that of Ellman *et al.* (1961) and is presented here in detail.

There are actually three determinations. One is the rate of acetylthiocholine (ATCh) hydrolysis by AChE. The second is the rate of ATCh hydrolysis by AChE that has been partially inhibited by a known concentration of inhibitor, e.g., Soman, and time of exposure. And the third is the rate of ATCh hydrolysis by AChE that has been partially inhibited by an unknown concentration of inhibitor – unknown in that some degradation has taken place. Since calculations require partial inhibition, the second and especially the third determinations may require repeating with greater or smaller dilutions.

What follows is typical and laboratory tested, but may be varied within limits. AChE (500 units) is dissolved in 20 ml of the previously described PIPES buffer containing 1 – 2 mg gelatin; 20 mg 5,5'-dithiobis-(2-nitrobenzoic acid) (DTNB) and 7.5 mg $NaHCO_3$ in 10 ml 0.1 M phosphate, pH 7; 18 mg ATCh Br in 10 ml water; all solutions are iced. Combine 3.8 ml PIPES and 0.2 ml AChE and let stand at room temperature for 30 min (constant temperature bath for greater accuracy). While waiting, prepare a blank of 0.2 ml DTNB, 0.2 ml ATCh, and 3.0 ml PIPES in a disposable cuvette (these are used throughout). At about 28 min put 0.2 ml of each of DTNB and ATCh in another cuvette, at exactly 30 min add 3.0 ml of the PIPES-AChE combination, mix rapidly, and record at once at 412 nm against the blank. What is wanted is an AChE preparation such that 0.2 ml will cause an absorbency change of about 1 in 5 – 10 min. When the original 20 ml is sufficiently diluted to do this, it can be frozen in 2 – 3 ml aliquots that will last indefinitely. The DTNB and ATCh can be kept iced and used for several days. The blank will do nicely for half a day.

For the second determination make 3 mM Soman in the buffer of your choice (1.5 ml 0.01 M Soman in water added to 3.5 ml buffer is convenient; the Soman in water can then be kept iced), and make sufficient dilutions to arrive at 10^{-8} M, all with ice-cold water and kept iced. Combine 3.6 ml PIPES, 0.2 ml AChE, and 0.2 ml Soman (now 5×10^{-10} M). Keep at room or bath temperature for exactly 30 min and add 3.0 ml to DTNB and ATCh as previously described. Suppose that 84% inhibition is found. An application of the second-order rate equation for irreversible inhibition (Aldridge, 1950),

$$k = \frac{1}{[I]t} \ln \frac{a}{a - x} \tag{1}$$

will give a k of 1.22×10^8 M^{-1} min^{-1}.

If 3 mM Soman is now incubated with, e.g., squid type OPA anhydrase for some period of time, and an aliquot removed, diluted ice-cold, incubated with AChE, and added to DTNB and ATCh, it may be found, let us say, to cause only 10% inhibition. A rearrangement of the rate equation, now with [I] as the unknown, will reveal that only 6% of the Soman remains, i.e., that the concentration had dropped from 3 mM to 0.18 mM. Extremes have been chosen for clarity only, although the inhibition constant for Soman and AChE is approximately correct (Chettur *et al.*, 1988).

The purification of squid type OPA anhydrase from optic ganglion and hepatopancreas by ammonium sulfate precipitation and column chromatographic procedures (Hoskin and Long, 1972; Garden *et al.*, 1975) is routinely performed in this laboratory. Improvements and *tricks of the trade*, mainly the work of Kathleen E. Steinmann (Steinmann, 1989), will be published in the near future. For example, initial homogenization need not be thorough; the enzyme is extremely soluble so that centrifuging out coarse debris actually improves the specific activity without lowering the yield. Also, the ammonium sulfate precipitation differentiates this enzyme from the Mazur type OPA anhydrases.

Finally, because the comparison of enzymatic hydrolysis rates of DFP and Soman is an important basis for characterization, some description of the safe handling of these compounds at substrate concentrations is given. Since Soman is orders of magnitude more hazardous than DFP, advice applying to Soman is more than adequate for DFP. The Materials and Methods section of Chettur *et al.* (1988) should be consulted. If Soman is synthesized or otherwise obtained, it will probably be held in an all-glass 1

Table I. Organophosphorus Acid Anhydrase (OPA Anhydrase) Activity of Components of the Squid, *Loligo pealei*

Component	*DFP* (μmoles per hr and per g, fresh wt)	*Soman* (μmoles per hr and per g, fresh wt)	*Soman/DFP* (ratio)	Mn^{2+} *Stimulation* (%)
Axoplasm	175	68	0.39	6
Optic ganglion	329	97	0.29	6
Hepatopancreas	848	371	0.44	34
Saliva	220	59	0.27	0
Purified*	23,600†	5,800†	0.25	−2
Gill	52	55	1.06	69
Heart	13.8	21	1.52	22
Mantle	1.8	3.6	2.00	230
Blood	0.12	0.24	2.00	140

* From optic ganglia or hepatopancreas.
† μmoles min^{-1} g protein $^{-1}$

ml ampoule. It is convenient to have this re-dispensed into 20 – 40 mg portions. This is best done on a day and in a lab of low humidity, into 1 × 100 mm pre-weighed capillary tubes. An individually vented fume-hood should contain a 2 liter beaker with about 200 ml 10% KOH, and a Bunsen burner. The fume hood is closed to the point where the following operations can still be done, but not to where a 10 ml beaker will be sucked inward. A *tell-tale* tissue will show that the hood is working. The ampoule is nicked with a diamond-point, cracked open with a molten Pasteur pipette tip, and set upright in a small beaker. The capillary is dipped into the Soman until the liquid rises about half way. It may be necessary to tip the beaker somewhat. The capillary is removed so that the liquid is retained in its mid-portion, moved directly to the edge of the flame, and sealed with twirling. The capillary is turned without touching the sealed end (which should be tilted slightly down at this point) and the other end is sealed. A slight bulging will assure that both ends have a positive seal. The capillary is laid in a numbered slot on an accordion-pleated sheet of paper. When about 25 capillaries are filled or the Soman supply is exhausted, the beaker and ampoule, and any wipes that may have been used, are dropped into the alkali. The filled capillaries are left for 24 hrs in the fume hood air stream and then wiped and re-weighed. In our laboratory the capillaries are stored in a small locked refrigerator located in a fume hood. On making an aqueous Soman solution, the correct amount of water in a glass-stoppered cylinder is ice chilled, the capillary is nicked near one end, and the tip is broken off and dropped into the cylinder. The capillary is reversed so that the open end is actually into the mouth of the cylinder, the closed end is nicked and broken, and the capillary (now draining) and the second tip are dropped into the cylinder, which is closed, shaken, and re-iced. At the end of an experimental period, a few pellets of alkali are dropped into the cylinder, which is shaken and left standing for 24 hrs. Any fume hood used for preparation or experimentation should be equipped with a water aspirator attached to a large alkali bottle with an attached tube and Pasteur pipette for aspirating out a cylinder or reaction vessel. Every few days the alkali-decomposed Soman can be safely disposed of with a large flush volume of tap water.

3. Results and Discussion

The two most distinguishing characteristics of squid type OPA anhydrase are that it hydrolyzes DFP several times faster than Soman (the original observations were with Tabun) and that the activity is not stimulated by Mn^{2+} in the 10^{-3} M concentration range. This is true of the squid components in which the enzyme is found, and especially of the purified enzyme. This is shown in the upper part of Table I wherein published and unpublished results from over a decade have been averaged and summarized. The Mn^{2+} stimulation of hepatopancreas homogenate should be considered in comparison to the ubiquitous (Mazur type) OPA anhydrase sources. For a wide selection of these, including sonicated or homogenized bacteria, yeast, *Tetrahymena*, lobster and *Spisula* (surf clam) nerve, vertebrate liver, kidney, and brain, and rat sciatic nerve (Hoskin and Dettbarn, 1989), Soman is hydrolyzed 5 to 50 times faster than DFP, and Mn^{2+} causes a 50% to 400% (= 5-fold) increase in activity. This is also true of the OPA anhydrase partially purified from hog kidney by the method of Storkebaum and Witzel (1975). If these two contrasting characteristics of squid type OPA anhydrase and Mazur type OPA anhydrase are valid, then the squid hepatopancreas may contain some of the Mazur type enzyme. Similarly, the data in the lower half of Table I suggest that the other parts of the squid may contain some of the squid type OPA anhydrase. Figure 1 of Hoskin *et al.* (1984) is illustrative of these distributions, but should be reconstructed with more highly purified Mazur type OPA anhydrase.

A notable OPA anhydrase has been partially purified from an obligate thermophilic (OT) bacterium (Chettur *et al.*, 1988). Its two most distinguishing characteristics are that its hydrolysis of Soman is stimulated 80-fold (8000% in the terminology of Table I) by Mn^{2+} in the 10^{-3} M concentration range, and that it does not hydrolyze DFP at all. In a volume devoted to the squid, this OT OPA anhydrase would not normally be included except that it may represent an ultimate Mazur type OPA anhydrase. In that sense, it will be used to emphasize another property of the squid type OPA anhydrase.

In a recent meeting (Reiner *et al.*, 1988) several papers described, in passing, the inclusion of ethylenediaminetetraacetate (EDTA) at 10^{-3} M, usually to inhibit the enzymatic hydrolysis of Paraoxon (see Figure 1) not hydrolyzed by squid type OPA anhydrase). We have also been given free access to the results of Professor Heinz Rüterjans (1989) of Johann Wolfgang Goethe University, Frankfurt, Germany, which suggest that squid type OPA anhydrase may be a Ca^{2+}-requiring enzyme. For our investigations we have used OT OPA anhydrase as a model Mazur type enzyme, and squid type OPA anhydrase, both purified without addition of Ca^{2+}, Mg^{2+}, or Mn^{2+}, but with no precautions other than the use of deionized and double-distilled water. We have also used 5 – 10 mg freshly extruded axoplasm which, on addition to a 5 ml system will be approximately 10^{-6} M in Ca^{2+} and 10^{-5} M in Mg^{2+}; the Mn^{2+} concentration will remain negligible. The effects of these three cations added so as to be at 10^{-3} M in contact with the three enzyme sources for 30 minutes prior to addition of substrate is shown in Table II.

EDTA is a powerful chelator of all three of the cations considered in Table II; ethylenebis(oxyethylenenitrilo)tetraacetate (EGTA), of Ca^{2+} and Mn^{2+}; 1,10-

Table II. Effects of 10^{-3} M Cations on OPA Anhydrases

Cation	*Squid Enzyme (Purified or Axoplasm) (DFP or Soman)*	*OT Enzyme (Soman)**
Ca^{2+}	−10	0
Mg^{2+}	−11	0
Mn^{2+}	−20	+8,000†

* DFP not hydrolyzed at all, hence cannot be tested.
† This is an 80-fold stimulation.
+ refers to % stimulation.
− refers to % inhibition.

Table III. Effects of 10^{-4} M Chelators on OPA Anhydrases

Chelator	*Squid Enzyme (Purified or Axoplasm) (DFP or Soman)*	*OT Enzyme (Soman)**
EDTA	−92	−96
EGTA	−75	−89
1,10-ϕ	−5	−99
8-OHQ-5-SA	−2	−100

* DFP not hydrolyzed at all, hence cannot be tested.
+ refers to % stimulation.
− refers to % inhibition.

phenanthroline (1,10-ϕ), only of Mn^{2+}; and 8-hydroxyquinoline-5-sulfonate (8-OHQ-5SA), mainly of Mn^{2+} (Martell and Smith, 1974). The effects of these four chelators at 10^{-4} M for 30 minutes are shown in Table III.

The evidence of Tables II and III suggests that squid-type OPA anhydrase is Ca^{2+}-requiring, but not further divalent cation stimulated. By contrast, the OT enzyme is powerfully Mn^{2+} stimulated, and probably Mn^{2+}-requiring. The question was then posed, if Ca^{2+} is pulled out of a presumed Ca^{2+}-requiring site, is it Ca^{2+} specifically that will go back in? To answer this, squid enzyme was incubated for 30 minutes with EGTA at 10^{-4} M, then a small volume of Ca^{2+} or Mg^{2+} was added to attain a concentration of 10^{-3} M, and after another 30 minutes substrate was added. The results are presented in Table IV. Again, the differences between purified squid-type OPA anhydrase and squid axoplasm are so small that the data have been averaged. The results of Tables II, III, and IV show unequivocally that squid-type OPA anhydrase is a Ca^{2+} enzyme. The same experiment as in Table IV but with OT enzyme as a

Table IV. Recovery of Squid-Type OPA Anhydrase Activity After Chelator Inhibition

First Treatment		*Followed by*		
Chelator	*Time**	*Cation*	*Time**	*% Recovery†*
10^{-4} M EGTA	30	10^{-3} M Ca^{2+}	30	84 ± 20
10^{-4} M EGTA	30	10^{-3} M Mg^{2+}	30	20 ± 10

* Time in minutes
† Number of each experiment = 8

model Mazur-type OPA anhydrase, and Mn^{2+}, gave results that were not readily interpretable. The difference between chelator-inhibited OT enzyme and unstimulated OT enzyme is small compared to the difference between unstimulated OT enzyme and Mn^{2+}-stimulated OT enzyme. Thus addition of 10^{-3} M Mn^{2+} after 10^{-4} M chelator gave a very large recovery, but it was not possible to decide between a refilled Mn^{2+} requiring site and a filled Mn^{2+} stimulatory site, or merely a filled Mn^{2+} stimulatory site. However, on the basis of the 1,10-ϕ inhibition of Table III it appears likely that the OT OPA anhydrase is Mn^{2+} requiring. Other Mazur-type OPA anhydrase sources show similar results, including the ambiguities. In addition there is evidence (Little *et al.*, 1989), confirmed in this laboratory, that rat and mouse liver contains an OPA anhydrase that is Mn^{2+} and Mg^{2+} stimulated and probably requiring.

The properties of squid-type OPA anhydrase that can now be given with a high degree of confidence are listed on the left-hand side of Table V. Those for the enzyme or enzymes termed Mazur-type OPA anhydrase are given, with less certainty, on the right. The narrow distribution of the squid enzyme, especially its presence in the squid axon, is a most provocative feature. We have speculated that the enzyme might be associated with the metabolism of 2-hydroxyethanesulfonate (isethionate) (Hoskin and Kordik, 1977; Hoskin and Noonan, 1980), a major anion only of cephalopod nerve. However, hepatopancreas remains a major exception.

Based on an ultimate activity of approximately 10^3 μmoles DFP per minute and per mg protein in a single electrophoresis band, 1.62 μmoles DFP per minute and per g axoplasm, and assuming that axoplasm is typical intracellular material, squid-type OPA anhydrase is about 0.002% of the intracellular protein, not too unlike the *average* level of *typical* proteins or enzymes.

The Ca^{2+} requirement and the stereoselectivity give the illusion of a relationship to physiological function, but as yet the relationship remains unknown. Unfortunately for the stereoselectivity approach, for DFP without a chiral center the K_M for the squid enzyme is in the 10^{-3} M range but ten times greater for Soman and another phosphonate, both with chiral centers (Chettur *et al.*, 1988). The active site also favors a branched side-chain (isopropyl) over ethyl or methyl (Gay and Hoskin, 1979). Squid-type OPA anhydrase is not an ATPase, nor a phosphatase, nor does fluorophosphate fit the active site (see footnote 4 of Hoskin and Long, 1972).

Perhaps the most that can be said at this time is that the squid encounters, in its external or internal environment, a molecule that resembles DFP, either in its structure or in its properties. In a more immediately practical context, it seems reassuring that

Table V. Properties of Organophosphorus Acid (OPA) Anhydrases

Squid Type	*Mazur Type*
Narrow distribution, squid nerve, saliva, hepatopancreas	Ubiquitous
Molecular weight, 30,000	Variable, 45–90,000
Soman/DFP ≈ 0.25	Soman/DFP, 5–50 and higher
Hydrolyzes all isomers of Soman; some stereoselectivity in rates	Stereoselectivity variable; often quite stereospecific
Mn^{2+} indifferent or slightly inhibited	Mn^{2+} stimulated 2- to 20-fold and as high as 80-fold
Ca^{2+} requiring, not Ca^{2+} stimulated	May be Mg^{2+} requiring and stimulated
Purification, 60% $(NH_4)_2SO_4$	$(NH_4)_2SO_4$ labile
Mipafox indifferent	Mipafox inhibited

environmentally hazardous compounds such as nerve gases and insecticides may be detoxified by an enzyme found in the squid. Although such a relatively limited source is probably not practical for large-scale detoxication, the potential exists for accomplishing this by a combination of the techniques of genetic manipulation and biochemical engineering.

ACKNOWLEDGMENTS. This work has been supported by a grant from Army Research Office and past grants from NIH. I am indebted to Professor Heinz Rüterjans of Johann Wolfgang Goethe University, Frankfurt, Germany, and to all my past and present colleagues, especially the co-authors on the most recent papers from this laboratory. Funds and facilities have also been provided by IIT Research Institute.

References

Aldridge, W. N., 1950, Some properties of specific cholinesterase with particular reference to the mechanism of inhibition by diethyl p-nitrophenyl thiophosphate (E605) and analogues, *Biochem. J.* **46**:451–460.

Chettur, G., DeFrank, J. J., Gallo, B. J., Hoskin, F. C. G., Mainer, S., Robbins, F. M., Steinmann, K. E., and Walker, J. E., 1988, Soman-hydrolyzing and -detoxifying properties of an enzyme from a thermophilic bacterium, *Fundam. Appl. Toxicol.* **11**:373–380.

Ellman, G. L., Courtney, K. D., Andres, V., Jr., and Featherstone, R. M., 1961, A new and rapid colorimetric determination of acetylcholinesterase activity, *Biochem. Pharmacol.* **7**:88–95.

Frant, M. S. and Ross, J. W., Jr., 1966, Electrode for sensing fluoride ion activity in solution, *Science* **154**:1553–1555.

Garden, J. M., Hause, S. K., Hoskin, F. C. G., and Roush, A. H., 1975, Comparison of DFP-hydrolyzing enzyme purified from head ganglion and hepatopancreas of squid (*Loligo pealei*) by means of isoelectric focusing, *Comp. Biochem. Physiol.* **52C**:95–98.

Good, N. E., Winget, G. D., Winter, W., Connolly, T. N., Izawa, S., and Singh, R. M. M., 1966, Hydrogen ion buffers for biological research, *Biochemistry* **5**:467–477.

Hoskin, F. C. G., 1969, Possible significance of "DFPase" in squid nerve, *Biol. Bull.* **137**:389–390.

Hoskin, F. C. G., 1971, Diisopropylphosphorofluoridate and Tabun. Enzymatic hydrolysis and nerve function, *Science* **172**:1243–1245.

Hoskin, F. C. G., 1976, Distribution of diisopropylphosphorofluoridate-hydrolyzing enzyme between sheath and axoplasm of squid giant axon, *J. Neurochem.* **26**:1043–1045.

Hoskin, F. C. G., 1985, Inhibition of a Soman- and diisopropyl phosphorofluoridate (DFP)-detoxifying enzyme by Mipafox, *Biochem. Pharmacol.* **34**:2069–2072.

Hoskin, F. C. G. and Brande, M., 1973, An improved sulfur assay applied to a problem of isethionate metabolism in squid axon and other nerves, *J. Neurochem.* **20**:1317–1327.

Hoskin, F. C. G. and Dettbarn W. D., 1989, *pers. commun.*

Hoskin, F. C. G. and Kordik, E. R., 1977, Hydrogen sulfide as a precursor for the synthesis of isethionate in the squid giant axon, *Arch. Biochem. Biophys.* **180**:583–586.

Hoskin, F. C. G. and Long, R. J., 1972, Purification of a DFP-hydrolyzing enzyme from squid head ganglion, *Arch. Biochem. Biophys.* **150**:548–555.

Hoskin, F. C. G. and Noonan, P. K., 1980, Taurine and isethionate in squid nerve, in: *Natural Sulfur Compounds*, (D. Cavallini, G. E. Gaull, and V. Zappia, eds.) pp. 253–263, Plenum, New York.

Hoskin, F. C. G. and Prusch, R. D., 1983, Characterization of a DFP-hydrolyzing enzyme in squid posterior salivary gland by use of Soman, DFP, and manganous ion, *Comp. Biochem. Physiol.* **75C**:17–20.

Hoskin, F. C. G. and Roush, A. H., 1982, Hydrolysis of nerve gas by squid-type diisopropyl phosphorofluoridate hydrolyzing enzyme on agarose resin, *Science* **215**:1255–1257.

Hoskin, F. C. G., Chettur, G., Mainer, S., Steinmann, K. E., DeFrank, J. J., Gallo, B. J., Robbins, F. M., and Walker, J. E., 1989, Soman hydrolysis and detoxification by a thermophilic bacterial enzyme, in: *Enzymes Hydrolyzing Organophosphorus Compounds*, (E. Reiner, F. C. G. Hoskin, and N. W. Aldridge, eds.) pp. 53–64, Ellis Horwood, Chichester, England.

Hoskin, F. C. G., Kirkish, M. A., and Steinmann, K. E., 1984, Two enzymes for the detoxication of organophosphorus compounds – Sources, similarities, and significance, *Fundam. Appl. Toxicol.* **4**(2): Part 2: S165–S172.

Hoskin, F. C. G., Landis, W. G., and Durst, H. D., (organizers), 1987, *DFPase Workshop*, Marine Biological Laboratory, Woods Hole, MA.

Hoskin, F. C. G., Rajan, K. S., and Steinmann, K. E., 1988, Organophosphorus acid (OPA) anhydrase from squid: a calcium-dependent P-F splitting enzyme, *Biol. Bull.* **175**:305–306.

Hoskin, F. C. G., Reese, E. T., and Smith, W. J., 1987, Mipafox – inhibitor of cholinesterase, neurotoxic esterase, and DFPase: is there a "Mipafox-ase"?, in: *Neurobiology of Acetylcholine*, (N. J. Dun and R. L. Perlman, eds.) pp. 451–458, Plenum, New York.

Hoskin, F. C. G., Rosenberg, P., and Brzin, M., 1966, Re-examination of the effect of DFP on electrical and cholinesterase activity of squid giant axon, *Proc. Nat. Acad. Sci. USA* **55**:1231–1234.

Katz, B., 1960, Book reviews, *Perspectives Biol. Med.* **3**:563–565.

Landis, W. G. and Hoskin, F. C. G., (organizers), 1988, *Second OPA Anhydrase Workshop*, National Academy of Sciences Study Center, Woods Hole, MA.

Landis, W. G., Savage, R. E., Jr., and Hoskin, F. C. G., 1985, An organofluorophosphate hydrolyzing activity in *Tetrahymena thermophila*, *J. Protozool.* **32**:517–519.

Little, J. S., Broomfield, C. A., Fox-Talbot, M. K., Boucher, L. J., MacIver, B., and Lenz, D. E., 1989, Partial characterization of an enzyme that hydrolyzes Sarin, Soman, Tabun, and DFP, *Biochem. Pharmacol.*, **38**:23–29.

Martell, A. E. and Smith, R. M., 1974, *Critical Stability Constants, Volumes 1 and 2*, Plenum, New York.

Mazur, A., 1946, An enzyme in animal tissue capable of hydrolyzing the phosphorus-fluorine bond of alkyl fluorophosphates, *J. Biol. Chem.* **164**:271–289.

Mounter, L. A., 1963, Metabolism of organophosphorus anticholinesterase agents, in: *Handbuch der Experimentallen Pharmakologie: Cholinesterases and Anticholinesterase Agents*, (G. B. Koelle, ed.) pp. 486–504, Springer-Verlag, Berlin.

Nachmansohn, D., 1959, *Chemical and Molecular Basis of Nerve Activity*, Academic Press, New York, (see also revised edition, 1975).

Rüterjans, H., 1989, *pers. commun.*

Rajan, K. S., Mainer, S., Hoskin, F. C. G., and Luskus, L., 1985, An enzyme electrode for the quantification of organic phosphonofluoridates, *Biophys. J.* **47**:9a.

Reiner, E., Hoskin, F. C. G., and Aldridge, N. W., (eds.), 1989, *Enzymes Hydrolyzing Organophosphorus Compounds*, Ellis Horwood, Chichester, England.

Steinmann, K. E., 1989, *pers. commun.*

Storkebaum, W. and Witzel, H., 1975, Study on the enzyme-catalyzed splitting of triphosphates, *Forschungsber. Landes Nordrhein-Westfalen No.* **2523**:1–22.

Umbreit, W. W., Burris, R. H., and Stauffer, J. F., 1957, *Manometric Techniques*, 3rd ed., Burgess, Minneapolis, MN.

Williams, L. W., 1909, *The Anatomy of the Common Squid, Loligo pealii, Leseur*, p. 34, E. J. Brill, Leiden.

Chapter 22

Squid as Elite Athletes: Locomotory, Respiratory, and Circulatory Integration

RON O'DOR, H. O. PÖRTNER, and R. E. SHADWICK

1. Introduction

The giant axons of squid are among the best characterized animal cells, but we know much more about how they work in isolation than in the intact animal. This is largely because squid are *high-strung* and easier to work with dead than alive. Recent studies suggest that it is the inherent inefficiency of jet propulsion relative to undulatory propulsion in fish competitors that caused squid to become so evolutionarily *strung-out* (O'Dor and Webber, 1986). Even jet propelled flight of the large squid, *Dosidicus gigas*, in the open ocean has been filmed (Cole and Gilbert, 1970). Extrapolations from locomotor studies on smaller squid suggest if these squid would reach the size of the squid, *Architeuthis* ought to be able to swim around the world in 80 days (O'Dor, 1988c). The power output of a swimming squid approaches that of an air breather at similar size and temperature, and its oxygen consumption is higher than anything else in the sea. Since the jet automatically produces a large flow of water over the gills, squids are not limited by oxygen availability like other aquatic species (O'Dor, 1988a), but their basically molluscan circulatory systems seem to be prone to failure while operating near capacity (Wells, 1983). This chapter will outline what little we know about the integration of nerve, muscle, respiratory and circulatory systems in exercising squid, and will focus on new techniques which now make it possible to learn more.

2. Physiology in active squid

Studies of the dynamic functions of squid respiratory, circulatory and even nerve-muscle systems require techniques for linking healthy squid to physiological transducers. Many animals, particularly relatively inactive ones, behave reasonably normally when constrained or anaesthetized, and most of the available information on squids comes from attempts to use such approaches. However, their highly tuned

R. O'DOR • Biology Department, Dalhousie University, Halifax, Nova Scotia, Canada B3H 4J1. H. O. PÖRTNER • Institut für Zoologie IV, Universität Düsseldorf, Universitätsstrasse 1 D-4000 Düsseldorf 1, F. R. Germany R. E. SHADWICK • Biology Department, University of Calgary, Calgary, Alberta, Canada T2N 1N4

machinery does not respond well to such situations. Control of cephalopod circulatory systems is diffuse, with a high degree of nervous regulation at all levels (Wells, 1983; Wells and Smith, 1987), and these nervous links tend to fail under the same influences that produce general anaesthesia. Thus, tissues in anaesthetized squid rapidly become anoxic, even when the animals are artificially ventilated. If a squid survives a lengthy surgical procedure, it still requires several hours to return to a normal physiological state and may well rip out the plumbing with its dexterous arms or bite through the wires (or the surgeon) with its powerful beak. Attempting to handle squid without anaesthesia is generally worse, since they use their full power output in attempts to escape; this may push their circulatory systems past their limits, and, in any case, wind up exhausted. Constrained squid struggle similarly and measurements from them can hardly be considered normal. Thus, much of the little we *know* about squid physiology must be considered cautiously.

Several new techniques have emerged in recent years which promise to improve, if not revolutionize, studies of squid. Techniques for long-term maintenance of healthy squids have improved dramatically (Hanlon *et al.*, 1983; Hanlon, 1987; Hanlon, 1990). Messenger *et al.* (1985) reviewed the use of various anesthetics in cephalopods and showed that magnesium chloride is more effective than those more commonly used. We have recently taken advantage of this anesthetic for simple chronic cannula implantation. Several groups (O'Dor, 1982; Freadman *et al.*, 1984; Webber and O'Dor, 1985; Wells *et al.*, 1988; Pörtner *et al.*, 1987, 1989) have demonstrated that squid swim stably in swim-tunnels and have used these aquatic treadmills to monitor various physiological variables in exercising squid. A less sophisticated, but perhaps more elegant and certainly cheaper approach to providing a stable platform is to *crazy-glue* squid to floating styrofoam blocks; surprisingly, they don't seem to mind (M. E. DeMont, 1989). An alternate solution is to let the squid swim freely while using ultrasonic telemetry to eliminate the physical link between the transducer on the squid and the recording system (Webber and O'Dor, 1986). These new techniques, their limits and potentials are outlined below.

2.1. Anesthetics

The most recent and thorough, although by no means exhaustive, study of cephalopod anesthetics rediscovers magnesium ions and tests them on several species over a wide range of sizes (Messenger *et al.*, 1985). An isotonic 7.5% (weight/volume) solution of $MgCl_2 \cdot 6H_2O$ in distilled water, mixed with an equal volume of seawater just prior to use is recommended (Pantin, 1946). 3% (w/v) urethane in seawater and 2% (v/v) ethanol in seawater, also commonly used, are more traumatic, and, because they initially induce activity, the animals rapidly accumulate an oxygen debt due to anaerobic metabolism (Best and Wells, 1983). Magnesium takes longer (ca. 13 min for anaesthesia and 8 min for recovery) in octopuses. Cold (3 – 5 °C for octopuses at 24 °C) can also be used without trauma, but is a less effective muscle relaxant (Andrews and Tansy, 1981).

We have used ethanol, cold and magnesium with *Loligo pealei*, *Loligo opalescens* and *Illex illecebrosus*, and our experience is generally similar. For large animals at lower temperatures (ca. 400 g and 10 – 15 °C) we use 1.5% ethanol for quick, light anaesthesia for procedures such as weighing or tattooing. Animals appear normal to us within a few minutes after recovery, but do not rejoin their schools for several hours; perhaps drunken squid are not allowed in school. Mortality is high when squid are left in ethanol long enough to abolish all reflexes. We now use Pantin's formula, aerated and chilled to maintain squid during relatively long surgical procedures. Chilling not only aids anaesthesia but also lowers metabolic rate and reduces the anoxia, which squid do not handle well. We are experimenting with reduced levels of magnesium to maintain anaesthesia while allowing circulatory recovery but have not found the magic formula.

For surgery lasting 10 – 12 min, a squid (*Illex illecebrosus*, *Loligo pealei*) was removed from the holding tank and kept in ca. 5 – 7 °C sea water bubbled with pure oxygen. After 10 – 15 min the animal was transferred to oxygenated anesthetic of the same temperature until muscle relaxation occurred and the animals no longer responded to handling. For surgery (see below) animals were placed on a plastic bag filled with ice and covered with a leather cloth wetted with seawater to prevent skin damage. The mantle cavity was perfused by recirculating oxygenated, pre-cooled anesthetic over both gills. After surgery animals were returned to oxygenated sea water at 10 – 15 °C and watched until mantle and heart contractions resumed. The period of recovery, which was critical for survival, could be shortened by gentle pressure on the mantle and heart, simulating the process and frequency of mantle contraction in the resting animal. Ventilation of the mantle cavity ensured oxygenated blood at the gills which facilitated the early onset of cardiac activity and oxygen transport to the tissues (Pörtner *et al.*, 1987, 1989).

Further work on anaesthesia is essential for the future of squid research. Cephalopods are receiving increasing attention as animal models for biomedical research but are also receiving increasing attention from animal welfare groups and agencies, world-wide. Studying the performance of animals quickly provides a feel for their sensitivities. Good anesthetics are essential for both the humanitarian treatment of these animals and for meaningful physiological observations.

2.2. Cannulation

Surgery on squid is complicated by the limited time available prior to the hypoxia which is consistently induced by anaesthesia. Cuttlefish (*Sepia officinalis*) have been worked on for 20 min (Johansen *et al.*, 1982) and octopuses (*Octopus dofleini*) for up to 60 min (Johansen and Martin, 1962) with good survival, but even with the procedure outlined above 10 – 12 min is the maximum to ensure full recovery in squid. Bourne (1982) noted that the short time limit only allowed placement of a single cannula per animal in describing techniques for making short-term blood pressure measurements from the ventricle, anterior aorta, anterior *vena cava* and the afferent and efferent branchial arteries. Access to the latter three vessels is possible through the mantle opening so that incisions through the mantle, which are almost impossible to reclose for chronic exercise experiments, are not required. We have successfully used this approach in *Illex illecebrosus* and *Loligo pealei* for chronic cannulations in the anterior *vena cava* to obtain blood samples from undisturbed, nonrestrained, resting and exercising squids more than 24 h after surgery (Pörtner *et al.*, 1987, 1989).

Surgery is most successful when the blood in the *vena cava* is oxygenated (blue) at the start. Prior to the cannulation the funnel must be *unsnapped* from the ventral mantle to expose the *vena cava* which is punctured with a needle (20 gauge) that marks the site of puncture with india ink. A PE (polyethylene) 50 cannula filled with filtered sea water, which had been bent (by 180°) 2 – 5 cm from the end, drawn to a fine tip with 3 added holes to prevent blockage, was fed into the lumen of the *vena cava*. The

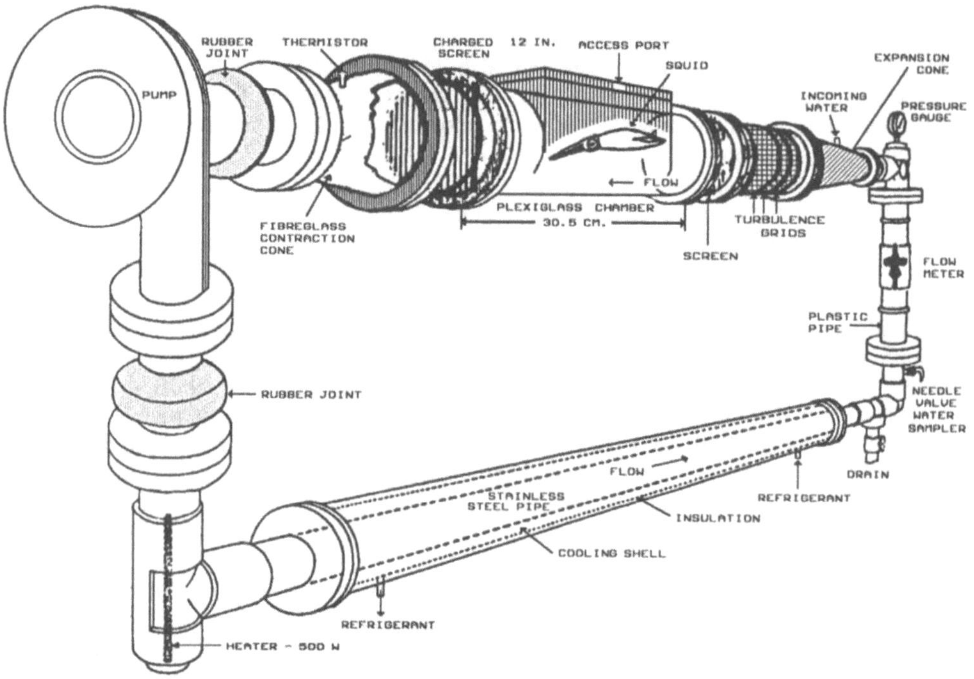

Figure 1. Design of the Brett tunnel-respirometer first used to swim the squid, *Loligo opalescens*. (Modified from Brett, 1964). The length of the plexiglass chamber is 30.5 cm. Oxygen consumption was measured using current speeds from 0 to 1 m/sec at 5, 10, and 15 °C (O'Dor, 1982).

cannula was tied to the wall of the *vena cava* with fine surgical suture and sealed using Bostik No. 7432 cyanoacrylate glue (Bostik GmbH, Oberursel, FRG). To prevent the animal grabbing and chewing on the cannula, it was led back into the mantle cavity and through the mantle wall close to the tip, where it was again secured by suture and glue. Surgery was finished by *resnapping* the funnel to the mantle.

2.3. Swim-tunnels

The wide variety of swimming chambers, beginning with the rotating annulus of Fry and Hart (1948), used to study fish during forced activity has been reviewed by Beamish (1978). One of the most flexible and successful designs (Brett, 1964) was a low volume, high power swim-tunnel respirometer used extensively in studies of salmon bioenergetics (Reviewed in Brett and Groves, 1980). This same machine was used in the first study of swimming performance in a squid (*Loligo opalescens*; O'Dor, 1982; Fig. 1), and machines of similar design have been used in numerous studies of fish locomotion (Webb, 1978), respiration and circulation (Jones and Randall, 1978), neuro-muscular organization (Bone, 1978), and muscle metabolism (Driedzic and Hochachka, 1978). Given a satisfactory anesthetic and the fact that squid will swim

Table I. Performance and metabolic rates of squid and fish during sustained swimming.

	Oncorhynchus[1] *nerka*	*Illex*[2] *illecebrosus*	*Loligo*[3] *pealei*	*Loligo*[4] *opalescens*
Temperature (°C)	15	15	22	14
Total length (m)	0.37	0.42	0.24	0.20
Mass (g)	500	400	100	40
Critical speed ($m{\cdot}s^{-1}$)	1.35	0.76	0.60	0.36
Active metabolism (ml $O_2{\cdot}kg^{-1}h^{-1}$)	480	1047	950	862
Standard metabolism (ml $O_2{\cdot}kg^{-1}h^{-1}$)	40	313	320	254
Resting metabolism (ml $O_2{\cdot}kg^{-1}h^{-1}$)	–	202	300	239
Scope for activity (ml $O_2{\cdot}kg^{-1}h^{-1}$)	440	734	630	608
Gross cost of transport ($J{\cdot}kg^{-1}m^{-1}$)	1.9	7.6	9.6	12.6
Net cost of transport ($J{\cdot}kg^{-1}m^{-1}$)	1.7	5.4	6.1	7.7
Oxygen debt ($ml{\cdot}kg^{-1}$)	329	167	–	73

Data from: [1] Brett 1965. [2] Webber and O'Dor 1985. [3] Freadman *et al.* 1984, 1989. [4] O'Dor 1982.

Definitions as given by Brett 1965 and O'Dor 1982. Metabolism for squid resting on the bottom is lower than extrapolated metabolism at zero speed (standard metabolism) because they are negatively buoyant. (From O'Dor and Webber 1986)

quite contentedly and reliably against a wide range of currents speeds, most of the experimental methods applied to fish in the last two decades can be applied to squid over the next two decades as well. It should be possible to publish twice as many articles, however, since squid will swim either tail-first (preferred) or head-first. A brief survey of experiments conducted to date on squid will give an idea of the potential of this approach, but the fish literature indicates that for every technique applied to squid, so far, there are at least ten which have not been tried.

When sealed from incurrent water and the atmosphere, a recirculating swim-tunnel becomes a respirometer, allowing continuous monitoring of a fixed volume of water for oxygen consumption and metabolic end-product excretion for normally swimming squid. Limited data on oxygen consumption at various speeds is available for several species of squid as shown in Table I. Ammonia production has been shown to parallel oxygen consumption, suggesting protein is the principal fuel at sustainable swimming speeds (Hoeger *et al.*, 1987). However, data on carbon dioxide accumulation to complete the information on respiratory quotients and fuels used is lacking. Anaerobic metabolism can be estimated from the oxygen debt which is paid off after swimming is stopped.

Well designed tunnels have non-refracting surfaces which allow the swimming animal to be filmed or video-taped for kinematic analysis and estimation of the drag forces acting at various speeds. Since the expansion and contraction of the mantle musculature is easily visualized or quantified using a video dimension analyzer (Gosline and Shadwick, 1983), detailed accounts of the frequency and volume of jetting cycles are possible. The same data also indicate the time course and extent of shortening of muscle contractions, as well as providing an indirect indication of nerve activity (Gosline *et al.*, 1983). These muscles work against the volume of water inside the mantle to produce the pressure which drives the jet, rather than dissipating their effort

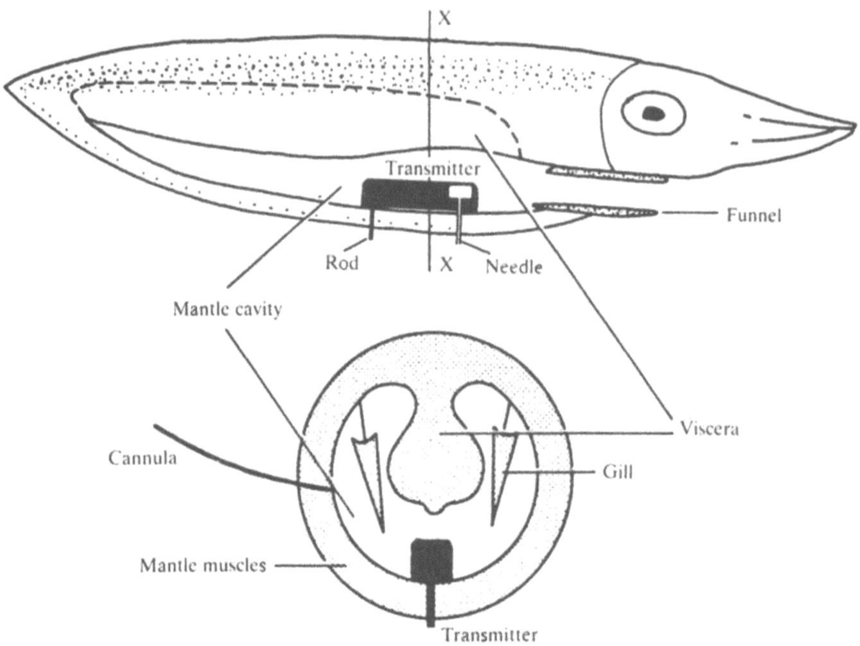

Figure 2. An integrating, differential pressure transducer with ultrasonic transmitter mounted *within* the mantle of an *Illex illecebrosus* to measure the work of jet propulsion in freely swimming squid (From Webber and O'Dor, 1986).

externally as fish muscles do. Estimating jet thrust of tethered squid from mantle cavity pressures measured through cannulae (Trueman and Packard, 1968; Johnson *et al.*, 1972) is much simpler than estimating the thrust of undulatory swimmers (Lighthill, 1960). Mantle pressures can also be measured with cannulae during graded exercise in tunnel-respirometers (Webber and O'Dor, 1985; Webber and O'Dor, 1986) and are well correlated with oxygen consumption.

These two variables are closely related to power output and power consumption, allowing direct estimates of locomotor efficiency (O'Dor, 1988a), which has not been possible with fish. Combining pressure data with kinematics allows a virtually complete analysis of squid locomotion (O'Dor, 1988b); actual equations of motion are quite complex, since squid can change fin activity, funnel diameter, angle of attack and jet cycle frequency and volume as they swim at increasing speeds. The integration of various mantle muscle types, funnel and fins produces various *gaits* which clearly require much more complex nervous control than has been attributed to squid in the past (Wilson, 1960; Packard and Trueman, 1974; Mackie, 1990). Detailed analysis of this activity is still lacking for any squid but should be possible using cuff-electrode techniques (Pinsker *et al.*, 1986) on squid swimming in a tunnel at various speeds.

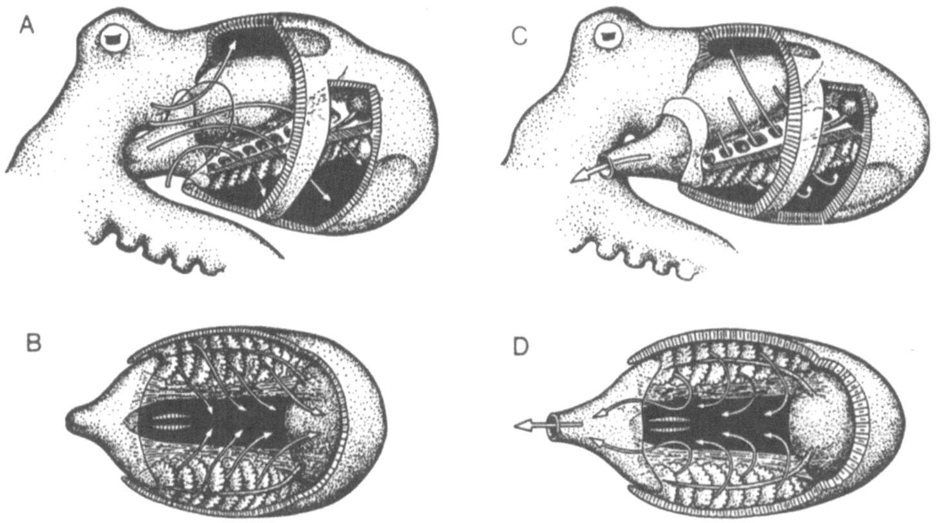

Figure 3. Continuous water flow during inhalation and exhalation in *Octopus vulgaris*. (*A*) and (*B*) show views from the side and from below during the brief expansion of the mantle; (*C*) and (*D*) show the flow during the longer, exhalant, phase (From Wells and Smith, 1985).

2.4. Ultrasonic transducer-transmitters

Webber and O'Dor (1986) used an integrating, differential pressure transducer coupled to an ultrasonic transmitter constructed by Vemco Ltd., Shad Bay, Nova Scotia, to compare the mantle cavity pressures of squids in a swim-tunnel at known speeds to those of the same animals swimming freely in static water. Squid almost seem to be designed for this, since the electronic package fits conveniently **inside** the mantle cavity producing minimal effects on drag and no external marks, allowing the squid to rejoin their schools and even feed (Fig. 2). The frequency limits on ultrasonics prevent rapid or complex physiological data from being transmitted, but electronic pre-processing can reduce such signals as electrocardiograms (EKGs) to heart rates before transmission. Similar packages for blood pressure, electromyograms (EMGs), and many other physiological variables should make experiments on freely moving and even free-living squid feasible.

3. Respiratory physiology

The studies of locomotor performance cited above indicate that, although squid can match or better fish in acceleration (Foyle and O'Dor, 1988), squid work twice as hard to go half as fast as comparable sized fish in steady swimming because of

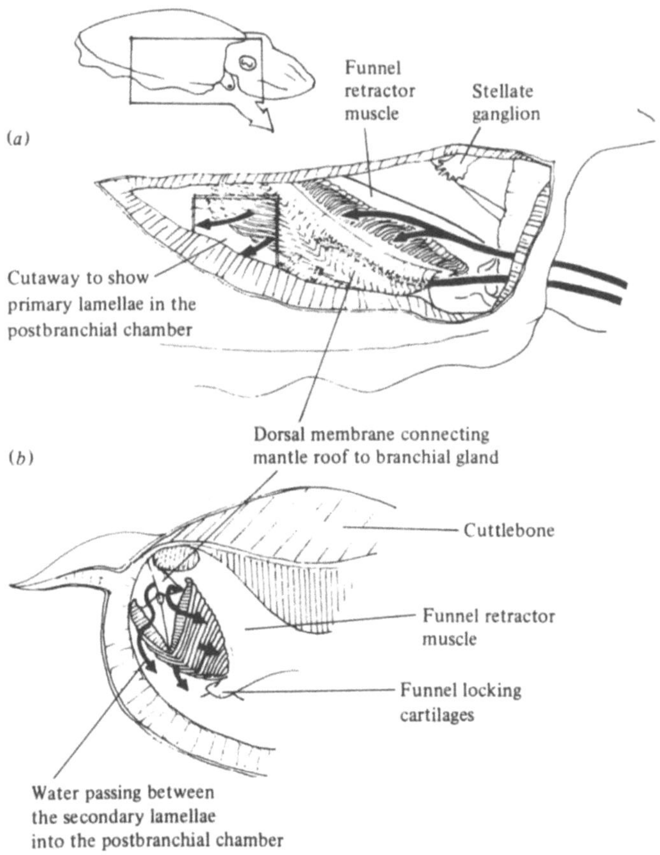

Figure 4. The position of the gill of *Sepia* in a freshly killed specimen. (*A*) From above and to one side. (*B*) In an animal cut transversely, viewed from behind looking forwards. Black arrows show the direction of water flow through the gill (From Wells and Wells, 1982).

the lower Froude efficiency of a high velocity jet. Following Packard's (1972) arguments that squid and fish evolved in parallel, as competitors, O'Dor and Webber (1986) examined the physiological constraints on cephalopods. The physiological and metabolic adaptations needed to provide power outputs twice those of fish at comparable temperatures were suggested to be responsible for the short lives and communal reproductive patterns of squid. Since the main locomotor system in squid *is* the ventilatory system as well, oxygen extraction from water is not a problem, even at the highest power outputs. Limiting factors are oxygen delivery to the tissues and adenosinetriphosphate (ATP) production by the tissues (O'Dor, 1988a).

3.1. Ventilatory flow

Wells and Wells (1982) showed, surprisingly recently, that octopuses and cuttlefish, more sedentary relatives of squids, force water over their gills in a complex pattern producing a countercurrent to the flow of blood and allowing oxygen to be extracted with typical efficiencies of 40%, occasionally rising as high as 75% in hypoxia. The gills do not simply hang into a bag which expands and contracts like a lung with a large dead space, as they appear to in dissected specimens. Differential

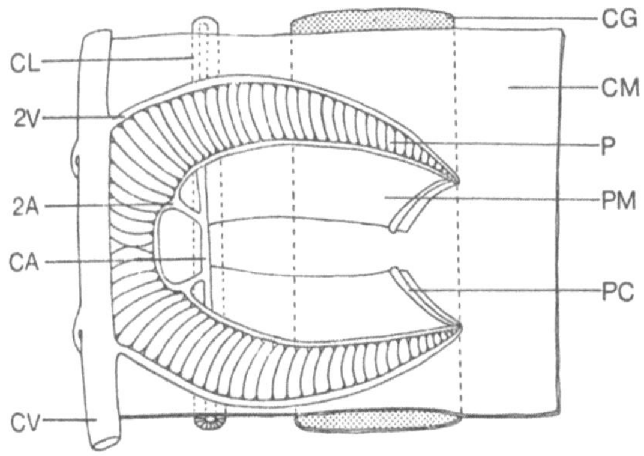

Figure 5. A section of the right gill of *Illex illecebrosus*, showing the organization of vessels and support structures. A second set of pinnae have been displaced distally and proximally: other pinnae form a mirror-image on the opposite side of the ctenidial membrane (See Fig. 4). *CA*, ctenidial artery; *CG*, ctenidial (branchial) gland; *CL*, ctenidial ligament; *CM*, ctenidial mesentary; *CV*, ctenidial vein; *P*, pinnal lamella; *PC*, pinnal cartilage; *PM*, pinnal mesentary; *2A*, pinnal artery; *2V*, pinnal vein (From Bradbury, 1970).

contraction rates in various parts of the mantle actually produce continuous, unidirectional flow over the gills of octopus (Fig. 3; Wells and Smith, 1985). The system shown in Figure 4 is from *Sepia*, a less pelagic decapod relative of squids which uses its fins more than its jet. As the mantle expands slowly, water is drawn in at its lip over the collapsed funnel. The inhalant water must pass between the gill lamellae to reach the large post-branchial chamber behind the gills, because each gill wedges between a mantle wall and a funnel retractor muscle. As the mantle contracts, deoxygenated water in this chamber is forced out medially through the funnel to produce the exhalant jet. The radial gill folds prevent backflow. Swimming squid probably do not produce a continuous flow, but gill ventilation occurs during the longer refilling phase, rather than the jet. There also appears to be partial ventilatory-circulatory coupling (Shadwick *et al.*, 1990).

3.2. Ventilatory regulation

No squid system has been described in such detail, but, while structures are similar (Bradbury, 1970; Fig. 5), squid are probably less rather than more sophisticated. Squid gills are smaller relative to body weight than those of octopus or *Sepia*, and the highest extraction efficiencies measured, even in hypoxia, are less than 15% (Shadwick *et al.*, 1990; Wells *et al.*, 1988). During maximum aerobic exercise, oxygen consumption increases to 3 times resting values, but extraction efficiency actually decreases to about 8%. The increases in ventilatory frequency (2.5 ×) and ventilatory stroke volume (1.5 ×) are dictated by the need for greater reaction mass for the jet, not by the oxygen requirement which is automatically met. It thus appears unlikely that any complex regulatory systems for squid respiration will be found. Wells *et al.* (1988) have shown that even during hypoxia the respiratory frequency is correlated with activity or temperature rather than the need to increase oxygen extraction. Extraction can increase to 20%, which may reflect changes in gill perfusion (Wells, 1990) or simply changes in the binding characteristics of hemocyanin (see below). In fact, strictly speaking, squid do not have a system for ventilating their gills; they use their principal locomotor system for the purpose. Moving the mantle, which may be up to 40% of a squid's body weight, to provide ventilation at rest is probably a major factor in the high resting metabolic rate of squid and may explain their *live fast, die young*

Table II. Respiratory and circulatory performance of cephalopods and fish.

	Nautilus[1] *pompilius*	*Octopus*[2] *vulgaris*	*Sepia*[3] *officinalis*	*Loligo*[4] *pealei*	*Salmo*[5] *gairdneri*
Temperature (°C)	17	22	22	22	10
Approximate weight (g)	500	1000	300	100	1200
Ventilation volume ($ml \cdot kg^{-1} min^{-1}$)	296	280*	770*	5000[20]	200
O_2 extraction (% from water)	7	63[6]	40[7]	14[21]	33
O_2 consumption ($ml \cdot kg^{-1} min^{-1}$) Rest	0.5[8]	0.7[9]	1.7[11]	5.0[13]	0.6
O_2 consumption ($ml \cdot kg^{-1} min^{-1}$) Active	2.2[8]	2.5[10]	4.9[12]	15.8[13]	4.3
Blood volume (% body weight)	–	5[14]	–	–	5[15]
Respiratory pigment (% w/v in blood)	–	9[16]	–	–	8[17]
Arterial O2 (volume %)	2.0	3.4	4.0	4.3	10.4
Bohr coefficient (log P_{50}/pH)	–0.2	–1.6	–1.5	–1.3	–05[17]
$P_{50}O_2$ (kPa at pH 7.4)	2.3	6.0	4.3	8.0	2.0[17]
Venous O_2 (kPa)	2.7	4.0	–	6.4	4.4
O_2 extraction (% from blood)	35	85	–	88	32
Cardiac output ($ml \cdot kg^{-1} min^{-1}$) Rest	5	42*	–	130*	18
Cardiac output ($ml \cdot kg^{-1} min^{-1}$) Active	–	150*	–	420*	53
Circulation time (s)	–	90	–	10[21]	64
Heart rate ($beats \cdot min^{-1}$)	12[18]	42[16]	40[19]	102[19]	38
Systolic pressure (kPa)	3.5[18]	3.9[16]	–	7.2[19]	5.6

* These values were calculated from other data in the table using the Fick principle. All other data from first reference given for species except as noted.

[1] Johansen *et al.*, 1978
[2] Houlihan *et al.*, 1982
[3] Brix *et al.*, 1981
[4] Redfield and Goodkind, 1929
[5] Kiceniuk and Jones, 1977
[6] Hazelhoff, 1938
[7] Wells and Wells, 1982
[8] Redmond *et al.*, 1978; minimum - 17 °C, maximum - 25 °C.
[9] Wells *et al.*, 1983a
[10] Wells *et al.*, 1983b
[11] Johansen *et al.*, 1982
[12] Montuori, 1913; maximum value
[13] Freadman *et al.*, 1984, Freadman, 1989
[14] O'Dor and Wells, 1985
[15] Stevens, 1968
[16] Wells, 1983
[17] Weber *et al.*, 1976
[18] Bourne *et al.*, 1978
[19] Bourne, 1982
[20] O'Dor, 1988
[21] Shadwick *et al.*, 1990. (Modified from O'Dor and Webber, 1986)

life-history strategy. It is obvious, however, that an animal evolving to maximize its power output for attack and escape cannot afford to carry around the large gills needed to provide efficient oxygen extraction at rest.

4. Circulatory physiology

Cephalopod circulatory systems are closed and highly regulated, but reflect their molluscan origins. Wells' (1983) review of the class is comprehensive to its date, and a recent multi-author review (Schipp, 1987) provides an update. Neither includes much information on squids; the most frequently cited works on them are still pre-1935, including Williams' 1909 anatomy, based on his work at the Marine Biological Laboratory in the last century. Work on the octopod and sepioid circulatory systems has, in contrast, been intense by cephalopod standards in the last decade, with over 50 articles published. A review of all this new literature is outside the scope of this volume (and these authors), but we have attempted to reflect the new findings insofar as they are pertinent to squids. In contrast to the situation in the respiratory system, we must anticipate that a squid circulatory system will be much *more* efficient and sophisticated than that of an octopus, since it must deliver nearly 7 times the oxygen at maximum aerobic scope (Table II).

4.1. Cellular elements

The hemocytes, the only cells in cephalopod blood, have no pigment and no apparent role in respiration, although Heming *et al.* (1988) have shown that they actively transport H^+. Their two principal roles are prevention of blood loss and phagocytosis (Cowden and Curtis, 1981). Cephalopod blood lacks clotting proteins, but the hemocytes adhere to damaged tissue and then *hold hands*, crenating to close a wounded vessel. Fortunately, they don't have to resist the full pressure of the circulation, since virtually all cephalopod blood vessels are contractile and independently innervated as discussed below. Thus central nervous control can provide a temporary mechanical seal to help the hemocytes until fibroblasts make a permanent repair – except when you are doing surgery and the animal is under anaesthesia!

The origin of hemocytes was debated for a century, but Johnson (1987) has recently reviewed and confirmed that the large spongy mass of *white body*, filling the blood sinuses behind the eyes, is the source. This tissue was one of the first cephalopod bits to be cultured *in vitro* (Necco and Martin, 1967), which should make further studies on the role of the hemocytes in defense against foreign material practical. There is no evidence for specific proteins binding to foreign material to make it a target for hemocytes (Bayne, 1983), but mammalian red cells exposed to *Eledone* serum were more rapidly phagocytized than unexposed cells (Stuart, 1968). The relationship between the roving hemocytes and sessile phagocytes which line vessels remains unclear, as well. Gray (1969) suggested that they were the same, but Bayne (1973) questioned (on equivocal grounds, in the authors' opinions) whether hemocytes are either related or phagocytic. Hemocytes make up 1 – 2% of blood volume as it is collected, but Wells (1983) suggests this may be an overestimate, if sessile cells are mobilized when tissue damage occurs. We have no quantitative measures, but motile hemocytes are readily observed in the circulation of transparent

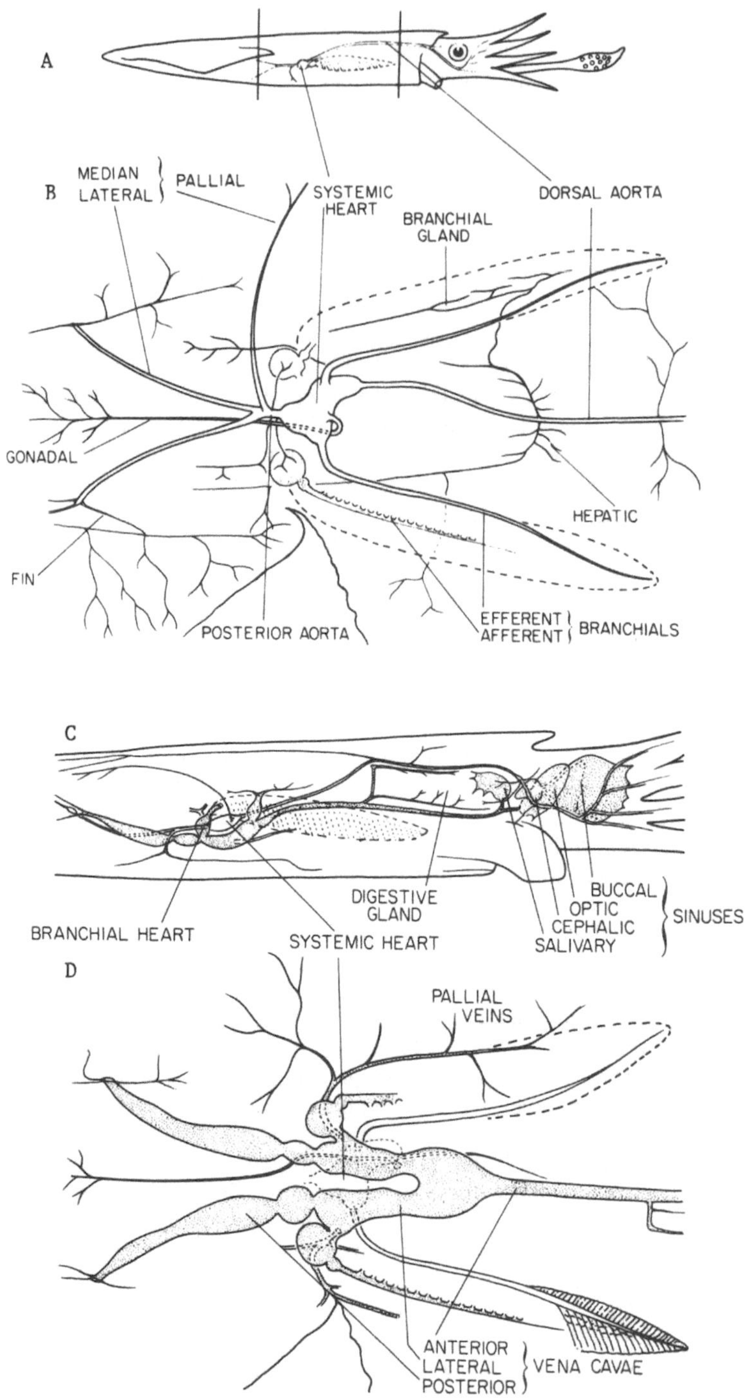
A
B
MEDIAN
LATERAL
PALLIAL
SYSTEMIC
HEART
DORSAL AORTA
BRANCHIAL
GLAND
GONADAL
HEPATIC
FIN
POSTERIOR AORTA
EFFERENT
AFFERENT
BRANCHIALS
C
DIGESTIVE
GLAND
BRANCHIAL HEART
SYSTEMIC HEART
BUCCAL
OPTIC
CEPHALIC
SALIVARY
SINUSES
D
PALLIAL
VEINS
ANTERIOR
LATERAL
POSTERIOR
VENA CAVAE

hatchling and juvenile squids with sufficient frequency to support this proportion in intact animals.

4.2. Oxygen carrier pigment

The most clearly molluscan aspect of the system is the oxygen carrying pigment hemocyanin, which is discussed in detail by Mangum (1990). Although the amount of oxygen carried per unit volume of blood is *appallingly inadequate* (Wells, 1983) relative to vertebrates with cellular hemoglobins, hemocyanin must not be a bad second best given the high oxygen consumptions squids manage. A 9 – 10% solution of this copper-bearing protein has about the same viscosity as human plasma but only 25% of carrying capacity of human blood (Wells, 1983). The formation of 4,000 kiloDalton (kD) complexes from 400 kD hemocyanin monomers reduces viscosity almost as effectively as concentrating it in cells, but the fact that it takes 50 kD of protein and two Cu^+ ions for each domain capable of binding an oxygen molecule, still gives a maximum carrying capacity for blood of only 4.5 vol% O_2. Although the low oxygen carrying capacity is a constraint on squids, there are compensations. Squid blood has a lower viscosity than whole vertebrate blood and must be easier to pump. The absence of cells also removes constraints on the delivery of blood to tissues. The details of mechanisms at this level are not clear, but the size of *capillaries* can be smaller and more variable (Browning, 1982). The number of capillaries and their proportion of tissue volume can be similar to those in rats in comparable tissues (Abbott and Bundgaard, 1987).

Hemocyanin's biochemical properties, as described by Mangum are, perhaps, even more adaptive than those of vertebrate hemoglobin's for allowing physiological differences to maximize O_2 loading at the gills and unloading at the tissues. The combination of Bohr effect and cooperativity permit over 90% of the transported O_2 to be released on each pass according to Redfield and Goodkind's (1929) calculations from constrained *Loligo pealei*; this is nearly 3 times as much as in fish (Table II). Their value has been criticized because of the unnatural conditions of the animals, but both Houlihan *et al.*'s (1986) recent report of 90% utilization in exercising *Octopus* and our experiments (Pörtner *et al.*, 1987, 1989) on freely swimming squids support their values. The placement of our cannulae allowed us to sample the relatively small flow of venous blood returning from the head directly, where utilization is only about 60%, but the low pH values (ca. 7.2) in the mantles of animals swum slowly (10 – 15 min) to exhaustion at supracritical speeds should produce nearly 100% unloading. We still know little about the mechanisms partitioning blood flow between the head and the high and variable demand of the mantle musculature, but it is clear that it requires the most sophisticated system in the invertebrate world.

Figure 6. (*A*) and (*B*), respectively, show the arterial system of *Loligo pealei* in relation to the body and gills, and the details around the systemic heart. (*C*) and (*D*) show the major anterior venous sinuses and the connections of the *vena cavae* (From Wells, 1983, after Williams, 1909).

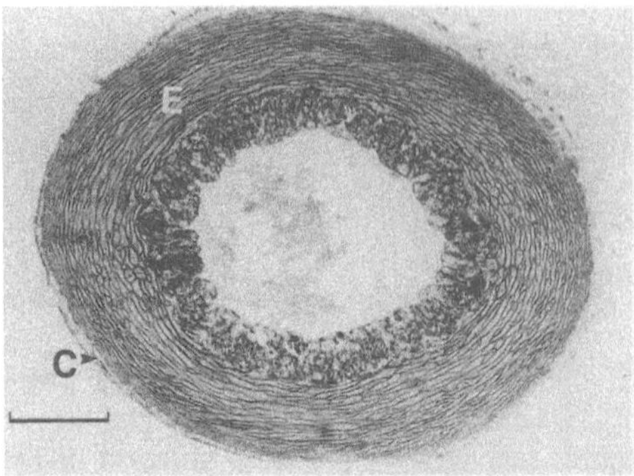

Figure 7. A transverse section of the dorsal aorta from the squid, *Nototodarus sloani*, stained by Gomori's aldehyde fuchsin technique to show the presence of concentric lamellae of elastic fibers (*E*) in the *tunica media*. Collagen fibers (*C*) are found in the *tunica adventitia*. Scale bar, 250 μM.

4.3. Circulatory system

Figure 6 shows the basic layout of a squid circulatory system. It is, arguably, as closed as any vertebrate system and more closed than any invertebrate one. The fluid it contains should, therefore, be called blood, not hemolymph, making cephalopods nature's only true bluebloods. Even the sinuses are not the open lacunae in direct contact with tissue found in other invertebrates, but may be considered enlarged veins lined with endothelial cells (Williams, 1902). No squid's blood volume has been determined, but, assuming it is 5% of body weight like *Octopus vulgaris* (O'Dor and Wells, 1985), means that to supply 1000 ml $O_2 \cdot kg^{-1}h^{-1}$ for exercise a squid must pump 22.2 $l \cdot kg^{-1}h^{-1}$ of blood with an average recirculation time of 10 sec. This is 8 times the maximum cardiac output of a fish and only slightly less than human limits.

These values should make it clear why three hearts are required, and, even then, all of the major capacitance vessels must be muscular and aid blood flow with peristaltic movements. The paired branchial hearts which are about 0.5% of body weight collect blood from the *vena cavae* and force it through the gill capillaries before the blood enters the main systemic heart, which is about 0.25% of body weight (Martin and Aldrich, 1970). People are usually surprised at the presence these *extra* gill hearts which are an innovation of advanced cephalopods, absent in *Nautilus* and other molluscs. However, it is really the systemic heart which is surprising, functionally. Fish hearts pump into their gills directly, and it is only in land vertebrates which must work against hydrostatic pressures produced by gravity that separate respiratory and systemic circuits develop fully. Presumably, in squid it is the need to work against the back-pressures of the hydroskeleton, which allows arms without joints and opposed muscles without bones, that makes this heart essential.

Many features of the squid circulatory system are surprisingly mammalian. In addition to the arrangement of gill and systemic hearts, which allows a dual circulation with respiratory and systemic circuits in series, the dorsal aorta of squid and other cephalopods have passive elasticity which contributes significantly to the dynamics of blood flow. They provide an elastic reservoir which helps smooth the pulsatile flow of blood from the heart. During systolic ejection blood swells the aorta, stretching the vessel wall which then recoils and moves blood passively during diastole. The elastic properties of cephalopod aortae increase with blood pressure (Gosline and Shadwick, 1982), with squid, the highest pressure cephalopods, having those most similar to the

Table III. A comparison of mechanical properties of rat and squid aortae

	E† *(kPa)*	*C‡* *($m \cdot s^{-1}$)*	*Z‡* *($MPa \cdot s \cdot m^{-3}$)*		*Heart Rate* *(min^{-1})*	*Temperature* *(°C)*
			Rest	*Active*		
Squid (400g)	350	4	600	60	120	12
	–	–	–	102	250	20
	–	–	–	420*	800*	37
Rat (400g)	400	4.5	500	300	ca 600	37

† Calculated for the mean blood pressure.
‡ From Eqns. 1 and 2.
* Projected, based on a Q_{10} of 2.9 from Dykens and Mangum, 1979.

mammalian aorta. Arterial elasticity in cephalopods demonstrates a remarkable evolutionary convergence with vertebrate circulatory systems. Structural similarities between aortae are also striking (Fig. 7); both contain rubber-like elastic fibers arranged in concentric lamellae throughout the *tunica media*, which is reinforced by stiff collagen fibers in the *tunica adventitia* (Shadwick and Gosline, 1983, 1985a).

The importance of arterial elasticity in squid can be appreciated when we consider that pressure wave velocity (C) and hydraulic impedance to blood flow (Z) are directly influenced by the elastic modulus of the vessel wall (E). These hemodynamic parameters can be calculated from the vessel dimensions and E:

$$C^2 = \frac{Eh}{2dR} \qquad (1)$$

$$Z = \frac{dC}{R^2} \qquad (2)$$

where h is the vessel wall thickness, R is the internal radius and d is the blood density. The modulus, E, can be calculated from stress/strain data derived from *in vitro* uniaxial or biaxial mechanical tests (See Shadwick and Gosline, 1985b for methods.). Because of the small diameter of the squid aorta (typically about 1 mm in a 40 g *Loligo opalescens* and 2 mm in a 350 g *Illex illecebrosus*) and difficulties experienced in surgery, no *in vivo* measurements of vessel elasticity have yet been made.

In mammals, E increases with pressure at a rate which is greatest over the physiological range. For the large pelagic squid *Nototoarus sloani*, Gosline and Shadwick (1982) found that the increase in E with pressure was most dramatic between about 10 and 20 kiloPascals (kPa) (1kPa = 1000 Newtons/m^2 = 1×10^4 dynes/cm^2) and they suggested that this range probably represents *in vivo* blood pressures in this species. The same method predicts a slightly lower physiological pressure range for *Loligo opalescens* of 5 to 11 kPa (Shadwick and Gosline, 1989), possibly reflecting a difference in activity levels in these two species. In *Loligo pealei* averaging 150 g,

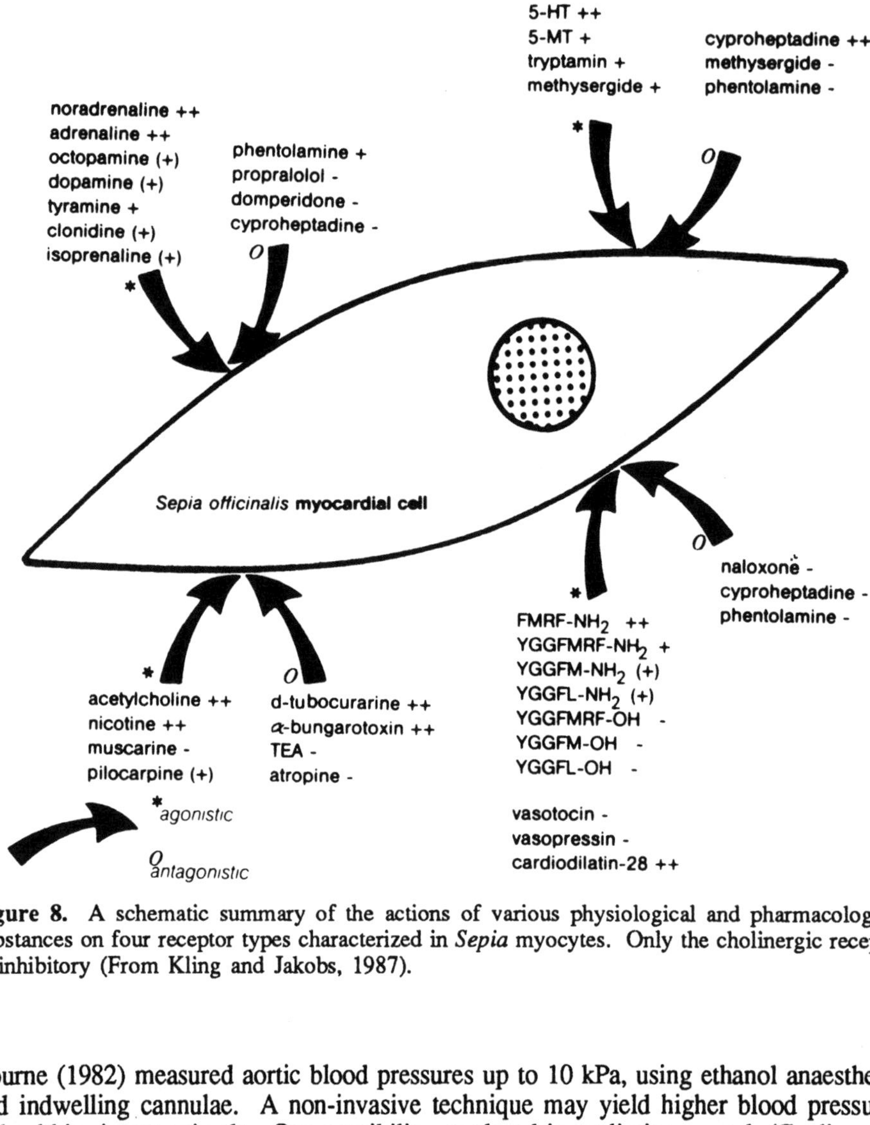

Figure 8. A schematic summary of the actions of various physiological and pharmacological substances on four receptor types characterized in *Sepia* myocytes. Only the cholinergic receptor is inhibitory (From Kling and Jakobs, 1987).

Bourne (1982) measured aortic blood pressures up to 10 kPa, using ethanol anaesthesia and indwelling cannulae. A non-invasive technique may yield higher blood pressures in healthier intact animals. One possibility, explored in preliminary work (Gosline and Shadwick, 1989), is using a video dimension analyzer to measure aortic diameter pulsations in undisturbed squid by looking through the transparent mantle wall after removal of a small patch of skin. By comparing these measurements with a post-mortem *in vitro* pressure/diameter curve for the aorta it should be possible to estimate the *in vivo* pressure range. This technique could potentially be used for determining squid blood pressure increases during exercise.

The available evidence, although scant, suggests that squid blood pressures are higher than other cephalopods and approach those of the most active vertebrates. Pressure wave velocity, aortic impedance (Equations 1 and 2) and heart rates are comparable to those of mammals (Table III). Thus, while the arterial circulation of less active cephalopods is a simple Windkessel (Shadwick *et al.*, 1987), the squid aorta may behave more like a hydraulic transmission line system as in mammals. If so, pressure wave propagation effects, such as reflection, may reduce the pulsatile work of the heart, particularly in large and active squid.

Squid heart muscle seems highly adapted to sustained aerobic power output compared to other molluscs, and even to vertebrate hearts (Dykens and Mangum, 1978; Kling and Schipp, 1987). Obliquely striated cardiac myofibers have relatively small diameters, a large volume of sarcoplasmic reticulum and a mitochondrial core which occupies 50 to 80% of the cross-sectional area. This ensures that the diffusion distance for O_2 delivery from blood to the mitochondria is not greater than about 2 μm (Dykens and Mangum, 1978). Overall, the squid circulatory system is structurally well designed for a highly active lifestyle, but the ultimate limitation on circulatory performance in squid remains the low O_2 carrying capacity of the blood.

In a recent study we measured heart and respiratory rates simultaneously by direct observation of *Loligo opalescens* swimming in a tunnel respirometer at 12 °C (Shadwick and O'Dor, 1988; Shadwick *et al.*, 1990). Assuming a blood volume of 5% as before, that blood contains 4.3 vol% O_2 and that extraction is 88% (Redfield and Goodkind, 1929), we made the following preliminary calculations using O_2 consumption data from O'Dor's previous study (1982). To supply the O_2 requirement of 11 $ml{\cdot}kg^{-1}min^{-1}$ at the critical swimming speed of 1.7 $l{\cdot}s^{-1}$ the heart must pump 300 $ml{\cdot}kg^{-1}min^{-1}$ of blood. With a heart rate of 100 min^{-1}, this yields a stroke volume of 0.1 ml (1.53 $ml{\cdot}g^{-1}$ heart) and a circulation time of only 10 seconds. We also calculated that the power output of the heart to deliver this flow (assuming a mean blood pressure of 5 kPa) would be 13 $W{\cdot}kg^{-1}$. If blood pressure actually increases during exercise then the power values will, too.

4.4. Circulatory regulation

There is evidence for both nervous and hormonal regulation of most parts of the circulatory systems in squid (Bacq, 1933a, 1933b, 1934), although the details of the control systems remain unknown. The cephalopod systemic heart is basically myogenic, but its inotropic and chronotropic responses to acetylcholine (negative) and to catecholamines (positive) resemble those of vertebrates (Kling and Jakobs, 1987; Fig. 8). It also responds to a variety of peptides and pharmacological agents. Several sites for the release of putative neurosecretory cardioregulatory materials have been characterized in cephalopods (Martin and Voight, 1987), but the situation in squid is less clear than in *Octopus* and *Sepia*.

Isolated heart preparations show a pronounced Starling response, beating more powerfully as refilling rate is increased (Smith, 1981). This response appears to be quite important in exercising squids, since heart rate can actually decrease as exercise begins, becoming approximately synchronous with mantle contractions (Shadwick *et al.*, 1990; Wells *et al.*, 1988). It is unclear how any of the chemical modulators described so far could decrease frequency and increase stroke volume and power, as is required in this circumstance. A possible explanation, anticipated by Bourne (1987, 1984), is a dramatic decline in peripheral resistance in response to acetylcholine. Perhaps the acetylcholine which initiates motor activity, also shifts blood flow to the active muscles and influences heart rate.

The control of the branchial hearts and gill circulation appears to be quite different. Neural influence from rhythmically firing ganglia clearly dominates the limited myogenic potential of these hearts, and the response to acetylcholine in the

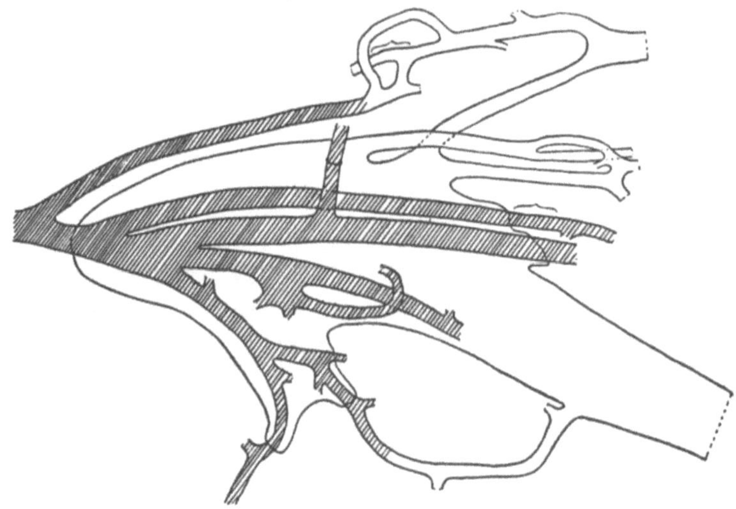

Figure 9. Some arterioles (shaded) in the mantle musculature and elsewhere pass **through** venous *end sinuses* before breaking up into a capillary plexus. Such structures may prevent backflow of venous blood when the mantle cavity is pressurized during jetting (From Williams, 1902).

gill vasculature appears to be an increase in resistance based on isolated gill preparations. However, isolated gills tend to close down regardless of what is done to them, and we really know very little about the regulation of gill perfusion in any cephalopod. Contraction of gill vessels may decrease diffusion distances and force blood through distal lamellae, thus increasing the efficiency of O_2 loading at the cost of increased resistance.

4.5. Auxiliary systems

It is doubtful that even three hearts are enough for the task. The human heart is proportionately twice as large as the systemic heart and has the advantage of working at a temperature 20 °C higher. It is hard to image how the systemic heart alone, even producing pressures over 10 kPa, can provide the required power to move all this blood. The mantle musculature itself may assist the heart which lies inside the mantle cavity. In actively swimming squid, pressures in the cavity become negative at the beginning of each refilling phase of the jet cycle (Webber and O'Dor, 1986), as a result of the recoil of collagen *springs* compressed during contraction (Gosline and Shadwick, 1983); this must also aid in venous return to the hearts. During contraction, cavity pressures may reach 50 kPa, creating a very large pressure gradient between the heart and the head, arms and, perhaps, even the outside of the mantle at some time in each cycle. Precisely how this system works is unknown, but Williams (1902) described arterioles which pass through venous lacunae before forming capillaries (Fig. 9); in such a system, venous pressure resulting from mantle contraction would collapse the arterioles, suggesting that these lacunae could function as one-way valves.

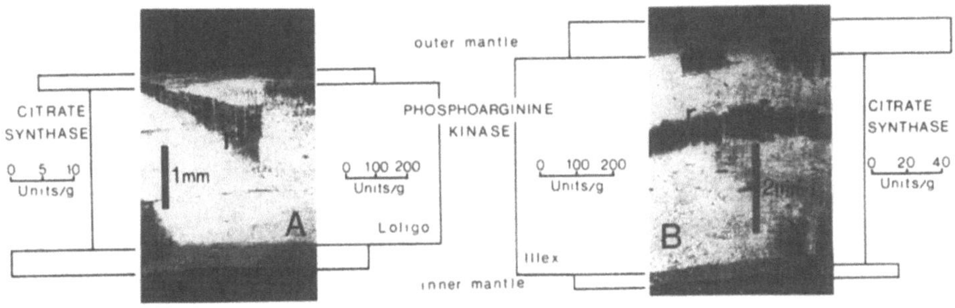

Figure 10. Both myopsid (*Loligo opalescens*, A) and oegopsid (*Illex illecebrosus*, B) squids have aerobic muscle fibers rich in mitochondria and citrate synthase, distributed in thin layers beneath their thin skins on the inner and outer surfaces of the mantle. The main mass of phosphoarginine kinase rich anaerobic muscle for escape jets is in the mantle core (From Mommsen *et al.*, 1981).

It is also possible that a significant portion of the oxygen consumed does not have to pass through the blood at all. Octopuses can apparently take 5 – 10% of their oxygen up across their skin (Wells and Wells, 1983). Squids have thinner skins (Hulet *et al.*, 1979), larger surface to volume ratios, more active muscles and much greater flows of water over both the inner and outer surfaces of the hollow tubular mantle. Figure 10 shows that the aerobic muscle fibers, containing up to 50% mitochondria by volume (higher even than insect flight muscles, Ellington, 1985) are localized close to either the inner or outer skin surface. We know that thin-skin is one of the principle causes of mortality of squid in captivity (Hulet *et al.*, 1979); perhaps oxygen transport is why evolution has forced squid to put up with this weakness.

References

Abbott, N. J. and Bundgaard, M., 1987, Microvessel surface area density and dimensions in brain and muscle of the cephalopod, *Proc. Roy. Soc. B* **230**:459–482.

Andrews, P. L. R. and Tansey, E. M., 1981. The effects of some anaesthetic agents in *Octop us vulgaris*. *Comp. Biochem. Physiol.* **70C**:241–247.

Bacq, Z. M., 1933a, Action d'adrenaline, d'ergotamine et de tyramine sur le ventricule median isolé de *Loligo pealii*, *C. R. Soc. Biol.* **114**:1358–1360.

Bacq, Z. M., 1933b, Réaction du ventricule median isolé de *Loligo pealii* à l'acetylcholine, à l'atropine et aux ions K, Ca et Mg, *C. R. Soc. Biol.* **114**:1360–1361.

Bacq, Z. M., 1934, Recherches sur la physiologie du système nerveux autonome, *Archs. Int. Physiol.* **38**:138–159.

Bayne, C. J., 1973, Internal defense mechanisms of *Octopus dofleini*, *Malacol. Rev.* **6**:13–17.

Bayne, C. J., 1983, Molluscan immunobiology, in: *The Mollusca*, Vol. **5** (A. S. M. Saleuddin and K. M. Wilber, eds.), Academic Press, London, pp. 407–486.

Beamish, F. W. H., 1978, Swimming capacity, in: *Fish Physiology*, Vol. **7** (W. S. Hoar and D. J. Randall, eds.), Academic Press, London, pp. 101–239.

Best, E. M. H. and M. J. Wells. 1983. The control of digestion in *Octopus* I: The anticipatory response and effects of severing the nerves to the gut. *Vie Milieu* **33**:135–142.

Bone, Q., 1978, Locomotor muscle, in: *Fish Physiology*, Vol. **7** (W. S. Hoar and D. J. Randall, eds.), Academic Press, London, pp. 361–424.

Bourne, G. B., 1982, Blood pressure in the squid, *Loligo pealei*, *Comp. Biochem. Physiol.* **72A**:23–27.

Bourne, G. B., 1984, Pressure-flow relations in the post-systemic circulation of the squid, *Loli go pealei*, *Comp. Biochem. Physiol.* **78A**:307–313.

Bourne, G. B., 1987, Hemodynamics in squid, *Experientia* **43**:500–502.

Bourne, G. B., Redmond, J. T., and Johansen, K. 1978, Some aspects of hemodynamics in *Nautilus pompilius*, *J. Exp. Zool.* **205**:63–70.

Bradbury, H. E., 1970, *Observations on the functional anatomy of the ommastrephid, Illex illecebrosus (LeSueur, 1821) (Coleoidea: Cephalopoda), with emphasis on musculature and the blood vascular system*, Ph.D. Thesis, Memorial University of Newfoundland, St. John's, Newfoundland, Canada.

Brett, J. R., 1964, The respiratory metabolism and swimming performance of young sockeye salmon, *J. Fish. Res. Board Can.* **21**:1183–1226.

Brett, J. R., 1965, The relation of size to rate of oxygen consumption and sustained swimming speed of sockeye salmon (*Oncorhynchus nerka*), *J. Fish. Res. Board Can.* **22**:1491–1501.

Brett, J. R. and Groves, T. D. D., 1980, Physiological energetics, in: *Fish Physiology*, Vol. **8** (W. S. Hoar, D. J. Randall and J. R. Brett, eds.), Academic Press, New York, pp. 279–352.

Bridges C. R., Bicudo, J. E. P. W., and Lykkeboe, G., 1979, Oxygen content measurement in blood containing hemocyanin, *Comp. Biochem. Physiol.* **62A**:457–462.

Brix, O., Lykkeboe, G., and Johansen, K., 1981, The significance of the linkage between the Bohr and Haldane effects in cephalopod bloods, *Resp. Physiol.* **44**:177–186.

Browning, J., 1982, The density and dimensions of exchange vessels in *Octopus pallidus*. *J. Zool. Lond.* **196**:569–579.

Cole, K. S. and Gilbert, D. L., 1970, Jet propulsion of squid, *Biol. Bull.* **138**:245–246.

Cowden, R. R. and Curtis, S. K., 1981, Cephalopods, in: *Invertebrate Blood Cells*, Vol. 1 (N. A. Ratcliffe and A. F. Rowley, eds.), Academic Press, London, pp. 301–323.

DeMont, M. E., 1989, *pers. commun.*

Driedzic, W. R. and Hochachka, P. W., 1978, Metabolism in fish during exercise, in: *Fish Physiology*, Vol. **8** (W. S. Hoar, D. J. Randall and J. R. Brett, eds.), Academic Press, London, pp. 503–543.

Dykens, J. A. and Mangum, C. P., 1979, The design of cardiac muscle and the mode of metabolism in mollusks, *Comp. Biochem. Physiol.* **62A**:549–554.

Ellington, C. P., 1985, Power and efficiency of insect flight muscles, *J. Exp. Biol.* **115**:293–306.

Foyle, T. P. and O'Dor, R. K., 1987, Predatory strategies of squid (*Illex illecebrosus*) attacking small and large fish, *Mar. Physiol. Behav.* **13**:155–168.

Freadman, M. A., Hernandez, L., and Scharold, J., 1984, Swimming biology of squid, *Loligo pealei*. *Am. Zool.* **24**:123A.

Freadman, M. A., 1989, *pers. commun.*

Fry, F. E. J. and Hart, J. S., 1948, Cruising speed of goldfish in relation to water temperature, *J. Fish. Res. Board Can.* **7**:169–175.

Gosline, J. M. and Shadwick, R. E., 1982, The biomechanics of the arteries of *Nautilus*, *Nototodarus* and *Sepia*, *Pacific Sci.* **36**:283–296.

Gosline, J. M. and Shadwick, R. E., 1983, The role of elastic energy storage mechanisms in swimming: an analysis of mantle elasticity in escape jetting in the squid, *Loligo opalesc ens*, *Can. J. Zool.* **61**:1421–1431.

Gosline, J. M. and Shadwick, R. E., 1989, *personal commun.*

Gosline, J. M., Steeves, J. D., Harman, A. D., and DeMont, M. E., 1983, Patterns of circular and radial mantle muscle activity in respiration and jetting of the squid, *Loligo opalescens*, *J . Exp. Biol.* **61**:1421–1431.

Gray, E. G., 1969, Electron microscopy of the glio-vascular organization of the brain of *Octo pus*, *Phil. Trans. Roy. Soc. B* **255**:13–32.

Hanlon, R. T., 1987, Mariculture, in: *Cephalopod Life Cycles*, Vol. **2** (P. Boyle, ed.) Academic Press, London, pp. 291–306.

Hanlon, R. T., 1990, Maintenance, rearing, and culture of teuthoid and sepioid squids, *this volume*.

Hanlon, R. T., Hixon, R. F., and Hulet, W. H., 1983, Survival, growth and behavior of the loliginid squids *Loligo pealei* and *Loliguncula brevis* (Mollusca: Cephalopoda) in closed sea water systems, *Biol. Bull.* **165**:637–685.

Hazelhoff, E. H., 1938, Uber de ausnutzung des Saurstoffs bei verschiedenen Wassertieren, *Z. Vergl. Physiol.* **26**:306–327.

Heming, T. A., Brown, S., Vanoye, C., and Bidani, A., 1988, Control of pH_i in squid hemocytes, *Proc. Int. Un. Biol. Sci.* 322 (abst.).

Hoeger, U., Mommsen, T. P., O'Dor, R. K., and Webber, D. M., 1987, Oxygen uptake and nitrogen in two cephalopods, octopus and squid, *Comp. Biochem. Physiol.* **87**:63–67.

Houlihan, D. F., Innes, A. J., Wells, M. J., and Wells, J., 1982, Oxygen consumption and blood gases of *Octopus vulgaris* in hypoxic conditions, *J. Comp. Physiol.* **148**:35–40.

Houlihan, D. F., Duthie, G., Smith, P. J., Wells, M. J., and Wells, J., 1986, Ventilation and circulation during exercise in Octopus vulgaris, *J. Comp. Physiol.* **148B**:683–689.

Hulet, W. H., Villoch, M. R., Hixon, R. F., and Hanlon, R. T., 1979, Fin damage in captured and reared squids, *Lab. Anim. Sci.* **29**:528–533.

Johansen, K., Redmond, J. R., and Bourne, G. B., 1978, Respiratory exchange and transport of oxygen in *Nautilus pompilius*, *J. Exp. Zool.* **205**:27–36.

Johansen, K., Brix, O., Kornerup, S., and Lykkeboe, G., 1982, Factors affecting O_2-uptake in the cuttlefish, *Sepia officinalis*, *J. Mar. Biol. Ass. U.K.* **62**:187–191.

Johnson, F. E., 1987, The vasculature of the white bodies in the ommastrephid squid *Illex illecebrosus* (LeSueur, 1821): a proposed route for the dissemination of newly formed hemocytes, *Can. J. Zool.* **65**:1607–1620.

Johnson, W., Soden, P. D., and Truman, E. R., 1972, A study in jet propulsion: an analysis of the motion of the squid *Loligo pealei*, *J. Exp. Biol.* **56**:155–165.

Jones, D. R. and Randall, D. J., 1978, The respiratory and circulatory systems during exercise, in: *Fish Physiology*, Vol. **7**, (W. S. Hoar and D. J. Randall, eds.) Academic Press, New York, pp. 425–501.

Kiceniuk, J. W. and Jones, D. R., 1977, The oxygen transport system in trout (*Salmo gairdneri*) during sustained exercise, *J. Exp. Biol.* **69**:247–260.

Kling, G. and Jakobs, P. M., 1987, Cephalopod myocardial receptors: Pharmacological studies on the isolated heart of *Sepia officinalis* (L.), *Experientia* **43**:511–525.

Kling, G. and Schipp, R., 1987, Comparative ultrastructural and cytochemical analysis of the cephalopod systemic heart and its innervation, *Experientia* **43**:502–511.

Lighthill, M. J., 1960, Note on the swimming of slender fish, *J. Fluid Mech.* **9**:305–317.

Mackie, G. O., 1990, The giant axon systems of squid and medusae, *Can. J. Zool.*, *in press.*

Mangum, C. P., 1990, Gas transport in the blood, *this volume.*

Martin, A. W. and Aldrich, F. A., 1970, Comparison of hearts and branchial heart appendages in some cephalopods, *Can. J. Zool.* **48**:751–756.

Martin, R. and Voight, K. H., 1987, The neurosecretory system of the octopus vena cava: A neurohemal organ, *Experientia* **43**:537–543.

Messenger, J. B., Nixon, M., and Ryan, K. P., 1985, Magnesium chloride as an anesthetic for cephalopods, *Comp. Biochem. Physiol.* **82C**:203–205.

Mommsen, T. P., Ballantyne, J., MacDonald, D., Gosline, J., and Hochachka, P. W., 1981, Analogs of red and white muscle in squid mantle, *Proc. Nat. Acad. Sci. U.S.A.* **78**:3274–3278.

Montuori, A., 1913, Les processus oxydatifs chez les animaux marins en rapport avec la loi de superficie, *Arch. Ital. Biol.* **59**:213–234.

Necco, A. and Martin, R., 1963, Behaviour and estimation of the mitotic activity of the white body cells in *Octopus vulgaris*, cultured *in vitro*, *Exp. Cell. Res.* **30**:588–590.

O'Dor, R. K., 1982, Respiratory metabolism and swimming performance of the squid, *Loligo opalescens*, *Can. J. Fish. Aquat. Sci.* **39**:580–587.

O'Dor, R. K., 1988a, Limitations on locomotor performance in squid, *J. Appl. Physiol.* **64**:128–134.

O'Dor, R. K., 1988b, The forces acting on swimming squid, *J. Exp. Biol.* **137**:421–442.

O'Dor, R. K., 1988c, The energetic limits on squid distributions, *Malacologia* **29**:113–119.

O'Dor, R. K. and Webber, D. M., 1986, The constraints on cephalopods: why squid aren't fish, *Can. J. Biol.* **64**:1591–1605.

O'Dor, R. K. and Wells, M. J., 1985, Circulation time, blood reserves and extracellular space in a cephalopod, *J. Exp. Biol.* **113**:461–464.

Packard, A., 1972, Cephalopods and fish: limits of convergence, *Biol. Rev.* **47**:241–307.

Packard, A. and Trueman, E. R., 1974, Muscular activity of the mantle of *Sepia* and *Loligo* (Cephalopoda) during respiration and jetting and its physiological interpretation, *J. Exp. Biol.* **61**:411–420.

Pantin, C. F. A., 1946, *Notes on microscopical techniques for zoologists*, Cambridge University Press, Cambridge.

Pinsker, H. M., Ferguson, G. P., Hanlon, R., Martini, F. M., and Long, T., 1986, Nonassociative learning and *in vivo* motor programs in the squid, *Proc. Int. Symp. Molluscan Neurobiol.*

Pörtner, H. O., Webber, D. M., Boutilier, R. G., and O'Dor, R. K., 1987, Acid-base regulation in unrestrained squid during and after swimming activity, *Bull. Can. Soc. Zool.* **18**:12(abst.).

Pörtner, H. O., Webber, D. M., Boutilier, R. G., and O'Dor, R. K., 1989, *unpublished data.*

Redfield, A. C. and Goodkind, R., 1929, The significance of the Bohr effect on the respiration and asphyxiation of the squid, *Loligo pealii*, *J. Exp. Biol.* **6**:340–349.

Redmond, J. R., Bourne, G. B., and Johansen, K., 1978, Oxygen uptake by *Nautilus pompilius*, *J. Exp. Zool.* **205**:45–50.

Schipp, R., 1987, General morphological and functional characteristics of the cephalopod circulatory system, *Experientia* **43**:474–477.

Shadwick, R. E. and Gosline, J. M., 1983, The structural organization of an elastic fiber network in the aortae of the cephalopod, *Octopus dofleini*, *Can. J. Zool.* **61**:1866–1879.

Shadwick, R. E. and Gosline, J. M., 1985a, Physical and chemical properties of rubber-like elastic fibers from the octopus aorta, *J. Exp. Biol.* **114**:239–257.

Shadwick, R. E. and Gosline, J. M., 1985b, Mechanical properties of the octopus aorta, *J. Ex p. Biol.* **114**:259–284.

Shadwick, R. E. and Gosline, J. M., 1989, *personal commun.*

Shadwick, R. E. and O'Dor, R. K., 1988, Respiratory and cardiac function during exercise in squid, *Bull. Can. Soc. Zool.* **19**(2):44(abst.)

Shadwick, R. E., Gosline, J. M., and Milson, W. K., 1987, Arterial hemodynamics in the cephalopod mollusc, *Octopus dofleini*, *J. Exp. Biol.*, **130**:87–106.

Shadwick, R. E., O'Dor, R. K., and Gosline, J. M., 1990, Respiratory and cardiac function during exercise in squid, *Can. J. Zool.*, *in press.*

Smith, P. J. S., 1981, The role of venous pressure in regulation of output from the heart of the octopus, *Eledone cirrosa* (Lam.), *J. Exp. Biol.* **93**:243–255.

Stevens, E. D., 1968, The effect of exercise on the distribution of blood to various organs in rainbow trout, *Comp. Biochem. Physiol.* **25**:615–625.

Stuart, A. E., 1968, The reticulo-endothelial system of the lesser octopus, *Eledone cirrosa*, *J. Pathol. Bacteriol.* **96**:401–412.

Trueman, E. R. and Packard, A., 1968, Motor performances of some cephalopods, *J. Exp. Bio l.* **49**:495–507.

Webb, P. W., 1978, Hydrodynamics: Nonscombroid fish, in: *Fish Physiology*. Vol. 7, (W. S. Hoar and D. J. Randall, eds.), Academic Press, London, pp. 190–239.

Webber, D. M. and O'Dor, R. K., 1985, Respiration and swimming performance of short-finned squid (*Illex illecebrosus*), *NAFO Sci. Coun. Studies* **9**:133–138.

Webber, D. M. and O'Dor, R. K., 1986, Monitoring the metabolic rate and activity of free-swimming squid with telemetered jet pressure, *J. exp. Biol.* **126**:205–224.

Weber, R. E., Wood, S. C., and Lamholt, J. P., 1976, Temperature acclimation and oxygen-binding properties of blood and multiple haemoglobins of rainbow trout, *J. Exp. Biol.* **65**:333–345.

Wells, M. J., 1983, Circulation in cephalopods, in: *The Mollusca*. Vol. **5**, (A. S.M. Saleuddin and K. M. Wilbur, eds.) Academic Press, London, pp. 239–290.

Wells, M. J., 1990, Oxygen extraction and jet propulsion in cephalopods, *Can. J. Zool.*, *in press.*

Wells, M. J. and Smith, P. J. S., 1985, The ventilation cycle in *Octopus*, *J. Exp. Biol.* **116**:375–383.
Wells, M. J. and Smith, P. J. S., 1987, The performance of the octopus circulatory system: A triumph of engineering over design, *Experientia* **43**:487–499.
Wells, M. J. and Wells, J., 1982, Ventilatory currents in the mantle of cephalopods, *J. Exp. Biol.* **99**:315–330.
Wells, M. J. and Wells, J., 1983, The circulatory response to acute hypoxia in *Octopus*, *J. Ex p. Biol.* **104**:59–71.
Wells, M. J., O'Dor, R. K., Mangold, K., and Wells, J., 1983a, Diurnal changes in activity and metabolic rate in *Octopus vulgaris*, *Mar. Behav. Physiol.* **9**:275–287.
Wells, M. J., O'Dor, R. K., Mangold, K., and Wells, J., 1983b, Oxygen consumption in movement by *Octopus vulgaris*, *Mar. Behav. Physiol.* **9**:289–303.
Wells, M. J., Duthie, G., Houlihan, D. F., Smith, P. J. S., and Wells, J., 1987, Blood flow and pressure changes in exercising octopuses (*Octopus vulgaris*), *J. Comp. Physiol.* **148B**:683–689.
Wells, M. J., Hanlon, R. T., Lee, P. G., and DiMarco, F. P., 1988, Respiratory and cardiac performance in *Loliguncula brevis* (Cephalopoda: Myopsida): The effects of activity, temperature and hypoxia, *J. Exp. Biol.*, **138**:17–36.
Williams, L. W., 1902, The vascular system of the common squid, *Loligo pealii*, *Am. Nat.* **36**:787–794.
Williams, L. W., 1909, *The anatomy of the common squid, Loligo pealii*, E. J. Brill, Leiden.
Wilson, D. M., 1960, Nervous control of movement in cephalopods, *J. Exp. Biol.* **37**:57–72.

Index

MIX
Papier aus verantwortungsvollen Quellen
Paper from responsible sources
FSC® C105338

If you have any concerns about our products,
you can contact us on
ProductSafety@springernature.com

In case Publisher is established outside the EU,
the EU authorized representative is:
Springer Nature Customer Service Center GmbH
Europaplatz 3, 69115 Heidelberg, Germany

Printed by Libri Plureos GmbH
in Hamburg, Germany